Lecture Notes in Computer Science 3400

Commenced Publication in 1973
Founding and Former Series Editors:
Gerhard Goos, Juris Hartmanis, and Jan van Leeuwen

James F. Peters Andrzej Skowron (Eds.)

Transactions on Rough Sets III

Volume Editors

James F. Peters
University of Manitoba
Department of Electrical and Computer Engineering
Winnipeg, Manitoba R3T 5V6, Canada
E-mail: jfpeters@ee.umanitoba.ca

Andrzej Skowron
Warsaw University
Institute of Mathematics
Banacha 2, 02-097 Warsaw, Poland
E-mail: skowron@mimuw.edu.pl

Library of Congress Control Number: 2005925573

CR Subject Classification (1998): F.4.1, F.1, I.2, H.2.8, I.5.1, I.4

ISSN 0302-9743
ISBN-10 3-540-25998-8 Springer Berlin Heidelberg New York
ISBN-13 978-3-540-25998-5 Springer Berlin Heidelberg New York

Springer is a part of Springer Science+Business Media

springeronline.com

Printed in Germany

Typesetting: Camera-ready by author, data conversion by Olgun Computergrafik
Printed on acid-free paper SPIN: 11427834 06/3142 5 4 3 2 1 0

Preface

Volume III of the Transactions on Rough Sets (TRS) introduces advances in the theory and application of rough sets. These advances have far-reaching implications in a number of research areas such as approximate reasoning, bioinformatics, computer science, data mining, engineering (especially, computer engineering and signal analysis), intelligent systems, knowledge discovery, pattern recognition, machine intelligence, and various forms of learning. This volume reveals the vigor, breadth and depth in research either directly or indirectly related to the rough sets theory introduced by Prof. Zdzisław Pawlak more than three decades ago. Evidence of this can be found in the seminal paper on data mining by Prof. Pawlak included in this volume. In addition, there are eight papers on the theory and application of rough sets as well as a presentation of a new version of the Rough Set Exploration System (RSES) tool set and an introduction to the Rough Set Database System (RSDS).

Prof. Pawlak has contributed a pioneering paper on data mining to this volume. In this paper, it is shown that information flow in a flow graph is governed by Bayes' rule with a deterministic rather than a probabilistic interpretation. A cardinal feature of this paper is that it is self-contained inasmuch as it not only introduces a new view of information flow but also provides an introduction to the basic concepts of flow graphs. The representation of information flow introduced in this paper makes it possible to study different relationships in data and establishes a basis for a new mathematical tool for data mining.

In addition to the paper by Prof. Pawlak, new developments in rough set theory are represented by five papers that investigate the validity, confidence and coverage of rules in approximation spaces (Anna Gomolińska), decision trees considered in the context of rough sets (Mikhail Ju. Moshkov), study of approximation spaces and information granulation (Andrzej Skowron, Roman Świniarski and Piotr Synak), a new interpretation of rough sets based on inverse probabilities and the foundations for a rough Bayesian model (Dominik Ślęzak), and formal concept analysis and rough set theory considered relative to topological approximations (Marcin Wolski). The theory papers in this volume are accompanied by four papers on applications of rough sets: knowledge extraction from electronic devices for power system substation event analysis and decision support (Ching-Lai Hor and Peter Crossley), processing of musical data using rough set methods, RSES and neural computing (Bożena Kostek, Piotr Szczuko, Paweł Żwan and Piotr Dalka), computational intelligence in bioinformatics (Sushmita Mitra), and an introduction to rough ethology, which is based on a biologically inspired study of collective behavior and reinforcement learning in intelligent systems using approximation spaces (James Peters).

This volume also celebrates two landmark events: a new version of RSES and the availability of a Rough Set Database System (RSDS). The introduction of a new version of the Rough Set Exploration System (RSES 2.2) is given in a paper by Jan G. Bazan and Marcin Szczuka. This paper gives an overview of the basic features of the new version of RSES: improved graphical user interface as well as production of decomposition trees and rules based on training samples. The decomposition tree and rules resulting from training can be used to classify unseen cases. The paper by Zbigniew Suraj and Piotr Grochowalski gives an overview of RSDS, which now includes over 1900 entries and over 800 authors. RSDS includes a number of useful utilities that make it possible for authors to update the database via the web, namely, append, search, download, statistics and help. In addition, RSDS provides access to biographies of researchers in the rough set community.

This issue of the TRS has been made possible thanks to the efforts of a great many generous persons and organizations. We express our thanks to the many anonymous reviewers for their heroic efforts in providing detailed reviews of the articles in this issue of the TRS. The editors and authors of this volume also extend an expression of gratitude to Alfred Hofmann, Ursula Barth, Christine Günther and the other LNCS staff members at Springer for their support in making this volume of the TRS possible. The Editors of this volume have been supported by the Ministry of Scientific Research and Information Technology of the Republic of Poland, Research Grant No. 3T11C00226, and the Natural Sciences and Engineering Research Council of Canada (NSERC), Research Grant No. 185986.

January 2005

James F. Peters
Andrzej Skowron

LNCS Transactions on Rough Sets

This journal subline has as its principal aim the fostering of professional exchanges between scientists and practitioners who are interested in the foundations and applications of rough sets. Topics include foundations and applications of rough sets as well as foundations and applications of hybrid methods combining rough sets with other approaches important for the development of intelligent systems.

The journal includes high-quality research articles accepted for publication on the basis of thorough peer reviews. Dissertations and monographs up to 250 pages that include new research results can also be considered as regular papers. Extended and revised versions of selected papers from conferences can also be included in regular or special issues of the journal.

Table of Contents

Regular Papers

Dissertations and Monographs

Flow Graphs and Data Mining

Zdzisław Pawlak[1,2]

[1] Institute for Theoretical and Applied Informatics,
Polish Academy of Sciences,
ul. Bałtycka 5, 44-100 Gliwice, Poland
[2] Warsaw School of Information Technology,
ul. Newelska 6, 01-447 Warsaw, Poland
zpw@ii.pw.edu.pl

Abstract. In this paper we propose a new approach to data mining and knowledge discovery based on information flow distribution in a flow graph. Flow graphs introduced in this paper are different from those proposed by Ford and Fulkerson for optimal flow analysis and they model flow distribution in a network rather than the optimal flow which is used for information flow examination in decision algorithms. It is revealed that flow in a flow graph is governed by Bayes' rule, but the rule has an entirely deterministic interpretation without referring to its probabilistic roots. Besides, a decision algorithm induced by a flow graph and dependency between conditions and decisions of decision rules is introduced and studied, which is used next to simplify decision algorithms.

Keywords: flow graph, data mining, knowledge discovery, decision algorithms.

Introduction

In this paper we propose a new approach to data analysis (mining) based on information flow distribution study in a flow graph.

Flow graphs introduced in this paper are different from those proposed by Ford and Fulkerson [4] for optimal flow analysis and they model rather flow *distribution* in a network, than the optimal flow.

The flow graphs considered in this paper are not meant to model physical media (e.g., water) flow analysis, but to model information flow examination in decision algorithms. To this end branches of a flow graph can be interpreted as decision rules. With every decision rule (i.e., branch) three coefficients are associated: the *strength, certainty* and *coverage factors.*

These coefficients have been used under different names in data mining (see, e.g., [14, 15]) but they were used first by Łukasiewicz [8] in his study of logic and probability.

This interpretation, in particular, leads to a new look at Bayes' theorem. Let us also observe that despite Bayes' rule fundamental role in statistical inference it has led to many philosophical discussions concerning its validity and meaning, and has caused much criticism [1, 3, 13].

J.F. Peters and A. Skowron (Eds.): Transactions on Rough Sets III, LNCS 3400, pp. 1–36, 2005.

This paper is a continuation of some of the authors' ideas presented in [10, 11], where the relationship between Bayes' rule and flow graphs has been introduced and studied (see also [6, 7]).

This paper consists of two parts. Part one introduces basic concepts of the proposed approach, i.e., flow graph and its fundamental properties. It is revealed that flow in a flow graph is governed by Bayes' rule, but the rule has an entirely deterministic interpretation that does not refer to its probabilistic roots. In addition, dependency of flow is defined and studied. This idea is based on the statistical concept of dependency but in our setting it has a deterministic meaning.

In part two many tutorial examples are given to illustrate how the introduced ideas work in data mining. These examples clearly show the difference between classical Bayesian inference methodology and the proposed one.

The presented ideas can be used, among others, as a new tool for data mining, and knowledge representation. Besides, the proposed approach throws new light on the concept of probability.

1 Flow Graphs

1.1 Overview

In this part the fundamental concepts of the proposed approach are defined and discussed. In particular flow graphs, certainty and coverage factors of branches of the flow graph are defined and studied. Next these coefficients are extended to paths and some classes of sub-graphs called connections. Further a notion of fusion of a flow graph is defined.

Further dependences of flow are introduced and examined. Finally, dependency factor (correlation coefficient) is defined.

Observe that in many cases the data flow order, represented in flow graphs, explicitly follows from the problem specification. However, in other cases the relevant order should be discovered from data. This latter issue will be discussed elsewhere.

1.2 Basic Concepts

A flow graph is a *directed, acyclic, finite* graph $G = (N, \mathcal{B}, \varphi)$, where N is a set of *nodes*, $\mathcal{B} \subseteq N \times N$ is a set of *directed branches*, $\varphi : \mathcal{B} \rightarrow R^+$ is a *flow function* and R^+ is the set of non-negative reals.

Input of a node $x \in N$ *is the set* $I(x) = \{y \in N : (y, x) \in \mathcal{B}\}$; *output* of a node $x \in N$ is defined by $O(x) = \{y \in N : (x, y) \in \mathcal{B}\}$.

We will also need the concept of *input* and *output* of a graph G, defined, respectively, as follows: $I(G) = \{x \in N : I(x) = \emptyset\}$, $O(G) = \{x \in N : O(x) = \emptyset\}$.

Inputs and outputs of G are *external nodes* of G; other nodes are *internal nodes* of G.

If $(x, y) \in \mathcal{B}$, then $\varphi(x, y)$ is a *throughflow* from x to y.

With every node x of a flow graph G we associate its *inflow*

$$\varphi_+(x) = \sum_{y \in I(x)} \varphi(y, x), \tag{1}$$

and *outflow*

$$\varphi_-(x) = \sum_{y \in O(x)} \varphi(x, y). \tag{2}$$

Similarly, we define an inflow and an outflow for the whole flow graph, which are defined by

$$\varphi_+(G) = \sum_{x \in I(G)} \varphi_-(x), \tag{3}$$

$$\varphi_-(G) = \sum_{x \in O(G)} \varphi_+(x). \tag{4}$$

We assume that for any internal node x we have $\varphi_+(x) = \varphi_-(x) = \varphi(x)$, where $\varphi(x)$ is a *throughflow* of node x.

Then, obviously, $\varphi_+(G) = \varphi_-(G) = \varphi(G)$, where $\varphi(G)$ is a *throughflow* of graph G.

The above formulas can be considered as *flow conservation equations* [4].

Example

We will illustrate the basic concepts of flow graphs by an example of a group of 1000 patients put to the test for certain drug effectiveness.

Assume that patients are grouped according to presence of the disease, age and test results, as shown in Fig. 1.

For example, $\varphi(x_1) = 600$ means that these are 600 patients suffering from the disease, $\varphi(y_1) = 570$ means that there are 570 old patients $\varphi(z_1) = 471$ means that 471 patients have a positive test result; $\varphi(x_1, y_1) = 450$ means that there are 450 old patients which suffer from disease etc.

Thus the flow graph gives clear insight into the relationship between different groups of patients.

Let us now explain the flow graph in more detail.

Nodes of the flow graph are depicted by circles, labeled by $x_1, x_2, y_1, y_2, y_3, z_1$, z_2. A branch (x, y) is denoted by an arrow from node x to y. For example, branch (x_1, z_1) is represented by an arrow from x_1 to z_1.

For example, inputs of node y_1 are nodes x_1 and x_2, outputs of node x_1 are nodes y_1, y_2 and y_3.

Inputs of the flow graph are nodes x_1 and x_2, whereas the outputs of the flow graph are nodes z_1 and z_2.

Nodes y_1, y_2 and y_3 are internal nodes of the flow graph. The throughflow of the branch (x_1, y_1) is $\varphi(x_1, y_1) = 450$. Inflow of node y_1 is $\varphi_+(y_1) = 450 + 120 = 570$. Outflow of node y_1 is $\varphi_-(y_1) = 399 + 171 = 570$. Inflow of the flow graph is $\varphi(x_1) + \varphi(x_2) = 600 + 400 = 1000$, and outflow of the flow graph is $\varphi(z_1) + \varphi(z_2) = 471 + 529 = 1000$.

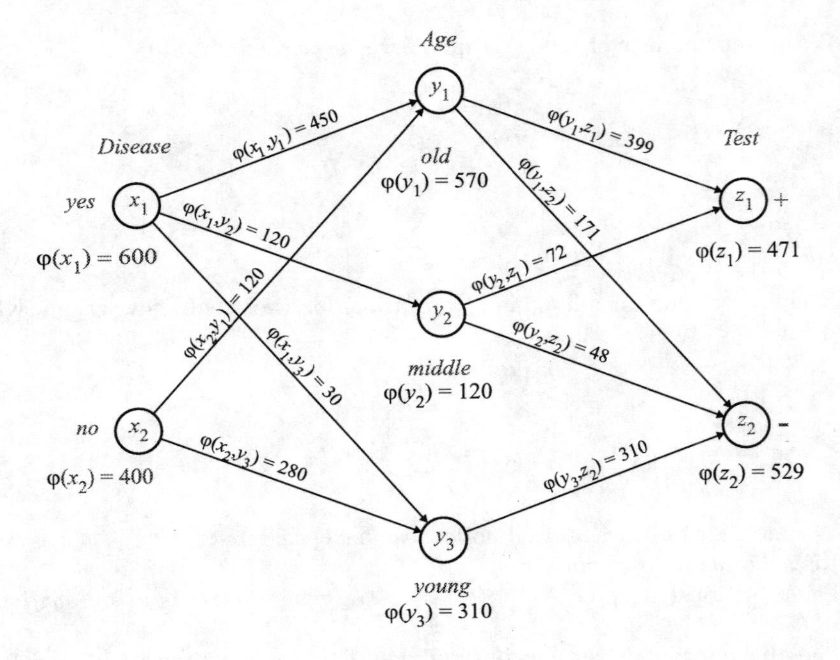

Fig. 1. Flow graph.

Throughflow of node y_1 is equal to $\varphi(y_1) = \varphi(x_1, y_1) + \varphi(x_2, y_1) = \varphi(y_1, z_1) + \varphi(y_2, z_2) = 570$. □

We will now define a *normalized flow graph.*

A normalized flow graph is a *directed, acyclic, finite* graph $G = (N, \mathcal{B}, \sigma)$, where N is a set of *nodes*, $\mathcal{B} \subseteq N \times N$ is a set of *directed branches* and $\sigma : \mathcal{B} \rightarrow \; < 0, 1 >$ is a *normalized flow* of (x, y) and

$$\sigma(x, y) = \frac{\varphi(x, y)}{\varphi(G)} \tag{5}$$

is a *strength* of (x, y). Obviously, $0 \leq \sigma(x, y) \leq 1$. The strength of the branch (multiplied by 100) expresses simply the percentage of a total flow through the branch.

In what follows we will use normalized flow graphs only, therefore by flow graphs we will understand normalized flow graphs, unless stated otherwise.

With every node x of a flow graph G we associate its *inflow* and *outflow* defined by

$$\sigma_+(x) = \frac{\varphi_+(x)}{\varphi(G)} = \sum_{y \in I(x)} \sigma(y, x), \tag{6}$$

$$\sigma_-(x) = \frac{\varphi_-(x)}{\varphi(G)} = \sum_{y \in O(x)} \sigma(x, y). \tag{7}$$

Obviously for any internal node x, we have $\sigma_+(x) = \sigma_-(x) = \sigma(x)$, where $\sigma(x)$ is a *normalized throughflow* of x.

Moreover, let

$$\sigma_+(G) = \frac{\varphi_+(G)}{\varphi(G)} = \sum_{x \in I(G)} \sigma_-(x), \tag{8}$$

$$\sigma_-(G) = \frac{\varphi_-(G)}{\varphi(G)} = \sum_{x \in O(G)} \sigma_+(x). \tag{9}$$

Obviously, $\sigma_+(G) = \sigma_-(G) = \sigma(G) = 1$.

Example (cont.) The normalized flow graph of the flow graph presented in Fig. 1 is given in Fig. 2.

In the flow graph, e.g., $\sigma(x_1) = 0.60$, that means that 60% of total inflow is associated with input x_1. The strength $\sigma(x_1, y_1) = 0.45$ means that 45% of total flow of x_1 flows through the branch (x_1, y_1) etc. □

Let $G = (N, \mathcal{B}, \sigma)$ be a flow graph. If we invert direction of all branches in G, then the resulting graph $G = (N, \mathcal{B}', \sigma')$ will be called an *inverted* graph of G. Of course, the inverted graph G' is also a flow graph and all inputs and outputs of G become inputs and outputs of G', respectively.

Example (cont.) The inverted flow graph of the flow graph from Fig. 2 is shown in Fig. 3. □

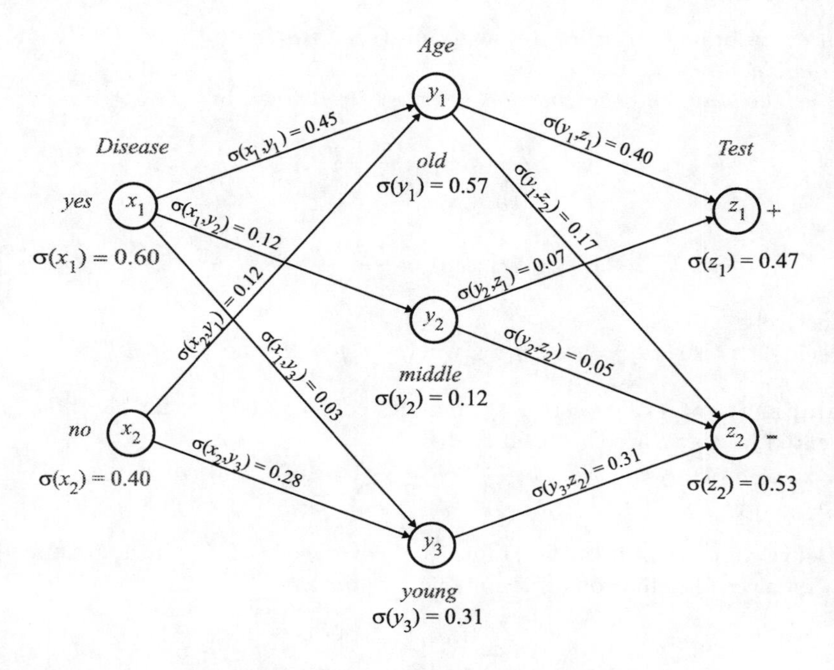

Fig. 2. Normalized flow graph.

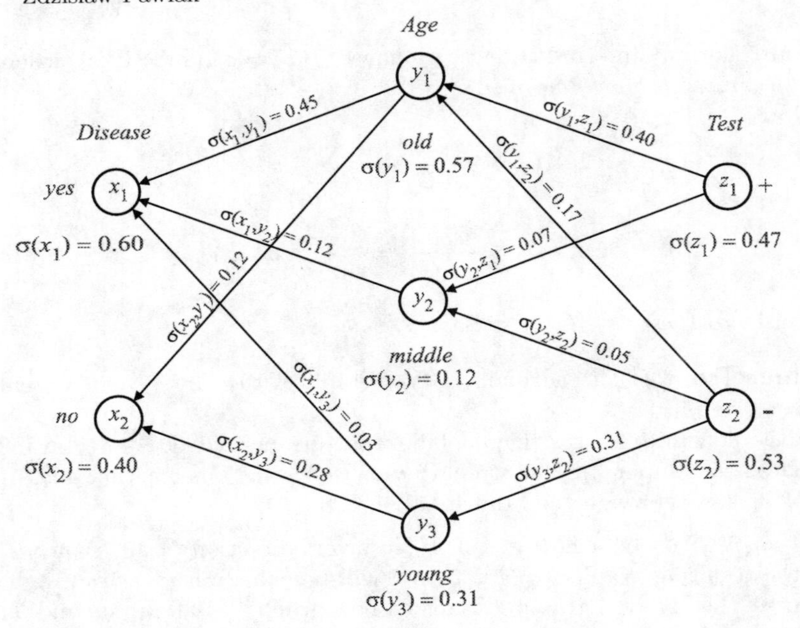

Fig. 3. Inverted flow graph.

1.3 Certainty and Coverage Factors

With every branch (x, y) of a flow graph G we associate the *certainty* and the *coverage factors.*

The *certainty* and the *coverage* of (x, y) are defined by

$$cer(x,y) = \frac{\sigma(x,y)}{\sigma(x)}, \tag{10}$$

and

$$cov(x,y) = \frac{\sigma(x,y)}{\sigma(y)}. \tag{11}$$

respectively.

Evidently, $cer(x, y) = cov(y, x)$, where $(x, y) \in \mathcal{B}$ and $(y, x) \in \mathcal{B}'$.

Example (cont.) The certainty and the coverage factors for the flow graph presented in Fig. 2 are shown in Fig. 4.

For example, $cer(x_1, y_1) = \frac{\sigma(x_1,y_1)}{\sigma(x_1)} = \frac{0.45}{0.60} = 0.75$, and $cov(x_1, y_1) = \frac{\sigma(x_1,y_1)}{\sigma(y_1)} = \frac{0.45}{0.57} \approx 0.79$. □

Below some properties of certainty and coverage factors, which are immediate consequences of definitions given above, are presented:

$$\sum_{y \in O(x)} cer(x,y) = 1, \tag{12}$$

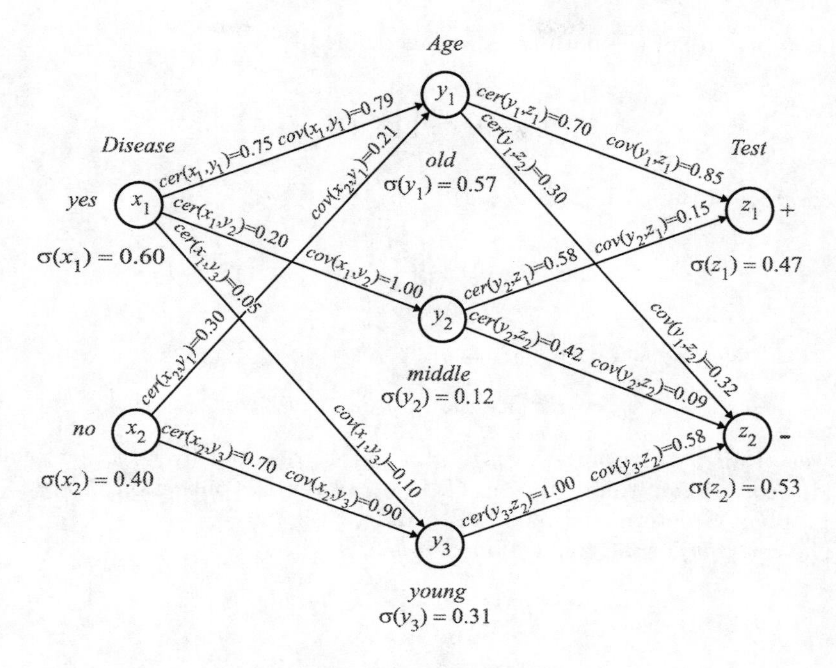

Fig. 4. Certainty and coverage.

$$\sum_{x \in I(y)} cov(x,y) = 1, \tag{13}$$

$$\sigma(x) = \sum_{y \in O(x)} cer(x,y)\sigma(x) = \sum_{y \in O(x)} \sigma(x,y), \tag{14}$$

$$\sigma(y) = \sum_{x \in I(y)} cov(x,y)\sigma(y) = \sum_{x \in I(y)} \sigma(x,y), \tag{15}$$

$$cer(x,y) = \frac{cov(x,y)\sigma(y)}{\sigma(x)}, \tag{16}$$

$$cov(x,y) = \frac{cer(x,y)\sigma(x)}{\sigma(y)}. \tag{17}$$

Obviously the above properties have a probabilistic flavor, e.g., equations (14) and (15) have a form of total probability theorem, whereas formulas (16) and (17) are Bayes' rules. However, these properties in our approach are interpreted in a deterministic way and they describe flow distribution among branches in the network.

1.4 Paths, Connections and Fusion

A (*directed*) *path* from x to y, $x \neq y$ in G is a sequence of nodes $x_1, \ldots, x_n$ such that $x_1 = x$, $x_n = y$ and $(x_i, x_{i+1}) \in \mathcal{B}$ for every i, $1 \leq i \leq n-1$. A path from x to y is denoted by $[x \ldots y]$.

The *certainty* of the path $[x_1 \ldots x_n]$ is defined by

$$cer[x_1 \ldots x_n] = \prod_{i=1}^{n-1} cer(x_i, x_{i+1}), \tag{18}$$

the *coverage* of the path $[x_1 \ldots x_n]$ is

$$cov[x_1 \ldots x_n] = \prod_{i=1}^{n-1} cov(x_i, x_{i+1}), \tag{19}$$

and the *strength* of the path $[x_1 \ldots x_n]$ is

$$\sigma[x_1 \ldots x_n] = \sigma(x_1) cer[x_1 \ldots x_n] = \sigma(x_n) cov[x_1 \ldots x_n]. \tag{20}$$

The set of all paths from x to y $(x \neq y)$ in G, denoted by $< x, y >$, will be called a *connection* from x to y in G. In other words, connection $< x, y >$ is a sub-graph of G determined by nodes x and y.

The *certainty* of the connection $< x, y >$ is

$$cer < x, y > = \sum_{[x \ldots y] \in < x, y >} cer[x \ldots y], \tag{21}$$

the *coverage* of the connection $< x, y >$ is

$$cov < x, y > = \sum_{[x \ldots y] \in < x, y >} cov[x \ldots y], \tag{22}$$

and the *strength* of the connection $< x, y >$ is

$$\sigma < x, y > = \sum_{[x \ldots y] \in < x, y >} \sigma[x \ldots y] =$$
$$= \sigma(x) cer < x, y > = \sigma(y) cov < x, y > . \tag{23}$$

If we substitute simultaneously any sub-graph $< x, y >$ of a given flow graph G, where x and y are input and output nodes of G respectively, by a single branch (x, y) such that $\sigma(x, y) = \sigma < x, y >$, then in the resulting graph G', called the *fusion* of G, we have $cer(x, y) = cer < x, y >$, $cov(x, y) = cov < x, y >$ and $\sigma(G) = \sigma(G')$.

Example (cont.) In the flow graph presented in Fig. 3 for the path $p = [x_1, y_1, z_1]$ we have $cer(p) = 0.75 \times 0.70 \approx 0.53$, $cov(p) = 0.85 \times 0.79 \approx 0.67$.

The connection $< x_1, z_1 >$ in the flow graph consists of paths $[x_1, y_1, z_1]$ and $[x_1, y_2, z_1]$. This connection is shown in Fig. 5 by bold lines.

For this connection we have $cer < x_1, z_1 > = 0.75 \times 0.70 + 0.20 \times 0.60 \approx 0.65$; $cov < x_1, z_1 > = 0.85 \times 0.79 + 0.15 \times 1.00 \approx 0.82$.

The strength of the connection x_1, z_1 is $0.68 \times 0.60 \approx 0.85 \times 0.47 \approx 0.40$. Connections $< x_1, z_2 >$, $< x_2, z_1 >$, and $< x_2, z_2 >$ are presented in Fig. 6, Fig. 7 and Fig. 8, respectively. □

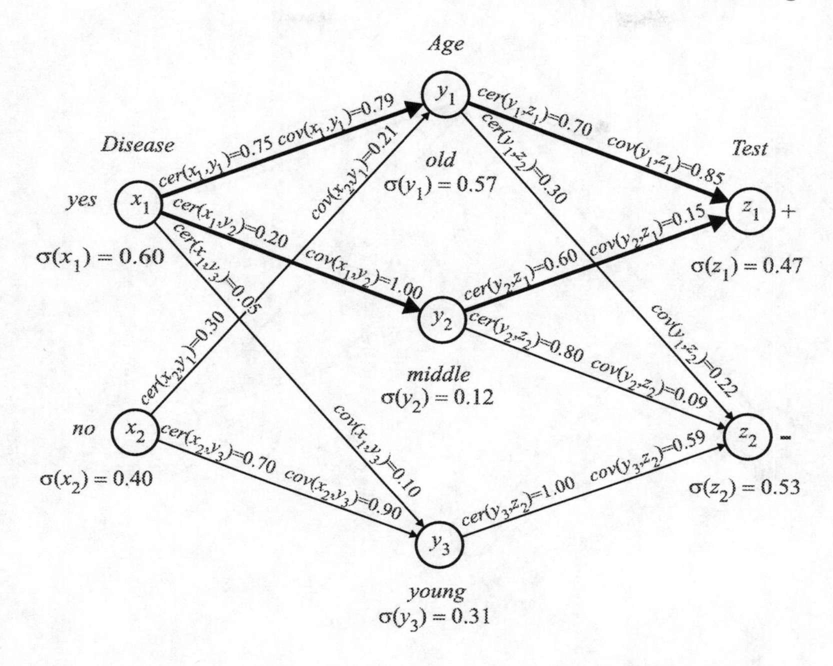

Fig. 5. Connection $< x_1, z_1 >$.

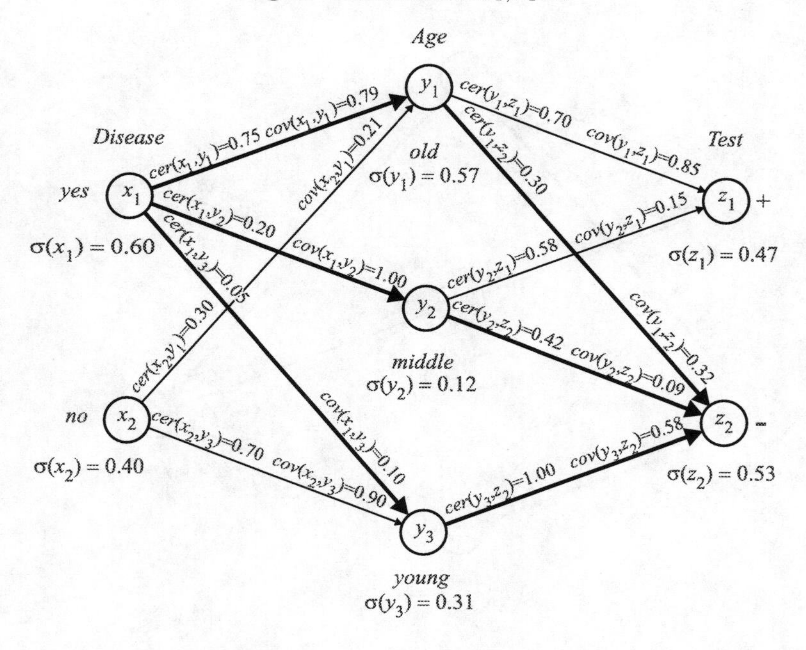

Fig. 6. Connection $< x_1, z_2 >$.

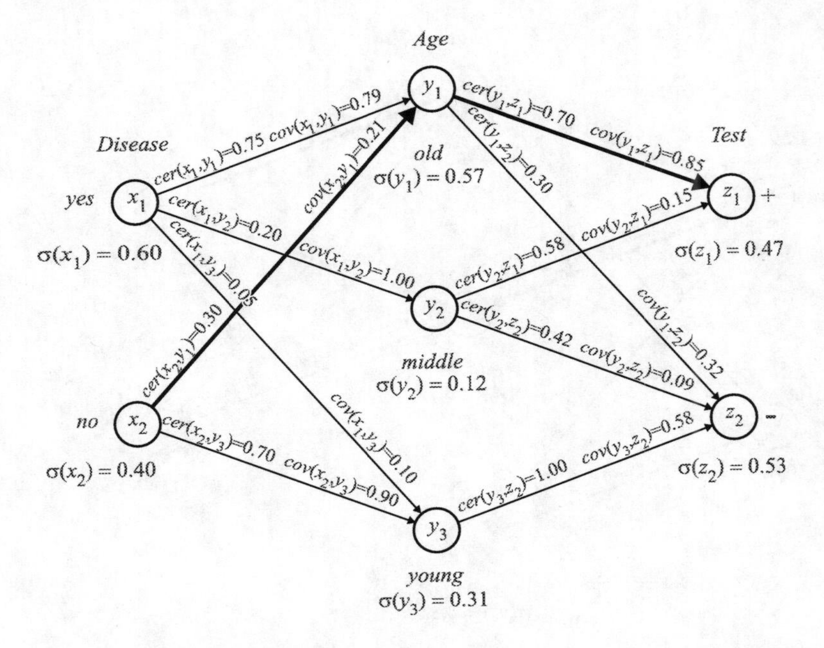

Fig. 7. Connection $< x_2, z_1 >$.

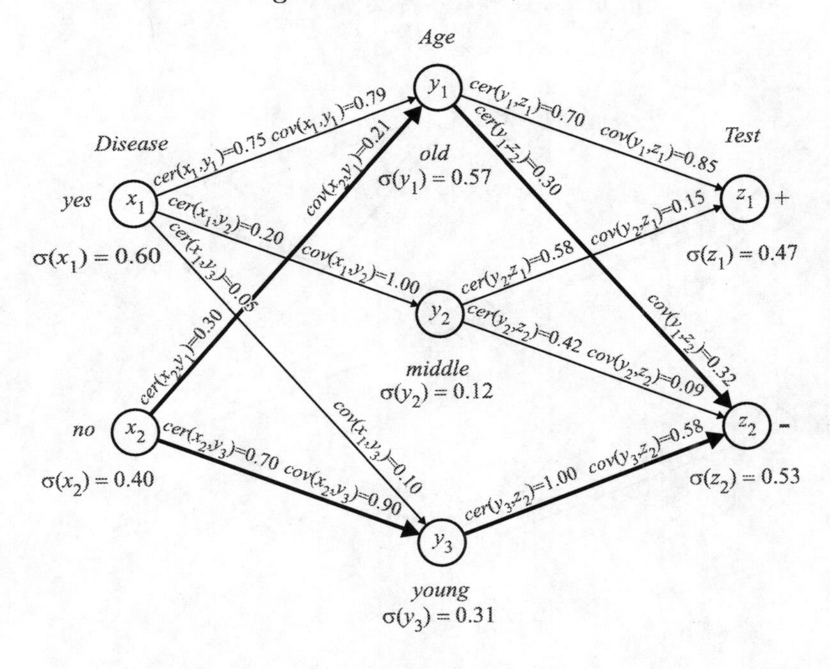

Fig. 8. Connection $< x_2, z_2 >$.

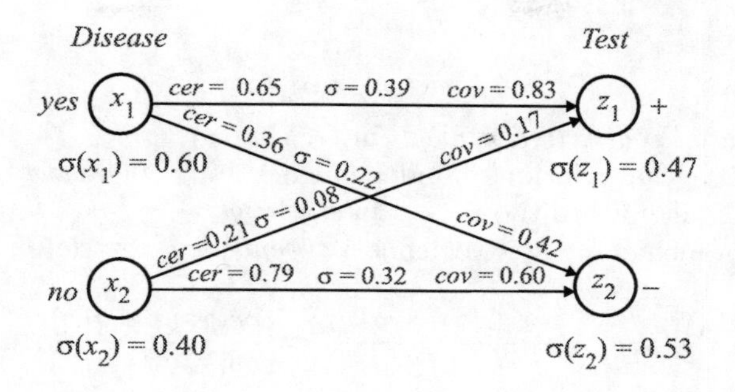

Fig. 9. Fusion of the flow graph.

Example (cont.) The fusion of the flow graph shown in Fig. 3 is given in Fig. 9.

The fusion of a flow graph gives information about the flow distribution between input and output of the flow graph, i.e., it leads to the following conclusions:

- if the disease is present then the test result is positive with certainty 0.65,
- if the disease is absent then the test result is negative with certainty 0.79.

Explanation of the test results is as follows:

- if the test result is positive then the disease is present with certainty 0.83,
- if the test result is negative then the disease is absent with certainty 0.60.

□

1.5 Dependences in Flow Graphs

Let x and y be nodes in a flow graph $G = (N, \mathcal{B}, \sigma)$, such that $(x, y) \in \mathcal{B}$. Nodes x and y are *independent* in G if

$$\sigma(x, y) = \sigma(x)\sigma(y). \tag{24}$$

From (24) we get

$$\frac{\sigma(x, y)}{\sigma(x)} = cer(x, y) = \sigma(y), \tag{25}$$

and

$$\frac{\sigma(x, y)}{\sigma(y)} = cov(x, y) = \sigma(x). \tag{26}$$

If

$$cer(x, y) > \sigma(y), \tag{27}$$

or

$$cov(x, y) > \sigma(x), \tag{28}$$

then x and y are *positively dependent* on x in G.

Similarly, if

$$cer(x, y) < \sigma(y), \tag{29}$$

or

$$cov(x, y) < \sigma(x), \tag{30}$$

then x and y are *negatively dependent* in G.

Let us observe that relations of independency and dependences are symmetric ones, and are analogous to those used in statistics.

For every branch $(x, y) \in \mathcal{B}$ we define a *dependency* (*correlation*) *factor* $\eta(x, y)$ defined by

$$\eta(x, y) = \frac{cer(x, y) - \sigma(y)}{cer(x, y) + \sigma(y)} = \frac{cov(x, y) - \sigma(x)}{cov(x, y) + \sigma(x)}. \tag{31}$$

Obviously $-1 \leq \eta(x, y) \leq 1$; $\eta(x, y) = 0$ if and only if $cer(x, y) = \sigma(y)$ and $cov(x, y) = \sigma(x)$; $\eta(x, y) = -1$ if and only if $cer(x, y) = cov(x, y) = 0$; $\eta(x, y) = 1$ if and only if $\sigma(y) = \sigma(x) = 0$.

It is easy to check that if $\eta(x, y) = 0$, then x and y are independent, if $-1 \leq \eta(x, y) < 0$ then x and y are negatively dependent and if $0 < \eta(x, y) \leq 1$ then x and y are positively dependent. Thus the dependency factor expresses a degree of dependency, and can be seen as a counterpart of the correlation coefficient used in statistics.

Example (cont.) Dependency factors for the flow graph shown in Fig. 9 are given in Fig. 10.

Thus, there is a positive dependency between the presence of the disease and the positive test result as well as between absence of the disease and negative test result. However, there is a much stronger negative dependency between presence of the disease and negative test result or similarly – between absence of the disease and positive left test result. More specifically:

- there is slight positive correlation between presence of the disease and positive test result ($\eta = 0.16$),
- there is low positive correlation between absence of the disease and negative test result ($\eta = 0.20$),
- there a negative correlation between presence of the disease and negative test result ($\eta = -0.19$),
- there is high negative correlation between absence of the disease and positive test result ($\eta = -0.38$). □

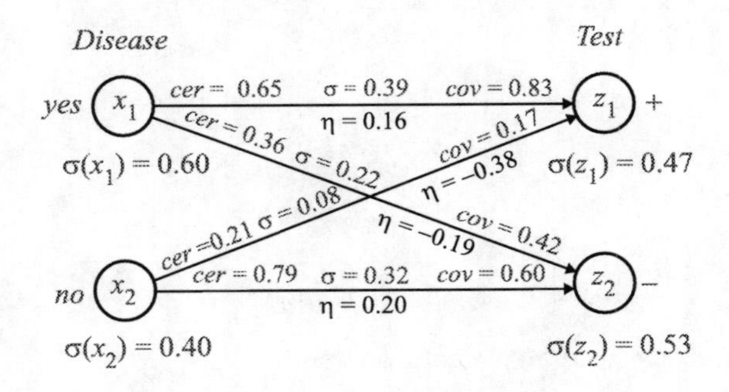

Fig. 10. Fusion of the flow graph.

1.6 Flow Graph and Decision Algorithms

Flow graphs can be interpreted as decision algorithms [5].

Let us assume that the set of nodes of a flow graph is interpreted as a set of logical formulas. The formulas are understood as propositional functions and if x is a formula, then $\sigma(x)$ is to be interpreted as a truth value of the formula. Let us observe that the truth values are numbers from the closed interval $[0, 1]$, i.e., $0 \leq \sigma(x) \leq 1$.

According to [3] these truth values can be also interpreted as probabilities. Thus $\sigma(x)$ can be understood as flow distribution ratio (percentage), truth value or probability. We will stick to the first interpretation.

With every branch (x, y) we associate a decision rule $x \rightarrow y$, read *if x then y*; x will be referred to as *condition*, whereas y – *decision* of the rule. Such a rule is characterized by three numbers, $\sigma(x, y)$, $cer(x, y)$ and $cov(x, y)$.

Let us observe that the inverted flow graph gives reasons for decisions.

Every path $[x_1 \dots x_n]$ determines a sequence of decision rules $x_1 \rightarrow x_2$, $x_2 \rightarrow x_3, \dots, x_{n-1} \rightarrow x_n$.

From previous considerations it follows that this sequence of decision rules can be interpreted as a single decision rule $x_1 x_2 \dots x_{n-1} \rightarrow x_n$, in short $x^* \rightarrow x_n$, where $x^* = x_1 x_2 \dots x_{n-1}$, characterized by

$$cer(x^*, x_n) = \frac{\sigma(x^*, x_n)}{\sigma(x^*)}, \tag{32}$$

$$cov(x^*, x_n) = \frac{\sigma(x^*, x_n)}{\sigma(x_n)}, \tag{33}$$

and

$$\sigma(x^*, x_n) = \sigma(x_1, \dots, x_{n-1}, x_n),\ \sigma(x^*) = \sigma(x_1, \dots, x_{n-1}). \tag{34}$$

From (32) we have

$$cer(x^*, x_n) = \frac{cer[x_1, \dots, x_{n-1}, x_n]}{cer[x_1, \dots, x_{n-1}]}.$$

The set of all decision rules $x_{i_1} x_{i_2} \dots x_{i_{n-1}} \rightarrow x_{i_n}$ associated with all paths $[x_{i_1} \dots x_{i_n}]$ such that x_{i_1} and x_{i_n} are input and output of the graph respectively will be called a *decision algorithm* induced by the flow graph.

If $x \rightarrow y$ is a decision rule, then we say that the condition and decision of the decision rule are independent if x and y are independent, otherwise the condition and decision of the decision rule are dependent (positively or negatively).

To measure the degree of dependency between the condition and decision of the decision rule $x \rightarrow y$, we can use the dependency factor $\eta(x, y)$.

Let us observe that if the conditions and decisions of a decision rule $x \rightarrow y$ are independent, then the decision rule is, in certain sense, useless, because such a decision rule indicates that there is no relationship between conditions and decisions and the decision can be eliminated from the decision algorithm.

On the other hand, the most important decision rules are those having the highest dependency factor and strength, for they indicate a strong relationship in substantial portion of the data. This property can be used to simplify the decision algorithms, because we can eliminate less relevant decision rules from the algorithm, at the cost of its lower classification power.

With every subset of decision rules $\delta_1, \ldots, \delta_n$ of the decision algorithm we can associate its strength equal to the sum of strengths of the decision rules, i.e., $\sum_{i=1}^{n} \sigma(\delta_i)$, which can be used as a measure of the classification power of the algorithm.

Example (cont.) The decision algorithm induced by the flow graph shown in Fig. 4 is given in the table:

	certainty	coverage	strength
$x_1, y_1 \to z_1$	0.71	0.67	0.32
$x_1, y_1 \to z_2$	0.31	0.25	0.14
$x_1, y_2 \to z_1$	0.58	0.15	0.07
$x_1, y_2 \to z_2$	0.42	0.09	0.05
$x_1, y_3 \to z_2$	1.00	0.06	0.03
$x_2, y_1 \to z_1$	0.40	0.18	0.08
$x_2, y_1 \to z_2$	0.20	0.01	0.04
$x_2, y_3 \to z_2$	1.00	0.53	0.28

The corresponding flow graph is presented in Fig. 11.

From the flow graph we can see, e.g., that 71% ill and old patients have a positive test result, whereas 100% young healthy patients have a negative test result. From the inverse flow graph we can conclude that positive test result have mostly (67%) ill and old patients and negative test result display mostly (53%) young healthy patients.

Consequently, for the decision rule $x_1, y_1 \to z_1$ (and the inverse decision rule $z_1 \to x_1, y_1$) we have the dependency factor $\eta \approx 0.19$,whereas for the decision rule $x_2, y_3 \to z_2$ (and its inverse decision rule), we have $\eta \approx 0.31$.

That means that the relationship between young healthy patients and negative test results is more substantial then – between ill old patients and positive test result.

The strength of the corresponding decision rules is 0.32 and 0.28, respectively. Thus they are rather strong decision rules. As the result if we drop all remaining decision rules from the decision algorithm, we obtain a very simple decision algorithm consisting of two decision rules, with strength $0.32 + 0.28 = 0.60$. This means that two decision rules instead eight suffices for previous proper classification of initial data in 60% cases. Adding the next strongest decision rule $x_1, y_2 \to z_2$ with $\sigma = 0.14$, we get a decision algorithm with strength $0.60 + 0.14 = 0.74$, which can classify properly of 74% cases.

1.7 Flow Graphs and Rough Sets

In this section we show that some flow graphs can be treated as representations of approximation spaces. To explain this let us consider an example based on

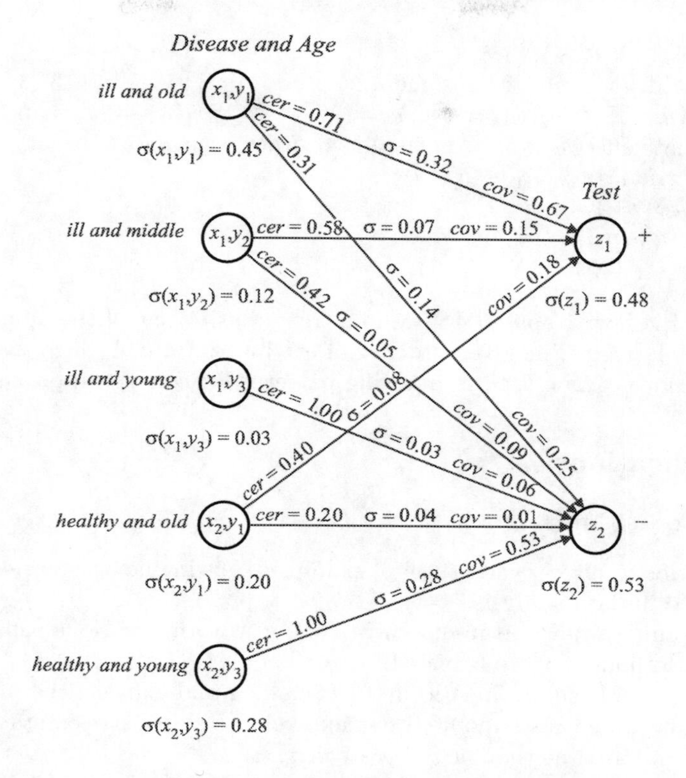

Fig. 11. Flow graph for the decision algorithm.

approximation spaces for information systems. Let us consider an information system $IS = (U, A)$ where U is the universe of objects and A is the set of attributes of the form $a : U \longrightarrow V_a$ [9]. Any such information system defines an approximation space $AS = (U, \mathcal{R}, \nu)$ [12] where $\mathcal{R}$ is a family of sets generated by descriptors over A, i.e.,

$$\mathcal{R} = \{X \subseteq U : X = \{u \in U : a(u) = v\} \text{ for some } a \in A,\ v \in V_a\} \tag{35}$$

and $\nu : P(U) \times P(U) \longrightarrow [0, 1]$ is the standard rough inclusion function defined by

$$\nu(X, Y) = \begin{cases} \frac{|X \cap Y|}{|X|} & \text{if } X \neq \emptyset \\ 1 & \text{if } X = \emptyset. \end{cases} \tag{36}$$

Hence, $\nu(X, Y)$ is a degree to which X is included in Y, for any $X, Y \in \mathcal{R}$.

Assume that $A = \{a_1, \ldots, a_m\}$ and $a_1 < \ldots < a_m$, i.e., A is linearly ordered by $<$.

Then one can construct a flow graph $G(AS) = (N, \mathcal{B}, \varphi)$ representing the approximation space $AS = (U, \mathcal{R}, \nu)$ where

1. $N = \{n_X : X \in \mathcal{R}\}$;
2. $n_X \mathcal{B} n_Y$ if and only if for some $a_i \in A, a_j \in A, v \in V_{a_i}, v' \in V_{a_j}$ we have $X = \{u \in U : a_i(u) = v\}$, $Y = \{u \in U : a_j(u) = v'\}$, and a_j is the immediate successor of a_i in the linear order $a_1, \ldots, a_m$;
3. For any nodes $n_X, n_Y \in N$:
   (a) $\varphi(n_X, n_Y) = |X \cap Y|/|U|$;
   (b) $cer(n_X, n_Y) = |X \cap Y|/|X|$;
   (c) $cov(n_X, n_Y) = |X \cap Y|/|Y|$.

Hence, the flow graph $G(AS)$ can be treated as a view of the approximation space AS relative to the given order $<$ of attributes from A. Such views as well as their fusions can be used in inducing patterns for concept approximations.

2 Applications

2.1 Introduction

In this section we give several tutorial examples showing how the presented ideas can be used in data analysis.

The examples have been chosen in such a way that various aspects of the proposed methodology are revealed.

In the example shown in section 2.2 (Smoking and Cancer) the probabilistic nature of data analysis is pointed out and relationship between statistical and flow diagram based methodology is revealed.

In the next example discussed in Section 2.3 (Hair, Eyes and Nationality) relationship between different sets of data is examined and the result need not to be necessarily interpreted in probabilistic terms.

Similar remark is valid for the next two examples.

Example given in Section 2.9 (Paint Demand and Supply) has entirely deterministic character and describes simply proportion between various ingredients.

In the remaining examples probabilistic character of data is rather immaterial and results can be understood as relationship between proportion of various features in the corresponding data sets.

Observe also that the numerical values of discussed coefficients may not satisfy exactly formulas given in the first chapter due to the round off errors in the computations.

2.2 Smoking and Cancer

In this section we show an application of the proposed methodology on the example taken from [5].

In Table 1 data concerning 60 people who do or do not smoke and do or do not have cancer are shown.

In Fig. 12 data given in Table 1 are presented as flow graph.

Normalized flow graph for the flow graph given in Fig. 12 is shown in Fig. 13.

From the flow graph we arrive at the following conclusions:

Table 1. Smoking and Cancer.

	Not smoke	Smoke	Total
Not cancer	40	10	50
Cancer	7	3	10
Total	47	13	60

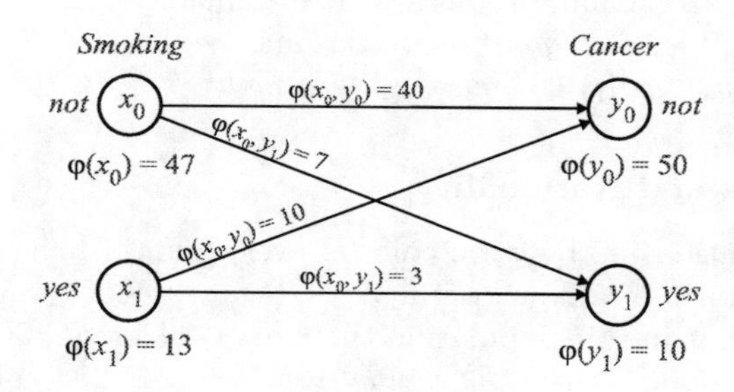

Fig. 12. Flow graph for Table 1.

- 85% non-smoking persons do not have cancer ($cer(x_0, y_0) = 40/47 \approx 0.85$),
- 15% non-smoking persons do have cancer ($cer(x_0, y_1) = 7/47 \approx 0.15$),
- 77% smoking persons do not have cancer ($cer(x_1, y_0) = 10/13 \approx 0.77$),
- 23% smoking persons do have cancer ($cer(x_1, y_1) = 3/13 \approx 0.23$).

From the flow graph we get the following reason for having or not cancer:

- 80% persons having not cancer do not smoke ($cov(x_0, y_0) = 4/5 = 0.80$),
- 20% persons having not cancer do smoke ($cov(x_1, y_0) = 1/5 = 0.20$),
- 70% persons having cancer do not smoke ($cov(x_0, y_1) = 7/10 = 0.70$),
- 30% persons having cancer do smoke ($cov(x_1, y_1) = 3/10 = 0.30$).

That means that not smoking persons mostly do not have cancer but smoking is mostly not associated with having cancer.

From the inverse flow graph, we conclude that the reason for having not cancer is not smoking but having cancer is not associated with smoking.

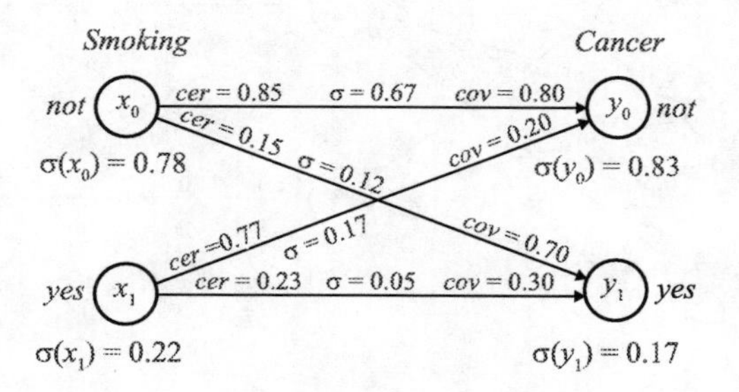

Fig. 13. Normalized flow graph for Table 1.

For the flow graph we have the following dependences: $\eta(x_0, y_0) = 0.01$, $\eta(x_0, y_0) = -0.06$, $\eta(x_1, y_0) = -0.09$ and $\eta(x_1, y_1) = 0.15$. These means that there is slight positive dependency between x_0 and y_0 and much stronger positive dependency between x_1 and y_1; x_0 and y_1 are negatively related and so are x_1 and y_0.

Let us also observe that in statistical terminology $\sigma(x_0), \sigma(x_1)$ are *priors*, $\sigma(x_0, y_0), \ldots, \sigma(x_1, y_1)$ are *joint distributions*, $cov(x_0, y_0), \ldots, cov(x_1, y_1)$ are *posteriors* and $\sigma(y_0)$, $\sigma(y_1)$ are *marginal probabilities.*

2.3 Hair, Eyes and Nationality

In Fig. 14 the relationship between color of eyes, color of hair and nationality is presented in the form of a flow graph.

That means that in this population there are 60% blond, 30% dark and 10% red haired; 80% blond haired have blue eyes whereas 20% blond haired have hazel eyes, etc. Similarly we see from the flow graph that 20% persons having blue eyes are Italian, and 80% persons with blue eyes are Swede, etc.

First let us compute "flow" in the graph and the result is shown in Fig. 15.

We can see from the flow graph that the strongest decision rules showing the relationship between color of hair and eyes are $x_1 \rightarrow y_1$ ($\sigma = 0.48$) and $x_2 \rightarrow y_2$ ($\sigma = 0.27$) with $\eta = 0.14$ and $\eta = 0.38$ respectively. These two decision rules have strength $0.48 + 0.27 = 0.75$. The dependency factors of these decision rules indicate that the relationship between dark hair and hazel eyes is much stronger then the dependency between blond hair and blue eyes.

Similarly the strongest decision rules showing the relationship between color of eyes and nationality are $y_1 \rightarrow z_2$ ($\sigma = 0.48$) and $x_2 \rightarrow z_1$ ($\sigma = 0.36$) with $\eta = 0.21$ and $\eta = 0.30$, respectively and strength 0.84. This shows that hazel eyes are more characteristic for Italians, then blue eyes for Swede.

The relationship between color of hair and nationality is computed using the concept of fusion and the result is shown in Fig. 16.

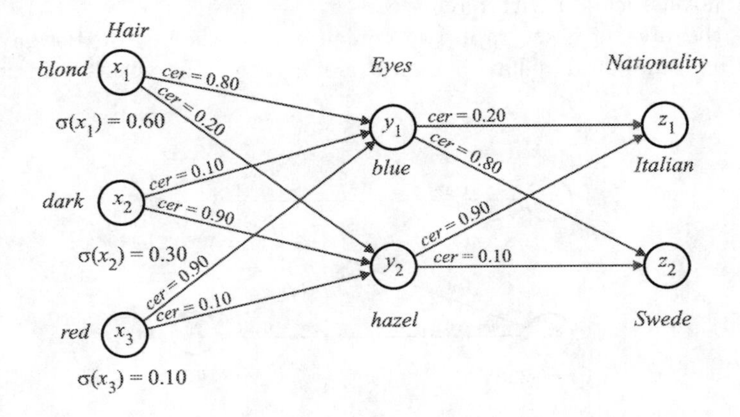

Fig. 14. Initial data.

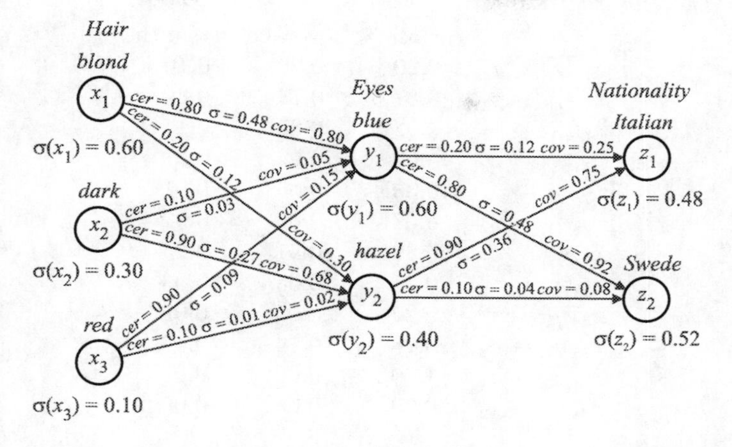

Fig. 15. Relationship between color of eyes, color of hair and nationality.

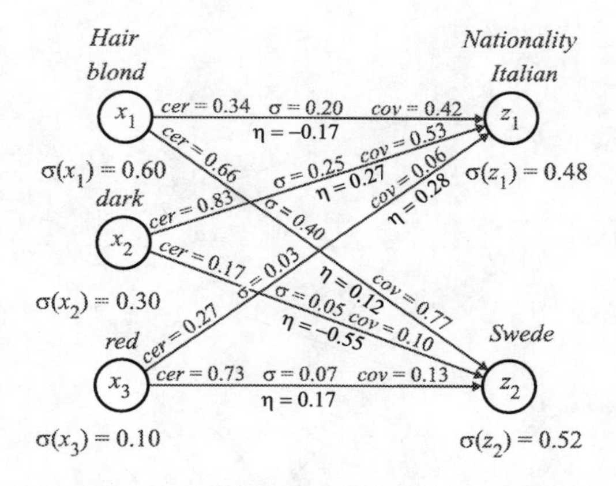

Fig. 16. Fusion of color of hair and nationality.

In this flow graph also degree of independence is given. We can see from the dependency coefficients that, e.g., there is a relatively strong negative dependency between dark hair and Swede ($\eta = -0.55$) blond hair and Italian ($\eta = -0.17$), but there is a positive dependency between dark hair and Italian ($\eta = 0.25$), however the first dependency has very low strength ($\sigma = 0.05$), whereas the second one has much higher strength ($\sigma = 0.20$). This means that in the first case 5% of the population display this property in contrast to the second case where 20% of the population support the dependency.

Let us also observe that the three decision rules $x_1 \to z_1$ ($\sigma = 0.20$), $x_2 \to z_1$ ($\sigma = 0.25$) and $x_1 \to z_1$ ($\sigma = 0.40$) have very high strength 0.85.

The decision algorithm induced by the flow graph shown in Fig. 15 is presented in table below:

	certainty	coverage	strength
$x_1, y_1 \to z_1$	0.20	0.20	0.10
$x_1, y_1 \to z_2$	0.81	0.74	0.39
$x_1, y_2 \to z_1$	0.92	0.23	0.11
$x_1, y_2 \to z_2$	0.08	0.02	0.01
$x_2, y_1 \to z_1$	0.33	0.05	0.01
$x_2, y_1 \to z_2$	0.67	0.04	0.02
$x_2, y_2 \to z_1$	0.89	0.51	0.24
$x_2, y_2 \to z_2$	0.11	0.05	0.03
$x_3, y_1 \to z_1$	0.22	0.05	0.02
$x_3, y_1 \to z_2$	0.78	0.14	0.07
$x_3, y_2 \to z_1$	1.00	0.02	0.01
$x_3, y_2 \to z_2$	0.00	0.00	0.00

Flow graph associated with the decision algorithm is shown in Fig. 17.

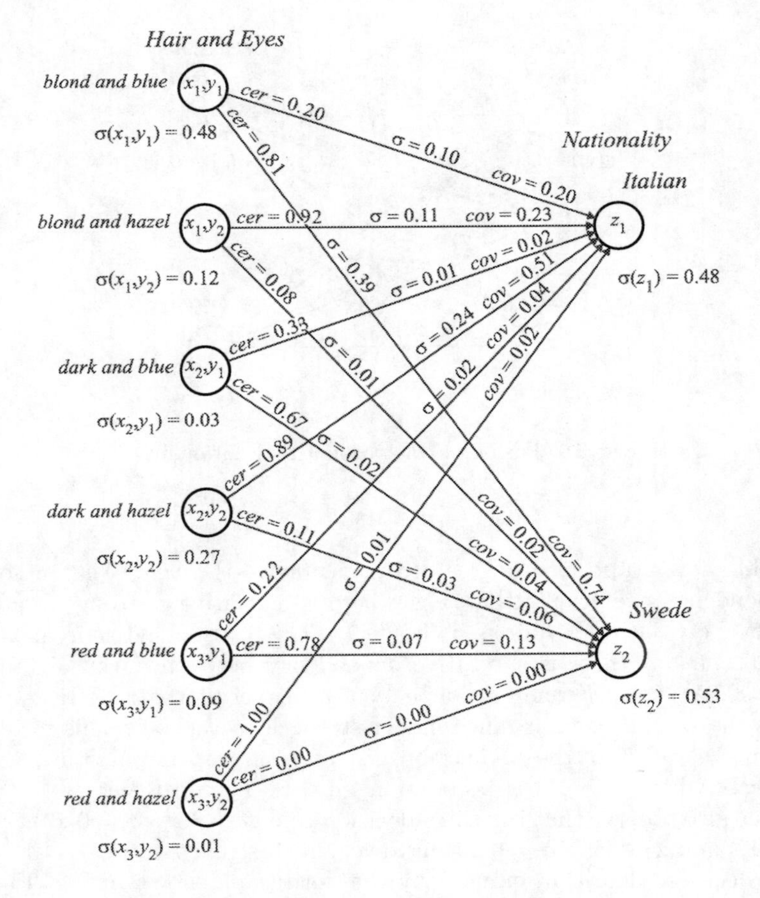

Fig. 17. Hair, eyes and nationality.

One can conclude from the flow graph that the most significant decision rules are $(x_1, y_1) \to z_2$ ($\sigma = 0.39$) and $(x_2, y_2) \to z_1$ ($\sigma = 0.24$) with the corresponding dependency factors $\eta = 0.10$ and 0.26, and strength $0.39 + 0.24 = 0.63$. That means that two decision rules enable us to classify correctly of the 63% cases.

Dependency factors indicate that dark hair and hazel eyes are more characteristic for Italians then blond hair and blue eyes for Swede.

Let us also mention that if the data are representative for a larger universe (form a proper sample of the universe), then the results can be also considered as promising hypotheses in this extended world. That is, we employ in this case inductive reasoning, i.e., induce from properties of a part of the universe properties of the whole universe.

2.4 Production Quality

Consider three industrial plants x_1, x_2 and x_3 producing three kinds of appliances y_1, y_2 and y_3. Some of the produced appliances are defective. The situation is presented in Fig. 18.

We want to find the relationship between plant and quality of products.

First we compute flow in the flow graph and the result is shown in Fig. 19.

Similarly as in the previous example we can find from the flow graph that the most significant decision rules describing the relationship between plant and product are $x_2 \to y_2$, $x_3 \to y_2$ and $x_3 \to y_3$ having together strength $0.18 + 0.10 + 0.40 = 0.68$, whereas the relationship between products and quality is best described by the decision rules $y_2 \to z_1$, $y_3 \to z_1$ and $y_3 \to z_2$ with strength $0.25 + 0.44 + 0.19 = 0.88$

In order to find relationship between producers and quality of their products, we compute the corresponding fusion and the result is given in Fig. 20. It is seen from the dependency coefficient that all decision rules except the rule $x_2 \to z_2$ have rather low rather low dependency factor. Because $\eta(x_2, z_2) = -0.17$ plant

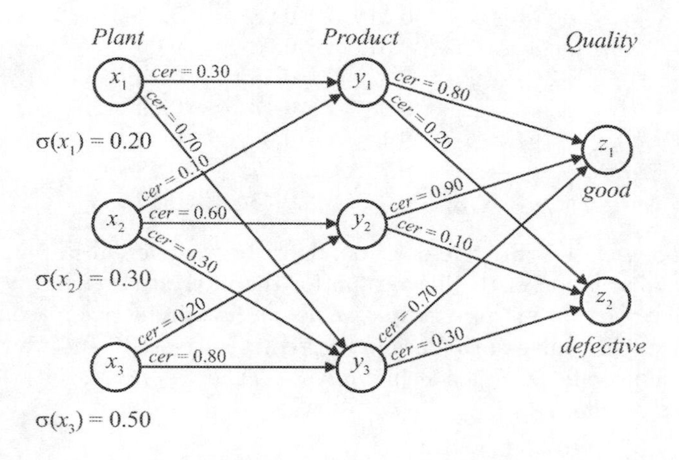

Fig. 18. Relationship between plant, product, and quality.

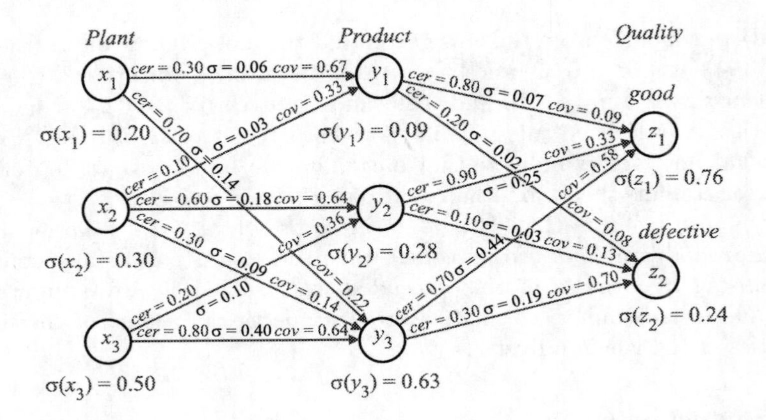

Fig. 19. Relationship between plants, products and quality.

x_2 produces the lowest ratio of defective products but the corresponding decision rule has a rather weak strength ($\sigma = 0.05$); that is, it has 5% support. In other words it is the best of all plants. It is interesting to observe that the four strongest decision rules provide 90% accuracy of classification.

The corresponding decision algorithm is shown below:

	certainty	coverage	strength
$x_1, y_1 \rightarrow z_1$	0.83	0.06	0.05
$x_1, y_1 \rightarrow z_2$	0.17	0.05	0.01
$x_1, y_3 \rightarrow z_1$	0.71	0.13	0.10
$x_1, y_3 \rightarrow z_2$	0.29	0.15	0.04
$x_2, y_1 \rightarrow z_1$	0.00	0.21	0.00
$x_2, y_1 \rightarrow z_2$	0.00	0.03	0.00
$x_2, y_2 \rightarrow z_1$	0.89	0.21	0.16
$x_2, y_2 \rightarrow z_2$	0.11	0.08	0.02
$x_2, y_3 \rightarrow z_1$	0.67	0.08	0.06
$x_2, y_3 \rightarrow z_2$	0.33	0.15	0.03
$x_3, y_2 \rightarrow z_1$	0.90	0.13	0.09
$x_3, y_2 \rightarrow z_2$	0.10	0.05	0.01
$x_3, y_3 \rightarrow z_1$	0.70	0.37	0.28
$x_3, y_3 \rightarrow z_2$	0.30	0.45	0.12

Flow graph associated with the decision algorithm is given in Fig. 21.

We leave discussion of the flow graph for the interested reader.

Let us observe that in this example we do not have inductive reasoning, whatsoever. The world (universe) we are interested in is "closed" and we search only for relationships valid in this specific universe. There is no reason to generalize the obtained results.

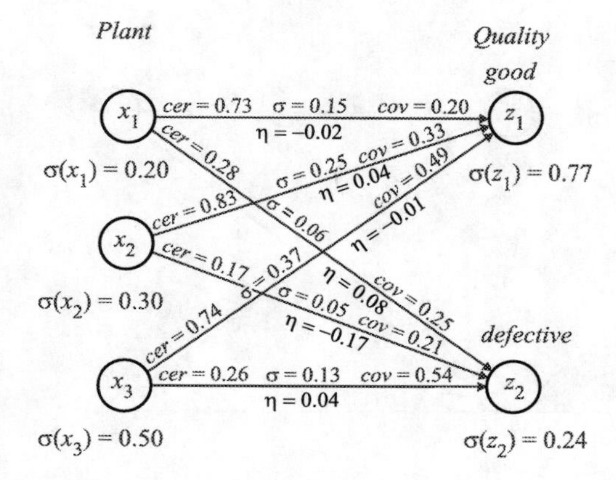

Fig. 20. Fusion between plant and quality.

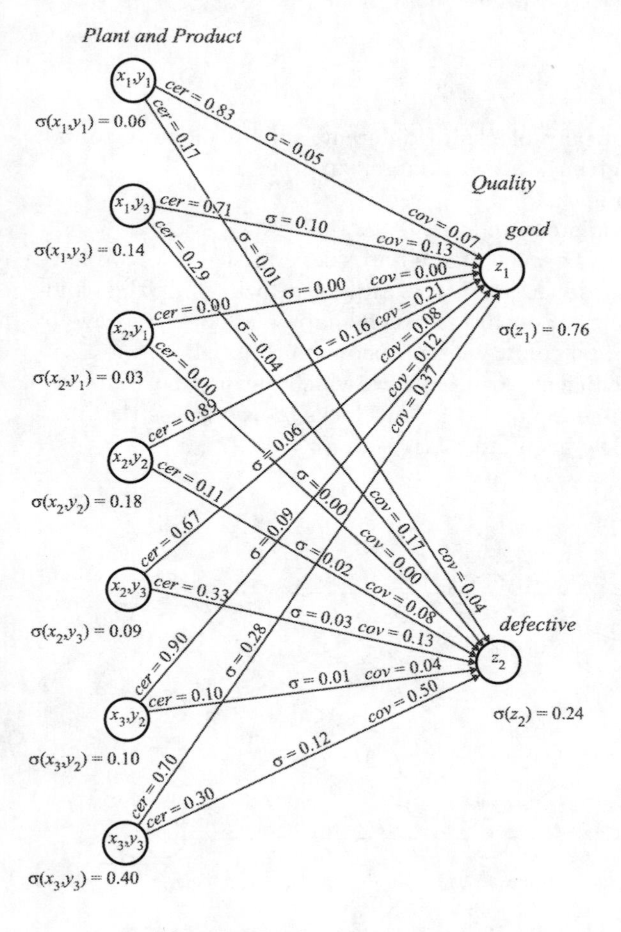

Fig. 21. Production quality.

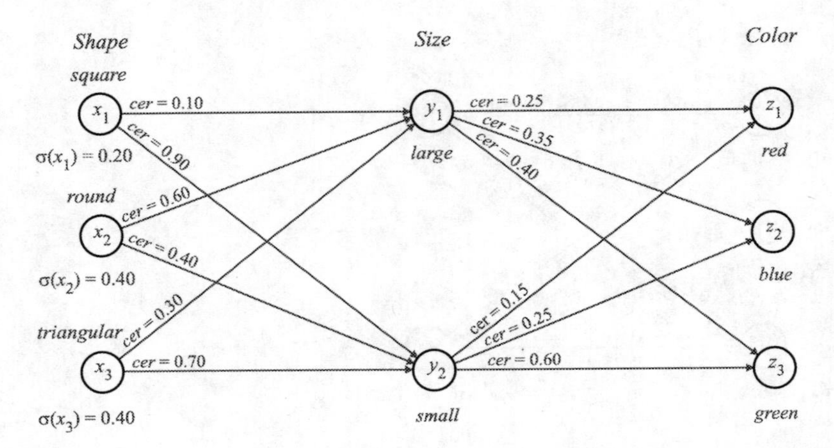

Fig. 22. Initial data.

2.5 Play Blocks

Suppose we are given a set of play blocks of different, shapes (square, round, triangular), colors (red, blue green) and size (large, small).

Initial data are shown in Fig. 22.

Corresponding flow graph is presented in Fig. 23.

In order to find relationship between shape and size, and size color we have to compute the corresponding dependency factors but we will omit this computation here. For finding the relationship between shape and color we have to compute first fusion of shape and color, which is shown in Fig. 24.

Almost all dependency coefficients are very low, which means that there is a very low relationship between shape and color of blocks, nevertheless there are strong decision rules in the flow graph, e.g., $x_3 \rightarrow z_3(\sigma = 0.22)$, $x_3 \rightarrow z_2(\sigma =$

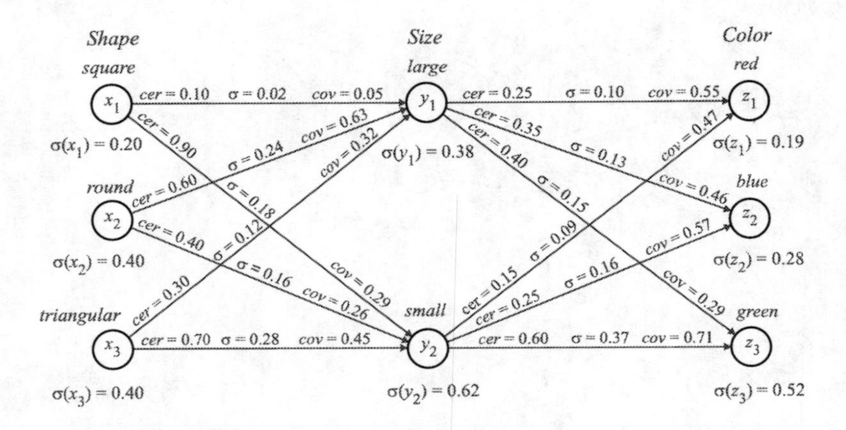

Fig. 23. Relationship between features of play blocks.

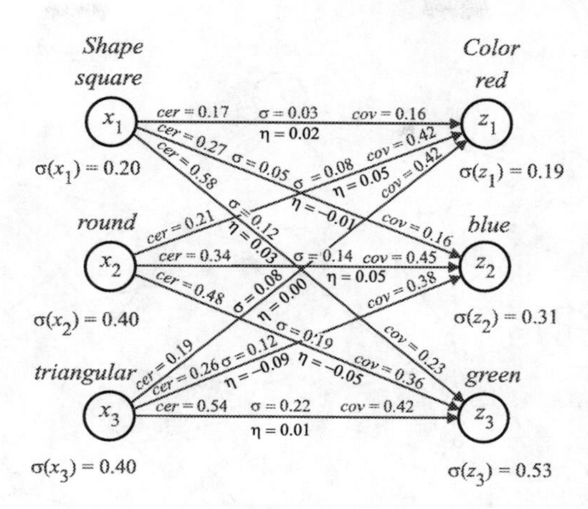

Fig. 24. Fusion of shape and color.

0.12), $x_2 \rightarrow x_3(\sigma = 0.19)$ and $x_2 \rightarrow z_2(\sigma = 0.14)$, which all together yields 67% accuracy of classification.

Analogously to the example discussed before (see Fig. 21) one can search for relationship between other features of play blocks.

Also in this example, similarly as in the previous one, we are not interested in inducing general rules. We are searching only here for relationships in a given data set, i.e., relations between various properties of a given set of objects.

2.6 Preference Analysis

Suppose that three models of cars x_1, x_2 and x_3 are sold to three disjoint groups of customers z_1, z_2 and z_3 through four dealers y_1, y_2, y_3 and y_4.

Moreover, let us assume that car models and dealers are distributed as shown in Fig. 25.

Computing strength and coverage factors for each branch we get results shown in Fig. 26.

In order to find consumer preferences in buying cars we have to compute fusion between car models and consumer group. The result is shown in Fig. 27.

From the flow graph we can see that consumer group z_1 mostly bought car x_3 (45%), consumer group x_2 mostly bought car x_3 (38%) and consumer group z_3 mostly bought cars x_3 too (69%). We can also conclude from the flow graph that car x_1 was mostly bought by consumer group z_2 (57%), car x_2 – by consumer group z_2 (60%) and car x_3 – by consumer group z_2 (39%).

The dependency coefficients reveal that the strangest negative dependency is between car model x_1 and consumer group $z_3(\eta = -0.37)$, whereas car model x_1 and consumer group z_1 shows the highest positive correlation ($\eta = 0.17$), with corresponding strengths $\sigma = 0.02$ and $\sigma = 0.06$.

Let us also notice that the five strongest decision rules provided 77% accuracy of classification.

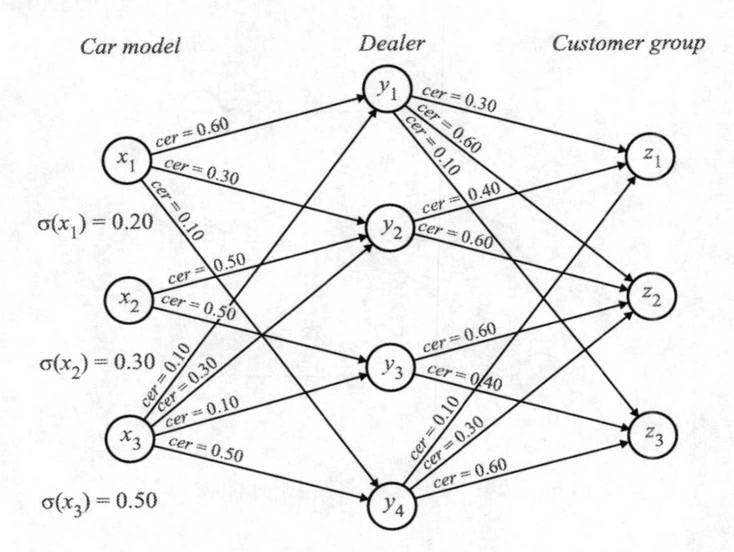

Fig. 25. Car and dealer distribution.

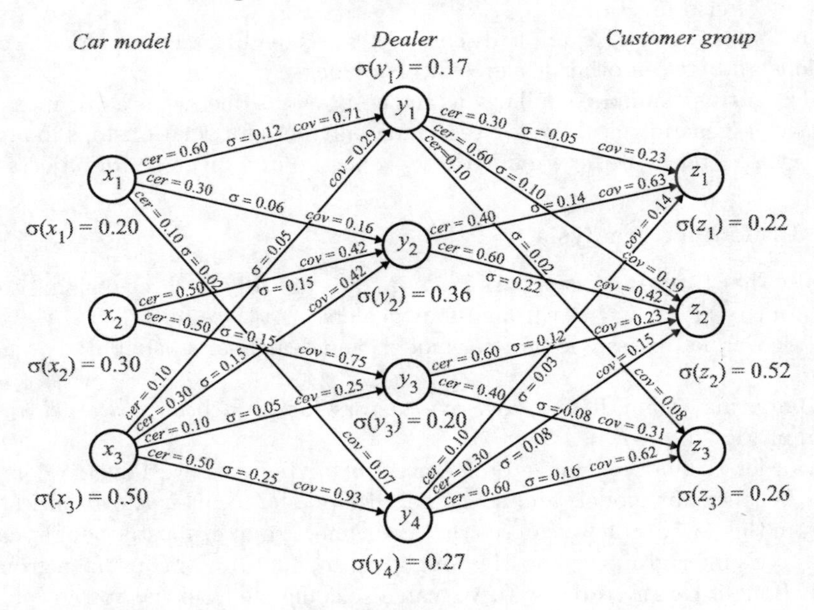

Fig. 26. Strength, certainty and coverage factors.

We can also ask how consumer preferences are related to car model and dealer. To this end we have to find the corresponding decision algorithm but we postpone this task here.

If this data set is representative for a greater universe then the obtained results can be induced for the whole universe.

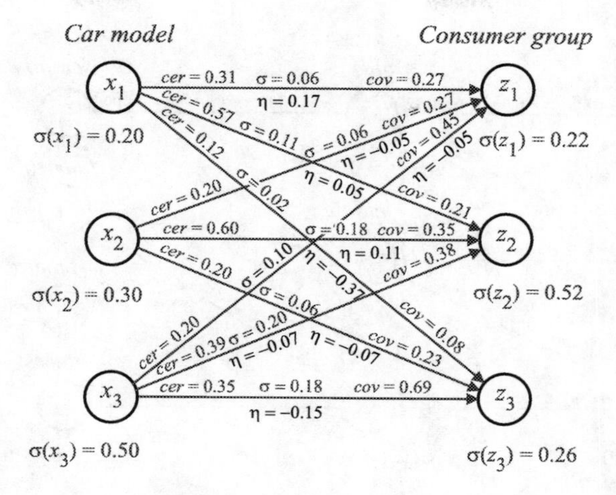

Fig. 27. Fusion of consumer preferences.

2.7 Voting Analysis

Consider three disjoint age groups of voters y_1 (*old*), y_2 (*middle aged*) and y_3 (*young*) – belonging to three social classes x_1 (*high*), x_2 (*middle*) and x_3 (*low*). The voters voted for four political parties z_1 (*Conservatives*), z_2 (*Labor*), z_3 (*Liberal Democrats*) and z_4 (*others*).

Social class and age group votes distribution is shown in Fig. 28.

First we want to find votes distribution with respect to age group. The result is shown in Fig. 29.

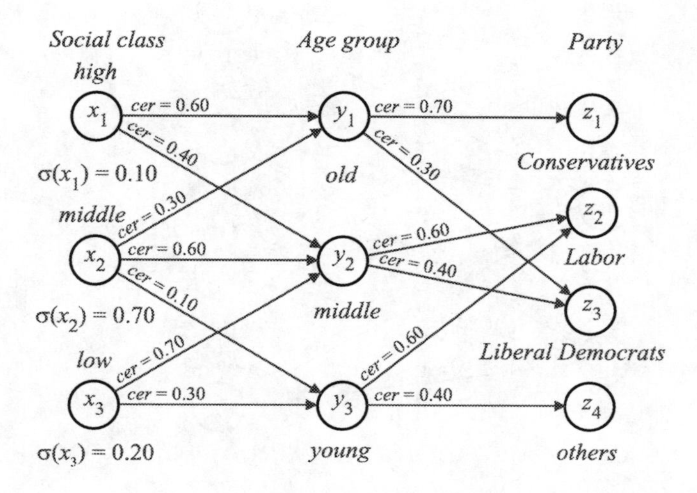

Fig. 28. Social class and age group votes distribution.

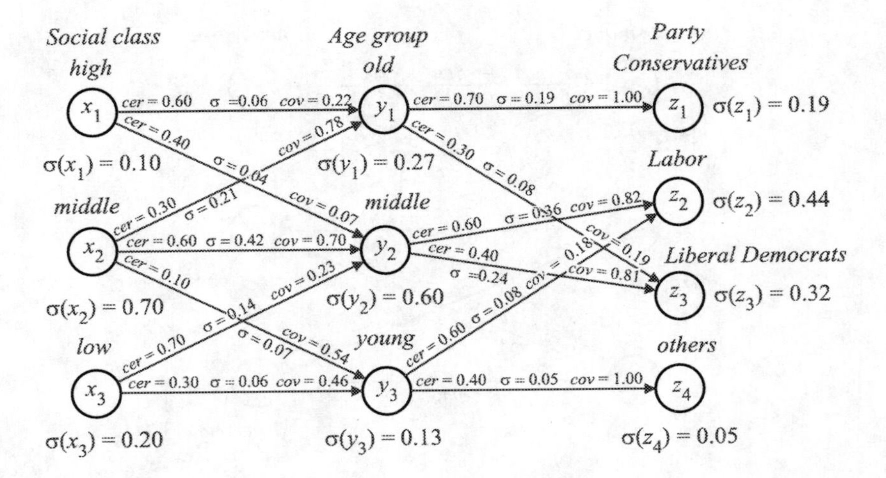

Fig. 29. Party votes distribution.

From the flow graph presented in Fig. 29 we can see that, e.g., party z_1 obtained 19% of total votes, all of them from age group y_1; party z_2 – 44% votes, which 82% are from age group y_2 and 18% – from age group y_3, etc.

If we want to know how votes are distributed between parties with respect to social classes, we have to compute fusion of the corresponding graph. Employing the algorithm presented previously we get the results shown in Fig. 30.

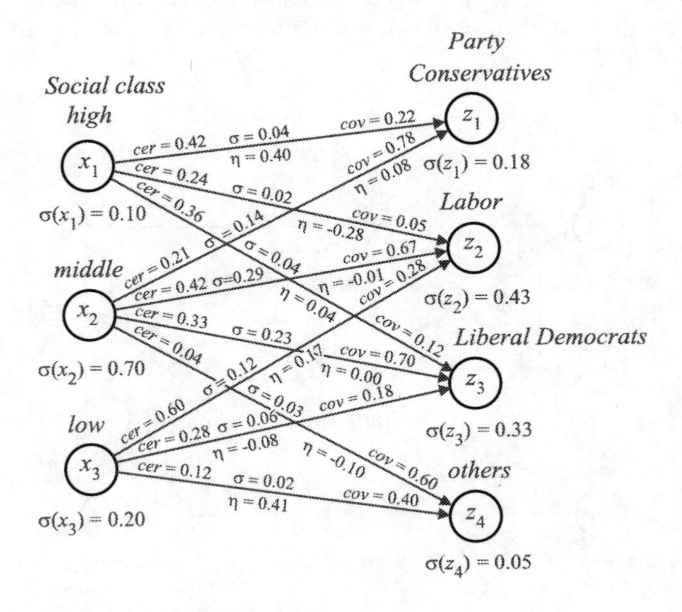

Fig. 30. Fusion of social class and party.

From the flow graph presented in Fig. 30, we can see that party z_1 obtained 22% votes from social class x_1 and 78% – from social class x_2, etc.

The strongest positive dependency occurs between social class x_1 (*high*) and party z_1 (*Conservatives*) $\eta = 0.40$ and social class x_3 (*low*) and party z_4 (*others*) $\eta = 0.41$, with corresponding strengths $\sigma = 0.04$ and $\sigma = 0.02$, which are rather low.

The highest negative correlation ($\eta = -0.28$) is between social class x_1 (*high*) and political party z_2 (*Labor*), with strength $\sigma = 0.02$, which is also low.

If we want to know how votes for political parties are distributed in relation to social class and age of voters, we have to derive the decision algorithm from the flow graph given in Fig. 30, but we will drop this here. Let us observe only, e.g., that old members of high social class voted mostly for Conservatives, middle aged members of middle social class voted mostly for Labor and young members of low social class voted mostly for Labor.

Let us also observe that the four strongest decision rules yields 0.78 strength, i.e., these four rules gives 78% accuracy of classification of party members with respect to social class.

A similar remark about induction as in the previous case of voting analysis applies here.

2.8 Promotion Campaign

Suppose we have three groups of customers classified with respect to age: *young* (*students*), *middle aged* (*workers*) and *old* (*pensioners*). Moreover, suppose we have data concerning place of residence of customers: *town, village* and *country.*

Let us assume that the customers are asked whether they will buy certain advertised product (e.g., a new tooth paste) in a promotion campaign.

The initial data are presented in Fig. 31.

That means that there are 25% young customers, 60% – middle aged and 15% old – in the data base. Moreover, we know that 75% of young customers live in towns, 20% – in villages and 5% – in the country, etc. We also have from the database that 75% town customers answered yes, 25% – no, etc.

We want to find a relationship between various customers' group and the final result of the promotion.

First, applying the ideas presented previously, we get the results shown in Fig. 32.

Fig. 32 shows the general structure of patterns between various customers groups and promotion results.

Suppose we are interested in finding the relationship between age group and final result of the promotion. To this end we have to compute fusion between age groups and the promotion result, or – the relationship between input and output of the flow graph. The result is shown in Fig. 33.

Fig. 33 contains also dependency factors between age groups and the promotion result.

It can be seen from the flow graph that all the dependency factors are very low and almost close to zero. That means, that in view of the data, practically,

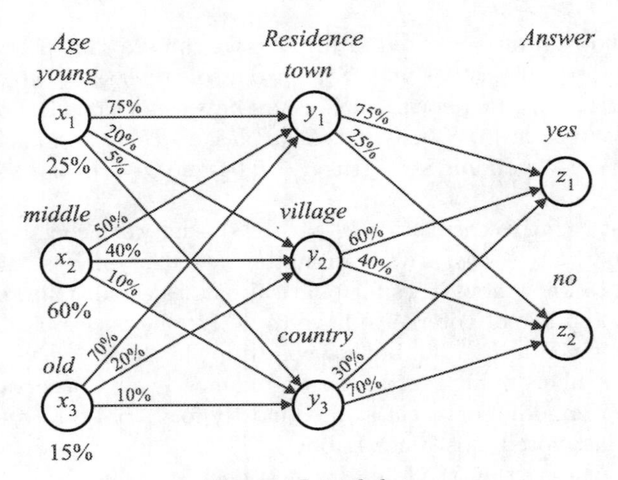

Fig. 31. Initial data.

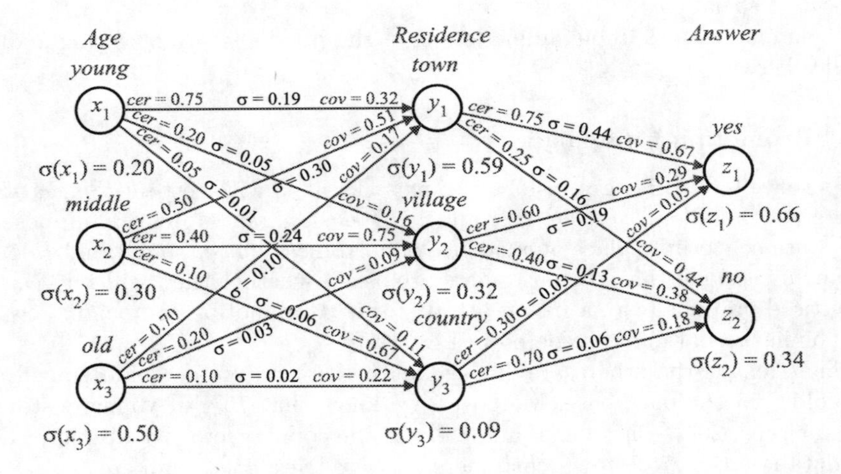

Fig. 32. Relationship between customers group and promotion.

there is no relationship between age group of customers and the final result of promotion, but there are three strong decision rules which provide all together 79% of classification accuracy.

We might be also interested in the relationship between customer's residence and promotion results. This relationship is shown in Fig. 34. We can see from the flow graph that there is relatively high negative dependency ($\eta = -0.38$) with strength $\sigma = 0.03$ between country customers group y_3 and answer z_1 (*yes*). Similarly there is high positive dependency ($\eta = 0.35$) with strength $\sigma = 0.16$ between country customers group y_3 and answer z_2 (*no*). There is also a substantial degree of negative dependency ($\eta = -0.16$) between town customers group y_1 and answer z_2 (*no*), with $\sigma = 0.16$.

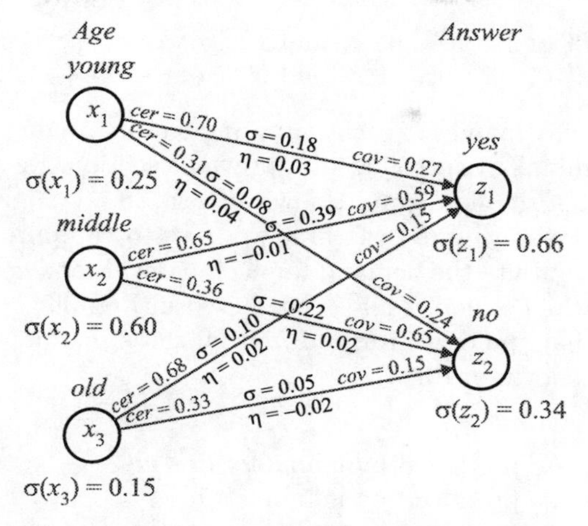

Fig. 33. Fusion of age group and promotion.

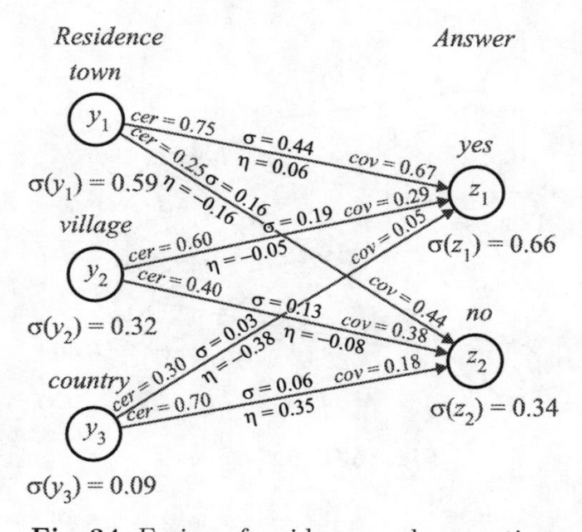

Fig. 34. Fusion of residence and promotion.

We can conclude from the flow graph in Fig. 34, e.g., that independently of age town customers mostly give positive answer to the promotion campaign and country customers give mostly negative answer to the promotion campaign.

Certainly, the results are valid only for the considered data. For another data (population), the results can be different.

2.9 Paint Demand and Supply

Suppose that cars are painted in two colors y_1 and y_2 and that 60% of cars have color y_1, whereas 40% cars have color y_2. Moreover, assume that colors y_1 and y_2 can be obtained by mixing three paints x_1, x_2 and x_3 in the following proportions:

– y_1 contains 20% of x_1, 70% of x_2 and 10% of x_3,
– y_2 contains 30% of x_1, 50% of x_2 and 20% of x_3.

We have to find the demand for each paint and supply among cars y_1 and y_2.

Employing terminology introduced in previous section, we can represent our problem by means of the flow graph shown in Fig. 35.

Thus, in order to solve our task, first we have to compute strength of each branch. Next, we compute the demand for each paint. Finally, we compute paint supply for each car. The final result is presented in Fig. 36.

Suppose now that the cars are produced by three manufacturers z_1, z_2 and z_3, in proportions shown in Fig. 37.

That means:

– 50% of cars y_1 are produced by manufacturer z_1
– 30% of cars y_1 are produced by manufacturer z_2
– 20% of cars y_1 are produced by manufacturer z_3

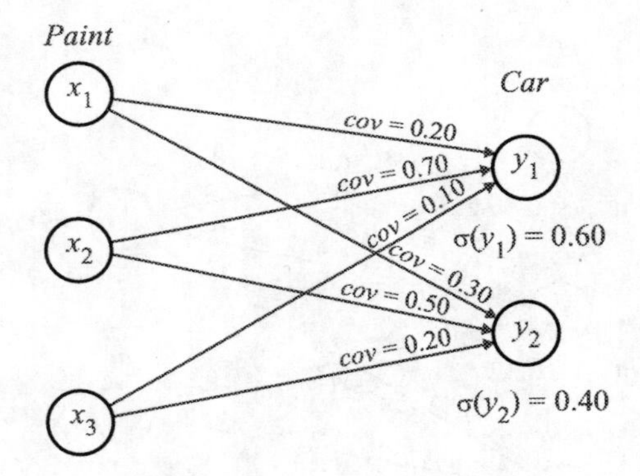

Fig. 35. Paint demand.

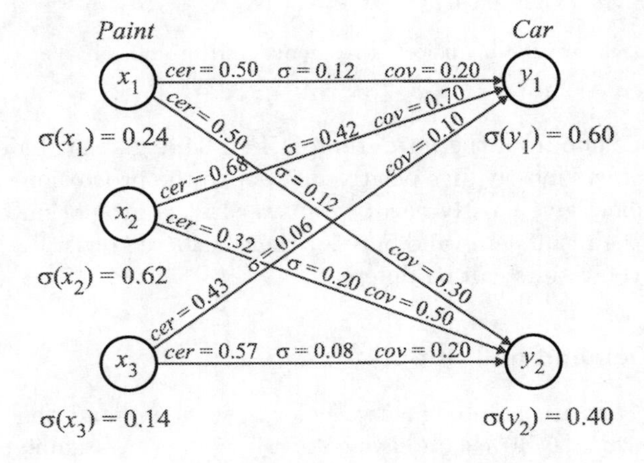

Fig. 36. Paint supply.

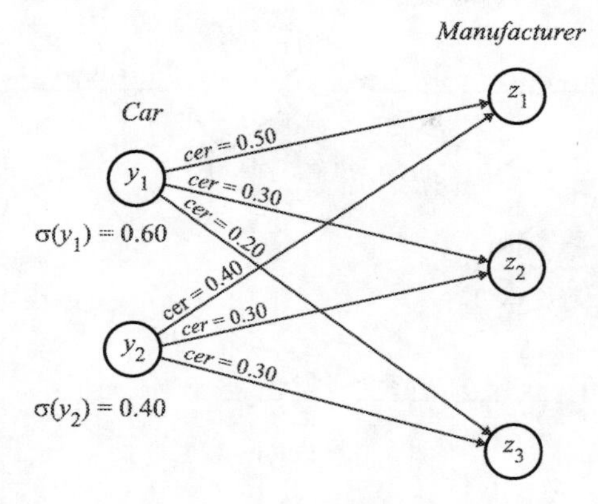

Fig. 37. Car production distribution.

and

- 40% of cars y_2 are produced by manufacturer z_1
- 30% of cars y_2 are produced by manufacturer z_2
- 30% of cars y_2 are produced by manufacturer z_3

Employing the technique used previously, we can compute car production distribution among manufacturers as shown in Fig. 38.

For example, manufacturer z_1 produces 65% of cars y_1 and 35% of cars y_2, etc. Finally, the manufacturer z_1 produces 46% cars, manufacturer z_2 – 30% cars and manufacturer z_3 – 24% of cars.

We can combine graphs shown in Fig. 36 and Fig. 38 and we obtain the flow graph shown in Fig. 39.

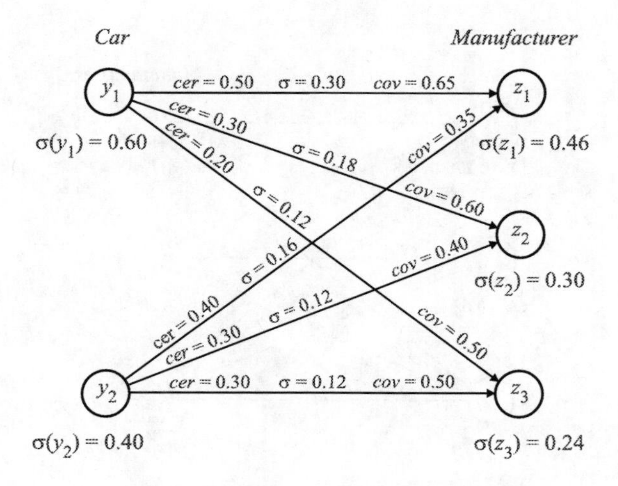

Fig. 38. Manufacturer distribution.

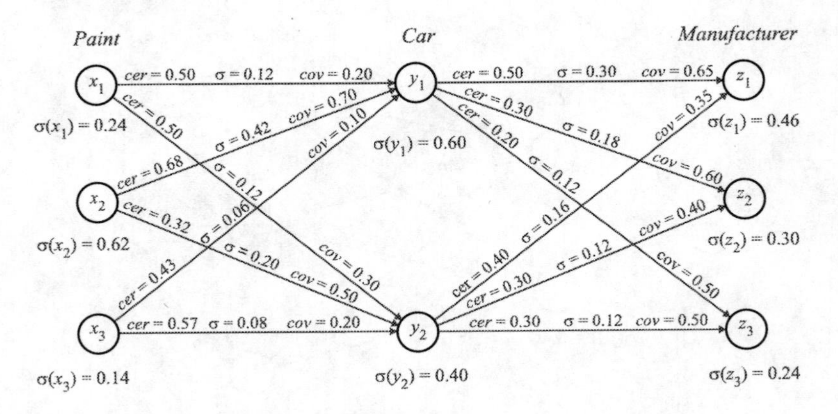

Fig. 39. Paint supply demand flow.

In order to find paint demand and supply by each manufacturer, we have to compute fusion between each paint and manufacture. The corresponding flow graph is presented in Fig. 40.

The meaning of the obtained results is the following.

Suppose that paints are delivered in the same units, say kg.

Thus manufacturer, e.g., z_1, demands 120 kg, 290 kg and 60 kg of paints x_1, x_2 and x_3, respectively. Whereas paint x_1 is delivered to manufacturer z_1, z_2 and z_3 in amounts 120 kg, 80 kg and 70 kg, respectively.

Consequently, we need 270 kg of paint x_1, 630 kg of paint x_2 and 140 kg of paint x_3.

Observe that this example has an entirely deterministic character and there is no probabilistic interpretation of the results needed whatsoever. Besides, we do not need to employ a decision algorithm to solve this task.

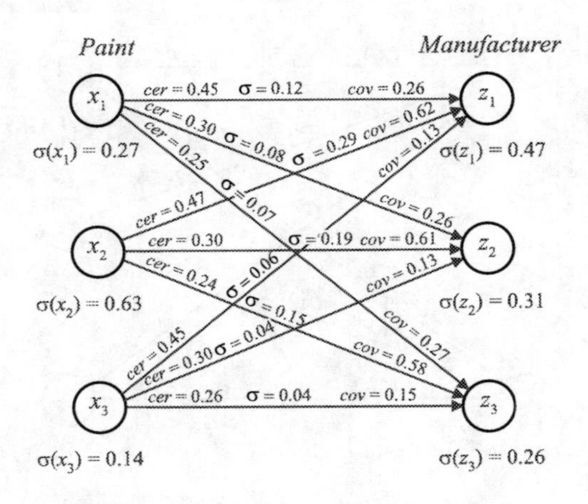

Fig. 40. Fusion of paint demand and supply flow.

In the examples in the previous sections, we considered sets of different objects, e.g., patients, customers, voters, play blocks, cars, etc. In this example we have an entirely different situation. We analyze various paints which are not sets but substances not consisting of elements but having various ingredients (which are also substances), e.g., blue, red paint etc. Thus we cannot use here set theoretical language, and define union, intersection or inclusion of sets (paints). Therefore we cannot say that blue paint is a subset of green paint, but that blue paint is an ingredient of green paint. Consequently, a flow graph can be in this case understood as a language for description of the relationship between various ingredients (substances), where $(x, y) \in \mathcal{B}$ means that x is ingredient of y. In this language $cer(x, y)$ expresses the ratio of substance y to substance x, in x, whereas $cov(x, y)$ is the ratio of x to y in y. This resembles somewhat the relation of being a part in a degree introduced in rough mereology by Polkowski and Skowron (see Section 1.7) but parts and ingredients are two different concepts. The concept of a part has set theoretical flavor, but ingredient has not.

Also inductive reasoning is not involved here. This example shows simply the relationship between demand and supply of some goods.

3 Conclusions

We propose in this paper a new approach to knowledge representation and data mining based on flow analysis in a new kind of flow network.

We advocate in this paper to represent relationships in data by means of flow graphs. Flow in the flow graph is meant to capture the structure of data rather than to describe any physical material flow in the network. It is revealed that the information flow in a flow graph is governed by Bayes' formula; however, the formula can be interpreted in an entirely deterministic way without referring to its probabilistic character. This representation allows us to study different relationships in data and can be used as a new mathematical tool for data mining.

Acknowledgments

Thanks are due to Professor Andrzej Skowron and Dr. Dominik Ślęzak for critical remarks.

References

1. J. M. Bernardo, A. F. M. Smith, Bayesian Theory. Wiley series in probability and mathematical statistics. John Wiley & Sons, Chichester, New York, Brisbane, Toronto, Singapore, 1994.
2. M. Berthold, D.J. Hand, Intelligent Data Analysis - An Introduction. Springer-Verlag, Berlin, Heidelberg, New York, 1999.
3. G.E.P. Box, G.C. Tiao, Bayesian Inference in Statistical Analysis. John Wiley and Sons, Inc., New York, Chichester, Brisbane, Toronto, Singapore, 1992.

4. L.R. Ford, D.R. Fulkerson, Flows in Networks. Princeton University Press, Princeton. New Jersey, 1962.
5. Ch. M. Grinstead, J. L. Snell, Introduction to Probability: Second Revised Edition American Mathematical Society, 1997.
6. S. Greco, Z. Pawlak, R. Słowiński, Generalized decision algorithms, rough inference rules and flow graphs. In: J.J. Alpigini, J.F. Peters, A. Skowron, N. Zhong (eds.), Rough Sets and Current Trends in Computing. Lecture Notes in Artificial Intelligence 2475, Springer-Verlag, Berlin, 2002, pp. 93-104.
7. S. Greco, Z. Pawlak, R. Słowiński, Bayesion confirmation measures within rough set approach, In: S. Tsumoto, R. Słowiński, J. Komorowski, J. Grzymała-Busse (eds.), Rough Sets and Current Trends in Computing (RSCTC 2004), Lecture Notes in Artificial Intelligence 3066, Springer Verlag, Berlin, 2004, pp. 261-270.
8. J. Łukasiewicz, Die logishen Grundlagen der Wahrscheinilchkeitsrechnung. Kraków (1913), in: L. Borkowski (ed.), Jan Łukasiewicz - Selected Works, North Holland Publishing Company, Amsterdam, London, Polish Scientific Publishers, Warsaw, 1970, pp. 16-63.
9. Z. Pawlak, Rough Sets: Theoretical Aspects of Reasoning about Data. Kluwer, Dordrecht, 1991.
10. Z. Pawlak, Rough sets, decision algorithms and Bayes' theorem. European Journal of Operational Research 136, 2002, pp. 181-189.
11. Z. Pawlak, Flow graphs, their fusion and data analysis, 2003, to appear.
12. A. Skowron, J. Stepaniuk, Tolerance approximation spaces, Fundamenta Informaticae 27(2-3), 1996, pp. 245-253.
13. R. Swinburne (ed.), Bayes' Theorem, Oxford University Press, 2002.
14. S. Tsumoto, H. Tanaka, Discovery of Functional Components of Proteins Based on PRIMEROSE and Domain Knowledge Hierarchy, Proceedings of the Workshop on Rough Sets and Soft Computing (RSSC-94), 1994: Lin, T.Y., and Wildberger, A.M. (Eds.), Soft Computing, SCS, 1995, pp. 280-285.
15. S.K.M. Wong, W. Ziarko, Algorithm for inductive learning. Bull. Polish Academy of Sciences 34, 5-6, 1986, pp. 271-276.

The Rough Set Exploration System

Jan G. Bazan[1] and Marcin Szczuka[2]

[1] Institute of Mathematics, University of Rzeszów,
Rejtana 16A, 35-310 Rzeszów, Poland
bazan@univ.rzeszow.pl
[2] Institute of Mathematics, Warsaw University,
Banacha 2, 02-097, Warsaw, Poland
szczuka@mimuw.edu.pl

Abstract. This article gives an overview of the Rough Set Exploration System (RSES). RSES is a freely available software system toolset for data exploration, classification support and knowledge discovery. The main functionalities of this software system are presented along with a brief explanation of the algorithmic methods used by RSES. Many of the RSES methods have originated from rough set theory introduced by Zdzisław Pawlak during the early 1980s.

1 Introduction

This paper introduces the latest edition of the *Rough Set Exploration System* (RSES) software system toolset that makes it possible to analyze tabular datasets utilizing various methods. In particular, many RSES methods are based on rough set theory (see, e.g., [20–28, 30–32]). The first version of RSES and its companion RSESlib became available over a decade ago. After a number of modifications, improvements, and removal of detected bugs, RSES has been used in many applications (see, e.g., [5, 11, 14, 15, 29, 34, 41]).

The RSESlib library of tools for rough set computations was successfully used for data analysis with encouraging results. Comparison with other classification systems (see, e.g., [2, 19, 29]) proves its value. The early version of RSESlib was also used in construction of the computational kernel of ROSETTA, an advanced system for data analysis (see,e.g, [12, 42]).

At the moment of this writing RSES version 2.2 is the most current. This version was prepared by the research team supervised by Professor Andrzej Skowron. Currently, the RSES R&D team consists of: Jan Bazan, Rafał Latkowski, Michał Mikołajczyk, Nguyen Hung Son, Nguyen Sinh Hoa, Dominik Ślęzak, Piotr Synak, Marcin Szczuka, Arkadiusz Wojna, Marcin Wojnarski, and Jakub Wróblewski.

The RSES ver. 2.2 software and its computational kernel – the RSESlib 3.0 library – maintains all advantages of previous versions. The algorithms have been redesigned to provide better flexibility, extended functionality and the ability to process massive data sets. New algorithms added to the library reflect the current state of our research in classification methods originating in rough sets

J.F. Peters and A. Skowron (Eds.): Transactions on Rough Sets III, LNCS 3400, pp. 37–56, 2005.

theory. Improved construction of library allows further extensions and supports augmentation of RSESlib into different data analysis tools.

Today RSES is freely distributed for non-commercial purposes. Anybody can download it from the Web site [40].

The RSES software and underlying computational methods have been successfully applied in many studies and applications. A system based on LTF-C (see Subsection 8.3) won the first prize in the EUNITE 2002 World Competition "Modeling the Bank's Client behavior using Intelligent Technologies" (see [41]). Approaches based on rule calculation have been successfully used in areas such as gene expression discovery (see, e.g., [34]), survival analysis in oncology (see, e.g., [5]), software change classification (see, e.g., [29]), classification of biomedical signal data (see, e.g., [15]), and classification of musical works and processing musical data (see, e.g, [11, 14]).

In this paper we attempt to provide quite a general description of capabilities of our software system. We also present a handful of basic facts about the underlying computational methods. The paper starts with introduction of basic notions (Section 2) that introduces the vocabulary for the rest of this article. Next, we describe, in general terms, main functionalities of RSES (Section 3) and architecture of the system (Section 4). Rest of the paper presents computational methods in the order introduced in Figure 1. Starting from input management (Section 5) we go through data preprocessing (Section 6) and data reduction (Section 7), to conclude with description of methods for classifier construction and evaluation (Section 8).

2 Basic Notions

In order to provide clear description further in the paper and avoid any misunderstandings we bring here some essential definitions from Rough Set theory. We will frequently refer to the set of notions introduced in this section. Quite comprehensive description of notions and concepts related to classical rough sets may be found in [13].

The structure of data that is central point of our work is represented in the form of *information system* [28] or, more precisely, the special case of information system called *decision table*.

Information system is a pair of the form $\mathbf{A} = (U, A)$ where U is a finite *universe* of *objects* and $A = \{a_1, ..., a_m\}$ is a set of *attributes*, i.e., mappings of the form $a_i : U \to V_a$, where V_a is called *value set* of the attribute a_i. The decision table is also a triple of the form $\mathbf{A} = (U, A, d)$ where the major feature that is different from the information system is the distinguished attribute d. In case of decision table the attributes belonging to A are called *conditional attributes* or simply *conditions* while d is called *decision* (sometimes *decision attribute*). The cardinality of the image $d(U) = \{k : d(s) = k \text{ for some } s \in U\}$ is called the *rank of d* and is denoted by $rank(d)$.

One can interpret the decision attribute as a kind of classifier on the universe of objects given by an expert, a decision-maker, an operator, a physician, etc. This way of looking at data classification task is directly connected to the general idea of *supervised learning* (learning with a "teacher").

Decision tables are usually called training sets or training samples in machine learning.

The i-th *decision class* is a set of objects $C_i = \{o \in U : d(o) = d_i\}$, where d_i is the i-th decision value taken from decision value set $V_d = \{d_1, ..., d_{rank(d)}\}$.

For any subset of attributes $B \subset A$ *indiscernibility relation* $IND(B)$ is defined as follows:

$$xIND(B)y \Leftrightarrow \forall_{a \in B} a(x) = a(y) \tag{1}$$

where $x, y \in U$. If a pair $(x, y) \in U \times U$ belongs to $IND(B)$ then we say that x and y are indiscernible by attributes from B.

Having indiscernibility relation we may define the notion of reduct. $B \subset A$ is a (global) *reduct* of information system if $IND(B) = IND(A)$ and no proper subset of B has this property. There may be many reducts for a given information system (decision table). We are usually interested only in some of them, in particular those leading to good classification models.

As the discernibility relation and reducts (reduction) are the key notions in classical rough sets we want to dwell a little on these notions and their variants used in RSES.

A *relative* (local) reduct for an information system $\mathbf{A} = (U, A)$ and an object $o \in U$ is an irreducible subset of attributes $B \subset A$ that suffices to discern o from all other objects in U. Such reduct is only concerned with discernibility relative to the preset object.

In case of decision tables the notion of *decision reduct* is handy. The decision reduct is a set $B \subset A$ of attributes such that it cannot be further reduced and $IND(B) \subset IND(d)$. In other words, decision reduct is a reduct that only cares about discernibility of objects that belong to different decision classes. This works under assumption that the table is consistent. If consistency is violated, i.e., there exist at least two objects with identical attribute values and different decision, the notion of decision reduct can still be utilized, but the notion *generalized decision* has to be used.

As in general case, there exists a notion of a decision reduct relative to an object. For an object $o \in U$ the relative decision reduct is a subset $B \subset A$ of attributes such that it cannot be further reduced and suffices to make o discernible from all objects that have decision value different from $d(o)$.

Dynamic reducts are reducts that are calculated in a special way. First, from the original information system a family of subsystems is selected. Then, for every member of this family (every subsystem) reducts are calculated. Reducts (subsets of attributes) that appear in results for many subtables are chosen. They are believed to carry essential information, as they are reducts for many parts of original data. For more information on dynamic reducts please refer to [2].

The set $\underline{B}X$ is the set of all elements of U which can be classified with certainty as elements of X, having the knowledge about them represented by attributes from B; the set $BN_B(X)$ is the set of elements which one can classify neither to X nor to $-X$ having knowledge about objects represented by B.

If $C_1, \ldots, C_{r(d)}$ are decision classes of $\mathbf{A}$ then the set $\underline{B}C_1 \cup \cdots \cup \underline{B}C_{rank(d)}$ is called *the B-positive region of* $\mathbf{A}$ and is denoted by $POS_B(d)$. The relative size of positive region, i.e., the ratio $\frac{|POS_B(d)|}{|U|}$, is an important indicator used in RSES.

Decision rule r is a formula of the form

$$(a_{i_1} = v_1) \wedge ... \wedge (a_{i_k} = v_k) \Rightarrow d = v_d \tag{2}$$

where $1 \leq i_1 < ... < i_k \leq m$, $v_i \in V_{a_i}$. Atomic subformulae $(a_{i_1} = v_1)$ are called *descriptors* or *conditions.* We say that rule r is *applicable* to object, or alternatively, the object *matches* rule, if its attribute values satisfy the premise of a rule.

With the rule we can associate some characteristics. $Supp_{\mathbf{A}}(r)$ denotes *Support,* and is equal to the number of objects from $\mathbf{A}$ for which rule r applies correctly, i.e., the premise of rule is satisfied and the decision given by rule is similar to the one preset in decision table. $Match_{\mathbf{A}}(r)$ is the number of objects in $\mathbf{A}$ for which rule r applies in general. Analogously the notion of matching set for a rule or collection of rules may be introduced (see [2–4]).

The notions of matching and supporting set are common to all classifiers, not only decision rules. For a classifier C we will denote by $Supp_{\mathbf{A}}(C)$ the set of objects that support classifier, i.e., the set of objects for which classifier gives the answer (decision) identical to that we already have. Similarly, $Match_{\mathbf{A}}(C)$ is a subset of objects in $\mathbf{A}$ that are recognized by C. Support and matching make it possible to introduce two measures that are used in RSES for classifier scoring. These are Accuracy and Coverage, defined as follows:

$$Accuracy_{\mathbf{A}}(C) = \frac{|Supp_{\mathbf{A}}(C)|}{|Match_{\mathbf{A}}(C)|}$$

$$Coverage_{\mathbf{A}}(C) = \frac{|Match_{\mathbf{A}}(C)|}{|\mathbf{A}|}$$

By *cut* for an attribute $a_i \in A$, such that V_{a_i} is an ordered set we will denote a value $c \in V_{a_i}$. Cuts mostly appear in the context of discretization of real-valued attributes. In such situation, the cut is a a value for an attribute such that it determines a split of attribute domain (interval) into two disjoint subintervals.

With the use of cut we may replace original attribute a_i by a new, binary attribute which tells as whether actual attribute value for an object is greater or lower than c (more in [17]).

Template of $\mathbf{A}$ is a propositional formula $\bigwedge(a_i = v_i)$ where $a_i \in A$ and $v_i \in V_{a_i}$. A generalized template is the formula of the form $\bigwedge(a_i \in T_i)$ where $T_i \subset V_{a_i}$. An object *satisfies* (matches) a template if for every attribute a_i occurring in the template the value of this attribute on considered object is equal to v_i (belongs to T_i in case of generalized template). The template induces in natural way the split of original information system into two distinct subtables. One of those subtables contains objects that satisfy the template, the other those that do not.

Decomposition tree is a binary tree, whose every internal node is labeled by some template and external node (leaf) is associated with a set of objects matching all templates in a path from the root to a given leaf (see [16] for more details).

3 Main Functionalities

RSES is a computer software system developed for the purpose of analyzing data. The data is assumed to be in the form of information system or decision table. The main step in the process of data analysis with RSES is the construction and evaluation of classifiers.

Classification algorithms, or classifiers, are algorithms that permit us to repeatedly make a forecast in new situation on the basis of accumulated knowledge. In our case the knowledge is embedded in the structure of classifier which itself is constructed (learned) from data (see [19]). RSES utilizes classification algorithms using elements of rough set theory, instance based learning, artificial neural networks and others. Types of classifiers that are available in RSES are discussed in Section 8.

The construction of classifier is usually preceded by several initial steps. First, the data for analysis has to be loaded/imported into RSES. RSES can accept several input formats as described in Section 5. Once the data is loaded, the user can examine it using provided visualization and statistics tools (see Figures 3F3 and 3F4).

In order to have a better chance for constructing (learning) a proper classifier, it is frequently advisable to transform the initial data set. Such transformation, usually referred to as *preprocessing* may consist of several steps. RSES supports preprocessing methods which make it possible to manage missing parts in data, discretize numeric attributes, and create new attributes. Preprocessing methods are further described in Section 6.

Once the data is preprocessed we may be interested in learning about its internal structure. By using classical rough set concepts such as reducts, dynamic reducts and positive region one may pinpoint dependencies that occur in our data set. Knowledge of reducts may lead to reduction of data by removing some of the redundant attributes. Reducts can also provide essential hints for the parameter setting during classifier construction. Calculation of reducts and their usage is discussed in Section 7.

The general scheme of data analysis process with use of RSES functionalities in presented in Figure 1.

4 The Architecture of RSES

To simplify the use of RSES algorithms and make it more intuitive the RSES graphical user interface was constructed. It is directed towards ease of use and visual representation of workflow. Project interface window consists of two parts. The visible part is the project workspace with icons representing objects created during blue computation. Behind the project window there is the history

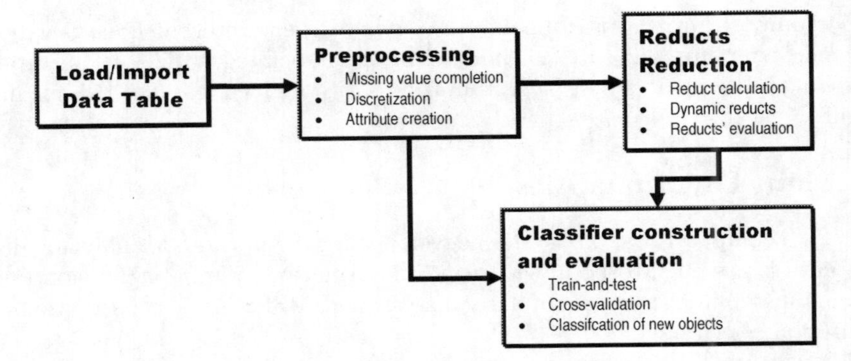

Fig. 1. RSES data analysis process.

window, reachable via tab, and dedicated to messages, status reports, errors and warnings. While working with multiple projects, each of them occupies a separate workspace accessible via tab at the top of workplace window.

It was designers' intention to simplify the operations on data within project. Therefore, the entities appearing in the process of computation are represented in the form of icons placed in the upper part of workplace. Such an icon is created every time the data (table, reducts, rules,...) is loaded from the file. Users can also place an empty object in the workplace and further fill it with results of operation performed on other objects. Every object appearing in the project has a set of actions associated with it. By right-clicking on the object the user invokes a context menu for that object. It is also possible to invoke an action from the general pull-down program menu in the main window. Menu choices make it possible to view and edit objects as well as include them in new computations. In

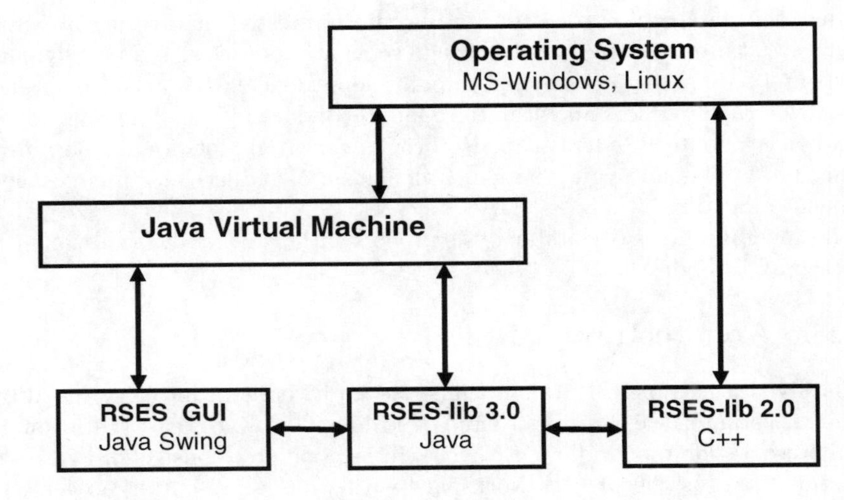

Fig. 2. The architecture of RSES.

many cases a command from context menu causes a new dialog box to open. In this dialog box the user can set values of parameters used in desired calculation. If the operation performed on the object leads to creation of a new object or modification of existing one then such a new object is connected with edge originating in object(s) which contributed to its current state. Placement of arrows connecting icons in the workspace changes dynamically as new operations are being performed. The user has the ability to align objects in workspace automatically, according to his/her preferences (eg. left, horizontal, bottom).

An important, recently added GUI feature is the possibility to display some statistical information about tables, rules and reducts in a graphical form. The look of various components of RSES interface is shown in Figure 3.

Behind the front-end that is visible to the user, there is RSES computational kernel. This most essential part of the system is built around the library of methods known as RSESlib ver. 3.0. The library is mostly written in Java but, it also uses a part that was implemented using C++. The C++ part is the legacy of previous RSESlib versions and contains those algorithms that could only lose optimality if re-implemented in Java. The layout of RSES components is presented in Figure 2. Currently, it is possible to install RSES in Microsoft Windows 95/98/2000/XP and in Linux/i386. The computer on which the RSES is installed has to be equipped with Java Runtime Environment.

5 Managing Input and Output

During operation certain functions in RSES may read and write information to/from files. Most of the files that can be read or written are regular ASCII text files. A particular sub-types can be distinguished by reviewing the contents or identifying file extensions (if used). Description of RSES formats can be found in [40].

As the whole system is about analyzing tabular data, it is equipped with abilities to read several tabular data formats. At the time of this writing the system can import text files formatted for old version of RSES (RSES1 format), Rosetta, and Weka systems. Naturally, there exists native RSES2 file format used to store data tables.

The old RSES1 format is just a text file that stores data table row by row. The only alternation is in the first row, which defines data dimension – number of rows and columns (attributes and objects). All the other input formats are more complex. The file in these formats consist of the header part and the data part. The header part defines structure of data, number and format of attributes, attribute names etc. Additional information from the header proved to be very useful during analysis, especially in interpretation of results. For detailed definition of these formats please refer to [42] for Rosetta, [43] for Weka, and [40] for RSES.

The RSES user can save and retrieve data entities created during experiment, such as rule sets, reduct sets etc. The option of saving the whole workspace (project) in a single file is also provided. The project layout together with underlying data structures is stored using dedicated, optimized binary file format.

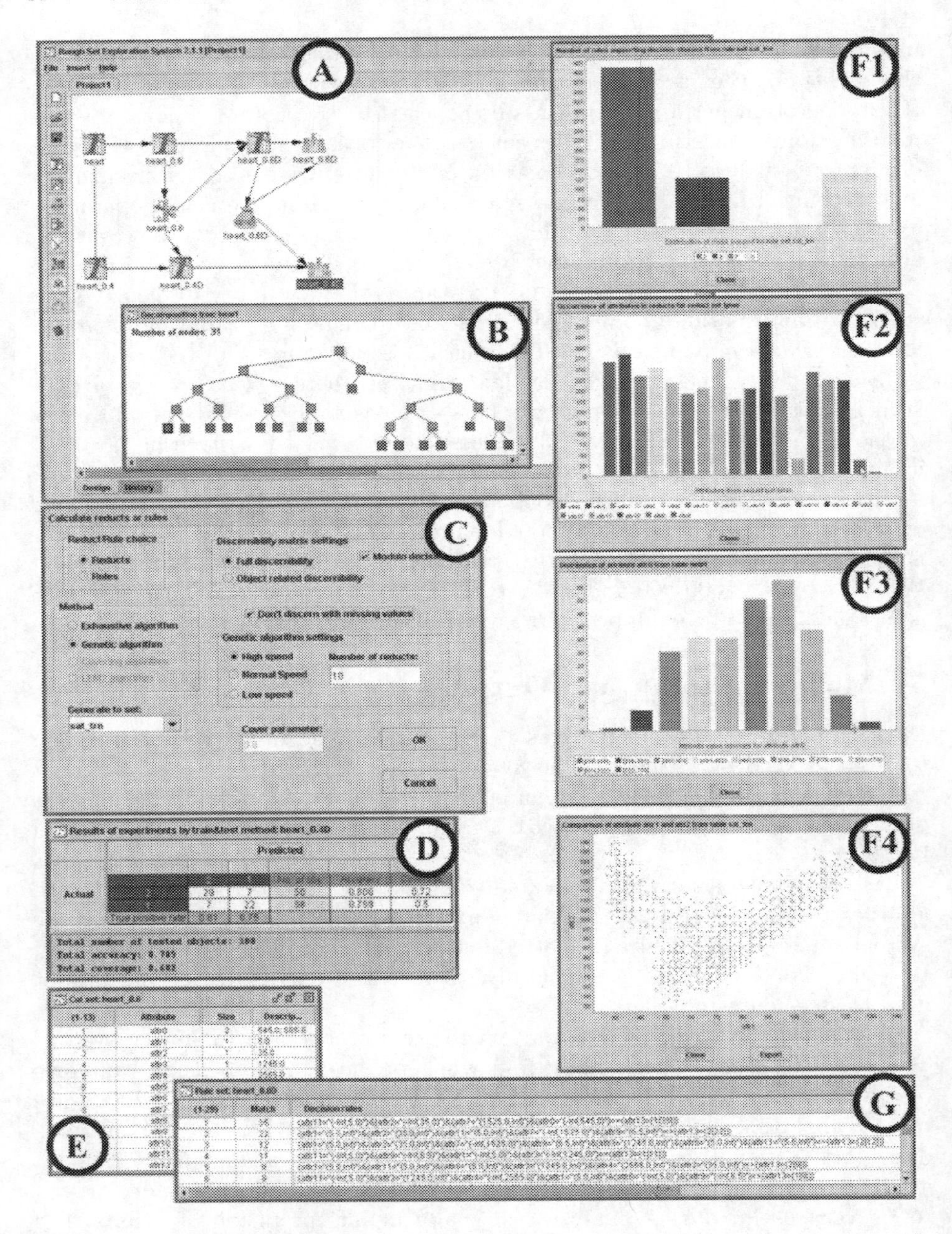

Fig. 3. RSES interface windows: A. Project window; B. Decomposition tree; C. Reduct/rule calculation; D. Classification results; E. Set of cuts; F1–F4. Graphical statistics for rules, reducts and attributes; G. Decision rules.

6 Data Preprocessing

The data that is an input to analysis with use of RSES can display several features that negatively influence quality and generality of classification model we want to construct. There may be several such problems with data, depending on classification paradigm we want to apply. Some of the RSES classifier construction methods require data (attributes) to be represented with use of specific value sets, e.g., only symbolic attributes are allowed. There may be other problems, such as missing information (missing values) or inconsistencies in the table. Finally, we may want to alter the original data by introducing new attributes that are better suited for classification purposes.

RSES provides users with several functionalities that make it possible to preprocess original data and prepare a training sample which more likely to lead to a good classifier.

6.1 Examining Data Statistics

Not a preprocessing method *per se*, RSES functions for producing data statistics may serve as a handy tool. By using them the user can get familiar with the data as well as compare some crucial data characteristics before and after data modification.

In RSES the user may examine distribution of a single attribute, in particular the decision distribution. The RSES system is capable of presenting numerical measurements for distribution of values of a single attribute as well as displaying the corresponding histogram (see Figure 3 F3). The information gathered for a single attribute includes range, mean value, and standard deviation. It is also possible to compare distribution of two attributes on a single plot as shown in Figure 3 F4.

6.2 Missing Value Completion

The missing elements in data table, so called `NULL` values, may pose a problem when constructing classifier. The lack of some information may be due to incomplete information, error or the constraints embedded in data collection process.

RSES offers four approaches to the issue of missing values. These are as follows:

- removal of objects with missing values,
- filling the missing part of data in one of two ways (see [10]):
  - filling the empty (missing) places with most common value in case of nominal attributes and filling with mean over all attribute values in data set in case of numerical attribute.
  - filling the empty (missing) places with most common value for the decision class in case of nominal attributes and filling with mean over all attribute values in the decision class in case of numerical attribute.

- analysis of data without taking into account those objects that have incomplete description (contain missing values). Objects with missing values (and their indiscernibility thereof) are disregarded during rule/reduct calculation. This result is achieved by activating corresponding options in the RSES dialog windows for reduct/rule calculation.
- treating the missing data as information (`NULL` is treated as yet another regular value for attribute).

6.3 Discretization

Suppose we have a decision table $\mathbf{A} = (U, A \cup \{d\})$ where $card(V_a)$ is high for some $a \in A$. Then there is a very low chance that a new object is recognized by rules generated directly from this table, because the attribute value vector of a new object will not match any of these rules. Therefore for decision tables with real (numerical) value attributes some discretization strategies are built into RSES in order to obtain a higher quality of classification.

Discretization in RSES is a two-fold process. First, the algorithm generates a set of cuts (see Figure 3 E). These cuts can be then used for transforming a decision table. As a result we obtain the decision table with the same set of attributes, but the attributes have different values. Instead of $a(x) = v$ for an attribute $a \in A$ and object $x \in U$ we rather get $a(x) \in [c_1, c_2]$, where c_1 and c_2 are cuts generated for attribute a by discretization algorithm. The cuts are generated in a way that the resulting intervals contain possibly most uniform sets of objects w.r.t decision.

The discretization method available in RSES has two versions, code-named *global* and *local*. Both methods are using a bottom-up approach which adds cuts for a given attribute one-by-one in subsequent iterations of algorithm. The difference between these two is in the way the candidate for new cut is scored. In the global method we consider in scoring all the objects in data table at every step. In the local method we only take part of objects that are concerned with this candidate cut, i.e., which have value of the currently considered attribute in the same range as the cut candidate. Naturally, the second (local) method is faster as less objects have to be examined at every step. In general, the local method is producing more cuts. The local method is also capable of dealing with nominal (symbolic) attributes. The grouping (quantization) of nominal attribute domain with use of local method always results in two subsets of attribute values.

6.4 Creation of New Attributes

RSES makes it possible to add an attribute to decision table. This new attribute is created as a weighted sum of selected existing (numerical) attributes. We may have several such attributes for different weight settings and different attributes participating in weighted sum. These attributes are carrying agglomerated information and are intended as a way of simplifying classifier construction for a given data set.

Linear combinations are created on the basis of collection of attribute sets consisting of k elements. These k-element attribute sets as well as parameters of combination (weights) are generated automatically by adaptive optimization algorithm implemented in RSES. As a measure for optimization we may use one of three possibilities. The measures take into account potential quality of decision rules constructed on the basis of newly created linear combination attribute. For details on these measures please turn to [33]. The user may specify the number (k) of new attributes to be constructed and the number of original attributes to be used in each linear combination.

7 Reduction and Reducts

As mentioned before (Section 2) there are several types of reducts that may be calculated in RSES for a given information system (decision table). The purposes for performing reduct calculation (reduction) may vary. One of advantages that reducts offer is better insight into data. By calculating reducts we identify this part of data (features) which carries most essential information. Reducts are also canvas for building classifiers based on decision rules.

Inasmuch as calculation of interesting, meaningful and useful reducts may be a complicated and computationally costly task, there is a necessity for larger flexibility when setting up an experiment involving reducts. For that reason there are several different methods for discovering reducts implemented in RSES.

7.1 Calculating Reducts

In Section 2 we mentioned four general types of reducts. All four of them can be derived in RSES by selecting appropriate settings for algorithms implemented (see Figure 3 C). It is important to mention that in most cases selecting object-related indiscernibility relation (local reducts) leads to creation of much larger set of reducts. On the other hand, selection of decision-dependant indiscernibility usually reduces computational cost.

There are two algorithms for reducts calculation available in RSES. First of them is an exhaustive algorithm. It examines subsets of attributes incrementally and returns those that are reducts of required type. This algorithm, although optimized and carefully implemented, may lead to very extensive calculations in case of large and complicated information systems (decision tables). Therefore, should be used with consideration.

To address problems with sometimes unacceptable cost of exhaustive, deterministic algorithm for reduct calculation, an alternative evolutionary method is implemented in RSES. This method is based on an order-based genetic algorithm coupled with heuristic. Theoretical foundations and practical construction of this algorithm are presented in [37] and [3].

The user, when invoking this method, has some control over its behavior since the population size and convergence speed may be set from the RSES interface (see Figure 3 C).

7.2 Calculating Dynamic Reducts

Dynamic reducts are reducts that remain to be such for many subtables of the original decision table (see [2], [3]). The process of finding the dynamic reduct is computationally more costly, as it requires several subtables to be examined in order to find the frequently repeating minimal subsets of attributes (dynamic reducts). Such dynamic reducts may be calculated for general or decision-related indiscernibility relation.

The purpose of creating dynamic reducts is that in many real-life experiments they proved to be more general and better suited for creation of meaningful and applicable decision rules. Examples of solutions and results obtained with use of dynamic reduct approach may be found in [2, 3].

The dynamic reduct calculation process involves sampling several subtables from original table and is controlled by number of options such as: number of sampling levels, number of subtables per sampling level, smallest and largest permissible subtable size, and so on. We also decide right on the start if we are interested in general or decision-related reducts.

As mentioned before, calculation of reducts may be computationally expensive. To avoid overly exhaustive calculation it is advisable to carefully select parameters for dynamic reduct calculation, taking into account size of data, number of samples, and size of attribute value sets.

7.3 From Reducts to Rules

Reducts in RSES are, above all, used to create decision rule sets. Equipped with collection of reducts (reduct set) calculated beforehand, the user may convert them into a set of decision rules. That may be achieved in two ways, as there are two methods for creating rules from reducts implemented in RSES.

First option is to calculate so called *global rules*. The algorithm scans the training sample object by object and produces rules by matching object against reduct. The resulting rule has attributes from reducts in conditional part with values of currently considered object, and points at decision that corresponds to the decision for this training object. Note, that for large tables and large reduct set the resulting set of rules may be quite large as well.

Another alternative is to generate all *local rules*. For each reduct a subtable, containing only the attributes present in this reduct, is selected. For this subtable algorithm calculates a set of minimal rules (rules with minimal number of descriptors in conditional part – see, e.g., [3]) w.r.t decision. Finally, the rule sets for all reducts are summed up to form result.

Sets of reducts obtained in RSES may be examined with use of included graphical module. This module makes it possible to find out how the attributes are distributed among reducts and how reducts overlap (see Figure 3 F2). In case of decision reduct it is also possible to verify the size of positive region.

Sets of decision rules obtained as a result may be quite large and some of the rules may be of marginal importance. This can be, among other things, the result of using reducts that are of mediocre quality.

The quality of reducts may be improved after they are calculated. One way to do that is by shortening. The user may specify the shortening threshold (between 0 and 1) and the algorithm will attempt to remove some attributes from reducts. The shortening is performed as long as the relative size of positive region for shortened reduct exceeds the threshold. Note, that in the process of shortening the resulting subset of attributes may no longer be a proper reduct but it is shorter and results in more general rules.

In case of dynamic reduct there is even more room for improvement here. We may perform reduct filtering, eliminating reducts of lower expected quality. To filter dynamic reducts in RSES the user has to set a value of *stability coefficient* (between 0 and 1). This coefficient is calculated for each dynamic reduct during the process of its creation. We will not bring here the entire theory behind the stability coefficient. Interested reader may find detailed explanation in [2] and [3]. Important thing to know is that stability coefficient keeps the record of appearances of a reduct for subtables sampled in reduct calculation process. The more frequent occurrence of the reduct (the greater stability) the higher stability coefficient. High stability (coefficient) of a dynamic reducts strongly suggest that it contains vital piece of information. Naturally, there is no point in considering stability coefficient filtering in case on non-dynamic (regular) reducts, as there in no sampling involved in their creation, and their stability coefficients always equal 1.

8 Construction and Utilization of Classifiers

Several types of classifiers are represented in RSES, and we present them in some detail in subsequent parts of this section. All of them follow the scheme of construction, evaluation and usage.

The classifier in RSES is constructed on the basis of training set consisting of labeled examples (objects with decisions). Such a classifier may be further used for evaluation with use of test/validation set or applied to new, unseen and unlabeled cases in order to establish the value of decision (classification) for them.

The evaluation of the classifier's performance in RSES may be conducted in two ways. We can either apply a *train-and-test* (also known as hold-out) or *cross-validation* procedure.

In train-and-test scenario the (labeled) data is split into two parts of which first becomes the training, second the testing/validation set. The classifier is build on the basis of the training set and then evaluated with use of testing one. The choice of method for splitting data into training and testing set depends on the task at hand. For some tasks this split is imposed by the task, the nature of data or the limitations of the methods to be applied. If there are no constraints on the data split, the training and testing sample is chosen by random. The responses given by the classifier for test table are compared with desired answers (known for our data set) and the classification errors are calculated. The results of such procedure are stored in dedicated object in RSES project interface. The set of results, when displayed (see Figure 3D), provide the user with values of accuracy and coverage of classifier for the entire testing data as well as for particular decision classes. The distribution of errors made by classifier on test data set is shown in detail using the typical confusion matrix (see, e.g., [19]).

In the k-fold cross-validation approach the data is split into k possibly equal parts (folds). Then the train-and-test procedure described above is applied repeatedly k times in such a way that each of k parts becomes the test set while the sum of remaining $k-1$ parts is used as a training set to construct classifier. In each run classification errors are calculated for the current test set. As a result the RSES returns the same set of results as in train-and-test approach but, the values of errors are averages over k iterations. The cross-validation approach to classifier evaluation is commonly used and has a good theoretical background (see [19]), especially for data sets with no more than 1000 objects. In RSES the application of cross-validation scheme is controlled with use of dedicated window which makes it possible to select number of folds and all important parameters for a classifier to be constructed and evaluated.

When using a previously constructed classifier for establishing decision for previously unseen, unlabeled objects, the user have to take care of the proper format of the examples. If during construction of classifier the training set was preprocessed (e.g., discretized) then the same procedure has to be repeated for the new data table. If the format of data and the classifier match, the result is created as a new column in the data table. This column contains the value of decision predicted by classifier for each object.

8.1 Decision Rules

Classifiers based on a set of decision rules are the most established methods in RSES. Several methods for calculation of the decision rule sets are implemented. Also, various methods for transforming and utilizing rule sets are available (see parts C, F1 and G of Figure 3).

The methods for retrieving rules, given a set of reducts, have been already described in Subsection 7.3. These methods produce set of rules by matching training objects against selected set of reducts. In RSES it is possible to calculate such rules instantly, without outputting the set of reducts. But, it has to be stated that the reduct calculation is performed in background anyway.

The two methods for rule calculation that use reducts, i.e., the exhaustive and GA algorithms, are accompanied with another two that are based on slightly different approach. These two are applying a *covering approach*. First of the two utilizes subsets of attributes that are likely to be local (relative) reducts. The details of this method are described in [38]. Second of the covering algorithm is a customized implementation of the LEM2 concept introduced by Jerzy Grzymała-Busse in [9]. In LEM2 a separate-and-conquer technique is paired with rough set notions such as upper and lower approximation. Both covering-based methods for rule calculation tend to produce less rules than algorithms based on explicit reduct calculation. They are also (on average) slightly faster. On the downside, the covering methods sometimes return too few valuable and meaningful rules.

In general, the methods used by RSES to generate rules may produce quite a bunch of them. Naturally, some of the rules may be marginal, erroneous or redundant. In order to provide better control over the rule-based classifiers some simple techniques for transforming rule sets are available in RSES. Note, that before any transformation of rule set it is advisable to examine the statistics produced in RSES for such an entity (see Figure 3 F1). The simplest way to alter the set of decision rules is by filtering them. It is possible to eliminate from rule set these rules that have insufficient support on training sample, or those that point at decision class other than desired.

More advanced operations on rule sets are *shortening* and *generalization.* Rule shortening is a method that attempts to eliminate descriptors from the premise of the rule. The resulting rule is shorter, more general (apply to more training objects) but, it may lose some of its precision. The shortened rule may be less precise, i.e., may give wrong answers (decision) for some of the matching training objects. The level to which we accept decrease of quality in favor of improved generality of rules is known as *shortening ratio* and may be set by the user of RSES. Generalization is the process which attempts to replace single descriptors (conditions) in the rule with more general ones. Instead of a unary condition of the form $a(x) = v$, where $a \in A$, $v \in V_a$, $x \in U$, the algorithm tries to use *generalized descriptors* of the form $a(x) \in V_c$, where $V_c \subset V_a$. Note, that in generalization process the implicit assumption about manageable size of V_a for each $a \in A$ is crucial for the algorithm to be computationally viable. A descriptor (condition) in a rule is replaced by its generalized version if such a change do not decrease size of positive region by the ratio higher than a threshold set by the user.

When we attempt to classify an object from test sample with use of generated rule set it may happen that various rules suggest different decision values. In such conflict situations we need a strategy to resolve controversy and reach a final result (decision). RSES provides a conflict resolution strategy based on voting among rules. In this method each rule that matches the object under consideration casts a vote in favor of the decision value it points at. Votes are summed up and the decision that got majority of votes is chosen. This simple method (present in RSES) may be extended by assigning weights to rules. Each rule then votes with its weight and the decision that has the highest total of weighted votes is the final one. In RSES this method (also known as *Standard voting*) assigns each rule the weight that is equal to the number of training objects supporting this rule.

8.2 Instance Based Method

As an instance based method we implemented the special, extended version of the k nearest neighbors (k-nn) classifier [7]. First the algorithm induces a distance measure from a training set. Then for each test object it assigns a decision based on the k nearest neighbors of this object according to the induced distance measure.

The distance measure ρ for the k-nn classifier is defined as the weighted sum of the distance measures ρ_a for particular attributes $a \in A$:

$$\rho(x,y) = \sum_{a \in A} w_a \cdot \rho_a(a(x), a(y)).$$

Two types of a distance measure are available to the user. The City-SVD metric [6] combines the city-block Manhattan metric for numerical attributes with the Simple Value Difference (SVD) metric for symbolic attributes.

The distance between two numerical values $\rho_a(a(x), a(y))$ is the difference $|a(x) - a(y)|$ taken either as an absolute value or normalized with the range $a_{\max} - a_{\min}$ or with the doubled standard deviation of the attribute a on the training set. The SVD distance $\rho_a(a(x), a(y))$ for a symbolic attribute a is the difference between the decision distributions for the values $a(x)$ and $a(y)$ in the whole training set. Another metric type is the SVD metric. For symbolic attributes it is defined as in the City-SVD metric and for a numerical attribute a the difference between a pair of values $a(x)$ and $a(y)$ is defined as the difference between the decision distributions in the neighborhoods of these values. The neighborhood of a numerical value is defined as the set of objects with similar values of the corresponding attribute. The number of objects considered as the neighborhood size is the parameter to be set by a user.

A user may optionally apply one of two attribute weighting methods to improve the properties of an induced metric. The distance-based method is an iterative procedure focused on optimizing the distance between the training objects correctly classified with the nearest neighbor in a training set. The detailed description of the distance-based method is described in [35]. The accuracy-based method is also an iterative procedure. At each iteration it increases the weights of attributes with high accuracy of the 1-nn classification.

As in the typical k-nn approach a user may define the number of nearest neighbors k taken into consideration while computing a decision for a test object. However, a user may use a system procedure to estimate the optimal number of neighbors on the basis of a training set. For each value k in a given range the procedure applies the leave-one-out k-nn test and selects the value k with the optimal accuracy. The system uses an efficient leave-one-out test for many values of k as described in [8].

When the nearest neighbors of a given test object are found in a training set they vote for a decision to be assigned to the test object. Two methods of nearest neighbors voting are available. In the simple voting all k nearest neighbors are equally important and for each test object the system assigns the most frequent decision in the set of the nearest neighbors. In the distance-weighted voting each nearest neighbor vote is weighted inversely proportional to the distance between a test object and the neighbor. If the option of filtering neighbors with rules is checked by a user, the system excludes from voting all the nearest neighbors that produce a local rule inconsistent with another nearest neighbor (see [8] for details).

The k-nn classification approach is known to be computationally expensive. The crucial time-consuming task is searching for k nearest neighbors in a training set. The basic approach is to scan the whole training set for each test object. To make it more efficient an advanced indexing method is used [35]. It accelerates searching up to several thousand times and allows to test datasets of a size up to several hundred thousand of objects.

8.3 Local Transfer Function Classifier

Local Transfer Function Classifier (LTF-C) is a neural network solving classification problems [36]. LTF-C was recently added to RSES as yet another classification paradigm. Its architecture is very similar to this of Radial Basis Function neural network (RBF) or Support Vector Machines (SVM) – the network has a hidden layer with gaussian neurons connected to an output layer of linear units. The number of inputs corresponds to the number of attributes while the number of linear neurons in output layers equals the number of decision classes. There are some additional restrictions on values of output weights that enable to use an entirely different training algorithm and to obtain very high accuracy in real-world problems.

The training algorithm of LTF-C comprises four types of modifications of the network, performed after every presentation of a training object:

1. changing positions (means) of gaussians,
2. changing widths (deviations) of gaussians, separately for each hidden neuron and attribute,
3. insertion of new hidden neurons,
4. removal of unnecessary or harmful hidden neurons.

As one can see, the network structure is dynamical. The training process starts with an empty hidden layer, adding new hidden neurons when the accuracy is insufficient and removing the units which do not positively contribute to the calculation of correct network decisions. This feature of LTF-C enables automatic choice of the best network size, which is much easier than setting the number of hidden neurons manually. Moreover, this helps to avoid getting stuck in local minima during training, which is a serious problem in neural networks trained with gradient-descend. The user is given some control over the process of network construction/trainig. In particular, it is for user to decide how rigid are the criteria for creating and discarding neurons in the hidden layer. Also, the user may decide whether to perform data (attribute) normalization or not.

8.4 Decomposition Trees

Decomposition trees are used to split data set into fragments not larger than a predefined size. These fragments, after decomposition represented as leafs in decomposition tree, are supposed to be more uniform and easier to cope with decision-wise.

The process of constructing a decomposition tree is fully automated, the user only has to decide about the maximal size of subtable corresponding to the leaf. The algorithm generates conditions one by one on subsequent levels of the tree. The conditions are formulated as a constraints for value of particular attribute. In this way, each node in the tree have an associated template as well as subset of training sample that corresponds to this template. It is possible to generate decomposition trees for data with numerical attributes. In this case discretization is performed during selection of conditions in tree nodes. A dedicated display method for presenting decomposition trees is implemented in RSES GUI, so that the user can examine interactively the resulting decomposition (see Figure 3B). For more information on underlying methods please turn to [16] and [18].

The decomposition tree is mostly used in RSES as a special form of classifier. Usually the subsets of data in the leafs of decomposition tree are used for calculation of decision rules (cf. [3]). The set of data in the leaf is selected by the algorithm in such a way, that (almost) all objects it contains belong to the same decision class. If such a set of objects is used to generate rules, there is a good chance of obtaining some significant decision rules for the class corresponding to the leaf.

The tree and the rules calculated for training sample can be used in classification of unseen cases. The rules originating in decomposition tree may be managed in the same manner as all other decision rules in RSES. It is possible to generalize and shorten them, although such modified rule sets may not be reinserted into original tree.

9 Conclusion

We have presented main features of the Rough Set Exploration system (RSES) hoping that this paper will attract more attention to our software. Interested reader, who wants to learn more about RSES, may download the software and documentation form the Internet (cf. [40]).

RSES will further grow, as we intend to enrich it by adding newly developed methods and algorithms. We hope that many researchers will find RSES an useful tool for experimenting with data, in particular using methods related to rough sets.

Acknowledgement

We would like to express our gratitude to all the current and previous members and supporters of RSES development team, in particular the creators of Rosetta [12, 42] – Aleksander Øhrn and Jan Komorowski. Special tribute should be paid to Professor Andrzej Skowron, who has for many years supervised this project.

Over the years development of our software was significantly supported by several national and international research grants. Currently RSES is supported by the grant 3T11C00226 from Polish Ministry of Scientific Research and Information Technology.

References

1. Alpigini J. J.,Peters J. F.,Skowron A.,Zhong N.(Eds.),Proceedings of 3rd Int. Conf. on Rough Sets and Current Trends in Computing (RSCTC2002), Malvern, PA, 14–16 Oct. 2002. LNAI 2475, Springer-Verlag, Berlin, 2002
2. Bazan J., A Comparison of Dynamic and non-Dynamic Rough Set Methods for Extracting Laws from Decision Tables. In: Skowron A., Polkowski L.(ed.), Rough Sets in Knowledge Discovery 1, Physica-Verlag, Heidelberg, 1998, pp. 321–365
3. Bazan J., Nguyen H. S., Nguyen S. H., Synak P., and Wróblewski J., Rough set algorithms in classification problem. In: Polkowski L., Tsumoto S., Lin T.Y. (eds.), Rough Set Methods and Applications, Physica-Verlag, Heidelberg, 2000, pp. 49–88.

4. Bazan J., Nguyen H. S., Nguyen T. T., Skowron A., Stepaniuk J., Decision rules synthesis for object classification. In: E. Orłowska (ed.), Incomplete Information: Rough Set Analysis, Physica - Verlag, Heidelberg, 1998, pp. 23-57.
5. Bazan J., Skowron A., Ślęzak D. Wróblewski J., Searching for the Complex Decision Reducts: The Case Study of the Survival Analysis. In: Zhong N. et al. (eds.), Proceedings of ISMIS 2003, LNAI 2871, Springer-Verlag, Berlin, 2003, pp. 160—168,
6. Domingos P., Unifying Instance-Based and Rule-Based Induction. Machine Learning, Vol. 24(2), 1996, pp. 141–168.
7. Duda R.O., Hart P.E., Pattern Classification and Scene Analysis. Wiley, New York, 1973.
8. Góra G., Wojna A.G., RIONA: a New Classification System Combining Rule Induction and Instance-Based Learning. Fundamenta Informaticae, Vol. 51(4), 2002, pp. 369–390.
9. Grzymała-Busse, J., A New Version of the Rule Induction System LERS. Fundamenta Informaticae, Vol. 31(1), 1997, pp. 27–39
10. Grzymała-Busse J., Hu M., A comparison of several approaches to missing attribute values in data mining. Proceedings of the Second International Conference on Rough Sets and Current Trends in Computing RSCTC'2000, LNAI 2005, Springer-Verlag, Berlin, 2000, pp. 378–385.
11. Hippe, M.: Towards the classification of musical works: A rough set approach. In: [1], 546-553
12. Komorowski J., Øhrn A., Skowron A., ROSETTA Rough Sets. In: Kloesgen W., Zytkow J. (eds.), Handbook of KDD, Oxford University Press, Oxford, 2002, pp. 554–559.
13. Komorowski J., Pawlak Z., Polkowski L., Skowron A., Rough Sets: A tutorial. In: Pal S.K., Skowron A., Rough Fuzzy Hybridization, Springer Verlag, Singapore, 1999, pp. 3–98
14. Kostek B., Szczuko P., Zwan P., Processing of musical data employing rough sets and artificial neural networks. In: Tsumoto, S., Slowinski, R., Komorowski, J., Grzymala-Busse, J.W.(Eds.), Rough Sets and Current Trends in Computing Lecture Notes in Artificial Intelligence, Vol. 3066, Springer Verlag, Berlin, 2004, pp. 539-548
15. Lazareck L., Ramanna S., Classification of swallowing sound signals: A rough set approach. In: Tsumoto, S., Slowinski, R., Komorowski, J., Grzymala-Busse, J.W.(Eds.), Rough Sets and Current Trends in Computing *Lecture Notes in Artificial Intelligence*, **3066** (2004) 679-684
16. Nguyen Sinh Hoa, Data regularity analysis and applications in data mining. Ph. D. Thesis, Department of Math., Comp. Sci. and Mechanics, Warsaw University, Warsaw, 1999
17. Nguyen S. H., Nguyen H. S., Discretization Methods in Data Mining. In: Skowron A., Polkowski L.(ed.), Rough Sets in Knowledge Discovery 1, Physica Verlag, Heidelberg, 1998, pp. 451–482
18. Nguyen S. H., Skowron A., Synak P., Discovery of data patterns with applications to decomposition and classfification problems. In: L. Polkowski and A. Skowron (eds.), Rough Sets in Knowledge Discovery 2, Physica-Verlag, Heidelberg, 1998, pp. 55–97.
19. Michie D., Spiegelhalter D. J., Taylor C. C., Machine Learning, Neural and Statistical Classification. Ellis Horwood, London, 1994
20. Pawlak Z., Rough sets. International J. Comp. Inform. Science, Vol. 11, 1982, pp. 341–356
21. Pawlak Z., Rough sets and decision tables. Lecture Notes in Computer Science, Vol. 208, Springer Verlag, Berlin, 1985, pp. 186–196

22. Pawlak Z., On rough dependency of attributes in information systems. Bulletin of the Polish Academy of Sciences, Vol. 33, 1985, pp. 551–599
23. Pawlak Z., On decision tables. Bulletin of the Polish Academy of Sciences, Vol. 34, 1986, pp. 553–572
24. Pawlak Z., Decision tables – a rough set approach. Bulletin of EATCS, Vol. 33, 1987, pp. 85–96
25. Pawlak Z., Skowron A., Rough membership functions. In: Yager R., et al.(Eds.), Advances in Dempster Shafer Theory of Evidence, Wiley, N.Y., 1994, pp. 251–271
26. Pawlak Z., In pursuit of patterns in data reasoning from data – the rough set way. In: [1], 2002, pp. 1–9
27. Pawlak, Z.: Rough sets and decision algorithms. In: [39], 2001, pp. 30–45
28. Pawlak Z., Rough Sets: Theoretical Aspects of Reasoning about Data. Kluwer, Dordrecht, 1991
29. Peters J.F., Ramanna S., Towards a software change classification system. Software Quality Journal, Vol. 11(2), 2003, pp. 121–148
30. Skowron A., Rauszer C., The discernibility matrices and functions in information systems. In: Slowinski R., (Ed.),Intelligent Decision Support: Handbook of Applications and Advances in Rough Set Theory, Kluwer Academic Publishers, Dordrecht, 1992, pp. 259–300
31. Skowron A., Polkowski L., Synthesis of decision systems from data tables. In: Lin T.Y, Cercone N. (Eds.), Rough Sets and Data Mining: Analysis for Imprecise Data, Kluwer Academic Publishers, Dordrecht, 1997, pp.331–362
32. Skowron A., Rough Sets in KDD (plenary talk). 16-th World Computer Congress (IFIP'2000). In: Zhongzhi Shi, Boi Faltings, Mark Musen (eds.) Proceedings of Conference on Intelligent Information Processing (IIP2000), Publishing House of Electronic Industry, Beijing, 2000, pp. 1–17.
33. Ślęzak D., Wróblewski J., Classification Algorithms Based on Linear Combinations of Features. Proceedings of PKDD'99, LNAI 1704, Springer-Verlag, Berlin, 1999, pp. 548–553.
34. Valdés J.J, Barton A.J, Gene Discovery in Leukemia Revisited: A Computational Intelligence Perspective. In: R. Orchard et al. (eds.), Proceedings of IEA/AIE 2004, LNAI 3029, Springer-Verlag, Berlin, 2004, pp. 118–127
35. Wojna A.G., Center-Based Indexing in Vector and Metric Spaces. Fundamenta Informaticae, Vol. 56(3), 2003, pp. 285-310.
36. Wojnarski M., LTF-C: Architecture, Training Algorithm and Applications of New Neural Classifier. Fundamenta Informaticae, Vol. 54(1), 2003, pp. 89–105
37. Wróblewski J., Genetic algorithms in decomposition and classification problem. In: Skowron A., Polkowski L.(ed.), Rough Sets in Knowledge Discovery 1, Physica Verlag, Heidelberg, 1998, pp. 471–487
38. Wróblewski J., Covering with Reducts - A Fast Algorithm for Rule Generation. Proceeding of RSCTC'98, LNAI 1424, Springer Verlag, Berlin, 1998, pp. 402-407
39. Ziarko, W., Yao, Y.,Eds.:Proceedings of 2nd Int. Conf. on Rough Sets and Current Trends in Computing (RSCTC2000), Banff, Canada, 16-19 Oct. 2000. *Lecture Notes in Artificial Intelligence*, Springer-Verlag, Berlin (2001)
40. The RSES Homepage, `http://logic.mimuw.edu.pl/~rses`
41. Report from EUNITE World competition in domain of Intelligent Technologies, `http://www.eunite.org/eunite/events/eunite2002/competitionreport2002.htm`
42. The ROSETTA Homepage, `http://rosetta.lcb.uu.se/general/`
43. The WEKA Homepage, `http://www.cs.waikato.ac.nz/~ml`

Rough Validity, Confidence, and Coverage of Rules in Approximation Spaces

Anna Gomolińska*

University of Białystok, Department of Mathematics,
Akademicka 2, 15267 Białystok, Poland
anna.gom@math.uwb.edu.pl

Abstract. From the granular computing perspective, the existing notions of validity, confidence, and coverage of rules in approximation spaces may be viewed as too crisp since granularity of the space is not, in general, taken into account in their definitions. In this article, an extension of the classical approach to a general rough case is discussed. We introduce and investigate graded validity, confidence, and coverage of rules as examples of rough validity, confidence, and coverage, respectively. The graded notions are based on the concepts of graded meaning of formulas and sets of formulas, studied in our earlier works. Among others, the notions of graded validitity, confidence, and coverage refine and extend the classical forms by taking into account granules of information drawn toward objects of an approximation space.

To Andrzej

1 Introduction

In this article, broadly speaking, we are interested in the relationships between satisfiability of premises (preconditions) and satisfiability of conclusions (postconditions) of rules in approximation spaces (ASs). Our main objectives are (i) to introduce rough granular versions of the notions of validity, confidence, and coverage of rules in approximation spaces as well as (ii) to grasp and emphasize various aspects and relationships between such purely logical concepts like entailment or validity of rules and the notions of partial validity of rules, confidence, and coverage which are much more practically oriented.

In logic, inference rules enjoying the property that every object satisfying the premises satisfies the conclusions of the rules are of high value for a practical use in reasoning. Such rules are called valid. Actually, many rules applied in practice are not valid in the logical sense. An index, called confidence, is used to estimate the degree of validity of rules. The standard confidence measures the fraction of

* The author expresses her gratitude to Andrzej Skowron and the anonymous referee for valuable and insightful remarks which helped to improve this article. The research has been partially supported by the grant 3T11C00226 from the Ministry of Scientific Research and Information Technology of the Republic of Poland.

J.F. Peters and A. Skowron (Eds.): Transactions on Rough Sets III, LNCS 3400, pp. 57–81, 2005.

objects satisfying both the premises and the conclusions of rules in the set of all objects satisfying the premises of rules. The higher the confidence of a rule, the more valid is the rule. Confidence is not the only characteristic used to measure the quality of rules. There are many such indices known to the KDD and soft computing communities [1–3]. Examples are coverage of rules working as the converse of confidence, a combination of both the confidence and the coverage, and the support of rules.

As suggested by the name itself, association rules, generated whether from the Pawlak information systems or from databases, are intended to describe formally the relationships (associations) between two finite collections of facts. For instance, an association rule may state that an occurrence of events $e_1, \ldots, e_m$ is associated with an occurrence of events $e'_1, \ldots, e'_n$, or that purchasing goods $g_1, \ldots, g_m$ is associated with purchasing goods $g'_1, \ldots, g'_n$. Decision rules, generated from decision tables, may be viewed as particular association rules which specify relationships between a finite set of conditions and a finite set of decisions. Confidence measures the degree to which the conclusions are associated with the premises, whereas coverage estimates the converse relationship. In an information system, the (standard) support of a rule is the set of objects of the universe, satisfying both the premises and the conclusions of the rule. In databases, the support of an association rule is the fraction of transactions containing both the premises and the conclusions of the rule. In decision (or classification) rules, the roles of premises and conclusions are not symmetric, and therefore both confidence as well as coverage are of interest. Nevertheless, premises and conlusions play symmetric roles in many cases of ordinary association rules. Since the coverage of a rule is the confidence of the converse rule, provided that the latter value is defined, the notion of coverage may be eliminated as being derivative in such cases. Usually, a great number of rules may be generated from an information system or a database. However, only rules having required properties such as a satisfactory quality are of importance in practice. The typical tasks concerning the confidence, the coverage, and the support of rules are (i) to compute all (association, decision, and other) rules, possibly with given premises or conclusions, such that the confidence and/or the coverage exceed some threshold values, and the support satisfies some conditions and (ii) to classify a new object by means of rules of a sufficient quality.

It is characteristic of an approximation space that one can measure the degree of inclusion of a set of objects in a set of objects of the universe by a rough inclusion function and that the universe is granulated into clusters of similar objects by an uncertainty mapping (see, e.g., [4, 5]). In the definitions of the classical notions of confidence and coverage of rules, only the first aspect is reflected. Thus, the classical approach is too crisp from the granular computing perspective. In [6–8], Polkowski and Skowron proposed a calculus of information granules, based on rough mereology [9–11]. According to their framework which comprises both the rough case and the fuzzy case, rules are labels for pairs of granules of information from the semantical perspective. Next, confidence and coverage of rules are obtained by measuring the degree of inclusion of a granule of information in a granule of information. Polkowski and Skowron's idea is profound, very general, yet simple.

In this article, in accordance with the Polkowski – Skowron framework, we discuss an extension of the classical individualistic approach to the concepts of validity, confidence, and coverage of rules to a general rough case. The general rough notions of satisfiability and meaning of sets of formulas, entailment, as well as applicability, validity, confidence, and coverage of rules, introduced in the paper, are pretty abstract. To be more concrete, we define rough graded forms of validity, confidence and coverage, and we investigate them in detail. These concepts, based on the notions of graded meaning of formulas and their sets which we introduced and studied in [12], are entirely grounded on the approximation space framework. The classical case is comprised by the graded one. From the technical point of view, the modification consists in replacement of the classical crisp meanings of formulas and sets of formulas by their graded versions and in dropping the assumption the rough inclusion function be standard. Unlike in the classical case, the rough graded meaning of a finite set of formulas X is, in general, different from the corresponding graded meaning of the conjunction of elements of X. Therefore, rules which are pairs of finite sets of formulas in our approach cannot be reduced to pairs of formulas in general.

In this paper, the cardinality of a set X is denoted by $\#X$, the power set of X by $\wp(X)$, and the Cartesian product of $n > 0$ copies of X by X^n, remembering that $X^1 = X$. Let $n > 0$, $i = 0, \ldots, n$, $(X_i, \leq_i)$ be non-empty partially ordered sets, $\geq_i$ – the converse relation of $\leq_i$, $<_i, >_i$ – the strict versions of $\leq_i, \geq_i$, respectively, and $s, t \in X_0 \times \ldots \times X_n$. Then, $\pi_i(t)$ is the i-th element of t, $\preceq$ is a partial ordering such that $s \preceq t$ if and only if $\forall i = 0, \ldots, n.\pi_i(s) \leq_i \pi_i(t)$, and $\succeq$ is its converse relation. For $0 \leq j \leq n-1$, a mapping $f : X_0 \times \ldots \times X_{n-1} \mapsto X_n$ is called *co-monotone* in the j-th variable if for any $s, t \in X_0 \times \ldots \times X_{n-1}$ such that $\pi_j(s) \leq_j \pi_j(t)$ and s, t being equal on the remaining places, it holds $f(t) \leq_n f(s)$. For example, $f : (\wp(U))^2 \mapsto \wp V$ is co-monotone in the second variable if and only if for any $X, Y, Z \subseteq U$, $Y \subseteq Z$ implies $f(X, Z) \subseteq f(X, Y)$. Given a relation $\varrho \subseteq X_0 \times X_1$ and $X \subseteq X_0$, the image of X is denoted by $\varrho^{\rightarrow}(X)$. The result of concatenation of arbitrary tuples t_1, t_2, composed in this order, is denoted by $t_1 t_2$. Thus, $(1, a, 3)(2, 3) = (1, a, 3, 2, 3)$. For any a, (a) denotes the "tuple" consisting of a only. For simplicity, parentheses will be dropped in formulas if no confusion results.

In Sect. 2, the Pawlak information systems and decision tables are briefly recalled. A concise overview of approximation spaces is given in Sect. 3. Section 4 is devoted to such notions as satisfiability, meaning of formulas and their sets, and entailment. We recall graded forms of these notions and introduce their formal rough versions. Applicability and validity of rules, in the classical, graded, and general rough forms are briefly discussed in Sect. 5. The standard confidence and the standard coverage of rules are overviewed in Sect. 6. Their general rough forms are introduced in Sect. 7. As a particular case, a graded confidence and a graded coverage are proposed and investigated. Section 8 is devoted to the related problem of confidence of association rules in databases. The last section contains concluding remarks.

2 The Pawlak Information Systems

An *information system* (IS) in the sense of Pawlak [13–15] may be viewed as a pair $\mathcal{A} = (U, A)$, where U (the universe of $\mathcal{A}$) is a non-empty finite set of objects and A is a non-empty finite set of attributes. With every attribute a, there is associated a set of values V_a and a is itself a mapping $a : U \mapsto V_a$ assigning values to objects. Along the standard lines, $\mathcal{A}$ is visualized by a table, where each and every object is described by a row of values of attributes. By assumption, objects are known by their description in terms of these values only. Objects having the same description are indiscernible. Every $B \subseteq A$ induces an *indiscernibility* relation on U, $\text{ind}_B \subseteq U^2$, which is an equivalence relation such that for any u, u',

$$(u, u') \in \text{ind}_B \text{ iff } \forall a \in B.a(u) = a(v). \tag{1}$$

For any object u, its equivalence class given by ind_B is called an *elementary granule* of information drawn toward u. Every equivalence relation ϱ on U induces a partition of U which is the family of all elementary granules given by ϱ. Sets of objects are viewed as concepts. Set-theoretical unions of families of elementary granules are called *definable* concepts. The remaining concepts may be approximated by definable concepts only. Let Γu denote the equivalence class of u given by ϱ. Any concept $X \subseteq U$ may be approximated in the rough sense [16, 17] by its *lower* and *upper rough approximations*, $\text{low}^{\cup}X$, $\text{upp}^{\cup}X$, respectively, given by

$$\text{low}^{\cup}X = \bigcup\{\Gamma u \mid \Gamma u \subseteq X\} \text{ and } \text{upp}^{\cup}X = \bigcup\{\Gamma u \mid \Gamma u \cap X \neq \emptyset\}. \tag{2}$$

If $\text{low}^{\cup}X = \text{upp}^{\cup}X$, then X is *exact*; and it is *rough* otherwise.

In practice, many attributes work as binary relations $a \subseteq U \times V_a$. For simplicity, let us consider the case, where each and every object is assigned a non-empty set of values $a^{\rightarrow}\{u\}$. It can happen that an attribute a is actually a mapping but due to some circumstances (uncertainty, incompleteness of information), its precise values are not available. For instance, the only information about $a(u)$ can be that $a(u) \in V \subseteq V_a$. Another variant is that several (or more) values of a on u are provided. This situation may be explained as either contradictory or non-deterministic. In the latter case, all values of a on u are collected into a set V, and $a(u) \in V$ is stated. To capture such and similar cases, the notion of an IS was generalized to a *multi-valued information system* (MVIS) [13–15]. Any MVIS may be viewed as a pair $\mathcal{A}$ as above except that every attribute a is perceived as a mapping $a : U \mapsto \wp V_a$, assigning sets of values to objects. If for every a and u, $\#a(u) = 1$, the MVIS becomes an ordinary IS.

A *decision table* (DT) [18–21] is an IS with the set of attributes split into two disjoint sets of condition attributes and decision attributes, respectively. Typically, the latter set consists of one decision attribute only. Multi-valued versions of DTs are called *multi-valued decision tables* (MVDTs). Broadly speaking, (MV)DTs are used in approximate classification of objects. They provide us with decision rules supporting classification of new objects.

Example 1. As an illustration, let us consider a fragment of a database on second-hand cars, represented in the form of an IS $\mathcal{A} = (U, A)$, where $U = \{1, \ldots, 57\}$ consists of 57 objects (cars) and $A = \{a_1, \ldots, a_6\}$ consists of 6 attributes (1 continuous, 1 integer, and 4 nominal). Every car (object) is specified in terms of such characteristics (attributes) as: make (a_1), model (a_2), body type (a_3), age (a_4), colour (a_5), and price (a_6). The range (the set of values) of a_1 is $V_{a_1} = \{\mathrm{C}, \mathrm{D}, \mathrm{F}, \mathrm{O}, \mathrm{P}\}$, where C is for citroën, D – daewoo, F – fiat, O – opel, P – peugeot; $V_{a_2} = \{\mathrm{as}, \mathrm{as2}, \mathrm{br}, \mathrm{la}, \mathrm{ne}, \mathrm{nu}, \mathrm{ve}, \mathrm{xs}, 306, 406\}$, where as is for astra, as2 – astra II, br– brava, la – lanos, ne – nexia, nu – nubira, ve – vectra, xs – xsara; $V_{a_3} = \{\mathrm{H}, \mathrm{S}, \mathrm{W}\}$, where H is for hatchback, S – sedan, W – station wagon; $V_{a_4} = \{1991, \ldots, 2003\}$; $V_{a_5} = \{\mathrm{bla}, \mathrm{bl}, \mathrm{bu}, \mathrm{gre}, \mathrm{gr}, \mathrm{re}, \mathrm{si}, \mathrm{wh}\}$, where bla is for black, bl – blue, bu – burgundy, gre – green, gr – grey, re – red, si – silver, wh – white; and $V_{a_6} = [10, 76]$ (prices are in thousands of the Polish monetary units (PLN)). System $\mathcal{A}$ is presented in the form of Tabl. 1.

Table 1. The information system $\mathcal{A}$.

u	a_1	a_2	a_3	a_4	a_5	a_6	u	a_1	a_2	a_3	a_4	a_5	a_6
1	C	xs	W	2003	gr	36.0	30	O	as2	H	1998	wh	31.5
2	C	xs	H	1998	bl	18.0	31	O	as2	W	2000	bl	36.2
3	C	xs	H	1998	gr	25.9	32	O	as	H	1998	bl	21.8
4	D	la	S	1998	gre	17.3	33	O	as	W	1996	re	19.2
5	D	la	H	2000	bl	17.4	34	O	as	W	1997	si	19.3
6	D	la	H	1999	gre	18.0	35	O	as2	H	1999	bl	33.8
7	D	la	H	1999	bl	18.0	36	O	as	H	1998	wh	17.2
8	D	la	S	2002	wh	23.6	37	O	as2	H	1998	bla	31.3
9	D	la	S	1999	gre	18.0	38	O	ve	S	2003	si	74.0
10	D	la	H	1998	si	16.4	39	O	ve	S	1996	gre	27.9
11	D	la	H	2000	gr	19.0	40	O	ve	S	1996	bla	27.2
12	D	la	S	1999	si	17.8	41	O	ve	S	1995	re	22.5
13	D	la	H	1999	bla	17.0	42	O	ve	S	1991	re	11.5
14	D	ne	S	1997	bla	13.4	43	O	ve	S	1996	gre	26.5
15	D	ne	H	1996	bu	12.3	44	O	ve	S	1998	si	16.0
16	D	ne	H	1996	re	13.2	45	O	ve	S	1997	gre	30.0
17	D	nu	W	1999	gre	16.8	46	P	306	H	1995	re	22.0
18	D	nu	S	1998	re	19.0	47	P	306	H	1997	bl	18.0
19	D	nu	W	1999	bl	23.5	48	P	306	H	1995	re	14.0
20	D	nu	S	1999	gr	23.5	49	P	306	W	1998	bu	22.4
21	D	nu	W	1999	re	21.0	50	P	306	W	1996	si	17.5
22	F	br	H	1997	gre	16.9	51	P	306	H	1994	bl	14.3
23	F	br	H	1997	bl	16.9	52	P	406	S	1996	re	23.5
24	F	br	H	2001	bl	14.9	53	P	406	S	1997	gr	27.4
25	O	as	W	2000	si	22.5	54	P	406	W	1997	re	26.5
26	O	as2	H	2001	bl	30.3	55	P	406	S	1997	bu	27.0
27	O	as	W	1999	re	22.1	56	P	406	S	1996	gre	24.7
28	O	as	W	2000	si	22.1	57	P	406	S	1996	bl	29.0
29	O	as	H	1999	bl	27.0							

3 Approximation Spaces

The notion of an *approximation space* (AS) was introduced by Skowron and Stepaniuk in [22], and elaborated in [4, 5, 23–25]. Any such space is a triple $\mathcal{M} = (U, \Gamma, \kappa)$, where U is a non-empty set, $\Gamma : U \mapsto \wp U$ is an *uncertainty mapping*, and $\kappa : (\wp U)^2 \mapsto [0, 1]$ is a *rough inclusion function* (RIF). In accordance with the original notation, Γ and κ should be equipped with lists of tuning parameters — dropped for simplicity here — to secure a satisfactory quality of approximation of concepts. The prototype of an AS is the rough approximation space induced by an IS (U, A), where for any $u, u' \in U$, $u' \in \Gamma u$ if and only if u, u' are indiscernible in a considered sense as mentioned in the preceding section. Since κ is not explicitly used, it may be omitted in this case.

Elements of U are called objects and are denoted by u with sub/superscripts if needed. Objects are known by their properties only. Some objects can be similar from an observer's perspective. Indiscernibility is viewed as a special case of similarity, hence every object is necessarily similar to itself. The notion of a *granule of information* was introduced by Zadeh [26]. Taking into account the modification proposed by Lin [27], we can say that a granule of information is a set of objects of some space, drawn together and/or toward some object on the basis of similarity or functionality. The uncertainty mapping Γ assigns to every object u, an elementary granule Γu of objects similar to u. By assumption, $u \in \Gamma u$. In this way, the universe U is covered by a family of elementary granules of information being clusters of objects assigned to elements of U by Γ.

Example 2 (Continuation of Example 1). We granulate the universe U of the system $\mathcal{A}$ by defining an uncertainty mapping Γ as follows[1]. First, for any u, u', let $u' \in \Gamma_{1,2}u$ if and only if $a_1(u), a_1(u') \in \{\mathrm{D}, \mathrm{F}\}$ and $a_2(u), a_2(u') \in \{\mathrm{br}, \mathrm{la}, \mathrm{ne}\}$ or $a_1(u), a_1(u') \in \{\mathrm{C}, \mathrm{D}, \mathrm{O}, \mathrm{P}\}$ and $a_2(u), a_2(u') \in \{\mathrm{as}, \mathrm{as2}, \mathrm{nu}, \mathrm{ve}, \mathrm{xs}, 306, 406\}$. Thus, $\Gamma_{1,2}$ generates a partition of U into two classes, one of which consists of such cars as fiat brava, daewoo lanos, and daewoo nexia, whereas the remaining class is formed of the rest of cars. Next, let $u' \in \Gamma_3 u$ if and only if $a_3(u), a_3(u') = \mathrm{W}$ or $a_3(u), a_3(u') \neq \mathrm{W}$. Also in this case, U is partitioned into two classes. According to Γ_3, the station wagons form one class and the remaining cars constitute the second class. In the sequel, let $u' \in \Gamma_4 u$ if and only if $a_4(u), a_4(u') \leq 1994$ or $a_4(u), a_4(u') \geq 2002$, or $1995 \leq a_4(u), a_4(u') \leq 2001$ and $|a_4(u) - a_4(u')| \leq 1$. $\Gamma_4^{\rightarrow}U$ is a covering but not a partition of U. The binary relation on U induced by Γ_4 is reflexive and symmetric but not transitive, i.e., it is a proper tolerance relation. For instance, car 22 is similar to car 18 which, in the sequel, is similar to car 21. However, car 22 is not similar to car 21. More generally, cars produced in 1994 and earlier are similar to one another, cars from 2002 – 2003 constitute another class of cars being mutually similar, cars from 1995 are similar to the cars from 1995 – 1996, cars from 1996 are similar to the cars from 1995 – 1997, ..., cars from 2000 are similar to the cars from 1999 – 2001, and finally, cars from 2001 are similar to the cars from 2000 – 2001. Next, let $u' \in \Gamma_{a6} u$ if and only if $|a_6(u) - a_6(u')| < 1.5$. Like in the preceding

[1] That is, we determine which objects of $\mathcal{A}$ are similar to one another.

case, $\Gamma_6^{\rightarrow}U$ is merely a covering of U, and the binary relation induced by Γ_6 is a tolerance relation on U. According to Γ_6, cars with prices differing by less than 1500 PLN are similar to one another. Finally, a global uncertainty mapping Γ is defined by

$$\Gamma u \stackrel{\text{def}}{=} \Gamma_{1,2}u \cap \Gamma_3 u \cap \Gamma_4 u \cap \Gamma_6 u, \tag{3}$$

and its values are presented in Tabl. 2.

The notion of a RIF goes back to Łukasiewicz [28] in connection with estimation of probability of implicative formulas. In line with Łukasiewicz's approach, for a finite U and any $X, Y \subseteq U$, the *standard* RIF, $\kappa^{\pounds}$, is defined by

$$\kappa^{\pounds}(X,Y) = \begin{cases} \frac{\#(X\cap Y)}{\#X} & \text{if } X \neq \emptyset \\ 1 & \text{otherwise.} \end{cases} \tag{4}$$

Given an arbitrary non-empty U, we call a RIF *quasi-standard* if it is defined as the standard one for finite first arguments. The idea underlying a general notion of RIF is that a RIF is a function, defined for pairs (X, Y) of sets of objects into the unit interval $[0, 1]$, measuring the degree of inclusion of X in Y. Polkowski and Skowron formulated a formal theory of being-a-part-in-degree, called *rough mereology*, which axiomatically describes a general notion of a RIF [8–11]. In our approach, every RIF $\kappa : (\wp U)^2 \mapsto [0, 1]$ is supposed to satisfy (A1)–(A3) for any $X, Y, Z \subseteq U$: (A1) $\kappa(X, Y) = 1$ if and only if $X \subseteq Y$; (A2) If $X \neq \emptyset$, then $\kappa(X, Y) = 0$ if and only if $X \cap Y = \emptyset$; (A3) If $Y \subseteq Z$, then $\kappa(X, Y) \leq \kappa(X, Z)$. In ASs, sets of objects may be approximated by means of

Table 2. Values of Γ.

u	Γu	u	Γu	u	Γu
1	1	20	20	39	39,40,43,53,55,57
2	2,18,36,47	21	21,27,28,49	40	39,40,43,53,55
3	3,29,55	22	4,10,22,23	41	41,46,52
4	4,6,7,9,10,12,13,22,23	23	4,10,22,23	42	42
5	5-7,9,12,13	24	24	43	39,40,43,53,55
6	4-7,9,11-13	25	19,25,27,28	44	36,44
7	4-7,9,11-13	26	26	45	37,45,57
8	8	27	19,21,25,27,28,49	46	41,46
9	4-7,9,11-13	28	19,21,25,27,28	47	2,18,36,47
10	4,10,12,13,22,23	29	3,29	48	48
11	6,7,9,11,12	30	30,37	49	19,21,27,49
12	4-7,9-13	31	31	50	50
13	4-7,9,10,12,13	32	32	51	51
14	14-16	33	33,34	52	41,52,56
15	14-16	34	33,34	53	39,40,43,53,55
16	14-16	35	35	54	54
17	17	36	2,36,44,47	55	3,39,40,43,53,55
18	2,18,47	37	30,37,45	56	52,56
19	19,25,27,28,49	38	38	57	39,45,57

uncertainty mappings and RIFs in a number of ways. For instance, the lower and upper rough approximations of a concept $X \subseteq U$ may be defined by (2), or in line with [4] by

$$\mathrm{low}X = \{u \mid \kappa(\Gamma u, X) = 1\} \text{ and } \mathrm{upp}X = \{u \mid \kappa(\Gamma u, X) > 0\}. \tag{5}$$

It follows from (A1)–(A3) that $\mathrm{low}^{\cup}X = \bigcup \Gamma^{\rightarrow}\mathrm{low}X$ and $\mathrm{upp}^{\cup}X = \bigcup \Gamma^{\rightarrow}\mathrm{upp}X$. In [29], we investigate various alternative ways of approximation of concepts in more detail.

4 The Meaning of Formulas and Entailment

We briefly recall the idea of introducing degrees into the notions of meaning of formulas and sets of formulas as well as the notion of entailment in an AS, described in detail in [12]. Given an approximation space $\mathcal{M} = (U, \Gamma, \kappa)$, consider a formal language L in which we can express properties of $\mathcal{M}$. Formulas of L are denoted by α, β, γ with subscripts if needed. All formulas of L form a set FOR. Assume that a commutative conjunction ($\wedge$) and a commutative disjunction ($\vee$) are among the connectives of L. For any non-empty finite $X \subseteq$ FOR, $\bigwedge X, \bigvee X$ denote the conjunction and the disjunction of all elements of X, respectively.

Suppose that a crisp relation of satisfiability of formulas for objects of U, denoted by $\models_c$, is given. We use the symbol "c" to denote "crisp" as a synonym of "precise" and as opposite to "soft" or "rough". As usual, $u \models_c \alpha$ reads as "α is *c-satisfied* for u" (see, e.g., [30–33])[2]. The *c-meaning* of α is understood as the set $||\alpha||_c = \{u \mid u \models_c \alpha\}$. Along the standard lines, α is *c-satisfiable* if its c-meaning is non-empty; otherwise α is *c-unsatisfiable*. For simplicity, c will be dropped if no confusion results.

Example 3. Given a DT $\mathcal{A} = (U, A)$, where $A = C \cup D$ and $C \cap D = \emptyset$. C, D are sets of condition and decision attributes, respectively. Values of attributes, i.e., elements of $W = \bigcup\{V_a \mid a \in A\}$ are denoted by v with subscripts if needed. A logical language to express properties of $\mathcal{A}$ may be defined as in [6, 7, 34]. For simplicity, entities (objects) are identified with their names. Constant symbols, being elements of $A \cup W$, are the only terms. The parentheses (,) and commas are auxiliary symbols. Atomic formulas are pairs of terms of the form (a, v), called *descriptors*, where $a \in A$ and $v \in W$. Primitive connectives are $\wedge$ (conjunction) and $\neg$ (negation). $\vee$ (disjunction), $\rightarrow$ (material implication), and $\leftrightarrow$ (double implication) are classically defined by means of $\wedge, \neg$. Compound formulas are formed from the atomic formulas along the standard lines. For any formula α and a set of formulas X, the sets of attribute symbols occurring in α and X are denoted by At_α and At_X, respectively. Notice that $At_X = \bigcup\{At_\alpha \mid \alpha \in X\}$. A formula α is a *template* if it is a conjunction of descriptors with distinct attribute symbols, i.e., if there is a natural number $n > 0$, $a_1, \ldots, a_n \in A$, and $v_1, \ldots, v_n \in W$ such that $\alpha = (a_1, v_1) \wedge \ldots \wedge (a_n, v_n)$ and $\forall i, j = 1, \ldots, n.(a_i = a_j \rightarrow v_i = v_j)$.

[2] Also, α is *c-true* of u or *c-holds* for u.

The relation of c-satisfiability (or, simply, satisfiability) of formulas for objects, $\models$, is defined for any formulas $(a,v), \alpha, \beta$ and an object u, as follows:

$$\begin{aligned} & u \models (a,v) \text{ iff } a(u) = v. \\ & u \models \alpha \wedge \beta \text{ iff } u \models \alpha \text{ and } u \models \beta. \\ & u \models \neg\alpha \text{ iff } u \not\models \alpha. \\ \text{Hence, } & ||(a,v)|| = \{u \mid a(u) = v\}; \\ & ||\alpha \wedge \beta|| = ||\alpha|| \cap ||\beta||; \\ & ||\neg\alpha|| = U - ||\alpha||. \end{aligned} \tag{6}$$

Example 4. Consider a MVDT $\mathcal{A}$ differing from the DT from the example above by that, instead of single values, attributes assign sets of values to objects. Sets of values of attributes are denoted by V with subscripts if needed. A logical language may be defined in lines with Example 3. In this case however, constant symbols are elements of $A \cup \wp W$. Atomic formulas are pairs of the form (a, V), called *(generalized) descriptors*, where $a \in A$ and $V \in \wp W$. A *(generalized) template* is a conjunction of generalized descriptors with distinct attribute symbols. Satisfiability of formulas is defined as in the preceding example except for atomic formulas, where for any descriptor (a, V) and an object u,

$$u \models (a,V) \text{ iff } a(u) \in V. \tag{7}$$

Finally, let us combine the language above with the language described in Example 3. Then, one can easily see that $u \models (a,v)$ if and only if $u \models (a,\{v\})$ or, in other words, $u \models (a,v) \leftrightarrow (a,\{v\})$. Moreover, if V is finite, then $u \models (a,V)$ if and only if $u \models \bigvee\{(a,v) \mid v \in V\}$. Next, for a finite family $X \subseteq \wp W$, $u \models \bigvee\{(a,V) \mid V \in X\}$ if and only if $u \models (a, \bigcup X)$. Similarly, $u \models \bigwedge\{(a,V) \mid V \in X\}$ if and only if $u \models (a, \bigcap X)$.

The crisp satisfiability and meaning are converted into rough graded forms by introducing degrees $t \in [0,1]$:

$$u \models_t \alpha \text{ iff } \kappa(\Gamma u, ||\alpha||) \geq t, \text{ and } ||\alpha||_t = \{u \mid u \models_t \alpha\}. \tag{8}$$

$u \models_t \alpha$ reads as "α is t-satisfied for u"[3], and $||\alpha||_t$ is called the *t-meaning* of α. As earlier, α is *t-satisfiable* if and only if $||\alpha||_t \neq \emptyset$. Notice that the t-satisfiability of a formula α for an object u entirely depends on the crisp satisfiability of α for objects similar to u.

Example 5 (Continuation of Example 1). Let the RIFs, considered in this example and its further extensions, be the standard ones. In this way, we arrive at an AS $\mathcal{M} = (U, \Gamma, \kappa^{\pounds})$, where Γ is defined by (3). Consider formulas $\alpha = (a_5, \mathrm{bl}) \vee (a_5, \mathrm{gr}) \vee (a_5, \mathrm{si})$, $\beta = (a_3, \mathrm{S})$, and $\gamma = (a_4, 1999) \vee \ldots \vee (a_4, 2003)$. Notice that α and γ may be rewritten in line with the MVIS-convention as $(a_5, \{\mathrm{bl}, \mathrm{gr}, \mathrm{si}\})$ and $(a_4, [1999, 2003])$, respectively. Formula α may informally

[3] Equivalently, α is *true* of u or *holds* for u *in degree* t.

read as "the colour (of a car) is blue, grey, or silver", β – "the body type is sedan", and γ – "a car was produced in 1999 – 2003". The crisp meaning $||\alpha||$ of α consists of all blue, grey, or silver cars from the database, $||\beta||$ is the set of all cars that their body type is sedan, and $||\gamma||$ is constituted by all cars of U, produced in 1999 or later. Thus,

$$\begin{aligned} ||\alpha|| &= \{1-3, 5, 7, 10-12, 19, 20, 23-26, 28, 29, 31, 32, 34, 35, 38, \\ &\quad 44, 47, 50, 51, 53, 57\}; \\ ||\beta|| &= \{4, 8, 9, 12, 14, 18, 20, 38-45, 52, 53, 55-57\}; \\ ||\gamma|| &= \{1, 5-9, 11-13, 17, 19-21, 24-29, 31, 35, 38\}. \end{aligned}$$

In the next step, the degrees of satisfiability of α, β, γ for objects of U are calculated (Tabl. 3). For instance,

$$\begin{aligned} ||\alpha||_{0.5} &= \{1-3, 5-7, 9-13, 18-20, 22-29, 31-36, 38, 44, 47, 50, 51\}; \\ ||\beta||_{0.5} &= \{8, 20, 38-46, 52, 53, 55-57\}; \\ ||\gamma||_{0.5} &= \{1, 4-9, 11-13, 17, 19-21, 24-29, 31, 35, 38, 49\}. \end{aligned}$$

The 0.5-meaning $||\alpha||_{0.5}$ of α contains every such car $u \in U$ that at least 50% of cars similar to u are blue, grey, or silver. For instance, car 10 is in $||\alpha||_{0.5}$ since from six cars similar to the car 10, two cars are silver and one is blue. Next, a car u is in $||\beta||_{0.5}$ if and only if at least 50% of cars similar to u are sedan. For instance, all cars similar to the car 39 are sedan and, hence, it belongs to $||\beta||_{0.5}$. Finally, a car u is in $||\gamma||_{0.5}$ if and only if at least 50% of cars similar to u were produced in 1999 or later. As a negative example take car 3 produced in 1998. Only one third of cars similar to 3 satisfies γ, viz., car 29.

Table 3. Degrees of inclusion of Γu in the crisp meanings of α, β, γ.

u	1	2	3	4	5	6	7	8	9	10	11	12	13	14	15
$\kappa^{\pounds}(\Gamma u, \|\|\alpha\|\|)$	1	$\frac{1}{2}$	$\frac{2}{3}$	$\frac{4}{9}$	$\frac{1}{2}$	$\frac{1}{2}$	$\frac{1}{2}$	0	$\frac{1}{2}$	$\frac{1}{2}$	$\frac{3}{5}$	$\frac{5}{9}$	$\frac{1}{2}$	0	0
$\kappa^{\pounds}(\Gamma u, \|\|\beta\|\|)$	0	$\frac{1}{4}$	$\frac{1}{3}$	$\frac{1}{3}$	$\frac{1}{3}$	$\frac{3}{8}$	$\frac{3}{8}$	1	$\frac{3}{8}$	$\frac{1}{3}$	$\frac{2}{5}$	$\frac{1}{3}$	$\frac{3}{8}$	$\frac{1}{3}$	$\frac{1}{3}$
$\kappa^{\pounds}(\Gamma u, \|\|\gamma\|\|)$	1	0	$\frac{1}{3}$	$\frac{5}{9}$	1	$\frac{7}{8}$	$\frac{7}{8}$	1	$\frac{7}{8}$	$\frac{1}{3}$	1	$\frac{7}{9}$	$\frac{3}{4}$	0	0
u	16	17	18	19	20	21	22	23	24	25	26	27	28	29	30
$\kappa^{\pounds}(\Gamma u, \|\|\alpha\|\|)$	0	0	$\frac{2}{3}$	$\frac{3}{5}$	1	$\frac{1}{4}$	$\frac{1}{2}$	$\frac{1}{2}$	1	$\frac{3}{4}$	1	$\frac{1}{2}$	$\frac{3}{5}$	1	0
$\kappa^{\pounds}(\Gamma u, \|\|\beta\|\|)$	$\frac{1}{3}$	0	$\frac{1}{3}$	0	1	0	$\frac{1}{4}$	$\frac{1}{4}$	0	0	0	0	0	0	0
$\kappa^{\pounds}(\Gamma u, \|\|\gamma\|\|)$	0	1	0	$\frac{4}{5}$	1	$\frac{3}{4}$	0	0	1	1	1	$\frac{5}{6}$	1	$\frac{1}{2}$	0
u	31	32	33	34	35	36	37	38	39	40	41	42	43	44	45
$\kappa^{\pounds}(\Gamma u, \|\|\alpha\|\|)$	1	1	$\frac{1}{2}$	$\frac{1}{2}$	1	$\frac{3}{4}$	0	1	$\frac{1}{3}$	$\frac{1}{5}$	0	0	$\frac{1}{5}$	$\frac{1}{2}$	$\frac{1}{3}$
$\kappa^{\pounds}(\Gamma u, \|\|\beta\|\|)$	0	0	0	0	0	$\frac{1}{4}$	$\frac{1}{3}$	1	1	1	$\frac{2}{3}$	1	1	$\frac{1}{2}$	$\frac{2}{3}$
$\kappa^{\pounds}(\Gamma u, \|\|\gamma\|\|)$	1	0	0	0	1	0	0	1	0	0	0	0	0	0	0
u	46	47	48	49	50	51	52	53	54	55	56	57			
$\kappa^{\pounds}(\Gamma u, \|\|\alpha\|\|)$	0	$\frac{1}{2}$	0	$\frac{1}{4}$	1	1	0	$\frac{1}{5}$	0	$\frac{1}{3}$	0	$\frac{1}{3}$			
$\kappa^{\pounds}(\Gamma u, \|\|\beta\|\|)$	$\frac{1}{2}$	$\frac{1}{4}$	0	0	0	0	1	1	0	$\frac{5}{6}$	1	1			
$\kappa^{\pounds}(\Gamma u, \|\|\gamma\|\|)$	0	0	0	$\frac{3}{4}$	0	0	0	0	0	0	0	0			

The crisp satisfiability and meaning as well as their graded forms are combined into one case, where $t \in T \stackrel{\text{def}}{=} [0,1] \cup \{c\}$. The natural ordering of reals, $\leq$, is extended to a partial ordering on T by assuming $c \leq c$. Thus, the set of all formulas t-satisfied for an object u is denoted by $|u|_t$:

$$|u|_t = \{\alpha \mid u \models_t \alpha\}. \tag{9}$$

Assume that the crisp satisfiability of formulas is generalized to sets of formulas along the classical lines, i.e., a set of formulas X is *satisfied* for u, $u \models X$, if and only if $\forall \alpha \in X.u \models \alpha$. Then, the crisp meaning of X is defined as the set $||X|| = \{u \mid u \models X\}$. Hence, $||X|| = \bigcap\{||\alpha|| \mid \alpha \in X\}$. As usual, X is *satisfiable* if $||X||$ is non-empty, and it is *unsatisfiable* in the remaining case. The crisp concept of meaning is refined by introducing degrees $t \in T_1 \stackrel{\text{def}}{=} T \times [0,1]$. Consider a RIF $\kappa^* : (\wp\text{FOR})^2 \mapsto [0,1]$. Then,

$$u \models_t X \text{ iff } \kappa^*(X, |u|_{\pi_1 t}) \geq \pi_2 t, \text{ and } ||X||_t = \{u \mid u \models_t X\}. \tag{10}$$

$u \models_t X$ reads as "X is *t-satisfied* for u", and $||X||_t$ is the *t-meaning* of X. Next, X is referred to as *t-satisfiable* if $||X||_t \neq \emptyset$; otherwise it is *t-unsatisfiable*. Observe that $||X||_{(c,1)} = ||X||$. It is worth noticing that the graded form of satisfiability of a set of formulas X is parameterized by two, formally independent parameters. The first one (i.e., $\pi_1 t$) may serve for switching between the basic (crisp) mode ($\pi_1 t = c$) and the graded mode ($\pi_1 t \in [0,1]$). In the latter case, $\pi_1 t$ may be tuned for achieving a satisfactory degree of satisfiability of single formulas. The second parameter (i.e., $\pi_2 t$) may be used in the control of the risk of error caused by the fact of disregarding of some elements of X. For instance, if κ^* is quasi-standard and $\pi_2 t = 0.9$, then at least 90% of formulas of X have to be satisfied in a required sense.

Example 6 (Continuation of Example 1). Consider sets of formulas $X = \{\alpha, \beta\}$ and $Y = \{\gamma\}$, where α, β, γ are as in Example 5. Observe that X is satisfied in the crisp sense by any sedan from the database which is blue, grey, or silver. The meaning of Y is the same as the meaning of γ. Thus, we arrive at the following sets of objects satisfying X, Y, respectively:

$$||X|| = ||\alpha|| \cap ||\beta|| = \{12, 20, 38, 44, 53, 57\}.$$
$$||Y|| = ||\gamma|| = \{1, 5-9, 11-13, 17, 19-21, 24-29, 31, 35, 38\}.$$

Now, we compute the graded $(0.5, t)$-meanings of X, Y for $t \in [0,1]$. Clearly, $||X||_{(0.5,0)} = ||Y||_{(0.5,0)} = U$. In the remaining cases, we obtain

$$||X||_{(0.5,t)} = ||\alpha||_{0.5} \cup ||\beta||_{0.5} = \{1-3, 5-13, 18-20, 22-29, 31-36, 38-47, 50-53, 55-57\} \text{ if } 0 < t \leq 0.5;$$
$$||X||_{(0.5,t)} = ||\alpha||_{0.5} \cap ||\beta||_{0.5} = \{20, 38, 44\} \text{ if } t > 0.5;$$
$$||Y||_{(0.5,t)} = ||\gamma||_{0.5} = \{1, 4-9, 11-13, 17, 19-21, 24-29, 31, 35, 38, 49\}.$$

Thus, if $0 < t \leq 0.5$, then X is satisfied by u in degree $(0.5, t)$ if and only if u satisfies either α or β, or both formulas in degree 0.5. If $t > 0.5$, then X is

$(0.5, t)$-satisfied by exactly these objects which satisfy both α and β in degree 0.5. The $(0.5, t)$-satisfiability of Y is the same as the 0.5-satisfiability of γ.

The graded meaning is an example of a soft (or, more precisely, rough) meaning. As a matter of fact, a large number of examples of rough meaning of formulas/sets of formulas may be defined [35]. For instance, a rough meaning of a formula α may be defined as $\mathrm{upp}||\alpha||$. In this case, α is satisfied for u if and only if α is c-satisfied for some object similar to u. In the case of sets of formulas, the meaning of X may be defined, e.g., as $\mathrm{upp}||X||_t$, where $t \in T_1$. Then, X is satisfied for u if and only if X is t-satisfied for some object similar to u. In yet another example, let the meaning of X be $\mathrm{low}||X_1|| \cap \mathrm{upp}||X_2||$, where $\{X_1, X_2\}$ is a partition of X. In this case, X is satisfied for u if and only if X_1 and X_2 are satisfied in the crisp sense for all objects similar to u and for some object similar to u, respectively. Consider a general case, where a set of relations of rough satisfiability of sets of formulas for objects of an AS $\mathcal{M}$ is given. Every such relation, $\models\!\approx$, uniquely determines a rough-meaning mapping $\nu_{\models\approx} : \wp\mathrm{FOR} \mapsto \wp U$ such that for any set of formulas X,

$$\nu_{\models\approx} X \stackrel{\mathrm{def}}{=} \{u \mid u \models\!\approx X\}. \tag{11}$$

On the other hand, given a rough-meaning mapping ν, the corresponding relation of rough satisfiability of sets of formulas for objects, $\models\!\approx_\nu$, may be defined for any set of formulas X and an object u, by

$$u \models\!\approx_\nu X \stackrel{\mathrm{def}}{\leftrightarrow} u \in \nu X. \tag{12}$$

Observe that

$$\models\!\approx_{\nu_{\models\approx}} = \models\!\approx \text{ and } \nu_{\models\approx_\nu} = \nu. \tag{13}$$

Let $\models\!\approx$ and ν correspond to each other. Then, $u \models\!\approx X$ reads as "X is *satisfied* for u in the sense of $\models\!\approx$", and νX is the *meaning* of X in the sense of ν[4]. As usual, X is *satisfiable* in the sense of $\models\!\approx$ (or ν) if and only if νX is non-empty.

The notion of entailment in an AS $\mathcal{M}$ is closely related to the notion of meaning of a set of formulas. In the crisp case, we can say along the standard lines that a set of formulas X *entails* a set of formulas Y in $\mathcal{M}$, $X \models Y$, if and only if satisfaction of X implies satisfaction of Y for every object u, i.e., $||X|| \subseteq ||Y||$. In [12], a graded form of entailment, relativized to an AS, was introduced. For $t = (t_1, t_2, t_3)$, $t_1, t_2 \in T_1$, and $t_3 \in [0, 1]$, X *t-entails* Y in $\mathcal{M}$[5], written $X \models_t Y$, if and only if for sufficiently many objects of U, a sufficient degree of satisfaction of X implies a sufficient degree of satisfaction of Y, where sufficiency is determined by t. Formally,

$$X \models_t Y \text{ iff } \kappa(||X||_{t_1}, ||Y||_{t_2}) \geq t_3. \tag{14}$$

[4] We can also say that X is *satisfied* for u in the sense of ν, and νX is the *meaning* of X in the sense of $\models\!\approx$.

[5] In other words, X *entails* Y *in degree* t.

Observe that $X \models_{((c,1),(c,1),1)} Y$ if and only if $X \models Y$. For simplicity, we can write $X \models_t \beta$ and $\alpha \models_t Y$ instead of $X \models_t \{\beta\}$ and $\{\alpha\} \models_t Y$, respectively. The above forms of entailment are merely special cases of a general notion of entailment in $\mathcal{M}$. Given rough-meaning mappings ν_1, ν_2 and $t \in [0,1]$, we say that X *t-entails* Y in $\mathcal{M}$ in the sense of ν_1, ν_2 if and only if $\kappa(\nu_1 X, \nu_2 Y) \geq t$.

Example 7 (Continuation of Example 1). Let X, Y be sets of formulas from Example 6. Observe that X does not entail Y in $\mathcal{M}$. The counterexample is car 44 being a silver sedan produced in 1998. Formally, $44 \models X$ but $44 \not\models Y$. Let $t_1 \in (0, 0.5]$, $t_2 \in (0,1]$, and $t_3 \in [0,1]$. In the graded case, X entails Y in degree $((0.5, t_1), (0.5, t_2), t_3)$ if $t_3 \leq 10/23$. That is, for any $t_3 \leq 10/23$, the fraction of cars, produced in 1999 or later in degree 0.5, in the class of all cars of U which are blue, grey, or silver in degree 0.5 and/or their body type is sedan in degree 0.5, is not less than t_3. The fact that Y entails X in degree $((0.5, t_2), (0.5, t_1), t_3)$, where $t_3 \leq 5/6$, may be described similarly.

5 Rules, Their Applicability and Validity

In our approach, a rule over L is a pair $r = (X, Y)$ of finite sets of formulas of L, where Y is non-empty in addition[6]. X is called the set of *premises (preconditions)* and Y – the set of *conclusions (postconditions)* of r. Rules without premises are *axiomatic*. The set of all rules over L is denoted by RUL. Whenever convenient, the sets of premises and conclusions of r will be denoted by P_r and C_r, respectively. If X is non-empty as well, the rule $r^{-1} = (Y, X)$ is the *converse* of r. For simplicity, rules $(X, \{\beta\})$ and $(\{\alpha\}, Y)$ may be written as (X, β) and (α, Y), respectively.

Example 8. Association rules, known also as *local dependencies* of attributes, are of interest within the framework of (MV)ISs. Association rules over the languages from Examples 3, 4 are of the form $r = (X, Y)$, where X, Y are non-empty sets of formulas such that $At_X \cap At_Y = \emptyset$. An association rule r expresses a local dependency between sets of attributes At_X, At_Y in a given (MV)IS. We call r *simple* if X, Y are sets of descriptors and for every distinct $\alpha, \beta \in X \cup Y$, $At_\alpha \neq At_\beta$[7]. In a given (MV)DT, an association rule r is called a *decision (classification)* rule if At_X consists of condition attributes and At_Y consists of decision attributes only.

Given an AS $\mathcal{M}$ and a rough-meaning mapping $\nu : \wp\mathrm{FOR} \mapsto \wp U$, we can say along the standard lines that r is *applicable* (resp., *applicable* to u) if and only if $\nu P_r \neq \emptyset$ (resp., $u \in \nu P_r$) which means satisfiability (satisfiability for u) of P_r in the sense of ν. In the graded case introduced and studied in [36], where $\nu = \|\cdot\|_t$ and $t \in T_1$, a rule r is said to be *t-applicable* (resp., *t-applicable* to u)

[6] Suppose that for some natural numbers m, n, $X = \{\alpha_1, \ldots, \alpha_m\}$ and $Y = \{\beta_1, \ldots, \beta_n\}$. r may be written equivalently in the "if – then" form as $\alpha_1, \ldots, \alpha_m \Rightarrow \beta_1, \ldots, \beta_n$.

[7] More generally, one might require $\bigwedge X, \bigwedge Y$ be templates.

if and only if $||P_r||_t \neq \emptyset$ (resp., $u \in ||P_r||_t$). Let us note that $(c,1)$-applicability coincides with the crisp applicability since $||\cdot||_{(c,1)} = ||\cdot||$.

In the crisp case, we say that a rule r is *valid (true)* in $\mathcal{M}$ if and only if r is applicable and the set of premises P_r entails the set of conclusions C_r, i.e., $\emptyset \neq ||P_r|| \subseteq ||C_r||$. In other words, r is valid in $\mathcal{M}$ if and only if there is at least one object of U satisfying all premises of r and for every such object u, u satisfies all conclusions as well. Thus, our notion of crisp validity of rules is obtained from the classical concept (see Example 9) by excluding the counterintuitive cases, where inapplicable rules are nevertheless true. If $\mathcal{M}$ is known, the reference to it will be omitted for simplicity. First, observe that an axiomatic rule r is valid if and only if $||C_r|| = U$, i.e., $\models C_r$. Recall that for any non-empty finite set of formulas X, $||X|| = ||\bigwedge X||$. Hence, r and the rule $(\bigwedge P_r, \bigwedge C_r)$ may be used interchangeably as long as the crisp form of validity of rules is regarded provided that r is non-axiomatic. In the remaining case, validity of r is equivalent with validity of $(\emptyset, \bigwedge C_r)$. In particular, if $\neg$ (negation) and $\rightarrow$ (implication), understood classically, occur in L, $P_r = \{\alpha\}$, and $C_r = \{\beta\}$, then r is valid if and only if $\emptyset \neq ||\alpha|| \subseteq ||\beta||$ if and only if $\not\models \neg\alpha$ and $\models \alpha \rightarrow \beta$.

Example 9. In the Hilbert-style logical systems, a rule r is defined as a non-empty, possibly infinite set of pairs of the form (X, β) called sequents of r. In the classical propositional logic [37], two kinds of validity of rules are considered: validity in a proper sense (or, simply, validity) and truth-preservingness[8]. Thus, r is *valid (true)* if and only if for every $(X, \beta) \in r$, X entails β, i.e., for every truth assignment f, $f \models X$ implies $f \models \beta$. On the other hand, r is *truth-preserving* if and only if for every $(X, \beta) \in r$, $\models X$ implies $\models \beta$. Every valid rule is also truth-preserving, but not vice versa. *Modus ponens* is a well-known example of a valid rule. The *uniform substitution* rule is truth-preserving, yet it is not valid.

From the perspective of approximate reasoning and decision making, the claim rules be valid in the crisp sense is usually too restrictive. Rules which are only true to some extent can be valuable as well. Having this motivation in mind and taking granulation of the universe of $\mathcal{M}$ into account, we arrive at the notion of graded validity with degrees of the form $t = (t_1, t_2, t_3)$, where $t_1, t_2 \in T_1$ and $t_3 \in [0,1]$. Thus, a rule r is *t-valid (true)*[9] in $\mathcal{M}$ if and only if r is t_1-applicable (i.e., $||P_r||_{t_1} \neq \emptyset$) and the set of premises P_r t-entails the set of conclusions C_r in $\mathcal{M}$ (i.e., $\kappa(||P_r||_{t_1}, ||C_r||_{t_2}) \geq t_3$). It is worthy to note that the $((c,1),(c,1),1)$-validity and the crisp validity coincide. In a general case, for $t \in [0,1]$, r is said to be *t-valid (true)* in the sense of ν_1, ν_2 as earlier if and only if r is applicable in the sense of ν_1 (i.e., $\nu_1 P_r \neq \emptyset$) and P_r *t-entails* C_r in the sense of ν_1, ν_2 (i.e., $\kappa(\nu_1 P_r, \nu_2 C_r) \geq t$).

Example 10 (Continuation of Example 1). Consider an association rule $r = (X, Y)$, where the set of premises X and the set of conlusions Y are from Example 6. Rule r formally expresses the intuition that the property of cars of being produced in years 1999 – 2003 is associated with the property of being

[8] Pogorzelski distinguishes *normal* and *unfailing* rules, respectively.

[9] Equivalently, r is *valid (true) in degree t*.

blue, grey, or silver and the property of being sedan. Let $t_2 \in (0,1]$. First, observe that r is both applicable as well as $(0.5, t_1)$-applicable for any $t_1 \in [0,1]$ since $||X||, ||X||_{(0.5,t_1)} \neq \emptyset$. That is, there is at least one car in our database which satisfies the premises of r in the crisp sense and at least one car which satisfies the premises of r in degree $(0.5, t_1)$. r is not valid in the crisp sense because $X \not\models Y$ (see Example 7). In other words, not every blue, grey, or silver sedan in U was produced in 1999 or later. In the graded case, r is valid in degree $((0.5, t_1), (0.5, t_2), t_3)$ if (i) $t_1 = 0$ and $t_3 \in [0, 8/19]$ or (ii) $t_1 \in (0, 0.5]$ and $t_3 \in [0, 10/23]$, or (iii) $t_1 \in (0.5, 1]$ and $t_3 \in [0, 2/3]$. For the comments on case (ii), which was presented equivalently in terms of the graded entailment, see Example 7.

6 Confidence and Coverage of Rules

The confidence (accuracy) and the coverage of a rule are examples of indices intended to measure the quality of rules in databases, (MV)ISs, and ASs [1–3]. Assume that U is finite, and the crisp satisfiability and the crisp meaning of formulas are defined as in Example 3 or Example 4. The set of objects satisfying both the premises and the conclusions of a rule r is called the *support* of r and it is denoted by $S(r)$. That is, $S(r) = ||P_r|| \cap ||C_r|| = ||P_r \cup C_r||$. The *confidence* and the *coverage* of r, $q_1^{\pounds} r$ and $q_2^{\pounds} r$, respectively, are defined as follows:

$$
\begin{aligned}
q_1^{\pounds} r &\stackrel{\text{def}}{=} \frac{\#S(r)}{\#||P_r||} \text{ if } ||P_r|| \neq \emptyset; \\
q_2^{\pounds} r &\stackrel{\text{def}}{=} \frac{\#S(r)}{\#||C_r||} \text{ if } ||C_r|| \neq \emptyset.
\end{aligned} \tag{15}
$$

Let us note that

$$
\begin{aligned}
q_1^{\pounds} r &= \kappa^{\pounds}(||P_r||, ||C_r||) \text{ if } ||P_r|| \neq \emptyset; \\
q_2^{\pounds} r &= \kappa^{\pounds}(||C_r||, ||P_r||) \text{ if } ||C_r|| \neq \emptyset.
\end{aligned} \tag{16}
$$

Henceforth, we shall call these forms of confidence and coverage *standard*. The notion of standard confidence of a rule may be traced back to Łukasiewicz [28] who first used this index to estimate the probability of implicative formulas. The standard confidence and the standard coverage of rules are commonly known in the rough-set and KDD communities. Among others, they were used by Tsumoto to investigate medical diagnostic rules [38]. As argued earlier, a non-axiomatic rule r may be represented by $(\bigwedge P_r, \bigwedge C_r)$. An interesting observation, noted already in [38], is that the confidence of r may serve as a measure of sufficiency of the condition $\bigwedge P_r$ for $\bigwedge C_r$, whereas the coverage of r may be a measure of necessity of the condition $\bigwedge C_r$ for $\bigwedge P_r$. It is worth emphasizing that confidence is defined for applicable rules only and coverage – for rules with satisfiable sets of conclusions. If for an arbitrary, unnecessarily applicable rule r, the confidence were defined as $\kappa^{\pounds}(||P_r||, ||C_r||)$, then inapplicable rules would be given the greatest degree of confidence. Indeed, if $||P_r|| = \emptyset$, then $\kappa^{\pounds}(||P_r||, ||C_r||) = 1$.

Such an approach would be in accordance with the philosophy of classical logic, where implication with a false antecedent is true and, hence, inapplicable rules are valid[10]. In our opinion, however, such a view is far from being intuitive. As regarding confidence, $q_1^{\pounds} r > 0$ is a minimal requirement to be satisfied by a "valuable" applicable rule r, which means that $S(r) \neq \emptyset$. Clearly, the higher the value $q_1^{\pounds} r$ the more trustworthy is the rule. In the limit case, where $q_1^{\pounds} r = 1$, r is simply valid in the sense defined in the preceding section.

Mollestad [39] proposed a framework for discovery of rough-set default rules in DTs, where the notion of a default rule is based on the concept of standard confidence.

Example 11. Reasoning by default in the absence of knowledge that a given case is exceptional is relatively common in real life. In [40], Reiter proposed a formalism called *default logic*, based on the notion of a *default rule*, to model such forms of commonsense reasoning. Serving as inference rules to derive conclusions under incomplete and/or uncertain knowledge, default rules have two sets of preconditions: one of them consists of ordinary premises and the other specifies exceptions making a given rule inapplicable. Within the rough set framework, Mollestad introduced another notion of a default rule, yet the general motivation was similar:

> *Default reasoning is a framework which is well suited to modelling [...] common-sense reasoning, making uncertain decisions, i.e. drawing conclusions in absence of knowledge. Indeterminism in information systems lends itself quite naturally to a model using default rules, that capture the* general patterns *that exist in the data, at the expense of making an incorrect decision in certain cases.*

A decision rule r is called a *default rule* with respect to a threshold value $0 < t \leq 1$ if and only if the standard confidence of r is at least t. Hence, an applicable rule r is a default rule with respect to t if and only if r is $((c, 1), (c, 1), t)$-valid.

Example 12 (Continuation of Example 1). Consider the rule r from Example 10. The support of r, $S(r)$, consists of all blue, grey, or silver cars of U of type sedan, produced in 1999 or later. That is, $S(r) = ||X|| \cap ||Y|| = \{12, 20, 38\}$. It also holds

$$\kappa^{\pounds}(||X||, ||Y||) = \frac{1}{2} \text{ and } \kappa^{\pounds}(||Y||, ||X||) = \frac{3}{22}.$$

Hence, the standard confidence of r equals $q_1^{\pounds} r = 0.5$. In other words, 50% of all cars of U which are blue, grey, or silver sedan were produced in 1999 or later. Next, the standard coverage of r equals $q_2^{\pounds} r = 3/22$, and this is the fraction of blue, grey, or silver cars of type sedan in the class of all cars of U produced in 1999 or later. Clearly, the standard coverage of r may be viewed as the standard confidence of the converse rule $r^{-1} = (Y, X)$.

---

[10] Recall remarks from Sect. 5.

7 Confidence and Coverage: A Rough Granular Approach

The characteristics $q_1^{\pounds}, q_2^{\pounds}$ estimate the quality of rules by measuring the degree of containment of a set of objects in a set of objects of an AS $\mathcal{M}$. In the crisp case, objects are treated as individuals, non-connected with one another. In other words, the semantical structure of the universe U of $\mathcal{M}$, granulated by Γ, is disregarded. Being motivated by the granular computing ideas [6–8], we propose to eliminate this apparent drawback by replacing the crisp meanings of sets of formulas in (16) by their rough counterparts, serving a particular purpose. Moreover, arbitrary (i.e., unnecessarily standard) RIFs may be assumed. In what follows, we first define general abstract notions of rough confidence, rough coverage, and rough support of rules. In the next step, we investigate a special, so-called "graded" case in detail. To start with, let us consider rough-meaning mappings ν_1, ν_2 and a RIF $\kappa : (\wp U)^2 \mapsto [0,1]$ as earlier. We define partial mappings of *rough confidence* and *rough coverage*, $q_1, q_2 : \mathrm{RUL} \overset{\circ}{\mapsto} [0,1]$, respectively, where for any rule r,

$$\begin{aligned} q_1 r &\overset{\text{def}}{=} \kappa(\nu_1 P_r, \nu_2 C_r) \text{ if } \nu_1 P_r \neq \emptyset; \\ q_2 r &\overset{\text{def}}{=} \kappa(\nu_1 C_r, \nu_2 P_r) \text{ if } \nu_1 C_r \neq \emptyset. \end{aligned} \tag{17}$$

$q_1 r$, called the *rough confidence (accuracy)* of r, is defined for the rules applicable in the sense of ν_1 only. On the other hand, $q_2 r$, the *rough coverage* of r, is defined only for the rules with sets of conclusions satisfiable in the sense of ν_1. Given $t \in [0,1]$ and a rule r applicable in the sense of ν_1, let us observe that $q_1 r \geq t$ if and only if r is t-valid in the sense of ν_1, ν_2. On the other hand, if r is non-axiomatic and its set of conclusions is satisfiable in the sense of ν_1, then $q_2 r \geq t$ if and only if the converse rule r^{-1} is t-valid in the sense of ν_1, ν_2. Finally, the *rough support* of r in the sense of ν_1, ν_2 may be defined as the set

$$S_{\nu_1,\nu_2} r = \nu_1 P_r \cap \nu_2 C_r. \tag{18}$$

That is, an object u supports the rule r in the sense of ν_1, ν_2 if and only if u satisfies the premises of r in the sense of ν_1 and the conclusions of r in the sense of ν_2. It is easy to show that for finite U and $\kappa = \kappa^{\pounds}$,

$$\begin{aligned} q_1 r &= \frac{\# S_{\nu_1,\nu_2} r}{\# \nu_1 P_r} \text{ if } \nu_1 P_r \neq \emptyset; \\ q_2 r &= \frac{\# S_{\nu_2,\nu_1} r}{\# \nu_1 C_r} \text{ if } \nu_1 C_r \neq \emptyset. \end{aligned} \tag{19}$$

A particular case of rough confidence, rough coverage, and rough support is obtained by taking $\nu_1 = ||\cdot||_{t_1}$ and $\nu_2 = ||\cdot||_{t_2}$, where $t_1, t_2 \in T_1$. In this case, the partial mappings of confidence and coverage are denoted by $q_{1,t}, q_{2,t}$, respectively, and the rough support of r is denoted by $S_t r$, where $t = (t_1, t_2)$. The *t-confidence* of r, $q_{1,t} r$, is defined for the t_1-applicable rules only, and

$$q_{1,t} r = \kappa(||P_r||_{t_1}, ||C_r||_{t_2}). \tag{20}$$

Table 4. The graded supports $S_t r$ of r.

$t_1 \backslash t_2$	$\{0\}$	$(0,1]$
$\{0\}$	U	$\|\|\gamma\|\|_{0.5}$
$(0,0.5]$	$\|\|\alpha\|\|_{0.5} \cup \|\|\beta\|\|_{0.5}$	$(\|\|\alpha\|\|_{0.5} \cup \|\|\beta\|\|_{0.5}) \cap \|\|\gamma\|\|_{0.5}$
$(0.5,1]$	$\|\|\alpha\|\|_{0.5} \cap \|\|\beta\|\|_{0.5}$	$\|\|\alpha\|\|_{0.5} \cap \|\|\beta\|\|_{0.5} \cap \|\|\gamma\|\|_{0.5}$

On the other hand, the *t-coverage* of r, $q_{2,t}r$, is defined for the rules with t_1-satisfiable sets of conclusions only, and

$$q_{2,t}r = \kappa(||C_r||_{t_1}, ||P_r||_{t_2}). \tag{21}$$

An object u *t-supports* a rule r[11], $u \in S_t r$, if and only if $u \in ||P_r||_{t_1} \cap ||C_r||_{t_2}$, i.e., if and only if the set of premises of r is t_1-satisfied and the set of conlusions of r is t_2-satisfied for u. It is easy to see that for finite U, $\kappa = \kappa^{\pounds}$, and $t = ((c,1),(c,1))$, the t-confidence, the t-coverage, and the t-support of r coincide with the standard confidence, coverage, and support of this rule, respectively.

Example 13 (Continuation of Example 1). For the rule r from Example 10, $t_1, t_2 \in [0,1]$, and $t = ((0.5,t_1),(0.5,t_2))$, we compute the graded t-supports and values of the t-confidence and the t-coverage of r[12]. In Tabl. 4, the t-supports of r (i.e., the intersections $||X||_{(0.5,t_1)} \cap ||Y||_{(0.5,t_2)}$) are given for various values of t_1, t_2. In particular, for any $t_2 > 0$,

$$S_{((0.5,0.5),(0.5,t_2))}r = \{1, 5-9, 11-13, 19, 20, 24-29, 31, 35, 38\};$$
$$S_{((0.5,1),(0.5,t_2))}r = \{20, 38\}.$$

In other words, the $((0.5,0.5),(0.5,t_2))$-support of r consists of cars $1, 5-9, 11-13, 19, 20, 24-29, 31, 35, 38$ being the only cars in our database which are blue, grey, or silver in degree 0.5 (i.e., for every car u, at least 50% of cars similar to u are blue, grey, or silver) AND/OR their type is sedan in degree 0.5 (i.e., for every car u, the body type is sedan for at least 50% of cars similar to u), AND they were produced in 1999 or later in degree 0.5 (i.e., for every car u, at least 50% of cars similar to u are from $1999-2003$). The $((0.5,1),(0.5,t_2))$-support of r consists of cars $20, 38$. This case differs from the previous one in that "AND/OR" is strengthened to "AND".

Next, we calculate degrees of the standard rough inclusion of $||X||_{(0.5,t_1)}$ in $||Y||_{(0.5,t_2)}$, and vice versa (Tabl. 5). Hence, we immediately obtain values of the graded t-confidence and t-coverage of r, $q_{1,t}r$ and $q_{2,t}r$, respectively (Tabl. 6). For example, for $t = ((0.5,0.7),(0.5,0.8))$, $q_{1,t}r = 2/3$ and $q_{2,t}r = 1/12$. The first result means that two third of cars of U being blue, grey, or silver in degree 0.5 and sedan in degree 0.5 were produced in 1999 or later in degree 0.5. The latter result is understood as "1/12 of cars of U, produced in 1999 or later in degree 0.5, are blue, grey, or silver in degree 0.5 and their type is sedan in degree 0.5".

[11] Equivalently, u *supports* r *in degree* t.

[12] In connection with our previous remarks on coverage of association rules which are not decision rules, the notion of graded coverage is of minor importance. We compute its values for the sake of illustration.

Table 5. Degrees of inclusion of $||X||_{(0.5,t_1)}$ in $||Y||_{(0.5,t_2)}$, and vice versa.

	$\kappa^{\pounds}(\lVert X\rVert_{(0.5,t_1)}, \lVert Y\rVert_{(0.5,t_2)})$		$\kappa^{\pounds}(\lVert Y\rVert_{(0.5,t_2)}, \lVert X\rVert_{(0.5,t_1)})$	
$t_1 \backslash t_2$	$\{0\}$	$(0,1]$	$\{0\}$	$(0,1]$
$\{0\}$	1	$\frac{8}{19}$	1	1
$(0,0.5]$	1	$\frac{10}{23}$	$\frac{46}{57}$	$\frac{5}{6}$
$(0.5,1]$	1	$\frac{2}{3}$	$\frac{1}{19}$	$\frac{1}{12}$

Table 6. Values of $q_{1,t}r$ and $q_{2,t}r$.

	$q_{1,t}r$		$q_{2,t}r$		
$t_1 \backslash t_2$	$\{0\}$	$(0,1]$	$\{0\}$	$(0,0.5]$	$(0.5,1]$
$\{0\}$	1	$\frac{8}{19}$	1	$\frac{46}{57}$	$\frac{1}{19}$
$(0,0.5]$	1	$\frac{10}{23}$	1	$\frac{5}{6}$	$\frac{1}{12}$
$(0.5,1]$	1	$\frac{2}{3}$	1	$\frac{5}{6}$	$\frac{1}{12}$

If the crisp meaning is classical, then for any non-empty finite set of formulas X, $||\bigwedge X|| = ||X||$. In the graded case, it holds $||\bigwedge X||_t \subseteq ||X||_{(t,1)}$, where $t \in T$ but the converse is not true in general. The situation does not improve if some $0 \leq s < 1$ is taken instead of 1, either. Thus, rules r and $(\bigwedge P_r, \bigwedge C_r)$ are not interchangeable as regarding the graded confidence and coverage. Below, we present properties of the rough graded form of confidence. In the sequel, properties of the rough graded coverage are given as well.

Theorem 1. *For any formulas α,β, finite sets of formulas X,Y,Z, where $Y \neq \emptyset$, a t_1-applicable rule r, $s = (s_1, s_2)$, $t = (t_1, t_2)$, $s, t \in T_1 \times T_1$, and $t_3 \in [0,1]$, we have:*

(*a*) $q_{1,t}r \geq t_3$ iff $P_r \models_{t(t_3)} C_r$.

(*b*) $q_{1,t}r > 0$ iff $\exists s \in (0,1].P_r \models_{t(s)} C_r$.

(*c*) If $\alpha \models \beta$, $||\alpha||_{\pi_1 t_1} \neq \emptyset$, and $\pi_2 t_2 = 0 \vee (\pi_2 t_1 > 0 \wedge \pi_1 t_2 \leq \pi_1 t_1)$, then $q_{1,t}(\alpha,\beta) = 1$.

(*d*) If $\alpha \models \beta$ and $||X||_{t_1} \neq \emptyset$, then $q_{1,t}(X,\alpha) \leq q_{1,t}(X,\beta)$.

(*e*) If $C_r \subseteq P_r$, $\pi_2 t_1 = 1$, and $\pi_1 t_2 \leq \pi_1 t_1$, then $q_{1,t}r = 1$.

(*f*) If $||Y||_{t_1} \neq \emptyset$ and $t_2 \preceq t_1$, then $q_{1,t}(Y,Y) = 1$.

(*g*) If $||X||_{t_1} \neq \emptyset$, $Y \subseteq Z$, and $\pi_2 t_2 = 1$, then $q_{1,t}(X,Z) \leq q_{1,t}(X,Y)$.

(*h*) If $X \subseteq Z$, $||Z||_{t_1} \neq \emptyset$, and $\pi_2 t_1 = 1$, then $q_{1,t}(X,Y) = 1$ implies $q_{1,t}(Z,Y) = 1$, and $q_{1,t}(X,Y) = 0$ implies $q_{1,t}(Z,Y) = 0$.

(*i*) If $\pi_2 t_2 = 1$, then $q_{1,t}r = 1$ iff $\forall \alpha \in C_r.q_{1,t}(P_r,\alpha) = 1$.

(*j*) If $s_1 = t_1$ and $s_2 \preceq t_2$, then $q_{1,t}r \leq q_{1,s}r$.

(*k*) If $||X||_{t_1} \neq \emptyset$, $||X \cap Z||_{t_1} = U$, and $\pi_2 t_1 = 1$, then $q_{1,t}(X - Z, Y) = q_{1,t}(X,Y)$.

(*l*) If $||X \cup Z||_{t_1} \neq \emptyset$, $||Z - X||_{t_1} = U$, and $\pi_2 t_1 = 1$, then $q_{1,t}(X \cup Z, Y) = q_{1,t}(X,Y)$.

Proof. We prove (b), (d), and (h) only. In case (b), $q_{1,t}r > 0$ if and only if $\kappa(||P_r||_{t_1}, ||C_r||_{t_2}) > 0$ (by the definition of $q_{1,t}$) if and only if there is $s \in (0,1]$ such that $\kappa(||P_r||_{t_1}, ||C_r||_{t_2}) \geq s$ if and only if there is $s \in (0,1]$ such that $P_r \models_{t(s)} C_r$ (by (14)). For (d) assume (d1) $\alpha \models \beta$ and (d2) $||X||_{t_1} \neq \emptyset$. By the latter assumption, $(X, \alpha), (X, \beta)$ are t_1-applicable rules for which

$$q_{1,t}(X, \alpha) = \kappa(||X||_{t_1}, ||\{\alpha\}||_{t_2}) \text{ and } q_{1,t}(X, \beta) = \kappa(||X||_{t_1}, ||\{\beta\}||_{t_2}).$$

If $\pi_2 t_2 = 0$, then $||\{\alpha\}||_{t_2} = ||\{\beta\}||_{t_2} = U$ by (10). Hence,

$$\kappa(||X||_{t_1}, ||\{\alpha\}||_{t_2}) = \kappa(||X||_{t_1}, ||\{\beta\}||_{t_2}) = 1.$$

In the remaining case, where $\pi_2 t_2 > 0$,

$$||\{\alpha\}||_{t_2} = ||\alpha||_{\pi_1 t_2} \text{ and } ||\{\beta\}||_{t_2} = ||\beta||_{\pi_1 t_2}$$

by (8)–(10). By (d1), $||\alpha|| \subseteq ||\beta||$. Hence, for any object u,

$$\kappa(\Gamma u, ||\alpha||) \leq \kappa(\Gamma u, ||\beta||)$$

in virtue of (A3). As a consequence, $||\alpha||_{\pi_1 t_2} \subseteq ||\beta||_{\pi_1 t_2}$ by (8). In the sequel,

$$\kappa(||X||_{t_1}, ||\alpha||_{\pi_1 t_2}) \leq \kappa(||X||_{t_1}, ||\beta||_{\pi_1 t_2})$$

by (A3). Thus, $q_{1,t}(X, \alpha) \leq q_{1,t}(X, \beta)$. For (h) assume that (h1) $X \subseteq Z$, (h2) $||Z||_{t_1} \neq \emptyset$, and (h3) $\pi_2 t_1 = 1$. We first show that (h4) $||Z||_{t_1} \subseteq ||X||_{t_1}$. Let u be an object such that $u \in ||Z||_{t_1}$. By (10) and (h3), $Z \subseteq |u|_{\pi_1 t_1}$. Hence, $X \subseteq |u|_{\pi_1 t_1}$ by (h1). As a consequence, $u \in ||X||_{t_1}$. By (h4), $||X||_{t_1} \subseteq ||Y||_{t_2}$ implies $||Z||_{t_1} \subseteq ||Y||_{t_2}$, and $||X||_{t_1} \cap ||Y||_{t_2} = \emptyset$ implies $||Z||_{t_1} \cap ||Y||_{t_2} = \emptyset$. By the definition of $q_{1,t}$ and (h2), $q_{1,t}(X, Y) = 1$ implies $q_{1,t}(Z, Y) = 1$, and $q_{1,t}(X, Y) = 0$ implies $q_{1,t}(Z, Y) = 0$. □

Some comments on the properties above are in order. Recall that $t(t_3)$ denotes the result of concatenation of t with the "tuple" consisting of t_3 only. The graded validity and the graded entailment are closely related to each other, viz., a t_1-applicable rule is $t(t_3)$-valid if and only if the set of premises of the rule $t(t_3)$-entails the set of its conclusions in virtue of (a). Properties (c), (e), and (f) provide us with sufficient conditions for the $t(1)$-validity in some cases of rules. By (d), the degree of t-confidence of a t_1-applicable rule with only one conclusion α will not decrease after replacing the conclusion by a formula entailed by α. It follows from (g) that the degree of t-confidence of a t_1-applicable rule will not decrease in case some conlusions are deleted and $\pi_2 t_2 = 1$ (which means that $t_2 = (s, 1)$ for some $s \in T$). Now assume that $\pi_2 t_1 = 1$. In virtue of (h), extending the set of premises of a t_1-applicable rule in such a way the resulting rule be still t_1-applicable does not change the degree of t-confidence if this degree equals 0 or 1. Where $\pi_2 t_2 = 1$, it follows from (i) that the sufficient and necessary condition for the $t(1)$-validity of a t_1-applicable rule r is that every rule, having the same premises as r and some conclusion of r as the only conclusion, is $t(1)$-valid. Taking the partial mapping of t-confidence $q_{1,t}$ as a partial mapping of three variables t_1, t_2, and r, property (j) may read as co-monotonicity of the graded confidence in the second variable (i.e., t_2). For $\pi_2 t_1 = 1$, properties (k), (l) may be concisely read as "the degree of t-confidence of a rule does not change by removing or, respectively, adding premises which are t_1-satisfiable for all objects".

Theorem 2. *For any formulas α, β, finite sets of formulas X, Y, Z, where $Y \neq \emptyset$, a rule r such that C_r is t_1-satisfiable, $s = (s_1, s_2)$, $t = (t_1, t_2)$, $s, t \in T_1 \times T_1$, and $t_3 \in [0, 1]$, it holds that:*

(*a*) $q_{2,t}r \geq t_3$ iff $C_r \models_{t(t_3)} P_r$.
(*b*) If $P_r, ||P_r||_{t_1} \neq \emptyset$, then $q_{1,t}r^{-1} = q_{2,t}r$ and $q_{2,t}r^{-1} = q_{1,t}r$.
(*c*) $q_{2,t}r > 0$ iff $\exists s \in (0, 1].C_r \models_{t(s)} P_r$.
(*d*) If $P_r \subseteq C_r$, $\pi_2 t_1 = 1$, and $\pi_1 t_2 \leq \pi_1 t_1$, then $q_{2,t}r = 1$.
(*e*) If $||Y||_{t_1} \neq \emptyset$ and $t_2 \preceq t_1$, then $q_{2,t}(Y, Y) = 1$.
(*f*) If $||Z||_{t_1} \neq \emptyset$, $Y \subseteq Z$, and $\pi_2 t_1 = 1$, then $q_{2,t}(X, Y) = 1$ implies $q_{2,t}(X, Z) = 1$, and $q_{2,t}(X, Y) = 0$ implies $q_{2,t}(X, Z) = 0$.
(*g*) If $||Y||_{t_1} \neq \emptyset$, $Z \subseteq X$, and $\pi_2 t_2 = 1$, then $q_{2,t}(X, Y) \leq q_{2,t}(Z, Y)$.
(*h*) If $\pi_2 t_2 = 1$, then $q_{2,t}r = 1$ iff $\forall \alpha \in P_r.q_{2,t}(\alpha, C_r) = 1$.
(*i*) If $s_1 = t_1$ and $s_2 \preceq t_2$, then $q_{2,t}r \leq q_{2,s}r$.
(*j*) If $||Y||_{t_1} \neq \emptyset$, $||X \cap Z||_{t_2} = U$, and $\pi_2 t_2 = 1$, then $q_{2,t}(X - Z, Y) = q_{2,t}(X, Y)$.
(*k*) If $||Y||_{t_1} \neq \emptyset$, $||Z - X||_{t_2} = U$, and $\pi_2 t_2 = 1$, then $q_{2,t}(X \cup Z, Y) = q_{2,t}(X, Y)$.

Observe that the degree of t-coverage of a rule r with a t_1-satisfiable set of conclusions equals 1 if and only if $||C_r||_{t_1} \subseteq ||P_r||_{t_2}$. Hence, if $P_r = \emptyset$ (i.e., r is axiomatic), then the t-coverage of r equals 1 since $||\emptyset||_{t_2} = U$. The graded coverage is closely related to the graded entailment, viz., it holds by (a) that the t-coverage of a rule with a t_1-satisfiable set of conclusions is equal or greater than t_3 if and only if the set of conclusions of the rule $t(t_3)$-entails the set of its premises. In virtue of (b), for a non-axiomatic t_1-applicable rule r with a t_1-satisfiable set of conclusions, the t-coverage and the t-confidence coincide with the t-confidence and the t-coverage of the converse rule, respectively. Except for decision rules, the roles of premises and conlusions are symmetric in the case of association rules. Therefore, the notion of coverage of an association rule may be treated less seriously, and the stress is laid on the concepts of support and confidence. Properties (d), (e) provide us with sufficient conditions for a rule be given the greatest degree of t-coverage in some cases. (f), (g) correspond to Theorem 1(h), (g), respectively. By (h), where $\pi_2 t_2 = 1$, the t-coverage of a rule r with a t_1-satisfiable set of conclusions equals 1 if and only if the t-coverage of every rule with the same conclusions as r and some premise of r as the only premise equals 1. Like in the case of Theorem 1(j), property (i) may be expressed as co-monotonicity of $q_{2,t}$ in the second variable if $q_{2,t}$ is viewed as a partial mapping of three variables t_1, t_2, and r. Thanks to (j), (k), the degree of t-coverage of a rule, where $\pi_2 t_2 = 1$, does not change by removing or, respectively, adding premises which are t_2-satisfiable for all objects.

8 Confidence of Association Rules in Databases

Confidence of association rules in databases is related to the topics of our article. The problem of mining association rules in large databases was introduced in [41] for sales transaction databases. Since then, it has attracted quite a lot of attention among the KDD community (see, e.g., [41–45]). Discovery of interesting association rules is of importance. For instance, association rules may be used for prediction of consumers' behaviour since they describe which bundles of commodies to purchase are associated with one another. An exemplary association rule may state that 30% of transactions which purchase sausage, mustard, and bread also purchase beer and charcoal.

The starting point is a finite non-empty set I of positive literals called *items*. Sets of items are referred to as *itemsets*. The itemsets observed actually are called *transactions*. A *database* D of size n on I consists of n transactions and may be defined as a pair $D = (U, \tau)$, where $U = \{1, \ldots, n\}$ and $\tau : U \mapsto \wp I$. Thus, elements of U may be viewed as labels (names) of transactions. An *association rule* is an expression of the form $X \Rightarrow Y$, where X, Y are non-empty disjoint itemsets. For any itemset $X \subseteq I$, the *support* of X, written $s(X)$, is defined as the fraction of transactions containing X. By the *support* of $X \Rightarrow Y$ we understand the support of $X \cup Y$. Where $s(X) > 0$, the *confidence* of $X \Rightarrow Y$ is then defined as the ratio $s(X \cup Y)/s(X)$. The main task concerning association rules consists in generating rules for which support and confidence are not less than some threshold values.

Association rules in D, defined above, may be viewed as simple association rules in an IS, where objects are elements of U. With every item i, we can associate an attribute $a_i : U \mapsto \{0, 1\}$ such that for any $u \in U$, $a_i(u) = 1$ if and only if $i \in \tau_u$ (i.e., item i belongs to the transaction τ_u). Every itemset X is described by its characteristic function $f_X : I \mapsto \{0, 1\}$ as usual, i.e., for any item i, $f_X(i) = 1$ if and only if $i \in X$. In our case, $a_i(u) = 1$ if and only if $f_{\tau_u}(i) = 1$. For any itemsets X, Y, it holds that $X \subseteq Y$ if and only if for every $i \in I$, $f_X(i) = 1$ implies $f_Y(i) = 1$. The crisp meaning of X, $||X||$, may be defined as the set of labels of transactions containing X, i.e., $||X|| \stackrel{\text{def}}{=} \{u \in U \mid X \subseteq \tau_u\}$. Observe that $||X|| \cap ||Y|| = ||X \cup Y||$ and, moreover,

$$s(X) = \frac{\#||X||}{n} \text{ and } s(X \cup Y) = \frac{\#(||X|| \cap ||Y||)}{n}. \tag{22}$$

Where $||X|| \neq \emptyset$, the confidence of $X \Rightarrow Y$ is exactly the standard confidence of $r = (X, Y)$ since

$$\frac{s(X \cup Y)}{s(X)} = \frac{\#(||X|| \cap ||Y||)}{\#||X||}. \tag{23}$$

We can go a step further and define a graded form of meaning of an itemset, e.g., by replacing the crisp inclusion by a RIF in the definition of $|| \cdot ||$. Next, rough graded forms of support and confidence of association rules may be introduced. Such soft association rules would take into account granulation of the space of items I. Clearly, it is important the semantical structure of I be considered since items are hardly ever independent from one another in real life.

9 Concluding Remarks

The classical notions of validity, confidence, and coverage of rules disregard the granular structure of ASs. Being motivated by ideas of rough granular computing, we extend the classical approach to a general rough case. As exemplary rough validity, confidence, and coverage of rules we take rough graded notions of validity, confidence, and coverage which we define and investigate in detail. From the technical point of view, the modification of the classical notions, resulting in the graded versions, consists in replacement of the crisp meaning of a set of formulas by its rough graded counterpart and in dropping the assumption the RIF be standard. As a consequence, granulation of the universe of an AS is more seriously taken into account in the definitions of the notions mentioned above.

In the future, the aim will be at practical applications of the graded and other rough forms of confidence and coverage in estimation of the quality of rules. Apart from the typical tasks like computation of association rules (and, in particular, decision rules), possibly given some premises or conclusions and rough forms of support, confidence, and coverage satisfying required conditions, a rough classifier will be constructed, where the quality of classification rules will be determined in terms of the rough support, confidence, and coverage of rules.

In this paper, the relationship between our approach and the fuzzy one has not been explored, and this may be another direction for the future research.

References

1. An, A., Cercone, N.: Rule quality measures for rule induction systems: Description and evaluation. J. Comput. Intelligence **17** (2001) 409–424
2. Bruha, I.: Quality of decision rules: Definitions and classification schemes for multiple rules. In Nakhaeizadeh, G., Taylor, C.C., eds.: Machine Learning and Statistics, The Interface, New York, John Wiley & Sons (1997) 107–131
3. Stepaniuk, J.: Knowledge discovery by application of rough set models. In Polkowski, L., Tsumoto, S., Lin, T.Y., eds.: Rough Set Methods and Applications: New Developments in Knowledge Discovery in Information Systems, Heidelberg New York, Physica (2001) 137–233
4. Skowron, A., Stepaniuk, J.: Tolerance approximation spaces. Fundamenta Informaticae **27** (1996) 245–253
5. Stepaniuk, J.: Approximation spaces, reducts and representatives. In Polkowski, L., Skowron, A., eds.: Rough Sets in Knowledge Discovery. Volume 2., Heidelberg, Physica (1998) 109–126
6. Polkowski, L., Skowron, A.: Rough sets: A perspective. In Polkowski, L., Skowron, A., eds.: Rough Sets in Knowledge Discovery. Volume 1., Heidelberg, Physica (1998) 31–56
7. Polkowski, L., Skowron, A.: Towards adaptive calculus of granules. In Zadeh, L.A., Kacprzyk, J., eds.: Computing with Words in Information/Intelligent Systems. Volume 1., Heidelberg, Physica (1999) 201–228

8. Polkowski, L., Skowron, A.: Rough mereological calculi of granules: A rough set approach to computation. J. Comput. Intelligence **17** (2001) 472–492
9. Polkowski, L., Skowron, A.: Rough mereology. In: LNAI. Volume 869., Berlin, Springer (1994) 85–94
10. Polkowski, L., Skowron, A.: Rough mereology: A new paradigm for approximate reasoning. Int. J. Approximated Reasoning **15** (1996) 333–365
11. Polkowski, L., Skowron, A.: Rough mereology in information systems. A case study: Qualitative spatial reasoning. In Polkowski, L., Tsumoto, S., Lin, T.Y., eds.: Rough Set Methods and Applications: New Developments in Knowledge Discovery in Information Systems, Heidelberg New York, Physica (2001) 89–135
12. Gomolińska, A.: A graded meaning of formulas in approximation spaces. Fundamenta Informaticae **60** (2004) 159–172
13. Pawlak, Z.: Mathematical Foundations of Information Retrieval. Volume 101 of CC PAS Report., Warsaw (1973)
14. Pawlak, Z.: Information systems – Theoretical foundations. Information Systems **6** (1981) 205–218
15. Pawlak, Z.: Information Systems. Theoretical Foundations. Wydawnictwo Naukowo-Techniczne, Warsaw (1983) In Polish.
16. Pawlak, Z.: Rough sets. Int. J. Computer and Information Sciences **11** (1982) 341–356
17. Pawlak, Z.: Rough Sets. Theoretical Aspects of Reasoning About Data. Kluwer, Dordrecht (1991)
18. Pawlak, Z.: Rough sets and decision tables. In: LNCS. Volume 208., Berlin, Springer (1985) 186–196
19. Pawlak, Z.: On decision tables. Bull. Polish Acad. Sci. Tech. **34** (1986) 553–572
20. Pawlak, Z.: Decision tables – A rough set approach. Bull. EATCS **33** (1987) 85–96
21. Pawlak, Z.: Bayes' theorem – The rough set perspective. In Inuiguchi, M., Hirano, S., Tsumoto, S., eds.: Rough Set Theory and Granular Computing. Volume 125 of Studies in Fuzziness and Soft Computing., Berlin Heidelberg, Springer (2003) 1–12
22. Skowron, A., Stepaniuk, J.: Generalized approximation spaces. In: Proc. 3rd Int. Workshop on Rough Sets and Soft Computing, San Jose, 1994, November, 10-12. (1994) 156–163
23. Peters, J.F.: Approximation space for intelligent system design patterns. Engineering Applications of Artificial Intelligence **17** (2004) 1–8
24. Skowron, A., Stepaniuk, J.: Generalized approximation spaces. In Lin, T.Y., Wildberger, A.M., eds.: Soft Computing, Simulation Councils, San Diego (1995) 18–21
25. Skowron, A., Stepaniuk, J.: Information granules and approximation spaces. In: Proc. 7th Int. Conf. on Information Processing and Management of Uncertainty in Knowledge-based Systems (IPMU'98), Paris, France, 1998, July, 8-10. (1998) 354–361
26. Zadeh, L.A.: Outline of a new approach to the analysis of complex system and decision processes. IEEE Trans. on Systems, Man, and Cybernetics **3** (1973) 28–44
27. Lin, T.Y.: Granular computing on binary relations I: Data mining and neighborhood systems. In Polkowski, L., Skowron, A., eds.: Rough Sets in Knowledge Discovery. Volume 1., Heidelberg, Physica (1998) 107–121
28. Łukasiewicz, J.: Die logischen Grundlagen der Wahrscheinlichkeitsrechnung. In Borkowski, L., ed.: Jan Łukasiewicz – Selected Works, Amsterdam London Warsaw, North-Holland, Polish Scientific Publ. (1970) 16–63 First published in Kraków, 1913.

29. Gomolińska, A.: A comparative study of some generalized rough approximations. Fundamenta Informaticae **51** (2002) 103–119
30. Barwise, J.: Information Flow. The Logic of Distributed Systems. Cambridge University Press, UK (1997)
31. Devlin, K.: Logic and Information. Cambridge University Press, UK (1991)
32. Kripke, S.A.: Semantical analysis of intuitionistic logic. In Crossley, J.N., Dummet, M.A.E., eds.: Formal Systems and Recursive Functions, Amsterdam, North-Holland (1965) 92–130
33. Skowron, A., Stepaniuk, J., Peters, J.F.: Rough sets and infomorphisms: Towards approximation of relations in distributed environments. Fundamenta Informaticae **54** (2003) 263–277
34. Pawlak, Z., Polkowski, L., Skowron, A.: Rough sets and rough logic: A KDD perspective. In Polkowski, L., Tsumoto, S., Lin, T.Y., eds.: Rough Set Methods and Applications: New Developments in Knowledge Discovery in Information Systems, Heidelberg New York, Physica (2001) 583–646
35. Gomolińska, A.: Satisfiability and meaning in approximation spaces. In Lindemann et al., G., ed.: Proc. Workshop Concurrency, Specification, and Programming (CS&P'2004), Caputh, Germany, 2004, September, 24-26. Volume 170 of Informatik-Berichte., Berlin, Humboldt-Universität zu Berlin (2004) 229–240
36. Gomolińska, A.: A graded applicability of rules. In Tsumoto, S., Słowiński, R., Komorowski, J., W., G.B.J., eds.: Proc. 4th Int. Conf. Rough Sets and Current Trends in Computing (RSCTC'2004), Uppsala, Sweden, 2004, June, 1-5. Volume 3066 of LNAI., Berlin Heidelberg, Springer (2004) 213–218
37. Pogorzelski, W.A.: Notions and Theorems of Elementary Formal Logic. Białystok Division of Warsaw University, Białystok (1994)
38. Tsumoto, S.: Modelling medical diagnostic rules based on rough sets. In Polkowski, L., Skowron, A., eds.: Proc. 1st Int. Conf. Rough Sets and Current Trends in Computing (RSCTC'1998), Warsaw, Poland, 1998, June, 22-26. Volume 1424 of LNAI., Berlin Heidelberg, Springer (1998) 475–482
39. Mollestad, T.: A Rough Set Approach to Data Mining: Extracting a Logic of Default Rules from Data, Ph.D. Dissertation. NTNU, Trondheim (1997)
40. Reiter, R.: A logic for default reasoning. Artificial Intelligence J. **13** (1980) 81–132
41. Agrawal, R., Imieliński, T., Swami, A.: Mining association rules between sets of items in large databases. In: Proc. ACM SIGMOD Int. Conf. on Management of Data, Washington, D.C. (1993) 207–216
42. Agrawal, R., Srikant, R.: Fast algorithms for mining association rules. In: Proc. 20th Conf. on Very Large Databases, Santiago, Chile (1994) 487–499
43. Kryszkiewicz, M.: Fast discovery of representative association rules. In Polkowski, L., Skowron, A., eds.: Proc. 1st Int. Conf. Rough Sets and Current Trends in Computing (RSCTC'1998), Warsaw, Poland, 1998, June, 22-26. Volume 1424 of LNAI., Berlin Heidelberg, Springer (1998) 214–221
44. Lin, T.Y., Louie, E.: Association rules with additional semantics modeled by binary relations. In Inuiguchi, M., Hirano, S., Tsumoto, S., eds.: Rough Set Theory and Granular Computing. Volume 125 of Studies in Fuzziness and Soft Computing., Berlin Heidelberg, Springer (2003) 147–156
45. Murai, T., Nakata, M., Sato, Y.: Association rules from a point of view of conditional logic. In Inuiguchi, M., Hirano, S., Tsumoto, S., eds.: Rough Set Theory and Granular Computing. Volume 125 of Studies in Fuzziness and Soft Computing., Berlin Heidelberg, Springer (2003) 137–145

Knowledge Extraction from Intelligent Electronic Devices

Ching-Lai Hor[1] and Peter A. Crossley[2]

[1] Centre for Renewable Energy Systems Technology
Loughborough University, Leicestershire, LE11 3TU,
United Kingdom
c.hor@lboro.ac.uk
http://www.crestuk.org

[2] Electric Power and Energy Systems,
Queen's University Belfast, Belfast, BT9 5AH,
United Kingdom
p.crossley@qub.ac.uk
http://www.ee.qub.ac.uk/power/

Abstract. Most substations today contain a large number of Intelligent Electronic Devices (IEDs), each of which captures and stores locally measured analogue signals, and monitors the operating status of plant items. A key issue for substation data analysis is the adequacy of our knowledge available to describe certain concepts of power system states. It may happen sometimes that these concepts cannot be classified crisply based on the data/information collected in a substation. The paper therefore describes a relatively new theory based on rough sets to overcome the problem of overwhelming events received at a substation that cannot be crisply defined and for detecting superfluous, conflicting, irrelevant and unnecessary data generated by microprocessor IEDs. It identifies the most significant and meaningful data patterns and presents this concise information to a network or regionally based analysis system for decision support. The operators or engineers can make use of the summary of report to operate and maintain the power system within an appropriate time. The analysis is based on time-dependent event datasets generated from a PSCAD/EMTDC simulation. A 132/11 kV substation network has been simulated and various tests have been performed with a realistic number of variables being logged to evaluate the algorithms.

1 Introduction

Advances in communication and microprocessor technologies have largely contributed to a significant increase in real time data and information that are now readily collected at various points in the network. An important requirement to fulfill our future information needs is our ability to extract knowledge from intelligent electronic devices (IEDs). Lacking this ability could lead to an ever-widening gap between what we understand and what we think we should understand. A better way to manage information is necessary in order to help

J.F. Peters and A. Skowron (Eds.): Transactions on Rough Sets III, LNCS 3400, pp. 82–111, 2005.

operators utilise the available data to monitor the status of the power system. An overabundance of data may cause a serious inconvenience to the operators at the critical moment. By improving how we retrieve information, our information overload and anxiety will be reduced and our confidence in making a correct decision increases. The motto of data analysis is to allow the data to speak for themselves. It is also very essential for us to understand how a *river* changes over time than to memorise facts about how and when it changed. The more we understand the process, the richer our learning experience would be and the more we find new ways to address our problem.

2 Intelligent Electronic Devices

An IED is a device incorporating one or more processors that has the capability to receive or send data/control from or to an external source. With its enhanced microprocessor and communication technology, this new unit not only provides a self and external circuit monitoring, real-time synchronisation for event reporting, but also increases the possibilities for remote and local control and data acquisition for use in network analysis [1]. In addition to the metering and status information, it also provides current and voltage phasors [2]. IEDs can be regarded as the eyes and ears of any remote power management systems and how they see or hear reality will have a significant impact on the system reliability, effectiveness and cost of a solution. Many utilities are discovering significant economic benefits through the use of IEDs, consequently they are rapidly becoming the mainstream product [3]. Figure 1 depicts the structure of a single package IED with all the functions associated with a conventional relaying scheme.

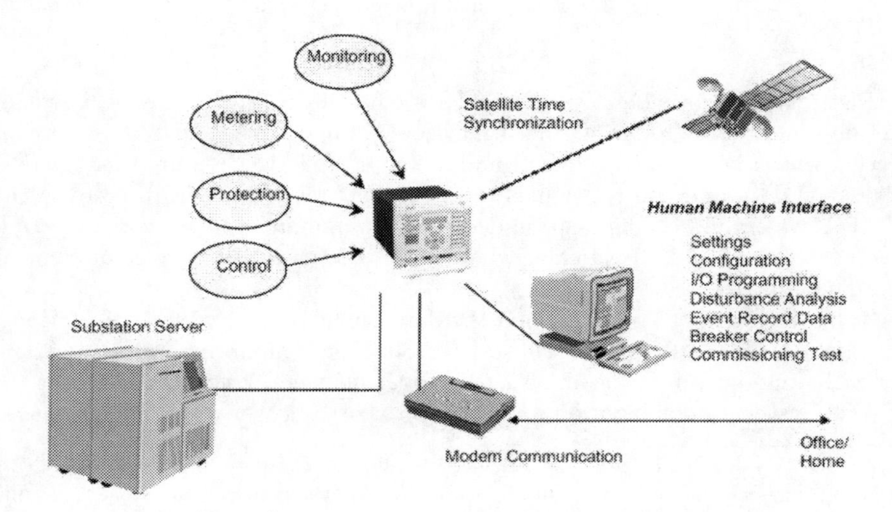

Fig. 1. Functional overview of an Intelligent Electronic Device.

3 Data Impact of IEDs

The growth in the number of IEDs has resulted a significant increase in the volume of substation data. This data is usually primitive and stored in a digital form. It has to be processed and analysed before any user is able to utilise the benefit of it.

Figure 2 shows a conventional protection system in which the data and control signal from the relay are sent via an RTU[1] to the SCADA[2] system. Extensive and costly cables may be required between various bays in the substation and the control room.

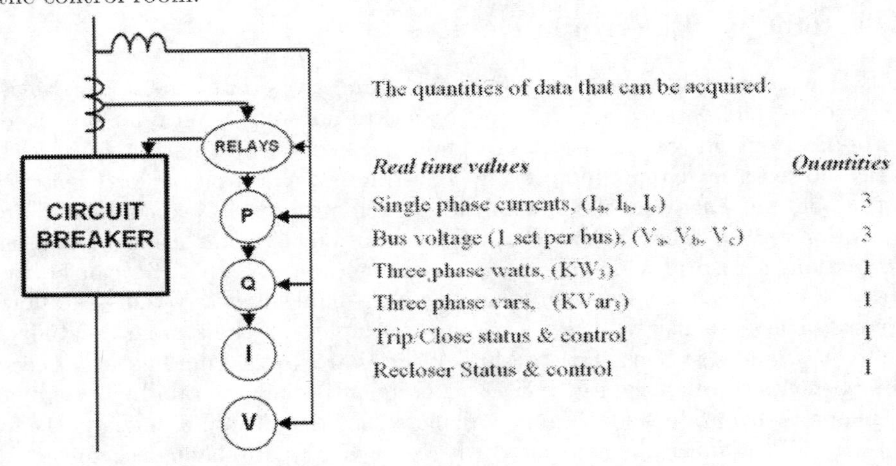

Fig. 2. Conventional protection system.

Figure 3 shows a modern protection system utilising an IED relay. The connection diagrams of Figure 2 and 3 may look similar except that the interconnection wiring between transducers and meters is no longer required for the IED relay. The data and control signals from the IED relay are sent directly to the SCADA system via the high-speed dedicated communication network. The volume of data increases drastically when an IED is used as the control element and data source.

Table 1 presents a comparison of the data quantities between a conventional RTU and an IED in a 10-feeder substation. Considering real time values only and not including data from incoming lines or transformers, it can be seen that the SCADA system and substation data is a factor of 10 larger when IEDs are used

[1] Remote Terminal Unit (RTU) is a device installed at a remote location that collects data, processes and codes the data into a format that is transmittable and transmits the data back to a central or master station.

[2] Supervisory Control and Data Acquisition (SCADA) is the system that collects data from IEDs in the network and utilises it for control and monitoring. It also provides a remote access to substation data and devices to the control centre.

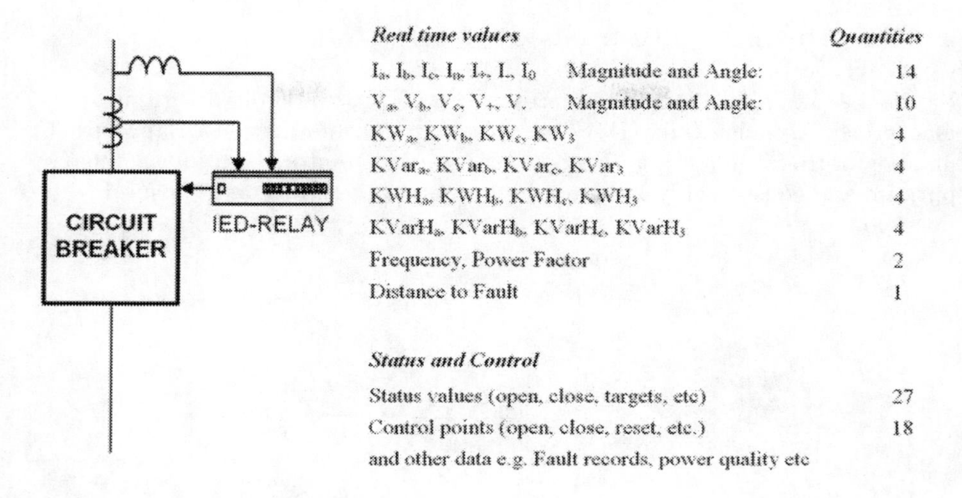

Fig. 3. Modern protection system.

Table 1. Comparison of RTU basis and IED basis.

Quantities	RTU	IED
Analogue	53	430
Status Quantities	20	270
Control Quantities	20	180

as a data source [4]. With the wealth of information that an IED can produce, engineers are no longer facing the lack of data problem [5].

In each IED, there are two types of data namely, operational and non-operational [6]. Operational data is instantaneous values of volts, amps, MW, MVAR and etc. It is typically conveyed to the SCADA system using the communications protocol. This operational data path from the substation to the SCADA system is continuous. Non-operational data, on the other hand, is the IED data that is used for analysis and historical archiving, and is not in a point format as operational data e.g. fault event logs, metering records and oscillography. This data is more difficult to extract from the IEDs since the IED vendor's proprietary ASCII commands are required for extraction of this non-operational data [7]. Operational data is critical for SCADA dispatchers to effectively monitor and control the utility's power system. However, the non-operational data has also tremendous value to the utility [8]. In this substation data analysis, only the operational data from IEDs has been considered. The proposed algorithm can however identify the certain IED relays that carry the crucial information. This not only reduces our attention span, but also saves our time in selecting which IEDs may contain the non-operational data we need for detailed post-fault analysis.

4 Future Information Needs

Figure 4 shows a future digital control system (DCS) integrated with an Information Management Unit (IMU) to deliver useful information to appropriate manpower groups in a utility. Each group uses the monitored data for a different purpose and consequently has varied information requirements that need to be satisfied.

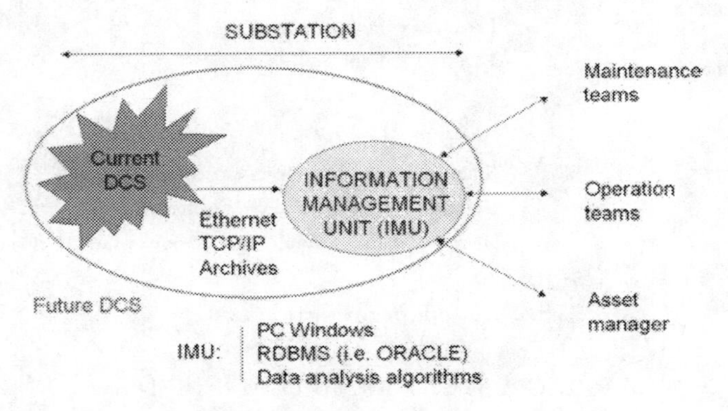

Fig. 4. Future DCS with Information Management Unit(IMU).

Our advances in communication and microprocessor technologies have largely contributed to a significant increase in real time data and information that are now readily collected at various points in the network. Therefore, an important requirement to fulfill our future information needs is the ability to extract knowledge from intelligent electronic devices (IEDs).

5 Discretisation

The power system operational state changes over time as the event evolves. It is thus important to determine the condition of the system based on the real time data collected from IEDs. This normally requires manipulating and processing a large volume of data/information before the status of the system could be verified. Therefore, to extract useful information, these numerical and continuous data values must be discretised into a range of thresholds [9].

The discretisation determines how coarsely we want to view the raw data. It can be formulated as $\mathrm{P} : \mathrm{D} \rightarrow \mathrm{C}$ assigning a class $\mathrm{c} \in \mathrm{C}$ (c is the member of class C) to each value $d \in \mathrm{D}$ (d is the member of attribute D) in the domain of the attribute being discretised. The classification engine effectively summarises key information about the power system condition and classifies the operational state in the decision attribute into four transition states: *normal*, *alert*, *emergency* and *safe*. Figure 5 shows the operational state of a power system and changing of operation points. At the operational point A, the system continues at the normal state whereas, at the operational point B, the system moves from the normal

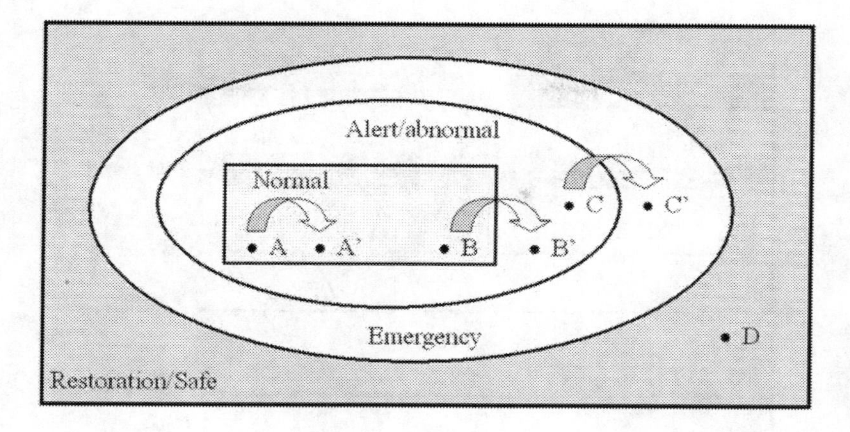

Fig. 5. The changing of operational points in power system.

to an alert/abnormal state [10]. The operational point C however indicates that the system, which is already at the alert condition transits to an emergency state. The operational point at B' may or may not be considered in an unsafe condition. C' is definitely in an unsafe operation. Both operational points A and D are in a safe operation.

6 Limit Supervision and Deadband Suppression

Sometimes the real time data received may oscillate a lot. If this happens close to the threshold, it will lead to a flood of events. The problem can be reduced by defining a hysteresis values represented as a *dotted line* and shaded area in Figure 6.

If the current threshold is exceeded, the classification engine will define the state as *abnormal* or *emergency* depending on the value of current and voltage drop on that feeder. This is because the magnitude of high current will justify different levels of instability and damage to the system. The state is classified as a normal condition if the value of data is lower than the hysteresis value below the threshold limit. If the feeder is taken out of operation, the current becomes zero and is below any low current limit. The engine will mistakenly classify the state as abnormal even though the system is still operating normally or safely. To overcome this problem in the classification, a zero dead band suppression has been used to exempt the value from the abnormal zone if the range is around zero [11].

The concept of hysteresis and zero deadband can also be applied in the voltage supervision. If the voltage sag threshold is exceeded, the classification engine will classify the state as *abnormal*. The state is only classified as a normal condition if the value of data is higher than the hysteresis value above the threshold limit. The system voltage is normally kept within a narrow range around the

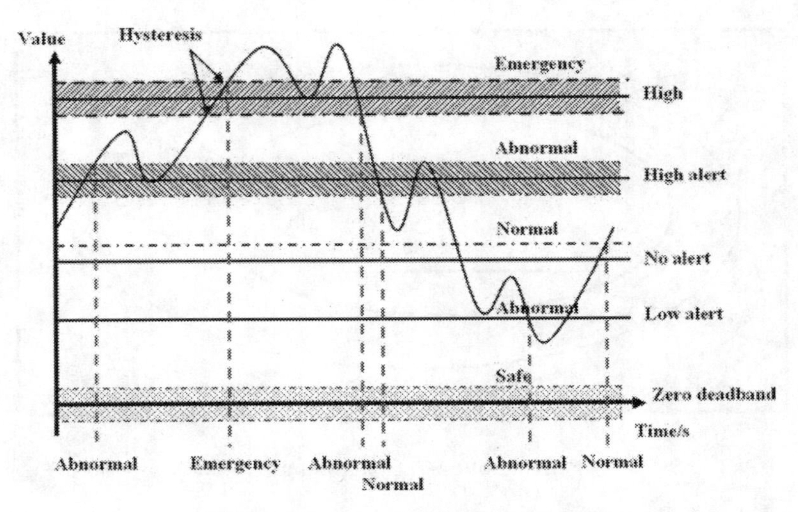

Fig. 6. Measurand limit supervision.

nominal value (90%). A zero dead band suppression is used to avoid any mistakes made by the classification engine when the feeder is taken out of operation and the voltage value becomes zero and is below any low voltage limit.

The normal operating range for the voltage in a power system is typically from 0.90pu to 1.10pu of the nominal value. Therefore, the threshold for each voltage and current signals are set as follows: –

1. The thresholds of the nominal voltage:
   - **Low (L) $\leq$ 0.90pu.**
     The voltage must drop to or below 0.90pu before it is classified as Low(L).
   - **0.90pu < Normal (N) < 1.10pu**
     The hysteresis value M is set at 0.91pu and 1.05pu ($\approx$1%). This means that if the voltage recovers from a sag, it must exceed at least 0.91pu before it can be safely classified as Normal(N). If the voltage drops from its high value, it must be below 1.05pu before it can be classified as Normal (N).
   - **High (H) $\geq$ 1.10pu.**
     The voltage must exceed or at least must be equal to 1.10pu before it is classified as High(H).
2. The thresholds of the nominal current:
   - **Low (L) $\leq$ 0.50pu.**
     The current must drop to or at least must be below 0.5pu before it is classified as Low(L).
   - **0.5pu < Normal (N) < 1.50pu**
     The hysteresis value M is set at 0.6pu and 1.40pu ($\approx$10%). This means that if the current recovers from its low value, it must exceed at least 0.6pu before it can be classified as Normal (N). If the current drops from its high value, it must be below 1.40pu before it can be classified as Normal (N).

- **High (H) $\geq$ 1.50pu.**
  The current must exceed or at least must be equal to 1.50pu before it is classified as High(H).

As the current varies significantly more than the voltage, a wider range of threshold is used. The nominal current range is considered to be between 0.50pu and 1.50pu. The pattern depends on how the threshold value has been set. If the current threshold is set reasonably high, then it may become less susceptible to the changes. This means that the voltage would respond earlier to a disturbance than current.

Figure 7 demonstrates how these continuous values are transformed into discrete data. Let $x_1 = (0.96, 5.10)$ and $x_2 = (0.88, 34.14)$.

The information from all relays are gathered in a common database and categorised into four different states that characterise the operational condition of the system seen by the relays. Normal (**N**) indicates that all the constraints and loads are satisfied, the voltages are close to the nominal values and the currents are either close to the nominal or low values. Alert (**A**) indicates one or more currents are high and the voltage is normal ($\approx$ nominal), or the currents are normal ($\approx$ nominal or low) but one or more voltages are high or low. Emergency (**E**) indicates at least two physical operating limits are violated (e.g. under voltages and over currents). Safe (**S**) is when the remaining partial system is operating in a normal state, but one or more loads are not satisfied – partial blackout after a breaker opened [12].

7 Types of Extraction

In this paper, an unsupervised extraction and a supervised extraction techniques based on rough sets are discussed. The former discovers patterns within the data without any pre-defined classes to improve the form in which the data and information are presented to network operators. It is self-organising and can perform data clustering to discover new relations between data. Such system is

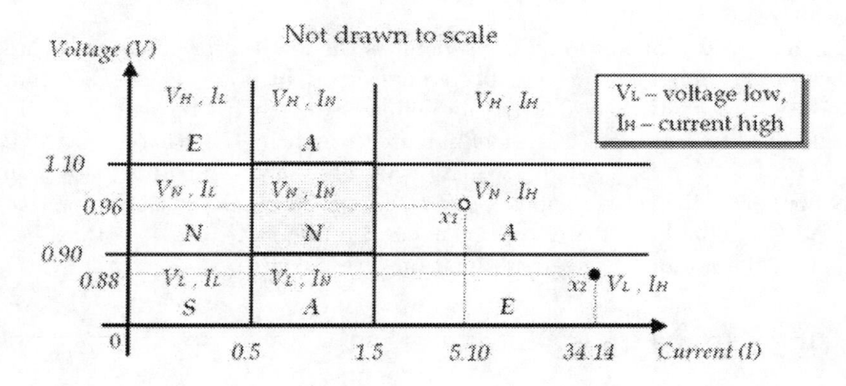

Fig. 7. Transform a continuous data into a range of intervals.

applied to an area in which the (event) classification is difficult to perform or simply unknown to human beings. It is useful where the relations governing a domain are yet to be discovered [13]. The later however relies on a pre-classified dataset (by expert) for training the algorithm. The dataset provides the system with a class of decisions and the number of time stamped events for each decision class. It is useful when the systems are intended to perform tasks that have previously been performed or encountered by human operators with a certain degree of success. This type of extraction suits most power system applications today particularly in the control centre because an outcome of state classification in the power system is usually known.

8 Rough Set Approach

Rough set approach hinges on two approximation concepts; lower and upper approximations, which defines a crisp and vague manner in the sets. If any concept of the universe can be formed as a union of some elementary sets, it is referred as crisp, otherwise it is vague. The fundamentals of rough sets are omitted as the knowledge is already available in the literatures found in the references: [14][15][16].

Let X denotes the subset of elements of the universe U ($X \subseteq U$), attribute $B \subseteq A$ and $[x]_B$ as the set of all objects x that are the equivalence class of the attribute B indiscernible with X. The lower approximation of X in B can be represented in Equation 1 as follows: –

$$B_*X = \{X \in U, [x]_B \in U/IND(B) : [x]_B \subseteq X\} \tag{1}$$

where $[x]_B \subseteq X$ is the lower approximation of X in B.

The upper approximation of X in B can be represented in Equation 2 as follows: –

$$B^*X = \{x \in U, [x]_B \in U/IND(B) : [x]_B \cap X \neq \emptyset\} \tag{2}$$

where $[x]_B \cap X \neq \emptyset$ means that some upper approximations of X are the element of X.

The upper approximation always includes the lower approximation. The difference of the upper and lower approximation, $BN_B(X) = B^*X - B_*X$ is a B-Boundary of X in U. Set X is crisply definable in U with respect to B, if and only if $BN_B(X) = 0$. Set X is indefinable or rough in U with respect to B, if and only if $B^*X \neq B_*X$ when $BN_B(X) \neq 0$. The lower and upper approximations further define three regions: *positive region*, $POS_B(X) = B_*X$, *negative region*, $NEG_B(X) = U - B^*X$ and *boundary region*, $BN_B(X) = B^*X - B_*X$. Figure 8 displays the notion of set approximations graphically.

8.1 Reducts and Core

Reducts and core are the two major concepts in Rough Set Theory used in the knowledge base reduction. A reduct is defined as a minimal set of attributes

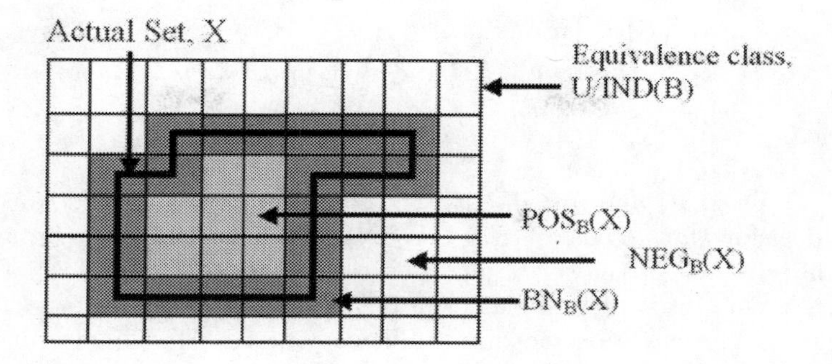

Fig. 8. Schematic view of the upper and lower approximations.

that preserves the indiscernibility relation computed on the basis of the full set of attributes. The core is defined as the set of relations that appears in all reducts, i.e. the set of all the indispensable relations required to characterise the equivalence relation [17].

9 Discernibility Matrix

The concept of a discernibility matrix is important when it comes to compute reducts and core. This matrix is a symmetric $n \times n$ where n denotes the number of elementary sets [18].

9.1 Information System

Let an information system be an ordered pair, $\mathcal{A} = (\mathrm{U}, \mathrm{A})$. Every subset of attributes $\mathrm{B} \subseteq \mathrm{A}$ defines a discernibility matrix M(B). Each entry $m(x_i, x_j)$ consists of the set of attributes that can be used to discern between objects x_i and x_j where $\{x_i, x_j\} \in \mathrm{U}$.

$$\begin{aligned} \mathrm{M}\left(\mathrm{B}\right) &= \left\{m_B\left(x_i, x_j\right)\right\}_{n \times n} \\ m_B\left(x_i, x_j\right) &= \left\{b \in B : b\left(x_i\right) \neq b\left(x_j\right)\right\} \end{aligned} \tag{3}$$

where $i, j = \{1, ..., n\}$ and $n = \left|\mathrm{U/IND}\left(\mathrm{B}\right)\right|$

If the $b(x_i)$ and $b(x_j)$ are symmetric and reflexive for all x_i and x_j, then $m_B\left(x_i, x_j\right) = m_B\left(x_j, x_i\right)$ and that $m_B\left(x_i, x_i\right) = \emptyset$. This means that only half the matrix entries is necessary for computing when constructing M(B) [18].

9.2 Decision System

A single decision attribute, d is used to represent a n-size decision attribute set D, i.e. $\mathrm{D} = \{d\}$. Assume the attribute $\mathrm{B} \subseteq \mathrm{A}$ and the decision table is represented as $\mathcal{D} = (\mathrm{U}, \mathrm{B} \cup \{d\})$. The discernibility matrix of a decision system, $M^d\left(B\right)$ can be defined as: –

$$M^d(B) = \{m_B(x_i, x_j)\}_{n \times n},$$
$$m_B^d(x_i, x_j) \begin{cases} \emptyset & if\ \forall\, d \in D\, [d(x_i) = d(x_j)] \\ \{r \in B : r(x_i) \neq r(x_j)\} \\ & if\ \exists d \in D\, [d(x_i) \neq d(x_j)] \end{cases} \tag{4}$$

where $i, j = \{1, ..., n\}$ and $n = |\text{U/IND (B)}|$. The notion $r\,(x)$ denotes the set of possible decisions for a given class $x \in$ U/IND (B). The entry $m_B^d\,(x_i, x_j)$ in the discernibility matrix is the set of all (condition) attributes from B that classify objects x_i and x_j into different classes in U/IND(B) if $r\,(x_i) \neq r\,(x_j)$. Empty set $\emptyset$ denotes that this case does not need to be considered. All disjuncts of minimal disjunctive form of this function define the reducts of B [16].

10 Discernibility Functions

A discernibility function $f\,(B)$ is a boolean function that expresses how an object (or a set of objects) can be discerned from a certain subset of the full universe of objects [19]. A boolean expression normally consists of Boolean variables and constants, linked by disjunction ($\bigvee$) operators.

Let $\bar{b}$ be a unique Boolean function of m Boolean variables $\{\bar{b}_1, \bar{b}_2, ..., \bar{b}_m\}$ associated with the corresponding attribute $b = \{b_1, b_2, ..., b_m\} \in$ B. Each element $m_B(x_i,x_j)$ of the discernibility matrix corresponds a boolean set $\bar{m}_B(x_i,x_j)$ $= \{\bar{b} : b \in m_B\,(x_i, x_j)\}$. For a set of Boolean variables, $\bar{S} = \{\bar{b}_1, \bar{b}_2, ..., \bar{b}_m\}$, then $\bigvee \bar{S} = \{\bar{b}_1 \vee \bar{b}_2 \vee ..... \vee \bar{b}_m\}$ [18].

10.1 Information System

Given an information system $\mathcal{A}$ = (U, A) for a set of attributes B $\subseteq$ A. Let $m = |\,\text{B}\,|$ and $n = |\text{U/IND (B)}\,|$. The discernibility function of attribute B is a Boolean function of m variables: –

$$f\,(B) = \bigwedge \left\{ \bigvee \bar{m}_B\,(x_i, x_j)\ : 1 \leq j \leq i \leq n \right\} \tag{5}$$

$\bigvee \bar{m}_B\,(x_i, x_j)$ is the disjunction taken over the set of Boolean variables $\bar{m}_B(x_i,x_j)$ corresponding to the discernibility matrix element $m_B\,(x_i, x_j)$ which is not equal to $\emptyset$.

10.2 Decision System

Given a decision system $\mathcal{D}$ = (U, B $\cup$ $\{d\}$), the discernibility function of $\mathcal{D}$ is: –

$$f_B^d\,(x_i) = \bigwedge \left\{ \bigvee \bar{m}_B^d\,(x_i, x_j)\ :\ 1 \leq j \leq i \leq n \right\} \tag{6}$$

where $n = |\text{U/IND (B)}|$, and $\bigvee \bar{m}_B^d\,(x_i, x_j)$ is the disjunction taken over the set of Boolean variables $\bar{m}_B^d\,(x_i, x_j)$ corresponding to the discernibility matrix $m_B^d\,(x_i, x_j)$ which is not equal to $\emptyset$ [16].

11 Knowledge Representation Systems

11.1 Protection Data

The substation data consists of both the protection data given in Table 2 and the measurement data given in Table 3 recorded by a number of intelligent electronic devices installed in the network. Some situations in the event dataset may occur in more than one decision class can specify. For instance, a new event might occur at the same time as an existing event and consequently the decision classes overlap [20]. To overcome such problem and prevent the loss of information, the dataset is split into two different tables. The measurement data table is reduced using rough sets and the final result is merged with the protection data to produce a summary of the events.

Table 2 shows that the IED1 relay has operated whilst the other relays remain stable. The auto-recloser (AR) has been disabled to simplify the example. The IED1 relay picked up the fault at 1.004s, tripped at 1.937s and the breaker BRK1 opened at 2.007s.

11.2 Information and Decision System

Table 3 can be regarded as a decision system. It consists of a time series data collected from the simulation and a set of pre-classified decision values. The

Table 2. Protection status of IED1.

Time	**IED1**				
t/s	Pickup	Trip	AR	52A	52B
0.139	0	0	0	0	1
1.004	1	0	0	0	1
1.937	1	1	0	0	1
2.007	1	1	0	1	0
2.010	0	0	0	1	0

IED: IED relay, Pickup: pickup time, Trip: trip time, 52A and 52B: Circuit breaker auxiliary contacts in which 52A and 52B: '01' close; '10' open.

Table 3. Decision system.

Time	**IED1**		**IED2**		**IED3**		**IED4**		**Decision**
t/s	V_1	I_1	V_2	I_2	V_3	I_3	V_4	I_4	d
0.139	N	N	N	N	N	N	N	N	N
1.003	N	H	N	N	N	N	N	N	A
1.004	L	H	N	N	N	N	N	H	E
1.005	L	H	N	L	N	H	L	H	E
1.006	L	H	L	L	L	H	L	H	E
2.007	L	N	L	L	L	H	L	H	E
2.011	L	N	L	N	L	H	L	N	E
2.012	L	L	N	N	N	N	N	N	S

information system is identical to the decision system except that it does not include the decision attribute $\boldsymbol{d}$. The table presents a simple dataset, which is composed of a set of discrete voltages and currents over a time period of 0.139s to 2.012s. The voltage and current are used because they are the fundamental components in the power systems. $\mathbf{U}$ is the universe of events and $\mathbf{B}$ is the set of condition attributes $\{\text{IED1}, \text{IED2}, \text{IED3}, \text{IED4}\}$.

12 Experiment

Real time data is the main asset for a substation control system but often very difficult to obtain for research purposes. Because of the complexity of a power network, it is almost impossible to anticipate or provide an infinite case of problems to investigate every scenario in a substation. To partially solve the problems, primary and secondary system of a 132/11kV substation given in Figure 9 and 10 have been modelled using PSCAD/EMTDC [21].

Observing the above set of events in Table 3, we may easily identify that the problem is actually within the supervised region of the IED1 relay. However, given that there are n number of events and m number of relays, this may become impractical in an actual problem. Additionally, the dataset is given in a perfect pattern, which may not be always the case with the real time data received from the control system. The irregularity in data pattern makes it much harder for a human to handle a complex situation. The following example utilises both the information system and decision system to extract knowledge from a substation. The data were collected from the 132/11kV substation model with a single bus arrangement given in Figure 9.

12.1 Supervised Extraction

Table 4 shows the discernibility matrix for Table 3. Due to the lack of space, the columns of the discernibility matrix in the table have been simplified such that IED1 is given as "1", IED2 as "2" and etc. The discernibility function is calculated using an absorption law in each column of Table 4 shown in Equation 6.

For better interpretation, the Boolean function attains the form '+' for the operator of disjunction ($\bigvee$) and '·' for the operator of conjunction ($\bigwedge$). The final discernibility function is: –

$$\begin{aligned} f(B) &= f(0.139) \cdot f(1.003) \cdot f(1.004) \cdot f(1.005) \cdot \ldots \cdot f(2.011) \\ &= 1 \cdot 1 \cdot (1+4) \cdot (1+2+3+4) \cdot \ldots \cdot (1+2+3+4) \\ &= 1 \Rightarrow \{\text{IED1}\} \end{aligned}$$

The example shows that the IED1 relay is the main source of information to justify the outcomes of interest. Table 3 can thus be reduced to Table 5.

Table 2 showed that the IED1 relay tripped at 1.937s and reset at 2.010s after the current has dropped below the threshold value. To generate a concise report that can assist the operators in their decision-making, the information given in Table 5 may be excessive for inclusion in the report. Thus, we have to condense the table. The solution is to retain the change of state information as shown in Table 6 since it provides useful information.

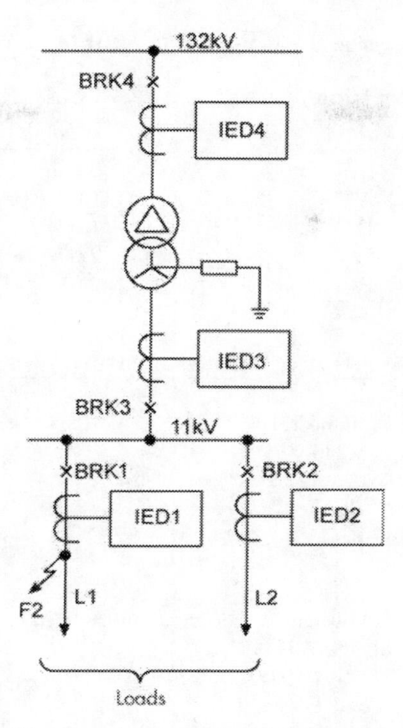

Fig. 9. 132/11kV substation model with a single bus arrangement.

Table 4. Discernibility matrix for the decision system.

Time	0.139	1.003	1.004	1.005	1.006	2.007	2.011	2.012
0.139	∅							
1.003	1	∅						
1.004	1,4	1,4	∅					
1.005	1,2,3,4	1,2,3,4	∅	∅				
1.006	1,2,3,4	1,2,3,4	∅	∅	∅			
2.007	1,2,3,4	1,2,3,4	∅	∅	∅	∅		
2.011	1,2,3,4	1,2,3,4	∅	∅	∅	∅	∅	
2.012	1	1	1,4	1,2,3,4	1,2,3,4	1,2,3,4	1,2,3,4	∅

Notation 1,2,3,4: $\{\text{IED1}, \text{IED2}, \text{IED3}, \text{IED4}\}$ and 1,4: $\{\text{IED1}, \text{IED4}\}$

The load feeder L1 experienced both a high current and voltage sag at approximately 1.003s. This indicates that a fault has occurred in that region, which can be further confirmed by the protection data from that region. Combining the Table 2 with the Table 6, various reports can be produced for respective operators, protection engineers and maintenance engineers. The reports are used mainly as an example. The format presented may not reflect the actual substation reports used in the power industries. The change of state within 5 ms should be combined as one event. For instance, the high current occurred at 1.003s and the voltage sag at 1.004s should be grouped as one event.

Table 5. Reduced decision table.

Time	IED1		Decision
t/s	V_1	I_1	d
0.139	N	N	N
1.003	N	H	A
1.004	L	H	E
2.007	L	N	E
2.012	L	L	S

Table 6. Change of states.

Time	IED1		Decision
t/s	V_1	I_1	d
0.139	N	N	N
1.003	•	H	A
1.004	L	•	E
2.007	•	N	E
2.012	•	L	S

Report for Operators

Location
Station name: Manchester
Station number: M111
Event date: 14th August 2003
Event time: 9:20:12pm
Event number: 001

Description
Load Feeder L1: voltage sag, current high ≈ 1.003s
IED1: tripped at 1.937s
Breaker status: BRK1 (2.007s – open)
Disconnection: Load Feeder L1
Faulted section: Load Feeder L1
System restored after 2.012s

Report for Protection Engineers

Location
Station name: Manchester
Station number: M111
Event date: 14th August 2003
Event time: 9:20:12pm
Event number: 001

Description
Load Feeder L1: voltage sag, current high ≈ 1.003s
IED1: picked up at 1.004s, tripped at 1.937s, Reset at 2.010s
Disconnection: Load Feeder L1
System restored after 2.012s
Faulted section: Load Feeder L1
Fault inception: 1.0s
Fault duration: 1.007s
Fault type: single phase to earth fault (A-G), permanent
Maximum fault magnitude: 17.78kA
Breaker status: BRK1 (2.007s – open)
Relay condition: healthy
Breaker condition: healthy

Report for Maintenance Engineers

Location
Station name: Manchester
Station number: M111
Event date: 14th August 2003
Event time: 9:20:12pm
Event number: 001

Description
Disconnection: Load Feeder L1
Faulted section: Load Feeder L1
Fault inception: 1.0s
Fault duration: 1.007s
Fault type: single phase to earth fault (A-G), permanent
Breaker status: BRK1 (2.007s – open)
Relay condition: healthy
Breaker condition: healthy

$$\text{The relay trip duration} = 2.007 - 1.004 = 1.003\text{s}.$$

12.2 Unsupervised Extraction

Table 7 shows the discernibility matrix for the information system in Table 3 (without the decision attribute d). Due to the lack of space, the columns of the discernibility matrix in the table have been simplified such that IED1 is given as "1", IED2 as "2" and etc. The discernibility function is calculated using an absorption law in each column of Table 7 using Equation 5.

Table 7 shows how the result can be computed in the discernibility matrix. The final discernibility function obtained is:

$$\begin{aligned} f(\mathrm{B}) &= f(0.139) \cdot f(1.003) \cdot f(1.004) \cdot \ldots \cdot f(2.011) \\ &= (1) \cdot (1) \cdot ((1+4) \cdot (2+3+4)) \cdot \ldots \cdot (1+2+3+4) \\ &= 1 \cdot (2+3) \cdot (2+4) \\ &= 1 \cdot (2 + (3 \cdot 4)) \end{aligned}$$

Table 7. Discernibility matrix for the information system.

Time	0.139	1.003	1.004	1.005	1.006	2.007	2.011	2.012
0.139	∅							
1.003	1	∅						
1.004	1,4	1,4	∅					
1.005	1,2,3,4	1,2,3,4	2,3,4	∅				
1.006	1,2,3,4	1,2,3,4	2,3,4	2,3	∅			
2.007	1,2,3,4	1,2,3,4	1,2,3,4	1,2,3	1	∅		
2.011	1,2,3,4	1,2,3,4	1,2,3,4	1,2,3,4	1,2,4	2,4	∅	

Notation 1,2,3,4: $\{\mathrm{IED1}, \mathrm{IED2}, \mathrm{IED3}, \mathrm{IED4}\}$ and 1,4: $\{\mathrm{IED1}, \mathrm{IED4}\}$

The example shows that the relays {IED1,IED2} or {IED1,IED3,IED4} are identified as the main source of information. We chose the solution of {IED1,IED2} as it contains the least number of IEDs, which appeared in Table 8 as the minimal set of relations or reducts. Like the previous case, to generate a concise report for a decision support, the information given in Table 8 have to be condensed. This can be done by retaining the change of state information as shown in Table 9.

Table 8. Reduced information system.

Time	IED1		IED2	
t/s	V_1	I_1	V_2	I_2
0.139	N	N	N	N
1.003	N	H	N	N
1.004	L	H	N	N
1.005	L	H	N	L
1.006	L	H	L	L
2.007	L	N	L	L
2.011	L	N	L	N
2.012	L	L	N	N

Table 9. Change of states.

Time	IED1		IED2	
t/s	V_1	I_1	V_2	I_2
0.139	N	N	N	N
1.003	•	H	•	•
1.004	L	•	•	•
1.005	•	•	•	L
1.006	•	•	L	•
2.007	•	N	•	•
2.011	•	•	•	N
2.012	•	L	N	•

The load feeder L1 experienced both a high current and voltage sag at approximately 1.003s. This indicates that a fault has occurred in that region which can be confirmed also by the feeder L2 data and the protection data from that region. Combining the Table 2 with the Table 9, various reports can be produced for respective operators, protection engineers and maintenance engineers.

Report for Operators

Location
Station name: Manchester
Station number: M111
Event date: 14th August 2003
Event time: 9:20:12pm
Event number: 001

Description
Load Feeder L1: voltage sag, current high $\approx$ 1.003s
Load Feeder L2: voltage sag, current low $\approx$ 1.005s
IED1: tripped at 1.937s
Breaker status: BRK1 (2.007s – open)
Disconnection: Load Feeder L1
Faulted section: Load Feeder L1
System restored after 2.012s

Report for Protection Engineers

Location
Station name: Manchester
Station number: M111
Event date: 14th August 2003
Event time: 9:20:12pm
Event number: 001

Description
Load Feeder L1: voltage sag, current high ≈ 1.003s
Load Feeder L2: voltage sag, current low ≈ 1.005s
IED1: picked up at 1.004s, tripped at 1.937s, Reset at 2.010s
Disconnection: Load Feeder L1
System restored after 2.012s
Faulted section: Load Feeder L1
Fault inception: 1.0s
Fault duration: 1.007s
Fault type: single phase to earth fault (A-G), permanent
Maximum fault magnitude: 17.78kA
Breaker status: BRK1 (2.007s – open)
Relay condition: healthy
Breaker condition: healthy

Report for Maintenance Engineers

Location
Station name: Manchester
Station number: M111
Event date: 14th August 2003
Event time: 9:20:12pm
Event number: 001

Description
Disconnection: Load Feeder L1
Faulted section: Load Feeder L1
Fault inception: 1.0s
Fault duration: 1.007s
Fault type: single phase to earth fault (A-G), permanent
Breaker status: BRK1 (2.007s – open)
Relay condition: healthy
Breaker condition: healthy

The IED1, IED3 and IED4 indicate that the fault is at the load feeder L1 because the IED1 has tripped. The upstream feeders 3 and 4 also experience a high current similar to the faulty feeder L1. Alternatively, since there are only two load feeders in the substation (see Figure 10), if IED1 relay detects a fault, the feeder L2 shall experience a voltage sag and current drop. Therefore, IED1 and IED2 tell us also that the fault is at the feeder L1. This means that the

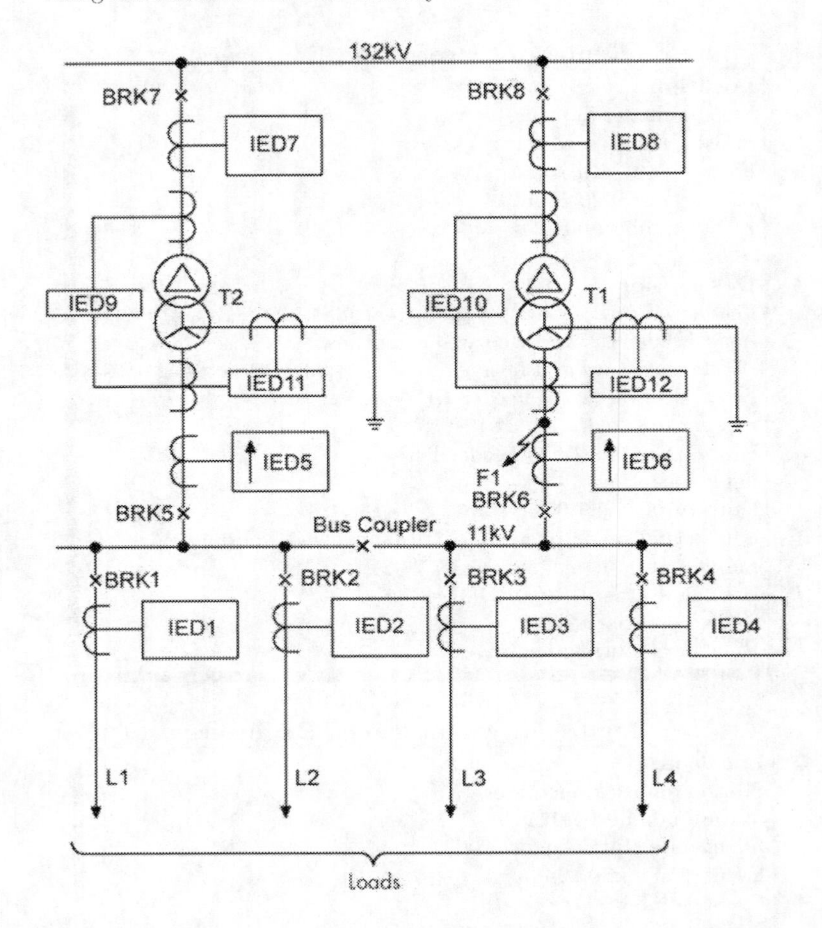

Fig. 10. 132/11kV substation model with a double bus arrangement.

reducts {IED1, IED2} and {IED1, IED3, IED4} both share the same information and one is considered redundant. {IED1, IED2} is chosen since it contains the least number of IEDs. This reduces the attention span because the operators can only concentrate on the IED1 and IED2 rather than all the IEDs to decide which one is useful for detailed analysis.

13 Substation Network Modelling

For further verification, a larger and more realistic network shown in Figure 10 has been developed using PSCAD/EMTDC [21]. A selection of fault scenarios were applied to the network and the operating response of the relays, circuit breakers, voltage and current sensors were collected and stored in an event database. The network model is protected by several types of relay models [22], each of which includes one or more of the protection functions in Table 10. The

Table 10. List of relay models in the 132/11kV double bus substation.

IED Plate numbers	**1,2,3,4**	**5,6**	**7,8**	**9,10**	**11,12**
Instantaneous Overcurrent, 50	$\surd$	$\surd$	$\surd$	$\times$	$\times$
Instantaneous Earth Fault, 50N	$\surd$	$\surd$	$\surd$	$\times$	$\times$
Time delayed Overcurrent, 51	$\surd$	$\surd$	$\surd$	$\times$	$\times$
Time delayed Earth Fault, 51N	$\surd$	$\surd$	$\surd$	$\times$	$\times$
Balanced Earth Fault, 51G	$\times$	$\times$	$\surd$	$\times$	$\times$
Standby Earth Fault, 51NB	$\times$	$\times$	$\times$	$\times$	$\surd$
Directional Phase Overcurrent, 67	$\times$	$\surd$	$\times$	$\times$	$\times$
Directional Earth Fault, 67N	$\times$	$\surd$	$\times$	$\times$	$\times$
Auto-recloser, 79	$\surd$	$\times$	$\times$	$\times$	$\times$
Restricted Earth Fault, 87N	$\times$	$\times$	$\times$	$\times$	$\surd$
Transformer Differential, 87T	$\times$	$\times$	$\times$	$\surd$	$\times$

$\surd$: available, $\times$: not available

Table 11. Protection trip status of IED6 and IED8.

Time	**IED6**				**IED8**		
t/s	67	50/51	52A	52B	50/51	52A	52B
0.139	0	0	0	1	0	0	1
1.039	1	1	0	1	0	0	1
1.119	1	1	1	0	0	1	0
1.133	0	1	1	0	0	1	0
1.139	0	0	1	0	0	1	0

set of collected situations for this example are approximately 7,000–10,000 cases and more than 300 conditions attributes are available but only 35 attributes are chosen. The case covers a wide range of typical voltage and current situations that occur in each scenario. Due to its large size, only change of state data is presented.

The bus-coupler BC is assumed closed prior to the fault. To prevent both transformers from tripping as the result of a fault on the 11kV terminal, the IED5 and IED6 are set to look into their respective transformers in accordance with IEEE nomenclature 67. Both 132/11kV transformers are protected by a differential unit protection, restricted earth fault protection and balanced earth fault protection [23]. The sensitive earth fault protection is not required since the neutral of the transformer is solidly earthed.

13.1 Supervised Extraction

The result obtained based on Table 12, 13 and 14, shows that there are two reduct sets in the considered set of attributes, which means that the decision table can be reduced and presented in two alternative ways; $\{\mathrm{IED1}, \mathrm{IED5}, \mathrm{IED6}\}$ and $\{\mathrm{IED1}, \mathrm{IED5}, \mathrm{IED8}\}$. We select the $\{\mathrm{IED1}, \mathrm{IED5}, \mathrm{IED6}\}$ as our main source of

Table 12. Directional A-B fault on the transformer T1 feeder 6 – Part I.

Time	IED1				IED2				IED3			
t/s	V_1	I_{A1}	I_{B1}	I_{C1}	V_2	I_{A2}	I_{B2}	I_{C2}	V_3	I_{A3}	I_{B3}	I_{C3}
0.139	N	N	N	N	N	N	N	N	N	N	N	N
1.002	L	N	N	N	L	N	N	N	L	N	N	N
1.003	L	N	N	N	L	N	N	N	L	N	N	N
1.005	L	N	N	N	L	N	N	N	L	N	N	N
1.007	L	N	N	N	L	N	N	N	L	N	N	N
1.008	L	N	N	N	L	N	N	N	L	N	N	N
1.015	L	N	L	N	L	N	L	N	L	N	L	N
1.019	L	L	L	N	L	L	L	N	L	L	L	N
1.129	L	N	L	N	L	N	L	N	L	N	L	N
1.133	L	N	L	N	L	N	L	N	L	N	L	N
1.134	L	N	N	N	L	N	N	N	L	N	N	N
1.135	L	N	N	N	L	N	N	N	L	N	N	N
1.137	L	N	N	N	L	N	N	N	L	N	N	N
1.138	L	N	N	N	L	N	N	N	L	N	N	N
1.139	L	N	N	N	L	N	N	N	L	N	N	N
1.140	L	N	N	N	L	N	N	N	L	N	N	N
1.142	L	N	N	N	L	N	N	N	L	N	N	N
1.153	N	N	N	N	N	N	N	N	N	N	N	N
1.172	N	N	N	N	N	N	N	N	N	N	N	N

Table 13. Directional A-B fault on the transformer T1 feeder 6 – Part II.

Time	IED4				IED5				IED6			
t/s	V_4	I_{A4}	I_{B4}	I_{C4}	V_5	I_{A5}	I_{B5}	I_{C5}	V_6	I_{A6}	I_{B6}	I_{C6}
0.139	N	N	N	N	N	N	N	N	N	N	N	N
1.002	L	N	N	N	N	N	N	N	N	N	N	N
1.003	L	N	N	N	L	N	N	N	L	N	N	N
1.005	L	N	N	N	L	H	H	N	L	N	N	N
1.007	L	N	N	N	L	H	H	N	L	H	H	N
1.008	L	N	N	N	L	H	H	N	L	H	H	N
1.015	L	N	L	N	L	H	H	N	L	H	H	N
1.019	L	L	L	N	L	H	H	N	L	H	H	N
1.129	L	N	L	N	L	H	H	N	L	H	H	N
1.133	L	N	L	N	L	H	H	N	L	H	H	L
1.134	L	N	N	N	L	H	H	N	L	H	H	L
1.135	L	N	N	N	L	H	H	N	L	H	H	L
1.137	L	N	N	N	L	H	H	N	L	H	H	L
1.138	L	N	N	N	L	H	H	N	L	H	H	L
1.139	L	N	N	N	L	H	N	N	L	N	N	L
1.140	L	N	N	N	L	N	N	N	L	N	N	L
1.142	L	N	N	N	L	N	N	N	L	L	L	L
1.153	N	N	N	N	L	N	N	N	L	L	L	L
1.172	N	N	N	N	N	N	N	N	L	L	L	L

Table 14. Directional A-B fault on the transformer T1 feeder 6 – Part III.

Time	IED7				IED8				T1	T2	Decision
t/s	V_7	I_{A7}	I_{B7}	I_{C7}	V_8	I_{A8}	I_{B8}	I_{C8}	NTRL	NTRL	d
0.139	N	N	N	N	N	N	N	N	L	L	N
1.002	N	N	N	N	N	N	N	N	L	L	A
1.003	N	N	N	N	N	N	N	N	L	L	A
1.005	N	N	H	N	N	N	H	N	L	L	E
1.007	N	N	H	H	N	N	H	H	L	L	E
1.008	N	H	H	H	N	H	H	H	L	L	E
1.015	N	H	H	H	N	H	H	H	L	L	E
1.019	N	H	H	H	N	H	H	H	L	L	E
1.129	N	H	H	H	N	H	H	H	L	L	E
1.133	N	H	H	H	N	H	H	H	L	L	E
1.134	N	H	H	H	N	H	H	H	L	L	E
1.135	N	H	H	H	N	H	H	N	L	L	E
1.137	N	H	H	N	N	H	H	N	L	L	E
1.138	N	H	H	N	N	N	H	N	L	L	E
1.139	N	H	H	N	L	N	N	N	L	L	E
1.140	N	N	N	N	L	N	N	L	L	L	A
1.142	N	N	N	N	L	L	L	L	L	L	A
1.153	N	N	N	N	L	L	L	L	L	L	A
1.172	N	N	N	N	L	L	L	L	L	L	S

IEDx: IED number X, V: voltage of the IEDx, I_A, I_B and I_C: the respective Phase A, Phase B and Phase C current recorded by the IEDx.

information for the fault F1 on the transformer feeder 6. Owing to the incoherent values caused by each changing phase current, we combine all phases of current e.g. I_A, I_B and I_C into one current magnitude e.g. I_x, in which $x = \{1, 2, ..., 7, 8\}$.

13.2 Reduct

Table 15 shows the reduct table computed from the indiscernibility functions. Identical situations happened at multiple times in the reducts table, consequently they are grouped into similar classes.

To generate a summary of report, the information given in Table 15 may still include some condition attributes, which can be considered redundant. It is necessary to identify a concise but informative message for the system operators at the emergency. To make the given table more condensed, we have to find out, which elements in the table that can safely be removed.

Change of states can provide us the crucial information and thus it is selected. The revised result is presented in Table 16 that describes the general overview of the events taken place.

In Table 16, any information about the normal operation is ignored as it is not important during the emergency. Combining with the information from the Table 11, various reports can be produced for respective operators, protec-

Table 15. Results from the decision system.

Time	IED1		IED5		IED6		Decision
t/s	V_1	I_1	V_5	I_5	V_6	I_6	d
0.139	N	N	N	N	N	N	N
1.002	L	N	N	N	N	N	A
1.003	L	N	L	N	L	N	A
1.005	L	N	L	H	L	N	E
1.007	L	N	L	H	L	H	E
1.015	L	L	L	H	L	H	E
1.134	L	N	L	H	L	H	E
1.139	L	N	L	H	L	L	E
1.140	L	N	L	N	L	L	A
1.153	N	N	L	N	L	L	A
1.172	N	N	N	N	L	L	S

Table 16. Change of states in the reduced decision system.

Time	IED1		IED5		IED6		Decision
t/s	V_1	I_1	V_5	I_5	V_6	I_6	d
0.139	N	N	N	N	N	N	N
1.002	L	•	•	•	•	•	A
1.003	•	•	L	•	L	•	A
1.005	•	•	•	H	•	•	E
1.007	•	•	•	•	•	H	E
1.015	•	L	•	•	•	•	E
1.134	•	N	•	•	•	•	E
1.139	•	•	•	•	•	L	E
1.140	•	•	•	N	•	•	A
1.153	N	•	•	•	•	•	A
1.172	•	•	N	•	•	•	S

tion engineers and maintenance engineers. System operators need a summary of report whereas protection engineers require detailed and specific information regarding the operation of protection system and its related equipment. Maintenance engineers however require a summary of the fault classification to diagnose the fault type and the cause of the event.

Report for Operators

Location
Station name: Manchester
Station number: M117
Event date: 16th August 2003
Event time: 11:12:11am
Event number: 002

Description
Feeder L1: voltage sag = 1.002s, current low = 1.015s
Feeder 5 and 6: voltage sag, current high ≈ 1.003s
IED6: tripped
Breaker status: BRK6, BRK8 (1.119s – open)
Disconnection: Transformer T1
Faulted section: Transformer T1 feeder 6
System restored after 1.172s

Report for Protection Engineers

Location
Station name: Manchester
Station number: M117
Event date: 16th August 2003
Event time: 11:12:11am
Event number: 002

Description
Feeder L1: voltage sag = 1.002s, current low = 1.015s.
Feeder 5 and 6: voltage sag, current high ≈ 1.003s
IED6: picked up = 1.007s, tripped = 1.049s, reset = 1.138s
Disconnection: Transformer T1
Faulted section: Transformer T1 feeder 6
Fault inception: 1.0s
Fault duration: 0.119s
Fault type: directional A-B, permanent
Maximum fault magnitude: 13.70kA
Breaker status: BRK6, BRK8 (1.119s – open)
Relay condition: healthy
Breaker condition: healthy

Report for Maintenance Engineers

Location
Station name: Manchester
Station number: M117
Event date: 16th August 2003
Event time: 11:12:11am
Event number: 002

Description
Disconnection: Transformer T1
Faulted section: Transformer T1 feeder 6
Fault inception: 1.0s
Fault duration: 0.119s
Fault type: directional A-B, permanent
Breaker status: BRK6, BRK8 (1.119s – open)
Relay condition: healthy
Breaker condition: healthy

13.3 Unsupervised Extraction

The result obtained based on Table 12, 13 and 14, indicated that apart of the redundant data sources e.g. IED2, IED3 and IED4, the algorithm selects all the remaining attributes i.e. $\{IED1, IED5, IED6, IED7, IED8\}$ as our main source of information for the fault F1 on the transformer load feeder 6. Owing to the incoherent values caused by each changing phase current, we combine all phases of current e.g. I_A, I_B and I_C into one current magnitude e.g. I_x, in which $x = \{1, 2, ..., 7, 8\}$. In this case study, a fault was applied to the transformer feeder and theoretically, we expected to see a significant pattern change recorded by IED1, IED5, IED6, IED7 and IED8 (as IED1, IED2, IED3 and IED4 are carrying the same information, only IED1 is selected).

13.4 Reduct

Table 17 shows the reduct table computed from the indiscernibility functions. Identical situations happened at the multiple and consecutive times are eliminated or grouped into similar classes. The events at t = 1.007s and t = 1.134s are similar but occurred at two different times. For better description, they are displayed chronologically. These two time events indicate the intermediate steps between the fault period and the recovery period.

Attribute V_x in Table 17 represents a three-phase r.m.s voltage. I_x merged all the phase A, B, C r.m.s current of IEDx together with the priority of information set to be in the order of High, Low and Normal. This is to reduce the incoherent pattern change for all the three phase currents. t = 0.139s is the time at which the simulation attains a steady state condition e.g. voltage = *normal*, current = *normal*.

The information given in Table 17 may be excessive for inclusion in the report. Thus, the table must be condensed. The solution is to retain the change of state information as it provides useful information. Table 18 shows the change of state derived from Table 17 that can be used to describe the overview of events in the substation. The message received about the normal operation is ignored because it is not important during the emergency. Combining with the information from the Table 11, a summary of report can be produced respectively

Table 17. Results from the information system.

Time	IED1		IED5		IED6		IED7		IED8	
t/s	V_1	I_1	V_5	I_5	V_6	I_6	V_7	I_7	V_8	I_8
0.139	N	N	N	N	N	N	N	N	N	N
1.002	L	N	N	N	N	N	N	N	N	N
1.003	L	N	L	N	L	N	N	N	N	N
1.005	L	N	L	H	L	N	N	H	N	H
1.007	L	N	L	H	L	H	N	H	N	H
1.015	L	L	L	H	L	H	N	H	N	H
1.134	L	N	L	H	L	H	N	H	N	H
1.139	L	N	L	H	L	L	N	H	L	N
1.140	L	N	L	N	L	L	N	N	L	L
1.153	N	N	L	N	L	L	N	N	L	L
1.172	N	N	N	N	L	L	N	N	L	L

Table 18. Change of states in the reduced information system.

Time	IED1		IED5		IED6		IED7		IED8	
t/s	V_1	I_1	V_5	I_5	V_6	I_6	V_7	I_7	V_8	I_8
0.139	N	N	N	N	N	N	N	N	N	N
1.002	L	•	•	•	•	•	•	•	•	•
1.003	•	•	L	•	L	•	•	•	•	•
1.005	•	•	•	H	•	•	•	H	•	H
1.007	•	•	•	•	•	H	•	•	•	•
1.015	•	L	•	•	•	•	•	•	•	•
1.134	•	N	•	•	•	•	•	•	•	•
1.139	•	•	•	•	•	L	•	•	L	N
1.140	•	•	•	N	•	•	•	N	•	L
1.153	N	•	•	•	•	•	•	•	•	•
1.172	•	•	N	•	•	•	•	•	•	•

for operators, protection engineers and maintenance engineers. These reports are used mainly as an example. The format presented may not reflect the actual substation reports used in the power industries.

Report for Operators

Location
Station name: Manchester, Station number: M117
Event date: 16th August 2003, Event time: 11:12:11am
Event number: 002

Description
Feeder L1: voltage sag = 1.002s, current low = 1.015s
Feeder 5 and 6: voltage sag, current high ≈ 1.005s
Feeder 7 and 8: current high = 1.005s
IED6: tripped = 1.049s
Breaker status: BRK6, BRK8 (1.119s – open)
Disconnection: Transformer T1
Faulted section: Transformer T1 feeder 6
System restored after 1.172s

Report for Protection Engineers

Location
Station name: Manchester
Station number: M117
Event date: 16th August 2003
Event time: 11:12:11am
Event number: 002

Description
Feeder L1: voltage sag = 1.002s, current low = 1.015s.
Feeder 5 and 6: voltage sag, current high ≈ 1.005s
Feeder 7 and 8: current high = 1.005s
IED6: pickup = 1.007s, tripped = 1.049s, reset = 1.138s
Disconnection: Transformer T1
Faulted section: Transformer T1 feeder 6
Fault inception: 1.0s
Fault duration: 0.119s
Fault type: directional A-B, permanent
Maximum fault magnitude: 13.70kA
Breaker status: BRK6, BRK8 (1.119s – open)
Relay condition: healthy
Breaker condition: healthy

Report for Maintenance Engineers

Location
Station name: Manchester
Station number: M117
Event date: 16th August 2003
Event time: 11:12:11am
Event number: 002

Description
Disconnection: Transformer T1
Faulted section: Transformer T1 feeder 6
Fault inception: 1.0s
Fault duration: 0.119s
Fault type: directional A-B, permanent
Relay condition: healthy
Breaker status: BRK6, BRK8 (1.119s – open)
Breaker condition: healthy

Comparing the two set of results obtained in the example, it shows that the unsupervised extraction is less efficient than the supervised extraction. It produces less concise result; 7 attributes are selected compare to only 2 attributes for the supervised extraction. However, the advantage is that it does not require any pre-defined classes and is entirely based on the relation between data. This helps minimise the human errors in classifying the events. Both techniques have their cons and pros, therefore it is difficult to generalise which method is better. It depends on the area of application in the power system.

13.5 Station Report

A station report can also be formed using the information available. During the emergency, the operators may not require a detailed report. Thus the information provided must be concise. An expert system can be used to process the facts given in Table 16 and 18 and the protection trip data in Table 11. A sample of station report can be generated as shown in Table 19.

The time of fault or disturbance inception can be estimated using the abnormal condition status flag or emergency condition status flag if the abnormal condition on the fault inception is not available. The duration of the fault is determined by measuring the time period from the relay pickup to breaker opening. The magnitude of fault current can be used to estimate the I^2t contact-wear and other conditions associated with circuit breakers. Evaluation of the time at which the trip is applied until all the main contacts are open gives a good check on the breaker operation with the knowledge of the circuit breaker type and its operating characteristics.

Table 19. Station report.

STATION REPORT				
Date of event	16th August 2003		Time of event	11:12:11am
Event Number	002		Sample rate	1ms
EVENT DESCRIPTION				
Both Transformer Feeders 5 and 6: Voltage sags and high current at approximately 1.003s Load Feeder L1: Voltage sag at 1.002s and low current at 1.015s Relay IED6: Tripped at 1.049s; Reset at 1.138s; Directional element (67) was set. Circuit Breakers: BRK6 and BRK8 opened at 1.119s.				
PROTECTION SYSTEM OPERATION ANALYSIS				
Directional (67)	Yes	Non-Directional		No
Primary relay	Start Time	1.049s	End Time	1.138s
Backup relay	Start Time	N/A	End Time	N/A
Auto Recloser	No	Number of operations		N/A
Recloser Duration	N/A	N/A	N/A	N/A
Relay First Pickup	1.039s	Avg. Trip Time		0.010s
Breaker Operation	1.119s	No. Breakers	2 (BRK6, BRK8)	
Breaker Time	0.080s	Breaker Status	52A(1)	52B(0)
Relay Condition	Healthy	Breaker Condition		Healthy
ESTIMATED FAULT DATA				
Fault Inception	1.000s	Fault Types	Phase A-B, Permanent	
Faulted Section	Transformer Feeder T1	Fault Duration	0.119s	
Maximum Fault Magnitude		Approximate: 13.700kA		
Breaker Contact Wear, I^2t		$(\text{Max } I_{\text{Fault}})^2 \times t_{\text{Fault Duration}}$		$24.21 \times 10^6 As$
LINE CURRENTS AND VOLTAGES ON FAULTED SECTION				
RMS Value	Pre-fault	Max. Fault	Post Fault	Unit
I_n	0.00	0.00	0.00	[kA]
I_a	1.03	12.24	0.00	[kA]
I_b	1.03	12.97	0.00	[kA]
I_c	1.03	1.03	0.00	[kA]
V_a	6.17	3.49	0.00	[kV]
V_b	6.14	3.49	0.00	[kV]
V_c	6.30	6.12	0.00	[kV]
V_{ab}	10.58	3.55	0.00	[kV]
V_{bc}	10.80	9.32	0.00	[kV]
V_{ca}	10.86	9.33	0.00	[kV]

14 Conclusion

The large quantities of data generated by processor-based relays and IEDs have created both a demand and opportunity for extracting knowledge. Superfluous data may confuse an operator and result in a slower response to the emergency. The challenge is to interpret the data correctly from IEDs. The essential consideration of achieving highly recallable and concise information is to determine the most relevant attributes in the dataset and eliminate irrelevant/unimportant attributes without losing crucial information.

This paper presented two approaches; supervised and unsupervised extraction based on rough sets to assist our substation event analysis and decision support. The simulation models are used to generate an event database for various fault scenarios. Though the model used in our experiment may look simplified, it is however believed to be adequate for a pilot study and the results obtained are consistent and satisfactory. Rough set approach allows us to explore about the data. It is generic and independent of substation topology. Thus, it can be applied to any form of substations for knowledge extraction. The summary of events identified by rough sets can yield significant benefits to utilities by helping engineers and operators respond to a fault condition correctly in a limited time frame.

The analysis of substation data using rough classification is a new research area that would certainly benefit energy utilities especially the threat from data/information overload during the emergency.

References

1. K. Behrendt and M. Dood. Substation Relay Data and Communication. In 27nd*nd Annual Western Protective Relay Conference*, Spokane, Washington, October 1995. [online] www.selinc.com/techpprs.
2. A. Apostolov. Distributed Intelligence in Integrated Substation Protection and Control Systems. In 13th *Conference of the Electric Power Supply Industry and Exhibition (CEPSI 2000)*, Manila, Philippines, October 2000.
3. E. Rick. IEDs: The eyes and ears of Transmission & Distribution. *Electric Light & Power Technology*. Intelligent Controls Inc., Saco, Maine, USA.
4. W. Ackerman. The Impact of IEDs on the Design of Systems used for Operation and Control of Power Systems. In *IEE* 5th *International Conference on Power System Management and Control*, London, UK, April 2002.
5. L. Smith. Requirements: Utility Perspective. *IEEE Tutorial on Artificial Intelligence Application in Fault Analysis: Section I*, July 2000.
6. J. McDonald. Substation Automation – IED Integration and Availability of Information. IEEE Power & Energy, March–April 2003.
7. D. Kreiss. Non-Operational Data: The Untapped Value of Substation Automation. Utility Automation, September/October 2003.
8. D. Kreiss. Utilities can enhance Bottom Line by leveraging Non-Operational Data. Utility Automation, November/December 2003.
9. C. Hor and P. Crossley. Substation Data Analysis with Rough Sets. In *Proceeding of 8th IEE Development in Power System Protection*, April 5th – 8th 2004.
10. G. Torres, "Application of rough sets in power system control center data mining," in *Proc. IEEE Power Engineering Society Winter Meeting*, New York, United States, Jan. 27th – 31st 2002.
11. K. Brand, V. Lohmann, and W. Wimmer. *Substation Automation Handbook*. Utility Automation Consulting Lohmann, Im Meyerhof 18, CH-5620 Bremgarten, Switzerland, first edition, 2003.
12. G. Gross, A. Bose, C. DeMarco, M. Pai, J. Thorp, and P. Varaiya. Real Time Security Monitoring and Control of Power Systems. Technical report, The Consortium for Electric Reliability Technology Solutions (CERTS) Grid of the Future White Paper, December 1999. [online] http://certs.lbl.gov/CERTS_Pubs.html.
13. Q. Shen and A. Chouchoulas. Rough Set-Based Dimensionality Reduction for Supervised and Unsupervised Learning. Technical report, Centre for Intelligent Systems and their Applications, Division of Informatics, University of Edinburgh, May 2001.

14. L. Polkowski. *Advances in Soft Computing: Rough Sets Mathematical Foundation.* Phyica-Verlag Publisher, A Springer Verlag Company, Berlin, 2002.
15. Z. Pawlak and A. Skowron. Rough Set Rudiments. *Bullentin International Rough Set Society,* 3(4):pp. 181–185, 1999.
16. J. Komorowski, Z. Pawlak, L. Polkowskis, and A. Skowron. *Rough Sets: A Tutorial.* In: Rough Fuzzy Hybridization – A New Trend in Decision Making, (S.K. Pal, A. Skowron, Eds.), Springer Verlag Publisher, Singapore, 1999.
17. J. Komorowski, Z. Pawlak, L. Polkowskis, and A. Skowron. *A Rough Set Perspective on Data and Knowledge.* In: Handbook of Data Mining and Knowledge Discovery, (W. Klösgen, J. Zytkow, Eds.), Oxford University Press, 2000.
18. A. Skowron and C. Rauszer. *The Discernibility Matrices and Functions in Information Systems,* volume pp. 311–362. R. Slowinski, ed., Intelligent Decision Support. Handbook of Applications and Advances of the Rough Set Theory, Kluwer, Dordrecht, 1992.
19. A. Øhrn. *Discernibility and Rough Sets in Medicine: Tools and Applications.* PhD thesis, Department of Computer Science and Information Science, Norwegian University of Science and Technology (NTNU), Trondheim, Norway, 2000.
20. G. Roed. Knowledge Extraction from Process Data: A Rough Set Approach to Data Mining on Time Series. Master's thesis, Department of Computer and Information Science, Norwegian University of Science and Technology, Trondheim, Norway, 1999.
21. C.L. Hor, A. Shafiu, P.A. Crossley, F. Dunand, *Modeling a Substation in a Distribution Network: Real Time Data Generation for Knowledge Extraction,* IEEE PES Summer Meeting 2002, Chicago, Illinois, USA, July 21st – 25th, 2002.
22. C.L. Hor, K. Kangvansaichol, P.A. Crossley, A. Shafiu, *Relays Models for Protection Studies,* IEEE Power Tech 2003, Bologna, Italy, June 23st – 26th, 2003.
23. Central Networks, *Long term development statement for Central Networks East & Central,* Summary Information, November 2003.
[online] http://www.centralnetworks.co.uk/Content/Service/serv_longterm.aspx

Processing of Musical Data Employing Rough Sets and Artificial Neural Networks

Bożena Kostek, Piotr Szczuko, Paweł Żwan, and Piotr Dalka

Gdańsk University of Technology, Multimedia Systems Department,
Narutowicza 11/12, 80-952 Gdańsk, Poland
{bozenka,szczuko,zwan,dalken}@sound.eti.pg.gda.pl
http://www.multimed.org

Abstract. This article presents experiments aiming at testing the effectiveness of the implemented low-level descriptors for automatic recognition of musical instruments and musical styles. The paper discusses first some problems in audio information analysis related to MPEG-7-based applications. A short overview of the MPEG-7 standard focused on audio information description is also given. System assumptions for automatic identification of music and musical instrument sounds are presented. A discussion on the influence of descriptor selection process on the classification accuracy is included. Experiments are carried out basing on a decision system employing Rough Sets (RS) and Artificial Neural Networks (ANNs).

1 Introduction

The aim of this study is to automatically classify musical instrument sounds or a musical style on the basis of a limited number of parameters, and to test the quality of musical sound parameters that are included in the MPEG-7 standard. Recently defined MPEG-7 standard is designed to describe files containing digital representations of sound, video, images and text information allowing the content to be automatically queried in multimedia databases that can be accessed via the Internet. MPEG-7 standard specifies the description of features related to the audio-video (AV) content as well as information related to the management of the AV content. In order to guarantee interoperability for some low-level features, MPEG-7 standard also specifies part of the extraction process. MPEG-7 descriptions take two possible forms: a textual XML form (high-level descriptors) suitable for editing, searching, filtering, and browsing and a binary form (low-level descriptors) suitable for storage, transmission, and streaming. For many applications, the mapping between low-level descriptions and high-level queries has to be done during the description process. The search engine or the filtering device have to analyze the low-level features, and on this basis, perform the recognition process. This is a very challenging task for audio analysis research [12]. The technology related to intelligent search and filtering engines using low-level audio features, possibly together with high-level features, is still

J.F. Peters and A. Skowron (Eds.): Transactions on Rough Sets III, LNCS 3400, pp. 112–133, 2005.

very limited. A major question remains open what is the most efficient set of low-level descriptors that have to be used to allow a certain class of recognition tasks to be performed on the description itself. Of importance is the fact that the parameters of MPEG-7 musical sounds were selected based on many but separately carried on experiments, examples of which may be found in many publications [3, 5–7, 11, 13, 15, 22, 32], hence the need to test the quality of the parameters in the process of automatic identification of musical instrument classes. Especially interesting is material presented by Wieczorkowska et al., in which authors try to identify the most significant temporal musical sound features and descriptors. Their main goal was to verify whether temporal patterns observed for particular descriptors would facilitate classification of new cases introduced to the decision system [32], however conclusions are not yet definite.

The experiments presented in this article are based on a decision-making system employing learning algorithms. The paper discusses a number of parametrization methods that are used for describing musical objects. The starting point in the experiments is engineering a music database. Among others, problems discussed in the paper are related to how musical objects are saved and searched in the constructed sound database. Despite the standard relatively clear-cut instructions on how musical objects should be described (e.g. using the feature vector), it mostly covers the indexing of files in databases rather than focusing on the algorithmic tools for searching these files.

In general, music can be searched using a so-called "query-by-example" scheme. Such a scheme includes musical signals (single, polyphonic sounds, audio files), human voice sounds (both speech and singing), music scores (graphical form), and MIDI (Musical Instrument Digital Interface) code (encoded music) or verbal description. Each signal comes as a different representation. The paper shows examples of feature vectors (FVs) based on time, spectral, time-frequency musical signals representations. In addition, the problem of how to query effectively multimedia contents applies to search tools as well. To address that, many research centers focus on this problem (see International Music Retrieval Symposia website [11]). Earlier research at the Department of Multimedia Systems (former Sound and Vision Department) of the Gdansk University of Technology showed that for automatic search of musical information, learning decision-making systems work much better than typical statistical or topological methods. For instance, rough set-based decision systems allow for searching databases with incomplete information and inconsistent entries. This theory, founded by Pawlak [25] is now extremely well developed, and one can find applications based on rough sets in many domains [1, 2, 4, 14, 15, 19, 24, 26–30]. The rough set theory proposed by Pawlak provides an effective tool for extracting knowledge from data tables. The rough set decision system is also very valuable in the feature selection process. Another example of learning algorithms is the Artificial Neural Network (ANN). ANNs are computing structures that adjust themselves to the specific application by training rather than by having a defined algorithm. The network is trained by minimizing the error of the network classification. When a specific threshold error is reached, the learning process is considered as completed. ANN may serve as both pattern classifier and feature selectors.

For the purpose of experiments the Rough Set Exploration System (RSES) system developed by the Warsaw University [1, 2, 8] is used. RSES has been created to enable multi-directional practical investigations and experimental verification of research in decision support systems and classification algorithms, in particular to applications of rough set theory [1, 2]. On the other hand, the ANN employed is a toolbox working in the Matlab environment. The use of rough sets and artificial neural networks allows for identifying musical objects even if during the training process only a limited number of examples is presented to the decision system.

2 Low-Level MPEG-7 Descriptors

Development of streaming media on the Internet has caused a huge demand for making audio-visual material searchable in the same way as text is. Audio and video contain a lot of information that can be used in indexing and retrieval applications. Many research efforts have been spent to describe the content of audio and video signals, using specific indexing parameters and extraction techniques. The MPEG-7 standard provides a uniform and standardized way of describing the multimedia content. A part of this standard is specifically dedicated to the description of the audio content. The MPEG-7 Audio framework consists of two description levels: low-level audio descriptors, and high-level ones related to semantic information and metadata. Metadata are compiled from collections of low-level descriptors that are the result of detailed analysis of the content actual data samples and signal waveforms. MPEG-7 expresses these descriptions in XML, thereby providing a method of describing the audio or video samples of the content in textual form [9, 12, 21]. The MPEG-7 standard defines six categories of low-level audio descriptors.In addition a so-called "silence" parameter is also defined. Generally, it may be said that such a categorization of parameters is related to the need to visualize sound, and to extract significant signal parameters regardless of the audio signal type (e.g. musical sounds, polyphonic signals, etc.). The MPEG-7 low-level descriptors can be stored as an XML file that serves as a compact representation of the analyzed audio. These six categories contain the following parameters [9]:

1. Basic: Audio Waveform, Audio Power
2. Basic Spectral: Audio Spectral Envelope, Audio Spectral Centroid, Audio Spectrum Spread, Audio Spectrum Flatness
3. Signal Parameters: Audio Fundamental Frequency, Audio Harmonicity
4. Timbral Temporal: Attack Time, Temporal Centroid
5. Timbral Spectral: Harmonic Spectrum Centroid, Harmonic Spectral Deviation, Harmonic Cepstral Spread, Harmonic Spectral Variation, Spectral Centroid
6. Spectral Basis: Audio Spectrum Basis, Audio Spectrum Projection

3 System Assumptions

A very interesting example of musical databases is CDDB that contains over 1 million CDs [10]. However, apart from world-wide known databases there exist many local databases. An example of such a musical database is the one created at the Multimedia Systems Department of the Gdansk University of Technology. This musical database consists of three sub-databases, each related to different format of music. Therefore, the first one contains musical sounds [15, 20], the second one – MIDI files [4], and the third one fragments of music from CDs [19]. Searching for music, MIDI-based recording or a sound signal (audio type) in a music database includes the following tasks:

- audio registrating – employing an audio card
- uploading the pattern file
- playing back the pattern file
- automatic parameterizing the pattern file
- defining query criteria (metadata description)
- searching in the parameter database based on the metadata description and vector parameter similarity with the following options:
  - searching for the specific file and not downloading it
  - script compiling in the server environment/downloading the file with parameters from the server
  - reviewing and playing query results
  - downloading the file selected

Fig. 1 shows the system operational principle presenting in which way new files are introduced into the system and how the query is done. Within this scheme all other categories of musical files can also be searched. When sound signals are introduced into the system, it applies preprocessing, then generates feature vectors by extracting low-level descriptors defined for this procedure [18].

A database record consists of the following components: category, [feature vector], metadata description and resource link. The parameter file for each database objectincludes a parametric description along with metadata. The query process does not require parameterizing all database objects, this is done

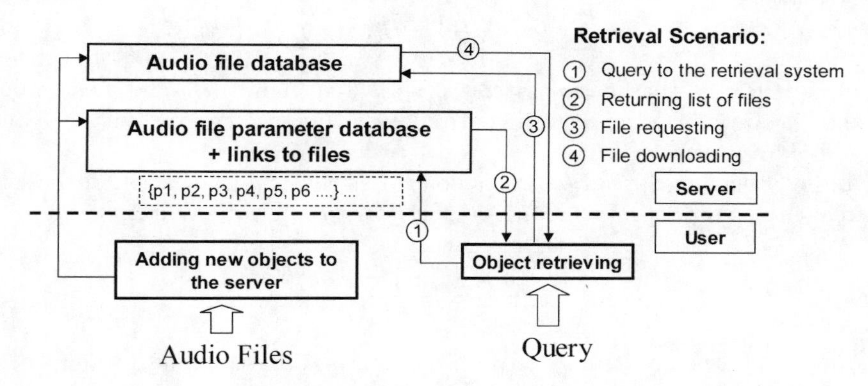

Fig. 1. System operational principle.

when introducing a new object to the database. The search is for a file that matches the description. In this way, an audio file can be specifically described in a standardized form that can be stored in a database. This provides a reliable way of identifying a particular item of content without actually analyzing its waveform or decoding its essence data, but rather by scanning its description.

4 MPEG-7 Based Parameters Applied to Musical Sounds

Before automatic parameter extraction of musical sounds can be done, the sound fundamental frequency must be estimated. A modified Schroeder's histogram is applied for this purpose [20]. As was earlier reported the effectiveness of detection of the musical sound fundamental frequency is very high for the selected instruments (98 %), thus the calculation of signal parameters do not suffer from the erroneous pitch detection [20]. In the next step parametrization process is performed. The following MPEG-7 descriptors are taken for this study [31]. It should be noted, that features, parameters, and features vectors are standard terminology in audio processing and refer to sound descriptors, thus these terms are often used interchangeably.

- Audio Spectrum Envelope (*ASE*) describes the short-term power spectrum of the waveform as a time series of spectra with logarithmic frequency axis. According to MPEG-7 recommendation, the spectrum consists of one coefficient representing power between 0 Hz and 62.5 Hz, a series of coefficients representing power in 1/4-octave resolution sized bands between 62.5 Hz and 16 kHz, and a coefficient representing power beyond 16 kHz. This results in 34 coefficients for each spectral frame. The mean values and variance of each coefficient over time are denoted as $ASE_1...ASE_{34}$ and $ASEv_1...ASEv_{34}$, respectively.
- Audio Spectrum Centroid (*ASC*) describes the center of gravity of the log-frequency power spectrum. Power spectrum coefficients below 62.5 Hz are replaced by a single coefficient, with power equal to their sum and a nominal frequency of 31.25 Hz. Frequencies of all coefficients are scaled to an octave scale with its zero point set to 1 kHz. The spectrum centroid is calculated as follows:
$$C = \sum_n (log_2(\frac{f(n)}{1000}) \cdot P_x(n) / \sum_n P_x(n)) \tag{1}$$
where $P_x(n)$ is the power associated with frequency *f(n)*. The mean value and the variance of spectrum centroid over time are denoted as ASC and $ASCv$, respectively.
- Audio Spectrum Spread (*ASS*) is defined as the *RMS* deviation of the log-frequency power spectrum with respect to its center of gravity:
$$S = \sqrt{\frac{\sum_n (log_2((\frac{f(n)}{1000} - C)^2 \cdot P_x(n)))}{\sum_n P_x(n)}} \tag{2}$$
where C is the spectrum centroid. The mean value and the variance of S over time are denoted as ASS and $ASSv$, respectively.

- Audio Spectrum Flatness (SFM) describes the properties of the short-term power spectrum of an audio signal. For each frequency band, the spectrum flatness measure is defined as the ratio of the geometric and the arithmetic mean of the power spectrum coefficients $c(i)$ within the band b (i.e. from coefficient index il to coefficient index ih, inclusive:

$$SFM_b = \frac{\sqrt[ih(b)-il(b)+1]{\prod_{i=il(b)}^{ih(b)} c(i)}}{\frac{1}{ih(b)-il(b)+1} \cdot \sum_{i=il(b)}^{ih(b)} c(i)} \tag{3}$$

The mean values and the variance of each SFM_b over time are denoted as $SFM_1 ... SFM_{24}$ and $SFMv_1 ... SFMv_{24}$, respectively.
- Log Attack Time (LAT) is defined as the logarithm (decimal basis) of the time difference between when the signal starts (T_0) and when it reaches its sustained part (T_1):

$$LAT = log_{10}(T_1 - T_0) \tag{4}$$

- Temporal Centroid (TC) is defined as the time averaged over the energy envelope SE:

$$TC = \frac{\sum_{n=1}^{length(SE)} (n/sr \cdot SE(n))}{\sum_{n=1}^{length(SE)} (SE(n))} \tag{5}$$

where sr is the sampling rate.
- Spectral Centroid (SC) is the average of the Instantaneous Spectral Centroid (ISC) values computed in each frame. They are defined as the power weighted average of the frequency of bins in the power spectrum:

$$ISC = \frac{\sum_{k=1}^{length(S)} (f(k) \cdot S(k))}{\sum_{k=1}^{length(S)} (S(k))} \tag{6}$$

where $S(k)$ is the kth power spectrum coefficient and $f(k)$ stands for the frequency of the kth power spectrum coefficient. The mean value and variance of $IHSC$ over time are denoted as SC and SCv, respectively.
- Harmonic Spectral Centroid (HSC) is the average of the Instantaneous Harmonic Spectral Centroid $IHSC$ values computed in each frame. They are defined as the amplitude (linear scale) weighted mean of the harmonic peaks of the spectrum:

$$IHSC = \frac{\sum_{h=1}^{nb_h} (f(h) \cdot A(h))}{\sum_{h=1}^{nb_h} (A(h))} \tag{7}$$

where nb_h is the number of harmonics taken into account, $A(h)$ is the amplitude of the harmonic peak number h and $f(h)$ is the frequency of the harmonic peak number h. The mean value and the variance of $IHSC$ over time are denoted as HSC and $HSCv$, respectively.
- Harmonic Spectral Deviation (HSD) is the average of the Instantaneous Harmonic Spectral Deviation $IHSD$ values computed in each frame. They

are defined as the spectral deviation of log-amplitude components from the global spectral envelope:

$$IHSD = \frac{\sum_{h=1}^{nb_h}(|log_{10}(A(h) - log_{10}(SE(h))|)}{\sum_{h=1}^{nb_h}(log_{10}A(h))} \quad (8)$$

where *SE(h)* is the local spectrum envelope (mean amplitude of the three adjacent harmonic peaks) around the harmonic peak number h. To evaluate the ends of the envelope (for $h = 1$ and $h = nb_h$) the mean amplitude of two adjacent harmonic peaks is used. The mean value and the variance of *IHSD* over time are denoted as *HSD* and *HSDv*, respectively. Mean values and variation denoted as *HSD* and *HSDv*, respectively.

- Harmonic Spectral Spread (*HSS*) is the average of the Instantaneous Harmonic Spectral Spread (*IHSS*) values computed in each frame. They are defined as the amplitude weighted standard deviation of the harmonic peaks of the spectrum, normalized by the harmonic spectral centroid:

$$IHSS = \frac{1}{IHSC} \cdot \sqrt{\frac{\sum_{h=1}^{nb_h} A^2(h) \cdot (f(h) - IHSC)^2}{\sum_{h=1}^{nb_h} A^2(h)}} \quad (9)$$

where *IHSC* is the harmonic spectrum centroid defined by Eq. 7. The mean value and the variance of harmonic spectrum spread over time are denoted as *HSS* and *HSSv*, respectively.

- Harmonic Spectral Variation (*HSV*) is the average of is Instantaneous Harmonic Spectral Variation (*IHSV*) values computed in each frame. They are defined as the normalized correlation between the amplitude of the harmonic peaks of two adjacent frames:

$$IHSV = 1 - \frac{\sum_{h=1}^{nb_h} A_{-1}(h) \cdot A(h)}{\sqrt{\sum_{h=1}^{nb_h} A_{-1}^2(h)} \cdot \sqrt{\sum_{h=1}^{nb_h} A^2(h)}} \quad (10)$$

where *A(h)* is the amplitude of the harmonic peak number h at the current frame and $A_{-1}(h)$ is the amplitude of the harmonic peak number h at the preceding frame. The mean value and the variance of harmonic spectrum variation over time are denoted as *HSV* and *HSVv*, respectively.

Besides MPEG-7-based descriptors, a few additional parameters were used:

- KeyNumber (*KeyNum*) expresses the pitch of a sound according to the MIDI standard:

$$KeyNum = 69 + 12 \cdot \log_2(\frac{f_0}{440}) \quad (11)$$

- Content of even harmonics in spectrum (h_{ev})

$$h_{ev} = \frac{\sqrt{\sum_{k=1}^{\lfloor N/2 \rfloor} A_{2k}^2}}{\sqrt{\sum_{n=1}^{N} A_n^2}} \quad (12)$$

The number of all attributes derived from MPEG-7 analysis was 139.

4.1 Sound Descriptor Analysis

A number of parameters describing a musical instrument sound should be as low as possible because of the limited computer system resources. The process of decreasing the feature vector length means removing redundancy from the set describing an audio signal. Therefore, evaluation criteria of the effectiveness of particular parameters for the sound classification have to be used.

Fisher statistic is often used as such a criterion. It is defined for a parameter A and two classes X and Y, in the case analyzed two classes of instruments [15, 16]:

$$V = \frac{\bar{A}_X - \bar{A}_Y}{\sqrt{S_{AX}^2/n_X + S_{AY}^2/n_Y}} \tag{13}$$

where AX and AY are mean values of parameter A for instruments X and Y; n_X, n_Y are the cardinalities of two sets of sound parameters; and S_{AX}^2 and S_{AY}^2 are variance estimators:

$$S_{AX}^2 = \frac{1}{(n_X - 1)} \cdot \sum_{i=1}^{n_X} (A_{Xi} - \bar{A}_X)^2 \tag{14}$$

$$S_{AY}^2 = \frac{1}{(n_Y - 1)} \cdot \sum_{i=1}^{n_Y} (A_{Yi} - \bar{A}_Y)^2 \tag{15}$$

The bigger the absolute values $|V|$ of Fisher statistics, the easier it is to divide a multidimensional parameter space into areas representing different classes. It is much easier to differentiate between two musical instruments based on the given parameter if its mean values for both instruments are clearly different, variances are small and the quantity of audio samples is large.

Values of the calculated Fisher statistic for selected parameters and for the selected pair of instruments are shown in Table 1. It was found for example that *HSD* and *HSS* parameters are useful for the separation of musical sounds of different groups (brass, woodwinds, strings). Figure 2 shows an example of distribution of these parameters values for instruments of similar musical scales.

High value of the Fisher statistic of h_{ev} parameter for the pair bassoon-clarinet proves its usefulness for separation of a clarinet sounds from other musical instruments. Initial experiments show that Timbre Descriptors are insufficient for the separation of musical instruments from the same group with satisfactory effectiveness. Therefore, the feature vector needs to be complemented by some additional descriptors of a musical signal.

Table 1. Analysis of parameter separability based on Fisher statistics.

instrument pairs	h_{ev}	LAT	SC	...	HSV	ASE_1	ASE_2	...	SFM_{24}
bassoon – clarinet	28.79	3.22	15.73	...	10.72	9.10	10.56	...	14.18
bassoon – oboe	3.02	0.13	53.78	...	1.36	8.31	8.75	...	17.71
bassoon – trombone	1.22	0.34	12.78	...	0.33	8.42	9.54	...	5.46
bassoon – F. horn	1.94	5.43	4.48	...	0.58	6.90	7.26	...	1.85
...	...	...	...	...	...	...	...	...	...
cello – tuba	4.31	15.82	12.51	...	5.22	4.72	0.72	...	22.55

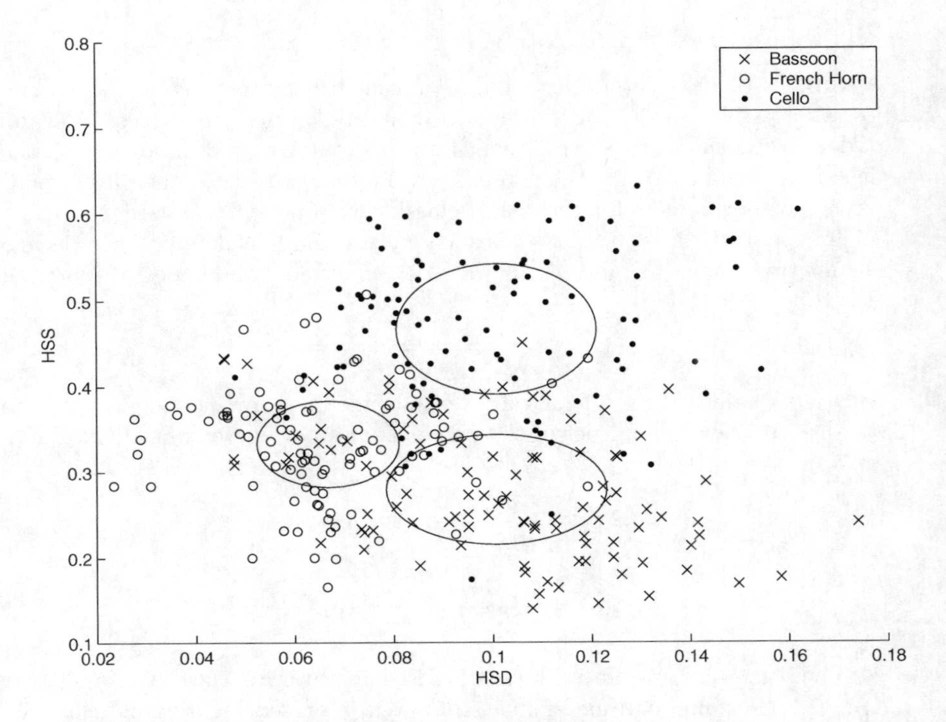

Fig. 2. Example of two parameter value distribution for three instruments.

More effective sound description is provided by the parameters connected directly with the power spectrum density: Audio Spectrum Descriptors, particularly Audio Spectrum Envelope (ASE) and Audio Spectrum Flatness (SFM). It has been noticed that ASE and $ASEv$ descriptors calculated for low- and high-frequency bands allows for distinguishing between instrument classes; mid-frequency band values are less useful for musical instrument classification. On the other hand, SFM descriptors are the most accurate in mid-frequency bands. $SFMv$ descriptors prove to be redundant, thus they have not been included in FVs. In addition, the Pearson's correlation analysis has been performed, and some other parameters have been eliminated. After a thorough analysis, a final content of the feature vector used in experiments with neural networks (31) attributes was as follows:

$$\{ASE_2, ..., ASE_5, ASE_8, ASE_9, ASE_{18}, ASE_{21}, ASE_{23}, ..., ASE_{31}, ASE_{33}, ASE_{34}, ASEv_5, ..., ASEv_9, ASEv_{21}, ASEv_{31}, ASEv_{34}, ASC, ASS, ASSv, SFM_{13}, ..., SFM_{19}, SFM_{21}, SFM_{22}, SFM_{24}, HSC, HSD, HSDv, HSS, HSSv, KeyNum, h_{ev}, LAT\}$$

4.2 ANN-Based Classification Results

A three-layer neural network (NN) of the feed-forward type was used in the experiments. Its structure has been defined as follows:

- number of neurons in the initial layer is equal to the number of elements of the feature vector
- number of neurons in the hidden layer is twice as large as the number of neurons in the initial layer
- each neuron in the output layer is matched to the different class of the instrument, thus the number of neurons in the output layer is equal to the number of classes of instruments being identified
- neurons in the initial and the output layers have log-sigmoid transfer functions, while neurons in the hidden layer have tan-sigmoid transfer functions.

The initial stage of experiments started with the training phase of the neural network. Sounds samples come from two sources, those recorded at the Multimedia Systems Department of the Gdansk University of Technology [15, 16], contained in the Catalogue, and those from MUMS (McGill University Master Samples) from the McGill University [23]. Vectors of parameters were randomly divided into two independent sets: training and testing. Such a technique of estimating error rates was used in all experiments presented in this article. Each set consisted in 50 % of vectors contained in the database. This means that it was possible for the training set to contain only a few FVs of some instruments while the test set might include much larger number of FVs of the same instrument. Error back-propagation algorithm was used to train the neural network. The process of training was considered as finished when the value of the cumulative error of network responses for the set of test vectors had dropped below the assumed threshold value or when the cumulative error of network responses for the test set of vectors had been rising for more than 10 consecutive cycles. The recognized class of the instrument was determined by the highest value of the output signals of neurons in the output layer. The training procedure was repeated 10 times and the best-trained network was chosen for further experiments.

Detailed results of the musical sound classification with the neural network algorithms are presented in Tables 2 and 3. Confusion Matrix of Neural Network Classifier is shown in Table 2. The ANN results for the MPEG-7-based representation are given by the diagonal of Tab. 2 – correctly recognized class of an instrument. The test set accuracy denotes the ratio of the number of correctly classified objects from the test set and all objects in the test set. Other values in Table 2 refer to errors done in the recognition process. Table 3 gives the number of sound samples tested in experiments.

As shown in Tables 2 and 3 tuba sound recognition effectiveness was 100%. Other instrument sounds were recognized with lower accuracy, for instance some mix-up between tenor trombone and French horn sounds occurred. Trombone sounds that were incorrectly recognized were classified by the system as French horn sounds, and vice versa; also clarinet and oboe sounds were confused with each other (a few cases of erroneous system answers). Therefore the total test set accuracy for these 10 musical instrument classes was 92.35%. The total test set accuracy is meant here as average of all test set accuracy values.

Table 2. Recognition effectiveness of the MPEG-7-based descriptors with the NN classifier.

Instrument	BA	CL	Oboe	TT	FH	AS	VI	TR	TU	CE
bassoon(BA)	**95.5**	0.0	1.5	0.5	.15	0.0	0.0	0.0	0.5	0.5
B flat clarinet (CL)	0.5	**93.0**	5.4	0.0	0.0	0.0	0.5	0.0	0.0	0.5
oboe	0.0	4.4	**91.2**	0.0	0.0	0.0	1.9	2.5	0.0	0.0
tenor trombone	3.2	0.5	0.0	**86.8**	6.9	0.5	0.0	1.1	1.1	0.0
French horn (FH)	1.2	1.2	0.0	7.4	**87.7**	0.0	0.6	0.6	0.0	1.2
alto saxophone (AS)	0.0	0.8	0.0	0.8	0.8	**93.2**	0.8	1.7	0.8	0.8
violin (VI)	0.0	0.6	1.7	1.1	0.0	0.6	**91.2**	0.0	1.7	3.3
trumpet (TR)	0.0	2.8	1.4	2.1	0.0	1.4	0.7	**91.6**	0.0	0.0
F flat tuba(TU)	0.0	0.0	0.0	0.0	0.0	0.0	0.0	0.0	**100.0**	0.0
cello (CE)	0.0	1.0	0.0	0.5	0.0	0.0	5.2	0.0	0.0	**93.3**

Table 3. Summary of results obtained by the neural network algorithm tested on the MPEG-7-based FVs.

Musical Instrument	No. of samples	No. of errors	Effectiveness [%]
bassoon	200	9	95.5
B fat clarinet	186	13	93.0
oboe	159	14	91.2
tenor trombone	189	25	86.8
French horn	162	20	87.7
alto saxophone	118	8	93.2
violin	181	16	91.2
trumpet	143	12	91.6
F Flat tuba	143	12	100
cello	143	12	93.3

4.3 MPEG-7 and Wavelet-Based Joint Representation Quality Testing

MPEG-7 parameters were calculated for the whole sound, however, a decision has been done to enhance the description by adding descriptors based on the sound transient phase, only. Earlier research on wavelet parameters for the transient state of musical sounds allowed for an easy calculation of such parameters [16]. One of the main advantages of wavelets is that they offer a simultaneous localization in time and frequency domain. Frames consisting of 2048 samples taken from the transient of a sound were analyzed. In earlier experiments performed by one of the authors [16], it was found that Daubechies filters (2nd order) have the computational load considerably lower than that of other types of filters, therefore they were used in the analysis.

For the purpose of this study several parameters were calculated. They were derived by observing both energy and time relations within the wavelet subbands. Energy-related parameters are based on energy coefficients computed for the wavelet spectrum subbands normalized over the whole energy of the parameterized frame corresponding to the starting transient. On the other hand, time-related wavelet parameters refer to the number of coefficients that have exceeded the given threshold. Such a threshold helps to differentiate between "tone-like" and "noise-like" characteristics of the wavelet spectrum [16]. Wavelet-based parameters were as follows: cumulative energy (E_{cn}) related to the nth subband, energy of nth subband E_n, energy ratio of nth subband to $(n-1)$th subband, e_n – time-related parameter allowing for characterization of the wavelet pattern, calculated for each wavelet spectrum subband, and referring to the number of coefficients that have exceeded the given threshold and f_n – variance of the first derivative of the absolute value of the wavelet coefficient sequence [16, 17]. By means of the Fisher statistic the number of wavelet coefficients was reduced from 50 to a few most important (13), thus forming the following feature vector:

$$\{E_5, E_6, E_8, (E_{10}/E_9), e_6, e_7, e_8, e_{10}, f_7, f_{10}, E_{c7}, E_{c8}, E_{c10}\}.$$

To assess the quality of the expanded feature vectors, musical instrument sounds underwent recognition for a combined representation of the steady and transient states (simultaneous MPEG-7 and wavelet parametrization resulting in 44 parameters). To that end, ANNs were used again. Confusion Matrix of Neural Network classifier resulting from the testing process is shown in Table 4. The system results for the wavelet-based representation are given by the diagonal of Tab. 4 (correctly recognized class of instrument). Other values denote errors done in the classification process.

As seen from Tab. 4 and 5 the system recognition effectiveness is better than while employing FVs containing only MPEG-7 descriptors. For example, accuracy of alto saxophone sounds increases up to 98.2%. Still, some musical instrument sounds are recognized incorrectly. It can be seen that the wavelet-based parametrization help to better differentiate between French horn and tenor trombone sounds, thus much better results for this instrument are obtained. Fewer errors occur for other instruments, while tuba does not do that well. In overall, the combined MPEG-7 and wavelet representation ensures better classification results, thus the total test set accuracy is equal to approx. 94%. The number of errors diminishes from 131 to 106, a reduction of 20%. This is because the parametrization process covers both the transient and steady state

Table 4. Recognition effectiveness of the MPEG-7+wavelet-based FVs.

Instrument	BA	CL	Oboe	TT	FH	AS	VI	TR	TU	CE
bassoon(BA)	**96.7**	1.1	0.0	1.1	0.5	0.0	0.0	0.0	0.0	0.5
B flat clarinet (CL)	0.5	**91.4**	6.5	0.0	0.5	0.0	1.1	0.0	0.0	0.0
oboe	0.0	3.6	**91.7**	0.0	0.0	0.0	3.0	1.2	0.0	0.6
tenor trombone	2.3	0.0	0.0	**86.8**	6.3	0.0	1.1	0.0	2.9	0.6
French horn (FH)	2.0	0.7	0.0	2.6	**94.8**	0.0	0.0	0.0	0.0	0.0
alto saxophone (AS)	0.0	0.0	0.9	0.0	0.0	**98.2**	0.0	0.9	0.0	0.0
violin (VI)	0.5	0.5	2.2	0.0	0.0	0.0	**92.4**	0.0	0.0	4.3
trumpet (TR)	0.0	1.8	0.6	0.6	0.0	1.2	0.6	**95.2**	0.0	0.0
F flat tuba(TU)	0.0	0.0	1.9	3.2	0.0	0.0	0.0	0.0	**94.8**	0.0
cello (CE)	0.0	0.0	0.0	0.5	0.0	0.0	2.8	0.5	0.0	**96.8**

Table 5. Effectiveness of the ANN for MPEG-7+wavelet-based FVs.

Musical Instr.	No. of samples	No. of errors	Effectiveness [%]
bassoon	183	6	96.7
B flat clarinet	185	16	91.4
oboe	169	14	91.7
tenor trombone	174	23	86.8
French horn	153	8	94.8
alto saxophone	112	2	98.2
violin	184	14	92.39
trumpet	165	8	95.15
F flat tuba	155	8	94.8
cello	216	7	96.76

Table 6. Recognition effectiveness for instrument groups based on NNs.

Instrument group	No. of sounds	Erroneous recognition	Effectiveness[%]
string	411	6	98.54
woodwinds	902	17	98.12
brass	712	18	97.47

of a sound, and the features resulting form this process of the parameter vector complement each other.

Further, another experiment based on ANNs, was carried out. Instruments were assigned to three classes and classification was performed once again. Results obtained were gathered in Table 6. As seen from Table 6 the system accuracy is very good (approx. 98/classified. Therefore, it can be said that the FV representation containing both MPEG-7- and wavelet-based parameters is more effective than while using MPEG-7 features only.

4.4 Rough Set-Based Feature Vector Quality Testing

For the purpose of experiments based on the rough set method, the RSES system was employed. The RSES is a system containing functions for performing various data exploration tasks such as: data manipulation and edition, discretization of numerical attributes, calculation of reducts. It allows for generation of decision rules with use of reducts, decomposition of large data into parts that share the same properties, search for patterns in data, etc. Data are represented as a decision table in RSES, which contains condition attribute values and decisions. A standard term used in the rough set theory is 'attribute' instead of feature, but in the research presented we use also 'feature vector' term since it may apply to both ANNs and rough sets. Since the system was presented very thoroughly at the rough set society forum, thus no details are to be shown here, however, a reader interested in the system may visit its homepage [8].

It is common to divide a database in two parts to create a training and a test set. During training phase the rough set system tries to extract knowledge from

Table 7. Format of the decision table.

KeyNum	h_{ev}	*LAT*	*HSC*	...	ASE_4	ASE_5	...	SFM_{22}	SFM_{24}	*Decision*
57.64	0.9272	0.1072	914	...	–0.1761	–0.1916	...	0.0971	0.0707	cello
57.68	0.9178	0.1140	818	...	–0.1634	–0.1727	...	0.0927	0.0739	cello
...	...	...	...	...	...	...	...	...	...	...
53.03	0.7409	–0.758	875	...	–0.2115	–0.2115	...	0.1108	0.0775	t. tromb.

data contained in the training set. The goal is to gain knowledge which will be valid not only in the database considered. Then, it is tested against a test set to check whether the knowledge acquired from the training set is of general nature.

In the case of experiments presented data were divided into 50%/50% training and test sets (1695 samples in each set). Since a rough set-based system allows for analyzing significance of attributes to the recognition process, therefore all 139 attributes derived from the MPEG-7 analysis were used in the decision table (Table 7). To generalize the problem and to classify unknown objects induced rules operate only on subintervals. Subinterval ranges were obtained as a result of discretization algorithm. MD-heuristic was used for finding cuts in the attribute domain, which discerned largest numbers of pairs of objects [2, 8]. The same cuts were then applied to discretize attribute values of test objects. A global discretization method was used first. In further experiments also local discretization was employed. Once the discretization process has been completed, rule generation is performed. The RSES system uses genetic algorithm to induce rules [8].

Each rule has an implication form, conditioning decision on attribute values:

$$\begin{aligned} &[ASE_9 \in (-\infty, -0.15525) \wedge ASE_{10} \in (-0.16285, +\infty) \wedge \\ &\wedge ASE_{11} \in (-0.16075, +\infty) \wedge ASE_{13} \in (-0.18905, +\infty) \wedge \\ &\wedge ASE_{26} \in (+\infty, -0.18935) + \infty)] \Rightarrow [decision = violin] \end{aligned}$$

Rules were used to classify unknown objects from the test set. All results were obtained during classification of test sets only. Attributes of a new object were discretized with required cuts, and then, all rules were applied. If more that one rule was matching object, then the final decision was based on a voting method. Test set and total test set accuracy measures are defined for the RSES system as follows:

$$A_{i\{\text{with } D=d1\}} = \frac{\#\text{ of correctly classified obj.}_{\{\text{with } D=d1\}}}{\text{Total }\#\text{ of classified obj.}_{\{\text{with } D=d1\}}\text{ in the test set}} \tag{16}$$

– where A_i denotes *Test Set Accuracy*

$$Total\ Test\ Set\ Accuracy = \sum_{i=1}^{D} A_i \cdot b_i \tag{17}$$

– where b_i is the total # of classified objects$_{\{\text{with } D=d1\}}$ in the test set.

The total test set classification accuracy was approximately 84%. The worst results were obtained for bassoon and strings, instruments belonging to this group were most often confused with each other. The best result obtained was for an oboe.

In the next step a local discretization method was employed, and results obtained are presented in Fig. 3 and Table 8. Over 27000 rules were induced in tests, the principle of which was to classify a set of ten musical instruments. Rules of the length equals 4 cover more than 12000 cases, rules of length equals 3 support more than 7000 cases, and 5 – correspondingly, 6400 cases. Maximum length of the induced rule (i.e. number of attributes in implication) is 7, minimum 2, the average being 4 (Fig. 3). Results of the system classification employing MPEG-7-based descriptors are given in Table 8. Denotations in Table 8 are as follows: bassoon (BA), B flat clarinet (CL), oboe (OB), tenor trombone (TT), French horn (FH), alto saxophone (AS), violin (VI), trumpet (TR), F flat tuba (TU), cello (CE), Accuracy (Acc.), coverage (Cov). The total test set accuracy is approx. 88.4%. The best system performance was found for sound samples of a tuba (98%).

Table 8. Effectiveness of the RSES system while testing MPEG-7 descriptors.

Mus. Instr.	BA	CL	OB	TT	FH	SA	VI	TR	TU	CE	No. of obj.	Acc.	Cov
BA	**160**	6	6	1	2	1	0	2	8	0	186	0.86	1
CL	0	**159**	13	0	3	1	8	2	0	1	187	0.85	1
OB	0	5	**147**	1	0	0	10	1	0	0	164	0.895	1
TT	6	0	1	**142**	16	1	0	3	8	0	177	0.802	1
FH	5	0	1	11	**133**	2	0	0	11	0	163	0.816	1
SA	0	1	0	0	0	**116**	3	5	0	0	125	0.928	1
VI	1	3	12	1	0	2	**167**	3	0	4	193	0.865	1
TR	0	1	1	1	0	2	3	**137**	0	0	145	0.945	1
TU	0	0	0	2	0	1	0	0	**145**	0	148	0.98	1
CE	0	3	3	0	1	3	12	0	0	**188**	210	0.895	1

Another classification scheme included wavelet-based parameters into feature vectors. Data were again divided into 50%/50% training and test sets (1695 samples contained in each set). The length of rules generated was limited to 10, in total 27321 rules were derived. The total test set accuracy was 89% (see Table 9), thus comparing with results obtained for FVs containing MEPG-7-based descriptors only, the system classification performance generally improved, however, not significantly. Denotations in Table 9 are the same as before. Table 9 represents the confusion matrix for the rough set classifier. Results for the combined representation are given by the diagonal of Tab. 9 (number of sound correctly recognized). Other values denote the erroneous system answer made in the recognition process. The classification accuracy for each musical instrument class is also shown in this table. As seen from Table 9, the maximum number of errors occurred again for French horn sounds, whereas the best identification was

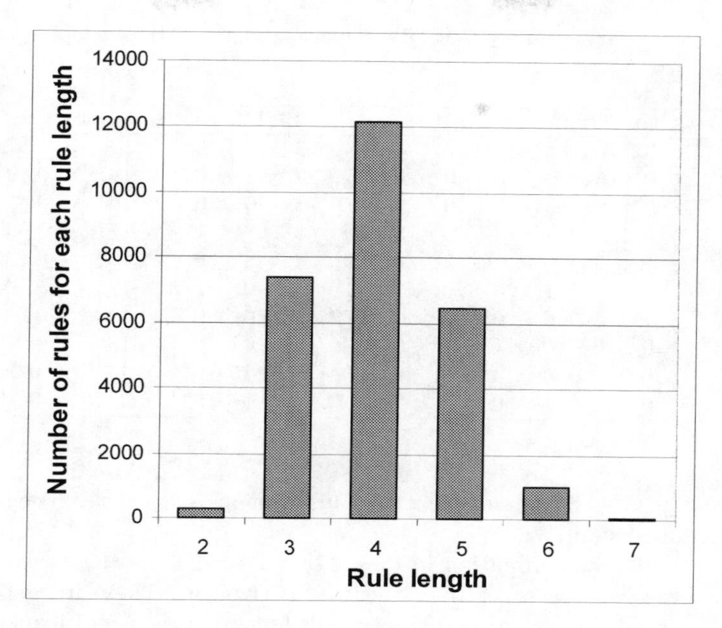

Fig. 3. Histogram of the rule length.

once more for tuba sounds. Overall, all results were better apart from the identification accuracy for French horn, which was lower than in the case while using MPEG-7 descriptors, only. French horn sound samples were often confused with trombone samples. This means that wavelet-based descriptors are not suitable for discerning between French horn and trombone sound samples.

It is also interesting to apply the rough set-based system in the context of feature vector redundancy search. As expected, reducts obtained in the analysis consist of a fraction of the whole feature vector (70 of 139 features). For the MPEG-7-based descriptors the following 70 features were employed in the analysis:

$\{KeyNum, Br, h_{ev}, LAT, TC, SC, SCv, HSC, HSCv, HSD, HSDv, HSS,$ $HSSv, HSV, HSVv, ASE_{1...5,7...9,11,13,17,19,21,24...28,31,33,34}, ASEv_{2,7,11,16,17,19},$ $ASEv_{20,22,23,28,31...34}, ASC, ASS, ASSv, SFM_{2,9,13,14,15,17,19,20,21,22,23},$ $SFMv_{4,6,14,18,23}.$

On the other hand, for the MPEG-7-and wavelet-based feature vectors, the following attributes (8) were found as the most significant: $\{HSD$, TC, E_8, $KeyNum$, SC, E_5, ASE_{14} and $Ec_{10}\}$ in the sense that they were the most frequent descriptors used in rules.

Comparing with the Fisher analysis it may be said that many descriptors used previously were also recognized by the RSES system as the significant ones. However, in the rough set-based analysis a smaller number of descriptors were eliminated from FVs, and this may be one of advantages of the rough system over ANNs. In the case, when instruments that were not seen by the system are

Table 9. Classification accuracy (local discretization method, 1695 samples in training and testing).

Mus. Instr.	BA	CL	OB	TT	FH	SA	VI	TR	TU	CE	No. of obj.	Acc.	Cov
BA	**162**	0	2	1	2	4	3	2	0	0	186	0.871	1
CL	0	**168**	12	0	0	1	1	3	2	0	187	0.898	1
OB	0	2	**147**	0	0	2	10	3	0	0	164	0.896	1
TT	4	0	1	**146**	12	3	0	2	8	1	177	0.825	1
FH	11	3	0	19	**120**	0	0	2	8	0	163	0.736	1
SA	0	0	1	0	0	**118**	1	4	0	1	125	0.944	1
VI	2	4	6	0	0	1	**172**	2	0	6	193	0.891	1
TR	0	0	0	1	0	6	0	**138**	0	0	145	0.952	1
TU	0	0	0	3	0	0	0	0	**145**	0	148	0.98	1
CE	0	3	1	0	0	4	11	0	0	**191**	210	0.91	1

tested, there is a better chance that this instrument is correctly recognized by such a decision system.

After redundancy elimination in FVs, MPEG-7- and wavelet-based resulted in 65 attributes that were used in the rule derivation. They are gathered in Table 10, other descriptors were disregarded while creating a set of rules for this classification task. For comparison, Table 11 provides most significant descriptors indicated by Fisher analysis.

In another experiment the division ratio of training and test samples was 2 to 3 (2370 training samples and 1020 test samples). The analysis resulted in 34508 rules, and classification accuracy reached 91.93%. Table 12 provides a summary of results obtained.

It was also decided that the experiment be extended to include 24 musical instrument classes. They were as follows: alto trombone (1), altoflute (2), Bach trumpet (3), bass clarinet (4), bass trombone (5), bassflute (6), bassoon (7), Bb clarinet (8), C trumpet, (9), CB (10), cello (11), contrabass clarinet (12), contrabassoon (13), Eb clarinet (14), English horn (15), flute (16), French horn (17), oboe (18), piccolo (19), trombone (20), tuba (21), viola (22), violin (23), violin ensemble (24) (see Table 13). Recognition effectiveness for the rough set-based decision system is shown in Table 13. Most errors were due to similarity in timbre of instruments, for example: such pairs of instruments as: clarinet and bass clarinet, trombone and bass trombone, and also contrabass (CB) and cello were often misclassified due to their timbre similarity. The worst results were obtained for flute sound samples. This may signify that some descriptors which were redundant for other instruments and were therefore skipped in the classification process are needed to be contained in FVs for flute sound samples. On the other hand, some instruments such as piccolo, tuba and violin ensemble were recognized perfectly (100% accuracy). In overall, in the case of 24 instruments the system accuracy was equal to 0.78.

Results of musical instrument classification based on rough sets are very satisfying. The classification accuracy on an average is greater than 90% for a dozen of instrument classes. It must be also stressed that the algorithm worked under very

Table 10. Most significant attributes generated by the RSES system.

Num.	Attribute	Signif.	Num.	Attribute	Signif.	Num.	Attribute	Signif.
1	SC	0.321	11	$ASSv$	0.085	21	SFM_{19}	0.060
2	TC	0.184	12	ASE_{17}	0.071	22	ASE_3	0.059
3	HSD	0.168	13	$HSVv$	0.071	23	ASE_1	0.055
4	$KeyNum$	0.146	14	ASE_{19}	0.070	24	$ASEv_{22}$	0.055
5	LAT	0.131	15	ASE_{34}	0.069	25	SFM_{23}	0.054
6	Br	0.118	16	ASE_2	0.068	26	HSV	0.054
7	h_{ev}	0.107	17	$SFMv_{14}$	0.065	27	$SFMv_{18}$	0.054
8	$ASEv_{23}$	0.094	18	ASE_4	0.064	28	ASE_{28}	0.053
9	ASC	0.087	19	ASE_{33}	0.062	29	ASE_{25}	0.051
10	HSC	0.085	20	SFM_{21}	0.062	30	$ASEv_{16}$	0.051

Table 11. Most significant descriptors derived from Fisher analysis; FS – the average value of the Fisher statistics of all instrument pairs.

No.	Attr. Name	FS	No.	Attr. Name	FS	No.	Attr. Name	FS
1	SFM_{16}	21.34	11	SFM_{17}	18.17	21	ASE_5	14.38
2	SFM_{15}	20.73	12	ASE_{26}	18.06	22	$KeyNum$	13.99
3	ASC	20.30	13	SFM_{19}	17.49	23	SFM_{13}	13.81
4	ASE_{24}	20.15	14	ASE_{27}	17.32	24	ASE_9	13.46
5	SFM_{14}	19.32	15	ASE_{28}	15.89	25	ASE_4	13.20
6	SC	19.06	16	HSD	14.64	26	SFM_{20}	13.19
7	HSC	19.02	17	ASE_8	14.58	27	ASE_{29}	12.90
8	ASE_{23}	18.75	18	ASE_{22}	14.57	28	SFM_{24}	12.67
9	SFM_{18}	18.65	19	ASE_6	14.51	29	ASE_{21}	12.57
10	ASE_{25}	18.61	20	ASE_7	14.41	30	SFM_{23}	12.13

demanding conditions: audio samples originated from two different sources and in most experiments only 50% of the samples were included in the training/test sets. It should be also remembered that these classes contained sound samples of differentiated articulation. Classification results are instrument-dependent. Instruments having very similar timbre (e.g. tuba – trombone) or the same scale range (e.g. trombone – bassoon) were most often confused with each other.

Also, results obtained are close to those acquired with artificial neural networks. However, a very essential feature of the rough set-based decision system is that the system supplies the researcher with a set of transparent rules.

5 Musical Style Classification

Apart from the experiments regarding musical instrument sound classification, an attempt to use parameters contained in the MPEG-7 standard for a musical style classification was carried out. A set of 162 musical pieces was used in the experiment. Each of them was categorized to be classical music, jazz or rock. Approximately 15 one-second-long samples, starting one minute from the

Table 12. Classification accuracy (local discretization method, 2370 training and 1020 test samples).

Musical Instrument	No. of samples	Accuracy
bassoon	112	0.964
B flat clarinet	112	0.938
oboe	99	0.889
tenor trombone	106	0.858
French horn	98	0.878
alto saxophone	75	0.907
violin	116	0.905
trumpet	87	0.92
F flat tuba	89	0.989
cello	126	0.944

Table 13. Recognition effectiveness for instruments based on rough sets.

No.	1	2	3	4	5	6	7	8	9	10	11	12
Eff.	0.67	0.71	0.88	0.57	0.5	0.83	0.88	0.8	0.88	0.79	0.81	0.83
No.	**13**	**14**	**15**	**16**	**17**	**18**	**19**	**20**	**21**	**22**	**23**	**24**
Eff.	0.75	0.63	0.88	0.44	0.74	0.75	1	0.59	1	0.87	0.81	1

beginning of a piece, were extracted from every piece, forming musical samples. Randomly chosen eight samples from every piece were added to the training set, depending on the classification algorithm; other samples were included in the test set. Parameters contained in the MPEG-7 standard that are applicable for such an analysis were included in the feature vector. Therefore, the feature vector consisted of only Audio Spectrum Descriptors. Results of the musical style classification are shown in Table 14. The results of musical style classification are lower by approximately 10% than the results of musical instrument identification. It is believed that an extension of the feature vector by the specialized parameters describing rhythm would improve the classification effectiveness significantly.

Nearly the same effectiveness was obtained by both the neural network (approx. 80/system (82.43/discretization was slightly better suited to classification tasks than a local discretization method. It can be noticed that the difference between the results obtained by these three algorithms is nearly negligible, however, results obtained are not yet satisfactory.

6 Conclusions

In the paper it is shown that it is possible to classify automatically musical instrument sounds and a musical style on the basis of a limited number of parameters. In addition, it is noticed that some MPEG-7-based low-level audio descriptors are better suited than others for the automatic musical sound classification. Parameters derived directly from the audio spectrum (i.e. Audio Spectrum Descriptors) seem to be the most significant ones in the recognition process.

Table 14. Effectiveness of musical style classification.

Classification system	Style	No. of samples	No. of errors	Effectiveness [%]
Neural network	classical	412	57	86.2
	jazz	505	99	80.4
	rock	176	48	72.7
TOTAL				79.77
RSES global discretization	classical	474	63	86.7
	jazz	573	147	74.3
	rock	306	42	86.3
Total				82.43
RSES local discretization	classical	474	88	81.4
	jazz	573	141	75.4
	rock	306	39	87.3
TOTAL				81.37

Moreover, they are much more universal than Timbre Descriptors because they may be used for the classification of musical signals of every type. Among Audio Spectrum Descriptors, the simplest parameter seems to be the most important one, such as for example Audio Spectrum Envelope descriptor which consists of coefficients describing power spectrum density in the octave bands. It has also been presented that combining the MPEG-7 and wavelet-based features has led, in general, to better classification system performance. This is because each set of features complemented the other, and the two combined covered both the transient and steady state of the sounds being analyzed. These conclusions are derived on the basis of experiments employing neural networks and rough sets as decision systems. Both classifiers are adequate for identification tasks, however, the rough set-based system gives information about tasks performed. This may be the starting point for more deep analysis of attributes that should be used in the description of musical signals and musical styles.

Rough Set classification is performing better than the ANN classification in terms of accuracy, although, RS uses a larger number of attributes to achieve good results. Smaller size of a feature vector means shorter analysis time, and faster classification, but at the same time lower accuracy. More attributes may also signify less sensitivity to any particular characteristics of musical instruments.

Classification system performance for automatic identification of a musical style is lower than for musical instrument classes, thus in such a case FVs should be extended by some additional attributes. Therefore, there is a continuing need for further work on the selection and effectiveness of parameters for the description of musical sounds, and audio signals in general.

Acknowledgements

The research is sponsored by the Committee for Scientific Research, Warsaw, Grant No. 4T11D 014 22, Poland, and the Foundation for Polish Science, Poland.

References

1. Bazan, J., Szczuka, M.: RSES and RSESlib – A Collection of Tools for Rough Set Computations, In *Proc. of RSCTC'2000*, LNAI 2005, Springer Verlag, Berlin (2001)
2. Bazan, J., Szczuka, M, Wroblewski, J..: A New Version of Rough Set Exploration System, In *Proc. RSCTC LNAI 2475*, J.J. Alpigini et al. (Eds), Springer Verlag, Heidelberg, Berlin (2002) 397–404
3. Brown, J. C.: Computer identification of musical instruments using pattern recognition with cepstral coefficients as features, *J. Acoust. Soc. of America*, Vol. 105 (1999) 1933–41
4. Czyzewski, A., Szczerba, M., Kostek, B.: Musical Phrase Representation and Recognition by Means of Neural Networks and Rough Sets, *Transactions on Rough Sets*, (2004) 259–284
5. Eronen, A., Klapuri, A.: Musical instrument recognition using cepstral coefficients and temporal features, In *Proc. IEEE International Conference on Acoustics, Speech, and Signal Processing ICASSP'00*, Vol. 2 (2000) 753–6
6. Eronen, A.: Comparison of Features for Musical instrument recognition, In *Proc. IEEE Workshop on the Applicationis of Signal Processing to Audio and Acoustics*, (2001) 19–22
7. Herrera, P., Peeters, G., Dubnov, S.: Automatic classification of musical instrument sounds, *J. of New Music Research*, Vol. 32, (19) (2003) 3–21
8. http://logic.mimuw.edu.pl/˜rses/ (RSES homepage)
9. http://www.meta-labs.com/mpeg-7-aud
10. http://www.gracenote.com
11. http://www.ismir.net
12. Hunter, J.: An overview of the MPEG-7 Description Definition Language (DDL), *IEEE Transactions on Circuits and Systems for Video Technology*, 11 (6), June (2001) 765–772
13. Kaminskyj, I.: Automatic Musical Vibrato Detection System, In *Proc. ACMA Conference* (1998) 9–15
14. Komorowski, J., Pawlak, Z., Polkowski, L., Skowron, A.: Rough Sets: A Tutorial, In *Rough Fuzzy Hybridization: A New Trend in Decision-Making*. Pal, S. K., Skowron, A. (Eds.), Springer-Verlag (1998) 3–98
15. Kostek, B.: Soft Computing in Acoustics, Applications of Neural Networks, Fuzzy Logic and Rough Sets to Musical Acoustics, Physica Verlag, Heidelberg, NY (1999).
16. Kostek, B., Czyzewski, A.: Representing Musical Instrument Sounds for their Automatic Classification, *J. Audio Eng. Soc.*, Vol. 49, 9, (2001) 768–785
17. Kostek, B., Zwan, P.: Wavelet-based automatic recognition of musical instruments, In *Proc. 142nd Meeting of the Acoustical Soc. of America*, Fort Lauderdale, Florida, USA, Dec. 3–7. (2001) 2754
18. Kostek, B., Szczuko P., Zwan, P.: Processing of Musical Data Employing Rough Sets and Artificial Neural Networks, In *RSCTC 2004*, Uppsala, LNAI 3066, Springer Verlag, Berlin, Heidelberg, (2004) 539–548
19. Kostek B., Czyzewski A., Processing of Musical Metadata Employing Pawlak's Flow Graphs, Rough Set Theory and Applications (RSTA), 1, LNCS 3100, *J. Transactions on Rough Sets*, 2004 285–305
20. Kostek B., Musical Instrument Classification and Duet Analysis Employing Music Information Retrieval Techniques, *Proc. of the IEEE*, vol. 92, No. 4, April (2004) 712–729

21. Lindsay, A.T., Herre, J.: MPEG-7 and MPEG-7 Audio – An Overview, *J. Audio Eng. Soc.*, Vol. 49, 7/8, (2001) 589–594
22. Martin, K. D., Kim, Y. E.: Musical instrument identification: A pattern-recognition approach, In *Proc. 136th Meeting, J. Acoust. Soc. of America*, Vol. 103 (1998) 1768
23. Opolko, F., Wapnick,J.: "MUMS – McGill University Master Samples," CD's, 1987.
24. Pal, S. K., Polkowski, L., Skowron, A.: Rough-Neural Computing. Techniques for Computing with Words. Springer Verlag, Berlin, Heidelberg, New York (2004)
25. Pawlak, Z.: Rough Sets, *International J. Computer and Information Sciences*, Vol. 11 (5) (1982)
26. Pawlak, Z.: Rough Sets: Theoretical Aspects of Reasoning About Data. Boston, MA: Kluwer (1991).
27. Pawlak, Z.: Probability, Truth and Flow Graph. *Electronic Notes in Theoretical Computer Science*, Elsevier, Vol. 82 (4) (2003)
28. Pawlak, Z.: Elementary Rough Set Granules: Towards a Rough Set Processor. In *Rough-Neural Computing. Techniques for Computing with Words.* Pal, S.K., Polkowski L., Skowron A. (Eds.). Springer Verlag, Berlin, Heidelberg, New York (2004) 5–13
29. Peters, J.F, Skowron, A., Grzymala-Busse, J.W., Kostek, B., Swiniarski, R.W., Szczuka, M.S.(eds.): Transactions on Rough Sets I Series : Lecture Notes in Computer Science , 3100 (2004)
30. Polkowski, L., Skowron, A.: Rough sets: A perspective. In: L. Polkowski and A. Skowron (eds.), Rough Sets in Knowledge Discovery 1: Methodology and Applications, Physica-Verlag, Heidelberg (1998) 31–56.
31. Szczuko, P., Dalka, P., Dabrowski, M., Kostek, B.: MPEG-7-based Low-Level Descriptor Effectiveness in the Automatic Musical Sound Classification, In *116 Audio Eng. Convention*, Preprint No. 6105, Berlin, (2004)
32. Wieczorkowska, A., Wróblewski, J., Synak, P., Ślęzak, D.: Application of temporal descriptors to musical instrument sound recognition, *J. of Intelligent Information Systems*, Vol. 21 (1) (2003), 71–93

Computational Intelligence in Bioinformatics

Sushmita Mitra

Machine Intelligence Unit, Indian Statistical Institute, Kolkata 700 108, India
sushmita@isical.ac.in

Abstract. Computational intelligence poses several possibilities in Bioinformatics, particularly by generating low-cost, low-precision, good solutions. Rough sets promise to open up an important dimension in this direction. The present article surveys the role of artificial neural networks, fuzzy sets and genetic algorithms, with particular emphasis on rough sets, in Bioinformatics. Since the work entails processing huge amounts of incomplete or ambiguous biological data, the knowledge reduction capability of rough sets, learning ability of neural networks, uncertainty handling capacity of fuzzy sets and searching potential of genetic algorithms are synergistically utilized.

Keywords: Rough sets, soft computing, biological data mining, proteins, gene expression, artificial neural networks, genetic algorithms.

1 Introduction

Bioinformatics [1, 2] involves the application of computer technology to the management of biological information, encompassing a study of the inherent genetic information, underlying molecular structure, resulting biochemical functions, and the exhibited phenotypic symptoms. One needs to decode, analyze and interpret the vast amount of genetic data that are available. *Biological data mining* is an emerging field of research and development for further progress in this direction [3], involving tasks like classification, clustering, rule mining and visualization.

Proteins constitute an important ingredient of living beings and are made up of a sequence of amino acids. There can be a large number of 3D states for a protein. The determination of an optimal conformation constitutes protein folding. It is a highly complex process, providing enormous information on the presence of active sites and possible drug interaction.

Proteins in different organisms, that are related to one another by evolution from a common ancestor, are called *homologues.* This relationship can be recognized by multiple sequence comparisons. Since the traditional dynamic programming method for local alignment is too slow, Basic Local Alignment Search Tool (BLAST) [4] is often found to be more efficient. BLAST is a heuristic method to find the highest locally optimal alignments between a query sequence and a database. Although BLAST does not allow the presence of gaps in between, its extension Gapped BLAST [5] allows insertions and deletions to be introduced into alignments. BLAST improves the overall speed of search while retaining good sensitivity, by breaking the query and database sequences into fragments

J.F. Peters and A. Skowron (Eds.): Transactions on Rough Sets III, LNCS 3400, pp. 134–152, 2005.

(words) and initially seeking matches between these fragments. An efficient extension to BLAST is Position-specific iterative BLAST (Psi-BLAST) [5], that includes gaps while searching for distant homologies by building a profile (general characteristics).

Unlike a genome, which provides only static sequence information, microarray experiments produce gene expression patterns that provide dynamic information about cell function. Gene expression data being typically high-dimensional, it requires appropriate data mining strategies like clustering for further analysis.

In addition to the combinatorial approach, there also exists scope for computational intelligence, especially for generating low-cost, low-precision, good solutions. Soft computing is another terminology, often used in lieu of computational intelligence. It entails a consortium of methodologies that works synergistically and provides flexible information processing capability for handling real life ambiguous situations [6]. The main constituents of computational intelligence (soft computing), at this juncture, include fuzzy logic, neural networks, genetic algorithms and rough sets.

Soft computing tools like artificial neural networks (ANNs), fuzzy sets, genetic algorithms (GAs) and rough sets have been used for analyzing the different genomic sequences, protein structures and folds, and gene expression microarrays [7]. Since the work involves processing huge amounts of incomplete or ambiguous data, the knowledge reduction capability of rough sets, learning ability of neural networks, uncertainty handling capacity of fuzzy sets and the searching potential of genetic algorithms are utilized in this direction. We do not go into the basics fuzzy sets, ANNs and GAs here, as these are widely available in literature [6, 8, 9]. Rough sets are described briefly in Section 5.

In this article we provide a survey on the role of computational intelligence in modeling various aspects of Bioinformatics involving genomic sequence, protein structure and gene expression microarray. Major tasks of pattern recognition and data mining, like clustering, classification and rule generation, are considered. Section 2 introduces the relevant basics from biology. The different tools of soft computing covered include artificial neural networks (ANNs), neuro-fuzzy computing and genetic algorithms (GAs), with particular emphasis on rough sets. These are described in Sections 3–5, categorized on the basis of their domain and function. Finally, Section 6 concludes the article.

2 Preliminaries

In this section we provide a basic understanding of the protein structure, folding, and microarray data. The nucleus of a cell contains chromosomes that are made up of the double helical deoxyribonucleic acid (DNA) molecules. The DNA consists of two strands, consisting of a string of four nitrogenous bases, *viz.*, adenine (A), cytosine (C), guanine (G), thymine (T). DNA in the human genome is arranged into 24 distinct chromosomes. Each chromosome contains many genes, the basic physical and functional units of heredity. However, genes comprise only about 2% of the human genome; the remainder consists of noncoding regions, whose functions may include providing chromosomal structural integrity and regulating where, when, and in what quantity proteins are made.

The DNA is *transcribed* to produce messenger (m)-RNA, which is then *translated* to produce protein. The m-RNA is a single-stranded chain of ribose groups (one for each base), which are linked together by phosphates. There exist '*Promoter*' and '*Termination*' sites in a gene, responsible for the initiation and termination of transcription. Translation consists of mapping from triplets (codons) of four bases to the 20 amino acids building block of proteins.

A gene is primarily made up of sequence of triplets of the nucleotides (exons). Introns (non coding sequence) may also be present within gene. The coding zone indicates that it is a template for a protein. As an example, in the human genome only 3%–5% of the sequence are used for coding, or constitute the gene. There are sequences of nucleotides within the DNA that are spliced out progressively in the process of transcription and translation. In brief, the DNA consists of three types of non-coding sequences. These are (i) intergenic regions between genes that are ignored during the process of transcription, (ii) intragenic (or introns) regions within the genes that are spliced out from the transcribed RNA to yield the building blocks of the genes, referred to as exons, and (iii) pseudogenes that are transcribed into the RNA and stay there, without being translated, due to the action of a nucleotide sequence.

An amino acid is an organic molecule consisting of an amine (NH) and a carboxylic (CO) acid group (backbone), together with a side-chain (hydrogen atom and residue R) that differentiates between them. Proteins are polypeptides, formed within cells as a linear chain of amino acids. Chemical properties that distinguish the 20 different amino acids cause the protein chains to fold up into specific three-dimensional structures that define their particular functions in the cell.

An alignment is a mutual arrangement of two or more sequences, that exhibits where the sequences are similar and where they differ. An optimal alignment is one that exhibits the most correspondence and the least difference. It is the alignment with the highest score, but may or may not be biologically meaningful. Basically there are two types of alignment methods, *viz.*, global alignment and local alignment. While global alignment maximizes the number of matches between the sequences along the entire length of the sequence, local alignment gives the highest score to a local match between a pair of sequences.

Given the *primary* structure of a protein, in terms of a linear sequence of amino acids, folding attempts to predict its stable 3D structure. However, considering all interactions governed by the laws of physics and chemistry to predict 3D positions of different atoms in the protein molecule, a reasonably fast computer would need one day to simulate 1 ns of folding.

The 2D *secondary* structure can involve an α helix (with the CO group of the ith residue hydrogen (H)-bonded to the NH group of the $(i+4)$th one) or a β sheet (corrugated or hairpin structure) formed by the H-bonds between the amino acids. The parts of the protein that are not characterized by any regular H-bonding patterns are called random coils or turns. Fig. 1(a) depicts the different secondary structures of proteins, generated using RasMol[1].

[1] http://www.umass.edu/microbio/rasmol/

The *tertiary* structure refers to the 3D conformation of the protein, as illustrated in Fig. 1(b). The objective is to determine the minimum energy state for a polypeptide chain folding. The process of protein folding involves minimization of an energy function, that is expressed in terms of several variables like bond lengths, bond angles and torsional angles. The major factors affecting folding include (i) hydrogen bonding, (ii) hydrophobic effect, (iii) electrostatic interactions, (iv) Van der Waals' forces, and (v) conformational entropy.

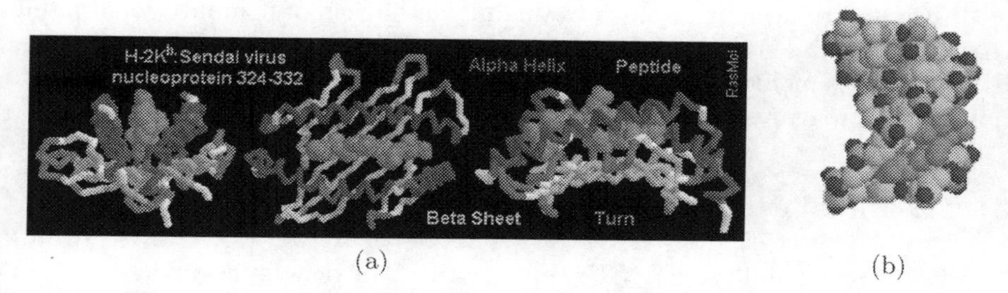

(a) (b)

Fig. 1. Protein structures: (a) 2D secondary, and (b) 3D tertiary.

Protein binding sites exhibit highly selective recognition of small organic molecules, utilizing features like complex three-dimensional *lock* (active sites) into which only specific *keys* (drug molecules or enzymes) will *dock*. Any solution to the docking problem requires a powerful search technique to explore the conformation space available to the protein and *ligand*, along with a good understanding of the process of molecular recognition to devise scoring functions for reliably predicting binding modes.

Reverse transcribed m-RNA or cDNA microarrays (gene arrays or gene chips) [2] usually consist of thin glass or nylon substrates containing specific DNA gene samples spotted in an array by a robotic printing device. This measures the relative m-RNA abundance between two samples, which are labeled with different fluorescent dyes *viz.* red and green. The *m*-RNA binds (hybridizes) with cDNA probes on the array. The relative abundance of a spot or gene is measured as the logarithmic ratio between the intensities of the dyes, and constitutes the gene expression data.

The data contains a high level of noise due to experimental procedures. Moreover, the expression values of single genes demonstrate large biological variance within tissue samples from the same class. Gene expression levels can be determined for samples taken (i) at multiple time instants of a biological process (different phases of cell division) or (ii) under various conditions (e.g., tumor samples with different histopathological diagnosis). Each gene corresponds to a high-dimensional row vector of its expression profile.

Microarrays provide a powerful basis to simultaneously monitor the expression of thousands of genes, in order to identify the complex mechanisms that govern the activation and interaction of genes in an organism. Short DNA patterns (or binding sites) near the genes serve as switches that control gene expression. Therefore, similar patterns of expression correspond to similar binding site patterns. A major cause of coexpression of genes is their sharing of the regulation mechanism (coregulation) at the sequence level. Clustering of coexpressed genes, into biologically meaningful groups, helps in inferring the biological role of an unknown gene that is coexpressed with a known gene(s). Cluster validation is essential, from both the biological and statistical perspectives, in order to biologically validate and objectively compare the results generated by different clustering algorithms.

In the following four sections we highlight the role of computational intelligence, *viz.*, ANNs [10–14], GAs [15–18] and rough sets [19], in different areas of Bioinformatics including genomic sequence prediction, protein structure prediction and microarrays.

3 Artificial Neural Networks

The learning capability of ANNs, typically in data-rich environments, come in handy when discovering regularities from large datasets. This can be unsupervised as in clustering, or supervised as in classification. The connection weights and topology of a trained ANN are often analyzed to generate a mine of meaningful (comprehensible) information about the learned problem in the form of rules.

Hybridization with fuzzy sets results in neuro-fuzzy computing encompasses the generic and application-specific merits of ANNs and fuzzy sets. This enables better uncertainty handling and rule generation in a more human-understandable form.

3.1 Microarray

Each DNA array contains the measures of the level of expression many genes, and distances are obtained from pairwise comparison of the patterns. Let $gene_j(e_{j1}, \ldots, e_{jn})$ denote the expression pattern for the jth gene. The *Euclidean distance* between the jth and kth genes, computed as

$$d_{j,k} = \sqrt{\sum_i (e_{ji} - e_{ki})^2}, \tag{1}$$

is suitable when the objective is to cluster genes displaying similar levels of expression. The *Pearson correlation coefficient* $-1 \le r \le 1$ measures the similarity in trend between two profiles. The distance is given as

$$d_{j,k} = (1 - r) = 1 - \frac{\sum_i((e_{ji} - \hat{e}_j)(e_{ki} - \hat{e}_k))/n}{\sigma_{e_j} * \sigma_{e_k}}, \tag{2}$$

where $\hat{e}_j$ and σ_{e_j} indicate the mean and standard deviation, respectively, of all points of the jth profile.

Kohonen's Self Organizing Map (SOM) has been applied to the clustering of gene expression data [20, 21]. It provides a robust and accurate approach to the clustering of large and noisy data. SOMs require a selected node in the gene expression space (along with its neighbors) to be rotated in the direction of a selected gene expression profile (pattern). However, the predefinition of a two-dimensional topology of nodes can often be a problem considering its *biological relevance*.

The Self-Organizing Tree Algorithm (SOTA) [13] is a dynamic binary tree that combines the characteristics of SOMs and divisive hierarchical clustering. As in SOMs, the gene expression profiles are sequentially and iteratively presented at the terminal nodes, and the mapping of the node that is closest (along with its neighboring nodes) is appropriately updated. Upon convergence, the node containing the most variable (measured in terms of distance) population of expression profiles is split into sister nodes, causing a growth of the binary tree. Unlike conventional hierarchical clustering, SOTA is linear in complexity to the number of profiles. The number of clusters need not be known in advance, as in c-means clustering. The algorithm starts from the node having the most heterogeneous population of associated input gene profiles. A statistical procedure is followed for terminating the growing of the tree, thereby eliminating the need for an arbitrary choice of cutting level as in hierarchical models. However, no validation is provided to establish the biological relevance.

Classification of acute leukemia, having highly similar appearance in gene expression data, has been made by combining a pair of classifiers trained with mutually exclusive features [14]. Gene expression profiles were constructed from 72 patients having acute lymphoblastic leukemia (ALL) or acute myeloid leukemia (AML), each constituting one sample of the DNA microarray[2]. Each pattern consists of 7129 gene expressions. A neural network combines the outputs of the multiple classifiers. Feature selection with nonoverlapping correlation (such as Pearson and Spearman correlation coefficients) encourages the classifier ensemble to learn different aspects of the training data in a wide solution space.

An evolving modular fuzzy neural network, involving dynamic structure growing (and shrinking), adaptive online learning and knowledge discovery in rule form, has been applied to the leukemia and colon cancer[3] gene expression data [22]. Feature selection improves classification by reducing irrelevant attributes that do not change their expression between classes. The Pearson correlation coefficient is used to select genes that are highly correlated with the tissue classes. Rule generation provides physicians, on whom the final responsibility for any decision in the course of treatment rests, with a justification about how a classifier arrived at a judgement. Fuzzy logic rules, extracted from the trained network, handle the inherent noise in microarray data while offering the knowledge in a human-understandable linguistic form. These rules point to genes (or their combinations) that are strongly associated with specific types of cancer, and may be used for the development of new tests and treatment discoveries.

[2] http://www.genome.wi.mit.edu/MPR

[3] http://microarray.princeton.edu/oncology

3.2 Primary Genomic Sequence

Eukaryotic[4] genes are typically organized as exons (coding regions) and introns (non-coding regions). Hence the main task of gene identification, from the primary genomic sequence, involves coding region recognition and splice junction[5] detection. Fig. 2 illustrates the exons and introns in genes.

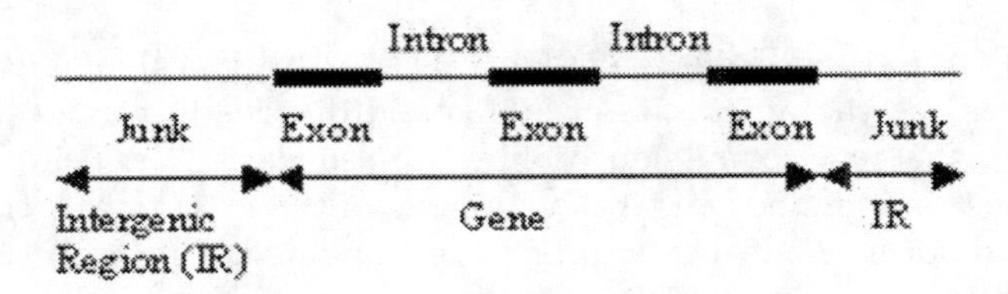

Fig. 2. Parts of a DNA.

A multilayer perceptron (MLP), with backpropagation learning, was used to identify exons in DNA sequences [23]. Thirteen input features used include sequence length, exon *GC* composition, Markov scores, splice site (donor/acceptor) strength, surrounding intron character, etc., calculated within a fixed 99-nucleotide sequence window and scaled to lie between 0 and 1. A single output indicated whether a specific base, central to the said window, was either coding or non-coding.

Prediction of the exact location of transcription initiation site has been investigated [24] in mammalian promoter regions, using MLP with different window sizes of input sequence. MLPs were also employed to predict the translation initiation sites [25], with better results being generated for bigger windows on the input sequence. Again, some of the limitations of MLPs, like convergence time and local minima, need to be appropriately handled in all these cases.

Promoter regions, in DNA and RNA sequences, were classified by using a knowledge-based ANN [26] that encodes and refines expert knowledge about the domain. This was extended by incorporating GAs to search through the topology space of neural net ensembles, for recognizing and predicting *E. coli*[6] promoter sequences [27]. Biosequence classification of DNA, using the *expectation maximization* algorithm[7] in conjunction with ANN, has been developed for recognizing *E. coli* promoters [28]. Binding sites are located on the DNA sequence, followed by their alignment, feature selection (based on information content) and feature representation using orthogonal encoding. These features are provided as input to an ANN for promoter recognition.

Identification of important binding sites, in a peptide involved in pain and depression, has been attempted [29] using feedforward ANNs. Rules in M-of-N

[4] Organisms (except viruses, bacteria and algae, that are prokaryotic) having well-developed subcellular compartments, including a discrete nucleus.

[5] Splice junctions are positions at which, after primary transcription of the DNA into RNA, the introns of a gene are excised to form edited m-RNA.

[6] Escherichia coli is a bacterium.

[7] An iterative refinement clustering algorithm that is model-based.

form are extracted by detecting positions in the DNA sequence where changes in the stereochemistry give rise to significant differences in the biological activity. Browne *et al.* also predict splice site junctions in human DNA sequences, that has a crucial impact on the performance of gene finding programs. Donor sites are nearly always located immediately preceding a GT sequence, while acceptor sites immediately follow an AG sequence. Hence GT and AG pairs within a DNA sequence are markers for potential splice junction sites, and the objective is to identify which of these sites correspond to real sites followed by prediction of likely genes and gene products. The resulting rules are shown to be reasonably accurate and roughly comparable to those obtained by an equivalent $C5$ decision tree, while being simpler at the same time.

Rules were also generated from a pruned MLP [30], using a penalty function for weight elimination, to distinguish donor and acceptor sites in the splice junctions from the remaining part of the input sequence. The pruned network consisted of only 16 connection weights. A smaller network leads to better generalization capability as well as easier extraction of simpler rules. Ten rules were finally obtained in terms of AG and GT pairs.

Kohonen's SOM has been used for the analysis of protein sequences [31], involving identification of protein families, aligned sequences and segments of similar secondary structure, in a highly visual manner. Other applications of SOM include prediction of cleavage sites in proteins [32] and prediction of beta-turns [33]. Clustering of human protein sequences were investigated [34] with a 15×15 SOM, and the performance was shown to be better than that using statistical non-hierarchical clustering.

SOTA has been employed for clustering protein sequences [35] and amino acids [36]. However, if the available training data is too small to be adequately representative of the actual dataset then the performance of the SOM is likely to get affected.

3.3 Protein Secondary Structure

A step on the way to a prediction of the full 3D structure of protein is predicting the local conformation of the polypeptide chain, called the secondary structure. The whole framework was pioneered by Chou and Fasmann [37]. They used a statistical method, with the likelihood of each amino acid being one of the three (alpha, beta, coil) secondary structures being estimated from known proteins.

Around 1988 the first attempts were made by Qian and Sejnowski [10], to use MLP with backpropagation to predict protein secondary structure. Three output nodes correspond to the three secondary structures. Performance is measured in terms of overall correct classification Q (64.3%) and Matthews Correlation Coefficient (MCC). We have

$$Q = \sum_{i=1}^{l} w_i Q_i = \frac{C}{N} \tag{3}$$

for an l-class problem, with Q_i indicating the accuracy for the ith class, w_i being the corresponding normalizing factor, N representing the total number of samples, and C being the total number of correct classifications.

$$MCC = \frac{(TP * TN) - (FP * FN)}{\sqrt{(TP + FP)(TP + FN)(TN + FP)(TN + FN)}}, \quad (4)$$

where TP, TN, FP and FN correspond to the number of true positive, true negative, false positive and false negative classifications, respectively. Here $N = TP + TN + FP + FN$ and $C = TP + TN$, and $-1 \leq MCC \leq +1$ with +1 (-1) corresponding to a perfect (wrong) prediction. The values for MCC for the α-helix, β-strand and random coil were found to be 0.41, 0.31, 0.41, respectively.

The performance of this method was improved by Rost and Sander [11], by using a cascaded three-level network with multiple-sequence alignment. The three levels correspond to a sequence-to-structure net, a structure-to-structure net, and a jury (combined output) decision, respectively. Correct classification increased to 70.8%, with the MCC being 0.60, 0.52, 0.51, respectively, for the three secondary classes.

Prediction of protein secondary structure has been further developed by Riis and Krogh [12], with ensembles of combining networks, for greater accuracy in prediction. The *Softmax* method is used to provide simultaneous classification of an input pattern into multiple classes. A normalizing function at the output layer ensures that the three outputs always sum to one. A logarithmic likelihood cost function is minimized, instead of the usual squared error. A window is selected from all the single structure networks in the ensemble. The output is determined for the central residue, with the prediction being chosen as the largest of the three outputs normalized by Softmax.

The use of ensembles of small, customized subnetworks is found to improve predictive accuracy. Customization involves incorporation of domain knowledge into the subnetwork structure for improved performance and faster convergence.

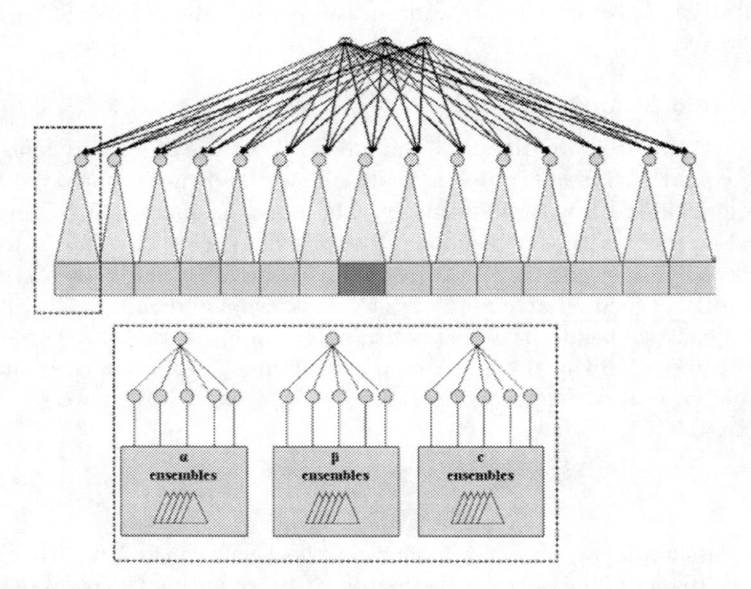

Fig. 3. Secondary protein structure prediction using ensemble of ANNs.

For example, the helix-network has a built-in period of three residues in its connections in order to capture the characteristic periodic structure of helices. Fig. 3 provides the schematic network structure. Overall accuracy increased to 71.3%, with the MCC becoming 0.59, 0.50, 0.41, respectively, for the three secondary classes.

3.4 Protein Tertiary Structure

One of the earliest ANN-based protein tertiary structure prediction in the backbone [38] used MLP, with binary encoding for a 61-amino acid window at the input. There were 33 output nodes corresponding to the three secondary structures, along with distance constraints between the central amino acid and its 30 preceding residues.

Interatomic C^{α} distances between amino acid pairs, at a given sequence separation, were predicted [39] to be above (or below) a given threshold corresponding to contact (or non-contact). The input consisted of two sequence windows, each with 9 or 15 amino acids separated by different lengths of sequence, and a single output indicated the contact (or non-contact) between the central amino acids of the two sequence windows.

4 Genetic Algorithms

Protein structure prediction typically uses experimental information stored in protein structural databases, like the Brookhaven National Laboratory Protein Data Bank. A common approach is based on sequence alignment with structurally known proteins. However, these techniques are likely to encounter difficulties for proteins with completely new sequences. The experimental approach involving X-ray crystallographic analysis and nuclear magnetic resonance (NMR) are very expensive and time-consuming. GAs [9] offer innovative approaches in overcoming these problems. In this section, we review the application of GAs to different aspects of Bioinformatics.

4.1 Primary Genomic Sequence

The simultaneous alignment of many amino acid sequences is one of the major research areas of Bioinformatics. Given a set of homologous sequences, multiple alignments can help predict secondary or tertiary structures of new sequences. GAs have been used for this purpose [40]. Fitness is measured by globally scoring each alignment according to a chosen objective function, with better alignments generating a higher fitness. The cost of multiple alignment A_c is expressed as

$$A_c = \sum_{i=1}^{N-1} \sum_{j=1}^{N} W_{i,j} cost(A_i, A_j), \tag{5}$$

where N is the number of sequences, A_i is the aligned sequence i, $cost(A_i, A_j)$ is the alignment score between two aligned sequences A_i and A_j, and $W_{i,j}$ is the weight associated with that pair of sequences. The cost function includes the sum of the substitution costs, as given by a substitution matrix, and the cost of insertions/deletions using a model with affine gap (gap-opening and gap-extension) penalties. Roulette wheel selection is carried out among the population of possible alignments, and insertion/deletion events in the sequences are modeled using a *gap insertion* mutation operator.

Given N aligned sequences $A_1 \ldots A_N$ in a multiple alignment, with $A_{i,j}$ being the pairwise projection of sequences A_i and A_j, $length(A_{i,j})$ the number of ungapped columns in this alignment, $score(A_{i,j})$ the overall consistency between $A_{i,j}$ and the corresponding pairwise alignment in the library, and $W'_{i,j}$ the weight associated with this pairwise alignment, the fitness function was modified [41] to

$$F = \frac{\sum_{i=1}^{N-1} \sum_{j=1}^{N} W'_{i,j} * score(A_{i,j})}{\sum_{i=1}^{N-1} \sum_{j=1}^{N} W'_{i,j} * length(A_{i,j})}. \tag{6}$$

The main difference with eqn. (5) is the library, that replaces the substitution matrix and provides position-dependent means of evaluation.

4.2 Protein Tertiary Structure and Folding

Tertiary protein structure prediction and folding, using GAs, has been reported in Ref. [15, 16]. The objective is to generate a set of *native-like* conformations of a protein based on a force field, while minimizing a fitness function based on potential energy. Proteins can be represented in terms of (a) three-dimensional Cartesian coordinates of its atoms and (b) the torsional angle Rotamers, which are encoded as bit strings for the GA. The Cartesian coordinates representation has the advantage of being easily convertible to and from the 3D conformation of a protein. Bond lengths, b, are specified in these terms. In the torsional angles representation, the protein is described by a set of angles under the assumption of constant standard binding geometries. The different angles involved are the

1. Bond angle θ,
2. Torsional angle ϕ, between N (amine group) and C_α,
3. angle ψ, between C_α and C' (carboxyl group),
4. Peptide bond angle ω, between C' and N, and
5. Side-chain dihedral angle χ.

The potential energy $U(r_1, \ldots, r_N)$ between N atoms is minimized. It is expressed as

$$\begin{aligned} U(r_1, \ldots, r_N) = &\textstyle\sum_i K_b(b_i - b_0^i)^2 + \sum_i K_\theta(\theta_i - \theta_0^i)^2 + \sum_i K_\phi[1 - \cos(n\phi_i - \delta)] \\ &+ \textstyle\sum_{i,j} \frac{q_i q_j}{4\pi\varepsilon_0\varepsilon_r r_{ij}} + \sum_{i,j} \varepsilon \left[\left(\frac{\sigma_{ij}}{r_{ij}}\right)^{12} - 2\left(\frac{\sigma_{ij}}{r_{ij}}\right)^{6} \right], \end{aligned} \tag{7}$$

where the first three harmonic terms on the right-hand side involve the bond length, bond angle, and torsional angle of covalent connectivity, with b_0^i and θ_0^i indicating the down-state (low energy) bond length and bond angle, respectively for the ith atom. The effects of hydrogen bonding and that of solvents (for nonbonded atom pairs i, j, separated by at least four atoms) is taken care of by the electrostatic Coulomb interaction and Van der Waals' interaction, modeled by the last two terms of the expression. Here K_b, K_θ, K_ϕ, σ_{ij} and δ are constants, q_i and q_j are the charges of atoms i and j, separated by distance r_{ij}, and ε indicates the dielectric constant.

Additionally, a protein acquires a folded conformation favorable to the solvent present. The calculation of the entropy difference between a folded and unfolded state is based on the interactions between a protein and solvent pair. Since it is not yet possible to routinely calculate an accurate model of these interactions, an *ad hoc* pseudo-entropic term E_{pe} is added to drive the protein to a globular state. E_{pe} is a function of its actual diameter, which is defined to be the largest distance between a pair of C_α carbon atoms in a conformation. We have

$$E_{pe} = 4^{(actual\_diameter - expected\_diameter)} \text{ kcal/mol}, \tag{8}$$

where $expected\_diameter/m = 8 * \sqrt[3]{len/m}$ is the diameter in its native conformation and len indicates the number of residues. This penalty term ensures that extended conformations have larger energy (or lower fitness) values than globular conformations. It constitutes the conformational entropy constituent of potential energy, in addition to the factors involved in eqn. (7).

An active site structure determines the functionality of a protein. A ligand (enzyme or drug) docks into an active site of a protein. Genetic Optimization for Ligand Docking (GOLD) [17] is an automated ligand docking program. It evaluates nonmatching bonds while minimizing the potential energy (fitness function), defined in terms of Van der Waals' internal and external (or ligand-site) energy, torsional (or dihedral) energy, and hydrogen bonds. Each chromosome in GOLD encodes the internal coordinates of both the ligand and active protein site, and a mapping between the hydrogen-bonding sites. Reproduction operators include crossover, mutation, and a migration operator to share genetic material between populations. The output is the ligand and protein conformations, associated with the fittest chromosome in the population, when the GA terminates. The files handled are the Cambridge Crystallographic Database, Brookhaven PDB, and the Rotamer library[8].

4.3 Microarray

The identification of gene subsets for classifying two-class disease samples has been modeled as a multiobjective evolutionary optimization problem [18], involving minimization of gene subset size to achieve reliable and accurate classification based on their expression levels. The Non-Dominated Sorting GA (NSGA-II) [42], a multiobjective GA, is used for the purpose. This employs elitist selection and an explicit diversity preserving mechanism, and emphasizes the non-dominated solutions. It has been shown that this algorithm can converge to the global Pareto front, while simultaneously maintaining the diversity of population. The main steps of NSGA-II are as follows.

[8] Provides the relationship between side-chain dihedral angles and backbone conformation.

1. Initialize the population.
2. Calculate the fitness.
3. Rank the population using the *dominance* criteria.
4. Calculate the *crowding distance*.
5. Do selection using *crowding selection operator*.
6. Do crossover and mutation to generate children population.
7. Combine parent and children population, and do *non-dominated sorting*.
8. Replace the parent population by the best members of the combined population.
   Initially, members of lower fronts replace the parent population. When it is not possible to accommodate all the members of a particular front, then that front is sorted according to the crowding distance. Then as many individuals are selected on the basis of higher crowding distance, which makes the population of new parent population same as the old one.

Results are provided on three cancer samples, *viz.*, Leukemia, Lymphoma[9] and Colon. An l-bit binary string, where l is the number of selected (filtered) genes in the disease samples, represents a solution. The major difficulties faced in solving the optimization problem include the availability of only a few samples as compared to the number of genes in each sample, and the resultant huge search space of solutions. Moreover many of the genes are redundant to the classification decision, and hence need to be eliminated. The three objectives minimized are (i) the gene subset size, (ii) number of misclassifications in training, and (iii) number of misclassifications in test samples.

5 Rough Sets

The theory of *rough sets* [43, 44] is a major mathematical tool for managing uncertainty that arises from granularity in the domain of discourse – that is, from the indiscernibility between objects in a set. The intention is to approximate a *rough* (imprecise) concept in the domain of discourse by a pair of *exact* concepts, called the lower and upper approximations. The lower approximation is the set of objects definitely belonging to the vague concept, whereas the upper approximation is the set of objects possibly belonging to the same. Figure 4 provides a schematic diagram of a rough set [7].

A basic issue related to many practical applications of knowledge databases is whether the whole set of attributes in a given information system is always necessary to define a given partition of the universe. Many of the attributes are superfluous, *i.e.*, we can have *'optimal'* subsets of attributes which define the same partition as the whole set of attributes. These subsets are called the *reducts* in rough set theory [43], and correspond to the minimal feature set that are sufficient to represent a decision.

Rough sets have been applied mainly to microarray gene expression data, in mining tasks like classification [45, 46] and clustering [19]. These are described in this section.

---
[9] http://llmpp.nih.gov/lymphoma/data/figure1/figure1.cdt

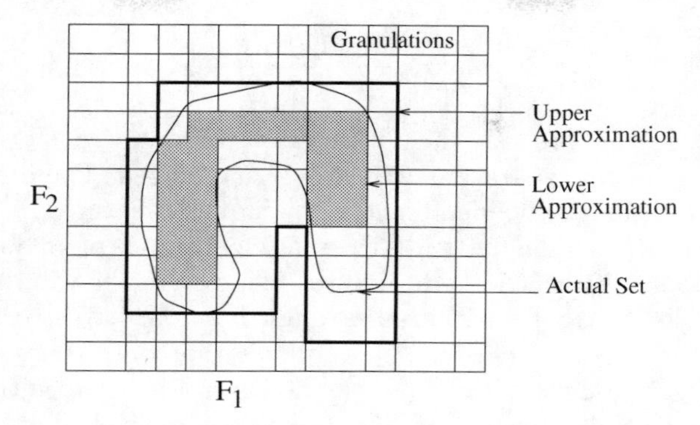

Fig. 4. Lower and upper approximations in a rough set.

5.1 Clustering

In the rough c-means clustering algorithm, the concept of c-means is extended by viewing each cluster as an interval or rough set [47]. A rough set Y is characterized by its lower and upper approximations $\underline{B}Y$ and $\overline{B}Y$ respectively. This permits overlaps between clusters. Here an object $\mathbf{X}_k$ can be part of at most *one* lower approximation. If $\mathbf{X}_k \in \underline{B}Y$ of cluster Y, then simultaneously $\mathbf{X}_k \in \overline{B}Y$. If $\mathbf{X}_k$ is not a part of any lower approximation, then it belongs to two or more upper approximations.

The centroid $\mathbf{m}_i$ of cluster U_i is computed as

$$\mathbf{m}_i = \begin{cases} w_{low} \frac{\sum_{\mathbf{X}_k \in \underline{B}U_i} \mathbf{X}_k}{|\underline{B}U_i|} + w_{up} \frac{\sum_{\mathbf{X}_k \in (\overline{B}U_i - \underline{B}U_i)} \mathbf{X}_k}{|\overline{B}U_i - \underline{B}U_i|} & \text{if } \overline{B}U_i - \underline{B}U_i \neq \emptyset \\ w_{low} \frac{\sum_{\mathbf{X}_k \in \underline{B}U_i} \mathbf{X}_k}{|\underline{B}U_i|} & \text{otherwise,} \end{cases} \tag{9}$$

where the parameters w_{low} and w_{up} correspond to the relative importance of the lower and upper approximations respectively. Here $|\underline{B}U_i|$ indicates the number of pattern points in the lower approximation of cluster U_i, while $|\overline{B}U_i - \underline{B}U_i|$ is the number of elements in the rough boundary lying between the two approximations. It is to be noted that a major disadvantage of this algorithm is the involvement of too many user-defined parameters.

An evolutionary rough c-means clustering algorithm has been applied to microarray gene expression data [19]. Rough sets are used to model the clusters in terms of upper and lower approximations. GAs are used to determine the optimal values of the parameters w_{low} and *threshold* for each c (number of clusters). Each parameter is encoded using ten bits in a chromosome. The value of the corresponding Davies-Bouldin index is chosen as the fitness function to be minimized while arriving at an optimal partitioning. Crossover and mutation probabilities of $p_c = 0.8$ and $p_m = 0.02$ were selected for a population size of 20 chromosomes.

The Davies-Bouldin index is a function of the ratio of the sum of within-cluster distance to between-cluster separation. The optimal clustering, for $c = c_0$, minimizes

$$\frac{1}{c}\sum_{k=1}^{c} \max_{l \neq k} \left\{ \frac{S(U_k) + S(U_l)}{d(U_k, U_l)} \right\}, \tag{10}$$

for $1 \leq k, l \leq c$. In this process, the within-cluster distance $S(U_k)$ is minimized and the between-cluster separation $d(U_k, U_l)$ is maximized. The distance can be chosen as the traditional Euclidean metric for numeric features.

The parameter *threshold* measures the relative distance of an object $\mathbf{X}_k$ from a pair of clusters having centroids $\mathbf{m}_i$ and $\mathbf{m}_j$. The larger the value of *threshold*, the more likely is $\mathbf{X}_k$ to lie within the rough boundary (between upper and lower approximations) of a cluster. This implies that only those points which definitely belong to a cluster (lie close to the centroid) occur within the lower approximation. A small value of *threshold* implies a relaxation of this criterion, such that more patterns are allowed to belong to any of the lower approximations.

The parameter w_{low} controls the importance of the objects lying within the lower approximation of a cluster in determining its centroid. A lower w_{low} implies a higher w_{up}, and hence an increased importance of patterns located in the rough boundary of a cluster towards the positioning of its centroid. It is observed that the performance of the algorithm is dependent on the choice of w_{low}, w_{up} and *threshold*. We allowed $w_{up} = 1 - w_{low}$, $0.5 < w_{low} < 1$ and $0 < threshold < 0.5$.

The main steps of the algorithm are provided below.

1. Choose the initial means $\mathbf{m}_i$ for the c clusters.
2. Initialize the population of chromosomes encoding parameters *threshold* and relative importance factor w_{low}.
3. Tune the parameters by minimizing the Davies-Bouldin index [eqn. (10)] as the fitness function for the GA, considering objects lying within the lower approximation of each cluster.
4. Assign each data object (pattern point) $\mathbf{X}_k$ to the lower approximation $|\underline{B}U_i|$ or upper approximation $|\overline{B}U_i|$ of cluster U_i, by computing the difference in its distance $d(\mathbf{X}_k, \mathbf{m}_i) - d(\mathbf{X}_k, \mathbf{m}_j)$ from cluster centroid pairs $\mathbf{m}_i$ and $\mathbf{m}_j$.
5. **If** the minimal distance pair $d(\mathbf{X}_k, \mathbf{m}_i) - d(\mathbf{X}_k, \mathbf{m}_j)$ is less than *threshold*
   then $\mathbf{X}_k \in \overline{B}U_i$ and $\mathbf{X}_k \in \overline{B}U_j$ and $\mathbf{X}_k$ cannot be a member of any lower approximation,
   else $\mathbf{X}_k \in \underline{B}U_i$ such that distance $d(\mathbf{X}_k, \mathbf{m}_i)$ is minimum over the c clusters.
6. Compute new mean for each cluster U_i using eqn. (9).
7. Repeat Steps 3)-6) until convergence.

It was found that the algorithm performed particularly well over the Colon cancer gene expression data, involving a collection of 62 measurements from colon biopsy (22 normal and 40 colon cancer samples) with 2000 genes (features).

Gene expression data typically consists of a small number of samples with very large number of features, of which many are redundant. We first did some initial clustering on the expression values, to detect those genes that were highly coexpressed (or correlated) in either of the two output cases. In this manner, we selected 29 out of the existing 2000 genes for further processing. This was followed by clustering on the samples. The optimum values of parameters generated by the evolutionary rough c-means algorithm was $w_{low} = 0.92$ and $threshold = 0.39$.

5.2 Classification

Classification rules (in *If–Then* form) have been extracted from microarray data [45], using rough sets with supervised learning. The underlying assumption is that the associated genes are organized in an ontology, involving super- and sub-classes. This biological knowledge is utilized while generating rules in terms of the minimal characteristic features (reducts) of temporal gene expression profiles. A rule is said to *cover* a gene if the gene satisfies the conditional part, expressed as a conjunction of attribute-value pairs. The rules do not discriminate between the super-and sub-classes of the ontology, while retaining as much detail about the predictions without losing precision.

Gastric tumor classification in microarray data is made using rough set based learning [46], implemented with ROSETTA involving genetic algorithms and dynamic reducts [48]. The fitness function incorporates measures involving the classification performance (discernibility) along with the size of the reduct. Thereby precedence is provided to solutions having less number of attributes. A major problem with microarray data being the smaller number of objects with a comparatively larger number of attributes, a preprocessing stage of feature selection based on bootstrapping is made. In the absence of appropriate feature selection, one may however end up with thousands of reducts that are simply artifacts of the data having neither satisfactory prediction performance nor effective generalization ability. The dataset consists of 2504 human genes corresponding to the conditional attributes, while the 17 tumor types are clubbed as six different clinical parameters or the decision attributes.

6 Conclusions and Discussion

Bioinformatics is a new area of science where a combination of statistics, molecular biology, and computational methods is used for analyzing and processing biological information like gene, DNA, RNA, and proteins. Proteins play a very important role in Bioinformatics. Improper folding of protein structure is responsible for causing many diseases. Therefore, accurate structure prediction of proteins is an important area of study. Microarrays, sequenced genomes, and the explosion of Bioinformatics research have led to astonishing progress in our understanding of molecular biology. With the availability of huge volume of high-dimensional data, it holds ample promise for the emergent field of biological data mining.

Soft computing tools, like ANNs, GAs and rough sets, have been used for analyzing the different protein sequences, structures and folds, as well as microarrays. Since the work entails processing huge amounts of incomplete or ambiguous data, the learning ability of neural networks, uncertainty handling capacity of rough sets, and the searching potential of GAs are utilized in this direction.

We have provided, in this article, a structured review on the role of computational intelligence in different aspects of Bioinformatics, mainly involving pattern recognition and data mining tasks. It is categorized based on the tool used, the domain of operation, and the function modeled. Protein structures have been considered at the primary, secondary and tertiary levels. Microarray data, involving gene expressions, has been dealt with separately.

References

1. Baldi, P., Brunak, S.: Bioinformatics: The Machine Learning Approach. Adaptive Computation and Machine Learning. The MIT Press, Cambridge, MA (2001)
2. Special Issue on Bioinformatics. IEEE Computer **35** (2002)
3. Special Issue on Bioinformatics, Part I: Advances and Challenges. Proceedings of the IEEE **90** (2002)
4. Altschul, S.F., Gish, W., Miller, W., Myers, E.W., Lipman, D.J.: Basic local alignment search tool. Journal of Molecular Biology **215** (1990) 403–410
5. Altschul, S.F., Madden, T.L., Schaffer, A.A., Zhang, J., Zhang, Z., Miller, W., Lipman, D.J.: Gapped BLAST and PSI-BLAST: A new generation of protein database search programs. Nucleic Acids Research **25** (1997) 3389–3402
6. Zadeh, L.A.: Fuzzy logic, neural networks, and soft computing. Communications of the ACM **37** (1994) 77–84
7. Mitra, S., Acharya, T.: Data Mining: Multimedia, Soft Computing, and Bioinformatics. John Wiley, New York (2003)
8. Haykin, S.: Neural Networks: A Comprehensive Foundation. Macmillan College Publishing Co. Inc., New York (1994)
9. Goldberg, D.E.: Genetic Algorithms in Search, Optimization and Machine Learning. Addison-Wesley, Reading, MA (1989)
10. Qian, N., Sejnowski, T.: Predicting the secondary structure of globular proteins using neural network models. Journal of Molecular Biology **202** (1988) 865–884
11. Rost, B., Sander, C.: Prediction of protein secondary structure at better than 70% accuracy. Journal of Molecular Biology **232** (1993) 584–599
12. Riis, S.K., Krogh, A.: Improving prediction of protein secondary structure using structured neural networks and multiple sequence alignments. Journal of Computational Biology **3** (1996) 163–183
13. Herrero, J., Valencia, A., Dopazo, J.: A hierarchical unsupervised growing neural network for clustering gene expression patterns. Bioinformatics **17** (2001) 126–136
14. Cho, S.B., Ryu, J.: Classifying gene expression data of cancer using classifier ensemble with mutually exclusive features. Proceedings of the IEEE **90** (2002) 1744–1753
15. Fogel, G., Corne, D., eds.: Evolutionary Computation in Bioinformatics. Morgan Kaufmann, San Francisco (2002)
16. Schulze-Kremer, S.: Genetic algorithms for protein tertiary structure prediction. In Männer, R., Manderick, B., eds.: Parallel Problem Solving from Nature II. North Holland, Amsterdam (1992) 391–400
17. Jones, G., Willett, P., Glen, R.C., Leach, A.R., Taylor, R.: Development and validation of a genetic algorithm for flexible docking. Journal of Molecular Biology **267** (1997) 727–748
18. Deb, K., Raji Reddy, A.: Reliable classification of two-class cancer data using evolutionary algorithms. BioSystems **72** (2003) 111–129
19. Mitra, S.: An evolutionary rough partitive clustering. Pattern Recognition Letters **25** (2004) 1439–1449
20. Torkkola, K., Gardner, R.M., Kaysser-Kranich, T., Ma, C.: Self-organizing maps in mining gene expression data. Information Sciences **139** (2001) 79–96
21. Tamayo, P., Slonim, D., Mesirov, J., Zhu, Q., Kitareewan, S., Smitrovsky, E., Lander, E.S., Golub, T.R.: Interpreting patterns of gene expression with self-organizing maps: Methods and applications to hematopoietic differentiation. Proceedings of National Academy of Sciences USA **96** (1999) 2907–2912

22. Futschik, M.E., Reeve, A., Kasabov, N.: Evolving connectionist systems for knowledge discovery from gene expression data of cancer tissue. Artificial Intelligence in Medicine **28** (2003) 165–189
23. Uberbacher, E.C., Xu, Y., Mural, R.J.: Discovering and understanding genes in human DNA sequence using GRAIL. Methods Enzymol **266** (1996) 259–281
24. Larsen, N.I., Engelbrecht, J., Brunak, S.: Analysis of eukaryotic promoter sequences reveals a systematically occurring CT-signal. Nucleic Acids Res **23** (1995) 1223–1230
25. Pedersen, A.G., Nielsen, H.: Neural network prediction of translation initiation sites in eukaryotes: Perspectives for EST and genome analysis. Ismb **5** (1997) 226–233
26. Towell, G.G., Shavlik, J.W.: Knowledge-based artificial neural networks. Artificial Intelligence **70** (1994) 119–165
27. Opitz, D.W., Shavlik, J.W.: Connectionist theory refinement: Genetically searching the space of network topologies. Journal of Artificial Intelligence Research **6** (1997) 177–209
28. Ma, Q., Wang, J.T.L., Shasha, D., Wu, C.H.: DNA sequence classification via an expectation maximization algorithm and neural networks: A case study. IEEE Transactions on Systems, Man, and Cybernetics, Part C: Applications and Reviews **31** (2001) 468–475
29. Browne, A., Hudson, B.D., Whitley, D.C., Ford, M.G., Picton, P.: Biological data mining with neural networks: Implementation and application of a flexible decision tree extraction algorithm to genomic problem domains. Neurocomputing **57** (2004) 275–293
30. Setiono, R.: Extracting rules from neural networks by pruning and hidden-unit splitting. Neural Computation **9** (1997) 205–225
31. Hanke, J., Reich, J.G.: Kohonen map as a visualization tool for the analysis of protein sequences: Multiple alignments, domains and segments of secondary structures. Comput Applic Biosci **6** (1996) 447–454
32. Cai, Y.D., Yu, H., Chou, K.C.: Artificial neural network method for predicting HIV protease cleavage sites in protein. J. Protein Chem **17** (1998) 607–615
33. Cai, Y.D., Yu, H., Chou, K.C.: Prediction of beta-turns. J. Protein Chem **17** (1998) 363–376
34. Ferran, E.A., Pflugfelder, B., Ferrara, P.: Self-organized neural maps of human protein sequences. Protein Sci **3** (1994) 507–521
35. Wang, H.C., Dopazo, J., de la Fraga, L.G., Zhu, Y.P., Carazo, J.M.: Self-organizing tree-growing network for the classification of protein sequences. Protein Sci **7** (1998) 2613–2622
36. Wang, H.C., Dopazo, J., Carazo, J.M.: Self-organizing tree-growing network for classifying amino acids. Bioinformatics **14** (1998) 376–377
37. Chou, P., Fasmann, G.: Prediction of the secondary structure of proteins from their amino acid sequence. Advances in Enzymology **47** (1978) 45–148
38. Bohr, H., Bohr, J., Brunak, S., Cotterill, R.M.J., Fredholm, H.: A novel approach to prediction of the 3-dimensional structures of protein backbones by neural networks. FEBS Letters **261** (1990) 43–46
39. Lund, O., Frimand, K., Gorodkin, J., Bohr, H., Bohr, J., Hansen, J., Brunak, S.: Protein distance constraints predicted by neural networks and probability distance functions. Protein Eng **10** (1997) 1241–1248
40. Notredame, C., Higgins, D.G.: SAGA: Sequence alignment by genetic algorithm. Nucleic Acids Research **24** (1996) 1515–1524

41. Notredame, C., Holm, L., Higgins, D.G.: COFFEE: An objective function for multiple sequence alignments. Bioinformatics **14** (1998) 407–422
42. Deb, K., Agarwal, S., Pratap, A., Meyarivan, T.: A fast elitist non-dominated sorting genetic algorithm for multi-objective optimization: NSGA-II. In: Proceedings of the Parallel Problem Solving from Nature VI Conferences. (2000) 849–858
43. Pawlak, Z.: Rough Sets, Theoretical Aspects of Reasoning about Data. Kluwer Academic, Dordrecht (1991)
44. Skowron, A., Rauszer, C.: The discernibility matrices and functions in information systems. In Słowiński, R., ed.: Intelligent Decision Support, Handbook of Applications and Advances of the Rough Sets Theory. Kluwer Academic, Dordrecht (1992) 331–362
45. Midelfart, H., Lægreid, A., Komorowski, J. In: Classification of gene expression data in an ontology. Volume 2199 of Lecture Notes in Computer Science. Springer Verlag, Berlin (2001) 186–194
46. Midelfart, H., Komorowski, J., Nørsett, K., Yadetie, F., Sandvik, A.K., Lægreid, A.: Learning rough set classifiers from gene expressions and clinical data. Fundamenta Informaticae **53** (2002) 155–183
47. Lingras, P., West, C.: Interval set clustering of Web users with rough k-means. Technical Report No. 2002-002, Dept. of Mathematics and Computer Science, St. Mary's University, Halifax, Canada (2002)
48. Wroblewski, J.: Finding minimal reducts using genetic algorithms. Technical Report 16/95, Warsaw Institute of Technology - Institute of Computer Science, Poland (1995)

Rough Ethology: Towards a Biologically-Inspired Study of Collective Behavior in Intelligent Systems with Approximation Spaces

James F. Peters

Department of Electrical and Computer Engineering, University of Manitoba,
Winnipeg, Manitoba R3T 5V6 Canada
jfpeters@ee.umanitoba.ca

The treatment of behavior patterns as organs
has not merely removed obstacles to analysis,
it has also positively facilitated causal analysis.
–N. Tinbergen, 1963.

Abstract. This article introduces an ethological approach to evaluating biologically-inspired collective behavior in intelligent systems. This is made possible by considering ethology (ways to explain agent behavior) in the context of approximation spaces. The aims and methods of ethology in the study of the behavior of biological organisms were introduced by Niko Tinbergen in 1963. The rough set approach introduced by Zdzisław Pawlak provides a ground for concluding to what degree a particular behavior for an intelligent system is a part of a set of behaviors representing a norm or standard. A rough set approach to ethology in studying the behavior of cooperating agents is introduced. Approximation spaces are used to derive action-based reference rewards for a swarm. Three different approaches to projecting rewards are considered as a part of a study of learning in real-time by a swarm. The contribution of this article is the introduction of an approach to rewarding swarm behavior in the context of an approximation space.

Keywords: Approximation space, behavior, ethology, intelligent systems, learning, rough sets, swarm.

1 Introduction

This paper introduces a biologically-inspired approach to observing the collective behavior of intelligent systems, which is part of growing research concerning intelligent systems in the context of rough sets (see,e.g., [16, 19], [25, 26, 29], [32–36], [52–54]) with particular emphasis on approximation spaces. Considerable work has been done on approximation spaces ([26, 32–35], [36, 44, 48, 51–53, 56, 58–60, 64]) based on generalized approximation spaces introduced in [52, 53]). This work on approximation spaces is an outgrowth of the original approximation space definition in [26]. Approximation spaces constructed relative to patterns gleaned from observations of the behavior of cooperating agents (e.g., swarming by bots [3, 5, 15, 36]) provide gateways to knowledge about how intelligent systems learn. An agent itself is something that has observable behavior. There

J.F. Peters and A. Skowron (Eds.): Transactions on Rough Sets III, LNCS 3400, pp. 153–174, 2005.

is growing interest in biology as a source of paradigms useful in the study of intelligent systems (see, e.g., [2, 3, 5, 7–9, 15, 36, 70]). For example, a number of the features of swarm behavior can be identified with methods used by ethologists in the study of the behavior of biological organisms [4, 12–14, 67, 68]. One can consider, for instance, the survival value of a response to a proximate cause (stimulus) at the onset of a behavior by an agent interacting with its environment as well as cooperating with other agents in an intelligent system. In addition, at any instance in time, behavior ontogeny (origin and development of a behavior) and behavior growth (short term evolution) can be considered as means of obtaining a better understanding of the interactions between agents and the environment in a complex system. In the study of swarm behavior, it is necessary for a swarm to choose its next action based on a projected reward if an action is chosen. Three different approaches to projecting rewards for swarm actions are considered in this article as part of a study of learning in real-time by a swarm. First, a projected action-reward based on the average number of times a swarm enters a state where it has sufficient energy (its high state) is considered. For each proximate cause, there is usually more than one possible response. Swarm actions with higher rewards tend to be favored. Second, two forms of projected action rewards that depend on what is known as reference rewards are considered. One form of reference reward is an average of all recent rewards independent of the actions chosen (see, e.g., [66]). The second form of reference reward is derived in the context of an approximation space. This form of reference reward is pattern-based and action-specific. Each action has its own reference reward which is computed within an approximation space that makes it possible to measure the closeness of action-based blocks of equivalent behaviors to a standard. The contribution of this article is the introduction of an approach to rewarding swarm behavior in the context of an approximation space. This paper has the following organization. Basic concepts from rough sets are briefly presented in Sect. 2. A distinction between universes containing models of behavior and universes of behaviors is also given in Sect. 2. An ethological perspective on intelligent system behavior is given in Sect. 3. A concise overview of approximation spaces is presented in Sect. 4. A testbed for studying swarmbot behavior is introduced in Sect. 5. A sample structure for what is known as an ethogram viewed in the context of an approximation space is considered in Sect. 6. Approximation space-based rewards are introduced in Sect. 7. Reflections on an ontology of behavior and hierarchical learning are given in Sect. 8.

2 Basic Concepts: Rough Sets

This section briefly presents some fundamental concepts in rough set theory that provide a foundation for projecting rewards for actions by collections of cooperating agents. This section also introduces a distinction between two types of universes that are relevant to the study of the collective behavior in intelligent systems. The rough set approach introduced by Zdzisław Pawlak [20] and elaborated in [21–31] provides a ground for concluding to what degree a set of equivalent behaviors are a part of a set of behaviors representing a standard.

For computational reasons, a syntactic representation of knowledge is provided by rough sets in the form of data tables. Informally, a data table is represented as a collection of rows each labeled with some form of input, and each column is labeled with the name of an attribute (feature) that computes a value using the row input. Traditionally, row labels have been used to identify a sample element commonly called an object (see, e.g., [26]) belonging to some universe (e.g., states, processes, entities like animals, persons, and machines). In this work, the term *sample element* (i.e., member of a population) is used instead of *object* because we want to consider universes with elements that are either models of behavior or behaviors themselves observed in real-time. This distinction is explained later in this section. Formally, a data (information) table IS is represented by a pair (U, A), where U is a non-empty, finite set of elements and A is a non-empty, finite set of attributes (features), where $a : U \longrightarrow V_a$ for every $a \in A$. For each $B \subseteq A$, there is associated an equivalence relation $Ind_{IS}(B)$ such that $Ind_{IS}(B) = \{(x, x') \in U^2 | \forall a \in B.a(x) = a(x')\}$. Let $U/Ind_{IS}(B)$ denote a partition of U, and let *B(x)* denote a set of B-indiscernible elements containing x. *B(x)* is called a block, which is in the partition $U/Ind_{IS}(B)$. For $X \subseteq U$, the sample X can be approximated from information contained in B by constructing a B-lower and B-upper approximation denoted by B_*X and B^*X, respectively, where $B_*X = \{x \in U | B(x) \subseteq X\}$ and $B^*X = \{x \in U | B(x) \cap X \neq \emptyset\}$. The B-lower approximation B_*X is a collection of sample elements that can be classified with full certainty as members of X using the knowledge represented by attributes in B. By contrast, the B-upper approximation B^*X is a collection of sample elements representing both certain and possible uncertain knowledge about X. Whenever B_*X is a proper subset of B^*X, i.e., $B_*X \subset B^*X$, the sample X has been classified imperfectly, and is considered a rough set [59].

2.1 Universes Containing Models of Behavior

In this section, we briefly consider universes containing models of behavior. A *model* is high-level description of something such as a system, process or behavior. A *model of behavior* is a set containing interacting objects that define the behavior of an agent. The term *behavior* in this work refers to the way an agent responds to a stimulus that results in a change of state. Sets of interacting objects are represented by what are known as collaboration diagrams (see, e.g., [11, 18]). This sense of the term of *object* differs from the term commonly used in rough set theory. In the context of a model of behavior, an *object* is an instance of a type or class. For example, let $X \in U$, where X represents a sample model of behavior and U is a collection of such models. This view of the world befits the case where the behavior models belonging to U can be spliced together to form a complex model of system behavior. In this sense, $X \in U$ is a higher-order object in the sense that each X is an instance of a type of model. Consider, for example, a model of behavior for a physical system such as an autobot (see Fig. 1).

Such a model of behavior would include objects of type *Receiver* (for receiving messages), *Auto* (for monitoring incoming messages and sensor measure-

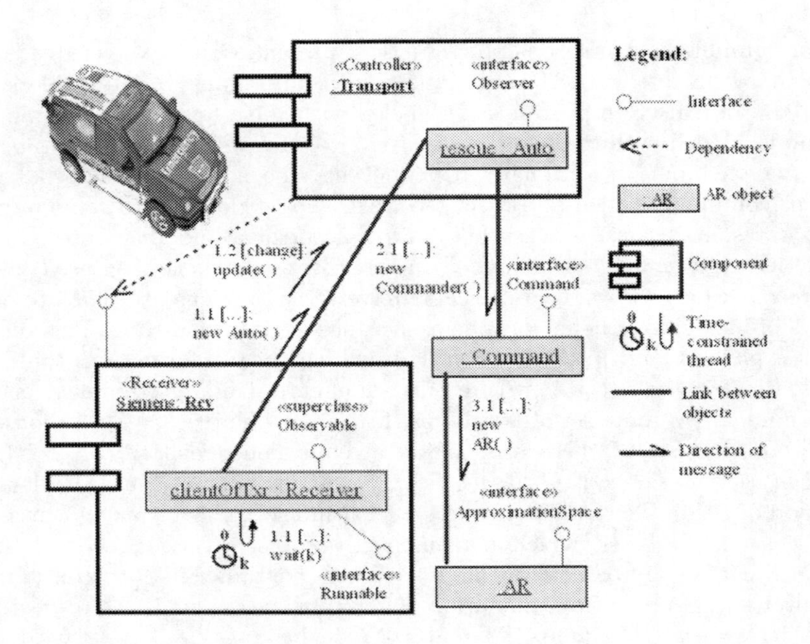

Fig. 1. Partial Collaboration Diagram for an AutoBot.

ments), *Command* (for communicating action requests), *AR* (for approximate reasoning). The *AR* object periodically derives pattern-based reference rewards during on-line learning. The *ApproximationSpace* interface prescribes methods (e.g., *lower* for a lower approximation, *upper* for an upper approximation, *include* for rough inclusion) that are implemented in the *AR*. The interaction between objects that constitutes the behavior of an agent is defined by exchanges of messages, where a message sent by one object to another object plays the role of a stimulus at an instant in time and closely resembles what Tinbergen calls a *proximate cause* [67]. In the social behavior of animals, for example, a proximate cause is some stimulus observed very close in time to an observed action by something (e.g., collision-avoidance by a starling that is part of a flock of starlings spiraling overhead [68], which is a common occurrence during feeding time at the bird sanctuary in Singapore). The response by an object that receives a message will be some observable action. Internal actions by an object are not observable.

For simplicity, consider, for example, the case of an AutoBot with a behavior partially represented by four objects spread across two components (see Fig. 1). The partial collaboration diagram for an AutoBot agent in Fig. 1 exhibits some of the possible messages between four objects. The notation "rescue : Auto" in Fig. 1 specifies that an object named "rescue" is an instance of an *Auto* type. The notation 1.1 [...] : $wait(k)$ in Fig. 1 specifies a message that repeatedly stimulates a $clientOfTxr$ (i.e., client of a transmitter) object of type Receiver. The $clientOfTxr$ object has been designed to receive messages from a transmitter

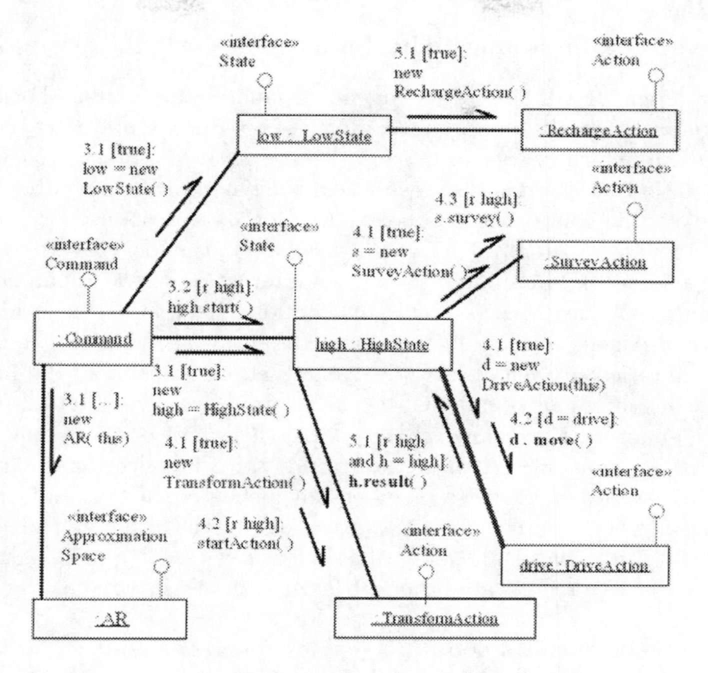

Fig. 2. Collaboration Diagram with Interacting Objects for an Autobot.

object in a wireless network. A link between a pair of objects describes a communication channel that makes message-passing between the objects possible. The diagram in Fig. 2 is a expansion of the model of behavior shown in Fig. 1. The object of type *Command* in Fig. 2 is linked to a state pattern (collections of objects representing state-action combinations). Two objects in a state pattern for a bot are shown in Fig. 2, namely, *high* and *low* of type *LowState* and *HighState*, respectively. An agent is in a high state, if its battery has sufficient charge for required actions. Otherwise an agent is in a *low* state. The heavy line in the diagram in Fig. 2 highlights a link between the object named *high* and the object named *drive*, which illustrates a possible interaction (exchange of messages) between two objects. The object named *high* in Fig. 2 passes a copy of itself when it instantiates the *drive* object in message 4.1. Then the *high* object stimulates the *drive* object with a d.move( ). The *drive* object responds to the *high* object with a *h.result(...)* message numbered 5.1, provided the action reward r is greater than threshold th (see, e.g., condition $[r > th]$ for message 5.1). In this work, the magnitude of reward r is a measure of survivability. Implicit in the behavior modeled in Fig. 2 is a characterization of an an autobot that learns based on action-rewards. A detailed explanation of models of behavior viewed in the context of intelligent system models is given in [32, 34, 36].

2.2 Universes Containing Behaviors

Intuitively, a behavior is the way an agent responds to a stimulus. Formally, the term *behavior* denotes the observable effects of a sequence of events in the form of observed stimulation of one or more entities (e.g., arrival of a message received by an agent) and observed action-response during an interaction. A behavior $x = (pc_t, a_t)$ is a tuple containing an observed action a_t in response to a proximate cause pc_t at time t. An *interaction* in a behavior can be viewed on different levels (see, e.g., [36]). On an object level, an interaction between pairs of objects is an exchange of messages (stimuli) and actions (responses) with identifiable features. On an agent level, an interaction is the result of a collaboration between collections of objects. Each observable agent is an instance of a type (a member of a species of agents) and can be viewed as a composite object. On the collective agent level, an interaction is an exchange of messages between cooperating agents and *concerted* actions in response ("swarming"). Each collection of cooperating agents is an instance of a type (a member of a species of society) and is, in effect, a higher-level, very complex object. Notice, again, that behavior at each level (object, agent, and swarm), has identifiable features (see, e.g., [36]). A record of an observed behavior is a tabulation of feature values in what is known as an ethogram.

Behavior is best understood in the context of a collaboration between entities in real-time. An *entity in real-time* belonging to a collaboration is a part of a realization of a model of behavior for either an individual agent (e.g., a bot) or for a collection of agents (e.g., cooperating bots in a swarm). Consider, for example, the realization in real-time of the behavior of an agent described by the collaboration diagram in Fig. 2. Notice that three message labeled 4.1 are sent concurrently by the object of type $HighState$. In real-time, one might observe a response (some action such as movement in some direction or change in speed, color or shape) from any or all of the three stimulated objects. In an environment where there is a certain amount of unpredictability (e.g., selection of the best camera angle needed to inspect some structure or the best angle needed to turn left or right to navigate around an obstacle), an observed behavior resulting from the implementation of a model of behavior for an agent engaged in learning in real-time may not conform to what a designer expects. In such an environment, an agent that learns in real-time considers action probability, action preference, and action reward, and the choice between two or more possible actions by an agent is not always clear, if one considers the results of recent actions. Similarly, a swarm that learns in real-time considers action probabilities, action preferences and action rewards as a guide in choosing an action in response to a stimulus. If one considers the collective behavior of cooperating agents (swarm), then the choice of actions by a swarm is influenced by actions that improve the survivability of the swarm (sustaining cooperation). The choice of the next action by a swarm is not always clear, and can vary from one situation to the next. It is this variability in observed behavior that motivates the consideration of what are known as ethograms in the context of rough sets.

3 Intelligent System Behavior: An Ethological Perspective

In a hierarchical model of the behavior of an intelligent system, features can be identified for each layer of the hierarchy (see, e.g., [26]). A number of features of the behavior of an agent in an intelligent system can be discovered with ethological methods. *Ethology* is the study of the behavior and interactions of animals (see, e.g., [4, 12–14, 67, 68]). The biological study of behavior provides a rich source of features useful in modeling and designing intelligent systems in general (see, e.g., [36]). It has been observed that animal behavior has patterns with a biological purpose and that these behavioral patterns have evolved. Similarly, patterns of behavior can be observed in various forms of intelligent systems, which respond to external stimuli and which evolve. In the search for features of intelligent system behavior, one might ask *Why does a system behave the way it does?* Tinbergen's four whys [67] are helpful in the discovery of some of the features in the behavior of intelligent systems, namely, (1) *proximate cause* (stimulus), (2) *action response* together with survival value, (3) recent behavior *growth* (evolution over short periods of time), and (4) *behavior ontogeny* (origin of current behavior).

A proximate cause (e.g., excitation, recognized pattern, arrival of a message) triggers a behavior occurring currently. The survival value of a behavior correlates with proximate cause. A principal proximate cause for a behavior is the availability of sufficient energy to sustain the life of an organism. In this work, behavior growth g_t at time t is measured using two reference values, namely, $\bar{r}_{t-1}$ (average reward) and $\bar{l}_{t-1}$ (average number of times a bot or a swarm enters a low state) at time t *-1*. Being in a high state is necessary for survival. Hence, it is reasonable to associate g_t with the frequency that a swarm is in a high state. A decline in growth level (i.e., bot population decline) serves as an indicator of a deterioration in a behavior (e.g., a bot failing to recharge its battery when its energy level is low, where low energy level defines a pattern that is part of a recovery strategy in the behavior of a bot). Behavior ontogeny (origin of a behavior) is associated with the action that has the highest frequency. The assumption made here is that a predominant action will have the greatest influence on the currect behavior.

4 Approximation Spaces

The classical definition of an approximation space given by Zdzisław Pawlak in [25] is represented as a pair (U, Ind), where the Indiscernibility relation Ind is defined on a universe of objects U. As a result, any subset X of U has an approximate characterization in an approximation space [64]. A generalized approximation space was introduced by Skowron and Stepaniuk in [52, 53]. A *generalized approximation space* is a system $GAS = (U, I, \nu)$ where

- U is a non-empty set of objects, and $\mathcal{P}(U)$ is the powerset of U.
- $I : U \rightarrow \mathcal{P}(U)$ is an uncertainty function.
- $\nu : \mathcal{P}(\mathrm{U}) \times \mathcal{P}(\mathrm{U}) \rightarrow [0, 1]$ denotes rough inclusion

The uncertainty function I defines a neighborhood of every sample element x belonging to the universe U. That is, $I(x)$ can be interpreted as a *neighborhood of* x (see,e.g., [41]). The neighborhood of a sample element x can be defined using the indiscernibility relation Ind, where $I(x) = [x]_{Ind}$. The sets computed with the uncertainty function $I(x)$ can be used to define a covering of U [46]. The rough inclusion function ν computes the degree of overlap between two subsets of U. Let $\mathcal{P}(U)$ denote the powerset of U. In general, rough inclusion $\nu : \mathcal{P}(\mathrm{U})$ x $\mathcal{P}(\mathrm{U}) \rightarrow [0, 1]$ can be defined in terms of the relationship between two sets where

$$\nu(X,Y) = \begin{cases} \frac{|X \cap Y|}{|Y|}, & \text{if Y } \neq \emptyset \\ 1 \quad, & \text{otherwise} \end{cases}$$

for any $X, Y \subseteq U$. In the case where $X \subseteq Y$, then $\nu(X, Y) = 1$. The minimum inclusion value $\nu(X, Y) = 0$ is obtained when $X \cap Y = \emptyset$ (i.e., X and Y have no elements in common). In a hierarchical model of an intelligent system, one or more approximation spaces would be associated with each layer [36], which is related to recent research on layered learning (see,e.g., [16, 65]), information granulation (see,e.g., [57, 61]) and informorphisms (see,e.g., [58]).

4.1 Example: Approximation Space for a Swarmbot

To set up an approximation space for a swarmbot, let $DT_{sbot} = (U_{beh}, A, \{d\})$ be a decision system, where U_{beh} is a non-empty set of behaviors, A is a set of swarmbot behavior features, and d is a distinguished attribute representing a decision. Assume that $I_B : U_{beh} \rightarrow \mathcal{P}(U_{beh})$ is an uncertainty function that computes a subset of U_{beh} relative to parameter B (subset of attributes in A). For example, $I_B(x)$ for $x \in U_{beh}$, $B \subseteq A$ can be used to compute B_*D of $D \subseteq U_{beh}$. Further, let B_*D represent a standard for swarmbot behaviors, and let $B_a(x)$ be a block in the partition of U_{beh} containing x relative to action a (i.e., $B_a(x)$ contains behaviors for a particular action a that are equivalent to x). The block $B_a(x)$ is defined in (1).

$$B_a(x) = \{y \in U_{beh} : xInd\,(B \cup \{a\})\, y\} \tag{1}$$

Then we can measure the closeness of $B_a(x)$ to B_*D as in (2):

$$\nu_B(B_a(x), B_*D) = \frac{|B_a(x) \cap B_*D|}{|B_*D|} \tag{2}$$

B_*D represents certain knowledge about the behaviors in D. For this reason, B_*D provides a useful behavior standard or behavior norm in gaining knowledge about the proximity of behaviors to what is considered normal. The term *normal* applied to a set of behaviors denotes forms of behavior that have been accepted. The introduction of some form of behavior standard makes it possible to measure the closeness of blocks of equivalent action-specific behaviors to those behaviors that are part of a standard.

5 Swarmbot Behavior Testbed

This section gives an overview of a testbed for studying the behavior of individual bots as well as the collective behavior of cooperating bots (called swarmbot or sbot). A bot such as the one shown in Fig. 3 that has been designed to crawl along powerlines during inspection of various structures (e.g., towers, conductors, vibration dampers, insulators) is given in this section. Cooperating bots form swarms for various reasons (e.g., obstacle navigation, inspection, security, and acquiring energy) as illustrated in Fig. 4.

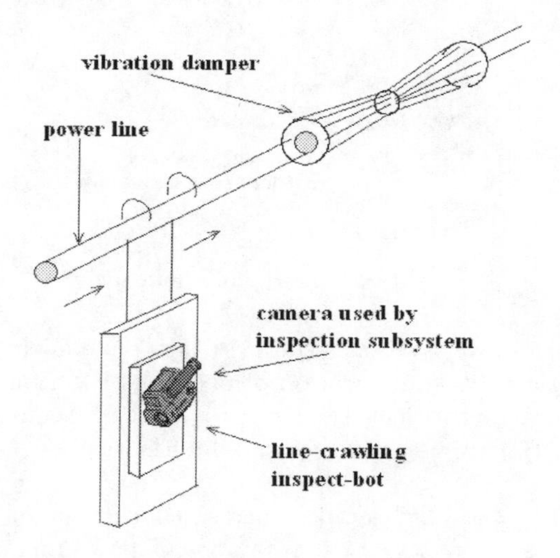

Fig. 3. Inspection bot.

The swarmbot paradigm has been chosen because it leads to a simplified design of individual bots. A snapshot showing part of a sample swarmbot testbed run is shown in Fig. 5. To facilitate the study of swarm behavior, the current version of the testbed generates ethograms that make it possible track learning by an sbot. The obstacle navigation and inspection capabilities are the result of cooperation between a number of bots. The cooperative behavior of a collection of bots working together give the bots the appearance of a super individual.

Various forms of on-line and off-line learning are commonly found in what is known as layered learning are currently being considered in the design of the testbed. *On-line learning* occurs in real-time during the operation of cooperating bots. For example, the survivability feature provides a basis for a form of on-line reinforcement learning in a swarmbot (i.e., as a result of exchange of information between bots in a swarmbot, the swarmbot learns to improve its survivability). The design of the swarmbot testbed is based on a rough mereological approach, which is built on the basis of an inclusion relation *to be a part to a degree* that generalizes the rough set approach (see,e.g., [44, 45, 49]) and provides a basis for a pattern-based form of on-line learning. *Off-line learning* is for fixed tasks that can be trained beforehand. For example, this form of learning is useful in

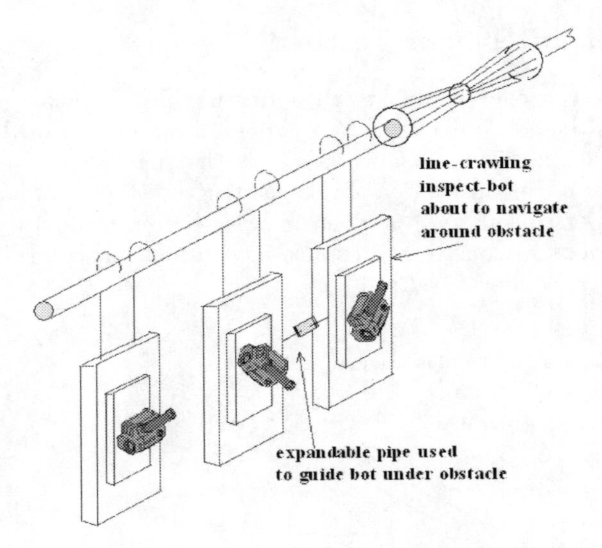

Fig. 4. Caterpillar sbot.

robotic inspection of power system structures that have fairly uniform patterns (e.g.,electromagnetic fields,configuration of obstacles). Various forms of off-line learning using neural networks (see, e.g., [6, 10, 19]) and C4.5 decision tree learning (see, e.g., [47]) are being incorporated in the testbed, but not considered in this article.

Each line-crawling bot has been designed so that it cannot navigate by itself around an obstacle such as a vibration damper. This simplifies the design of a bot. Such a bot must have help from another bot. In other words, inspection and complete navigation is only possible by cooperating bots. A sample snapshot of the individual and collective behavior of inspect bots in a line-crawling swarmbot testbed is shown in Fig. 5. The principal testbed display symbols are shown in Fig. 6. Briefly, this testbed makes it possible to experiment with various off- and on-line learning algorithms and various swarm behaviors such as cooperation between bots where one bot requires the help of another bot to crawl past an obstacle, message-passing between bots using wireless communication, and responding to various adversaries such as wind, rain, lightning, hunters and birds.

Also included in the sbot testbed is a provision for tracking feature values for individual bots as well as swarms (see sample feature window in Fig. 5) over time (e.g., energy level). Feature tracking by the testbed is useful in behavior survivability estimates. A bot that encounters an obstacle enters standby mode, and attempts to send a call for help to neighboring bots. The appendages of an inspection bot give the bot the ability to hang onto and roll along an unobstructed length of line. The survivability of a bot in standby mode decreases over time (the longer it waits, the less chance it has to renew its energy).

A bot in standby mode learns to cope with its decreasing energy problem and no response from another bot, by exiting standby mode and searching for a means to renewing its energy (this usually means the bot adjusts its angle of rotation

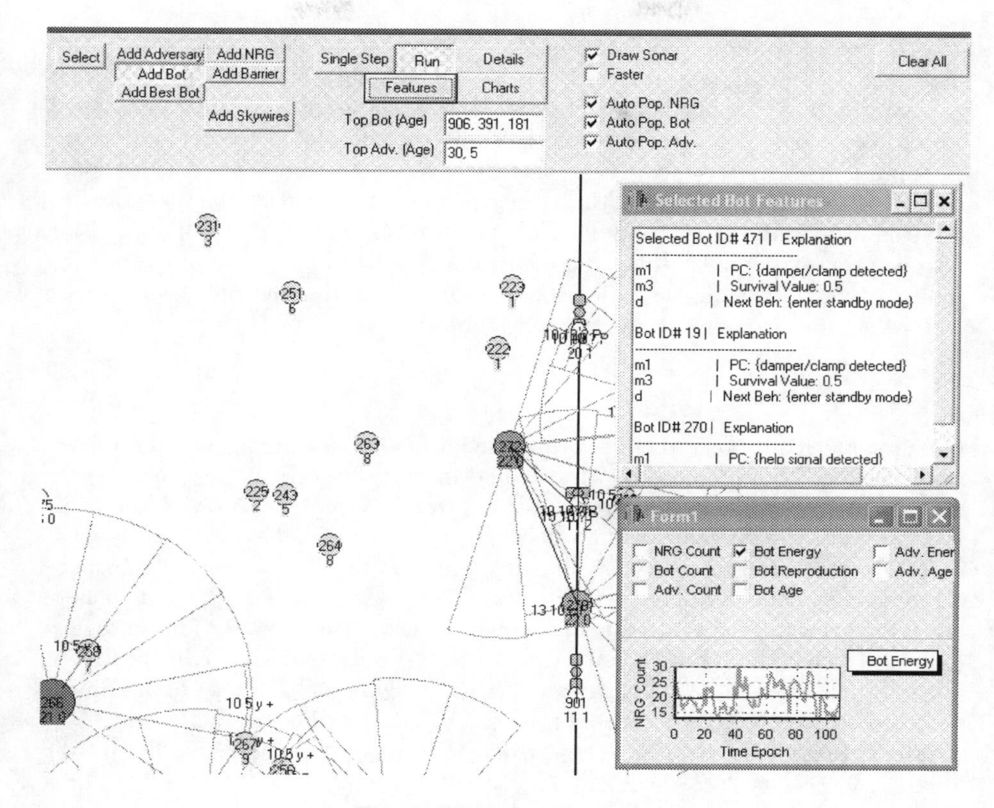

Fig. 5. Swarmbot Testbed.

Symbol	Meaning	Symbol	Meaning
	Vibration damper.		Bot with ID 45, energy level 110.
	Tower clamp attached to wire.		Swarm (line shows wireless connection between bots).
	Energy source, ID 263, level 8		Energy radiating from bot proximity sensors.

Fig. 6. Testbed Symbols.

so that its solar panels can absorb the sun's rays). Cooperation between bots begins by one bot responding to a wireless communication from another bot. This marks the birth of a swarmbot. A responding bot docks or connects itself to a bot asking for help. Then the responding bot provides stability and support while the other bot opens its calipers at the tips of its appendages (closed calipers hold onto a line). After this, the responding bot pushes the other bot past an obstacle. Such cooperation between bots is one of the hallmarks of swarmbot behavior (see, e.g., [15]), and in multiagent systems containing independent agents that exhibit some degree of coordination (see, e.g., [65]). Many of the details concerning this sbot have been already been reported (see, e.g., [39, 40]). It should be noted that the work on line-crawling sbots with a vision system used to monitor and detect faults in power system structures is to some extent related to recent research on unmanned aerial vehicles that classify vehicles from image sequence data and track moving vehicles over time (see,e.g., [17, 71]). That is, the line-crawling sbots and the UAVs are autonomous, above-ground vehicles and both have vision systems. However, the UAV has more freedom of movement and monitors moving vehicles, whereas the line-crawling sbots are restricted to movement along a wire and are limited to monitoring stationary objects.

6 Ethograms

Feature values of observed collective behavior of cooperating agents in an intelligent system are tabulated in what is known as an ethogram (see Table 1). In this work, an *ethogram* is represented by a table, where each row of the table contains the feature values for an observed patterns of behavior. At this writing, the production of ethograms has been automated for swarms. Let U_{beh} be a universe of behaviors, and let $x \in U_{beh}$. Let s_t, pc_t, a_t, o_t, g_t, r_t, d_t denote behavior features, namely, state, proximate cause (stimulus), action, ontogeny (origin of behavior), growth, reward and action decision at time t, respectively. For example, let $s_t, d_t \in \{0, 1\}$, and let $pc_t \in PC$, where PC is a set of proximate cause codes (see Fig. 7, e.g.,where $p_t \in \{3, 9, 6, 13\}$) and $a_t \in AC$, which AC is a set of action codes (see Fig. 7, e.g., where $a_t \in \{4, 5, 10, 11, 12, 7, 8, 16, 14, 15\}$). Further, let $r_t, g_t \in \Re$ (reals). It is also the case that $PC \cap AC = \oslash$. For a swarm, $s = 0$ indicates that the average charge on bot batteries in a swarm is low, whereas $s = 1$ indicates high average charge. The projected action reward $\bar{r}_{t+1}$ at time $t + 1$ is computed in different ways. In this section, $\bar{r}_{t+1}$ equals the average number of number of times the swarm enters a high state up to the current state. The decision $d_t = 1$ indicates that a swarm chooses an action a_t at time t, and $d_t = 0$ indicates that the swarm rejects (refuses) action a_t. A set of sample codes for proximate causes and actions is given in Fig. 7.

A partial list of proximate causes and corresponding responses are also given in Fig. 7. For example, a swarm is endowed with curiosity (proximate cause code 3), which means that when a swarm has no special task to perform, it wanders in search of features of its environment and learns to recognize environmental patterns. This proximate cause triggers either a *wander* (action code 4) if it has sufficient charge on its batteries or a *fast wander* (action code 5) if it has a near maximum charge on its batteries. A *low average charge* for a swarm (proximate cause code 9) means that a swarm enters an alarm state, where two or more of its members have low batteries. The response to this proximate cause is some form of battery recovery operation: *get battery* (action code 10) if a swarm requests a battery replacement, or *recharge* (action code 11) if a swarm begins recharging its batteries, or *wait for sunshine* (action code 12), if a swarm with solar panels begins waiting for sunshine to charge its batteries. The *adversary* proximate cause (proximate cause code 6) occurs whenever the existence of a

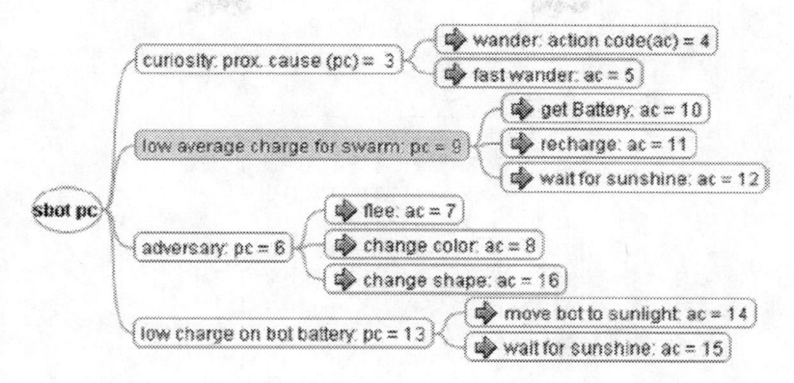

Fig. 7. Proximate Cause and Action-Response Codes.

swarm is threatened by some identified source (e.g., high electromagnetic field, high wind, rain, attacking birds, lightning and so). The response by a swarm to an adversary will be, for example, to $flee$ from or to avoid the adversary (action code 7) or to *change color* (action code 8) or to *change shape* (action code 16). For example, if a swarmbot is threatened by high wind, its individual bots will change shape by doing such things as retracting their arms (code 16). During normal operation, the bots in an sbot hanging from a wire have an oblong, parallelepiped shape. When each bot retracts its arms and partially folds its camera platform inward, it is possible for the shape of the bots in a swarm to approximate either a sphere or an ellipsoid (ideal shapes during the buffeting of high wind). A proximate cause in the form of *low charge on a bot battery* (proximate cause code 13) is a threat to the survival of a swarm. The response to proximate cause 13 is some form of a rescue operation by an sbot. This has different forms such as the one suggested in Fig. 7, namely, *moving a bot to sunlight* (action code 14) or *waiting for sunshine* (action code 15). Behavior ontogeny o_t at time t equals the code for the recent action with the highest frequency. If all recent actions have the same frequency, $o_t = 0$. Let g_t, $\overline{g}_{t-1}$, $\overline{r}_{t-1}$, $\overline{l}_{t-1}$, γ be behavior growth at time t (current state), average behavior growth, average action reward, average number of times recent behaviors are in a low state at time $t-1$ and step size, respectively. Then a proposed model for behavior growth g_t is given in (3).

$$g_t = \overline{g}_{t-1} + \gamma(\overline{r}_{t-1} - \overline{l}_{t-1}) \tag{3}$$

The model for behavior growth g_t in (3) is based on the intuition that growth tends to diminish as $\overline{l}_{t-1}$ (average number of times a behavior enters its low state) increases. In effect, $\overline{l}_{t-1}$ plays the role of a reference value for growth. An ethogram will have the form shown in Table 1. A trial value of *step size* γ equal to 0.01 has been used. In choosing its response to a proximate cause, an agent learns to adapt its behavior so that its behavior growth does not decline and its projected reward is acceptable. In the sample behaviors represented by Table 1, g_t values tend not to change after a certain point in time while behavior rewards tend to fluctuate.

Table 1. Sample Ethogram for a swarm ($\gamma = 0.01$).

$X \backslash A$	s_t	pc_t	a_t	o_t	g_t	r_t	d_t
x0	0	9	11	0	0.1	0.0	0
x1	0	9	11	11	0.1	0.0	1
x2	0	9	11	11	0.1	0.0	1
x3	1	3	4	11	0.2	0.3	1
x4	0	9	10	0	0.6	0.2	0
x5	0	9	11	11	0.6	0.2	1
x6	1	3	4	11	0.6	0.3	1
x7	1	3	5	11	0.6	0.4	0
x8	0	9	10	11	0.6	0.3	1
x9	0	9	11	11	0.6	0.3	1
x10	1	3	4	11	0.6	0.4	1
x11	1	3	5	11	0.6	0.4	1
x12	0	9	10	11	0.6	0.4	1
x13	0	9	11	11	0.6	0.4	1
x14	0	9	10	11	0.6	0.3	0
x15	0	9	11	11	0.6	0.3	0
x16	1	3	4	11	0.6	0.4	0
x17	1	3	5	11	0.6	0.4	1
x18	1	3	4	11	0.6	0.5	1
x19	1	3	4	11	0.6	0.5	1

As this swarm ages, the rewards for its actions tend to diminish. One might wonder how well the overall sample behavior of a swarm matches what one would expect. A step towards the solution to this problem is examined in this article where one considers sample behaviors in the context of an approximation space.

6.1 Example: Approximation Space for an Ethogram

Consider an approximation space (U_{beh}, I_B, ν_B), where U_{beh} is a universe of cooperating agent (swarm) behaviors, and B is a behavior feature set. The mapping $I_B : U_{beh} \longrightarrow \mathcal{P}(U_{beh})$ where $I_B(x) = [x]_{IND(B)}$ for $x \in U_{beh}$ is used to derive a lower approximation B_*D, where D is a decision class. In effect, B_*D contains those blocks of equivalent behaviors which are subsets of D. The set B_*D will serve as a norm or standard useful in measuring the acceptability of blocks of equivalent behaviors relative to a specific action. The lower approximation has been chosen as a standard because it provides certain knowledge about blocks of equivalent behaviors contained in D. Next, consider the reduct B, decision class D, and lower approximation B_*D extracted from Table 1 and shown in (4).

$$\begin{aligned}
&B = \{a_t, o_t, g_t, r_t\} \\
&Decision\ D = \{x \in U_{beh} : d_t(x) = 1\} = \\
&\quad \{\text{x1}, \text{x10}, \text{x11}, \text{x12}, \text{x13}, \text{x17}, \text{x18}, \text{x19}, \text{x2}, \text{x3}, \text{x5}, \text{x6}, \text{x8}, \text{x9}\} \\
&B_*D = \{\text{x1}, \text{x12}, \text{x13}, \text{x18}, \text{x19}, \text{x2}, \text{x3}, \text{x5}, \text{x6}\}
\end{aligned} \tag{4}$$

Recall that a reduct is a minimal set of attributes $B \subseteq A$ that can be used to discern (i.e., distinguish) all objects obtainable by all of the attributes of an information system. In a decision system $DT = (U, A \cup \{d\})$, it is the case that $Ind_{DT}(B) = Ind_{DT}(A)$ for reduct B. In effect, a reduct is a subset B of attributes A of a decision system that preserves the partitioning of the universe U. For the sample DT in Table 1, we have chosen reduct B to highlight the importance of certain sbot features such as action a_t and reward r_t in discerning the sample elements of the decision system. This is consistent with what we have found in other decision tables representing sample behaviors of a swarm. The action-blocks extracted from the ethogram represented by Table 1 as well as rough inclusion values are given in Table 2.

Table 2. Action Block Rough Inclusion Values.

Action Blocks	$\nu_B(B_a(x), B_* D)$
$a_t = 10$(get battery)	inclusion value
$[0, 9, 10, 0, 0.6, 0.2] : B_{(10)}(x4) =$ x4	0.0
$[0, 9, 10, 11, 0.6, 0.3] : B_{(10)}(x14) = \{x14, x8\}$	0.0
$[0, 9, 10, 11, 0.6, 0.4] : B_{(10)}(x12) = \{x12\}$	0.11
$a_t = 11$(recharge)	inclusion value
$[0, 9, 11, 0, 0.1, 0.0] : B_{(11)}(x0) = \{x0\}$	0.0
$[0, 9, 11, 11, 0.1, 0.0] : B_{(11)}(x1) = \{x1, x2\}$	0.22
$[0, 9, 11, 11, 0.6, 0.2] : B_{(11)}(x5) = \{x5\}$	0.11
$[0, 9, 11, 11, 0.6, 0.3] : B_{(11)}(x15) = \{x15, x9\}$	0.0
$[0, 9, 11, 11, 0.6, 0.4] : B_{(11)}(x13) = \{x13\}$	0.11
$a_t = 4$(wander)	inclusion value
$[1, 3, 4, 11, 0.2, 0.3] : B_{(4)}(x3) = \{x3\}$	0.11
$[1, 3, 4, 11, 0.6, 0.3] : B_{(4)}(x6) = \{x6\}$	0.11
$[1, 3, 4, 11, 0.6, 0.4] : B_{(4)}(x10) = \{x10, x16\}$	0.0
$[1, 3, 4, 11, 0.6, 0.5] : B_{(4)}(x18) = \{x18, x19\}$	0.22
$a_t = 5$(fast wander)	inclusion value
$[1, 3, 5, 11, 0.6, 0.4] : B_{(5)}(x11) = \{x11, x17, x7\}$	0.0

7 Approximation Space-Based Rewards

In this section, two approaches to deriving a reference reward $\overline{r}_t$ needed to estimate a projected action reward $\overline{r}_{t+1}$ are considered. A reference reward serves as a basis for measuring the difference between the reward in the current state and rewards for recent actions. One approach to deriving $\overline{r}_t$ is to compute the average action reward up to time t [66]. A second approach is to derive $\overline{r}_t$ periodically (e.g., every 500 ms) within an approximation space and to compute the average degree of inclusion of action blocks in a standard. In both cases, $\overline{r}_t$ is computed periodically on-line (in real-time). The computation of reference rewards is part of a framework for reinforcement learning by either an individual or by a swarm. In this paper, the focus is on reinforcement learning by a swarm.

Let α, r_t, $\overline{r}_t$, denote step size in (0, 1], reward and average reference reward at time t, respectively, and let $\overline{r}_{t+1}$ denote the projected reference reward at time $t + 1$. Then a model for projected reference reward $\overline{r}_{t+1}$ at time $t + 1$ is given in (5).

$$\overline{r}_{t+1} = \overline{r}_t + \alpha(r_t - \overline{r}_t) \tag{5}$$

The projected reward $\overline{r}_{t+1}$ at time $t + 1$ suggested by [66] does not take into account the patterns associated with each of the action rewards leading up to the current state. If we consider the computation of average rewards in the context of an approximation space, then it is possible to arrive at action-based reference rewards relative to action blocks belonging to a decision class. Let $U_{beh}/Ind_{DT}(B)$ denote a partition of U_{beh} relative to the set of attributes B in the decision system DT. Let $B_a(x)$ denote a block in partition $U_{beh}/Ind_{DT}(B)$ with respect to action a. In what follows, the feature set B would always include feature a (action). In addition, notice that the inclusion values for each block $B_a(x)$ are independent of the decision class. This is the case in Table 2, where, for example, the inclusion values for $B_{(a=11)}(x0)$ with $d(x0) = 0$ are $B_{(a=11)}(x1)$ with $d(x1) = 1$ are computed. The degree of inclusion of each of the blocks for a particular action yields useful information in reinforcement learning by a swarm. For example, the rough inclusion values of the action blocks for $a_t = 11(recharge)$ relative to B_*D in Table 2 give us an indication of the closeness of the recharge behaviors to the standard. We are interested in viewing this information collectively as a means of guiding the distribution of near-future rewards for recharge-behavior. One way to do this is by averaging the closeness values for each of the recharge blocks represented in Table 2. In general, in the context of an approximation space, a reference reward $\overline{r}_t$ for each action a is computed as shown in (6).

$$\overline{r}_t = \frac{\sum_{i=1}^{n} \nu_B(B_a(x)_i, B_*D)}{n} \tag{6}$$

where n is the number of blocks $B_a(x)$ in $U_{beh}/Ind_{DT}(B)$. The action-based reference rewards derived from Table 2 are given in Table 3.

The values in Table 3 match the intuition that the *recharge* and *get battery* actions have low reference rewards because a swarm is most vulnerable at times when it has low energy. By contrast, wandering actions by a swarm occur when a swarm has higher energy. Hence the reference rewards in Table 3 also match

Table 3. Action-Based Reference Rewards (d = 1).

Action	reference reward $\overline{r}_t$
$a_t = 4$ (*wander*)	0.11
$a_t = 5$ (*fast wander*)	0.0
$a_t = 10$ (*get battery*)	0.0367
$a_t = 11$ (*recharge*)	0.088

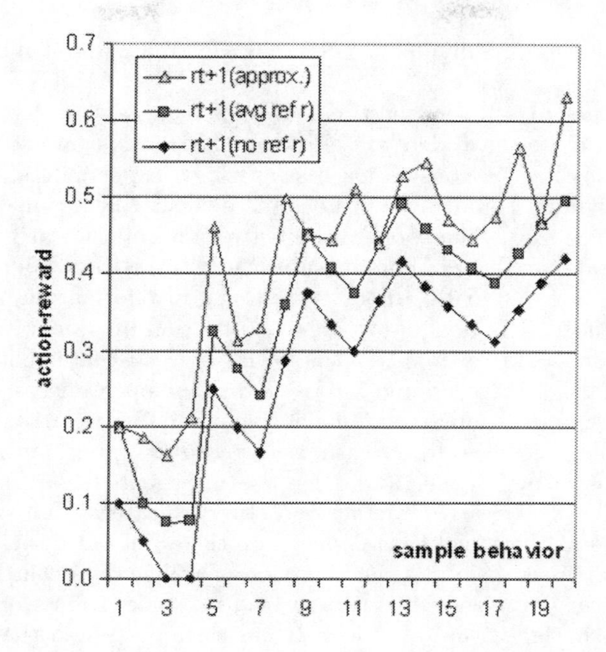

Fig. 8. Comparison of Behavior Action Rewards.

our intuition. That is, one can expect that projected rewards will increase if we use the behavior standard represented by the lower approximation derived from Table 1. A comparison of the three types of rewards computed with and without a reference reward is shown in Fig. 8.

Over time, one can expect that projected rewards will gradually increase while a swarm is learning. This is the case for all three types of rewards shown in Fig. 8. However, for either of the projected rewards with a reference reward for the choice of an action, the two sample behaviors follow each other closely. Swarm behavior rewards are slightly better when reference rewards are derived within the framework of an approximation space. The drawback to the approximation space approach to deriving reference rewards is that it is necessary to recompute the reference reward periodically to reflect changes in the environment of a swarm. The way to combat this problem is to identify a standard drawn from a comprehensive representation of proximate causes of swarm behavior.

8 Ontology of Behavior and Hierarchical Learning

In this section, we briefly consider the basis for an ontology of behavior. An ontology of behavior is a study of what it means for a behavior to be a part of the life of an entity. A complete ontology of behavior for a collection of cooperating agents would consider a hierarchy of behaviors. Families, societies, insect colonies, flocks of birds, and collections of cooperating bots (swarms) are sources of examples of hierarchies of behavior. There are obvious dependencies between the behavior of individual agents that contribute to social behavior (i.e., behavior on a higher level).

It is beneficial to ask the question, "How is a hierarchy of behaviors possible?" Until now, we have deliberately focused attention on swarm behavior. This is made possible by considering a swarm as a super individual. The idea of a super individual comes from studies of the social behavior of animals. It has been observed that when cooperation between agents is nearly perfect, then such cooperation by a team of bots in a robocup competition [65] or by dancing Orio bots from Sony [8] or by a flock geese flying in v-formation has a singular character. In effect, it is possible to view a collection of cooperating agents as a single individual. For example, it has been observed that flight manoeuvers of a flock of Starlings flying round a roost have the appearance of a super individual [68]. In that case, observed behavior is understood relative to proximate causes and action-responses by a swarm as opposed to observing the behavior of a member of a swarm. A collection of cooperating agents (swarm) survives and is "held together" by effective interaction between members of a swarm.

Cooperation at the social level depends on the ability of a society or swarm to learn from the rewards of its own behavior as well as the rewards of behavior of individual behavior. Hence, it is reasonable to consider frameworks for behavior knowledge-sharing within a society. It has been suggested that this can be accomplished by introducing approximation spaces at each level in a behavior hierarchy (see, e.g., [36] as well as [50, 61]). A principal advantage in considering behavior in the context of an approximation space is that one can then consider what it means for blocks of equivalent behaviors to be a part of a set of behaviors that provide a standard for learning. In effect, approximation spaces provide the basis for an ontology of behavior.

It is understood that whenever two agents begin communicating with each other, learning occurs at the individual level (action choices based on local behavior rewards) as well as at the social level (concerted action choices based on social behavior rewards). In a behavior hierarchy, a model of exchange of knowledge in the context of approximation spaces of agents has been proposed (see, e.g., [58]). A basic idea in learning on a particular level is deciding when an action in response to a stimulus belongs to a particular decision class. This means that either an individual agent or a collection of cooperating agents (swarm) must recognize a behavior pattern relative to a decision class. One way to do this in real time is to consider learning in the context of an approximation space, where a judgment about membership of a behavior in a decision class is influenced by pattern-based rewards.

9 Conclusion

The end result of viewing intelligent systems behavior within an approximation space is to derive a norm or standard that can be used to compute projected action rewards. Ultimately, it is important to consider ways to evaluate the behavior of an intelligent system as it unfolds. Rather than take a rigid approach where a system behavior is entirely predictable based on its design, there is some benefit in relaxing the predictability requirement and considering how one might gain approximate knowledge about evolving patterns of behavior in an intelligent system in real-time. The studies of animal behavior by ethologists provide a number of features useful in the study of changing intelligent systems behavior in response to environmental changes (sources of stimuli) as well as internal influences (e.g., image classification results, battery energy level, response to behavior pattern recognition, various forms of learning). Behavior decision tables (called ethograms in this article) constantly change during the life of an intelligent system because of changing proximate causes and changing rewards of corresponding action responses. As a result, there is a need for a cooperating system of agents to gain, measure, and share knowledge about changing behavior patterns. Part of the future work in this research is a consideration of an ontology of behavior that takes into account knowledge-sharing as well as learning on different levels in a hierarchy of behaviors. This future work will include a study of various forms of action preferences, action-probabilities and action-rewards in the context of approximation spaces useful in on-line learning by either a swarm or an individual.

Acknowledgements

This research has been supported by Natural Sciences and Engineering Research Council of Canada (NSERC) grant 185986 and grants T209, T217, T137, and T247 from Manitoba Hydro. Many thanks to Andrzej Skowron for his incisive comments and for suggesting many helpful ways to improve this article. Many thanks also to Sheela Ramanna, Christopher Henry, Dan Lockery and Maciej Borkowski for their helpful suggestions.

References

1. Alpigini, J. J.,Peters, J. F.,Skowron, A.,Zhong, N.,Eds.:Proceedings of 3rd Int. Conf. on Rough Sets and Current Trends in Computing (RSCTC2002), Malvern, PA, 14–16 Oct. 2002. *Lecture Notes in Artificial Intelligence*, Springer-Verlag, Berlin (2002)
2. Applewhite, A.: The view from the top. *IEEE Spectrum* (Nov. 2004), 36–51
3. Bonabeau, E., Dorigo, M., Theraulaz, G. : *Swarm Intelligence. From Natural to Artificial Systems*, Oxford University Press, UK (1999)
4. Cheng, K.: Generalization and Tinbergen's four whys. *Behavioral and Brain Sciences* **24** (2001) 660–661
5. Dorigo, M: Swarmbots, *Wired* (Feb. 2004) 119
6. Duda, R. O., Hart, P.E., Stork, D. G.: *Pattern Classification*, Wiley-Interscience, Toronto (2001)
7. Fahle, M, Poggio, T. (Eds.): *Perceptual Learning*, The MIT Press, Cambridge, MA (2002)
8. Geppert, L.: Sony's Orio. *IEEE Spectrum* (Feb. 2004)
9. Harnad, S. (Ed.): *Categorical Perception. The Groundwork of cognition*, Cambridge University Press, UK (1987)
10. Hastie,T.,Tibshirani,R.,Friedman,J.: *The Elements of Statistical Learning. Data Mining, Inference, and Prediction*, Springer–Verlag, Berlin (2001)

11. Holt, J.: *UML for Systems Engineering. Watching the Wheels.* The Institute of Electrical Engineers, Herts, UK (2001)
12. Kruuk, H.: *Niko's Nature. A life of Niko Tinbergen and his science of animal behavior.* Oxford University Press, London (2003)
13. Lehner, P. N.: *Handbook of Ethological Methods*, 2nd Ed. Cambridge University Press, UK (1996)
14. Martin, P., Bateson, P.: *Measuring Behavior.* Cambridge University Press, Cambridge, MA (1993)
15. Mondada, F., Bonani, M., Magnenat, S., Guignard, A., Floreano, D.: Physical connections and cooperation in swarm robotics. In: Frans Groen, Nancy Amato, Andrea Bonarini, Eiichi Yoshida and Ben Kröse editors, Proceedings of the 8th Conference on Intelligent Autonomous Systems (IAS8), Amsterdam, NL, March 10-14 (2004) 53–60
16. Nguyen, S. H., Bazan, J., Skowron, A., Nguyen, H. S.: Layered learning for concept synthesis, *Transactions on Rough Sets* **I** (2004) 187–208
17. Son, N. H.,Skowron, A.,Szczuka,M. S.: Analysis of image sequences for the Unmanned Aerial Vehicle. In: Hirano, S. Inuiguchi, M., Tsumoto S. (Eds.), *Bulletin of the International Rough Set Society* **5**, No. 1/2 (2001) 185–184
18. OMG Unified Modeling Language (UML) Specification. Object Management Group, `http://www.omg.org`
19. Pal, S. K., Polkowski, L., Skowron, A. (eds.): *Rough-Neural Computing. Techniques for Computing with Words.* Springer-Verlag, Heidelberg (2004)
20. Pawlak, Z.: Rough sets. *International J. Comp. Inform. Science.* **11** (1982)341–356
21. Pawlak, Z.: Rough sets and decision tables. *Lecture Notes in Computer Science*, **208**, Springer Verlag, Berlin (1985) 186–196
22. Pawlak, Z.: On rough dependency of attributes in information systems. *Bulletin Polish Acad. Sci. Tech.* **33** (1985) 551–599
23. Pawlak, Z.: On decision tables. *Bulletin Polish Acad. Sci. Tech.*, **34** (1986) 553–572
24. Pawlak, Z.: Decision tables—a rough set approach, *Bulletin ETACS*, **33** (1987) 85–96
25. Pawlak, Z.: Elementary rough set granules: Toward a rough set processor. In: [19], (2004) 5–14
26. Pawlak, Z.: *Rough Sets. Theoretical Reasoning about Data.* Kluwer, Dordrecht (1991)
27. Pawlak,Z., Skowron,A.: Rough membership functions. In: Yager,R., et al.,Eds., *Advances in Dempster Shafer Theory of Evidence.* Wiley, N.Y.(1994) 251–271
28. Pawlak, Z.: Some issues on rough sets. *Transactions on Rough Sets* **I** (2004) 1–58
29. Pawlak, Z.: In pursuit of patterns in data reasoning from data–The rough set way. In: [1], (2002) 1–9
30. Pawlak, Z.: Rough sets and decision algorithms. In: [72], (2001) 30–45
31. Pawlak, Z.: Flow graphs and decision algorithms. In: [69], (2003) 1–10
32. Peters, J. F.: Design patterns in intelligent systems. *Lecture Notes in Artificial Intelligence*, **2871**, Springer-Verlag, Berlin (2003) 262–269
33. Peters, J. F., Ramanna, S.: Intelligent systems design and architectural patterns. In: Proceedings IEEE Pacific Rim Conference on Communication, Computers and Signal Processing (PACRIM'03) (2003) 808–811
34. Peters, J. F. : Approximation space for intelligent system design patterns. *Engineering Applications of Artificial Intelligence* **17**, No. 4, (2004) 1–8
35. Peters, J. F., Ramanna, S.: Measuring acceptance of intelligent system models. In: M. Gh. Negoita et al. (Eds.), Knowledge-Based Intelligent Information and Engineering Systems, *Lecture Notes in Artificial Intelligence* **3213**, Part I, (2004) 764–771

36. Peters, J. F. : Approximation spaces for hierarchical intelligent behavioral system models. In: B.D.-Keplicz, A. Jankowski, A. Skowron, M. Szczuka (Eds.), Monitoring, Security and Rescue Techniques in Multiagent Systems. *Advances in Soft Computing*. Physica-Verlag, Heidelberg (2004) 13–30
37. Peters, J. F., Ramanna, S.: Approximation space for software models, *Transactions on Rough Sets* **I** (2004) 338–355
38. Peters, J. F., Skowron, A., Stepaniuk, J., Ramanna, S.: Towards an ontology of approximate reason. *Fundamenta Informaticae* **51**, Nos. 1, 2 (2002) 157–173
39. Peters, J. F., Ahn, T. C., Borkowski, M., Degtyaryov, V., Ramanna, S.: Line-crawling robot navigation: A rough neurocomputing approach. In: C. Zhou, D. Maravall, D. Ruan (Eds.), Autonomous Robotic Systems. *Studies in Fuzziness and Soft Computing* **116**. Physica-Verlag, Heidelberg, (2003) 141–164
40. Peters, J.F., Ahn, T.C., Borkowski, M.: Object-classification by a line-crawling robot: A rough neurocomputing approach. In: J.J. Alpigini, J.F. Peters, A. Skowron, N. Zhong (Eds.), Rough Sets and Current Trends in Computing, *Lecture Notes in Artificial Intelligence* **2475**. Springer-Verlag, Berlin (2002) 595–601
41. Peters, J. F., Skowron, A., Synak, P., Ramanna, S. : Rough sets and information granulation. In: Bilgic, T., Baets, D., Kaynak, O. (Eds.), Tenth Int. Fuzzy Systems Assoc. World Congress IFSA, Instanbul, Turkey, *Lecture Notes in Artificial Intelligence* **2715**, Springer-Verlag, Heidelberg, (2003) 370–377
42. Polkowski, L. and Skowron, A. (Eds.): Rough Sets in Knowledge Discovery 1. *Studies in Fuzziness and Soft Computing* **18**. Physica-Verlag, Heidelberg (1998)
43. Polkowski, L. and Skowron, A. (Eds.): Rough Sets in Knowledge Discovery 2. *Studies in Fuzziness and Soft Computing* **19**. Physica-Verlag, Heidelberg (1998)
44. Polkowski, L. and Skowron, A.: Rough mereology: A new paradigm for approximate reasoning. *Int. J. Approximate Reasoning* **15/4** (1996) 333–365
45. Polkowski, L. and Skowron, A.: Rough meriological calculi of granules: A rough set approach to computation. *Computational Intelligence* **17**(3) (2001) 472–492
46. Polkowski, L.: *Rough Sets. Mathematical Foundations.* Physica–Verlag, Heidelberg (2002)
47. Quinlan, J. R.: *C4.5: Programs for Machine Learning.* Morgan Kaufmann, Amsterdam (1988)
48. Skowron, A.: Toward intelligent systems: Calculi of information granules. In: Hirano, S. Inuiguchi, M., Tsumoto S. (Eds.), *Bulletin of the International Rough Set Society* **5**, No. 1/2 (2001) 9–30
49. Skowron, A., Peters, J. F.: Rough sets: Trends and Challenges. In: [69](2003) 25-34
50. Skowron, A., Stepaniuk, J.: Information systems in hierarchical modeling. In: G. Lindemann, H.-D. Burkhard, L. Czaja. A. Skowron. H. Schlingloff, Z. Suraj (eds.), Proceedings of the Workshop on Concurrency, Specification and Programming (CSP 2004), vol. **1-3**, Caputh, Germany, September 24-26, 2004, Informatik-Bericht Nr. **170**, Humboldt Universität (2004) 378–389
51. Skowron, A., Synak, P., Peters, J. F.: Spacio-temporal approximate reasoning over hierarchical information maps. In: G. Lindemann, H.-D. Burkhard, L. Czaja. A. Skowron. H. Schlingloff, Z. Suraj (eds.), Proceedings of the Workshop on Concurrency, Specification and Programming (CSP 2004), vol. 1–3, Caputh, Germany, September 24–26, 2004, Informatik-Bericht Nr. **170**, Humboldt Universität (2004) 358–371
52. Skowron, A. Stepaniuk, J.: Generalized approximation spaces. In: Proceedings of the Third International Workshop on Rough Sets and Soft Computing, San Jose (1994) 156–163

53. Skowron, A. Stepaniuk, J.: Generalized approximation spaces. In: Lin, T. Y., Wildberger, A. M. (Eds.), Soft Computing, Simulation Councils, San Diego (1995) 18–21
54. Skowron, A., Stepaniuk, J.: Tolerance approximation spaces. *Fundamenta Informaticae*, 27 (1996) 245–253
55. Skowron, A., Stepaniuk, J.: Information granules and approximation spaces. In: Proc. of the 7th Int. Conf. on Information Processing and Management of Uncertainty in Knowledge-based Systems (IPMU'98), Paris (1998) 1354–1361
56. Skowron, A., Stepaniuk, J.,: Information granules and rough neural computing. In: [19], (2004) 43–84.
57. Skowron, A. Stepaniuk, J.: Information granules: Towards Foundations of Granular Computing. *Int. Journal of Intelligent Systems* **16**, (2001) 57–85
58. Skowron, A., Stepaniuk, J. Peters, J. F.: Rough sets and infomorphisms: Towards approximation of relations in distributed environments. *Fundamenta Informaticae* **54**, Nos. 2, 3 (2003) 263–277
59. Skowron, A. and Swiniarski, R. W.: Information granulation and pattern recognition. In: [19], (2004) 599–636
60. Skowron, A., Swiniarski, R., Synak, P.: Approximation spaces and information granulation. In: S. Tsumoto, R. Slowinski, J. Komorowski, J.W. Grzymala-Busse (Eds.), Rough Sets and Current Trends in Computing, *Lecture Notes in Artificial Intelligence* **3066**. Springer, Berlin (2004) 116–126
61. Skowron, A., Stepaniuk, J.: Information granules in distributed environment. In: N. Zhong, A. Skowron, S. Ohsuga (Eds.), New Directions in Rough Sets, Data Mining, and Granular-Soft Computing, *Lecture Notes in Artificial Intelligence* **1711**. Berlin: Springer (1999) 357–365
62. Skowron, A., Stepaniuk, J.: Towards discovery of information granules. In: Proc. of the 3rd European Conf. on Principles and Practice of Knowledge Discovery in Databases, *Lecture Notes in Artificial Intelligence* **1704**, Berlin, Springer-Verlag (1999) 542–547
63. Skowron, A., Stepaniuk, J.: Constraint sums in information systems. In: S. Tsumoto, R. Slowinski, J. Komorowski, J.W. Grzymala-Busse (Eds.), Rough Sets and Current Trends in Computing, *Lecture Notes in Artificial Intelligence* **3066**. Springer, Berlin (2004) 300–309
64. Stepaniuk, J.: Approximation spaces, reducts and representatives. In: [43], 109–126
65. Stone, P.: *Layered Learning in Multiagent Systems. A Winning Approach to Robotic Soccer*. The MIT Press, Cambridge, MA, 2000.
66. Sutton, R. S., Barto, A. G.: *Reinforcement Learning: An Introduction*. The MIT Press, Cambridge, MA (1998)
67. Tinbergen N.: On aims and methods of ethology, *Zeitschrift für Tierpsychologie* **20** (1963) 410–433
68. Tinbergen N.: *Social Behavior in Animals with Special Reference to Vertebrates*. The Scientific Book Club, London, 1953.
69. Wang, G. Liu, Q., Yao, Y., Skowron, A. (Eds.), Proceedings 9th Int. Conf. on Rough Sets, Fuzzy Sets, Data Mining, and Granular Computing (RSFDGrC2003). *Lecture Notes in Artificial Intelligence* **2639**. Springer-Verlag, Berlin (2003)
70. Watanabe, S.: *Pattern Recognition: Human and Mechanical*, Wiley, Toronto (1985)
71. WITAS project. `http://www.ida.liu.se/ext/witas/eng.html`. 2001.
72. Ziarko, W., Yao, Y.,Eds.:Proceedings of 2nd Int. Conf. on Rough Sets and Current Trends in Computing (RSCTC2000), Banff, Canada, 16-19 Oct. 2000. *Lecture Notes in Artificial Intelligence*, Springer-Verlag, Berlin (2001)

Approximation Spaces and Information Granulation

Andrzej Skowron[1], Roman Świniarski[2,3], and Piotr Synak[4]

[1] Institute of Mathematics, Warsaw University,
Banacha 2, 02-097 Warsaw, Poland
skowron@mimuw.edu.pl
[2] Institute of Computer Science, Polish Academy of Sciences,
Ordona 21, 01-237 Warsaw, Poland
[3] Department of Mathematical and Computer Sciences,
San Diego State University,
5500 Campanile Drive San Diego, CA 92182, USA
rswiniar@sciences.sdsu.edu
[4] Polish-Japanese Institute of Information Technology,
Koszykowa 86, 02-008 Warsaw, Poland
synak@pjwstk.edu.pl

Abstract. In this paper, we discuss approximation spaces in a granular computing framework. Such approximation spaces generalise the approaches to concept approximation existing in rough set theory. Approximation spaces are constructed as higher level information granules and are obtained as the result of complex modelling. We present illustrative examples of modelling approximation spaces that include approximation spaces for function approximation, inducing concept approximation, and some other information granule approximations. In modelling of such approximation spaces we use an important assumption that not only objects but also more complex information granules involved in approximations are perceived using only partial information about them.

1 Introduction

The rough set approach is based on the concept of approximation space. Approximation spaces for information systems [1] are defined by partitions or coverings defined by attributes of a pattern space. One can distinguish two basic components in approximation spaces: an uncertainty function and an inclusion function [2]. This approach has been generalised to the rough mereological approach (see, e.g., [3–5]). The existing approaches are based on the observation that the objects are perceived via information about them and due to the incomplete information they can be indiscernible. Hence, with each object one can associate an indiscernibility class, called also (indiscernibility) neighbourhood [6]. In the consequence, testing if a given object belongs to a set is substituted by checking a degree to which its neighbourhood is included into the set. Such an approach covers several generalisations of set approximations like those based on the tolerance relation or the variable precision rough set model [7].

J.F. Peters and A. Skowron (Eds.): Transactions on Rough Sets III, LNCS 3400, pp. 175–189, 2005.

In real-life applications approximation spaces are complex information granules that are not given directly with data but they should be discovered from available data and domain knowledge by some searching strategies (see, e.g., [5,8]). In the paper we present a general approach to approximation spaces based on granular computing. We show that the existing approaches to approximations in rough sets are particular cases of our approach. Illustrative examples include approximation spaces with complex neighbourhoods, approximation spaces for function approximation and for inducing concept approximations. Some other aspects of information granule construction, relevant for approximation spaces, are also presented. Furthermore, we discuss one more aspect of approximation spaces based on the observation that the definition of approximations does not depend only on perception of partial information about objects but also of more complex information granules.

The presented approach can be interpreted in a multi-agent setting [5, 9]. Each agent is equipped with its own relational structure and approximation spaces located in input ports. The approximation spaces are used for filtering (approximating) information granules sent by other agents. Such agents are performing operations on approximated information granules and sending the results to other agents, checking relationships between approximated information granules, or using such granules in negotiations with other agents. Parameters of approximation spaces are analogous to weights in classical neuron models. Agents are performing operations on information granules (that approximate concepts) rather than on numbers. This analogy has been used as a starting point for the rough-neural computing paradigm [10] of computing with words [11].

2 Concept Approximation

In this section we consider the problem of concepts approximation over a universe U^∞ (concepts that are subsets of U^∞). We assume that the concepts are perceived only through some subsets of U^∞, called samples. This is a typical situation in machine learning, pattern recognition, and data mining approaches [12–14]. We explain the rough set approach to induction of concept approximations.

Let $U \subseteq U^\infty$ be a finite sample. By Π_U we denote a perception function from $P(U^\infty)$ into $P(U)$ defined by $\Pi_U(C) = C \cap U$ for any concept $C \subseteq U^\infty$. The problem we consider is how to extend the approximations of $\Pi_U(C)$ to approximation of C over U^∞. In the rough set approach the approximation of a concept is defined by means of a so called approximation space $AS = (U, I, \nu)$, where $I : U \to P(U)$ is an uncertainty function such that $x \in I(x)$ for any $x \in U$, and $\nu : P(U) \times P(U) \to [0,1]$ is a rough inclusion function (for details see Section 4). We show that the problem can be described as searching for an extension $AS_C = (U^\infty, I_C, \nu_C)$ of the approximation space AS, relevant for approximation of C. This makes it necessary to show how to extend the inclusion function ν from U to relevant subsets of U^∞ that are suitable for the approximation of C. Observe (cf. Definition 5) that for the approximation of C it is enough to induce the necessary values of the inclusion function ν_C without knowing the exact value of $I_C(x) \subseteq U^\infty$ for $x \in U^\infty$.

Let AS be a given approximation space for $\Pi_U(C)$ and let us consider a language L in which the neighbourhood $I(x) \subseteq U$ is expressible by a formula $pat(x)$, for any $x \in U$. It means that $I(x) = \|pat(x)\|_U \subseteq U$, where $\|pat(x)\|_U$ denotes the meaning of $pat(x)$ restricted to the sample U. In the case of rule based classifiers patterns of the form $pat(x)$ are defined by feature value vectors.

We assume that for any new object $x \in U^\infty \setminus U$ we can obtain (e.g., as a result of sensor measurement) a pattern $pat(x) \in L$ with semantics $\|pat(x)\|_{U^\infty} \subseteq U^\infty$. However, the relationships between information granules over U^∞ like sets: $\|pat(x)\|_{U^\infty}$ and $\|pat(y)\|_{U^\infty}$, for different $x, y \in U^\infty$ (or between $\|pat(x)\|_{U^\infty}$ and $y \in U^\infty$), are, in general, known only if they can be expressed by relationships between the restrictions of these sets to the sample U, i.e., between sets $\Pi_U(\|pat(x)\|_{U^\infty})$ and $\Pi_U(\|pat(y)\|_{U^\infty})$.

The set of patterns $\{pat(x) : x \in U\}$ is usually not relevant for approximation of the concept $C \subseteq U^\infty$. Such patterns are too specific or not general enough, and can directly be applied only to a very limited number of new objects. However, by using some generalisation strategies, one can search, in a family of patterns definable from $\{pat(x) : x \in U\}$ in L, for such new patterns that are relevant for approximation of concepts over U^∞. Let us consider a subset $PATTERNS(AS, L, C) \subseteq L$ chosen as a set of pattern candidates for relevant approximation of a given concept C. For example, in the case of a rule based classifier one can search for such candidate patterns among sets definable by subsequences of feature value vectors corresponding to objects from the sample U. The set $PATTERNS(AS, L, C)$ can be selected by using some quality measures checked on meanings (semantics) of its elements restricted to the sample U (like the number of examples from the concept $\Pi_U(C)$ and its complement that support a given pattern). Then, on the basis of properties of sets definable by those patterns over U we induce approximate values of the inclusion function ν_C on subsets of U^∞ definable by any of such pattern and the concept C.

Next, we induce the value of ν_C on pairs (X, Y) where $X \subseteq U^\infty$ is definable by a pattern from $\{pat(x) : x \in U^\infty\}$ and $Y \subseteq U^\infty$ is definable by a pattern from $PATTERNS(AS, L, C)$.

Finally, for any object $x \in U^\infty \setminus U$ we induce the approximation of the degree $\nu_C(\|pat(x)\|_{U^\infty}, C)$ applying a conflict resolution strategy $Conflict\_res$ (a voting strategy, in the case of rule based classifiers) to two families of degrees:

$$\{\nu_C(\|pat(x)\|_{U^\infty}, \|pat\|_{U^\infty}) : pat \in PATTERNS(AS, L, C)\}, \tag{1}$$

$$\{\nu_C(\|pat\|_{U^\infty}, C) : pat \in PATTERNS(AS, L, C)\}. \tag{2}$$

Values of the inclusion function for the remaining subsets of U^∞ can be chosen in any way – they do not have any impact on the approximations of C. Moreover, observe that for the approximation of C we do not need to know the exact values of uncertainty function I_C – it is enough to induce the values of the inclusion function ν_C. Observe that the defined extension ν_C of ν to some subsets of U^∞ makes it possible to define an approximation of the concept C in a new approximation space AS_C by using Definition 5.

In this way, the rough set approach to induction of concept approximations can be explained as a process of inducing a relevant approximation space.

3 Granule Approximation Spaces

Using the granular computing approach (see, e.g., [5]) one can generalise the approximation operations for sets of objects, known in rough set theory, to arbitrary information granules. The basic concept is the rough inclusion function ν [3–5].

First, let us recall the definition of an information granule system [5].

Definition 1. *An information granule system is any tuple $GS = (E, O, G, \nu)$ where E is a set of elements called elementary information granules, O is a set of (partial) operations on information granules, G is a set of information granules constructed from E using operations from O, and $\nu : G \times G \longrightarrow [0, 1]$ is a (partial) function called rough inclusion.*

The main interpretation of rough inclusion is to measure the inclusion degree of one granule in another.

In the paper we use the following notation: $\nu_p(g, g')$ denotes that $\nu(g, g') \geq p$ holds; $Gran(GS) = G$; $\mathcal{G}$ denotes a given family of granule systems. For every non-empty set X, let $P(X)$ denote the set of all subsets of X.

We begin with the general definition of approximation space in the context of a family of information granule systems.

Definition 2. *Let $\mathcal{G}$ be a family of information granule systems. A granule approximation space with respect to $\mathcal{G}$ is any tuple*

$$\mathcal{AS}_{\mathcal{G}} = (GS, G, Tr), \tag{3}$$

where GS is a selected (initial) granule system from $\mathcal{G}$; $G \subseteq Gran(GS)$ is some collection of granules; a transition relation Tr is a binary relation on information granule systems from $\mathcal{G}$, i.e., $Tr \subseteq \mathcal{G} \times \mathcal{G}$.

Let $GS \in \mathcal{G}$. By $Tr[GS]$ we denote the set of all information granule systems reachable from GS:

$$Tr[GS] = \{GS' \in \mathcal{G} : GS\ Tr^*\ GS'\}, \tag{4}$$

where Tr^* is the reflexive and transitive closure of the relation Tr. By $Tr[GS, G]$ we denote the set of all Tr-terminal granule systems reachable from GS that consist of information granules from G:

$$Tr[GS, G] = \{GS' : (GS, GS') \in Tr^* \text{ and } G \subseteq Gran(GS') \text{ and } Tr[GS'] = \{GS'\}\}. \tag{5}$$

The elements of $Tr[GS, G]$ are called the candidate granule systems for approximation of information granules from G, generated by Tr from GS (G-candidates, for short). For any system $GS^* \in Tr[GS, G]$ we define approximations of granules from G by information granules from $Gran(GS^*) \setminus G$. Searching for granule systems from $Tr[GS, G]$ that are relevant for approximation of given information granules is one of the most important tasks in granular computing.

By using granule approximation space $\mathcal{AS}_{\mathcal{G}} = (GS, G, Tr)$, for a family of granule systems $\mathcal{G}$, we can define approximation of a given granule $g \in G$ in terms of its lower and upper approximations[1]. We assume that there is additionally

[1] If there is no contradiction we use $\mathcal{AS}$ instead of $\mathcal{AS}_{\mathcal{G}}$.

given a "make granule" operation $\oplus : P(G^*) \longrightarrow G^*$, where $G^* = Gran(GS^*)$, for any $GS^* \in Tr(GS, G)$, that constructs a single granule from a set of granules. A typical example of $\oplus$ is set theoretical union, however, it can be also realised by a complex classifier. The granule approximation is thus defined as follows:

Definition 3. *Let $0 \leq p < q \leq 1$, $\mathcal{AS} = (GS, G, Tr)$ be a granule approximation space, and $GS^* \in Tr[GS, G]$. The $(\mathcal{AS}, GS^*, \oplus, q)$-lower approximation of $g \in G$ is defined by*

$$LOW(\mathcal{AS}, GS^*, \oplus, q)(g) = \oplus(\{g' \in Gran(GS^*) \setminus G : \nu^*(g', g) \geq q\}) \quad (6)$$

where ν^* denotes the rough inclusion of GS^*.

The $(\mathcal{AS}, GS^*, \oplus, p)$-upper approximation of g is defined by

$$UPP(\mathcal{AS}, GS^*, \oplus, p)(g) = \oplus(\{g' \in Gran(GS^*) \setminus G : \nu^*(g', g) > p\}) \quad (7)$$

where ν^* denotes the rough inclusion of GS^*.

The numbers p, q can be interpreted as inclusion degrees that make it possible to control the size of the lower and upper approximations. In the case when $p = 0$ and $q = 1$ we have the case of full inclusion (lower approximation) and any non-zero inclusion (upper approximation).

One can search for optimal approximations of granules from G defined by $GS^* \in Tr[GS, G]$ using some optimisation criteria or one can search for relevant fusion of approximations of granules from G defined by $GS^* \in Tr[GS, G]$.[2]

In the following sections we discuss illustrative examples showing that the above scheme generalises several approaches to approximation spaces and set approximations. In particular, we include examples of information granules G and their structures, the rough inclusion ν as well as the $\oplus$ operation.

4 Approximation Spaces

Let us recall the definition of an approximation space from [1, 2]. For simplicity of reasoning we omit parameters that label components of approximation spaces.

Definition 4. *An approximation space is a system $AS = (U, I, \nu)$, where*

- *U is a non-empty finite set of objects,*
- *$I : U \rightarrow P(U)$ is an uncertainty function such that $x \in I(x)$ for any $x \in U$,*
- *$\nu : P(U) \times P(U) \rightarrow [0, 1]$ is a rough inclusion function.*

A set $X \subseteq U$ is *definable in AS* if and only if it is a union of some values of the uncertainty function.

The standard rough inclusion function ν_{SRI} defines the degree of inclusion between two subsets of U by

$$\nu_{SRI}(X, Y) = \begin{cases} \frac{card(X \cap Y)}{card(X)} & \text{if } X \neq \emptyset \\ 1 & \text{if } X = \emptyset. \end{cases} \quad (8)$$

[2] This problem will be investigated elsewhere.

The lower and the upper approximations of subsets of U are defined as follows.

Definition 5. *For any approximation space* $AS = (U, I, \nu)$, $0 \leq p < q \leq 1$, *and any subset* $X \subseteq U$ *the q-lower and the p-upper approximation of* X *in* AS *are defined by*

$$LOW_q(AS, X) = \{x \in U : \nu(I(x), X) \geq q\}, \tag{9}$$

$$UPP_p(AS, X) = \{x \in U : \nu(I(x), X) > p\}, \tag{10}$$

respectively.

Approximation spaces can be constructed directly from information systems or from information systems enriched by some similarity relations on attribute value vectors. The above definition generalises several approaches existing in the literature like those based on equivalence or tolerance indiscernibility relation as well as those based on exact inclusion of indiscernibility classes into concepts [1, 7].

Let us observe that the above definition of approximations is a special case of Definition 3. The granule approximation space $\mathcal{AS} = (GS, G, Tr)$ corresponding to $AS = (U, I, \nu)$ can be defined by

1. GS consisting of information granules being subsets of U (in particular, containing neighbourhoods that are values of the uncertainty function I) and of rough inclusion $\nu = \nu_{SRI}$.
2. $G = P(U)$.
3. $Tr[GS, G]$ consisting of exactly two systems: GS and GS^* such that
   - $Gran(GS^*) = G \cup \{(x, I(x)) : x \in U\}$;
   - the rough inclusion ν is extended by $\nu((x, I(x)), X) = \nu(I(x), X)$ for $x \in U$ and $X \subseteq U$.

Suppose the "make granule" operation $\oplus$ is defined by

$$\oplus(\{(x, \cdot) : x \in Z\}) = Z \text{ for any } Z \subseteq U.$$

Then for the approximation space AS and granule approximation space $\mathcal{AS}$ we have the following:

Proposition 1. *Let* $0 \leq p < q \leq 1$. *For any* $X \in P(U)$ *we have:*

$$LOW_q(AS, X) = LOW(\mathcal{AS}, GS^*, \oplus, q)(X), \tag{11}$$

$$UPP_p(AS, X) = UPP(\mathcal{AS}, GS^*, \oplus, p)(X). \tag{12}$$

5 Approximation Spaces with Complex Neighbourhoods

In this section we present approximation spaces that have more complex uncertainty functions. Such functions define complex neighbourhoods of objects, e.g., families of sets. This aspect is very important from the point of view of complex concepts approximation. A special case of complex uncertainty functions is such with values in $P^2(U)$, i.e., $I : U \longrightarrow P^2(U)$. Such uncertainty functions appear, e.g., in the case of the similarity based rough set approach. One can define $I(x)$ to be a family of all maximal cliques defined by the similarity relation which x belongs to (for details see Section 8).

We obtain the following definition of approximation space:

Definition 6. *A k-th order approximation space is any tuple* $AS = \left(U, I^k, \nu\right)$, *where*

- U *is a non-empty finite set of objects,*
- $I^k : U \rightarrow P^k(U)$ *is an uncertainty function,*
- $\nu : P(U) \times P(U) \rightarrow [0,1]$ *is a rough inclusion function.*

Let us note that up to the above definition there can be given different uncertainty functions for different levels of granulation. The inclusion function can be also defined in this way, however, in most cases we induce it recursively from ν. For example, in the case of set approximation by means of given approximation space $AS = (U, I^k, \nu)$ we are interested in an inclusion function $\nu^k : P^k(U) \times P(U) \rightarrow [0,1]$, defined recursively by the corresponding relation ν_p^k as follows

- $\nu_p^k(\mathcal{Y}, X)$ iff $\exists Y \in \mathcal{Y}\ \nu_p^{k-1}(Y, X)$,
- $\nu_p^1(Y, X)$ iff $\nu_p(Y, X)$,

for $X \subseteq U$ and $\mathcal{Y} \subseteq P^k(U)$.

The definition of set approximations for the k-th order approximation spaces depends on the way the values of uncertainty function are perceived. To illustrate this point of view we consider the following two examples: the complete perception of neighbourhoods and the perception defined by the intersection of the family $I(x)$. In the former case we consider a new definition of set approximations.

Definition 7. *Let* $0 \leq p < q \leq 1$. *For any k-th order approximation space* $AS = \left(U, I^k, \nu\right)$, ν^k *induced from* ν, *and any subset* $X \subseteq U$ *the q-lower and the p-upper approximation of* X *in* AS *are defined by*

$$LOW_q(AS, X) = \left\{x \in U : \nu^k \left(I^k(x), X\right) \geq q\right\}, \tag{13}$$

$$UPP_p(AS, X) = \left\{x \in U : \nu^k \left(I^k(x), X\right) > p\right\}, \tag{14}$$

respectively.

We can observe, that the approximation operations for those two cases are, in general, different.

Proposition 2. *Let us denote by* $AS^\cap = (U, I^\cap, \nu)$ *the approximation space obtained from the second order approximation space* $AS = \left(U, I^2, \nu\right)$ *assuming* $I^\cap(x) = \cap I^2(x)$ *for* $x \in U$. *We also assume that* $x \in Y$ *for any* $Y \in I^2(x)$ *and* $x \in U$. *Then, for* $X \subseteq U$ *we have*

$$LOW(AS, X) \subseteq LOW(AS^\cap, X) \subseteq X \subseteq UPP(AS^\cap, X) \subseteq UPP(AS, X). \tag{15}$$

One can check (in an analogous way as in Section 4) that the above definition of approximations is a special case of Definition 3.

6 Relation and Function Approximation

One can directly apply the definition of set approximation to relations. For simplicity, but without loss of generality, we consider binary relations only. Let $R \subseteq U \times U$ be a binary relation. We can consider approximation of R by an approximation space $AS = (U \times U, I, \nu)$ in an analogous way as in Definition 5:

$$LOW_q (AS, R) = \{(x, y) \in U \times U : \nu (I (x, y)), X) \geq q\}, \quad (16)$$

$$UPP_p (AS, R) = \{(x, y) \in U \times U : \nu (I (x, y), X) > p\}, \quad (17)$$

for $0 \leq p < q \leq 1$. This definition can be also easily extended to the case of complex uncertainty function as in Definition 7. However, the main problem is how to construct relevant approximation spaces, i.e., how to define uncertainty and inclusion functions. One of the solutions is the following uncertainty function

$$I(x, y) = I(x) \times I(y), \quad (18)$$

(assuming that one dimensional uncertainty function is given) and the standard inclusion, i.e., $\nu = \nu_{SRI}$.

Now, let us consider an approximation space $AS = (U, I, \nu)$ and a function $f : Dom \longrightarrow U$, where $Dom \subseteq U$. By $Graph(f)$ we denote the set $\{(x, f(x)) : x \in Dom\}$. We can easily see that if we apply the above definition of relation approximation to f (it is a special case of relation) then the lower approximation is almost always empty. Thus, we have to construct the relevant approximation space $AS^* = (U \times U, I^*, \nu^*)$ in different way, e.g., by extending the uncertainty function as well as the inclusion function on subsets of $U \times U$. We assume that the value $I^*(x, y)$ of the uncertainly function, called the neighbourhood (or the window) of (x, y), for $(x, y) \in U \times U$, is defined by

$$I^*(x, y) = I(x) \times I(y). \quad (19)$$

Next, we should decide how to define values of the inclusion function on pairs $(I^*(x, y), Graph(f))$, i.e., how to define the degree r to which the intersection

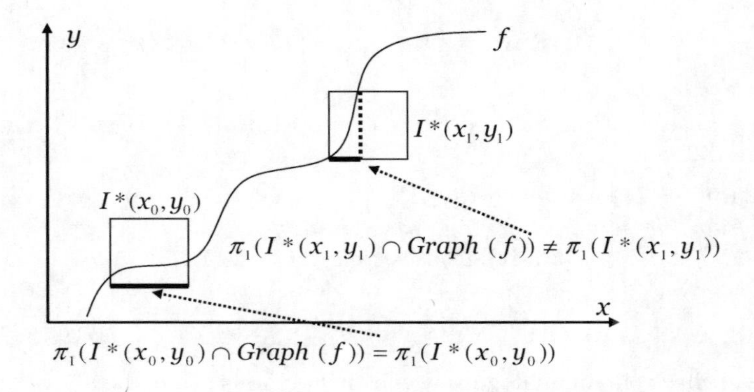

Fig. 1. Function approximation.

$I^*(x, y) \cap Graph(f)$ is included into $Graph(f)$. If $I(x) \cap Dom \neq \emptyset$, one can consider a ratio r of the fluctuation in $I(y)$ of the function $f \upharpoonright I(x)$ to the fluctuation in $I(x)$, where by $f \upharpoonright I(x)$ we denote the restriction of the function f to $I(x)$. If $r = 1$ then the window is in the lower approximation of $Graph(f)$; if $0 < r \leq 1$ then the window $I^*(x, y)$ is in the upper approximation of $Graph(f)$. If $I(x) \cap Dom = \emptyset$ then the degree r is equal to zero. Using the above intuition, we assume that the inclusion holds to degree one if the domain of $Graph(f)$ restricted to $I(x)$ is equal to $I(x)$. This can be formally defined by the following condition:

$$\pi_1(I^*(x, y) \cap Graph(f)) = \pi_1(I^*(x, y)) \tag{20}$$

where π_1 denotes the projection on the first coordinate. Condition (20) is equivalent to:

$$\forall x' \in I(x)\ \exists y' \in I(y)\ y' = f(x'). \tag{21}$$

Thus, the inclusion function ν^* for subsets $X, Y \subseteq U \times U$ is defined by

$$\nu^*(X, Y) = \begin{cases} \frac{card(\pi_1(X \cap Y))}{card(\pi_1(X))} & \text{if } \pi_1(X) \neq \emptyset \\ 1 & \text{if } \pi_1(X) = \emptyset. \end{cases} \tag{22}$$

Hence, the relevant inclusion function in approximation spaces for function approximations is such a function that does not measure the degree of inclusion of its arguments but their perceptions, represented in the above example by projections of corresponding sets. Certainly, one can chose another definition based, e.g., on the density of pixels (in the case of images) in the window that are matched by the function graph.

We have the following proposition:

Proposition 3. *Let $AS^* = (U \times U, I^*, \nu^*)$ be an approximation space with I^*, ν^* defined by (19), (22), respectively, and let $f : Dom \longrightarrow U$ where $Dom \subseteq U$. Then we have*

1. $(x, y) \in LOW_1(AS^*, Graph(f))$
   if and only if $\forall x' \in I(x)\ \exists y' \in I(y)\ y' = f(x')$;
2. $(x, y) \in UPP_0(AS^*, Graph(f))$
   if and only if $\exists x' \in I(x)\ \exists y' \in I(y)\ y' = f(x')$.

In the case of arbitrary parameters p, q satisfying $0 \leq p < q \leq 1$ we have

Proposition 4. *Let $AS^* = (U \times U, I^*, \nu^*)$ be an approximation space with I^*, ν^* defined by* (19), (22), *respectively, and let $f : Dom \longrightarrow U$ where $Dom \subseteq U$. Then we have*

1. $(x, y) \in LOW_q(AS^*, Graph(f))$ *if and only if*
   $card(\{x' \in I(x) : \exists y' \in I(y) : y' = f(x')\}) \geq q \cdot card(I(x))$;
2. $(x, y) \in UPP_p(AS^*, Graph(f))$ *if and only*
   $card(\{x' \in I(x) : \exists y' \in I(y) : y' = f(x')\}) > p \cdot card(I(x))$.

In our example, we define the inclusion degree between two subsets of Cartesian product using, in a sense, the inclusion degree between their projections. Hence, subsets of Cartesian products are perceived by projections.

Again, one can consider the definition of approximation space for function approximation as a special case of the granule approximation space introduced in Definition 2 with the non standard rough inclusion introduced in this section.

7 Concept Approximation by Granule Approximation Space

The granule approximation space $\mathcal{AS} = (GS, G, Tr)$ modelling the described process of concept approximations under fixed U^∞, $C \subseteq U^\infty$, sets of formulas (patterns) $\{pat(x) : x \in U\}$, $PATTERNS(AS, L, C)$ and their semantics $\|\cdot\|_{U^\infty}$ can be defined by

1. GS consisting of the following granules: $C \in P(U^\infty)$, the sample $U \subseteq U^\infty$, $C \cap U$, $U \setminus C$, sets $\|pat(x)\|_U$, defined by $pat(x)$ for any $x \in U$, and the rough inclusion $\nu = \nu_{SRI}$.
2. $G = \{C\}$.
3. The transition relation Tr extending GS to GS' and GS' to GS^*. $Gran(GS')$ is extended from $Gran(GS)$ by the following information granules: the sets $\|pat(x)\|_{U^\infty}$, defined by $pat(x)$ for any $x \in U^\infty$, sets $\|pat\|_{U^\infty}$, for $pat \in PATTERNS(AS, L, C)$. The rough inclusion is extended using estimations described above. GS^* is constructed as follows:
   - $Gran(GS^*) = G \cup$
     $\cup\{(x, \|pat(x)\|_{U^\infty}, \|pat\|_{U^\infty}) : x \in U^\infty \wedge pat \in PATTERNS(AS, L, C)\}$;
   - The rough inclusion ν is extended by:
     $$\nu((x, X, Y), C) = Conflict\_res(\{\nu_C(X, Y) : Y \in \mathcal{Y}\}, \{\nu_C(Y, C) : Y \in \mathcal{Y}\}) \quad (23)$$
     where $X, Y \subseteq U^\infty$, $\mathcal{Y} \subseteq P(U^\infty)$ are sets and the family of sets on which values of ν_C have been estimated in (1) and (2);
   - The operation "make granule" $\oplus$ satisfies the following constraint:
     $$\oplus\{(x, \cdot, \cdot) : x \in Z\} = Z \text{ for any } Z \subseteq C^\infty.$$

8 Relational Structure Granulation

In this section we discuss an important role that the relational structure granulation [5], [8] plays in searching for relevant patterns in approximate reasoning, e.g., in searching for relevant approximation patterns (see Section 2 and Figure 2).

Let us recall, that the uncertainty (neighbourhood) function of an approximation space forms basic granules of knowledge about the universe U. Let us consider the case where the values of neighbourhood function are from $P^2(U)$.

Assume that together with an information system $\mathbb{A} = (U, A)$ [1] there is also given a similarity relation τ defined on vectors of attribute values. This relation can be extended to objects. An object $y \in U$ is similar to a given object $x \in U$ if the attribute value vector on x is τ-similar to the attribute value vector on y. Now, consider a neighbourhood function defined by $I_{\mathbb{A},\tau}(x) = \{[y]_{IND(A)} : x\tau y\}$.

Neighbourhood functions cause a necessity of further granulation. Let us consider granulation of a relational structure R by neighbourhood functions. We would like to show that due to the relational structure granulation we obtain new information granules of more complex structure and, in the consequence, more general neighbourhood functions than those discussed above. Hence, basic granules of knowledge about the universe corresponding to objects become more complex.

Assume that a relational structure R and a neighbourhood function I are given. The aim is to define a new relational structure R_I called the I–granulation[3] of R. This is done by granulation of all components of R, i.e., relations and functions (see Section 6), by means of I.

The relational structure granulation defines new patterns that are created for pairs of objects together with some inclusion and closeness measures defined among them. Such patterns can be used for approximation of a target concept (or a concept on an intermediate level of hierarchical construction) over objects composed from pairs (x, y) interpreted, e.g., as parts of some more compound objects. Such compound objects are often called structured or complex objects.

Certainly, to induce approximations of high quality it is necessary to search for relevant patterns for concept approximation expressible in a given language. This problem, known as feature selection, is widely discussed in machine learning, pattern recognition, and data mining (see, e.g., [12–14]).

Let us consider an exemplary degree structure $D = ([0,1], \leq)$ and its granulation $D_{I_0} = (P([0,1]), \leq_{I_0})$ by means of an uncertainty function $I_0 : [0,1] \rightarrow P([0,1])$ defined by $I_0(x) = \{y \in [0,1] : \lfloor 10^k x \rfloor = \lfloor 10^k y \rfloor\}$, for some integer k, where for $X, Y \subseteq [0,1]$ we assume $X \leq_{I_0} Y$ iff $\forall x \in X, \forall y \in Y\ x \leq y$. Let $\{X_s, X_m, X_l\}$ be a partition of $[0,1]$ satisfying $x < y < z$ for any $x \in X_s$, $y \in X_m$, $z \in X_l$. Let $AS_0 = ([0,1], I_0, \nu)$ be an approximation space with the standard inclusion function ν. We denote by S, M, L the lower approximations of X_s, X_m, X_l in AS_0, respectively, and by S–M, M–L the boundary regions between X_s, X_m and X_m, X_l, respectively. Moreover, we assume $S, M, L \neq \emptyset$. In this way we obtain restriction of D_{I_0} to the structure $(Deg, \leq_{I_0})$, where $Deg = \{S, S\text{–}M, M, M\text{–}L, L\}$. Now, for a given (multi-sorted) structure $(U, P(U), [0,1], \leq, I_0, \nu)$, where $\nu : P(U) \times P(U) \rightarrow [0,1]$ is an inclusion function, we can define its I_0–granulation by

$$(U, P(U), Deg, \leq_{I_0}, \{\nu_d\}_{d \in Deg}) \tag{24}$$

where $Deg = \{S, S\text{–}M, M, M\text{–}L, L\}$ and $\nu_d(X, Y)$ iff $\nu_p(X, Y)$ for some p, d', such that $p \in d'$ and $d \leq_{I_0} d'$.

[3] In general, granulation is defined using the uncertainty function and the inclusion function from a given approximation space AS. For simplicity, we restrict our initial examples to I–granulation only.

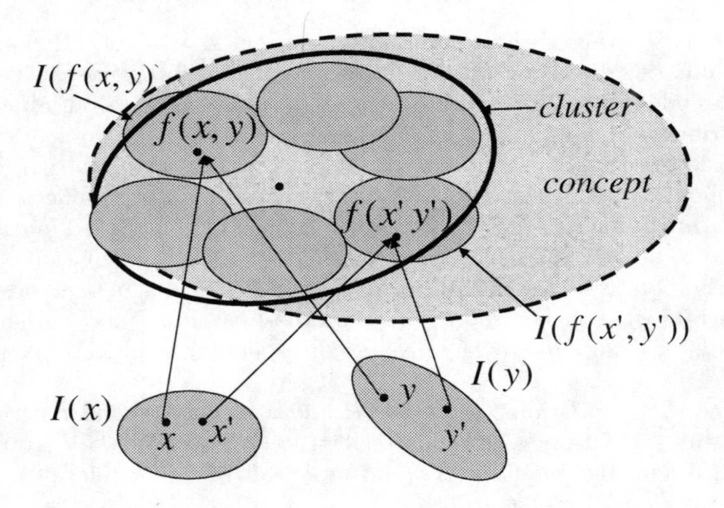

Fig. 2. Relational structure granulation.

Thus, for any object there is defined a neighbourhood specified by the value of uncertainty function from an approximation space. From those neighbourhoods some more relevant ones (e.g., for the considered concept approximation), should be discovered. Such neighbourhoods can be extracted by searching in a space of neighbourhoods generated from values of uncertainty function by applying to them some operations like generalisation operations, set theoretical operations (union, intersection), clustering, and operations on neighbourhoods defined by functions and relations in an underlying relational structure[4]. Figure 2 illustrates an exemplary scheme of searching for neighbourhoods (patterns, clusters) relevant for concept approximation. In this example, f denotes a function with two arguments from the underlying relational structure. Due to the uncertainty, we cannot perceive objects exactly but only by using available neighbourhoods defined by the uncertainty function from an approximation space. Hence, instead of the value $f(x, y)$ for a given pair of objects (x, y) one should consider a family of neighbourhoods $\mathcal{F} = \{I(f(x', y')) : (x', y') \in I(x) \times I(y)\}$. From this family $\mathcal{F}$ a subfamily $\mathcal{F}'$ of neighbourhoods can be chosen that consists of neighbourhoods with some properties relevant for approximation (e.g., neighbourhoods with sufficiently large support and/or confidence with respect to a given target concept). Next, the subfamily $\mathcal{F}'$ can be generalised to clusters that are relevant for the concept approximation, i.e., clusters sufficiently included into the approximated concept (see Figure 2). Observe also that the inclusion degrees can be measured by granulation of the inclusion function from the relational structure.

Now, let us present some examples illustrating information system granulation on searching for concept approximation.

Let $\mathbb{A} = (U, A)$ be an information system with universe U of objects described by some features from an attribute set A. In many cases, there are also given

[4] Relations from such structure may define relations between objects or their parts.

some additional relational structures on V_a, e.g., relations r_a defined for each attribute $a \in A$. Using $\{r_a\}_{a \in A}$ one can define relational structures over U in many different ways. For example, $r_a \subseteq V_a \times V_a$ can be a similarity relation for any $a \in A$. Such relations can be used to define similarity between objects $Sim \subseteq U \times U$, e.g., by $xSimy$ iff $r_a(a(x), a(y))$ for any $a \in A$. Then, for each $x \in U$ one can consider a relational structure R_x defined by a tolerance class $Sim(x) = \{y \in U : xSimy\}$ with relation Sim reduced to $Sim(x)$. In this way we obtain a new universe $U_{Rel} = \{R_x : x \in U\}$.

The trajectories of objects in time, $o(t)$, are basic objects in spatio-temporal reasoning and time series analysis. By restricting $o(t)$ to some time window of fixed length one can construct the basic relational structures forming objects of a new information system.

In the case of decision problems, i.e., when the initial information system $\mathbb{A}$ is a decision table, the task is to define relevant decisions for the considered problems and to define conditions making it possible to approximate new decision classes. These new decisions can be related to different tasks, e.g., to prediction in time series [13, 14], decision class approximations (robust with respect to deviations defined by similarities) [13, 14], and preference analysis [15]. For solving such tasks, the methods searching for relevant granulation of relational structures representing objects are very important.

Relational structures also arise in many pattern recognition problems as the result of (perception) representation of the object structure or data dimension reduction. Information granules considered in such applications are equal to elementary granules (indiscernibility classes) of information systems determined by some relational structures. Below we discuss this kind of granulation in more detail.

Let $\mathbb{A} = (U, A)$ be an information system where $a : U \longrightarrow V_a$ for any $a \in A$. Assume that $f : X \longrightarrow U$ and $b_a(x) = a(f(x))$ for any $x \in X$ and $a \in A$. Any such pair $(\mathbb{A}, f)$ defines a relational structure $\mathcal{R}$ on $Y = X \cup U \cup \bigcup_{a \in A} V_a$ with unary relations r_U, r_X, r_{V_a} and binary relations r_f and r_a, for $a \in A$, where $y \in r_X$ iff $y \in X$, $y \in r_U$ iff $y \in U$, $y \in r_{V_a}$ iff $y \in V_a$, for any $y \in Y$ and $a \in A$; $r_f \subseteq Y \times Y$ is a partial function defined by f, $r_a \subseteq Y \times Y$ is a partial function defined by a for any $a \in A$. Information granules over such a relational structure $\mathcal{R}$ are B-indiscernibility classes (elementary granules) of the information system $\mathcal{B} = (X, B)$ where $B = \{b_a : a \in A\}$. Elementary granules $[x]_{IND(B)}$ for $x \in X$, where $IND(B)$ is the B-indiscernibility relation, have the following property:

$$\begin{aligned} [x]_{IND(B)} &= f^{-1}(Inf_A^{-1}(Inf_A(f(x)))) \\ &= \bigcup \{y \in [f(x)]_{IND(A)} : f^{-1}(\{y\})\} \end{aligned} \tag{25}$$

where $f^{-1}(Z)$ denotes the counter-image of the set Z with respect to the function f.

The function f is used in pattern recognition [12, 14] applications to extract relevant parts of classified objects or to reduce the data dimension. Searching for relevant (with respect to some target concepts) granules of the form defined in (25) is performed by tuning f and A.

9 Conclusions

We discussed the problems of approximation space modelling for concept approximation. We presented consequences of the assumption that information granules involved in concept approximations are perceived by partial information about them. Illustrative examples of approximation spaces were also included. We emphasised the role of relational structure granulation in searching for relevant approximation spaces.

In our further work we would like to use the presented approach for modelling of searching processes for relevant approximation spaces using data and domain knowledge represented, e.g., in a natural language.

Acknowledgements

The research has been supported by the grant 3 T11C 002 26 from the Ministry of Scientific Research and Information Technology of the Republic of Poland.

References

1. Pawlak, Z.: Rough Sets: Theoretical Aspects of Reasoning about Data. Volume 9 of System Theory, Knowledge Engineering and Problem Solving. Kluwer Academic Publishers, Dordrecht, The Netherlands (1991)
2. Skowron, A., Stepaniuk, J.: Tolerance approximation spaces. Fundamenta Informaticae **27** (1996) 245–253
3. Polkowski, L., Skowron, A.: Rough mereology: A new paradigm for approximate reasoning. International Journal of Approximate Reasoning **15** (1996) 333–365
4. Polkowski, L., Skowron, A.: Towards adaptive calculus of granules. In Zadeh, L.A., Kacprzyk, J., eds.: Computing with Words in Information/Intelligent Systems. Physica-Verlag, Heidelberg, Germany (1999) 201–227
5. Skowron, A., Stepaniuk, J.: Information granules and rough-neural computing. In [10] 43–84
6. Lin, T.Y.: The discovery, analysis and representation of data dependencies in databases. In Polkowski, L., Skowron, A., eds.: Rough Sets in Knowledge Discovery 1: Methodology and Applications. Volume 18 of Studies in Fuzziness and Soft Computing. Physica-Verlag, Heidelberg, Germany (1998) 107–121
7. Ziarko, W.: Variable precision rough set model. Journal of Computer and System Sciences **46** (1993) 39–59
8. Peters, J.F., Skowron, A., Synak, P., Ramanna, S.: Rough sets and information granulation. In Bilgic, T., Baets, D., Kaynak, O., eds.: Tenth International Fuzzy Systems Association World Congress IFSA, Istanbul, Turkey, June 30-July 2. Volume 2715 of Lecture Notes in Artificial Intelligence. Springer-Verlag, Heidelberg, Germany (2003) 370–377
9. Skowron, A.: Toward intelligent systems: Calculi of information granules. Bulletin of the International Rough Set Society **5** (2001) 9–30
10. Pal, S.K., Polkowski, L., Skowron, A., eds.: Rough-Neural Computing: Techniques for Computing with Words. Cognitive Technologies. Springer-Verlag, Heidelberg, Germany (2003)

11. Zadeh, L.A.: A new direction in AI: Toward a computational theory of perceptions. AI Magazine **22** (2001) 73–84
12. Duda, R.O., Hart, P.E., Stork, D.G.: Pattern Classification. John Wiley & Sons, New York, NY (2001)
13. Hastie, T., Tibshirani, R., Friedman, J.: The Elements of Statistical Learning: Data Mining, Inference, and Prediction. Springer-Verlag, Heidelberg, Germany (2001)
14. Kloesgen, W., Żytkow, J., eds.: Handbook of Knowledge Discovery and Data Mining. Oxford University Press, New York, NY (2002)
15. Słowiński, R., Greco, S., Matarazzo, B.: Rough set analysis of preference-ordered data. In Alpigini, J.J., Peters, J.F., Skowron, A., Zhong, N., eds.: Third International Conference on Rough Sets and Current Trends in Computing RSCTC, Malvern, PA, October 14-16, 2002. Volume 2475 of Lecture Notes in Computer Science. Springer-Verlag, Heidelberg, Germany (2002) 44–59

The Rough Set Database System: An Overview

Zbigniew Suraj[1] and Piotr Grochowalski[2]

[1] Chair of Computer Science Foundations,
University of Information Technology and Management, Rzeszów, Poland
zsuraj@wsiz.rzeszow.pl
[2] Institute of Mathematics, Rzeszów University, Rzeszów, Poland
piotrg@univ.rzeszow.pl

Abstract. The paper describes the "Rough Sets Database System" (called in short the RSDS system) for the creation of a bibliography on rough sets and their applications. This database is the most comprehensive online rough sets bibliography currently available and is accessible from the RSDS website at http://www.rsds.wsiz.rzeszow.pl. This service has been developed to facilitate the creation of a rough sets bibliography for various types of publications. At the moment the bibliography contains over 1900 entries from more than 815 authors. It is possible to create the bibliography in *HTML* or *BibTeX* format. In order to broaden the service contents it is possible to append new data using a specially dedicated online form. After appending data online the database is updated automatically. If one prefers sending a data file to the database administrator, please be aware that the database is updated once a month. In the present version of the RSDS system, we have broadened information about the authors as well as the Statistics sections, which facilitates precise statistical analysis of the service. In order to widen the abilities of the RSDS system we added new features including:

- Detailed information concerning the software connected with the rough sets methodology.
- Scientific biographies of the outstanding people who work on rough sets.

Keywords: rough sets, fuzzy systems, neural networks, evolutionary computing, data mining, knowledge discovery, pattern recognition, machine learning, database systems.

1 Introduction

Rough set theory introduced by Professor Zdzisław Pawlak in 1981 [4] is a rapidly developing discipline of theoretical and applied computer science. It has become apparent during recent years that a bibliography on this subject is urgently needed as a tool for both the efficient research, and the use of the rough set theory.

The aim of this paper is to present the RSDS system for the creation of a bibliography on the rough sets and their applications; papers on other topics

J.F. Peters and A. Skowron (Eds.): Transactions on Rough Sets III, LNCS 3400, pp. 190–201, 2005.

have been included whenever the rough sets play a decisive role for the presented matters, or in case outstanding applications of the rough set theory are discussed. In compiling the bibliography for the database we faced the fact that many important ideas and results are contained in reports, theses, memos, etc.; we have done our best to arrive at a good compromise between the completeness of the bibliography and the restriction to generally available publications.

Another difficulty we had to cope with was sometimes a completely different spelling of the authors' names. The following served among others as the sources for the bibliography database:

- The publications in the journal Fundamenta Informaticae and others.
- Books on the rough set theory and applications as well as proceedings of the international conferences on the rough sets mentioned in the references at the end of this article.
- Other materials available at the International Rough Set Society.
- Queries for the "rough sets" on the website of the database.

The service has been developed in order to facilitate the creation of the rough sets bibliography, for various types of publications. At present, it is possible to create the bibliography in *HTML* or *BibTeX* format. In order to broaden the service contents, it is possible to append new data using a specially dedicated form. After appending data online, the database is updated automatically. If one prefers sending a data file to the database administrator, please be aware that the database is updated once a month.

The following types of publications are available in the service: an article, book, booklet, inbook, incollection, inproceedings, manual, mastersthesis, phdthesis, proceedings, techreport, unpublished.

This paper is organized as follows. Section 2 presents an overview of information used to characterize the RSDS system. The future plans for the RSDS system are discussed in section 3. Conclusions are given in section 4.

2 A Description of the RSDS System

2.1 Home Page

Having the system activated, the English version of the home page appears on a display. The service menu comprises several options making it possible to move around the whole system. The menu includes the following: *Home page*, *Login*, *Append*, *Search*, *Download*, *Send*, *Write to us*, *Statistics*, *Help*, *Software*, *People*.

2.2 Appending Data

In order to append new data to the bibliographic database at first one should go to the *Append* section. Before appending new data, a user must log into the system using a special form. That form includes the fields allowing to insert a user's *id* and user's *password*. If a user inserts a wrong user's *id* or *password* then a message describing the mistake displays on the screen. If a user wants

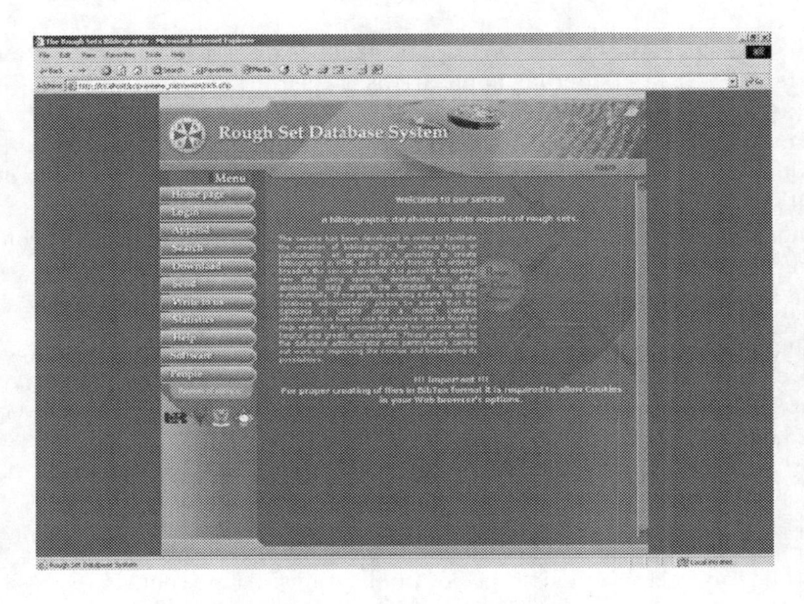

Fig. 1. The starting page of the RSDS system.

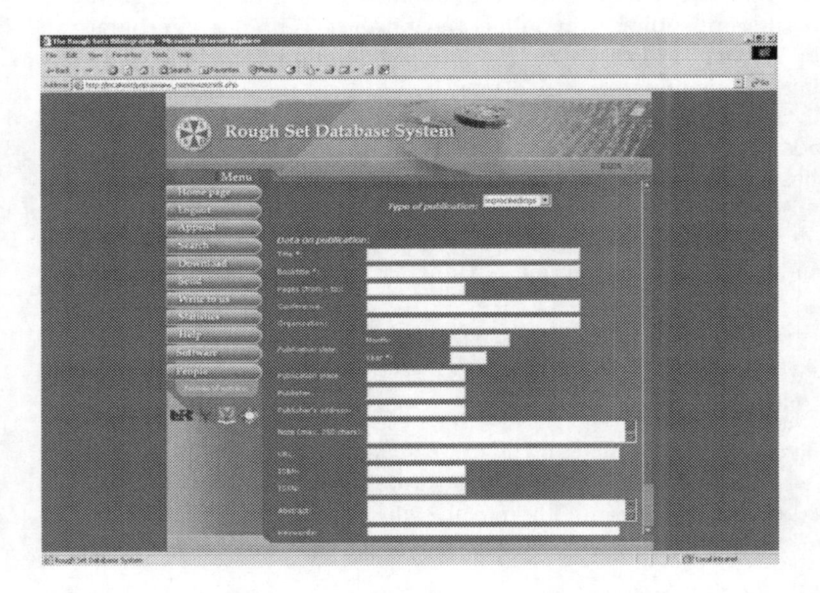

Fig. 2. A screenshot for appending (online) new data to the RSDS system.

to log in at first, then one must use another special form, by clicking the *First login* button. That form includes the fields allowing to insert: a user's *name* and user's *surname*, *e-mail*, user's *id* and user's *password*. Next, the entered data is verified in the database. If all data is correct, the account for the user is created at once, and then one is logged into the system automatically with new data number in the database. This information helps to implement the existing data changes. After logging in, a special form displays, and it is then possible to type new data (excluding data about the authors; another form is used for to entering the authors' data). After providing information concerning the publication type, the form is updated with fields which require inputting specific data. The fields required for proceeding with data input are marked with a star symbol (*). The required fields are described by the *BibTeX* format specification. After entering the required data, it is possible to proceed with the next step – which is inputting the authors' or editors' data. The authors' data inputting form will be reloaded until the last author's data is entered. A user decides when to stop entering the authors' data by clicking the *End* button. For the entered data verification, all the data is displayed prior to sending to the database. After accepting, the data is sent.

There follows the list concerning the publication types together with descriptions of the fields required.

Publication	Description
article	An article from a journal. *Fields required:* author, title, journal, year. *Optional fields:* volume, number, pages, month, note.
book	A book with the known, given publisher. *Fields required:* author or editor, title, publisher, year. *Optional fields:* volume, series, address, edition, month, note.
booklet	Printed and bound matter, whilst the publisher is unknown. *Fields required:* title. *Optional fields:* author, address, month, year, note.
inbook	A part of a book, could be a chapter or given pages. *Fields required:* author or editor, title, chapter or pages, publisher, year. *Optional fields:* volume, series, address, edition, month, note.
incollection	A part of a book with its own title. *Fields required:* author, title, book title, publisher, year. *Optional fields:* editor, chapter, pages, address, month, note.
inproceedings	An article published in the conference proceedings. *Fields required:* author, title, book title, year. *Optional fields:* author, organization, publisher, address, month, note.
manual	Manual or documentation. *Fields required:* title. *Optional fields:* author, organization, address, edition, month, year, note.

mastersthesis	M.Sc. thesis. *Fields required:* author, title, school, year. *Optional fields:* address, month, note.
phdthesis	Ph.D. thesis. *Fields required:* author, title, school, year. *Optional fields:* address, month, note.
proceedings	Proceedings. *Fields required:* title, year. *Optional fields:* editor, publisher, organization, address, month, note.
techreport	Report, usually with a given number, being periodically issued. *Fields required:* author, title, institution, year. *Optional fields:* number, address, month, note.
unpublished	A document with a given author and title data, unpublished. *Fields required:* author, title, note. *Optional fields:* month, year.

Explanation on existing fields.

address	Publisher's address.
author	Forename and surname of an author (or authors).
booktitle	Title of a quoted in part book.
chapter	The chapter number.
edition	Issue, edition.
editor	Forenames and surnames of editors. If there also exists the field "author", the "editor" denotes the editor of a larger entity, of which the quoted work is a part.
institution	Institution publishing the printed matter.
journal	Journal's name.
month	Month of issue or completion of the manuscript.
note	Additional information useful to a reader.
number	The journal or the report number. Usually journals are being identified by providing their year and a number within the year of issue. A report, in general, has only a number.
organization	Organization supporting a conference.
pages	One or more page numbers; for example 42–11, 7,41,73–97.
publisher	Publisher's name.
school	University, college, where the thesis be submitted.
series	The name of a book series. If one quotes a book from a given series, then the "title" field denotes the title of a book, whilst the "series" field should contain the entire series' name.

title	The title of the work.
volume	The periodical's or the book's volume.
year	Year of issue. In case of unpublished work, the year of completing writing. Year only in number format e.g. 1984.
URL	The WWW Universal Resource Locator that points to the item being referenced. This often is used for technical reports to point to the ftp site where the postscript source of the report is located.
ISBN	The International Standard Book Number.
ISSN	The International Standard Serial Number. Used to identify a journal.
abstract	An abstract of a publication.
keywords	Key words attached to a publication. This can be used for searching a publication.

Note: All data must be appended in the Latin alphabet – without national marks.

2.3 Searching Data

In order to search the database search go to the *Search* section. An alphabetical search and advanced search options are possible. An advanced search allows to find the bibliographic data according to different combinations of the following fields: a title, author, editor, journal's name, conference's name, publisher, keywords and abstract. The searched data can be even more precisely defined, using the options of narrowing the search according to a year and type of a publication. When using an alphabetic search (according to authors) we can see three icons, next to the author's name, which mean:

- An icon representing a magnifier – information about an author.
- An icon representing an envelope – e-mail address of an author.
- An icon representing a house – www site of an author.

The required data can be sent to a user in two formats: at first *HTML* format data is displayed and then, after clicking the *BibTeX* link, the *BibTeX* format file is created. It is then possible to download the created file with the *.*tex* extension (with an entered file's name). There are two file downloading methods that have been applied for a user's comfort:

- Saving directly to a user's local hard drive.
- Sending the file as an e-mail attachment.

Before editing data into the database, a user must log in the system and then, using the *Search* option, display *HTML* format chosen data on the screen. After clicking the *Edit* button, a special form displays with existing data and it is then possible to edit this data. A user decides when to stop editing the data by clicking the *Submit entry* button. After that, the data is sent to a database administrator. If a user logs in as an *administrator*, then there is possibility of deleting the redundant data from the database.

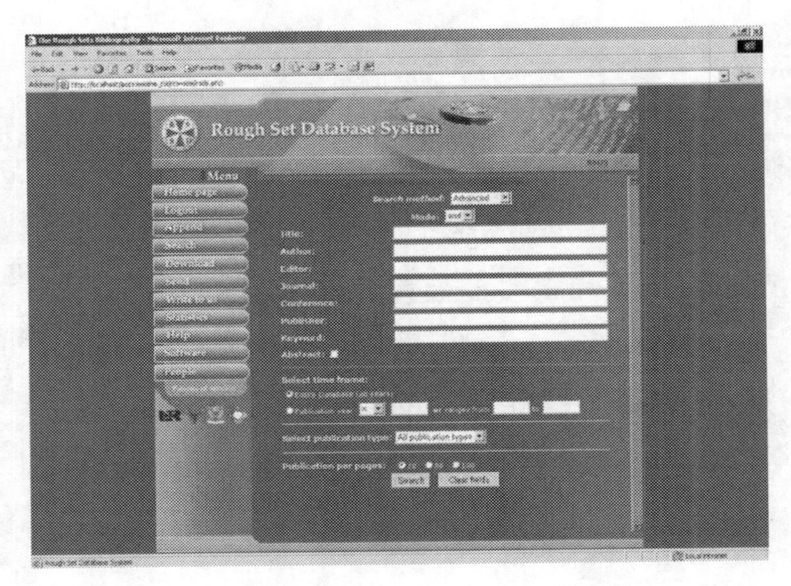

Fig. 3. A screenshot for an advanced search data.

2.4 Downloading a File

Before saving the data to a file, one must specify the operating system for which the file with the entered file's name and the **.tex* extension should be created. The two methods for downloading the file in the RSDS system have been implemented:

- Save to a user's local hard drive.
- Send as an e-mail attachment.

2.5 Sending a File

It is possible to submit a file with the bibliographic data to the database administrator, who has the software allowing to append automatically large data to the database. In order to do it, one can use a special dedicated form. Submissions in the form of *BibTeX* files are preferred. Please note that submissions are not immediately available, as the database is updated in batches once a month.

2.6 Write to Us

This section allows to write and send the comments concerning the service to us by using a special dedicated form. This form includes a field for comments and the *Send* button. Any comments about our service will be helpful and greatly appreciated. Please post them to the database administrator who permanently carries out work on improving the service and broadening its possibilities.

2.7 Statistics

The section *Statistics* includes statistical information concerning the system. This section has been divided into three pages:

- Page 1 contains information describing: the number of users' visits to the site, number of the authors in a database, as well as the dynamic diagrams related to: the number and types of publications in a database, number and years of published works.
- Page 2 contains the statistics depicting the countries from which the users of the service come.
- Page 3 contains the monthly and yearly statistics of visits to the service.

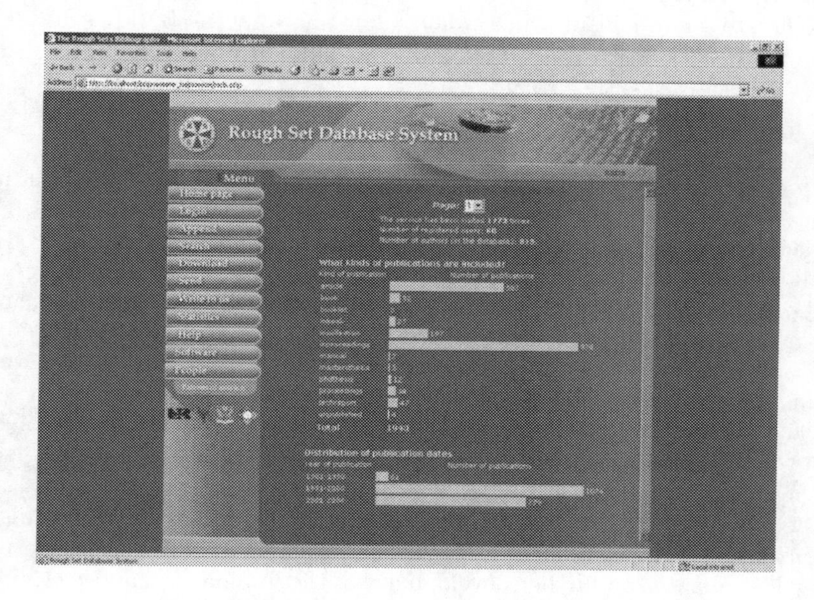

Fig. 4. Some statistics of the RSDS system.

2.8 Software

In this section there is an opportunity to search for information concerning the software connected with the rough sets. There are two ways of searching demanded information:

- A search through submitting an application's name.
- An alphabetic search.

Apart from a description of the searched software, the RSDS system allows to download a searched application.

2.9 People

This section, entitled *People*, allows to find the biographies of outstanding people concerned with the rough sets methodology. After having found a person, this person's biography, e-mail, the name and address of the academy the person works at, is available.

3 The Future Plans for the RSDS System

We plan to extend the RSDS system possibilities to the following, among others:

- Implementation of new methods of a search data.
- Adding the database FAQ.
- Updating the bibliographic database.
- To create an "intelligent" advisor for the users.
- To create a mechanism which would allow to search the Internet in order to gain new data for the base, using different mechanisms.

4 Conclusions

We have created the RSDS system by applying some of the basic possibilities of computer tools, which are needed in the bibliography database systems. These tools support a user in searching for the rough sets publications as well as downloading files in a natural and very effective way.

The main point of the RSDS system is its extensibility: it is easy to connect other methods and tools to the system.

It seems that our system presented in the paper is a professional database system which offers a stable platform for extensions.

Using the RSDS system is an opportunity for an information exchange between the scientists and practitioners who are interested in the foundations and applications of the rough sets.

The developers of the RSDS system hope that the increase in the dissemination of the results, methods, theories and applications based on the rough sets, will stimulate the further development of the foundations and methods for real-life applications in the intelligent systems.

For future updating of the bibliography we will appreciate receiving all forms of help and advice. In particular, we would like to become aware of any relevant contributions which are not referred to in this bibliography database. All submitted material will also be included in the RSDS system.

The RSDS system has been designed and implemented at Rzeszow University, and installed at University of Information Technology and Management in Rzeszow.

The RSDS system runs on any computer with any operating system connected to the Internet. The service is based on the Internet Explorer 6.0, Opera 7.03 as well as Mozilla 1.3 (correct operation requires the web browser with the accepting cookie option enabled).

Acknowledgments

We are grateful to Professor Andrzej Skowron from Warsaw University (Poland) for stimulating discussions about this work and providing the bibliographic data for the RSDS system. We wish to thank our colleagues from the Logic Group of Warsaw University for their help in searching for data. Our deepest thanks go to the staff of the Chair of Computer Science Foundations of University of Information Technology and Management in Rzeszow as well as the staff of the Computer Science Laboratory of Rzeszow University for their support and their infinite patience.

The research has been partially supported by the grant 3 T11C 005 28 from Ministry of Scientific Research and Information Technology of the Republic of Poland.

References

1. J.J. Alpigini, J.F. Peters, A. Skowron, N. Zhong (Eds.): Rough Sets and Current Trends in Computing. Third International Conference, RSCTC 2002, Malvern, PA, USA, October 14–16, 2002, Lecture Notes in Artificial Intelligence 2475, Springer-Verlag, Berlin 2002.
2. Cios, K.J., Pedrycz, W., Swiniarski, R.W.: Data Mining. Methods for Knowledge Discovery. Kluwer Academic Publishers, Dordrecht 1998.
3. Demri, S.P., Orlowska, E.,S.: Incomplete Information: Structure, Inference, Complexity. Springer-Verlag, Berlin 2002.
4. L. Czaja (Ed.): Proceedings of the Workshop on Concurrency, Specification and Programming, CS&P'2003, Vol. 1–2, Czarna, Poland, September 25–27, 2003, Warsaw University, 2003.
5. S. Hirano, M. Inuiguchi, S. Tsumoto (Eds.): Proceedings of International Workshop on Rough Set Theory and Granular Computing (RSTGC'2001), Matsue, Shimane, Japan, May 20–22, 2001. Bulletin of International Rough Set Society 5/1–2 (2001).
6. M. Inuiguchi, S. Miyamoto (Eds.): Proceedings of the First Workshop on Rough Sets and Kansei Engineering in Japan, December 14–15, 2002, Tokyo, Bulletin of International Rough Set Society 7/1–2 (2003).
7. M. Inuiguchi, S. Hirano, S. Tsumoto (Eds.): Rough Set Theory and Granular Computing, Studies in Fuzziness and Soft Computing, Vol. 125, Springer-Verlag, Berlin 2003.
8. T.Y. Lin (Ed.): Proceedings of the Third International Workshop on Rough Sets and Soft Computing (RSSC'94). San Jose State University, San Jose, California, USA, November 10–12, 1994.
9. T.Y. Lin, A.M. Wildberger (Eds.): Soft Computing: Rough Sets, Fuzzy Logic, Neural Networks, Uncertainty Management, Knowledge Discovery. Simulation Councils, Inc., San Diego, CA, 1995.
10. T.Y. Lin (Ed.): Proceedings of the Workshop on Rough Sets and Data Mining at 23rd Annual Computer Science Conference, Nashville, Tenessee, March 2, 1995.
11. T.Y. Lin (Ed.): Journal of the Intelligent Automation and Soft Computing 2/2 (1996) (special issue).
12. T.Y. Lin (Ed.): International Journal of Approximate Reasoning 15/4 (1996) (special issue).
13. T.Y. Lin, N. Cercone (Eds.): Rough Sets and Data Mining. Analysis of Imprecise Data. Kluwer Academic Publishers, Dordrecht 1997.
14. E. Orlowska (Ed.): Incomplete information: Rough set analysis. Physica-Verlag, Heidelberg, 1997.
15. S.K. Pal, A. Skowron (Eds.): Rough Fuzzy Hybridization: A New Trend in Decision-Making. Springer-Verlag, Singapore 1999.

16. Pawlak, Z.: Rough Sets – Theoretical Aspects of Reasoning about Data. Kluwer Academic Publishers, Dordrecht 1991.
17. S.K. Pal, L. Polkowski, A. Skowron (Eds.): Rough-Neural Computing. Techniques for Computing with Words. Springer-Verlag, Berlin 2004.
18. W. Pedrycz, J.F. Peters (Eds.): Computational Intelligence in Software Engineering. World Scientific Publishing, Singapore 1998.
19. Polkowski, L.: Rough Sets. Mathematical Foundations. Springer-Verlag, Berlin 2002.
20. L. Polkowski, A. Skowron (Eds.): Rough Sets in Knowledge Discovery 1. Methodology and Applications. Physica-Verlag, Heidelberg 1998.
21. L. Polkowski, A. Skowron (Eds.): Rough Sets in Knowledge Discovery 2. Applications, Case Studies and Software Systems. Physica-Verlag, Heidelberg 1998.
22. L. Polkowski, A. Skowron (Eds.): Proceedings of the First International Conference on Rough Sets and Current Trends in Computing (RSCTC'98), Warsaw, Poland, 1998, Lecture Notes in Artificial Intelligence 1424, Springer-Verlag, Berlin 1998.
23. L. Polkowski, S. Tsumoto, T.Y. Lin (Eds.): Rough Set Methods and Applications. New Developments in Knowledge Discovery in Information Systems. Physica-Verlag, Heidelberg, 2000.
24. A. Skowron, S.K. Pal (Eds.): Pattern Recognition Letters 24/6 (2003) (special issue).
25. A. Skowron, M. Szczuka (Eds.): Proceedings of an International Workshop on Rough Sets in Knowledge Discovery and Soft Computing, RSDK, Warsaw, Poland, April 5–13, 2003, Warsaw University, 2003.
26. R. Slowinski, J. Stefanowski (Eds.): Proceedings of the First International Workshop on Rough Sets: State of the Art. And Perspectives. Kiekrz – Poznan, Poland, September 2–4, 1992.
27. R. Slowinski (Ed.): Intelligent Decision Support – Hanbook of Applications and Advances of the Rough Sets Theory. Kluwer Academic Publishers, Dordrecht 1992.
28. R. Slowinski, J. Stefanowski (Eds.), Foundations of Computing and Decision Sciences 18/3–4 (1993) 155–396 (special issue).
29. Z. Suraj (Ed.): Proceedings of the Sixth International Conference on Soft Computing and Distributed Processing (SCDP 2002), June 24–25, 2002, Rzeszow, Poland, University of Information Technology and Management Publisher, Rzeszow 2002.
30. Z. Suraj, P. Grochowalski: The Rough Sets Database System: An Overview in: S. Tsumoto, Y.Y. Yao (Eds.): Bulletin of International Rough Set Society 1/2 (2003). First Workshop on Rough Sets and Kansei Engineering, Tokyo, Japan, December 14–15, 2002.
31. Z. Suraj, P. Grochowalski: The Rough Sets Database System: An Overview in: S. Tsumoto, R. Slowinski, J. Komorowski, J.W. Grzymala-Busse (Eds.): Rough Sets and Current Trends in Computing. Fourth International Conference, RSCTC 2004, Uppsala, Sweden, June 1–5, 2004, Lecture Notes in Artificial Intelligence 3066, Springer-Verlag, Berlin 2004.
32. S. Tsumoto, S. Kobayashi, T. Yokomori, H. Tanaka, and A. Nakamura (Eds.): Proceedings of the Fourth International Workshop on Rough Sets, Fuzzy Sets and Machine Discovery (RSFD'96). The University of Tokyo, November 6–8, 1996.
33. S. Tsumoto (Ed.): Bulletin of International Rough Set Society 1/1 (1996).
34. S. Tsumoto (Ed.): Bulletin of International Rough Set Society 1/2 (1997).
35. S. Tsumoto, Y.Y. Yao, and M. Hadjimichael (Eds.): Bulletin of International Rough Set Society 2/1 (1998).

36. S. Tsumoto, R. Slowinski, J. Komorowski, J.W. Grzymala-Busse (Eds.): Rough Sets and Current Trends in Computing. Fourth International Conference, RSCTC 2004, Uppsala, Sweden, June 1–5, 2004, Lecture Notes in Artificial Intelligence 3066, Springer-Verlag, Berlin 2004.
37. P.P. Wang (Ed.): Proceedings of the International Workshop on Rough Sets and Soft Computing at Second Annual Joint Conference on Information Sciences (JCIS'95), Wrightsville Beach, North Carolina, 28 September – 1 October, 1995.
38. P.P. Wang (Ed.): Proceedings of the Fifth International Workshop on Rough Sets and Soft Computing (RSSC'97) at Third Annual Joint Conference on Information Sciences (JCIS'97). Duke University, Durham, NC, USA, Rough Set & Computer Science 3, March 1–5, 1997.
39. G. Wang, Q. Liu, Y.Y. Yao, A. Skowron (Eds.). Rough Sets, Fuzzy Sets, Data Mining, ad Granular Computing. 9th International Conference, RSFDGrC 2003, Chongqing, China, May 26–29, 2003, Lecture Notes in Artificial Intelligence 2639, Springer-Verlag, Berlin 2003.
40. W. Ziarko (Ed.): Proceedings of the Second International Workshop on Rough Sets and Knowledge Discovery (RSKD'93). Banff, Alberta, Canada, October 12–15, 1993.
41. W. Ziarko (Ed.): Rough Sets, Fuzzy Sets and Knowledge Discovery (RSKD'93). Workshops in Computing, Springer-Verlag & British Computer Society, London, Berlin 1994.
42. W. Ziarko (Ed.): Computational Intelligence: An International Journal 11/2 (1995) (special issue).
43. W. Ziarko (Ed.): Fundamenta Informaticae 27/2–3 (1996) (special issue)
44. W. Ziarko, Y.Y. Yao (Eds.): Rough Sets and Current Trends in Computing. Second International Conference, RSCTC 2000, Banff, Canada, October 16–19, 2000, Lecture Notes in Artificial Intelligence 2005, Springer-Verlag, Berlin 2001.

Rough Sets and Bayes Factor

Dominik Ślęzak

Department of Computer Science, University of Regina,
Regina, SK, S4S 0A2 Canada
slezak@uregina.ca

Abstract. We present a novel approach to understanding the concepts of the theory of rough sets in terms of the inverse probabilities derivable from data. It is related to the Bayes factor known from the Bayesian hypothesis testing methods. The proposed Rough Bayesian model (RB) does not require information about the prior and posterior probabilities in case they are not provided in a confirmable way. We discuss RB with respect to its correspondence to the original Rough Set model (RS) introduced by Pawlak and Variable Precision Rough Set model (VPRS) introduced by Ziarko. We pay a special attention on RB's capability to deal with multi-decision problems. We also propose a method for distributed data storage relevant to computational needs of our approach.

Keywords: Rough Sets, Probabilities, Bayes Factor.

1 Introduction

The theory of *rough sets*, introduced by Pawlak in 1982 (see [10] for references), is a methodology of dealing with uncertainty in data. The idea is to approximate the *target concepts* (*events*, *decisions*) using the classes of *indiscernible* objects (in case of qualitative data – the sets of records with the same values for the features under consideration). Every concept X is assigned the *positive*, *negative*, and *boundary regions* of data, where X is *certain*, *impossible*, and *possible but not certain*, according to the data based information.

The above principle of rough sets has been extended in various ways to deal with practical challenges. Several extensions have been proposed as related to the data based probabilities. The first one, *Variable Precision Rough Set* (*VPRS*) model [24] proposed by Ziarko, softens the requirements for certainty and impossibility using the grades of the *posterior probabilities.* Pawlak [11, 12] begins research on the connections between rough sets and *Bayesian reasoning*, in terms of operations on the *posterior*, *prior*, and *inverse probabilities.* In general, one can observe a natural correspondence between the fundamental notions of rough sets and statistics, where a hypothesis (target concept X_1) can be verified *positively*, *negatively* (in favor of the *null hypothesis*, that is a *complement* concept X_0), or *undecided*, under the given evidence [4, 15].

Decision rules resulting from the rough set algorithms can be analyzed both with respect to the data derived posterior probabilities (certainty, accuracy)

J.F. Peters and A. Skowron (Eds.): Transactions on Rough Sets III, LNCS 3400, pp. 202–229, 2005.

and the inverse probabilities (coverage), like in machine learning methods [8, 9]. However, only the posterior probabilities decide about membership of particular cases to the positive/negative/boundary regions – the inverse probabilities are usually used just as the optimization parameters, once the posterior probabilities are good enough (cf. [16, 21]). Several rough set approaches to evaluation of "goodness" of the posterior probabilities were developed, like, for example, the above-mentioned parameter-controlled grades in VPRS. In [25] it is proposed to relate those grades to the prior probabilities of the target concepts. This is, actually, an implicit attempt to relate the rough set approximations with the Bayesian hypothesis testing, where comparison of the posterior and prior probabilities is crucial [1, 20].

In [18] a simplified probabilistic rough set model is introduced, where a given new object, supported by the evidence E, is in the positive region of X_1, if and only if the posterior probability $Pr(X_1|E)$ is greater than the prior probability $Pr(X_1)$. It is equivalent to inequality $Pr(E|X_1) > Pr(E|X_0)$, which means that the observed evidence is more probable assuming hypothesis X_1 than its complement X_0 (cf. [19]). This is the first step towards handling rough sets in terms of the inverse probabilities. Its continuation [17] points at relevance to the *Bayes factor* [4, 6, 15, 20], which takes the form of the following ratio:

$$B_0^1 = \frac{Pr(E|X_1)}{Pr(E|X_0)} \tag{1}$$

The Bayes factor is a well known example of comparative analysis of the inverse probabilities, widely studied not only by philosophers and statisticians but also within the domains of machine learning and data mining. Such analysis is especially important with regards to the rule confirmation and interestingness measures (cf. [5, 7]), considered also in the context of the rough set based decision rules [3, 12, 21]. In this paper we develop the foundations for *Rough Bayesian* (*RB*) model, which defines the rough-set-like positive/negative/boundary regions in terms of the Bayes factor. In this way, the inverse probabilities, used so far in the analysis of the decision rules obtained from the rough set model, become to be more directly involved in the specification of this model itself.

Operating with B_0^1 provides two major advantages, similar to those related to its usage in Bayesian reasoning and probabilistic data mining methods. Firstly, the posterior probabilities are not always derivable directly from data, in a reliable way (see e.g. Example 3 in Subsection 2.2). In such cases, information is naturally represented by means of the inverse probabilities corresponding to the observed evidence conditioned by the states we want to verify, predict, or approximate. Within the domain of statistical science, there is a discussion whether (and in which cases) the inverse probabilities can be combined with the prior probabilities using the *Bayes rule*. If it is allowed, then the proposed RB model can be rewritten in terms of the posterior probabilities and starts to work similarly as VPRS. However, such translation is impossible in case we can neither estimate the prior probabilities from data nor define them using background knowledge. Then, the data based inverse probabilities remain the only basis for constructing the rough-set-like models.

The second advantage of basing a rough set model on the Bayes factor is that the inverse probabilities provide clearer ability of comparing likelihoods of concepts. In the probabilistic rough set extensions proposed so far, the posterior probabilities $Pr(X_1|E)$ are compared to constant parameters [24, 25] or to the prior probabilities $Pr(X_1)$ of *the same* target concepts [18, 19]. A direct comparison of probabilities like $Pr(X_1|E)$ and $Pr(X_0|E)$ would not have too much sense, especially when the prior probabilities of X_1 and X_0 differ significantly. Comparison of the inverse probabilities $Pr(E|X_1)$ and $Pr(E|X_0)$ is more natural, as corresponding to relationship between the ratios of the posterior and prior probabilities for different concepts:

$$\frac{Pr(E|X_1)}{Pr(E|X_0)} = \frac{Pr(X_1|E)/Pr(X_1)}{Pr(X_0|E)/Pr(X_0)} \tag{2}$$

It shows that the analysis of the Bayes factor is equivalent to comparison of the ratios of the gain in probabilistic belief for X_1 and X_0 under the evidence E (cf. [18]). Therefore, the RB model can be more data sensitive than the approaches based on the posterior probabilities, especially for the problems with more than two target concepts to be approximated. RB is well comparable to Bayesian hypothesis testing methods, where B_0^1 is regarded as a summary of the evidence for X_1 against X_0 provided by the data, and also as the ratio of the posterior and prior odds. Finally, RB may turn out to be applicable to the problems where the prior probabilities are dynamically changing, remain unknown, or simply undefinable. Although we do not discuss such situations, we refer to the reader's experience and claim that it may be really the case for real-life data sets.

The article is organized as follows: Section 2 presents non-parametric probabilities derivable from data, with their basic intuitions and relations. It also contains basic information about the way of applying the Bayes factor in decision making. Section 3 presents the original rough set approach in terms of the posterior and, what is novel, the inverse data based probabilities. Then it focuses on foundations of the VPRS model and corresponding extensions of rough sets. Section 4 introduces the Rough Bayesian approach related to the Bayes factors calculated for the pairs of decision classes. The proposed model is compared with VPRS, both for the cases of two and more target concepts. In particular, it requires extending the original formulation of VPRS onto the multi-target framework, which seems to be a challenging task itself. Section 5 includes a short note on an alternative, distributed way of representing the data for the needs of the Rough Bayesian model. Section 6 summarizes the article and discusses directions for further research.

2 Data and Probabilities

2.1 Data Representation

In [10] it was proposed to represent the data as an *information system* $\mathbb{A} = (U, A)$, where U denotes the *universe* of *objects* and each *attribute* $a \in A$ is identified with function $a : U \to V_a$, for V_a denoting the set of values of a.

U	a_1	a_2	a_3	a_4	a_5	d
u_1	1	1	0	1	2	0
u_2	0	0	0	2	2	0
u_3	2	2	2	1	1	1
u_4	0	1	2	2	2	1
u_5	2	1	1	0	2	0
u_6	2	2	2	1	1	1
u_7	0	1	2	2	2	0
u_8	2	2	2	1	1	1
u_9	2	2	2	1	1	1
u_{10}	0	0	0	2	2	0

U	a_1	a_2	a_3	a_4	a_5	d
u_{11}	1	2	0	0	2	0
u_{12}	1	1	0	1	2	1
u_{13}	0	1	2	2	2	1
u_{14}	2	1	1	0	2	0
u_{15}	2	2	2	1	1	0
u_{16}	1	1	0	1	2	1
u_{17}	1	1	0	1	2	0
u_{18}	2	1	1	0	2	0
u_{19}	2	2	2	1	1	1
u_{20}	2	2	2	1	1	1

Fig. 1. Decision system $\mathbb{A} = (U, A \cup \{d\})$, $U = \{u_1, \ldots, u_{20}\}$, $A = \{a_1, \ldots, a_5\}$. Decision d induces classes $X_0 = \{u_1, u_2, u_5, u_7, u_{10}, u_{11}, u_{14}, u_{15}, u_{17}, u_{18}\}$ and $X_1 = \{u_3, u_4, u_6, u_8, u_9, u_{12}, u_{13}, u_{16}, u_{19}, u_{20}\}$.

Each subset $B \subseteq A$ induces a partition over U with classes defined by grouping together the objects having identical values of B. We obtain the partition space U/B, called the *B-indiscernibility relation* $IND_{\mathbb{A}}(B)$, where elements $E \in U/B$ are called the *B-indiscernibility classes* of objects.

Information provided by $\mathbb{A} = (U, A)$ can be applied to approximate the *target events* $X \subseteq U$ by means of the elements of U/B, $B \subseteq A$. We can express such targets using a distinguished attribute $d \notin A$. Given $V_d = \{0, \ldots, r-1\}$, we define the sets $X_k = \{u \in U : d(u) = k\}$. We refer to such extended information system $\mathbb{A} = (U, A \cup \{d\})$ as to a *decision system*, where d is called the *decision attribute*, and the sets X_k are referred to as the *decision classes*.

Elements of U/B correspond to *B-information vectors* $w \in V_B$ – collections of *descriptors* (a, v), $a \in B$, $v \in V_a$. They are obtained using *B-information function* $B : U \to V_B$ where $B(u) = \{(a, a(u)) : a \in B\}$.

Example 1. Consider $\mathbb{A} = (U, A \cup \{d\})$ in Fig. 1 and $B = \{a_1, a_3\}$. B-information vector $\{(a_1, 2), (a_3, 2)\}$ corresponds to conjunction of conditions $a_1 = 2$ and $a_3 = 2$, which is satisfied by the elements of $E = \{u_3, u_6, u_8, u_9, u_{15}, u_{19}, u_{20}\}$. In other words, $B(u_i) = \{(a_1, 2), (a_3, 2)\}$ holds for $i = 3, 6, 8, 9, 15, 19, 20$. □

2.2 Types of Probabilities

Let us assume that events X_k are labelled with the prior probabilities $Pr(X_k)$, $\sum_{l=0}^{r-1} Pr(X_l) = 1$, $r = |V_d|$. It is reasonable to claim that any X_k is likely to occur and that its occurrence is not certain – otherwise, we would not consider such an event as worth dealing with. The same can be assumed about indiscernibility classes $E \in U/B$, $B \subseteq A$, in terms of probabilities of their occurrence in data $\mathbb{A} = (U, A \cup \{d\})$. We can express such requirements as follows:

$$0 < Pr(X_k) < 1 \text{ and } 0 < Pr(E) < 1 \tag{3}$$

Let us also assume that each class E is labelled with the posterior probabilities $Pr(X_k|E)$, $\sum_{l=0}^{r-1} Pr(X_l|E) = 1$, which express beliefs that X_k will occur under the evidence corresponding to E.

Remark 1. We can reconsider probabilities in terms of the attribute-value conditions. For instance, if $k = 1$ and $E \in U/B$ groups the objects satisfying conditions $a_1 = 2$ and $a_3 = 2$, then we can write $Pr(d = 1)$ instead of $Pr(X_1)$, and $Pr(d = 1|a_1 = 2, a_3 = 2)$ instead of $Pr(X_1|E)$. □

In machine learning and data mining [5, 8, 9], the posterior probabilities correspond to the certainty (accuracy, precision) factors. One can also compare prior and posterior knowledge to see whether a new evidence (satisfaction of conditions) increases or decreases the belief in a given event (membership to a given decision class). This is, actually, the key idea of Bayesian reasoning [1, 20], recently applied also to rough sets [18, 19]. The easiest way of the data based prior and posterior probability estimation is the following:

$$Pr(X_k|E) = \frac{|X_k \cap E|}{|E|} \quad \text{and} \quad Pr(X_k) = \frac{|X_k|}{|U|} \tag{4}$$

Example 2. In case of Fig. 1, we get $Pr(d = 1|a_1 = 2, a_3 = 2) = 6/7$, which estimates our belief that objects satisfying $a_1 = 2$ and $a_3 = 2$ belong to X_1. It seems to increase the belief in X_1 with respect to $Pr(d = 1) = 1/2$. □

One can also use the inverse probabilities $Pr(E|X_k)$, $\sum_{E \in U/B} Pr(E|X_k) = 1$, which express a likelihood of the evidence E under the assumption about X_k [20]. The posterior probabilities are then derivable by using the Bayes rule. For instance, in case of $\mathbb{A} = (U, A \cup \{d\})$ with two decision classes, we have:

$$Pr(X_1|E) = \frac{Pr(E|X_1)Pr(X_1)}{Pr(E|X_0)Pr(X_0) + Pr(E|X_1)Pr(X_1)} \tag{5}$$

Remark 2. If we use estimations $Pr(E|X_k) = |X_k \cap E|/|X_k|$, then (5) provides the same value of $Pr(X_k|E)$ as (4). For instance, $Pr(d = 1|a_1 = 2, a_3 = 2) = (3/5 \cdot 1/2)/(1/10 \cdot 1/2 + 3/5 \cdot 1/2) = 6/7$. □

In some cases estimation (4) can provide us with invalid values of probabilities. According to the Bayesian principles, it is then desirable to combine the inverse probability estimates with the priors expressing background knowledge, not necessarily derivable from data. We can see it in the following short study:

Example 3. Let us suppose that X_1 corresponds to a rare but important target event like, e.g., some medical pathology. We are going to collect the cases supporting this event very accurately. However, we are not going to collect information about all the "healthy" cases as X_0. In the medical data sets we can rather expect the 50:50 proportion between positive and negative examples. It does not mean, however, that $Pr(X_1)$ should be estimated as $1/2$. It is questionable whether the posterior probabilities $Pr(X_1|E)$ should be derived from such data using estimation with $|E|$ in denominator – it is simply difficult to accept that $|E|$ is calculated as the non-weighted sum of $|E \cap X_0|$ and $|E \cap X_1|$. □

2.3 Bayes Factor

Example 3 shows that in some situations the posterior probabilities are not derivable from data in a credible way. In this paper, we do not claim that this is a frequent or infrequent situation and we do not focus on any specific real-life data examples. We simply show that it is still possible to derive valuable knowledge basing only on the inverse probabilities, in case somebody cannot trust or simply does not know priors and posteriors.

Our idea to refer to the inverse probabilities originates from the notion of *Bayes factor*, which compares the probabilities of the observed evidence E (indiscernibility class or, equivalently, conjunction of conditional descriptors) under the assumption concerning a hypothesis X_k (decision class) [4, 6, 15, 20]. In case of systems with two decision classes, the Bayes factor takes the form of B_0^1 defined by equation (1). It refers to the posterior and prior probabilities, as provided by equation (2). However, we can restrict to the inverse probabilities, if we do not know enough about the priors and posteriors occurring in (2).

The Bayes factor can be expressed in various ways, depending on the data type [15]. In case of decision table real valued conditional attributes, it would be defined as the ratio of probabilistic densities. For symbolic data, in case of more than two decision classes, we can consider pairwise ratios

$$B_l^k = \frac{Pr(E|X_k)}{Pr(E|X_l)} \tag{6}$$

for $l \neq k$, or ratios of the form

$$B_{\not k}^k = \frac{Pr(E|X_k)}{Pr(E|\neg X_k)} \qquad \text{where } \neg X_k = \bigcup_{l \neq k} X_l \tag{7}$$

In [6], it is reported that twice of the logarithm of B_0^1 is on the same scale as the deviance test statistics for the model comparisons. The value of B_0^1 is then used to express a degree of belief in hypothesis X_1 with respect to X_0, as shown in Fig. 2. Actually, the scale presented in Fig. 2 is quite widely used by statisticians while referring to the Bayes factors. We can reconsider this way of hypothesis verification by using the significance threshold $\varepsilon_1^0 \geq 0$ in the following criterion:

$$X_1 \text{ is verified } \varepsilon_1^0\text{-positively, if and only if } Pr(E|X_0) \leq \varepsilon_1^0 Pr(E|X_1) \tag{8}$$

For lower values of $\varepsilon_1^0 \geq 0$, the positive hypothesis verification under the evidence $E \in U/B$ requires more significant advantage of $Pr(E|X_1)$ over $Pr(E|X_0)$. Actually, it is reasonable to assume that $\varepsilon_1^0 \in [0, 1)$. This is because for $\varepsilon_1^0 = 1$, we simply cannot decide between X_1 and X_0 (cf. [18]) and for $\varepsilon_1^0 > 1$ one should rather consider X_0 instead of X_1 (by switching X_0 with X_1 and using possibly different $\varepsilon_0^1 \in [0, 1)$ in (8)). Another special case, $\varepsilon_1^0 = 0$, corresponds to *infinitely strong* evidence for hypothesis X_1, yielding $Pr(E|X_0) = 0$. This is the reason why we prefer to write $Pr(E|X_0) \leq \varepsilon_1^0 Pr(E|X_1)$ instead of $B_0^1 \geq 1/\varepsilon_1^0$ in (8).

The B_0^1 ranges proposed in [6]	Corresponding $2\log B_0^1$ ranges	Corresponding ε_1^0 ranges based on (8)	Evidence for X_1 described as in [6]
less than 1	less than 0	more than 1	negative (support X_0)
1 to 3	0 to 2	0.3 to 1	barely worth mentioning
3 to 12	2 to 5	0.1 to 0.3	positive
12 to 150	5 to 10	0.01 to 0.1	strong
more than 150	more than 10	less than 0.01	very strong

Fig. 2. The Bayes factor significance scale proposed in [6], with the corresponding ranges for $\varepsilon_1^0 \geq 0$ based on criterion (8). The values in the third column are rounded to better express the idea of working with inequality $Pr(E|X_0) \leq \varepsilon_1^0 Pr(E|X_1)$.

3 Rough Sets

3.1 Original Model in Terms of Probabilities

In Subsection 2.1 we mentioned that decision systems can be applied to approximation of the target events by means of indiscernibility classes. A method of such data based approximation was proposed in [10], as the theory of *rough sets*. Given $\mathbb{A} = (U, A \cup \{d\})$, $B \subseteq A$, and $X_k \subseteq U$, one can express the main idea of rough sets in the following way: The *B-positive*, *B-negative*, and *B-boundary rough set regions* (abbreviated as *RS-regions*) are defined as

$$\begin{aligned} \mathcal{POS}_B(X_k) &= \bigcup\{E \in U/B : Pr(X_k|E) = 1\} \\ \mathcal{NEG}_B(X_k) &= \bigcup\{E \in U/B : Pr(X_k|E) = 0\} \\ \mathcal{BND}_B(X_k) &= \bigcup\{E \in U/B : Pr(X_k|E) \in (0,1)\} \end{aligned} \tag{9}$$

$\mathcal{POS}_B(X_k)$ is an area of the universe where the occurrence of X_k is *certain*. $\mathcal{NEG}_B(X_k)$ covers an area where the occurrence of X_k is *impossible*. Finally, $\mathcal{BND}_B(X_k)$ defines an area where the occurrence of X_k is *possible* but *uncertain*. The boundary area typically covers large portion of the universe, if not all. If $\mathcal{BND}_B(X_k) = \emptyset$, then X_k is *B-definable*. Otherwise, X_k is a *B-rough set*.

Example 4. For $\mathbb{A} = (U, A \cup \{d\})$ from Fig. 1 and $B = \{a_1, a_3\}$, we obtain

$$\begin{aligned} \mathcal{POS}_B(X_1) &= \emptyset \\ \mathcal{NEG}_B(X_1) &= \{u_2, u_5, u_{10}, u_{14}, u_{18}\} \\ \mathcal{BND}_B(X_1) &= \{u_1, u_3, u_4, u_6, u_7, u_8, u_9, u_{11}, u_{12}, , u_{13}, u_{15}, u_{16}, u_{17}, u_{19}, u_{20}\} \end{aligned}$$

As we can see, X_1 is a B-rough set in this case. □

The following basic result emphasizes the decision-making background behind rough sets. To be sure (enough) about X_k we must be convinced (enough) against any alternative possibility X_l, $l \neq k$. This is a feature we would like to keep in mind while discussing extensions of the original rough set model, especially when the word "*enough*" becomes to have a formal mathematical meaning.

Proposition 1. *Let $\mathbb{A} = (U, A \cup \{d\})$, $X_k \subseteq U$, and $B \subseteq A$ be given. We have equality*

$$\mathcal{POS}_B(X_k) = \bigcap_{l:l \neq k} \mathcal{NEG}_B(X_l) \tag{10}$$

Proof. This is because $Pr(X_k|E) = 1$ holds, if and only if for every X_l, $l \neq k$, there is $Pr(X_l|E) = 0$. □

The RS-regions can be interpreted also by means of the inverse probabilities, which was not discussed so far in the literature. We formulate it as a theorem to emphasize its intuitive importance, although the proof itself is trivial. In particular, this result will guide us towards drawing a connection between rough sets and the Bayes factor based testing described in Subsection 2.3.

Theorem 1. *Let $\mathbb{A} = (U, A \cup \{d\})$ and $B \subseteq A$ be given. Let the postulate (3) be satisfied. Consider the k-th decision class $X_k \subseteq U$. For any $E \in U/B$ we obtain the following characteristics:*

$$\begin{array}{ll} E \subseteq \mathcal{POS}_B(X_k) \Leftrightarrow \forall_{l:\, l \neq k} Pr(E|X_l) = 0 \\ E \subseteq \mathcal{NEG}_B(X_k) \Leftrightarrow Pr(E|X_k) = 0 \\ E \subseteq \mathcal{BND}_B(X_k) \Leftrightarrow Pr(E|X_k) > 0 \wedge \exists_{l:\, l \neq k} Pr(E|X_l) > 0 \end{array} \tag{11}$$

Proof. Beginning with the positive region, we have

$$\forall_{l:\, l \neq k} Pr(E|X_l) = 0 \Leftrightarrow \forall_{l:\, l \neq k} Pr(X_l|E) = \frac{Pr(E|X_l) Pr(X_l)}{Pr(E)} = 0$$

Since $\sum_{l=0}^{r-1} Pr(X_l|E) = 1$, the above is equivalent to $Pr(X_k|E) = 1$. For the negative region we have

$$Pr(E|X_k) = 0 \Leftrightarrow Pr(X_k|E) = \frac{Pr(E|X_k) Pr(X_k)}{Pr(E)} = 0$$

Finally, for the boundary region, we can see that

$$\begin{array}{l} Pr(E|X_k) > 0 \Leftrightarrow Pr(X_k|E) > 0 \\ \exists_{l:\, l \neq k} Pr(E|X_l) > 0 \Leftrightarrow \exists_{l:\, l \neq k} Pr(X_l|E) > 0 \Leftrightarrow Pr(X_k|E) < 1 \end{array}$$

All the above equivalences follow from the postulate (3), the Bayes rule, and the fact that probability distributions sum up to 1. □

Remark 3. The formula for $\mathcal{POS}_B(X_k)$ can be also rewritten as follows:

$$E \subseteq \mathcal{POS}_B(X_k) \Leftrightarrow Pr(E|X_k) > 0 \wedge \forall_{l:\, l \neq k} Pr(E|X_l) = 0 \tag{12}$$

$Pr(E|X_k) > 0$ is redundant since conditions (3) and equalities $Pr(E|X_l) = 0$, $l \neq k$, force it anyway. However, the above form including $Pr(E|X_k) > 0$ seems to be more intuitive. □

Theorem 1 enables us to think about the rough set regions as follows (please note that interpretation of the positive region is based on characteristics (12)):

1. Object u belongs to $\mathcal{POS}_B(X_k)$, if and only if the vector $B(u) \in V_B$ is *likely* to occur under the assumption that u supports the event X_k and *unlikely* to occur under the assumption that it supports any alternative event X_l, $l \neq k$.

2. Object u belongs to $\mathcal{NEG}_B(X_k)$, if and only if the vector $B(u) \in V_B$ is *unlikely* to occur under the assumption that u supports the event X_k.
3. Object u belongs to $\mathcal{BND}_B(X_k)$, if and only if the vector $B(u) \in V_B$ is *likely* to occur under the assumption that u supports X_k but this is also the case for some alternative events X_l, $l \neq k$.

As a conclusion, the rough set model can be formulated without using the prior and posterior probabilities. It means that in case of rough sets we do not need any kind of background knowledge even if the only probabilities reasonably represented in data are the inverse ones. The rough set regions are not influenced by the changes of the prior probabilities. We do not even need the existence of those probabilities – postulate (3) could be then read as a requirement that every decision class under consideration is supported by some objects and that some alternative decisions are supported as well.

3.2 Variable Precision Rough Set Model

Although presented by means of probabilities in the previous subsection, the rough set regions were originally defined using simple set theoretic notions, namely inclusion (for positive regions) and empty intersection (for negative regions). Probabilities then occurred in various works [2, 13, 18, 21–24] to enable the initial rough set model to deal more flexibly with the indiscernibility classes *almost included* and *almost excluded* from the target events. In other words, one can use the probabilities to soften the requirements for certainty and impossibility in the rough set model. It provides better applicability to practical problems, where even a slight increase or decrease of probabilities can be as important as expecting them to equal 1 or 0.

The first method using non-0-1 posterior probabilities in rough sets is the *Variable Precision Rough Set* (*VPRS*) model [24]. It is based on parameter-controlled grades of the posterior probabilities in defining the approximation regions. The most general asymmetric VPRS model definition relies on the values of the *lower* and *upper limit certainty thresholds* α and β[1]. To deal with systems with many decision classes, we will understand α and β as vectors

$$\alpha = (\alpha_0, \ldots, \alpha_{r-1}) \text{ and } \beta = (\beta_0, \ldots, \beta_{r-1}) \tag{13}$$

where α_k and β_k refer to decision classes X_k, $k = 0, \ldots, r-1$, $r = |V_d|$. Let system $\mathbb{A} = (U, A \cup \{d\})$ and $B \subseteq A$ be given. The *VPRS-regions* are defined as follows:

$$\begin{array}{ll} \mathcal{POS}_B^{\beta}(X_k) & = \bigcup\{E \in U/B : Pr(X_k|E) \geq \beta_k\} \\ \mathcal{NEG}_B^{\alpha}(X_k) & = \bigcup\{E \in U/B : Pr(X_k|E) \leq \alpha_k\} \\ \mathcal{BND}_B^{\alpha,\beta}(X_k) & = \bigcup\{E \in U/B : Pr(X_k|E) \in (\alpha_k, \beta_k)\} \end{array} \tag{14}$$

[1] Originally, the notation l, u was proposed for the lower and upper certainty thresholds. We use α, β instead to avoid coincidence with notation for decision classes, where l may occur as the index, and with notation for elements of the universe, often denoted by $u \in U$.

The β-positive region $\mathcal{POS}_B^\beta(X_k)$ is defined by the upper limit parameter β_k, which reflects the *least acceptable* degree of $Pr(X_k|E)$. Intuitively, β_k represents the desired level of improved prediction accuracy when predicting the event X_k based on the information that event E occurred. The α-negative region $\mathcal{NEG}_B^\alpha(X_k)$ is controlled by the lower limit parameter α_k. It is an area where the occurrence of the set X_k is *significantly*, as expressed in terms of α_k, less likely than *usually*. Finally, the (α,β)-boundary region $\mathcal{BND}_B^{\alpha,\beta}(X_k)$ is a "gray" area where there is no sufficient bias towards neither X_k nor its complement.

As proposed in [25], we suggest the following inequalities to be satisfied while choosing the VPRS parameters for particular decision systems:

$$0 \leq \alpha_k < Pr(X_k) < \beta_k \leq 1 \tag{15}$$

The reason lays in interpretation of the VPRS-regions. In case of $\mathcal{POS}_B^\beta(X_k)$, the improvement of prediction accuracy is possible only if $\beta_k > Pr(X_k)$. In case of $\mathcal{NEG}_B^\alpha(X_k)$, the word "*usually*" should be understood as the prior probability $Pr(X_k)$. Therefore, we should be sure to choose $\alpha_k < Pr(X_k)$ to obtain practically meaningful results.

Another explanation of (15) is that without it we could obtain $E \in U/B$ contained in negative or positive VPRS-regions of *all* decision classes in the same time. This is obviously an extremely unwanted situation since we should not be allowed to verify negatively all hypotheses in the same time. We could be uncertain about all the decision classes, which would correspond to the boundary regions equal to U for all decision classes, but definitely not negatively (positively) convinced about all of them.

Remark 4. The above explanation of postulate (15) should be followed by recalling the meaning of Proposition 1 in the previous subsection. Here, it should be connected with the following *duality* property of the VPRS regions [24, 25]: For $\mathbb{A} = (U, A \cup \{d\})$, $V_d = \{0,1\}$, let us consider the limits satisfying equalities

$$\alpha_0 + \beta_1 = \alpha_1 + \beta_0 = 1 \tag{16}$$

Then we have the following identities:

$$\mathcal{POS}_B^\beta(X_0) = \mathcal{NEG}_B^\alpha(X_1) \text{ and } \mathcal{POS}_B^\beta(X_1) = \mathcal{NEG}_B^\alpha(X_0) \tag{17}$$

Further, equations (16) can be satisfied consistently with (15). This is because $0 \leq \alpha_0 < Pr(X_0) < \beta_0 \leq 1$ is equivalent to $1 \geq \beta_1 > Pr(X_1) > \alpha_1 \geq 0$. □

Identities (17) are important for understanding the nature of rough-set-like decision-making and its correspondence to the statistical hypothesis testing. It would be desirable to extend them onto the case of more than two decision classes, although it is not obvious how to approach it.

3.3 Further Towards the Inverse Probabilities

In the context of machine learning, the VPRS model's ability to flexibly control approximation regions' definitions allows for efficient capturing probabilistic relations existing in data. However, as we discussed before, the estimates of the posterior probabilities are not always reliable. Below we rewrite VPRS in terms of the inverse probabilities, just like we did in case of the original RS-regions.

Proposition 2. *Let $\mathbb{A} = (U, A \cup \{d\})$, $V_d = \{0,1\}$, and $B \subseteq A$ be given. Consider parameters $\alpha = (\alpha_0, \alpha_1)$, $\beta = (\beta_0, \beta_1)$ such that conditions (15) and (16) are satisfied. Then we have inequalities*

$$\begin{array}{l} \mathcal{POS}_B^{\beta}(X_0) = \mathcal{NEG}_B^{\alpha}(X_1) = \bigcup\{E \in U/B : Pr(E|X_1) \leq \varepsilon_0^1 Pr(E|X_0)\} \\ \mathcal{POS}_B^{\beta}(X_1) = \mathcal{NEG}_B^{\alpha}(X_0) = \bigcup\{E \in U/B : Pr(E|X_0) \leq \varepsilon_1^0 Pr(E|X_1)\} \end{array} \tag{18}$$

where coefficients $\varepsilon_0^1, \varepsilon_1^0$ defined as

$$\varepsilon_0^1 = \frac{\alpha_1 Pr(X_0)}{\beta_0 Pr(X_1)} \quad \textit{and} \quad \varepsilon_1^0 = \frac{\alpha_0 Pr(X_1)}{\beta_1 Pr(X_0)} \tag{19}$$

belong to the interval $[0,1)$.

Proof. Consider α_0 and β_1 such that $\alpha_0 + \beta_1 = 1$ (the case of α_{01} and β_0 can be shown analogously). We want to prove

$$Pr(X_1|E) \geq \beta_1 \Leftrightarrow \alpha_0 \geq Pr(X_0|E) \Leftrightarrow Pr(E|X_0) \leq \frac{\alpha_0 Pr(X_1)}{\beta_1 Pr(X_0)} Pr(E|X_1) \tag{20}$$

We know that two first above inequalities are equivalent. By combining them together, we obtain the third equivalent inequality (its equivalence to both $Pr(X_1|E) \geq \beta_1$ and $Pr(X_0|E) \leq \alpha_0$ can be easily shown by contradiction):

$$Pr(X_1|E) \geq \beta_1 \Leftrightarrow \alpha_0 \geq Pr(X_0|E) \Leftrightarrow \alpha_0 Pr(X_1|E) \geq \beta_1 Pr(X_0|E) \tag{21}$$

It is enough to apply identity (2) to realize that the third inequalities in (20) and (21) are actually the same ones. It remains to show that $\varepsilon_0^1, \varepsilon_1^0 \in [0,1)$. It follows from the assumption (15). For instance, we have inequality $\varepsilon_0^1 < 1$ because $\alpha_1 < Pr(X_1)$ and $Pr(X_0) < \beta_0$. □

The above correspondence can be used to draw a connection between VPRS and the statistical reasoning models. It is possible to refer inequality $Pr(E|X_0) \leq \varepsilon_1^0 Pr(E|X_1)$, rewritten as

$$B_0^1 \geq \frac{\beta_1 Pr(X_0)}{\alpha_0 Pr(X_1)} \tag{22}$$

to the Bayes factor based statistical principles discussed e.g. in [20]. However, the remaining problem is that we need to use $Pr(X_0)$ and $Pr(X_1)$ explicitly in (22), which is often too questionable from practical point of view.

In [18], another version of VPRS is considered. The idea is to detect *any* decrease/increase of belief in decision classes. The rough set region definitions proposed in [18] look as follows:

$$\begin{array}{ll} \mathcal{POS}_B^*(X_1) & = \bigcup\{E \in U/B : Pr(X_1|E) > Pr(X_1)\} \\ \mathcal{NEG}_B^*(X_1) & = \bigcup\{E \in U/B : Pr(X_1|E) < Pr(X_1)\} \\ \mathcal{BND}_B^*(X_1) & = \bigcup\{E \in U/B : Pr(X_1|E) = Pr(X_1)\} \end{array} \tag{23}$$

which can be equivalently expressed as follows (cf. [19]):

$$\begin{aligned}\mathcal{POS}^*_B(X_1) &= \bigcup\{E \in U/B : Pr(E|X_1) > Pr(E|X_0)\}\\ \mathcal{NEG}^*_B(X_1) &= \bigcup\{E \in U/B : Pr(E|X_1) < Pr(E|X_0)\}\\ \mathcal{BND}^*_B(X_1) &= \bigcup\{E \in U/B : Pr(E|X_1) = Pr(E|X_0)\}\end{aligned} \quad (24)$$

This simple interpretation resembles the VPRS characteristics provided by Proposition 2, for ε_0^1 and ε_1^0 tending to 1. It also corresponds to the limit $\varepsilon_1^0 \to 1$ applied to inequality $Pr(E|X_0) \leq \varepsilon Pr(E|X_1)$ in the Bayes factor criterion (8). We could say that according to the scale illustrated by Fig. 2 in Subsection 2.3 the region $\mathcal{POS}^*_B(X_1)$ gathers any, even *barely worth mentioning but still positive*, evidence for X_1. It is completely opposite to the original rough set model. Indeed $\mathcal{POS}_B(X_1)$ gathers, according to Theorem 1, only *infinitely strong* evidence for X_1. Let us summarize it as follows:

1. Object u belongs to $\mathcal{POS}_B(X_1)$ (to $\mathcal{POS}^*_B(X_1)$), if and only if X_1 can be *positively* verified under the evidence of $B(u)$ at the maximal (minimal) level of statistical significance, expressed by (8) for $\varepsilon_1^0 = 0$ ($\varepsilon_1^0 \to 1$).
2. Object u belongs to $\mathcal{NEG}_B(X_1)$ (to $\mathcal{NEG}^*_B(X_1)$), if and only if X_1 can be *negatively* verified under the evidence of $B(u)$ at the maximal (minimal) level of significance (we replace X_0 and X_1 and use ε_0^1 instead of ε_1^0 in (8)).
3. Object u belongs to $\mathcal{BND}_B(X_1)$ (to $\mathcal{BND}^*_B(X_1)$), if and only if it is not sufficient to verify X_1 *neither positively nor negatively* at the maximal (minimal) level of significance under the evidence of $B(u)$.

As a result, we obtain two models – the original rough set model and its modification proposed in [18] – which refer to the Bayes factor scale in two marginal ways. They also correspond to special cases of VPRS, as it is rewritable by means of the inverse probabilities following Proposition 2. They both do not need to base on the prior or posterior probabilities, according to characteristics (11) and (24). From this perspective, the main challenge of this article is to fill the gap between these two opposite cases of involving the Bayes factor based methodology into the theory of rough sets. An additional challenge is to extend the whole framework to be able to deal with more than two target events, as it was stated by Theorem 1 in case of the original RS-regions.

4 Rough Bayesian Model

4.1 RB for Two Decision Classes

After recalling basic methods for extracting probabilities from data and the VPRS-like extensions of rough sets, we are ready to introduce a novel extension based entirely on the inverse probabilities and the Bayes factor. To prepare the background, let us still restrict to systems with two decision classes. Using statistical terminology, we interpret classes X_1 and X_0 as corresponding to the *positive* and *negative verification* of some *hypothesis*.

Let us refer to the above interpretation of the RS-regions originating from substitution of $\varepsilon_1^0 = 0$ to the criterion (8). By using positive values of ε_1^0, we can soften the requirements for the positive/negative verification. In this way

$\varepsilon_1^0 \in [0,1)$ plays a role of a *degree of the significance approximation*. We propose the following model related to this degree. We will refer to this model as to the *Rough Bayesian* model because of its relationship to Bayes factor (cf. [17]).

Definition 1. *Let* $\mathbb{A} = (U, A \cup \{d\})$, $V_d = \{0,1\}$, *and* $B \subseteq A$ *be given. For any parameters* $\varepsilon = (\varepsilon_0^1, \varepsilon_1^0)$, $\varepsilon_0^1, \varepsilon_1^0 \in [0,1)$, *we define the* B-positive, B-negative, *and* B-boundary rough Bayesian regions *(abbreviated as* RB-regions*) as follows (the regions for* X_0 *are defined analogously):*

$$\begin{aligned} \mathcal{BAYPOS}_B^{\varepsilon}(X_1) &= \bigcup\{E \in U/B : Pr(E|X_0) \leq \varepsilon_1^0 Pr(E|X_1)\} \\ \mathcal{BAYNEG}_B^{\varepsilon}(X_1) &= \bigcup\{E \in U/B : Pr(E|X_1) \leq \varepsilon_0^1 Pr(E|X_0)\} \\ \mathcal{BAYBND}_B^{\varepsilon}(X_1) &= \bigcup\{E \in U/B : Pr(E|X_0) > \varepsilon_1^0 Pr(E|X_1) \wedge \\ &\qquad\qquad\qquad\quad Pr(E|X_1) > \varepsilon_0^1 Pr(E|X_0)\} \end{aligned} \tag{25}$$

Remark 5. The choice of ε_0^1 and ε_1^0 is a challenge comparable to the case of other parameter-controlled models, e.g. VPRS based on the threshold vectors α and β. It is allowed to put $\varepsilon_0^1 = \varepsilon_1^0$ and use a common notation $\varepsilon \in [0,1)$ for both coefficients. It obviously simplifies (but does not solve) the problem of parameter tuning. Further discussion with that respect is beyond the scope of this particular article and should be continued in the nearest future. □

Remark 6. As in Subsection 2.3, we prefer not to use the Bayes factor ratio explicitly because of the special case of zero probabilities. However, if we omit this case, we can rewrite the RB positive/negative/boundary regions using inequalities $B_0^1 \geq 1/\varepsilon_1^0$, $B_1^0 \geq 1/\varepsilon_0^1$, and $\max\{B_1^0\varepsilon_0^1, B_0^1\varepsilon_1^0\} < 1$, respectively, where $B_1^0 = Pr(E|X_0)/Pr(E|X_1)$ and $B_1^0 = Pr(E|X_1)/Pr(E|X_0)$. □

Proposition 3. *For* $\varepsilon = (0,0)$, *the RB-regions are identical with the RS-regions.*

Proof. Derivable directly from Theorem 1[2]. □

Below we provide possibly simplest way of understanding the RB-regions:

1. Object u belongs to $\mathcal{BAYPOS}_B^{\varepsilon}(X_1)$, if and only if $B(u)$ is *significantly* (up to ε_1^0) *more likely* to occur under X_1 than under alternative hypothesis X_0.
2. Object u belongs to $\mathcal{BAYNEG}_B^{\varepsilon}(X_1)$, if and only if the alternative hypothesis X_0 makes $B(u)$ *significantly more likely* (up to ε_0^1) than X_1 does.
3. Object u belongs to $\mathcal{BAYBND}_B^{\varepsilon}(X_1)$, if and only if it is not *significantly more likely* under X_1 than under X_0 but also X_0 does not make $B(u)$ *significantly more likely* than X_1 does.

Another interpretation refers to identity (2). It shows that by using condition (8) we actually require that the increase of belief in X_0 given E, expressed by $Pr(X_0|E)/Pr(X_0)$, should be ε-*negligibly small* with respect to the increase of belief in X_1, that is that $Pr(X_0|E)/Pr(X_0) \leq \varepsilon_1^0 Pr(X_1|E)/Pr(X_1)$. According

[2] Although we refer here to the special case of two decision classes, the reader can verify that this proposition is also true for more general case discussed in the next subsection.

to yet another, strictly Bayesian interpretation, we are in $\mathcal{BAYPOS}_B^{\varepsilon}(X_1)$, if and only if the *posterior odds* $Pr(X_1|E)/Pr(X_0|E)$ are ε_1^0-significantly greater than the *prior odds* $Pr(X_1)/Pr(X_0)$. Identity (2) also shows that we do not need neither $Pr(X_k|E)$ nor $Pr(X_k)$ while comparing the above changes in terms of the belief gains and/or the prior and posterior odds.

The Rough Bayesian model enables us to test the target events directly against each other. For ε_1^0 tending to 1, we can replace $Pr(E|X_0) \leq \varepsilon_1^0 Pr(E|X_1)$ by $Pr(E|X_0) < Pr(E|X_1)$, as considered in Subsection 3.3. Also, across the whole range of $\varepsilon_1^0 \in [0,1)$, we obtain the following characteristics, complementary to Proposition 2:

Proposition 4. *Let* $\varepsilon = (\varepsilon_0^1, \varepsilon_1^0)$, $\varepsilon_0^1, \varepsilon_1^0 \in [0,1)$, *and* $\mathbb{A} = (U, A \cup \{d\})$ *with* $V_d = \{0,1\}$ *be given. The RB-regions are identical with the VPRS-regions for the following parameters:*

$$\alpha_0^{\varepsilon} = \frac{\varepsilon_1^0 Pr(X_0)}{\varepsilon_1^0 Pr(X_0) + Pr(X_1)} \quad \textit{and} \quad \beta_0^{\varepsilon} = \frac{Pr(X_0)}{Pr(X_0) + \varepsilon_0^1 Pr(X_1)}$$

$$\alpha_1^{\varepsilon} = \frac{\varepsilon_0^1 Pr(X_1)}{\varepsilon_0^1 Pr(X_1) + Pr(X_0)} \quad \textit{and} \quad \beta_1^{\varepsilon} = \frac{Pr(X_1)}{Pr(X_1) + \varepsilon_1^0 Pr(X_0)}$$

Proof. Let $B \subseteq A$ and $E \in U/B$ be given. We have to show the following:

$$\begin{array}{l} Pr(E|X_0) \leq \varepsilon_1^0 Pr(E|X_1) \Leftrightarrow Pr(X_1|E) \geq \beta_1^{\varepsilon} \Leftrightarrow Pr(X_0|E) \leq \alpha_0^{\varepsilon} \\ Pr(E|X_1) \leq \varepsilon_0^1 Pr(E|X_0) \Leftrightarrow Pr(X_1|E) \leq \alpha_1^{\varepsilon} \Leftrightarrow Pr(X_0|E) \geq \beta_0^{\varepsilon} \end{array} \tag{26}$$

Let us show, for example (the rest is analogous), that

$$Pr(E|X_0) \leq \varepsilon_1^0 Pr(E|X_1) \Leftrightarrow Pr(X_1|E) \geq \frac{Pr(X_1)}{Pr(X_1) + \varepsilon_1^0 Pr(X_0)} \tag{27}$$

Using the Bayes rule we rewrite the right above inequality as follows:

$$\frac{Pr(E|X_1)Pr(X_1)}{Pr(E|X_1)Pr(X_1) + Pr(E|X_0)Pr(X_0)} \geq$$

$$\geq \frac{Pr(E|X_1)Pr(X_1)}{Pr(E|X_1)Pr(X_1) + \varepsilon_1^0 Pr(E|X_1)Pr(X_0)}$$

The only difference is now between the term $Pr(E|X_0)$ at the left side and the term $\varepsilon_1^0 Pr(E|X_1)$ at the right side. Hence, (27) becomes clear. □

Example 5. Let us consider the data table from Fig. 1, for $B = \{a_1, a_3\}$ and $\varepsilon_0^1 = \varepsilon_1^0 = 1/5$. We obtain the following RB-regions:

$$\begin{array}{l} \mathcal{BAYPOS}_B^{1/5}(X_1) = \{u_3, u_6, u_8, u_9, u_{15}, u_{19}, u_{20}\} \\ \mathcal{BAYNEG}_B^{1/5}(X_1) = \{u_2, u_5, u_{10}, u_{14}, u_{18}\} \\ \mathcal{BAYBND}_B^{1/5}(X_1) = \{u_1, u_4, u_7, u_{11}, u_{12}, u_{13}, u_{16}, u_{17}\} \end{array}$$

In comparison to the original RS-regions, the case of $a_1 = 2$ and $a_3 = 2$ starts to support the B-positive RB-region of X_1. If we can assume that $Pr(X_1) = 1/2$, as derivable from the considered data table, then we obtain analogous result in terms of the VPRS-regions for

$$\alpha_1^{1/5} = \frac{1/5 \cdot 1/2}{1/5 \cdot 1/2 + (1 - 1/2)} = \frac{1}{6} \text{ and } \beta_1^{1/5} = \frac{1/2}{1/2 + 1/5(1 - 1/2)} = \frac{5}{6}$$

In particular, for $E = \{u \in U : a_1(u) = 2 \wedge a_3(u) = 2\}$, we get $Pr(X_1|E) = 6/7$ which is more than $\beta_1^{1/5} = 5/6$. □

In this way, the Rough Bayesian model refers to the VPRS idea of handling the posterior probabilities. We can see that coefficients $\alpha_k^\varepsilon, \beta_k^\varepsilon$, $k = 0, 1$, satisfy assumption (15). For instance we have

$$0 \leq \frac{\varepsilon_1^0 Pr(X_0)}{\varepsilon_1^0 Pr(X_0) + Pr(X_1)} < Pr(X_0) < \frac{Pr(X_0)}{Pr(X_0) + \varepsilon_0^1 Pr(X_1)} \leq 1$$

where inequalities hold, if and only if $\varepsilon_0^1, \varepsilon_1^0 \in [0, 1)$. The property (17) is satisfied as well, e.g.:

$$\frac{\varepsilon_1^0 Pr(X_0)}{\varepsilon_1^0 Pr(X_0) + Pr(X_1)} + \frac{Pr(X_1)}{Pr(X_1) + \varepsilon_1^0 Pr(X_0)} = 1$$

We can summarize the obtained results as follows:

Theorem 2. *Let $\mathbb{A} = (U, A \cup \{d\})$, $V_d = \{0, 1\}$, and $B \subseteq A$ be given. The VPRS and RB models are equivalent in the following sense:*

1. *For any $\alpha = (\alpha_0, \alpha_1)$ and $\beta = (\beta_0, \beta_1)$ satisfying (15) and (16), there exists $\varepsilon(\alpha, \beta) \in [0, 1) \times [0, 1)$ such that for $k = 0, 1$ we have*

$$\begin{aligned} \mathcal{BAYPOS}_B^{\varepsilon(\alpha,\beta)}(X_k) &= \mathcal{POS}_B^{\beta}(X_k) \\ \mathcal{BAYNEG}_B^{\varepsilon(\alpha,\beta)}(X_k) &= \mathcal{NEG}_B^{\alpha}(X_k) \\ \mathcal{BAYBND}_B^{\varepsilon(\alpha,\beta)}(X_k) &= \mathcal{BND}_B^{\alpha,\beta}(X_k) \end{aligned}$$

2. *For any $\varepsilon \in [0, 1) \times [0, 1)$, there exist $\alpha(\varepsilon)$ and $\beta(\varepsilon)$ satisfying (15) and (16) such that for $k = 0, 1$ we have*

$$\begin{aligned} \mathcal{POS}_B^{\beta(\varepsilon)}(X_k) &= \mathcal{BAYPOS}_B^{\varepsilon}(X_k) \\ \mathcal{NEG}_B^{\alpha(\varepsilon)}(X_k) &= \mathcal{BAYNEG}_B^{\varepsilon}(X_k) \\ \mathcal{BND}_B^{\alpha(\varepsilon),\beta(\varepsilon)}(X_k) &= \mathcal{BAYBND}_B^{\varepsilon}(X_k) \end{aligned}$$

Proof. Derivable directly from Propositions 2 and 4. □

It is important to remember that Theorem 2 holds only for $V_d = \{0, 1\}$. We will address more general case in the next subsections. For now, given $V_d = \{0, 1\}$, let us note that the advantage of the RB model with respect to VPRS is that any change of $Pr(X_1)$ results in automatic change of the lower and upper VPRS thresholds. It can be illustrated as follows:

Example 6. Let us continue the previous example but for $Pr(X_1) = 1/1000$, as if X_1 corresponded to a rare medical pathology discussed in Example 3. There is no sense to keep the upper limit for $Pr(X_1|E)$ equal to $5/6$, so the VPRS parameters should be changed. However, there is no change required if we rely on the RB-regions. With the same $\varepsilon_0^1 = \varepsilon_1^0 = 1/5$ we simply get different interpretation in terms of the posterior probabilities. Namely, we recalculate the VPRS degrees as

$$\alpha_1^{1/5} = \frac{1/5 \cdot 1/1000}{1/5 \cdot 1/1000 + (1 - 1/1000)} \approx \frac{1}{5000} \quad \text{and} \quad \beta_1^{1/5} \approx \frac{1}{200}$$

One can see that this time $\beta_1^{1/5}$ would have nothing in common with previously calculated $Pr(X_1|E) = 6/7$. However, using standard estimation $Pr(X_1|E) = |X_k \cap E|/|E|$ is not reasonable in this situation. We should rather use the Bayes rule leading to the following result:

$$Pr(d = 1|a_1 = 2, a_3 = 2) = \frac{3/5 \cdot 1/1000}{1/10 \cdot (1 - 1/1000) + 3/5 \cdot 1/1000} = \frac{2}{335}$$

The posterior probability becomes then to be referrable to $\beta_1^{1/5}$. □

As a result, we obtain a convenient method of defining the rough-set-like regions based on the inverse probabilities, which – if necessary – can be translated onto the parameters related to more commonly used posterior probabilities. However, the Rough Bayesian model can be applied also when such translation is impossible, that is when the prior probabilities are unknown or even undefinable. The RB-regions have excellent statistical interpretation following from their connections with the Bayes factor. Actually, we obtain a kind of *variable significance rough set model*, as it is parameterized by the significance thresholds $\varepsilon = (\varepsilon_0^1, \varepsilon_1^0)$. The choice of ε refers to the choice of significance levels illustrated by Fig. 2, Subsection 2.3. We can draw a direct connection between the RB-regions and particular states of the statistical verification process. We can also base on statistical apparatus while tuning $\varepsilon = (\varepsilon_0^1, \varepsilon_1^0)$, with two important special cases – the original rough set model for $\varepsilon_0^1 = \varepsilon_1^0 = 0$ and the model introduced in [18] for $\varepsilon_0^1, \varepsilon_1^0$ tending to 1.

4.2 RB for More Decision Classes

The way of comparing the inverse probabilities in Definition 1 has a natural extension onto the case of more decision classes. Below we reconsider the Rough Bayesian model for such a situation. Please note that the regions from Definition 1 are the special cases of the following ones.

Definition 2. *Let* $\mathbb{A} = (U, A \cup \{d\})$, $V_d = \{0, \ldots, r-1\}$, *and* $B \subseteq A$ *be given. Consider matrix*

$$\varepsilon = \begin{bmatrix} * & \varepsilon_1^0 & \cdots & \varepsilon_{r-1}^0 \\ \varepsilon_0^1 & * & & \vdots \\ \vdots & & * & \varepsilon_{r-1}^{r-2} \\ \varepsilon_0^{r-1} & \cdots & \varepsilon_{r-2}^{r-1} & * \end{bmatrix} \tag{28}$$

where $\varepsilon_k^l \in [0,1)$, *for* $k \neq l$. *We define the* B-positive, B-negative, *and* B-boundary RB-regions *as follows:*

$$\begin{array}{ll}\mathcal{BAYPOS}_B^{\varepsilon}(X_k) &= \bigcup\{E \in U/B : \forall_{l:\,l\neq k} Pr(E|X_l) \leq \varepsilon_k^l Pr(E|X_k)\} \\ \mathcal{BAYNEG}_B^{\varepsilon}(X_k) &= \bigcup\{E \in U/B : \exists_{l:\,l\neq k} Pr(E|X_k) \leq \varepsilon_l^k Pr(E|X_l)\} \\ \mathcal{BAYBND}_B^{\varepsilon}(X_k) &= \bigcup\{E \in U/B : \exists_{l:\,l\neq k} Pr(E|X_l) > \varepsilon_k^l Pr(E|X_k) \wedge \\ & \qquad \forall_{l:\,l\neq k} Pr(E|X_k) > \varepsilon_l^k Pr(E|X_l)\}\end{array} \tag{29}$$

Remark 7. As in Remark 6, we could use respectively conditions $\min_{l:l\neq k} B_l^k \geq 1/\varepsilon_k^l$, $\max_{l:l\neq k} B_k^l \geq 1/\varepsilon_l^k$, and $\max\left\{\min_{l:l\neq k} B_l^k \varepsilon_k^l, \max_{l:l\neq k} B_k^l \varepsilon_l^k\right\} < 1$, where the ratios B_l^k are defined by (6). The only special case to address would correspond to the zero inverse probabilities. □

Let us generalize the previous interpretation of the RB-regions as follows:

1. Object u belongs to $\mathcal{BAYPOS}_B^{\varepsilon}(X_k)$, if and only if $B(u)$ is *significantly more likely* to occur under X_k than under any other hypothesis X_l, $l \neq k$.
2. Object u belongs to $\mathcal{BAYNEG}_B^{\varepsilon}(X_k)$, if and only if there is an alternative hypothesis X_l, which makes $B(u)$ *significantly more likely* than X_k does.
3. Object u belongs to $\mathcal{BAYBND}_B^{\varepsilon}(X_k)$, if and only if $B(u)$ is not *significantly more likely* under X_k than under all other X_l but there is also no alternative hypothesis, which makes $B(u)$ *significantly more likely* than X_k does.

Remark 8. As in case of two decision classes, we can consider a simplified model with $\varepsilon_k^l = \varepsilon$, for every $k, l = 0, \ldots, r-1$, $k \neq l$. Appropriate tuning of many different parameters in the matrix (28) could be difficult technically, especially for large $r = |V_d|$. The examples below show that the multi-decision RB model has a significant expressive power even for one unified $\varepsilon \in [0,1)$. Therefore, we are going to put a special emphasis on this case in the future applications. □

Example 7. Fig. 3 illustrates decision system $\mathbb{A} = (U, A \cup \{d\})$ with $V_d = \{0,1,2\}$. Actually, it results from splitting the objects supporting X_0 in Fig. 1 onto two 5-object parts, now corresponding to decision classes X_0 and X_2. For $B = \{a_1, a_3\}$, we have five B-indiscernibility classes. We list them below, with the corresponding inverse probabilities.

U/B	Conditions	$P(E_i\|X_0)$	$P(E_i\|X_1)$	$P(E_i\|X_2)$
E_1	$a_1 = 1$, $a_3 = 0$	2/5	1/5	1/5
E_2	$a_1 = 0$, $a_3 = 0$	1/5	0	1/5
E_3	$a_1 = 2$, $a_3 = 2$	0	3/5	1/5
E_4	$a_1 = 0$, $a_3 = 2$	0	1/5	1/5
E_5	$a_1 = 2$, $a_3 = 1$	2/5	0	1/5

Let us start with the RS-regions. We obtain the following characteristics:

Decisions	$\mathcal{POS}_B(X_k)$	$\mathcal{NEG}_B(X_k)$	$\mathcal{BND}_B(X_k)$
X_0	$\emptyset$	$E_3 \cup E_4$	$E_1 \cup E_2 \cup E_5$
X_1	$\emptyset$	$E_2 \cup E_5$	$E_1 \cup E_3 \cup E_4$
X_2	$\emptyset$	$\emptyset$	U

U	a_1	a_2	a_3	a_4	a_5	d
u_1	1	1	0	1	2	0
u_2	0	0	0	2	2	2
u_3	2	2	2	1	1	1
u_4	0	1	2	2	2	1
u_5	2	1	1	0	2	0
u_6	2	2	2	1	1	1
u_7	0	1	2	2	2	2
u_8	2	2	2	1	1	1
u_9	2	2	2	1	1	1
u_{10}	0	0	0	2	2	0

U	a_1	a_2	a_3	a_4	a_5	d
u_{11}	1	2	0	0	2	0
u_{12}	1	1	0	1	2	1
u_{13}	0	1	2	2	2	1
u_{14}	2	1	1	0	2	0
u_{15}	2	2	2	1	1	2
u_{16}	1	1	0	1	2	1
u_{17}	1	1	0	1	2	2
u_{18}	2	1	1	0	2	2
u_{19}	2	2	2	1	1	1
u_{20}	2	2	2	1	1	1

Fig. 3. $\mathbb{A} = (U, A \cup \{d\})$, $U = \{u_1, \ldots, u_{20}\}$, $A = \{a_1, \ldots, a_5\}$. Decision classes: $X_0 = \{u_1, u_5, u_{10}, u_{11}, u_{15}\}$, $X_1 = \{u_3, u_4, u_6, u_8, u_9, u_{12}, u_{13}, u_{16}, u_{19}, u_{20}\}$, and $X_2 = \{u_2, u_7, u_{14}, u_{17}, u_{18}\}$.

Now, let us consider $\varepsilon = 1/3$. We obtain the following RB-regions:

Decisions	$\mathcal{BAYPOS}_B^{1/3}(X_k)$	$\mathcal{BAYNEG}_B^{1/3}(X_k)$	$\mathcal{BAYBND}_B^{1/3}(X_k)$
X_0	$\emptyset$	$E_3 \cup E_4$	$E_1 \cup E_2 \cup E_5$
X_1	E_3	$E_2 \cup E_5$	$E_1 \cup E_4$
X_2	$\emptyset$	E_3	$U \setminus E_3$

While comparing to the RS-regions, we can see that:

1. $\mathcal{BAYPOS}_B^{1/3}(X_1)$ and $\mathcal{BAYNEG}_B^{1/3}(X_2)$ start to contain E_3.
2. $\mathcal{BAYBND}_B^{1/3}(X_1)$ and $\mathcal{BAYBND}_B^{1/3}(X_2)$ do not contain E_3 any more.

This is because we have both $P(E_3|X_0) \leq 1/3 * P(E_3|X_1)$ and $P(E_3|X_2) \leq 1/3 * P(E_3|X_1)$. It means that E_3 at least three times more likely given hypothesis X_1 than given X_0 and X_2. According to the scale proposed in [4] and presented in Subsection 2.3, we could say that E_3 is a *positive* evidence for X_1. □

As a conclusion for this part, we refer to Proposition 1 formulated for the original rough set model as an important decision-making property. We can see that the Rough Bayesian model keeps this property well enough to disallow intersections between the positive and negative RB-regions of different decision classes. We will go back to this topic in the next subsection, while discussing the VPRS model for more than two decision classes.

Proposition 5. *Let $\mathbb{A} = (U, A \cup \{d\})$, $X_k \subseteq U$, and $B \subseteq A$ be given. Consider matrix ε given by (28) for $\varepsilon_k^l \in [0, 1)$, $k \neq l$. We have the following inclusion:*

$$\mathcal{BAYPOS}_B^{\varepsilon}(X_k) \subseteq \bigcap_{l:l \neq k} \mathcal{BAYNEG}_B^{\varepsilon}(X_l) \tag{30}$$

Moreover, if inequalities

$$\varepsilon_k^l \geq \varepsilon_k^m \varepsilon_m^l \tag{31}$$

hold for every mutually different k, l, m, then the equality holds in (30).

Proof. Assume $E \subseteq \mathcal{BAYPOS}_B^{\varepsilon}(X_k)$ for a given $k = 0, \ldots, r-1$. Consider any X_m, $m \neq k$. We have $E \subseteq \mathcal{BAYNEG}_B^{\varepsilon}(X_m)$ because there exists $l \neq m$ such that $Pr(E|X_m) \leq \varepsilon_l^m Pr(E|X_l)$. Namely, we can choose $l = k$.

Now, let us assume that $E \subseteq \mathcal{BAYNEG}_B^{\varepsilon}(X_l)$ for every X_l, $l \neq k$. Then, for any X_{l0}, $l0 \neq k$ there must exist X_{l1}, $l1 \neq l0$, such that $Pr(E|X_{l0}) \leq \varepsilon_{l1}^{l0} Pr(E|X_{l1})$. There are two possibilities: If $l1 = k$, then we reach the goal – we wanted to show that $Pr(E|X_{l0}) \leq \varepsilon_k^{l0} Pr(E|X_k)$ for any X_{l0}, $l0 \neq k$. If $l1 \neq k$, then we continue with $l1$. Since $l1 \neq k$, there must exist X_{l2}, $l2 \neq l1$, such that $Pr(E|X_{l1}) \leq \varepsilon_{l2}^{l1} Pr(E|X_{l2})$. Given inequalities (31), we get

$$Pr(E|X_{l0}) \leq \varepsilon_{l1}^{l0} Pr(E|X_{l1}) \leq \varepsilon_{l1}^{l0} \varepsilon_{l2}^{l1} Pr(E|X_{l2}) \leq \varepsilon_{l2}^{l0} Pr(E|X_{l2})$$

Therefore, we can apply the same procedure to every next $l2$ as we did with $l1$ above. At each next step we must select a brand new decision class – this is because the ε-matrix takes the values within $[0, 1)$. Since the number of decisions is finite, we must eventually reach the moment when a new $l2$ equals k. □

Corollary 1. *Let $\mathbb{A} = (U, A \cup \{d\})$ and $B \subseteq A$ be given. Consider the RB model with unified parameter $\varepsilon \in [0, 1)$, that is $\varepsilon_k^l = \varepsilon$ for every $k, l = 0, \ldots, r-1$, $k \neq l$. Then we have always $\mathcal{BAYPOS}_B^{\varepsilon}(X_k) = \bigcap_{l:l \neq k} \mathcal{BAYNEG}_B^{\varepsilon}(X_l)$.*

Proof. Directly from Proposition 5. □

4.3 VPRS for More Decision Classes

The question is whether the Rough Bayesian model is still rewritable in terms of the posterior probabilities, similarly to the case of two decision classes described by Theorem 2. Let us first discuss requirements for a posterior probability based rough set model in such a case. In Subsection 3.2, we used the parameter vectors $\alpha = (\alpha_0, \ldots, \alpha_{r-1})$ and $\beta = (\beta_0, \ldots, \beta_{r-1})$ satisfying condition (15). One would believe that if a unique X_k is supported strongly enough, then the remaining classes cannot be supported in a comparable degree. However, this is the case only for two complementary decision classes. If $|V_d| > 2$, then there might be two different classes X_k and X_l, $k \neq l$, satisfying inequalities $Pr(X_k|E) \geq \beta_k$ and $Pr(X_l|E) \geq \beta_l$. It would lead to supporting two decisions in the same time, which is an unwanted situation.

Example 8. Consider the decision system illustrated in Fig. 3. Please note that $Pr(X_0) = Pr(X_2) = 1/4$ and $Pr(X_1) = 1/2$. Let us choose parameters $\alpha = (1/10, 1/4, 1/10)$ and $\beta = (13/20, 3/4, 13/20)$. One can see that inequalities (15) are then satisfied. For $B = \{a_1, a_3\}$, we have five B-indiscernibility classes, as in Example 7. Their corresponding posterior probabilities look as follows:

U/B	Conditions	$P(X_0\|E_i)$	$P(X_1\|E_i)$	$P(X_2\|E_i)$
E_1	$a_1 = 1$, $a_3 = 0$	2/5	2/5	1/5
E_2	$a_1 = 0$, $a_3 = 0$	1/2	0	1/2
E_3	$a_1 = 2$, $a_3 = 2$	0	6/7	1/7
E_4	$a_1 = 0$, $a_3 = 2$	0	2/3	1/3
E_5	$a_1 = 2$, $a_3 = 1$	2/3	0	1/3

We obtain the following characteristics, if we keep using conditions (14):

Decisions	$\mathcal{POS}_B^{\beta}(X_k)$	$\mathcal{NEG}_B^{\alpha}(X_k)$	$\mathcal{BND}_B^{\alpha,\beta}(X_k)$
X_0	E_5	$E_3 \cup E_4$	$E_1 \cup E_2$
X_1	E_3	$E_2 \cup E_5$	$E_1 \cup E_4$
X_2	$\emptyset$	$\emptyset$	U

Luckily enough, we do not obtain non-empty intersections between positive regions of different decision classes. However, there is another problem visible: E_5 (E_3) is contained in the positive region of X_0 (X_1) but it is not in the negative region of X_2. It is a lack of a crucial property of the rough-set-like regions, specially emphasized by Propositions 1 and 5. □

Obviously, one could say that α and β in the above example are chosen artificially to yield the described non-empty intersection situation. However, even if this is a case, it leaves us with the problem how to improve the requirements for the VPRS parameters to avoid such situations. We suggest embedding the property analogous to those considered for the RS and RB models directly into the definition. In this way, we can also simplify the VPRS notation by forgetting about the upper grades β. This is a reason why we refer to the following model as to the *simplified* VPRS model.

Definition 3. *Let $\mathbb{A} = (U, A \cup \{d\})$ and $B \subseteq A$ be given. Consider vector $\alpha = (\alpha_0, \ldots, \alpha_{r-1})$ such that inequalities*

$$0 \leq \alpha_k < Pr(X_k) \tag{32}$$

are satisfied for every $k = 0, \ldots, r-1$. The simplified VPRS-regions *are defined as follows:*

$$\begin{array}{l} \mathcal{POS}_B^{\alpha}(X_k) = \bigcup\{E \in U/B : \forall_{l:\, l \neq k} Pr(X_l|E) \leq \alpha_l\} \\ \mathcal{NEG}_B^{\alpha}(X_k) = \bigcup\{E \in U/B : Pr(X_k|E) \leq \alpha_k\} \\ \mathcal{BND}_B^{\alpha}(X_k) = \bigcup\{E \in U/B : Pr(X_k|E) > \alpha_k \wedge \exists_{l:\, l \neq k} Pr(X_l|E) > \alpha_l\} \end{array} \tag{33}$$

Proposition 6. *Let $\mathbb{A} = (U, A \cup \{d\})$, $X_k \subseteq U$, $B \subseteq A$, and $\alpha = (\alpha_0, \ldots, \alpha_{r-1})$ be given. We have the following equality:*

$$\mathcal{POS}_B^{\alpha}(X_k) = \bigcap_{l:l \neq k} \mathcal{NEG}_B^{\alpha}(X_l) \tag{34}$$

Proof. Directly from (33). □

The form of (33) can be compared with the way we expressed the original RS-regions by (11). There, we defined the positive region by means of conditions for the negative regions of all other classes, exactly like for simplified VPRS above. Further, we can reformulate the meaning of Remark 3 as follows:

Proposition 7. *Let $\mathbb{A} = (U, A \cup \{d\})$, $B \subseteq A$, $E \in U/B$, and $\alpha = (\alpha_0, ..., \alpha_{r-1})$ satisfying (32) be given. Let us define vector $\beta = (\beta_0, \ldots, \beta_{r-1})$ in the following way, for every $k = 0, \ldots, r-1$:*

$$\beta_k = 1 - \sum_{l:l \neq k} \alpha_k \tag{35}$$

Then, for any X_k, we have the following equivalence:

$$E \subseteq \mathcal{POS}_B^\alpha(X_k) \Leftrightarrow Pr(X_k|E) \geq \beta_k \wedge \forall_{l:\, l \neq k} Pr(X_l|E) \leq \alpha_l\}$$

Moreover, vectors α and β satisfy together the assumption (15).

Proof. Directly based on the fact that $\sum_{l=0}^{r-1} Pr(X_l|E) = 1$. □

In this way one can see that by removing $\beta = (\beta_0, \ldots, \beta_{r-1})$ from the definition of VPRS we do not change its meaning. Vector β is fully recoverable from α using equations (35), which actually generalize postulate (16). In particular, for a special case of two decision classes, we obtain the following result.

Proposition 8. *Let $V_d = \{0, 1\}$ and let equalities (16) be satisfied. Then Definition 3 is equivalent to the original VPRS model.*

Proof. For two decision classes, given (16), conditions $Pr(X_k|E) \geq \beta_k$ and $\forall_{l:\, l \neq k} Pr(X_l|E) \leq \alpha_l$ are equivalent – there is only one different $l = 0, 1$ and one of equalities (17) must take place. Therefore, $\mathcal{POS}_B^\alpha(X_k)$ takes the same form as in (14). Negative regions are formulated directly in the same way as in (14). Hence, the boundary regions must be identical as well. □

Example 9. Let us go back to the three-decision case illustrated by Fig. 3 and consider parameters $\alpha = (1/10, 1/4, 1/10)$, as in Example 8. Let us notice that the vector $\beta = (13/20, 3/4, 13/20)$ from that example can be calculated from α using (35). Now, let us compare the previously obtained regions with the following ones:

Decisions	$\mathcal{POS}_B^\alpha(X_k)$	$\mathcal{NEG}_B^\alpha(X_k)$	$\mathcal{BND}_B^\alpha(X_k)$
X_0	$\emptyset$	$E_3 \cup E_4$	$E_1 \cup E_2 \cup E_5$
X_1	$\emptyset$	$E_2 \cup E_5$	$E_1 \cup E_3 \cup E_4$
X_2	$\emptyset$	$\emptyset$	U

Although, on the one hand, the crucial property (34) is now satisfied, we do not get any relaxation of conditions for the positive regions with respect to the original RS-regions analyzed for the same system in Example 7. The problem with Definition 3 seems to be that even a very good evidence for X_k can be ignored (put into boundary) because of just one other X_l, $l \neq k$, supported by E to a relatively (comparing to X_k) low degree. The RB-regions presented in Example 7 turn out to be intuitively more flexible with handling the data based probabilities. We try to justify it formally below. □

After introducing a reasonable extension (and simplification) of VPRS for the multi-decision case, we are ready to compare it – as an example of the posterior probability based methodology – to the Rough Bayesian model. Since it is an introductory study, we restrict ourselves to the simplest case of RB, namely to the unified ε-matrix (28), where $\varepsilon_k^l = \varepsilon$ for every $k, l = 0, \ldots, r-1$, $k \neq l$, for some $\varepsilon \in [0, 1)$. It refers to an interesting special case of simplified VPRS, where

$$\frac{\alpha_0(1 - Pr(X_0))}{(1 - \alpha_0)Pr(X_0)} = \ldots = \frac{\alpha_{r-1}(1 - Pr(X_{r-1}))}{(1 - \alpha_{r-1})Pr(X_{r-1})} \tag{36}$$

According to (36) the parameters for particular decision classes satisfy inequalities (32) in a proportional way. Its advantage corresponds to the problem of tuning vectors $\alpha = (\alpha_0, \ldots, \alpha_{r-1})$ for large values of $r = |V_d|$. Given (36) we can handle the whole α using a single parameter:

Proposition 9. *Let $\mathbb{A} = (U, A \cup \{d\})$, $B \subseteq A$, and $\alpha = (\alpha_0, \ldots, \alpha_{r-1})$ satisfying both (32) and (36) be given. There exists $\varepsilon \in [0, 1)$, namely*

$$\varepsilon = \frac{\alpha_k(1 - Pr(X_k))}{(1 - \alpha_k)Pr(X_k)} \quad \text{for arbitrary } k = 0, \ldots, r-1 \tag{37}$$

such that for every $k = 0, \ldots, r-1$ the value of α_k is derivable as

$$\alpha_k = \frac{\varepsilon Pr(X_k)}{\varepsilon Pr(X_k) + (1 - Pr(X_k))} \tag{38}$$

Proof. It is enough to substitute the right side of (37) as ε to the right side (38) and check that it indeed equals α_k. □

The following result shows that at the one-parameter level the Rough Bayesian model can be potentially more data sensitive than the simplified VPRS model. Obviously, similar comparison of more general cases is a desired direction for further research.

Theorem 3. *Let $\mathbb{A} = (U, A \cup \{d\})$ and $B \subseteq A$ be given. Consider vector $\alpha = (\alpha_0, \ldots, \alpha_{r-1})$ satisfying (32) and (36). Consider $\varepsilon \in [0, 1)$ given by (37) as the unified parameter for the RB model, that is $\varepsilon_k^l = \varepsilon$ for every $k, l = 0, \ldots, r-1$. Then we have the following inclusions, for every $k = 0, \ldots, r-1$:*

$$\begin{array}{l} \mathcal{POS}_B^\alpha(X_k) \subseteq \mathcal{BAYPOS}_B^\varepsilon(X_k) \\ \mathcal{NEG}_B^\alpha(X_k) \subseteq \mathcal{BAYNEG}_B^\varepsilon(X_k) \\ \mathcal{BND}_B^\alpha(X_k) \supseteq \mathcal{BAYBND}_B^\varepsilon(X_k) \end{array} \tag{39}$$

Proof. Using the same technique as in the proof of Proposition 4, we can show

$$Pr(X_k|E) \le \alpha_k \Leftrightarrow \varepsilon Pr(E|\neg X_k) \ge Pr(E|X_k)$$

where $\neg X_k = \bigcup_{l \neq k} X_l$. Further, using a simple translation, we can observe that

$$Pr(E|\neg X_k) = \frac{\sum_{l \neq k} Pr(X_l) Pr(E|X_l)}{\sum_{l \neq k} Pr(X_l)}$$

Now, we are ready to show inclusions (39). Let us begin with the second one. Assume that a given $E \in U/B$ is not in $\mathcal{BAYNEG}_B^\varepsilon(X_k)$, that is

$$\forall_{l: l \neq k}\, \varepsilon Pr(E|X_l) < Pr(E|X_k)$$

Then we get $\varepsilon \sum_{l:l \neq k} Pr(X_l) Pr(E|X_l) < \sum_{l:l \neq k} Pr(X_l) Pr(E|X_k)$, further equivalent to

$$Pr(E|X_k) > \varepsilon \cdot \frac{\sum_{l:l \neq k} Pr(X_l) Pr(E|X_l)}{\sum_{l:l \neq k} Pr(X_l)}$$

Hence, E is outside $\mathcal{NEG}_B^\alpha(X_k)$ and the required inclusion is proved. To show the first inclusion in (39), assume $E \subseteq \mathcal{POS}_B^\alpha(X_k)$. According to (34), we then have $E \subseteq \mathcal{NEG}_B^\alpha(X_l)$, for every X_l, $l \neq k$. Using just proved inclusion, we get $E \subseteq \mathcal{BAYNEG}_B^\alpha(X_l)$. By Corollary 1 we then obtain $E \subseteq \mathcal{BAYPOS}_B^\alpha(X_k)$, what we wanted to prove. The third inclusion in (39) is now derivable directly from the other two ones. □

Example 10. Let us recall the decision system from Fig. 3, where $Pr(X_0) = Pr(X_2) = 1/4$ and $Pr(X_1) = 1/2$. It turns out that the parameters $\alpha = (1/10, 1/4, 1/10)$ considered in Example 8 are derivable using (38) for $\varepsilon = 1/3$. According to Theorem 3, the RB-regions presented in Example 7 for $\varepsilon = 1/3$ are referrable to the simplified VPRS-regions from Example 9. It is an illustration for (39) – one can see that we should expect strict inclusions in all those inclusions. The specific problem with putting E_3 to the positive simplified VPRS-region of X_1 is that it is blocked by too high value of $Pr(X_2|E_3)$ although this value seems to be much lower than $Pr(X_1|E_3)$. We should avoid comparing these two posterior probabilities directly because it would be unfair with respect to X_2 for its prior probability is twice lower than in case of X_1. However, direct comparison of the inverse probabilities $Pr(E_3|X_1)$ and $Pr(E_3|X_2)$ shows that we can follow X_1 since E_3 is three times more likely given X_1 than given X_2. □

An interesting feature of the Rough Bayesian model is that it can use a single parameter $\varepsilon \in [0, 1)$ to produces valuable results, as illustrated by the above example. On the other hand, asymmetric extensions of RB are possible, even for $r(r-1)$ different parameters ε_k^l corresponding to comparison of $\varepsilon_k^l Pr(E|X_k)$ and $Pr(E|X_l)$. Further research is needed to understand expressive power of such extensions and their relevance to the other rough set approaches.

5 Distributed Decision Systems

In the examples considered so far, we referred to decision systems gathering objects supporting all decision classes together. On the other hand, while dealing with the Bayes factors and the RB-regions, we calculate only the inverse probabilities, which do not require putting the whole data in a single table. We propose a data storage framework, where the objects supporting the target concepts corresponding to different decision classes are stored in separate data sets. It emphasizes that in some situations data supporting particular events are *uncombinable* and the only probability estimates we can use are of the inverse character, that is they are naturally conditioned by particular decisions.

Definition 4. *Let the set of r mutually exclusive target events be given. By a* distributed decision system $\mathcal{A}$ *we mean the collection of r information systems*

$$\mathcal{A} = \{\mathbb{A}_0 = (X_0, A), \ldots, \mathbb{A}_{r-1} = (X_{r-1}, A)\} \tag{40}$$

where X_k denotes the set of objects supporting the k-th event, $k = 0, \ldots, r-1$, and A is the set of attributes describing all the objects in $X_0, \ldots, X_{r-1}$.

Any information derivable from $\mathbb{A}_k$ is naturally conditioned by X_k, for $k = 0, \ldots, r-1$. Given B-information vector $w \in V_B$, $B \subseteq A$, we can set up

$$Pr_k(B = w) = \frac{|\{u \in X_k : B(u) = w\}|}{|X_k|} \tag{41}$$

as the probability that a given object will have the values described by w on B *conditioned* by its membership to X_k.

Example 11. Let us consider $\mathcal{A}$ consisting of two information systems illustrated in Fig. 4. For instance, $Pr_0(a_1 = 2, a_3 = 2) = 1/10$ and $Pr_1(a_1 = 2, a_3 = 2) = 3/5$ are estimates of probabilities that a given object will satisfy $a_1 = 2$ and $a_3 = 2$, if it supports the events X_0 and X_1, respectively.

One can see that if we use estimation

$$Pr(B = w|d = k) = Pr_k(B = w) \tag{42}$$

then the inverse probabilities derivable from Fig. 4 are identical with those derivable from Fig. 1. Actually, we created Fig. 1 artificially by doubling the objects from $\mathbb{A}_1$ and merging them with $\mathbb{A}_0$ from Fig. 4. Therefore, if we assume that due to our knowledge we should put $Pr(X_0) = Pr(X_1) = 1/2$, then systems illustrated by Figures 1 and 4 will provide the same posterior probabilities. □

X_0	a_1	a_2	a_3	a_4	a_5
u_1	1	2	0	0	2
u_2	1	1	0	1	2
u_3	0	0	0	2	2
u_4	2	1	1	0	2
u_5	2	1	1	0	2
u_6	2	2	2	1	1
u_7	0	1	2	2	2
u_8	1	1	0	1	2
u_9	2	1	1	0	2
u_{10}	0	0	0	2	2

X_1	a_1	a_2	a_3	a_4	a_5
o_1	2	2	2	1	1
o_2	0	1	2	2	2
o_3	1	1	0	1	2
o_4	2	2	2	1	1
o_5	2	2	2	1	1

Fig. 4. Distribute decision system $\mathcal{A} = \{\mathbb{A}_0 = (X_0, A), \mathbb{A}_1 = (X_1, A)\}$, where $A = \{a_1, \ldots, a_5\}$, and $X_0 = \{u_1, \ldots, u_{10}\}$, $X_1 = \{o_1, \ldots, o_5\}$.

Distributed decision systems do not provide a means for calculation of the posterior probabilities unless we know the priors of all decision classes. On the other hand, we get full flexibility with respect to the changes of the prior probabilities, which can be easily combined with the estimates (42). For instance, let us go back to the case study from the end of Subsection 2.2 and assume that the objects in $\mathbb{A}_1 = (X_1, A)$ are very carefully chosen cases of a rare medical pathology while the elements of X_0 describe a representative sample of human beings not suffering from this pathology. Let us put $Pr(X_1) = 1/1000$. Then, as in Example 6, we get $Pr(d = 1|a_1 = 2, a_3 = 2) = 2/335$. It shows how different posterior probabilities can be obtained from the same distributed decision system for various prior probability settings. Obviously, we could obtain identical results from appropriately created classical decision systems (like in case of the system in Fig. 1). However, such a way of data translation is unnecessary or even impossible, if the prior probabilities are not specified.

From technical point of view, it does not matter whether we keep the data in the form of distributed or merged decision system, unless we use estimations (4). However, we find Definition 4 as a clearer way to emphasize the nature of the data based probabilities that we can really believe in. Indeed, the inverse probabilities (42) are very often the only ones, which can be reasonably estimated from real-life data sets. This is because the process of the data acquisition is often performed in parallel for various decisions and, moreover, the experts can (and wish to) handle the issue of the information representativeness only at the level of separate decision classes. Following this argumentation, let us reconsider the original RS-regions for distributed data, without the need of merging them within one decision system.

Definition 5. *Let the system* $\mathcal{A} = \{\mathbb{A}_0 = (X_0, A), \ldots, \mathbb{A}_{r-1} = (X_{r-1}, A)\}$ *be given. For any* X_k *and* $B \subseteq A$*, we define the* B-positive, B-negative, *and* B-boundary distributed rough set regions *(abbreviated as* DRS-regions*) as follows:*

$$\begin{array}{ll} \mathcal{DPOS}_B(X_k) &= \{w \in V_B : \forall_{l:\, l \neq k} Pr_l(B = w) = 0\} \\ \mathcal{DNEG}_B(X_k) &= \{w \in V_B : Pr_k(B = w) = 0\} \\ \mathcal{DBND}_B(X_k) &= \{w \in V_B : Pr_k(B = w) > 0 \wedge \exists_{l:\, l \neq k} Pr_l(B = w) > 0\} \end{array} \quad (43)$$

The difference between (43) and (9) is that the distributed rough set regions are expressed in terms of B-information vectors, regarded as the conditions satisfiable by the objects. Besides, both definitions work similarly if they refer to the same inverse probabilities.

Example 12. The DRS-regions obtained for $B = \{a_1, a_3\}$ from Fig. 4 look as follows:

$$\begin{array}{ll} \mathcal{DPOS}_B(X_1) &= \emptyset \\ \mathcal{DNEG}_B(X_1) &= \{\{(a_1, 0), (a_3, 0)\}, \{(a_1, 2), (a_3, 1)\}\} \\ \mathcal{DBND}_B(X_1) &= \{\{(a_1, 0), (a_3, 2)\}, \{(a_1, 1), (a_3, 0)\}, \{(a_1, 2), (a_3, 2)\}\} \end{array} \quad (44)$$

One can see that the supports of the above B-information vectors within the decision system from Fig. 1 correspond to the RS-regions in Example 4. □

The rough set extensions referring in a non-trivial way to the posterior and prior probabilities, like e.g. VPRS, cannot be rewritten in terms of distributed decision systems. However, it is possible for the Rough Bayesian model. Actually, it emphasizes that RB does not need to assume *anything* about the prior and posterior probabilities. We believe that in this form our idea of combining rough sets with the Bayes factor based approaches is possibly closest to the practical applications.

Definition 6. *Let* $\mathcal{A} = \{\mathbb{A}_0 = (X_0, A), \ldots, \mathbb{A}_{r-1} = (X_{r-1}, A)\}$ *be given. Consider matrix* ε *given by (28) for* $\varepsilon_k^l \in [0, 1)$, $k \neq l$. *For any* $k = 0, \ldots, r-1$ *and* $B \subseteq A$*, we define the* B-positive, B-negative, *and* B-boundary distributed rough Bayesian regions *(abbreviated as* DRB-regions*) as follows:*

$$\begin{array}{rl} \mathcal{DBAYPOS}_B^{\varepsilon}(X_k) = & \{w \in V_B : \forall_{l:\, l \neq k} Pr_l(B=w) \leq \varepsilon_k^l Pr_k(B=w)\} \\ \mathcal{DBAYNEG}_B^{\varepsilon}(X_k) = & \{w \in V_B : \exists_{l:\, l \neq k} Pr_k(B=w) \leq \varepsilon_l^k Pr_l(B=w)\} \\ \mathcal{DBAYBND}_B^{\varepsilon}(X_k) = & \{w \in V_B : \exists_{l:\, l \neq k} Pr_l(B=w) > \varepsilon_k^l Pr_k(B=w) \\ & \wedge \forall_{l:\, l \neq k} Pr_k(B=w) > \varepsilon_l^k Pr_l(B=w)\} \end{array} \quad (45)$$

Example 13. Fig. 5 illustrates a distributed system for three target events. They result from splitting $\mathbb{A}_0 = (X_0, A)$ from Fig. 4 onto equally large $\mathbb{A}_0 = (X_0, A)$ and $\mathbb{A}_2 = (X_2, A)$, similarly as we did in the previous sections with our exemplary non-distributed decision system. Let us start with calculation of the regions introduced in Definition 5. As usual, consider $B = \{a_1, a_3\}$. The DRS-regions for X_1 do not change with respect to (44). The remaining regions look as follows:

$$\begin{array}{ll} \mathcal{DPOS}_B(X_0) = \emptyset & \mathcal{DPOS}_B(X_2) = \emptyset \\ \mathcal{DNEG}_B(X_0) = \{\{(a_1,0),(a_3,2)\},\{(a_1,2),(a_3,2)\}\} & \mathcal{DNEG}_B(X_2) = \emptyset \\ \mathcal{DBND}_B(X_0) = V_B \setminus \mathcal{DNEG}_B(X_0) & \mathcal{DBND}_B(X_2) = V_B \end{array} \quad (46)$$

X_0	a_1	a_2	a_3	a_4	a_5
u_1	1	2	0	0	2
u_2	1	1	0	1	2
u_3	0	0	0	2	2
u_4	2	1	1	0	2
u_5	2	1	1	0	2

X_1	a_1	a_2	a_3	a_4	a_5
o_1	2	2	2	1	1
o_2	0	1	2	2	2
o_3	1	1	0	1	2
o_4	2	2	2	1	1
o_5	2	2	2	1	1

X_2	a_1	a_2	a_3	a_4	a_5
e_1	2	2	2	1	1
e_2	0	1	2	2	2
e_3	1	1	0	1	2
e_4	2	1	1	0	2
e_5	0	0	0	2	2

Fig. 5. System $\mathcal{A} = \{\mathbb{A}_0 = (X_0, A),\ \mathbb{A}_1 = (X_1, A),\ \mathbb{A}_2 = (X_2, A)\}$, where $A = \{a_1, \ldots, a_5\}$, $X_0 = \{u_1, \ldots, u_5\}$, $X_1 = \{o_1, \ldots, o_5\}$, $X_2 = \{e_1, \ldots, e_5\}$.

The B-boundary DRS-region for X_2 corresponds to the whole V_B. It means that any so far recorded B-information vector is *likely* to occur for a given object under the assumption that that object supports X_2, as well as under the assumption that it supports X_0 and/or X_1. Now, consider the DRB-regions for $\varepsilon = 1/3$. We obtain the following changes with respect to the (44) and (46):

1. $\mathcal{DBAYPOS}_B^{1/3}(X_1)$, $\mathcal{DBAYNEG}_B^{1/3}(X_2)$ start to contain $\{(a_1, 2), (a_3, 2)\}$.
2. $\mathcal{DBAYBND}_B^{1/3}(X_k)$, $k = 1, 2$, do not contain $\{(a_1, 2), (a_3, 2)\}$ any more.

The obtained DRB-regions are comparable with the RB-regions obtained previously for the corresponding non-distributed decision system from Fig. 3. □

Introduction of distributed decision systems has rather a technical than theoretical impact. It illustrates possibility of handling decomposed data, which can be especially helpful in case of many decision classes with diversified or unknown prior probabilities. Distributed systems provide the exact type of information needed for extracting the RB-regions from data. Hence, we plan implementing the algorithms referring to the Rough Bayesian model mainly for such systems.

6 Final Remarks

We introduced the Rough Bayesian model – a parameterized extension of rough sets, based on the Bayes factor and the inverse probabilities. We compared it with other probabilistic extensions, particularly with the VPRS model relying on the data based posterior probabilities. We considered both the two-decision and multiple-decision cases, where the direct comparison of the inverse probabilities conditioned by decision classes turns out to be more flexible than handling their posterior probabilities. Finally, we proposed distributed decision systems as a new way of storing data, providing estimations for the Rough Bayesian regions.

We believe that the framework based on the Rough Bayesian model is well applicable to the practical data analysis problems, especially if we cannot rely on the prior/posterior probabilities derivable from data and/or background knowledge. The presented results are also helpful in establishing theoretical foundations for correspondence between the theory of rough sets and Bayesian reasoning. Several basic facts, like, e.g., the inverse probability based characteristics of the original rough set model, support an intuition behind this correspondence.

Acknowledgments

Supported by the research grant from the Natural Sciences and Engineering Research Council of Canada.

References

1. Box, G.E.P., Tiao, G.C.: Bayesian Inference in Statistical Analysis. Wiley (1992).
2. Duentsch, I., Gediga, G.: Uncertainty measures of rough set prediction. Artificial Intelligence 106 (1998) pp. 77–107.
3. Greco, S., Pawlak, Z., Słowiński, R.: Can Bayesian confirmation measures be useful for the rough set decision rules? Engineering Application of Artificial Intelligence, 17 (2004) pp. 345–361.
4. Good, I.J.: The Interface Between Statistics and Philosophy of Science. Statistical Science 3 (1988) pp. 386–397.
5. Hilderman, R.J., Hamilton, H.J.: Knowledge Discovery and Measures of Interest. Kluwer (2002).
6. Jeffreys, H.: Theory of Probability. Clarendon Press, Oxford (1961).
7. Kamber, M., Shingal, R.: Evaluating the interestingness of characteristic rules. In: Proc. of the 2nd International Conference on Knowledge Discovery and Data Mining (KDD'96), Portland, Oregon (1996) pp. 263–266.
8. Kloesgen, W., Żytkow, J.M. (eds): Handbook of Data Mining and Knowledge Discovery. Oxford University Press (2002).
9. Mitchell, T.: Machine Learning. Mc Graw Hill (1998).
10. Pawlak, Z.: Rough sets – Theoretical aspects of reasoning about data. Kluwer Academic Publishers (1991).
11. Pawlak, Z.: New Look on Bayes' Theorem – The Rough Set Outlook. In: Proc. of JSAI RSTGC'2001, pp. 1–8.

12. Pawlak, Z.: Decision Tables and Decision Spaces. In: Proc. of the 6th International Conference on Soft Computing and Distributed Processing (SCDP'2002). June 24-25, Rzeszów, Poland (2002).
13. Pawlak, Z., Skowron, A.: Rough membership functions. In: Advances in the Dempster Shafer Theory of Evidence. Wiley (1994) pp. 251–271.
14. Polkowski, L., Tsumoto, S., Lin, T.Y. (eds): Rough Set Methods and Applications. Physica Verlag (2000).
15. Raftery, A.E.: Hypothesis testing and model selection. In: W.R. Gilks, S. Richardson, and D.J. Spiegelhalter (eds), Markov Chain Monte Carlo in Practice. Chapman and Hall, London (1996) pp. 163–187.
16. Ślęzak, D.: Approximate Entropy Reducts. Fundamenta Informaticae, 53/3-4 (2002) pp. 365–390.
17. Ślęzak, D.: The Rough Bayesian Model For Distributed Decision Systems. In Proc. of RSCTC'2004 (2004) pp. 384–393.
18. Ślęzak, D., Ziarko, W.: Bayesian Rough Set Model. In: Proc. of FDM'2002 (2002) pp. 131–135.
19. Ślęzak, D., Ziarko, W.: The Investigation of the Bayesian Rough Set Model. International Journal of Approximate Reasoning (2005) to appear.
20. Swinburne, R. (ed.): Bayes's Theorem. Proc. of the British Academy 113 (2003).
21. Tsumoto, S.: Accuracy and Coverage in Rough Set Rule Induction. In: Proc. of RSCTC'2002 (2002) pp. 373–380.
22. Wong, S.K.M., Ziarko, W.: Comparison of the probabilistic approximate classification and the fuzzy set model. International Journal for Fuzzy Sets and Systems, 21 (1986) pp. 357–362.
23. Yao, Y.Y.: Probabilistic approaches to rough sets. Expert Systems, 20/5 (2003) pp. 287–297.
24. Ziarko, W.: Variable precision rough sets model. Journal of Computer and Systems Sciences, 46/1 (1993) pp. 39–59.
25. Ziarko, W.: Set approximation quality measures in the variable precision rough set model. Soft Computing Systems, Management and Applications, IOS Press (2001) pp. 442–452.

Formal Concept Analysis and Rough Set Theory from the Perspective of Finite Topological Approximations

Marcin Wolski

Department of Logic and Methodology of Science,
Maria Curie-Skłodowska University, Poland
marcin.wolski@umcs.lublin.pl

Abstract. The paper examines Formal Concept Analysis (FCA) and Rough Set Theory (RST) against the background of the theory of finite approximations of continuous topological spaces. We define the operators of FCA and RST by means of the specialisation order on elements of a topological space X which induces a finite approximation of X. On this basis we prove that FCA and RST together provide a semantics for tense logic S4.t. Moreover, the paper demonstrates that a topological space X cannot be distinguished from its finite approximation by means of the basic temporal language. It means that from the perspective of topology S4.t is a better account of approximate reasoning then unimodal logics, which have been typically employed.

1 Introduction

Formal Concept Analysis (FCA) and Rough Set Theory (RST), introduced in the early 80's by Wille [14] and Pawlak [13] respectively, have become today leading theories of knowledge acquisition from data tables. This fact has resulted in a rapid growth of interest in their formal relationships and possible unifications. Generally, both theories are based on Galois connections and cluster data into coherent and meaningful entities called concepts. Concepts express knowledge about a given domain, which is represented by means of a data table. Some sets of objects may be directly defined by concepts but others may be only approximated. Basically, FCA is concerned with the formal structure of concepts whereas RST is engaged with approximations. Although we know today quite a lot about both theories and even about their relationships [3, 4, 15, 17], there is serious lack of results on logics reflecting their formal connexions. This paper aims at providing such logic, which – as we demonstrate – should be considered as a good tool for approximate reasoning.

Going into detail, the present article establishes certain formal relationships between FCA and RST on the ground of the theory of finite topological approximations. The idea of finite approximations (aslo called finitary substitutes) preserving important topological features of continuous topologies (such as manifolds) has been introduced by Sorkin in [12]. We are concerned here only with pure mathematical content of this idea, leaving aside its physical interpretations.

J.F. Peters and A. Skowron (Eds.): Transactions on Rough Sets III, LNCS 3400, pp. 230–243, 2005.

The article shows that the composition of derivation operators from FCA and the upper approximation as well as the lower approximation operators of RST naturally emerge from the so called specialisation order, which expresses incompletness of information about a given domain. It brings us a few new insights into the nature of these operators and relationships between FCA and RST. Notably, the composition of derivation operators of FCA is the opposite operator with respect to the specialisation order to the upper approximation of RST. It follows that FCA and RST taken together bring a semantics to tense logic S4.t. To our best knowledge it is the first strictly logical result concerning relationships between FCA and RST.

As we have said, a finite approximation preserves important topological features of an approximated space. Hence a logic of approximate reasoning should be sufficiently strong to guarantee that these important features will be maintained. RST is expressed by unimodal normal logics [2, 16], what in case of finite models means that RST is as strong as bisimilarity. But – as we prove it – a topological space and its finitary substitute are temporally bisimilar. Therefore, in the light of topology, S4.t interpreted by RST and FCA is a better account of approximate reasoning than unimodal logics provided with RST semantics.

2 Finitary Approximations

We begin this section with recalling a few basic definitions from general topology. Having that done we shall introduce the idea of finite topological approximations [12].

Definition 1 (Topological Space). *A* topological space *is a pair* $(X, \Im)$ *where* X *is a nonempty set and* $\Im$ *is a family of subsets of* X*, which contains the empty set* $\emptyset$ *and* X*, and is closed under finite intersections and infinite unions.*

Elements of $\Im$ are called *open sets*. A set A is called *closed* if its complement is open. It is often convenient to single out some special subsets of $\Im$ which are more flexible than the topology itself.

Definition 2 (Basis and Subbasis). *A collection* $\mathcal{B}$ *of open sets is a* basis *of a topological space* $(X, \Im)$ *if each open set in* X *is a union of some elements of* $\mathcal{B}$*. A collection* $\mathcal{E}$ *of open sets is a* subbasis *of a topological space* $(X, \Im)$ *if the family* $\mathcal{B}_{\mathcal{E}}$ *of all elements of* $\mathcal{E}$ *together with* $\emptyset$ *and* X *closed under finite intersections is a basis of* $(X, \Im)$*.*

Both concepts have many formal advantages; for example a topology $\Im$ might be uncountable, but have a countable base allowing the use of countable procedures. Their main advantage from the perspective of data analysis follows from the fact that elements of a basis or a subbasis might admit very concrete and desired description.

Definition 3 (Continuous Map, Open Map and Closed Map). *A map* $f : X_1 \mapsto X_2$ *between two topological spaces* $(X_1, \Im_1)$ *and* $(X_2, \Im_2)$ *is* continuous *if* $A \in \Im_2$ *implies* $f^{-1}(A) \in \Im_1$*. The continuous map* f *is* open *if it preserves open sets and* closed *if it preserves closed sets.*

Let $(X, \mathcal{F})$ be a topological space, where the family $\mathcal{F}$ is finite. We introduce the following equivalence:

$$xRy \text{ iff } x \in U \Leftrightarrow y \in U, \text{ for all } U \in \mathcal{F} \ . \tag{1}$$

Let $X_{\mathcal{F}} = X/\mathcal{F}$ be a set of equivalence classes of R and $f : X \mapsto X_{\mathcal{F}}$ be the natural projection taking $x \in X$ to the equivalence class to which x belongs. The space $X_{\mathcal{F}}$ is given the quotient topology $\Im_{\mathcal{F}}$, where $A \in \Im_{\mathcal{F}}$ iff $f^{-1}(A) \in \mathcal{F}$. Since $\mathcal{F}$ is finite, the topological space $(X_{\mathcal{F}}, \Im_{\mathcal{F}})$ consists of a finite number of points; in Sorkin's terminology $\Im_{\mathcal{F}}$ would be a *finitary approximation* of $\mathcal{F}$. Finiteness of $\Im_{\mathcal{F}}$ reflects the fact that we have only coarse information (expressed by $\Im$) about X and therefore we must work with a (finite) number of "information granules".

By definition a finite space is a finite set X equipped with a topology $\Im$. Since X is finite, $\Im$ is closed under arbitrary intersections and arbitrary unions. A space which satisfies this condition is called in the literature an *Alexandroff space*. A good deal of what applies to finite spaces applies to Alexandroff spaces as well and all results presented in this paper may be easily proved for Alexandroff spaces too. However, finite spaces are more essential in computer science (for example they admit matrix representations [8]) and even in physics (one usually starts with finite approximations representing finite observations and seeks their inverse limit to understand an infinite space [12]). That is why we put emphasis on finite topological spaces and finite topological approximations.

If topology $\Im$ is closed under arbitrary intersections then for each set $A \subseteq X$ there exists the smallest (with respect to $\subseteq$) member of $\Im$ which includes A. In consequence, each $x \in X$ has the smallest neighbourhood defined as follows:

$$\triangle\, x = \bigcap\{A \in \Im : x \in A\}. \tag{2}$$

It allows us to convert the relation of set inclusion on $\Im$ into a preorder on elements of X:

$$x \preceq y \text{ iff } \triangle\, y \subseteq \triangle x. \tag{3}$$

It is often convenient to express $\preceq$ by means of *neighbourhood systems*; let $\mathcal{O}_x$ denote the neighbourhood system of x (i.e. the family of all open sets which includes x), then

$$x \preceq y \text{ iff } \mathcal{O}_x \subseteq \mathcal{O}_y. \tag{4}$$

In any case it holds that $x \in \triangle(x)$ and therefore (3) means that every open set containing x contains y as well. Consequently, $x \preceq y$ iff the constant sequence $(y, y, y, \ldots)$ converges to x or $x \in \overline{\{y\}}$, where $\overline{\{y\}}$ denotes the closure of $\{y\}$, i.e. $\overline{\{y\}}$ is the smallest closed set which includes $\{y\}$.

There is a bijective correspondence between preorders (i.e. reflexive and transitive relations) and Alexandroff topologies. Given a topological space $(X, \Im)$ by (3) we get a preorder $\preceq$. Given a preorder $\preceq$ on X, we may produce a topology $\Im_{\preceq}$ induced by a basis consisting of the following sets:

$$\triangle\, x = \{y \in X : x \preceq y\} \text{ for all } x \in X \ . \tag{5}$$

Therefore a set A is open iff A is a union of sets of the form (5). It is also easy to define a closure by means of $\preceq$:

$$\triangledown x = \{y \in X : y \preceq x\} \text{ for all } x \in X. \tag{6}$$

A set A is closed iff A is a sum of sets of the form (6). It follows that

$$x \preceq y \text{ iff } x \in \triangledown y. \tag{7}$$

Let us emphasize again, that as long as the space $(X, \Im)$ is Alexandroff, locally finite (i.e. for every point $x \in X$ the family $\mathcal{O}_x$ is finite) or finite, its topology will be equivalent to a preorder.

3 Data Analysis and Modal Logics

In [15] we have introduced the theory of Galois connections as a general framework for data analysis, within which both FCA and RST may be interpreted. Now we examine FCA and RST together with their Galois connections within a topological setting. Firstly, we recall some basic definitions and then show how FCA and RST may be interpreted by means of specialisation order. There is another correspondence, studied in [6], which is related to our discussion. Namely, the correspondence between preorders (Alexandroff topologies) and RST. As we have said there is on-to-one correspondence between Alexandroff spaces and preorders. We have associated every preordered set $(X, \preceq)$ with a topological space $(X, \Im_{\preceq})$ whose open sets are exactly the up-sets of $(X, \preceq)$, i.e. invariant sets under $\triangle$. In [6] authors take another approach, which is based on down-sets as opens. Although these structures are mathematically very much the same, the modal models built on them are quite different. In consequence only up-sets allow one to prove the (standard) topological completness theorem of S4.

Most results concerning Galois connections are well-known and may be found in course books. Our presentation is based on Erne et al. [5]. At the end of this section we show how RST and FCA are related to modal logics. Surprisingly, the topological setting of a finitary substitute brings tense logic S4.t as the account of relationships between RST and FCA.

The context (or data table) is a triple $\langle A, B, R\rangle$, where A is a set of objects, B is a set of attributes or properties and $R \subseteq A \times B$. $\langle a, b\rangle \in R$, for $a \in A$ and $b \in B$, reads as a has a property b. Any context gives rise to a Galois connection, which in turn induces a complete lattice of concepts.

Definition 4 (Galois Connection). *Let $(P, \leq)$ and $(Q, \preceq)$ be partially ordered sets (posets). If $\pi_* : P \mapsto Q$ and $\pi^* : Q \mapsto P$ are functions such that for all $p \in P$ and $q \in Q$, $p \leq \pi^* q$ iff $\pi_* p \preceq q$, then the quadruple $\pi = \langle (P, \leq), \pi_*, \pi^*, (Q, \preceq)\rangle$ is called a* Galois connection, *where π_* and π^* are called the coadjoint and adjoint part of π, respectively.*

Since $\pi_*(p) = inf\{q \in Q : p \leq \pi^*(q)\}$ and $\pi^*(q) = sup\{p \in P : \pi_*(p) \preceq q\}$ we often call π_* the lower adjoint of π^*, whereas π^* is called the upper adjoint of π_* – that is why we use * as a subscript or a superscript, respectively.

Given two posets $(\mathbf{P}A, \subseteq)$ and $(\mathbf{P}B, \subseteq)$, where $\mathbf{P}A$ is the power set of A, a polarity is a (contravariant) Galois connection π between $(\mathbf{P}A, \subseteq)$ and $(\mathbf{P}B, \subseteq)^{op}$ $= (\mathbf{P}B, \supseteq)$. It follows that both π_* and π^* reverse order, and both $\pi^* \circ \pi_*$ and $\pi_* \circ \pi^*$ form (algebraic) closure operators on $(\mathbf{P}A, \subseteq)$ and $(\mathbf{P}B, \subseteq)$, respectively. The basic proposition is as follows[1]:

Proposition 1. *Any relation $R \subseteq A \times B$ induces a contravariant Galois connection – polarity ${R_+}^+ = \langle(\mathbf{P}A, \subseteq), R_+, R^+, (\mathbf{P}B, \subseteq)\rangle$, where R_+ and R^+ are defined as follows: for any $U \subseteq A$ and $V \subseteq B$,*

$$R_+U = \{b \in B : (\forall a \in U)\langle a, b\rangle \in R\},$$
$$R^+V = \{a \in A : (\forall b \in V)\langle a, b\rangle \in R\}.$$

Polarities give rise to complete lattices, which are often called Galois lattices. Galois lattices constitute the formal foundation of Formal Concept Analysis (FCA), a mathematical theory of concepts and conceptual hierarchies introduced by Wille [14] in the early 80's. Elements of a Galois lattice – called concepts – consist of two parts: extension and intension. The extension of a concept consists of all objects belonging to the concept, whereas the intension consists of attributes which belong to all these objects.

Definition 5 (Concept). *A* concept *of a given context $\langle A, B, R\rangle$ is a pair (U, V), where $U \subseteq A$ and $V \subseteq B$ such that $U = R^+V$ and $V = R_+U$.*

Operators R_+ and R^+ are called by Wille derivation operators. The collection of all concepts of a given context is ordered by a subconcept – superconcept relation defined as follows: $(U_1, V_1) \leq (U_2, V_2)$ iff $U_1 \subseteq U_2$ (equivalently, $V_1 \supseteq V_2$). The set of all concepts of a given context $\langle A, B, R\rangle$ together with the defined order $\leq$ is denoted by $\mathbb{C}\langle A, B, R\rangle = \{(U, V) : U = R^+V \wedge V = R_+U\}$. The fundamental theorem of FCA states that:

Proposition 2 (Wille). *For any formal context $\langle A, B, R\rangle$, $\mathbb{C}\langle A, B, R\rangle$ is a complete lattice, called the concept lattice of $\langle A, B, R\rangle$, for which infima (meet) and suprema (join) are respectively:*

$$\bigwedge_{t\in T}(U_t, V_t) = (\bigcap_{t\in T} U_t, R_+R^+ \bigcup_{t\in T} V_t),$$
$$\bigvee_{t\in T}(U_t, V_t) = (R^+R_+ \bigcup_{t\in T} U_t, \bigcap_{t\in T} V_t).$$

A covariant Galois connection between power sets is called an axiality.

Proposition 3. *Any relation $R \subseteq A \times B$ induces a covariant Galois connection – axiality ${R_\exists}^\forall = \langle(\mathbf{P}A, \subseteq), R_\exists, R^\forall, (\mathbf{P}B, \subseteq)\rangle$, where $R_\exists$ and $R^\forall$ are defined as follows: for any $U \subseteq A$ and $V \subseteq B$,*

$$R_\exists U = \{b \in B : (\exists a \in A)\langle a, b\rangle \in R \wedge a \in U\},$$
$$R^\forall V = \{a \in A : (\forall b \in B)\langle a, b\rangle \in R \Rightarrow b \in V\}.$$

The theoretical dual of ${R_\exists}^\forall$, defined as ${R^\exists}_\forall = \langle R^\exists, R_\forall\rangle = {(R^{-1})_\exists}^\forall$, is also an axiality but from $(\mathbf{P}B, \subseteq)$ to $(\mathbf{P}A, \subseteq)$. R^{-1} means the converse relation of R, that is, $bR^{-1}a$ iff aRb. Now, we recall basic concepts of RST.

[1] A detailed presentation may be found in [4, 14].

Definition 6 (Approximation Operators). *Let A be a set, E an equivalence relation on A, and $[a]_E$ – the equivalence class containing a. With each $U \subseteq A$, we can associate its E-lower and E-upper* approximations, $\underline{E}U$ *and* $\overline{E}U$, *respectively, defined as follows:*

$$\underline{E}U = \{a \in A : [a]_E \subseteq U\},$$
$$\overline{E}U = \{a \in A : [a]_E \cap U \neq \emptyset\}.$$

Definition 7 (Information System). *An* information system *is a structure $I = \langle A, \Omega, \{B_\omega : \omega \in \Omega\}, f \rangle$, where:*

- *A is a finite set of objects,*
- *Ω is a finite set of attributes,*
- *For each $\omega \in \Omega$, B_ω is a set of attributes values of attribute ω; we let $B = \bigcup_{\omega \in \Omega} B_\omega$,*
- *$f : A \times \Omega \mapsto B$ is a function such that $f(a, \omega) \in B_\omega$ for all $a \in A$ and $\omega \in \Omega$,*
- *If $Q = \{\omega_1, ..., \omega_n\} \subseteq \Omega$ then $f_Q(a) = \langle f(a, \omega_1), ..., f(a, \omega_n) \rangle$, i.e. $f_Q : A \mapsto B_Q = B_{\omega_1} \times ... \times B_{\omega_n}$.*

Intuitively $f_Q(a)$ represents a description of an object a, i.e. a certain amount of knowledge about the object a, whereas $f_\Omega(a)$ may be viewed as the full description of this object. Let $R \subseteq A \times B_\Omega$ be a relational version of f_Ω (i.e. aRd iff $d \in f_\Omega(a)$) and let θ be a kernel of f_Ω (i.e. $a_1 \theta a_2$ iff $f_\Omega(a_1) = f_\Omega(a_2)$), then (A, θ) is an approximation space. The following result – dressed differently – was proved by Düntsch and Gediga [3, 4].

Proposition 4 (Düntsch and Gediga). $R^\forall R_\exists U = \overline{\theta}U$ *and* $R^\exists R_\forall U = \underline{\theta}U$.

Düntsch and Gediga [3, 4] have also proved that the operators of FCA and RST are mutually definable: theoretically, it suffices to have only one theory, for example FCA, because the second one, in this case RST, may be easily derived. However, we advocate here and in [15] that both theories should be kept in their original form not only for practical reasons (what Düntsch and Gediga explicitly admit as well) but also for theoretical one. As we have pointed out in [15] covariant Galois connections give rise to a number of interesting lattices. Especially, in the fashion of FCA two complete lattices of concepts might be formed: the lattice of upper-concepts and the lattice of lower concepts. Recently Yao in [17] has introduced another interpretation of concept lattices of FCA and RST, which shows that concepts in FCA and RST have different meanings. On the other hand FCA might produce its own approximation operators by taking $X \setminus R^+ R_+ X \setminus U$ as the FCA-lower approximation of U and $R^+ R_+ U$ as the FCA-upper approximation of U. Hence FCA gives rise to an alternative theory of rough set with respect to RST and RST allows building a theory of formal concepts alternative to FCA. The present paper provides another and much stronger argument: it proves on the base of the theory of finite approximations that well-known temporal system S4.t may be supplied with a new semantics given by FCA and RST together. Moreover, from the perspective of general topology S4.t is much better account for approximate reasoning then unimodal logics with RST semantics. Therefore we need the operators of FCA and RST simultaneusly!

Firstly, we convert a context $\langle A, B, R\rangle$ into a topological space. Each attribute $b \in B$ may be interpreted as a set of objects which possess it. Under natural assumption that each object $a \in A$ is described at least by one attribute, our knowledge forms a covering of A. The covering (i.e. the set of attributes) is usually finite, so is the topology generated by it. According to Sect. 2 we can relax this condition and assume that the covering is locally finite. In this setting R means $\in$. Hence a context $\langle A, B, R\rangle$ may be viewed as the set A equipped with a family of its subsets B. In a very natural way, it brings us a topological space $(A, \Im_B)$, where $\Im_B$ is generated by the subbasis B.

In Sect. 2 the specialisation order $\preceq$ has been defined by: $x \preceq y$ iff $\mathcal{O}_x \subseteq \mathcal{O}_y$. It means that we have better knowledge about y than about x. It is possible only in the case of incomplete information about a given domain. In the case of complete information any two objects x and y are incomparable by $\preceq$ or $\mathcal{O}_x = \mathcal{O}_y$. Therefore the following theorem relates RST to incomplete information expressed by means of $\preceq$.

Proposition 5. *For any context $\langle A, \Im_B, R\rangle$ and its topological space $(A, \Im_B)$,*

(*i*) $R^\forall R_\exists U = \{a \in A : (\exists a' \in U) a \preceq a'\} = \nabla U,$
(*ii*) $R^\exists R_\forall U = \{a \in U : (\forall a' \in A) a \preceq a' \Rightarrow a' \in U\}.$

Proof. (*i*) $R_\exists U = \{b \in \Im_B : (\exists a \in A) a \in b \wedge a \in U\}$, and hence $R_\exists U = \bigcup_{a \in U} \mathcal{O}_a$. $R^\forall V = \{a \in A : (\forall b \in \Im_B) a \in b \Rightarrow b \in V\}$. It follows that $R^\forall V = \{a \in A : \mathcal{O}_a \subseteq V\}$. Thus, $R^\forall R_\exists U = \{a \in A : (\forall \mathcal{O}_a)(\exists a' \in U) a' \in \mathcal{O}_a\} = \{a \in A : (\exists a' \in U) a' \in \Delta a\}$. Since in our case the intersection of any open sets is open, $R^\forall R_\exists U = \{a \in A : \Delta a' \subseteq \Delta a\}$, and by definition $R^\forall R_\exists U = \{a \in A : (\exists a' \in U) a \preceq a'\} = \nabla U$.

(*ii*) $R_\forall U = \{b \in \Im_B : (\forall a \in A) a \in b \Rightarrow a \in U\}$ and $R^\exists V = \{a \in A : (\exists b \in \Im_B) a \in b \wedge b \in V\}$. Since Δa is the smallest open set containing a, $R^\exists R_\forall U = \{a \in U : (\forall a' \in A) a' \in \Delta a \Rightarrow a' \in U\}$. By definition we get that $R^\exists R_\forall U = \{a \in U : (\forall a' \in A) a \preceq a' \Rightarrow a' \in U\}$. □

Obviously, $R^\forall R_\exists U$ is the topological closure of U, whereas $R^\exists R_\forall U$ is its topological interior. It is well known result. More important here is the fact that the upper approximation operator may be defined by means of ∇. Now we prove a similar result for FCA.

Proposition 6. *Let be given a context $\langle A, \Im_B, R\rangle$ and its topological space $(A, \Im)$, then*

$$R^+ R_+ U = \{a \in A : (\exists a' \in U) a' \preceq a\} = \Delta U$$

Proof. $R_+ U = \{b \in \Im_B : (\forall a \in U) a \in b\}$, and $R^+ V = \{a \in A : (\forall b \in \Im_B) a \in b\}$. Hence $R^+ R_+ U = \bigcap\{b \in \Im_B : U \subseteq b\}$. It is the smallest open set containing U and therefore $R^+ R_+ U = \{a \in A : a \in \Delta a'$ for some $a' \in U\} = \{a \in A : a' \preceq a$ for some $a' \in U\} = \Delta U$. □

From both theorems it follows that the composition of derivation operators of FCA and the upper approximation operator of RST are mutually opposite with respect to the specialization order. It is a crucial fact for supplying temporal logic S4.t with a new semantics, which is the main objective of the present section.

There is a well known connection between RST and modal logics. Before we state it, let us recall a few basic definitions. The standard modal language is given by the following rule:

$$\phi = p \mid \top \mid \neg\phi \mid \phi \wedge \varphi \mid \Box\phi \mid \Diamond\phi$$

where p ranges over the set of proposition letters. A *frame* for the standard modal language is a pair $\mathcal{F}r = (W, \theta)$ where W is a nonempty set and θ is a binary relation on W. A *model* for the standard modal language is a pair $(\mathcal{F}r, V)$ where $\mathcal{F}r$ is a frame and V is a valuation function assigning to each proposition letter p a subset $V(p)$ of W. A valuation V can be extended to the set of all modal formulas. The part for pure Boolean formulas is as usual:

$$V(\top) = W,$$
$$V(\phi \wedge \varphi) = V(\phi) \cap V(\varphi),$$
$$V(\neg\phi) = W \setminus V(\phi).$$

The key part (for modal operators) may be given by means of approximation operators of RST.

$$V(\Diamond\phi) = \overline{\theta}(V(\phi)),$$
$$V(\Box\phi) = \underline{\theta}(V(\phi)).$$

Hence RST provides semantics for normal modal systems. For example, when θ is an equivalence relation the corresponding logic is S5. There have been extensive studies concerned with non-classical logics and RST. In consequence, modal logics with modal operators interpreted by the lower and upper approximations have been regarded as leading logics of approximate reasoning [2, 10].

Now we focus on modal logic S4 and its topological model.

Definition 8 (Topological Model). *A topological model is a triple $(X, \Im, V)$ where $(X, \Im)$ is a topological space and the valuation V assigns propositional letters subsets of X. The definition of truth is as follows:*

$$x \models p \ \textit{ iff } \ x \in V(p)$$
$$x \models \top \ \textit{ always}$$
$$x \models \phi \wedge \psi \ \textit{ iff } \ x \models \phi \ \textit{ and } \ x \models \psi$$
$$x \models \neg\phi \ \textit{ iff } \ x \not\models \phi$$
$$x \models \Box\phi \ \textit{ iff } \ (\exists O \in \Im) x \in O \models \phi$$

The last part of the truth definition says that $\Box\phi$ is true at x when ϕ is true in some open neighbourhood of x. Given that a finite (or Alexandroff) topological space $(X, \Im)$ gives rise to a preordered set $(X, \preceq_{\Im})$, it follows that $\Box\phi$ is true at x just when ϕ is true at all $\preceq$-successors of x. That is why we have used up-sets of $\preceq$ as opens of $(X, \Im_{\preceq})$ in (3). Let us recall now, that

Proposition 7. S4 *is strongly complete with respect to the class of all topological spaces.*

In contrast to the standard modal language, the basic temporal language has two primitive unary operators **F** and **P**. Each primitive operator has its dual operator $\mathbf{F} = \neg\mathbf{G}\neg$ and $\mathbf{P} = \neg\mathbf{H}\neg$. These traditional *a la Prior* temporal modalities express temporal aspects of truth:

$\mathbf{F}\varphi$ reads as "sometime in the **F**uture it will be the case that φ ",
$\mathbf{G}\varphi$ reads as "always in the future it is **G**oing to be the case that φ ",
$\mathbf{P}\varphi$ reads as "sometime in the **P**ast it was the case that φ ",
$\mathbf{H}\varphi$ reads as "it always **H**as been the case that φ ".

The basic temporal logic $K.t$ is axiomatized by the following schemata:

$$(\mathrm{K_G}) \quad \mathbf{G}(\phi \to \psi) \to (\mathbf{G}\phi \to \mathbf{G}\psi),$$
$$(\mathrm{K_H}) \quad \mathbf{H}(\phi \to \psi) \to (\mathbf{H}\phi \to \mathbf{H}\psi),$$
$$(\mathrm{GP}) \quad \phi \to \mathbf{GP}\phi,$$
$$(\mathrm{HF}) \quad \phi \to \mathbf{HF}\phi,$$

and the rules of inference

$$(\mathrm{NEC_G}) \quad \text{if } \vdash \phi \text{ then } \vdash \mathbf{G}\phi,$$
$$(\mathrm{NEC_H}) \quad \text{if } \vdash \phi \text{ then } \vdash \mathbf{H}\phi.$$

Temporal semantics is given by a *bidirectional frame* $\mathcal{F} = (W, \theta, \theta^{-1})$ where θ^{-1} is the converse of θ. Since the converse relation θ^{-1} is directly given by θ, one typically writes only (W, θ) instead of (W, θ, θ^{-1}). Adding to $K.t$ the transivity axiom we get tense logic S4.t.

$$(\mathrm{TRAN}) \quad \mathbf{G}\phi \to \mathbf{GG}\phi.$$

Let us come back to a finite (or Alexandroff) topological space $(X, \Im)$ and its preordered set $(W, \preceq_\Im)$. Of course $(W, \preceq_\Im, \succeq_\Im)$ is a bidirectional frame. According to the standard Kripke semantics of temporal logics $\mathbf{G}\phi$ is true at x when ϕ is true at all $\preceq_\Im$-successors of x, whereas $\mathbf{H}\phi$ is true at x just when ϕ is true at all $\succeq_\Im$-successors of x. It follows from (3) and (7) that

$$x \models \mathbf{G}(\phi) \text{ iff there exists an open set } \mathrm{X}' \subseteq \mathrm{X} \text{ such that } \mathrm{x} \in \mathrm{X}' \models \phi \quad (8)$$
$$x \models \mathbf{H}(\phi) \text{ iff there exists a closed set } \mathrm{X}' \subseteq \mathrm{X} \text{ such that } \mathrm{x} \in \mathrm{X}' \models \phi \quad (9)$$

However, all temporal intuitions in this case should be left aside. Hence we get the following proposition:

Proposition 8. S4.t *is strongly complete with respect to the class of all preordered bidirectional frames and their topological spaces. Definition of truth is given by (8) and (9). If V is a valuation function then:*

(i) $V(\mathbf{P}\phi) = R^+R_+V(\phi) = \triangle V(\phi)$,
(ii) $V(\mathbf{F}\phi) = R^\forall R_\exists V(\phi) = \bigtriangledown V(\phi)$,
(iii) $V(\mathbf{G}\phi) = R^\exists R_\forall V(\phi)$.

The proof of Proposition 8 is immediate from propositions 5, 6 and 7. However, let us take a closer look at the valuation function. According to Kripke semantics $V(\mathbf{G}\phi)$ is defined as a set of points whose all $\preceq$-successors make ϕ true. That is nothing than the set of points having an open neighbourhood $O \subseteq V(\phi)$. It is the interior of $V(\phi)$ and hence $V(\mathbf{G}\phi) = R^{\exists}R_{\forall}V(\phi)$. $V(\mathbf{F}\phi)$ is just a set of points, which have a $\preceq$-successor satisfying ϕ. By (7) it is the set of points belonging to the closure of some element of $V(\phi)$. Since the underlying space is Alexandroff it follows that $V(\mathbf{G}\phi)$ is the closure of $V(\phi)$. Hence $V(\mathbf{F}\phi) = \bigtriangledown V(\phi)$. Similarly $V(\mathbf{H}\phi)$ is defined as a set of points whose all $\succeq$-successors make ϕ true. By (7) it is a set of points whose closure is contained in $V(\phi)$. That is the biggest closed subset of $V(\phi)$. Since R^{+} and R_{+} form a contravariant Galois connection the operator $\mathbf{H}$ cannot be expressed directly by FCA derivation operators, but only can be defined as the dual operator to $\mathbf{P}$. $V(\mathbf{P}\phi)$ is defined as a set of points, which have a $\succeq$-successor making ϕ true. By (3) it is the set of points belonging to the smallest neighbourhood of some element of $V(\phi)$. That isnothing but the smallest open set containing $V(\phi)$. Hence $V(\mathbf{P}\phi) = \bigtriangleup V(\phi)$.

Both FCA and RST are very popular and attract a lot of scientific attention today. Many efforts have been made in order to establish their relationships and dependencies. But there is a lack of results on logics reflecting connexions between FCA and RST. Let us recall that such logical investigations into rough set aspects has been very fruitful [7, 10, 11, 16] what should enqurage us to make similar effort in the case of FCA . Proposition 9 brings not only the first logic of relationships between RST and FCA but also provides a modal unification of both theories. In the next section we prove that the modal fusion of RST and FCA is – from the perspectiove of general topology – a better account for approximate reasoning than so popular unimodal logics interpreted by RST.

4 Modal View of a Finitary Approximation

As we have just said, this section aims at proving that S4.t is a better account of approximate reasoning than unimodal logics interpreted by means of RST. Firstly, an approximation is closely related with its approximated space. In what follows, we prove that they are homotopy equivalent. Secondly, each language has its own expressive power. Given these two premises we look for a language which has expressive power as close to the concept of homotopy equivalence as possible.

As it is well known the crucial semantic concept for modal logics is a *bisimilarity*:

Definition 9 (Bisimulation). *Let* $\boldsymbol{M} = (W, \theta, V)$ *and* $\boldsymbol{M'} = (W', \theta', V')$ *be two models, a nonempty relation* $Z \subseteq W \times W'$ *is called a* bisimulation *between* $\boldsymbol{M}$ *and* $\boldsymbol{M'}$ *iff*

(*i*) *if* wZw' *then* $w \in V(p)$ *iff* $w' \in V(p)$ *for any proposition letter* p,
(*ii*) *if* wZw' *and* $w\theta v$ *then there exists* $v' \in W'$ *such that* $w'\theta' v'$ *and* vZv' *(the forth condition),*
(*iii*) *if* wZw' *and* $w'\theta' v'$ *then there exists* $v \in W$ *such that* $w\theta v$ *and* vZv' *(the back condition).*

Bisimulations may be viewed as relational generalizations of *bounded morphisms*, which are more convenient for us, since we work with topological spaces and continuous maps. However a bounded morphism is a stronger concept than a bisimulation: if there exists a bounded morphism between two models then these models are bisimilar.

Definition 10 (Bounded Morphism). *Let* $\boldsymbol{M}$*=*(W,θ,V)*and*$\boldsymbol{M'}$*=*(W',θ',V') *be two models, a map* $f : \boldsymbol{M} \mapsto \boldsymbol{M'}$ *is a* bounded morphism *iff*

(*i*) $w \in V(p)$ *iff* $f(w) \in V(p)$ *for any proposition letter* p,
(*ii*) f *is a homomorphism with respect to* θ,
(*iii*) *if* $f(w)\theta'v'$ *then there exists* $v \in W$ *such that* $w\theta v$ *and* $f(v) = v'$ *(the back condition).*

Let us recall that there is a bijective correspondence between preorders and Alexandroff spaces. Given a finite topological space $(X, \Im)$ by (4) we get a preorder $\preceq$. Given a preorder $\preceq$ on X we may produce a topology $\Im_{\preceq}$. Given this correspondence we get that the condition (ii) of Definition 10 translates to continuity while the condition (iii) means that f takes open sets to open sets. Hence a bounded morphism is just a continuous open map. The case of bisimulation is a bit more complicated.

Definition 11 (Topological Bisimulation). *Let be given two topological models* $(X, \Im, V)$ *and* $(X', \Im', V')$ *then a* topological bisimulation *is a nonempty relation* $Z \subseteq X \times X'$ *such that*

(*i*) *if* wZw' *then* $w \in V(p)$ *iff* $w' \in V(p)$ *for any proposition letter* p,
(*ii*) *if* wZw' *and* $w \in \mathcal{O} \in \Im$ *then there exists* $\mathcal{O}' \in \Im'$ *such that* $w' \in \mathcal{O}'$ *and for all* $v' \in \mathcal{O}'$ *there exists* $v \in \mathcal{O}$ *such that* vZv' *(the forth condition),*
(*iii*) *if* wZw' *and* $w' \in \mathcal{O}' \in \Im'$ *then there exists* $\mathcal{O} \in \Im$ *such that* $w \in \mathcal{O}$ *and for all* $v \in \mathcal{O}$ *there exists* $v' \in \mathcal{O}'$ *such that* vZv' *(the back condition).*

The notion of topological bisimulation has been introduced by Aiello and van Benthem in [1], where it has been proved that modal formulas are invariant under continuous open maps and topological bisimulations:

Proposition 9 (Aiello, van Benthem). *Let* $(X, \Im)$ *and* $(X', \Im')$ *be topological spaces and let* $Z \subseteq X \times X'$ *be a topological bisimulation between them. If wZw' then for any modal formula* α, $(X, \Im), w \models \alpha$ *iff* $(X', \Im'), w' \models \alpha$.

Proposition 10 (Aielo, van Benthem). *Let* $(X, \Im)$ *and* $(X', \Im')$ *be topological spaces and* $f : (X, \Im) \mapsto (X', \Im')$ *a continuous open map. For any modal formula* α, $(X, \Im) \models \alpha$ *iff* $(X', \Im') \models \alpha$.

In the case of the basic temporal language we work with bidirectional frames and models. Therefore a *temporal bisimulation* needs take into account also θ^{op}.

Definition 12 (Temporal Bisimulation). *Let* $\boldsymbol{M} = (W, \theta, V)$ *and* $\boldsymbol{M'} = (W', \theta', V')$ *be two models, a nonempty relation* $Z \subseteq W \times W'$ *is called a* temporal bisimulation *between* $\boldsymbol{M}$ *and* $\boldsymbol{M'}$ *iff it satisfies the conditions of Definition 9 and*

(iv) if wZw' and $v\theta w$ then there exists $v' \in W'$ such that $v'\theta' w'$ and vZv' (the forth condition),

(v) if wZw' and $v'\theta' w'$ then there exists $v \in W$ such that $v\theta w$ and vZv' (the back condition).

The similar change must be made in Definition 10 so as to get a *temporal bounded morphism.*

Definition 13 (Temporal Bounded Morphism). *Let $\boldsymbol{M} = (W, \theta, V)$ and $\boldsymbol{M}' = (W', \theta', V')$ be two bidirectional models, a map $f : \boldsymbol{M} \mapsto \boldsymbol{M}'$ is a* temporal bounded morphism *iff it satisfies the conditions of Definition 10 and*

(iv) f is a homomorphism with respect to θ^{op},

(v) if $v'\theta' f(w)$ then there exists $v \in W$ such that $v\theta w$ and $f(v) = v'$ (the back condition).

It means that a temporal bounded morphism must be additionally a closed map. Similarly we must change Definition 11.

Definition 14 (Topological Temporal Bisimulation). *Given two topological models $(X, \Im, V)$ and $(X', \Im', V')$ a* topological temporal bisimulation *is a nonempty relation $Z \subseteq X \times X'$ such that Z satsfies conditions of Definition 11 and*

(iv) if wZw' and $w \in S$, where S is a closed subset of X, then there exists a closed subset S' of W' such that $w' \in S'$ and for all $v' \in S'$ there exists $v \in S$ such that vZv' (the forth condition),

(v) if wZw' and $w' \in S'$, where S' is a closed subset of X' then there exists a closed subset S of X such that $w \in S$ and for all $v \in S$ there exists $v' \in S'$ such that vZv' (the back condition).

Proposition 11. *Let $(X, \Im)$ and $(X', \Im')$ be topological spaces and let $Z \subseteq X \times X'$ be a topological temporal bisiumulation between them. If wZw' then for any temporal formula α, $(X, \Im), w \models \alpha$ iff $(X', \Im'), w' \models \alpha$.*

Proposition 12. *Let $(X, \Im)$ and $(X', \Im')$ be topological spaces and $f : (X, \Im) \mapsto (X', \Im')$ a continuous open and closed map. For any temporal formula α, $(X, \Im) \models \alpha$ iff $(X', \Im') \models \alpha$.*

Both theorems are simple generalizations of propositions 9 and 10 respectively and therefore we leave them without proofs. Now let us come back to a finitary approximation. The following theorem connects it with the basic temporal language.

Proposition 13. *Let $(X, \mathcal{F})$ be a (locally) finite topological space and let $(X_{\mathcal{F}}, \Im_{\mathcal{F}})$ be its finitary substitution, then*

(i) The natural projection $f : (X, \mathcal{F}) \mapsto (X_{\mathcal{F}}, \Im_{\mathcal{F}})$ is a temporal bounded morphism.

(ii) $(X, \mathcal{F})$ and $(X_{\mathcal{F}}, \Im_{\mathcal{F}})$ are (topologically) temporally bisimilar.

(iii) For any temporal formula α, $(X, \mathcal{F}) \models \alpha$ iff $(X_{\mathcal{F}}, \Im_{\mathcal{F}}) \models \alpha$.

Proof. Obviously (i) implies (ii), which in turn implies (iii). So it suffices to prove only (i). Firstly the natural projection as a quotient map takes saturated open sets into opens and saturated closed sets into closed sets. Since, by the method of construction, each open or closed set is saturated, f is a continuous open and closed map. It means that f is a temporal bounded morphism. □

If we consider modal logics as reasoning about spaces of objects and their properties 13 claims that S4.t is much better account of approximate reasoning then unimodal systems. Also in spatial reasoning, which has attracted much of attention in computer science, S4.t should find a lot of applications. However, when considering modal logics as reasoning about binary relations, S4.t is as good as any other modal system – it applies to a certain class of relations.

Although a topological space and its finite approximation are temporally bisimilar, it does not mean that S4.t is sufficently strong.

Definition 15 (Homotopy). *Let X and X' be topological spaces and let $f, g : X \mapsto X'$ be continuous maps. A* homotopy *from f to g is a continuous map $H : X \times [0,1] \mapsto X'$ such that $H(x, 0) = f(x)$ and $H(x, 1) = g(x)$ for all $x \in X$.*

Definition 16 (Homotopy Type). *Let X and X' be a topological spaces and let $f : X \mapsto X'$ and $g : X' \mapsto X$ be continuous maps. If $g \circ f$ is homotopic to identity on X and $f \circ g$ i homotopic to identity X' then we say that X and X' are* homotopy equivalent *or have the same* homtopy type.

Proposition 14. *A topological space $(X, \mathcal{F})$ and its finitary substitute (approximation) $(X_{\mathcal{F}}, \Im_{\mathcal{F}})$ are homotopy equivalent.*

Proof. The proof is exactly the same as the proof of McCord's theorem [9], which claims that if X is a finite space then there exists a quotient T_0 space such that the quotient map is a homotopy equivalence. The proof is valid also in the case when X is an Alexandroff space.

Homotopy is much more flexible than homeomorphism (two spaces may be homotopy equivalent though not homeomorphic), yet it preserves many algebraic-topological properties (for example homology groups). However, neither a homotopy nor conditions which imply a homotopy may be defined by means of (standard) modal concepts such as open, closed and continuous maps. The author consulted this problem with topologists who said that there is no theorem about homotopy, open and closed map since there is no direct connection among them. However, none has provided a proof. McCord's theorem suggests that two spaces X and Y are homotopic if there is a closed and open map from X to Y and additionaly Y has at least as many points as there are elements of the minimal basis of X. Of course the latter condition is not modally definable.

5 Summary

In the course of this paper, we have established relationships between topological approximations on one hand and theories of data analysis, namely RST and FCA, on the other hand. Both RST and FCA have been interpreted by means of the specialisation order and proved together to provide a semantics to tense logic S4.t. As it is well-known, tense logics are stronger than unimodal systems. The semantic counterpart of their expressive power is the concept of temporal bisimilarity. We have proved that a (finite) topological space and its finite approximation are indistinguishable by means of the standard temporal language. It means that in the case of (topological or spatial) approximate reasoning we should employ (at least) S4.t interpreted by RST and FCA instead of popular unimodal systems with RST semantics.

References

1. Aiello, M., van Benthem, J., Logical patterns in space, in Barker-Plummer, D. et al. (Eds.), Logic Unleashed: Language, Diagrams, and Computation, CSLI Publ., Stanford.
2. Demri, S. P., Orłowska, E. S., Incomplete Information: Structure, Inference, Complexity, Monographs in Theoretical Computer Science, Springer, Heidelberg.
3. Düntch, I., Gediga, G. (2002), Modal-style operators in qualitative data analysis, in Proc. of the 2002 IEEE Conference on Data Mining (ICDM'2002), 155-162.
4. Düntch, I., Gediga, G. (2003), Approximation operators in qualitative data analysis, in de Swart, H. et al. (Eds.), Theory and Applications of Relational Structures as Knowledge Instruments, LNCS, 216-233, Springer, Heidelberg.
5. Erne, M. et al. (1991), A primer on Galois connections. http://www.math.ksu.edu/~strecker/primer.ps
6. Järvinen, J., Kortelainen, J., A note on definability in Rough Set Theory,
7. Lin, T. Y., Liu, Q. (1996), First order rough logic I: approximate reasoning via rough sets, Fundamenta Informaticae 27, 137-153.
8. May, J. P., Finite topological spaces, http://www.math.uchicago.edu/~ may/
9. McCord, M. C. (1966), Singular homology groups and homotopy groups of finite topological spaces, Duke Mathematical Journal 33, 465-474.
10. Orłowska, E. (1994), Rough set semantics for non-classical logics, in Ziarko, W. P. (ed.) Rough Sets, Fuzzy Sets and Knowledge Discovery, Springer Verlag, London, 143-148.
11. Rasiowa, H., Skowron, A. (1985), Approxiamtion logic, in Proceedings of the International Spring School, Mathematical Methods of Specification and Synthesis of Software Systems.
12. Sorkin, R. D. (1991), Finitary substitute for continuous topology, Int. J. Theoretical Physics 30, 923.
13. Pawlak, Z. (1982), Rough sets, Int. J. Computer and Information Sci., 11, 341-356.
14. Wille, R. (1982), Reconstructing lattice theory: An approach based on hierachies of concepts, in Rival, I. (Ed.), Ordered Sets, NATO Advanced Studies Institute, 445-470, Dordrecht, Reidel.
15. Wolski, M. (2004), Galois connection and data analysis, Fundamenta Informaticae 60(1-4), 401-415.
16. Yao, Y. Y., Lin, T. Y. (1996), Generalizations of rough sets using modal logic, Journal of the Intelligent Automation and Soft Computing 2, 103-120.
17. Yao, Y. Y., A comparative study of Formal Concept Analysis and Rough Set Theory in data analysis, Rough Stes and Current Trends in Computing 2004, 59-68.

Time Complexity of Decision Trees

Mikhail Ju. Moshkov[1,2]

[1] Faculty of Computing Mathematics and Cybernetics,
Nizhny Novgorod State University,
23, Gagarina Ave., Nizhny Novgorod, 603950, Russia
[2] Institute of Computer Science, University of Silesia
39, Będzińska St., Sosnowiec, 41-200, Poland
moshkov@unn.ac.ru

Abstract. The research monograph is devoted to the study of bounds on time complexity in the worst case of decision trees and algorithms for decision tree construction. The monograph is organized in four parts. In the first part (Sects. 1 and 2) results of the monograph are discussed in context of rough set theory and decision tree theory. In the second part (Sect. 3) some tools for decision tree investigation based on the notion of decision table are described. In the third part (Sects. 4–6) general results about time complexity of decision trees over arbitrary (finite and infinite) information systems are considered. The fourth part (Sects. 7–11) contains a collection of mathematical results on decision trees in areas of rough set theory and decision tree theory applications such as discrete optimization, analysis of acyclic programs, pattern recognition, fault diagnosis and probabilistic reasoning.

Keywords: decision tree, rough set theory, test theory, time complexity

1 Introduction

Decision trees are widely used in different applications. The theory of decision trees continues to be a source of rich mathematical problems. This theory is closely connected with rough set theory created by Z. Pawlak (cf. [153]–[162]), and developed by many authors (cf. [37, 150, 168, 169, 194–196, 201]).

1.1 Crisp and Rough Classification Problems

To better explain the role of decision trees among the models of algorithms in rough set theory, and to place the results discussed in this monograph in the context of rough set theory, let us consider the notion of decision system which is a special kind of information system [44]. Any decision system is specified by a number of conditional attributes that divide a set of objects into domains on which these attributes have fixed values. Our aim is to find the value of a decision attribute using only values of conditional attributes. If the decision attribute is constant on each domain, this classification problem is called crisp, otherwise the classification problem is considered rough. Rough set theory gives us some

J.F. Peters and A. Skowron (Eds.): Transactions on Rough Sets III, LNCS 3400, pp. 244–459, 2005.

tools to work with rough problems, for example, tools for measurement of the degree of the roughness.

In the case of a crisp classification problem we can always find the exact value of the decision attribute using only values of conditional attributes. In the case of the rough classification problem we can sometimes find only some information on the decision attribute value. For example, instead of the value of the decision attribute on an object from a domain we can find the set of values of the decision attribute on all objects from the domain. In this case, we have in some sense a reduction of a rough problem to a crisp problem. Various types of such reductions are considered in [193, 195, 199, 200].

1.2 Three Approaches to Classification Problem Solving

Typically, in rough set theory, three approaches to solving the problem of classification are used. These approaches are based on the notions of: relative reduct, complete (applicable to any object) decision rule system and decision tree. A relative reduct is a subset of conditional attributes which gives the same information on decision attribute value as a whole set of conditional attributes. The first two approaches are usual for rough set theory [146, 193, 199, 200]. Several efficient methods for construction of reducts and rule systems have been developed [5, 145, 147, 197]. The third approach is used in rough set theory investigations to a smaller degree.

There are three main types of characteristics of relative reducts, complete decision rule systems and decision trees that are of interest to us: complexity of description, precision and time complexity.

Two sources of decision systems are known: experimental data (in this case, usually we know only a part of the set of objects) and data derived from completely described problems in areas such as discrete optimization and fault diagnosis [77, 115].

If we have a decision system connected with results of experiments we can see on reducts, rule systems and trees as on ways for knowledge representation or as on predictors of the decision attribute value. In the first case the complexity of description of reducts, rule systems and trees is the most important for us. In the second case we try to find a reduct, a rule system or a tree which will work with existing and new objects in the most precise way.

If we have a decision system corresponding to a completely described problem we see usually on reducts, rule systems and trees as on algorithms for this problem solving. In this case we are interesting in algorithms with small time complexity.

This research monograph is devoted mainly to the study of the role of decision trees as algorithms for completely described problems. We consider only crisp problems of classification and assume that we know values of all attributes on all objects. This setting is very close to the setting of classification problems in test theory created by S.V. Yablonskii and I.A. Chegis [14, 216]. However, most of the results presented in this monograph can be used for investigation of rough problems, especially if we consider a reduction of rough problems to crisp ones.

1.3 Comparison of Approaches

Now we compare possibilities of relative reducts, complete decision rule systems and decision trees as algorithms for solving problems involving classification.

Any partition of the set of objects generated by a relative reduct can be simulated by the decision tree which computes sequentially, values of all attributes from the considered reduct. In general, the class of partitions generated by decision trees is essentially broader than the class of partitions generated by relative reducts. This makes it possible sometimes to construct decision trees which are faster than algorithms based on sequential computation of values of attributes from a relative reduct.

For example, for any natural n there exists a decision system with binary conditional attributes such that the minimal cardinality of relative reduct is equal to $2^n - 1$ and there exists a decision tree for which the depth (the maximal length of a path from the root to a terminal node) is equal to n.

Furthermore, any partition of the set of objects generated by a decision tree can be simulated by the complete system of decision rules corresponding to all paths from the root to terminal nodes in the considered tree. In general, the class of coverings generated by complete decision rule systems is essentially broader than the class of partitions generated by decision trees. This makes it possible sometimes to construct algorithms based on complete decision rule systems which are faster than decision trees, especially in a multi-processor computing environment.

For example, for any natural n there exists a decision system with binary conditional attributes such that the minimal depth of decision tree is equal to n and there exists a complete decision rule system for which the number of conditions in the left-hand side of each rule is equal to 1. In order to realize this advantage of the decision rule system we must have sufficient number of processors.

Thus, decision trees as algorithms for solving of completely described problems in some sense lie between algorithms based on relative reducts and algorithms based on complete decision rule systems. The consideration of decision trees is completely justified if we use either a uni-processor computing or multi-processor computing (with a small number of processors) environment.

Of course, there are also another important distinctions among reducts, trees and rule systems, especially if we consider partially described problems. For example, let we have new object which was not used under the construction of a decision tree and a decision rule system. For this object the decision tree will give us at most one value of the decision attribute, but the decision rule system can give us several values, even if the considered classification problem is crisp. This peculiarity allows to create new types of classification algorithms based on the use of decision rule systems and conflict resolution strategies.

1.4 On Contents of Monograph

This monograph is devoted to study of depth and weighted depth of deterministic decision trees over both finite and infinite information systems. Decision trees over finite information systems are investigated in rough set theory, test theory, theory of questionnaires, in machine learning, etc. The notion of infinite information system is useful in discrete optimization, pattern recognition, computational geometry. However, decision trees over infinite information systems are investigated to a lesser degree than over finite information systems.

The monograph consists of 11 sections and appendix. In Sect. 2 results of the monograph are discussed in context of decision tree theory. In Sect. 3 bounds on time complexity and algorithms for construction of decision trees for decision tables are considered. Sects. 4 and 5 are devoted to the development of local and global approaches to the investigation of the decision trees over arbitrary (finite and infinite) information systems. In Sect. 6 decision trees over quasilinear information systems are studied. Some applications of results from Sect. 6 are considered in Sects. 7 and 8. In Sect. 7 six classes of problems of discrete optimization, sorting and recognition over quasilinear information systems are studied. In Sect. 8 the complexity of acyclic programs in the basis $\{x + y, x - y, 1; \text{sign}(x)\}$ is investigated. In Sect. 9 the depth of decision trees for recognition of words of regular languages is studied. In Sect. 10 the problem of diagnosis of constant faults in combinatorial circuits is considered. Sect. 11 is devoted to study of decision trees for computation of values of observable variables in Bayesian networks. In the appendix, the structure of all classes of Boolean functions, closed relatively the substitution operation and the operations of insertion and deletion of unessential variable, is described. Definitions, notation and results contained in appendix are used in Sects. 3 and 10.

The major part of this monograph consists of the author's own results. The monograph is essentially revised and extended version of [79] containing many new results and two new sections devoted to problems of fault diagnosis and probabilistic reasoning.

2 Results of Monograph in Context of Decision Tree Theory

This section consists of two subsections. In the first subsection a review of several parts of decision tree theory is given. This will allow the reader to understand the nature of the results presented in this monograph. The outline of these results is contained in the second subsection.

We denote by $\mathbb{N}$ the set $\{0, 1, 2, \ldots\}$ of natural numbers including 0. The set of integers, the set of rational numbers and the set of real numbers are denoted by $\mathbb{Z}$, $\mathbb{Q}$ and $\mathbb{R}$.

Let f and g be partial functions from $\mathbb{N}$ to $\mathbb{N}$. Later we will use the following notation.

The equality $g(n) = O(f(n))$ means that there exist positive constants c and n_0 such that for any integer $n \geq n_0$ the values of $f(n)$ and $g(n)$ are definite and the inequality $g(n) \leq cf(n)$ holds.

The equality $g(n) = \Omega(f(n))$ means that there exist positive constants c and n_0 such that for any integer $n \geq n_0$ the values of $f(n)$ and $g(n)$ are definite and the inequality $g(n) \geq cf(n)$ holds.

The equality $g(n) = \Theta(f(n))$ means that there exist positive constants c_1, c_2 and n_0 such that for any integer $n \geq n_0$ the values $f(n)$ and $g(n)$ are definite and the inequalities $c_1 f(n) \leq g(n) \leq c_2 f(n)$ hold.

2.1 On Decision Tree Theory

Basic notions of decision tree theory and also several of its structural parts are discussed in this subsection. These parts comprise the investigation of decision trees over finite and infinite information systems and some applications.

Basic Notions. Let A be a nonempty set, B be a finite nonempty set of integers with at least two elements and F be a nonempty set of functions from A to B. Functions from F are called *attributes* and the triple $U = (A, B, F)$ is called *an information system*. If F is a finite set then U is called *a finite* information system. If F is an infinite set then U is called *an infinite* information system.

We will consider problems over the information system U. *A problem over* U is an arbitrary $(n+1)$-tuple $z = (\nu, f_1, \ldots, f_n)$ where $\nu : B^n \to \mathbb{Z}$, and $f_1, \ldots, f_n \in F$. The tuple $(\nu, f_1, \ldots, f_n)$ is called *the description* of the problem z, and the number $\dim z = n$ is called *the dimension* of the problem z. The problem z may be interpreted as a problem of searching for the value $z(a) = \nu(f_1(a), \ldots, f_n(a))$ for an arbitrary $a \in A$. Note that one can interpret $f_1, \ldots, f_n$ as conditional attributes and z as a decision attribute. Different problems of pattern recognition, discrete optimization, fault diagnosis and computational geometry can be represented in such form. The set of all problems over the information system U is denoted by Probl_U.

A decision tree over U is a finite tree with the root in which each terminal node is labelled by a number from $\mathbb{Z}$ (a result of the tree work); each nonterminal node is labelled by an attribute from F; each edge is labelled by a number from B (the value of the attribute for which the jump is realized along this edge). Edges starting in a nonterminal node are labelled by pairwise different numbers. *A complete path* in a decision tree is an arbitrary directed path from the root to a terminal node of the tree.

A weight function for the information system U is a function $\psi : F \to \mathbb{N} \setminus \{0\}$. The value $\psi(f)$ for an attribute $f \in F$ is called *the weight* of the attribute f, and can be interpreted as the complexity of the computation of the attribute value. Let Γ be a decision tree over the information system U. The weight of a complete path in Γ is the sum of weights of attributes attached to nodes of this path. We denote by $\psi(\Gamma)$ the maximal weight of a complete path in Γ. The number $\psi(\Gamma)$ is called *the weighted depth* of Γ. We denote by h the weight function such that $h(f) = 1$ for any attribute $f \in F$. The number $h(\Gamma)$ is called *the depth* of Γ. Note that $h(\Gamma)$ is the maximal length of a complete path in Γ.

The investigation of decision trees solving the problem $z = (\nu, f_1, \ldots, f_n)$ and using only attributes from the set $\{f_1, \ldots, f_n\}$ is based on the study of *the decision table* $T(z)$ associated with the problem z. The table $T(z)$ is a rectangular table with n columns which contains elements from B. The row $(\delta_1, \ldots, \delta_n)$ is contained in the table $T(z)$ if and only if the equation system

$$\{f_1(x) = \delta_1, \ldots, f_n(x) = \delta_n\}$$

is compatible (has a solution) on the set A. This row is labelled by the number $\nu(\delta_1, \ldots, \delta_n)$. For $i = 1, \ldots, n$ the i-th column is labelled by the attribute f_i.

The notions introduced above may be defined in variety of ways and some alternatives will be mentioned here. Researchers in the area of rough set theory [162] investigate not only exact (crisp) but also approximate (rough) settings of the problem z. In some works, for example in [166], the definition of the information system $U = (A, B, F)$ includes a probability distribution on a class of subsets of the set A. Not only deterministic decision trees such as those discussed in this monograph, but also different types of nondeterministic decision trees are studied in [80, 81, 83, 84, 90, 93, 100, 101, 107, 109, 113, 118, 120, 127]. Different representations of decision trees are considered, for instance, branching programs computing Boolean functions [149, 213]. Besides the weighted depth considered in this paper, different types of worst-case-time complexity measures [70, 78, 84, 93, 97, 136] and average-time complexity measures [16–19, 35, 129–133, 136, 166] have been studied. In addition, space complexity measures such as the number of nodes in a tree or in its representation [20, 127, 149, 176] have been investigated.

Example 2.1. (Problem on tree cups.) Consider the three inverted cups and a small ball under one of these cups shown in Fig. 1. The problem under consideration is to find the number of the cup under which the ball lies.

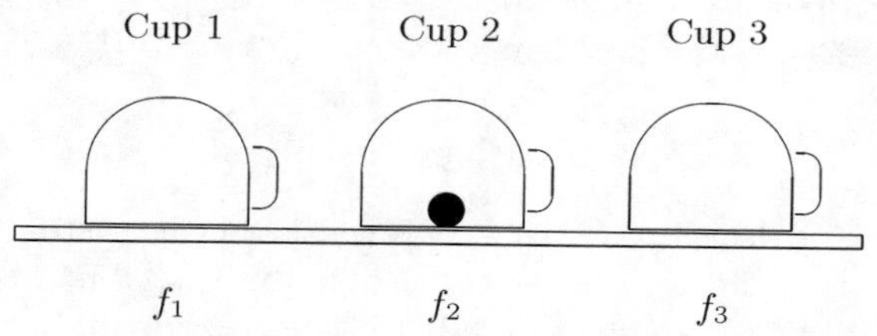

Fig. 1. Three cups and a small ball.

To solve this problem, we use three attributes f_1, f_2, f_3. These attributes are defined on the set $\{a_1, a_2, a_3\}$ where a_i is i-th cup which gives the location of the ball. If the ball lies under the i-th cup then the value of f_i is 1, otherwise the value of f_i is equal to 0.

We can represent our problem in the following form: $z = (\nu, f_1, f_2, f_3)$ where $\nu(1,0,0) = 1$, $\nu(0,1,0) = 2$, $\nu(0,0,1) = 3$, and $\nu(\delta_1, \delta_2, \delta_3) = 0$ for each $(\delta_1, \delta_2, \delta_3) \in \{0,1\}^3 \setminus \{(1,0,0),(0,1,0),(0,0,1)\}$. This is a problem over the information system $U = (A, B, F)$ where $A = \{a_1, a_2, a_3\}$, $B = \{0,1\}$ and $F = \{f_1, f_2, f_3\}$.

The decision tree represented in Fig. 2(a) solves the considered problem. This is a decision tree over U with a depth equal to 2. The decision table $T(z)$ is represented in Fig. 2(b).

Finite Information Systems. Considerable work has been done in the investigation of finite information systems. These include: rough set theory [37, 150], [153, 154, 162], [164, 165, 168, 169], and [193–195, 198, 201], test theory [14, 45],

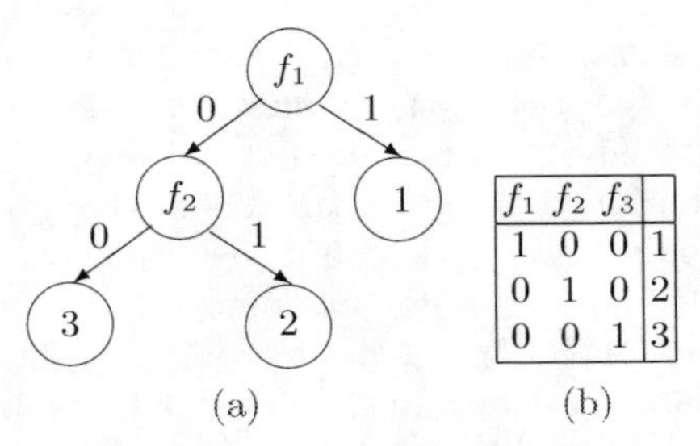

f_1	f_2	f_3	
1	0	0	1
0	1	0	2
0	0	1	3

Fig. 2. Decision tree and decision table for the three-cup problem.

[68, 70, 74, 79], [202, 203], and [214, 216, 220], theory of questionnaires [151, 166, 167], theory of decision tables [35, 170], machine learning [3, 21, 33, 34, 119, 124], and search theory [1, 206].

In particular, numerous papers are devoted to heuristic methods for decision tree construction [11–13, 25, 32, 50, 141–144, 174–177], and to feature selection and construction [48, 49, 179, 205].

In papers relating to finite information systems not only deterministic decision trees over the information system $U = (A, B, F)$ which solve a problem z over U are studied, but also certain objects like them are investigated. Among them there are finite subsets of the set F sufficient for the problem z solving (known as relative reducts and tests [14, 162, 197, 199]), decision rules [5, 36, 58, 147, 162] for the problem z which are true in U expressions of the kind

$$f_{i_1}(x) = \delta_1 \wedge \ldots \wedge f_{i_p}(x) = \delta_p \Rightarrow z(x) = \delta \ ,$$

and nondeterministic decision trees [84, 127] (decision rule systems is most simple kind of nondeterministic decision trees). Not only exact but also approximate settings of the problem z are considered [162], both time and space complexity measures are investigated [18, 102, 149, 166].

It is our view that these threads of research have so far been developed in isolation. The distinctions between them lie not only in nature of the problems under consideration but also in the character of the mathematical methods applied. The comparative analysis of the accumulated stock of problems, methods and results obtained in these areas would be useful for the progress of the decision tree theory.

Bounds on complexity and algorithms for construction of decision trees over finite information systems (represented in the form of decision tables) will be considered in Sect. 3. So-called Shannon functions for finite information systems will be studied in Sects. 4.3 and 5.2.

Infinite Information Systems. The principal results in decision tree theory related to infinite information systems have been achieved in the study of linear and algebraic decision trees.

Let $A \subseteq \mathbb{R}^n$, $D \subseteq \mathbb{R}$ and $F_n(D) = \{\text{sign}(\sum_{i=1}^{n} d_i x_i + d_{n+1}) : d_i \in D, 1 \leq i \leq n+1\}$ where for any $r \in \mathbb{R}$

$$\text{sign}(r) = \begin{cases} -1, \text{ if } r < 0 \ , \\ 0, \text{ if } r = 0 \ , \\ +1, \text{ if } r > 0 \ . \end{cases}$$

The information system $(A, E, F_n(D))$, where $E = \{-1, 0, +1\}$, will be called *linear* information system. Decision trees over $(A, E, F_n(D))$ are called *linear* decision trees.

In [23, 24, 60, 61] for certain problems $z = (\nu, f_1, \ldots, f_m)$ over $(\mathbb{R}^n, E, F_n(\mathbb{R}))$ the lower bounds near to $n \log_2 m$ have been obtained for the minimal depth of decision trees over $(\mathbb{R}^n, E, F_n(\mathbb{R}))$ solving these problems. Subsequently, the methods of proofs of the lower bounds have been generalized on *algebraic* decision trees which use attributes of the kind $\text{sign}(p(x_1, \ldots, x_n))$ where p is a polynomial with real coefficients [6, 31, 204, 218, 219].

In [22] the upper bound $(3 \cdot 2^{n-2} + n - 2)(\log_2 m + 1)$ for the minimal depth of a decision tree over $(\mathbb{R}^n, E, F_n(\mathbb{R}))$ solving a problem $z = (\nu, f_1, \ldots, f_m)$ over $(\mathbb{R}^n, E, F_n(\mathbb{R}))$ have been found for $n \geq 2$. In [68] the upper bound $(2(n+2)^3 \log_2(m + 2n + 2))/ \log_2(n+2)$ of the minimal depth for decision trees and for problems over $(\mathbb{Q}^n, E, F_n(\mathbb{Z}))$ is contained. The complete proof of weaker bound of the kind $4(n+2)^3 \ln(m + 2n + 2)$ is contained in [69]. Similar upper bound had been obtained in [56] for decision trees over $(\mathbb{R}^n, E, F_n(\mathbb{R}))$ and for problems over $(\mathbb{R}^n, E, F_n(\mathbb{Z}))$. In [77] decision trees over quasilinear information systems, which are generalization of linear and algebraic decision trees, were studied. The upper bounds on complexity were obtained for these which are similar to the bounds for the linear decision trees contained in [68]. This yields, in particular, positive solution of the problem set up in [57] of the existence of medium-value upper bounds on the minimum depth of decision trees over $(\mathbb{R}^n, E, F_n(\mathbb{R}))$ solving problems over $(\mathbb{R}^n, E, F_n(\mathbb{R}))$.

In our view, the problems of complexity of decision trees over arbitrary infinite information systems were not considered prior to [78, 79]. These studies develop the two approaches to decision trees over arbitrary information system $U = (A, B, F)$ based on the methods of test theory and rough set theory. One of the approaches is a local approach, where for a problem $z = (\nu, f_1, \ldots, f_n)$ over U, the decision trees are considered using only attributes from the set $\{f_1, \ldots, f_n\}$. The second approach is a global one where arbitrary attributes from F are used by decision trees.

Bounds on minimal depth and weighted depth of decision trees over arbitrary information systems will be considered in Sects. 4 and 5 in the framework of both local and global approaches. Decision trees over quasilinear information systems will be considered in Sect. 6. Also in these sections, some problems will be studied connected with algorithms for decision tree construction.

Applications of Decision Tree Theory. Applications of decision tree theory can be classified into four groups: applications in the field of the non-procedural programming languages; applications to analysis of algorithms which are not decision trees; applications in fields of research with their own specific mathematical models for infinite classes of problems and, at lastly, applications in fields where mathematical models for infinite classes of problems have not yet been developed.

The intrinsic feature of non-procedural programming is that the definition of the order of execution of the program instructions is not the programmer's but the compiler's duty. In [35] the use of decision tables and decision trees in non-procedural programming is studied in detail. The essence of the suggested approach consists of transformation of a non-procedural program from the decision table form into the decision tree by compiling system.

The results obtained in investigations of decision trees can be applied to the analysis of algorithms of different types. For example, the above-mentioned upper bounds on the complexity of linear decision trees are used in [73] to analyze the relationship between the depth of deterministic and nondeterministic acyclic programs in the basis $\{x + y, x - y, 1; \mathrm{sign}(x)\}$. Analogous upper bounds are obtained in [57] for simulation of parallel acyclic programs in the similar basis with the help of decision trees. In [8–10] high values of lower bounds on the depth of linear decision trees of special kind are used in proofs of non-polynomial lower bounds on time complexity of several known methods for solving NP-complete problems. In [96, 99] time complexity of deterministic and nondeterministic tree-programs in an arbitrary finite basis is considered.

In decision tree theory at least five application fields have been developed which have given rise to their own mathematical models for some infinite classes of problems. These are discrete optimization, computational geometry, discrete geometric pattern recognition, diagnosis of faults in circuits, and probabilistic reasoning. For the considered classes of problems specific research methods can be developed.

Different mathematical models are applied to different classes of discrete optimization problems. For example, the class of problems of a linear form minimization on finite subset of the set $\mathbb{R}^n$ is often considered. The research in the field of the linear decision trees has resulted in non-linear lower bounds [23, 60] and polynomial upper bounds [56, 68] on the depth of linear decision trees for some NP-complete problems of fixed dimension. Non-polynomial lower bounds on complexity of some known methods for solving NP-complete problems have also been found [9].

In the area of computational geometry, one of the most thoroughly explored problems is the localization problem. Let we have a partition of the geometrical space into finite number of domains. For a given point in the space it is required to find the domain containing this point. To define the class of localization problems it is sufficient to introduce the class of surfaces bounding the space domains. A systematic discussion of computational geometry can be found in [173].

The family of problems of fixed-length word recognition in some infinite language can be placed among known models for classes of problems of discrete geometric pattern recognition. Each attribute in a problem of this class entails the recognition of certain letter in a word. For example, the word of the length n in the k-letter alphabet corresponds to the k-color image on a screen containing n cells, and the recognition of the value of the i-th letter corresponds to the recognition of the color of the i-th screen cell [203, 214].

The field of the circuit fault diagnosis is the oldest application area initiated by [4, 14, 27, 59, 180, 216]. To describe a class of problems in this field it is sufficient to define the circuit class and the fault class. Many researchers have explored these problems studying both circuits with memory and memory-less circuits [52, 152, 178, 211, 212, 215].

Bayesian networks [39, 163] are useful tools for representation of joint probability distribution of variables. Some of these variables are hidden (unobservable). Using values of open (observable) variables and information about probability distribution from Bayesian network we can draw some conclusions about values of hidden variables. Use of decision trees can accelerate the process of recognition of all open variable values. For the description of a class of such problems the definition of a class of Bayesian networks is sufficient.

There are several fields of research in which decision trees are applied as algorithms, as a way of knowledge representation, or as predictors, but little or nothing is known of specific models for infinite classes of problems. These are fields such as medical diagnosis, geology, marketing, sociology, ecology, etc. Either common research schemes or specific ad hoc methods for solving separate problems would be appropriate in these areas.

This monograph does not deal directly with the problems concerning non-procedural programming languages, computational geometry and those areas without developed mathematical models for infinite classes of problems. For other mentioned application areas in Sects. 7–11 some problems of interest will be studied which illustrate the research features and instruments intrinsic to these areas.

2.2 On Results of Monograph

In this subsection the results from Sects. 3–11 of the monograph are discussed briefly.

Decision Trees for Decision Tables. Decision table is a rectangular table filled by integers. Rows of the table are pairwise different, and each row is labelled by an integer (a decision). We associate a two-person game to illustrate the concept of a decision table. The first player would choose a row of the table, and the second player would guess the decision corresponding to this row. In order to guess the decision, the second player can choose columns of the table and ask what is in the intersection of the chosen row and these columns. The strategies of the second player can be represented in the form of decision trees.

In Sect. 3 we consider decision tables as an independent object for investigations, and study bounds on complexity and algorithms for construction of decision trees for decision tables.

Lower bounds on complexity of decision trees based on different parameters of decision table (such as number of different decisions or minimal cardinality of relative reduct) are considered. An approach to the proof of lower bounds based on the "proof-tree" notion is considered. A proof-tree can be interpreted as a fragment of a strategy of the first player in the game which is modified in the following way: first player does not choose a row at the beginning of the game, but at least one row must satisfy his answers on questions of the second player.

Upper bounds on minimal complexity and algorithms for construction of decision trees considered in this section, are based on the use of uncertainty measures of decision tables such as the number of rows in the table or the number of pairs of rows with different decisions. To construct a decision tree, we either choose the question (attribute) which reduces the uncertainty to the greatest extent under certain constraints on the weight of the attribute ("greedy" algorithms), or minimize the total weight of the sequence of attributes (questions) which either reduces the uncertainty by half or gives the solution of the problem ("dichotomous" approach for decision tree construction). Upper bounds on complexity of decision trees based on "dichotomous" approach are studied in Sect. 3. In the case of depth these bounds can not be improved essentially. For the greedy algorithms the complexity and the precision are estimated. Based on results from [28, 51] we will show that under some assumptions on the class NP, one of algorithms performs nearly as well as best approximate polynomial algorithms for minimization of decision tree depth. In addition, we consider an algorithm which allows to construct for a given decision table a decision tree with minimal weighted depth. In Sect. 4.8 we describe all infinite information systems such that the algorithm has polynomial time complexity depending on number of columns in tables over the information system.

A decision table can be interpreted as a tabular representation of a partial function, and decision trees for this table can be interpreted as algorithms for this function computation. For decision tables corresponding to functions from arbitrary closed class of Boolean functions [217], unimprovable upper bounds were achieved for the minimal depth of decision trees depending on the number of variables of the functions. Proofs of these bounds illustrate methods of decision table study.

Sect. 3 contains mostly the results of [68, 70, 79, 80, 85, 89, 92]. Exact algorithm for minimization of decision tree weighted depth was considered in [133, 134] (see [19, 20, 136] for similar algorithms for another complexity measures, and [135] for some applications). It is impossible to establish the authorship of the lower bound on decision tree complexity depending on the number of different decisions in decision table. The approach to the proof of lower bounds based on the notion of proof-tree is similar to methods of analysis of search problems developed in [206]. The precision of "greedy" algorithms for decision tree construction have been investigated in [68, 70, 97, 106, 111, 126]. Apparently, the first publication suggesting the similar algorithm for decision tree construction was [174]. The early publications [38, 45, 148, 181] should be mentioned in which certain "greedy" algorithms for set cover problems of different type had been studied (see also [111, 125] for additional comments). Upper and lower bounds on decision tree complexity were used for study of problems of discrete optimization, pattern recognition, and fault diagnosis. Note that in [33] the "dichotomous" approach created in [70] was used for investigation of machine learning problems (see also [34] where results similar to [33] were obtained independently).

Two Approaches to the Study of Decision Trees. Let $U = (A, B, F)$ be an information system and ψ be a weight function for U. In Sects. 4 and 5 two approaches are developed to study the arbitrary pair of the kind (U, ψ): the local approach in Sect. 4 where for problem $z = (\nu, f_1, \dots, f_m)$ over U the decision trees are considered using only attributes from the set $\{f_1, \dots, f_m\}$ and the global one in Sect. 5 where for problem z solving all attributes from the set F can be used.

The main difficulty in the global approach is the necessity to choose appropriate attributes in large or infinite set F. However, in the framework of the global approach we can often construct more simple decision trees rather than in the framework of the local approach.

The first group of problems studied in Sects. 4 and 5 is associated with complexity bounds for decision trees. Let us define three parameters of a problem $z = (\nu, f_1, \dots, f_m)$ over U. Denote $\psi_U^l(z)$ the minimal weighted depth of a decision tree over U which solves the problem z and uses only attributes from the set $\{f_1, \dots, f_m\}$. Denote $\psi_U^g(z)$ the minimal weighted depth of a decision tree over U which solves the problem z. Denote $\psi(z) = \sum_{i=1}^{m} \psi(f_i)$. The value $\psi(z)$ is the complexity of the decision tree which solves the problem z in trivial way by computing sequentially the values of the attributes $f_1, \dots, f_m$. Let us consider the local and the global Shannon functions which allow to compare the values $\psi_U^l(z)$ and $\psi_U^g(z)$ with the value $\psi(z)$ which is the obvious upper bound on the former two. For $n \in \mathbb{N}$ the values of local and global Shannon functions are defined in the following way:

$$H_{U,\psi}^l(n) = \max\{\psi_U^l(z) : z \in \mathrm{Probl}_U, \psi(z) \le n\} ,$$
$$H_{U,\psi}^g(n) = \max\{\psi_U^g(z) : z \in \mathrm{Probl}_U, \psi(z) \le n\} .$$

It is shown that either the local Shannon function $H_{U,\psi}^l$ is bounded from above by a constant, or $H_{U,\psi}^l(n) = \Theta(\log_2 n)$, or $H_{U,\psi}^l(n) = n$ for infinitely many $n \in \mathbb{N}$.

The variety of ways in which the global Shannon function $H_{U,\psi}^g$ may behave is much wider. Let $\varphi : \mathbb{N} \to \mathbb{N}$ be a nondecreasing function such that for any n, $n \ge 7$, the inequalities

$$\lfloor \log_2 n \rfloor + 2 \le \varphi(n) \le n - 3$$

hold. Then there exists a pair (U, ψ) such that for any n, $n \ge 7$, the value of $H_{U,\psi}^g(n)$ is definite and the inequalities

$$\varphi(n) \le H_{U,\psi}^g(n) \le \varphi(n) + 2$$

hold.

The information system U is called *ψ-compressible* if $H_{U,\psi}^g(n) < n$ for large enough n. We describe all pairs (U, ψ) such that U is ψ-compressible.

If the depth h is taken in the context of complexity measure then either the global Shannon function $H_{U,h}^g$ is bounded from above by a constant, or

$H^g_{U,h}(n) = \Omega(\log_2 n)$ and $H^g_{U,h}(n) = O((\log_2 n)^\varepsilon)$ for any $\varepsilon > 0$, or $H^g_{U,h}(n) = n$ for any natural n.

For finite information systems and for the depth the behavior of local and global Shannon functions is studied in more detail.

The problems of the second group which are considered in Sects. 4 and 5 are connected with the conditions of solvability of decision tree optimization problems. We assume here that $F = \{f_i : i \in \mathbb{N}\}$ and ψ is a computable function which is defined not on attributes from the set F but on the set $\mathbb{N}$ of their numbers. So ψ is a general recursive function which does not take the value 0.

Three algorithmic problems can be defined:

Problem of Local Optimization for the Pair (U, ψ)*:* for a given problem $z = (\nu, f_{i_1}, \dots, f_{i_m})$ over U it is required to construct a decision tree over U which solves the problem z, uses only attributes from the set $\{f_{i_1}, \dots, f_{i_m}\}$ and which complexity is equal to $\psi^l_U(z)$.

Problem of Global Optimization for the Pair (U, ψ)*:* for a given problem z over U it is required to construct a decision tree over U which solves the problem z and which complexity is equal to $\psi^g_U(z)$.

Compatibility Problem for the Information System U*:* for a given equation system of the kind

$$\{f_{i_1}(x) = \delta_1, \dots, f_{i_m}(x) = \delta_m\} ,$$

where $f_{i_1}, \dots, f_{i_m} \in F$ and $\delta_1, \dots, \delta_m \in B$, it is required to determine whether this system is compatible on the set A.

The local optimization problem for the pair (U, ψ) is solvable if and only if the compatibility problem for the information system U is solvable.

If the compatibility problem for the information system U is unsolvable, then the global optimization problem for the pair (U, ψ) is also unsolvable. The inverse statement is true not for all pairs (U, ψ). The necessary and sufficient conditions on weight function ψ are found, under which for any information system U the compatibility problem for U and the global optimization problem for the pair (U, ψ) are solvable or unsolvable simultaneously. These conditions imply that for any $i \in \mathbb{N}$ the set $\mathbb{N}_\psi(i) = \{j : j \in \mathbb{N}, \psi(j) = i\}$ should be finite and the algorithm should exist which computes for an arbitrary $i \in \mathbb{N}$ the cardinality of the set $\mathbb{N}_\psi(i)$. Such complexity measures are called *proper*.

The third group of considered problems is concerned with study of ways of constructing decision trees. In the case of local approach, when the compatibility problem for the information system U is solvable, the following method of the decision tree construction has been studied in depth. For a given problem z we construct the decision table $T(z)$ and apply to it algorithms of decision tree construction described in Sect. 3. We consider bounds on complexity and precision of this method. In the case of global approach, when the compatibility problem for the information system U is solvable, one of ways of decision tree construction is to use proper weight functions. Other methods can be extracted from proofs of upper bounds on decision tree complexity (see Sects. 5 and 6).

Essential part of results of Sects. 4 and 5 had been first published in [79]. Other results can be found in [66, 75], [86]–[88], [93]–[95], [98], [103]–[105], [108], [112, 114], [117, 123], [127, 128, 133, 134].

Decision Trees over Quasilinear Information Systems and Related Topics. Quasilinear information systems are studied in Sect. 6. Let A be a nonempty set, $\varphi_1, \ldots, \varphi_n$ be functions from A to $\mathbb{R}$, and K be a subset of $\mathbb{R}$ containing 1 and closed relative to the operations of addition, subtraction and multiplication. Denote $F(A, K, \varphi_1, .., \varphi_n) = \{\text{sign}(\sum_{i=1}^{n} a_i\varphi_i(x) + a_{n+1}) : a_1, .., a_{n+1} \in K\}$, where sign is the function defined in Sect. 2.1. For any attribute $f = \text{sign}(\sum_{i=1}^{n} a_i\varphi_i(x) + a_{n+1})$ from $F(A, K, \varphi_1, \ldots, \varphi_n)$ let

$$r(f) = \max\{0, \max\{\log_2 |a_i| \, a_i \neq 0, 1 \leq i \leq n+1\}\}$$

(if $a_1 = \ldots = a_{n+1} = 0$, then $r(f) = 0$). The information system $U = (A, \{-1, 0, +1\}, F(A, K, \varphi_1, \ldots, \varphi_n))$ will be called *a quasilinear* information system. The main result of Sect. 6 is that for any problem $z = (\nu, f_1, \ldots, f_m)$ over U there exists a decision tree over U solving z for which the depth is at most $(2(n+2)^3 \log_2(m+2n+2))/\log_2(n+2)$, and for any attribute f used by this decision tree the inequality $r(f) \leq 2(n+1)^2(1+\log_2(n+1)+\max\{r(f_i) : 1 \leq i \leq m\})$ holds. If the set K coincides with the set $\mathbb{Z}$ then there exists an algorithm which for arbitrary problem z over U constructs a decision tree over U which solves z and possesses the mentioned properties. Also in this section, for some information systems and weight functions the behavior of the global Shannon function is studied.

In Sect. 7 six classes of problems over quasilinear information systems are considered: three classes of discrete optimization problems and three classes of recognition and sorting problems. For each class, examples and corollaries of the results of Sect. 6 are given. It is shown, for example, that for traveling salesman problem with $n \geq 4$ cities there exists a decision tree over linear information system solving it and satisfying the following conditions: the depth of the decision tree is at most $n^7/2$ and any attribute f used in it has integer coefficients and satisfies the inequality $r(f) \leq n^4 \log_2 n$.

The idea of application of test theory to the analysis of linear decision trees for discrete optimization problems can be attributed to Al.A. Markov. The idea was further developed by [69], whose tutors were Al.A. Markov and S.V. Yablonskii. Proof of the main result of Sect. 6 is like the proof of similar bound on complexity of the linear decision trees in [69]. Sects. 6 and 7 describe the results of [68, 69, 77, 79, 102, 116].

Acyclic programs in the basis $B_0 = \{x+y, x-y, 1; \text{sign}(x)\}$, which recognize membership of an element to a set, are studied in Sect. 8 based on the results obtained in Sect. 6. The depth $h(P)$, which is the maximal number of the operations made by program P, is used as the complexity measure. The comparison of the depth of deterministic and nondeterministic programs is made. It is shown that for any nondeterministic acyclic program P_1 with n input variables there exists a deterministic acyclic program P_2 which recognizes the same set as P_1 and for which the inequality

$$h(P_2) \leq 8(n+2)^7(h(P_1)+2)^2$$

holds. Sect. 8 contains results of [73].

As it was mentioned in Sect. 2.1, results similar to some results from Sects. 6–8 were obtained in [56, 57].

Applications to Pattern Recognition, Fault Diagnosis and Probabilistic Reasoning. In this section, we briefly discuss the results presented in Sects. 9–11 and concerned with three important application areas having their own store of mathematical models for corresponding infinite classes of problems.

In Sect. 9 the problem of recognition of words of fixed length in a regular language is considered. The word under consideration can be interpreted as a description of certain screen image in the following way: the i-th letter of the word encodes the color of the i-th screen cell. Let $\mathcal{L}$ be a regular language and $\mathcal{L}(n)$ be the set of words in the language $\mathcal{L}$ of the length n. The minimal depth of a decision tree which recognizes the words from $\mathcal{L}(n)$ and uses only attributes which recognize the i-th letter of the word, $i \in \{1,\ldots,n\}$, will be denoted by $h_{\mathcal{L}}(n)$. If $\mathcal{L}(n) = \emptyset$ then $h_{\mathcal{L}}(n) = 0$. We will consider a "smoothed" analog of the function $h_{\mathcal{L}}(n)$ which is the function $H_{\mathcal{L}}(n)$ defined in the following way:

$$H_{\mathcal{L}}(n) = \max\{h_{\mathcal{L}}(m) : m \leq n\} .$$

In Sect. 9 all regular languages are classified according to the complexity of the word recognition problem. It is shown that either $H_{\mathcal{L}}(n) = O(1)$, or $H_{\mathcal{L}}(n) = \Theta(\log_2 n)$, or $H_{\mathcal{L}}(n) = \Theta(n)$. Results of Sect. 9 had been published in [79, 82, 110].

Similar results for languages generated by some types of linear grammars and context-free grammars were obtained in [26, 41–43]. In [100] the classification of all regular languages depending on the depth of nondeterministic decision trees recognizing words of the language is obtained.

Different lines of investigation of applications of decision trees to constant fault diagnosis in combinatorial circuits are considered in Sect. 10. The faults under consideration are represented in the form of Boolean constants on some inputs of circuit gates. The diagnosis problem consists of recognition of the function realized by the circuit with a fixed tuple of constant faults from a given set of tuples. Each attribute is the result of observing the output of the circuit when the input (which is a binary tuple) is given.

Let B be a nonempty finite set (a basis) of Boolean functions and let $\mathrm{Circ}(B)$ be the set of one-output combinatorial circuits in the basis B. The number of gates in a circuit $S \in \mathrm{Circ}(B)$ will be denoted by $L(S)$, and the minimal depth of a decision tree which solves the problem of diagnosis of the circuit S relative to the set of all possible tuples of constant faults on gate inputs will be denoted by $h(S)$.

The first line of investigation comprises the study of complexity of fault diagnosis algorithms for arbitrary circuits in the basis B. Let us consider for this purpose the function $h_B^{(1)}$ which characterizes the worst-case dependency of $h(S)$ on $L(S)$ on the set $\mathrm{Circ}(B)$ of circuits:

$$h_B^{(1)}(n) = \max\{h(S) : S \in \mathrm{Circ}(B), L(S) \le n\} .$$

The basis B will be called *primitive* if at least one of the following conditions holds:

a) every function from B is either a disjunction $x_1 \vee \ldots \vee x_n$ or a constant;
b) every function from B is either a conjunction $x_1 \wedge \ldots \wedge x_n$ or a constant;
c) every function from B is a linear function or a constant.

We will show that $h_B^{(1)}(n) = O(n)$ for primitive basis B, and $\log_2 h_B^{(1)}(n) = \Omega(n^{1/2})$ for non-primitive basis B.

As opposed to the first one, the second line of research explores complexity of diagnostic algorithms for best (from the point of view of solution for the diagnosis problem) circuits realizing the Boolean functions given as formulas over B. Let $\Phi(B)$ be the set of formulas over the basis B. For a formula $\varphi \in \Phi(B)$ we will denote by $L(\varphi)$ the number of functional symbols in φ. Let $h(\varphi) = \min h(S)$, where the minimum is taken over all possible combinatorial circuits S (not necessarily in the basis B) which realize the same function as the formula φ. We will study the behavior of a function $h_B^{(2)}$ which characterizes the worst-case dependency of $h(\varphi)$ on $L(\varphi)$ on the set of formulas over B and is defined as follows:

$$h_B^{(2)}(n) = \max\{h(\varphi) : \varphi \in \Phi(B), L(\varphi) \le n\} .$$

We will show that $h_B^{(2)}(n) = O(n)$ for primitive basis B, and $\log_2 h_B^{(2)}(n) = \Omega(n^c)$ for non-primitive basis B, where c is a positive constant depending only on B.

The third line of research is to study the complexity of algorithms for the following problem $\mathrm{Con}(B)$: for a given circuit S from $\mathrm{Circ}(B)$ and an arbitrary set W of tuples of constant faults on inputs of gates of the circuit S, it is required to construct a decision tree which solves the diagnosis problem for the circuit S relative to the faults from W. Note that there exists a decision tree with at most $2\,|W|-1$ nodes which solves the diagnosis problem for the circuit S relative to the faults from W. If B is a primitive basis then there exists an algorithm which solves the problem $\mathrm{Con}(B)$ with polynomial time complexity. If B is a non-primitive basis then the problem $\mathrm{Con}(B)$ is NP-hard.

From the point of view of the diagnosis problem solving for arbitrary tuples of constant faults on inputs of gates of arbitrary circuits, only primitive bases seem to be admissible. The extension of the set of such bases is possible by the substantial restriction on the class of the circuits under consideration. The fourth line of research is the study of complexity of fault diagnosis algorithms for iteration-free circuits in the basis B. A combinatorial circuit is called *iteration-free* if each node (input or gate) of it has at most one issuing edge. Denote by $\mathrm{Circ}^1(B)$ the set of iteration-free circuits in the basis B with only one output. Consider the function $h_B^{(3)}$ which characterizes the worst-case dependency of $h(S)$ on $L(S)$ for circuits from $\mathrm{Circ}^1(B)$ and is defined as follows:

$$h_B^{(3)}(n) = \max\{h(S) : S \in \mathrm{Circ}^1(B), L(S) \le n\} .$$

A Boolean function $f(x_1, \ldots, x_n)$ is *quasimonotone* if there exist numbers $\sigma_1, \ldots, \sigma_n \in \{0, 1\}$ and a monotone Boolean function $g(x_1, \ldots, x_n)$ such that

$$f(x_1, \ldots, x_n) = g(x_1^{\sigma_1}, \ldots, x_n^{\sigma_n})$$

where $x^\sigma = x$ if $\sigma = 1$, and $x^\sigma = \neg x$ if $\sigma = 0$. The basis B will be called *quasiprimitive* if at least one of the following conditions holds:

a) each function from B is a linear function or a constant;
b) each function from B is a quasimonotone function.

We will show that $h_B^{(3)}(n) = O(n)$ if B is quasiprimitive, and $\log_2 h_B^{(3)}(n) = \Omega(n)$ if B is not quasiprimitive.

Note that there exist decision trees satisfying the bounds for quasiprimitive bases and possessing an effective description of the work.

The fifth line of research deals with circuit construction and effective diagnosis of faults based on the results obtained for the iteration-free circuits. Two functions are *equal* if one of them can be obtained from the other by operations of insertion and deletion of unessential variable. Based on the results of [209] one can show for each basis B_1, the existence of a quasiprimitive basis B_2 with the following properties:

a) the set of functions realized by circuits in the basis B_2 coincides with the set of functions realized by circuits in the basis B_1;
b) there exists a polynomial p such that for each formula $\varphi_1 \in \Phi(B_1)$ there exists a formula $\varphi_2 \in \Phi(B_2)$ which realizes the function equal to that realized by φ_1, and such that $L(\varphi_2) \leq p(L(\varphi_1))$.

Our approach to circuit construction and fault diagnosis is as follows. Let $\varphi_1 \in \Phi(B_1)$ be a formula realizing certain function f, $f \notin \{0, 1\}$, and let us construct a formula $\varphi_2 \in \Phi(B_2)$ realizing the function equal to f and satisfying the inequality $L(\varphi_2) \leq p(L(\varphi_1))$. Next a circuit S in the basis B_2 is constructed according to the formula φ_2 realizing the function f, satisfying the equality $L(S) = L(\varphi_2)$ and the condition that at most one edge results from each gate of the circuit S. In addition to the usual work mode of the circuit S there exists the diagnostic mode in which the inputs of the circuit S are "split" so that it becomes the iteration-free circuit $\tilde{S}$. The inequality $h(\tilde{S}) \leq cp(L(\varphi_1))$, where c is a constant depending only on the basis B_2, holds for the circuit $\tilde{S}$.

The results of Sect. 10 are taken from [72, 76, 91, 115]. Problems connected with complexity of algorithms for diagnosis of constant faults in combinatorial circuits have been studied by different authors [14, 178, 203, 215]. As a rule, the dependency of the algorithm's complexity on the number of inputs of the circuit were studied. Consider three series of publications which are most similar to approach taken in this section. From the results obtained in [30, 40] the bound $h_B^{(3)}(n) = O(n)$ can be derived immediately for arbitrary basis B with the following property: each function from B is realized by some iteration-free circuit in the basis $\{x \wedge y, x \vee y, \neg x\}$. In [184–191] for circuits over an arbitrary finite basis and faults of different type (not only the constant) the dependence is investigated of the minimal depth of a decision tree, which diagnoses circuit faults, on total number of inputs and gates in the circuit. In [137–140, 192] effective methods for diagnosis of faults of different types are considered.

In Sect. 11 the depth of decision trees which recognize values of all open variables from Bayesian Network (BN for short) is considered. BN ia a convenient tool for representation of joint probability distribution of variables. Some of these variables are hidden (unobservable). Using values of open (observable) variables and information about probability distribution from BN we can draw some conclusions about values of hidden variables.

The investigation of decision trees for recognition of all open variable values lends itself to the use of BN. Assume that the process of computation of open variable values is rather expensive (it may be connected with use of remote sensors, carrying out of experiments, etc.), there exists a decision tree whose depth is essentially less than the number of open variables, and there exists an efficient algorithm for simulation of the decision tree work. In such a case, it is appropriate to use this decision tree instead of the sequential computation of all open variable values.

We consider $(1, 2)$-BN in which each node has at most 1 entering edge, and each variable has at most 2 values. For an arbitrary $(1, 2)$-BN we obtain lower and upper bounds on minimal depth of decision tree that differ by not more than a factor of 4, and can be computed by an algorithm which has polynomial time complexity. The number of nodes in such decision trees can grow exponentially depending on number of open variables in BN. We will develop a polynomial algorithm for simulation of the decision trees whose depth lies between the stated bounds. Results discussed in this section are from [121, 122].

3 Decision Trees for Decision Tables

In this section upper and lower bounds on complexity of decision trees for decision tables and algorithms of decision tree construction are considered.

This section consists of seven subsections. The first subsection contains definitions of basic notions. In the second subsection, lower bounds on complexity of decision trees are considered. The third subsection contains upper bounds on decision tree complexity. In the fourth subsection greedy algorithm for decision tree construction is discussed. The fifth subsection contains an algorithm for construction of an optimal decision tree. Problems related to the complexity of decision tree optimization are considered in the sixth subsection. In the seventh subsection, the depth of decision trees for computation of Boolean functions from an arbitrary closed class is studied.

3.1 Basic Definitions and Notation

The notions of signature, decision table, decision tree, weight function, weighted depth and depth are introduced in this subsection.

For an arbitrary nonempty set (alphabet) D, the set of all finite words over D containing the empty word λ will be denoted by D^*.

Let $\mathbb{N} = \{0, 1, 2, \ldots\}$ and $\mathbb{Z}$ denote the set of integers. We define two partial functions min and max from $2^{\mathbb{N}}$ to $\mathbb{N}$ as follows. Let $C \subseteq \mathbb{N}$. If $C = \emptyset$ then the value of $\min C$ is indefinite, otherwise $\min C$ is the minimal number from C. If C is an empty or an infinite set then the value of $\max C$ is indefinite, otherwise $\max C$ is the maximal number from C.

A pair $\rho = (F, B)$, where F is a nonempty set and B is a finite nonempty set of integers with at least two elements, will be called *a signature*. Elements of the set F will be called *attributes*. We will consider decision tables filled by numbers from B, and we will associate columns of decision tables with attributes from F. Later we will assume that a signature $\rho = (F, B)$ is fixed, and $|B| = k$.

Let $\Omega_\rho = \{(f, \delta) : f \in F, \delta \in B\}^*$. A pair (f, δ) will be interpreted as the following condition for rows of a decision table: a row must have the number δ on the intersection with the column labelled by the attribute f. A word from Ω_ρ will be interpreted as a system (conjunction) of such conditions.

Decision Tables. *A decision table of the signature* ρ is a rectangular table filled by numbers from the set B. Rows of the table are pairwise different. Each row is labelled by a number from $\mathbb{Z}$. This number is interpreted as the decision. The columns of the table are labelled by attributes from F. If some columns are labelled by the same attribute then these columns coincide.

It is possible that a decision table does not contain rows. Such tables will be called *empty*.

Let T be a decision table of the signature ρ. Denote by $\mathrm{Row}(T)$ the set of rows of the decision table T. For any row $\bar{\delta}$ of the table T we denote by $\nu_T(\bar{\delta})$ the number (decision) corresponding to this row. Denote by $\mathrm{At}(T)$ the set of attributes from F which are labels of the table T columns.

If a table T' can be obtained from the table T by deletion of some rows then we will say that T' is *a subtable* of T.

Let $\Omega_\rho(T) = \{(f, \delta) : f \in \mathrm{At}(T), \delta \in B\}^*$. We will use words from the set $\Omega_\rho(T)$ for description of subtables of the table T. Let $u \in \Omega_\rho(T)$. Let us define the subtable Tu of the table T. If $u = \lambda$ then $Tu = T$. Let $u \neq \lambda$ and $u = (f_1, \delta_1) \ldots (f_m, \delta_m)$. Then Tu consists of such and only such rows of T which on the intersection with columns labelled by $f_1, \ldots, f_m$ have numbers $\delta_1, \ldots, \delta_m$ respectively. It is clear that Tu is a decision table of the signature ρ. It is possible that Tu is an empty table. A nonempty subtable T' of the table T will be called *separable* if there exists a word $u \in \Omega_\rho(T)$ such that $T' = Tu$.

Let Tab_ρ be the set of all decision tables of the signature ρ, and let Dtab_ρ be the set of all decision tables $T \in \mathrm{Tab}_\rho$ possessing the following property: either T is an empty table or all rows of T are labelled by the same number (decision). Decision tables from the set Dtab_ρ will be called *degenerate*.

The special signature $\rho_0 = (F_0, \{0, 1\})$, where $F_0 = \{f_i : i \in \mathbb{N}\}$, will be used in several of the examples.

Example 3.1. Consider the decision table T of the signature ρ_0 depicted in Fig. 3(a). For this table $\mathrm{Row}(T) = \{\bar{\delta}_i : i = 1, \ldots, 5\}$ where $\bar{\delta}_1 = (1, 1, 1)$, $\bar{\delta}_2 = (0, 1, 0)$, $\bar{\delta}_3 = (1, 1, 0)$, $\bar{\delta}_4 = (0, 0, 1)$, $\bar{\delta}_5 = (1, 0, 0)$, and $\nu_T(\bar{\delta}_1) = 1$, $\nu_T(\bar{\delta}_2) = 2$, $\nu_T(\bar{\delta}_3) = 2$, $\nu_T(\bar{\delta}_4) = 3$, $\nu_T(\bar{\delta}_5) = 3$.

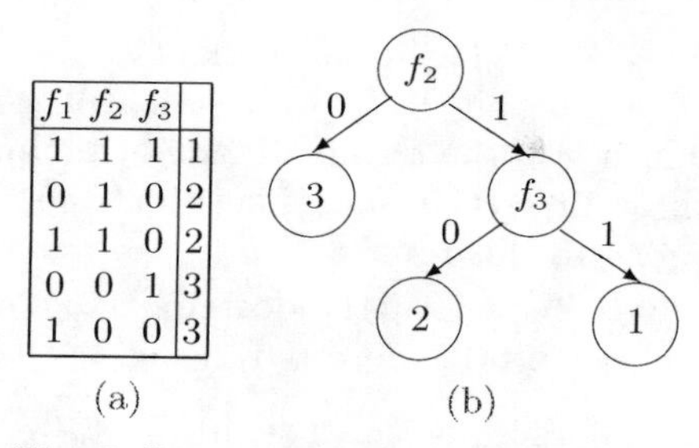

f_1	f_2	f_3	
1	1	1	1
0	1	0	2
1	1	0	2
0	0	1	3
1	0	0	3

Fig. 3. Decision table and decision tree.

Decision Trees. A finite directed tree containing exactly one node with no entering edges will be called *a finite rooted directed tree.* This singular node is called *the root* of the finite rooted directed tree. The nodes of the tree having no issuing edges will be called *terminal* nodes. The nonterminal nodes of the tree will be called *working* nodes. Let $\xi = w_1, d_1, \ldots, w_m, d_m, w_{m+1}$ be a sequence of nodes and edges of finite rooted directed tree G such that w_1 is the root, w_{m+1} is a terminal node, and for $i = 1, \ldots, m$ the edge d_i issues from the node w_i and enters the node w_{i+1}. Then ξ will be called *a complete path* in finite rooted directed tree G.

A labelled finite rooted directed tree will be called *a decision tree of the signature* ρ if it satisfies the following conditions:

a) every working node is labelled by an attribute from F;
b) every edge is labelled by a number from B, while the edges issuing from one and the same node are labelled by distinct numbers;
c) every terminal node is labelled by a number from $\mathbb{Z}$.

Denote by Tree_ρ the set of all decision trees of the signature ρ. For a decision tree $\Gamma \in \mathrm{Tree}_\rho$ denote $\mathrm{At}(\Gamma)$ the set of attributes from F used as labels at the working nodes of Γ, $\Omega_\rho(\Gamma) = \{(f, \delta) : f \in \mathrm{At}(\Gamma), \delta \in B\}^*$, and $\mathrm{Path}(\Gamma)$ the set of all complete paths in Γ.

Let us put into correspondence to a complete path ξ in Γ a word $\pi(\xi)$ from $\Omega_\rho(\Gamma)$. If the path ξ does not contain working nodes then $\pi(\xi) = \lambda$. Let the path ξ contain $m > 0$ working nodes, $\xi = w_1, d_1, \ldots, w_m, d_m, w_{m+1}$, and for $i = 1, \ldots, m$ the node w_i be labelled by the attribute f_i, and the edge d_i be labelled by the number δ_i. Then $\pi(\xi) = (f_1, \delta_1) \ldots (f_m, \delta_m)$.

Example 3.2. In Fig. 3(b) a decision tree of the signature ρ_0 is depicted. Denote this tree by Γ. For $i = 1, 2, 3$ denote by ξ_i the complete path in the thee Γ terminating at the node labelled by the number i. Then $\mathrm{Path}(\Gamma) = \{\xi_1, \xi_2, \xi_3\}$, $\pi(\xi_1) = (f_2, 1)(f_3, 1)$, $\pi(\xi_2) = (f_2, 1)(f_3, 0)$ and $\pi(\xi_3) = (f_2, 0)$.

Let $T \in \mathrm{Tab}_\rho$ and $\Gamma \in \mathrm{Tree}_\rho$. We will say that Γ is *a decision tree for the decision table* T if Γ satisfies the following conditions:

a) $\mathrm{At}(\Gamma) \subseteq \mathrm{At}(T)$;
b) if T is a nonempty table then for any row $\bar{\delta}$ of the table T there exists a complete path ξ in Γ such that $\bar{\delta}$ is a row of the subtable $T\pi(\xi)$, and the terminal node of the path ξ is labelled by the number $\nu_T(\bar{\delta})$.

The condition a) means that during the work of Γ we can ask questions only about columns of T. The subtable $T\pi(\xi)$ consists of all rows of the table T for which the work of Γ finishes in the terminal node of the path ξ. The condition b) means that for each row of T the result of decision tree Γ work coincides with the decision corresponding to this row.

Example 3.3. It is easily to show that the decision tree depicted in Fig. 3(b) is a decision tree for the decision table depicted in Fig. 3(a).

Weight Functions. A function $\psi : F \to \mathbb{N} \setminus \{0\}$ will be called *a weight function of the signature* ρ. The value $\psi(f)$ sometimes will be called *the weight* of the attribute $f \in F$. We denote by h the weight function of the signature ρ for which $h(f) = 1$ for any $f \in F$.

Extend a weight function ψ of the signature ρ on the sets F^*, Ω_ρ and Tree_ρ.

Let $\alpha \in F^*$. If $\alpha = \lambda$ then $\psi(\alpha) = 0$. Let $\alpha \neq \lambda$ and $\alpha = f_1 \dots f_m$. Then $\psi(\alpha) = \sum_{i=1}^{m} \psi(f_i)$.

Let $\beta \in \Omega_\rho$. If $\beta = \lambda$ then $\psi(\beta) = 0$. Let $\beta \neq \lambda$ and $\beta = (f_1, \delta_1) \dots (f_m, \delta_m)$. Then $\psi(\beta) = \sum_{i=1}^{m} \psi(f_i)$.

For $\Gamma \in \text{Tree}_\rho$ let $\psi(\Gamma) = \max\{\psi(\pi(\xi)) : \xi \in \text{Path}(\Gamma)\}$. The value $\psi(\Gamma)$ will be called *the weighted depth* of the decision tree Γ. The value $h(\Gamma)$ will be called *the depth* of the decision tree Γ. Note that $h(\Gamma)$ is the maximal length of a complete path in Γ.

Put into correspondence to a weight function ψ the function $\psi_\rho : \text{Tab}_\rho \to \mathbb{N}$. Let $T \in \text{Tab}_\rho$. Then $\psi_\rho(T) = \min\{\psi(\Gamma) : \Gamma \in \text{Tree}_\rho^{\text{all}}(T)\}$ where $\text{Tree}_\rho^{\text{all}}(T)$ is the set of all decision trees of the signature ρ for the table T. In other words, $\psi_\rho(T)$ is the minimal weighted depth of a decision tree for the table T. A decision tree Γ for the table T such that $\psi(\Gamma) = \psi_\rho(T)$ will be called *optimal.*

Example 3.4. Let T be the decision table depicted in Fig. 3(a), and Γ be the decision tree depicted in Fig. 3(b). It is easily to see $h(\Gamma) = 2$. Taking into account that Γ is a decision tree for T we obtain $h_{\rho_0}(T) \leq 2$.

Diagnostic Tables. A nonempty decision table $T \in \text{Tab}_\rho$ will be called *diagnostic* if its rows are labelled by pairwise different numbers (decisions). The study of diagnostic tables is a question of special interest in the present section for the reason that the diagnostic tables are, in a sense, the tables of maximal complexity. Namely, if two tables $T_1, T_2 \in \text{Tab}_\rho$ are differ only by numbers assigned to rows then $\psi_\rho(T_1) \leq \psi_\rho(T_2)$ if T_2 is diagnostic (see Lemma 3.6). The other reason is that diagnostic tables are frequently met in applications.

3.2 Lower Bounds on Complexity of Decision Trees

Lower bounds on complexity of decision trees for decision tables are studied in this subsection. An approach to proof of lower bounds based on the "proof-tree" notion is considered.

Lower Bounds on Complexity. Let $\rho = (F, B)$, $|B| = k$ and ψ be a weight function of the signature ρ. For any word $\alpha \in \Omega_\rho$ we denote by $\text{Alph}(\alpha)$ the set of letters from the alphabet $\{(f, \delta) : f \in F, \delta \in B\}$ contained in α.

Define a mapping $M_{\rho,\psi}$: $\text{Tab}_\rho \to \mathbb{N}$ as follows. Let T be a table from Tab_ρ with n columns which are labelled by attributes $f_1, \ldots, f_n$. For any $\bar{\delta} = (\delta_1, \ldots, \delta_n) \in B^n$ denote $M_{\rho,\psi}(T, \bar{\delta}) = \min\{\psi(\alpha) : \alpha \in \Omega_\rho(T), \text{Alph}(\alpha) \subseteq \{(f_1, \delta_1), \ldots, (f_n, \delta_n)\}, T\alpha \in \text{Dtab}_\rho\}$. Then $M_{\rho,\psi}(T) = \max\{M_{\rho,\psi}(T, \bar{\delta}) : \bar{\delta} \in B^n\}$.

Let $\bar{\delta} = (\delta_1, \ldots, \delta_n) \in B^n$. Denote $\alpha = (f_1, \delta_1) \ldots (f_n, \delta_n)$. Obviously, $\alpha \in \Omega_\rho(T)$, $\text{Alph}(\alpha) \subseteq \{(f_1, \delta_1), \ldots, (f_n, \delta_n)\}$ and $T\alpha \in \text{Dtab}_\rho$, i.e. T is a degenerate table. Therefore the value of $M_{\rho,\psi}(T, \bar{\delta})$ is definite. Hence the value of $M_{\rho,\psi}(T)$ is also definite.

Consider one more definition of the value $M_{\rho,\psi}(T, \bar{\delta})$. The weight of a column in T is the weight of the corresponding attribute. Let $\bar{\delta}$ be a row of T. Then $M_{\rho,\psi}(T, \bar{\delta})$ is the minimal total weight of columns which distinguish the row $\bar{\delta}$ from all rows with other decisions. Let $\bar{\delta}$ be not a row of T. Then $M_{\rho,\psi}(T, \bar{\delta})$ is the minimal total weight of columns which distinguish the row $\bar{\delta}$ from all rows with the exception, possibly, of some rows with the same decision.

Theorem 3.1. *Let ψ be a weight function of the signature ρ, and $T \in \text{Tab}_\rho$. Then $\psi_\rho(T) \geq M_{\rho,\psi}(T)$.*

Example 3.5. Let T be the table depicted in Fig. 3(a). Evidently, $T(f_i, 1)$ is a nondegenerate table for any $i \in \{1, 2, 3\}$. Hence $M_{\rho_0,h}(T, (1, 1, 1)) \geq 2$. One can show that $T(f_2, \delta_2)(f_3, \delta_3)$ is a degenerate table for any triple $(\delta_1, \delta_2, \delta_3) \in \{0, 1\}^3$. Therefore for any triple $\bar{\delta} \in \{0, 1\}^3$ we have $M_{\rho_0,h}(T, \bar{\delta}) \leq 2$. Thus, $M_{\rho_0,h}(T) = 2$. From Theorem 3.1 it follows that $h_{\rho_0}(T) \geq 2$. Using this inequality and the inequality $h_{\rho_0}(T) \leq 2$ from Example 3.4 we obtain $h_{\rho_0}(T) = 2$.

Define mappings S : $\text{Tab}_\rho \to \mathbb{N}$ and N : $\text{Tab}_\rho \to \mathbb{N}$. Let $T \in \text{Tab}_\rho$. Then $S(T) = \left|\{\nu_T(\bar{\delta}) : \bar{\delta} \in \text{Row}(T)\}\right|$ and $N(T) = |\text{Row}(T)|$. In other words, $S(T)$ it the number of different decisions corresponding to rows of T, and $N(T)$ is the number of rows in the table T. For any diagnostic table T (see Sect. 3.1) we have $\text{Row}(T) \neq \emptyset$ and $S(T) = N(T)$.

For real number a we denote by $\lceil a \rceil$ the minimal integer which is at least a. We denote by $\lfloor a \rfloor$ the maximal integer which is at most a.

Recall that $\rho = (F, B)$ is a signature for which $|B| = k$.

Theorem 3.2. *Let ψ be a weight function of the signature ρ, and T be a nonempty table from Tab_ρ. Then $\psi_\rho(T) \geq \lceil \log_k S(T) \rceil$.*

Corollary 3.1. *Let ψ be a weight function of the signature ρ, and T be a diagnostic table from Tab_ρ. Then $\psi_\rho(T) \geq \lceil \log_k N(T) \rceil$.*

Example 3.6. For the table T depicted in Fig. 3(a) the equality $S(T) = 3$ holds. Using Theorem 3.2 we obtain $h_{\rho_0}(T) \geq \lceil \log_2 3 \rceil = 2$.

Let $T \in \text{Tab}_\rho$ and let D be a subset of the set $\text{At}(T)$. The set D will be called *a test for the table T* if it satisfies the following conditions:

a) if $D = \emptyset$, then $T \in \text{Dtab}_\rho$, i.e. T is a degenerate table;
b) if $D \neq \emptyset$ and $D = \{f_{i(1)}, \ldots, f_{i(m)}\}$ then $T(f_{i(1)}, \delta_1) \ldots (f_{i(m)}, \delta_m) \in \text{Dtab}_\rho$ for any $\delta_1, \ldots, \delta_m \in B$.

In other words, a subset D of the set $\mathrm{At}(T)$ is a test for the table T if and only if any two rows of T with different decisions are distinct on columns of T corresponding to attributes from D.

Note that each relative reduct [162] of the table T is a test for the table T, and a test D for the table T is a relative reduct of the table T if and only if each proper subset of the set D is not a test for T.

Denote by $\mathrm{Test}_\rho(T)$ the set of all tests of the table T.

Let D be a finite subset of the set F. Define the value $\psi(D)$ as follows. If $D = \emptyset$, then $\psi(D) = 0$. Let $D \neq \emptyset$ and $D = \{f_1, \ldots, f_m\}$. Then $\psi(D) = \sum_{i=1}^{m} \psi(f_i)$.

Next define a mapping $J_\psi : \mathrm{Tab}_\rho \to \mathbb{N}$. Let $T \in \mathrm{Tab}_\rho$. Then $J_\psi(T) = \min\{\psi(D) : D \in \mathrm{Test}_\rho(T)\}$. Obviously $\mathrm{At}(T) \in \mathrm{Test}_\rho(T)$. Hence the value $J_\psi(T)$ is definite. It is clear that $J_h(T)$ is the minimal cardinality of a test for the table T, and it is the minimal cardinality of a relative reduct of the table T. Later we will write $J(T)$ instead of $J_h(T)$.

Theorem 3.3. *Let ψ be a weight function of the signature ρ, and $T \in \mathrm{Tab}_\rho$. Then $\psi_\rho(T) \geq \lceil \log_k((k-1)J(T)+1) \rceil$.*

Example 3.7. It is easily to show that the table T depicted in Fig. 3(a) has exactly two tests: $\{f_1, f_2, f_3\}$ and $\{f_2, f_3\}$. Hence $J(T) = 2$. Using Theorem 3.3 we obtain $h_{\rho_0}(T) \geq \lceil \log_2(2+1) \rceil = 2$.

Approach to Proof of Lower Bounds on Complexity. Let C be a nonempty subset of the set F and $C = \{f_1, \ldots, f_n\}$. A labelled finite rooted directed tree will be called *(C,ρ)-tree* if it satisfies the following conditions. The nodes of the tree are not labelled. Every edge of the tree is labelled by a pair from the set $\{(f_i, \delta) : f_i \in C, \delta \in B\}$. The root of the tree either is a terminal node or has n edges issuing from it and labelled by pairs of the kind $(f_1, \delta_1), \ldots, (f_n, \delta_n)$ respectively. Let w be a node of the tree which is not the root, and let $C(w)$ be the set of elements from C, not present in pairs used as labels at the edges in the path connecting the root with the node w. If $C(w) = \emptyset$ then w is a terminal node. Let $C(w) \neq \emptyset$ and $C(w) = \{f_{i(1)}, \ldots, f_{i(m)}\}$. Then either w is a terminal node or there are m edges issuing from w which are labelled by pairs of the kind $(f_{i(1)}, \sigma_1), \ldots, (f_{i(m)}, \sigma_m)$ respectively.

Let G be an (C,ρ)-tree. For every node w of the tree G we define a word $\zeta(w) \in \{(f_i, \delta) : f_i \in C, \delta \in B\}^*$. If w is the root of the tree G then $\zeta(w) = \lambda$. Let an edge issuing from a node w_1 and entering a node w_2 be labelled by a pair (f_i, δ). Then $\zeta(w_2) = \zeta(w_1)(f_i, \delta)$.

Let ψ be a weight function of the signature ρ, $T \in \mathrm{Tab}_\rho$, $r \in \mathbb{Z}$ and G be an $(\mathrm{At}(T), \rho)$-tree. The tree G will be called *a proof-tree for the bound $\psi_\rho(T) \geq r$* if $\psi(\zeta(w)) \geq r$ and $T\zeta(w)$ is a nonempty table for any terminal node w of G, and $T\zeta(w)$ is a nondegenerate table for any nonterminal node w of G.

A proof-tree can be interpreted as a fragment of a strategy of the first player in the game which is modified in the following way: first player does not choose a row at the beginning of the game, but at least one row must satisfy his answers on questions of the second player.

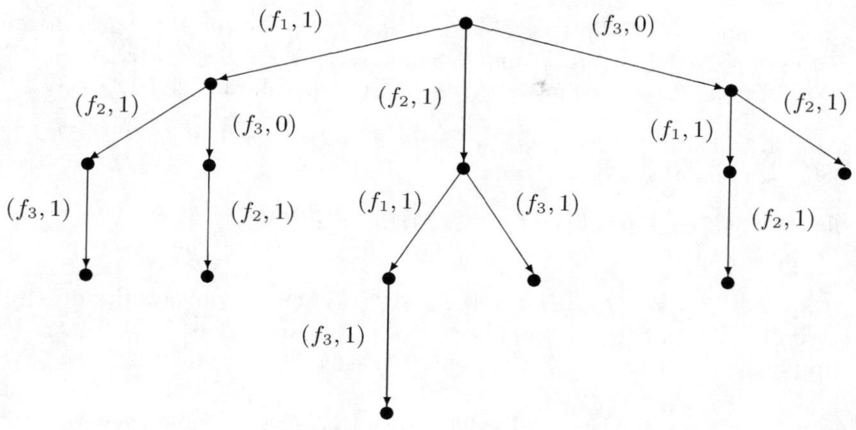

Fig. 4. $(\{f_1, f_2, f_3\}, \rho_0)$-tree.

Theorem 3.4. *Let ψ be a weight function of the signature ρ, T be a nonempty table from* Tab_ρ, *and $r \in \mathbb{Z}$. Then a proof-tree for the bound $\psi_\rho(T) \geq r$ exists if and only if the inequality $\psi_\rho(T) \geq r$ holds.*

Example 3.8. Denote by G the $(\{f_1, f_2, f_3\}, \rho_0)$-tree depicted in Fig. 4. Let T be the decision table depicted in Fig. 3(a), and Γ be the decision tree depicted in Fig. 3(b). Define the weight function ψ of the signature ρ_0 in the following way: $\psi(f_0) = 1$ and $\psi(f_i) = i$ for any $i \in \mathbb{N} \setminus \{0\}$. One can show that G is a proof-tree for the bound $\psi_{\rho_0}(T) \geq 5$. From Theorem 3.4 it follows that $\psi_{\rho_0}(T) \geq 5$. Taking into account that Γ is a decision tree for the table T, and $\psi(\Gamma) = 5$ we obtain $\psi_{\rho_0}(T) \leq 5$. Hence $\psi_{\rho_0}(T) = 5$.

Auxiliary Statements. This subsubsection contains lemmas which will be used later in proofs of Theorems 3.1–3.4.

Lemma 3.1. *Let ψ be a weight function of the signature ρ, and $T \in \mathrm{Tab}_\rho$. Then the value of $\psi_\rho(T)$ is definite and satisfies the inequality $\psi_\rho(T) \leq J_\psi(T)$.*

Proof. Let D be a test for the table T such that $\psi(D) = J_\psi(T)$. Consider a decision tree Γ from Tree_ρ described as follows. If $D = \emptyset$ then Γ contains only one node. Let $D \neq \emptyset$ and $D = \{f_1, \ldots, f_m\}$. In this case the set of nodes of the tree Γ is divided into $m + 1$ layers, where the j-th layer contains k^{j-1} nodes, $j = 1, \ldots, m + 1$. For $j = 1, \ldots, m$ every node of the j-th layer is labelled by the attribute f_j, and from every node of the j-th layer k edges are issuing each of which enters a node of the $(j + 1)$-th layer. These edges are labelled by pairwise different numbers from the set B. Let w be an arbitrary terminal node of the tree Γ, and let ξ be the complete path in Γ terminating at the node w. Taking into account that D is a test for the table T we conclude that $T\pi(\xi) \in \mathrm{Dtab}_\rho$. If $\mathrm{Row}(T\pi(\xi)) = \emptyset$ then node w is labelled by 0. Let $\mathrm{Row}(T\pi(\xi)) \neq \emptyset$. Then the node w is labelled by the number $r \in \mathbb{N}$ such that $\nu_T(\bar{\delta}) = r$ for any $\bar{\delta} \in \mathrm{Row}(T\pi(\xi))$.

One can show that the tree Γ is a decision tree for the table T. Therefore the value of $\psi_\rho(T)$ is definite and, obviously, $\psi_\rho(T) \leq \psi(\Gamma)$. It is clear that $\psi(\Gamma) = \psi(D)$. Using the equality $\psi(D) = J_\psi(T)$ we obtain $\psi_\rho(T) \leq J_\psi(T)$. □

Lemma 3.2. *Let $T \in \mathrm{Tab}_\rho$ and Γ be a decision tree for the table T. Then following statements hold:*

(a) for any $\xi_1, \xi_2 \in \mathrm{Path}(\Gamma)$ if $\xi_1 \neq \xi_2$ then $\mathrm{Row}(T\pi(\xi_1)) \cap \mathrm{Row}(T\pi(\xi_2)) = \emptyset$;
(b) for any $\xi \in \mathrm{Path}(\Gamma)$ the table $T\pi(\xi)$ is degenerate.

Proof. a) Let $\xi_1, \xi_2 \in \mathrm{Path}(\Gamma)$ and $\xi_1 \neq \xi_2$. Then, obviously, there exist an attribute $f \in \mathrm{At}(\Gamma)$ and numbers $\delta_1, \delta_2 \in B$ such that $\delta_1 \neq \delta_2$, $(f, \delta_1) \in \mathrm{Alph}(\pi(\xi_1))$ and $(f, \delta_2) \in \mathrm{Alph}(\pi(\xi_2))$. Hence $\mathrm{Row}(T\pi(\xi_1)) \cap \mathrm{Row}(T\pi(\xi_2)) = \emptyset$.

b) Assume that there exists $\xi \in \mathrm{Path}(\Gamma)$ such that $T\pi(\xi)$ is nondegenerate. Then there exist $\bar\delta_1, \bar\delta_2 \in \mathrm{Row}(T\pi(\xi))$ such that $\nu_T(\bar\delta_1) \neq \nu_T(\bar\delta_2)$. It follows from the statement (a) that ξ is the only path in $\mathrm{Path}(\Gamma)$ such that $\bar\delta_1 \in \mathrm{Row}(T\pi(\xi))$. Since Γ is a decision tree for the table T, we conclude that the terminal node of the path ξ is labelled by the number $\nu_T(\bar\delta_1)$. Considering the row $\bar\delta_2$ in the same way we see that the terminal node of the path ξ is labelled by the number $\nu_T(\bar\delta_2)$ which is impossible. Hence $T\pi(\xi) \in \mathrm{Dtab}_\rho$. □

Lemma 3.3. *Let $T \in \mathrm{Tab}_\rho$, and Γ be a decision tree for the table T. Then the set $\mathrm{At}(\Gamma)$ is a test for the table T.*

Proof. If $T \in \mathrm{Dtab}_\rho$ then, obviously, the set $\mathrm{At}(\Gamma)$ is a test for the table T. Consider the case when $T \notin \mathrm{Dtab}_\rho$. Using Lemma 3.2 one can show that $\mathrm{At}(\Gamma) \neq \emptyset$. Let $\mathrm{At}(\Gamma) = \{f_1, \ldots, f_m\}$. Assume the set $\mathrm{At}(\Gamma)$ is not a test for the table T. Then there exist numbers $\sigma_1, \ldots, \sigma_m \in B$ such that $T(f_1, \sigma_1) \ldots (f_m, \sigma_m)$ is a nondegenerate table. Denote $\beta = (f_1, \sigma_1) \ldots (f_m, \sigma_m)$. Choose certain row $\bar\delta \in \mathrm{Row}(T\beta)$. Since Γ is a decision tree for the table T, there exists a complete path $\xi \in \mathrm{Path}(\Gamma)$ such that $\bar\delta \in \mathrm{Row}(T\pi(\xi))$. It is easily to show that $\mathrm{Alph}(\pi(\xi)) \subseteq \mathrm{Alph}(\beta)$. Hence $\mathrm{Row}(T\beta) \subseteq \mathrm{Row}(T\pi(\xi))$. From Lemma 3.2 it follows that $T\pi(\xi) \in \mathrm{Dtab}_\rho$. Hence $T\beta \in \mathrm{Dtab}_\rho$ too which is impossible. Therefore $\mathrm{At}(\Gamma)$ is a test for the table T. □

For $\Gamma \in \mathrm{Tree}_\rho$ denote $L_t(\Gamma)$ the number of terminal nodes in the tree Γ, and $L_w(\Gamma)$ the number of working nodes in Γ.

Lemma 3.4. *Let $\Gamma \in \mathrm{Tree}_\rho$. Then $L_t(\Gamma) \leq k^{h(\Gamma)}$ and $L_w(\Gamma) \leq (k^{h(\Gamma)} - 1)/(k-1)$.*

Proof. If $h(\Gamma) = 0$ then the considered inequalities hold. Let $h(\Gamma) \geq 1$. Denote by G a decision tree from Tree_ρ in which exactly k edges issue from every working node and every complete path contains exactly $h(\Gamma) + 1$ nodes. It is easily to show that $L_t(\Gamma) \leq L_t(G) = k^{h(\Gamma)}$ and $L_w(\Gamma) \leq L_w(G) = \sum_{i=0}^{h(\Gamma)-1} k^i = (k^{h(\Gamma)} - 1)/(k-1)$. □

Let T be a nonempty table from Tab_ρ. Denote by $\mathrm{Tree}_\rho(T)$ the set of all decision trees Γ from Tree_ρ satisfying the following conditions:

a) $\mathrm{At}(\Gamma) \subseteq \mathrm{At}(T)$;
b) in every complete path in Γ working nodes are labelled by pairwise different attributes;
c) terminal nodes of the tree Γ are labelled by numbers from the set $\{\nu_T(\bar{\delta}) : \bar{\delta} \in \mathrm{Row}(T)\}$.

Lemma 3.5. *Let T be a nonempty table from* Tab_ρ. *Then there exists a decision tree Γ for the table T such that $\Gamma \in \mathrm{Tree}_\rho(T)$ and $\psi(\Gamma) = \psi_\rho(T)$.*

Proof. By Lemma 3.1, there exists a decision tree Γ_1 for the table T such that $\psi(\Gamma_1) = \psi_\rho(T)$. Remove from the tree Γ_1 all the edges and nodes not contained in at least one complete path ξ in the tree Γ_1 such that $\mathrm{Row}(T\pi(\xi)) \neq \emptyset$. As a result obtain certain tree from Tree_ρ which will be denoted by Γ_2. It is easily to show that Γ_2 is a decision tree for the table T and $\psi(\Gamma_2) \leq \psi(\Gamma_1)$. Let w be a working node in the tree Γ_2. If at least two edges issue from the node w then the node w is left untouched. Let exactly one edge (entering the node u) issue from w. We remove from the tree the node w together with the edge issuing from it. Edge entering node w is at that re-oriented so as to enter the node u. Having overlooked in such a way all the working nodes in tree Γ_2 we produce certain tree from Tree_ρ which will be denoted by Γ. As is easily to show Γ is a decision tree for the table T and $\psi(\Gamma) \leq \psi(\Gamma_2)$. Taking into account that $\psi(\Gamma_2) \leq \psi(\Gamma_1)$ and $\psi(\Gamma_1) = \psi_\rho(T)$ obtain $\psi(\Gamma) = \psi_\rho(T)$. Obviously, $\mathrm{At}(\Gamma) \subseteq \mathrm{At}(T)$. It is easily to show that in the tree Γ_2 every terminal node is labelled by a number from the set $\{\nu_T(\bar{\delta}) : \bar{\delta} \in \mathrm{Row}(T)\}$, and in every complete path if any two working nodes are labelled by the same attribute then at least from one of these nodes exactly one edge is issuing. From these properties of the tree Γ_2 one can easily deduce $\Gamma \in \mathrm{Tree}_\rho(T)$. □

Lemma 3.6. *Let T_1, T_2 be tables from* Tab_ρ *which are differ only by numbers assigned to rows, and T_2 be a diagnostic table. Then $\psi_\rho(T_1) \leq \psi_\rho(T_2)$.*

Proof. By Lemma 3.1, there exists a decision tree Γ_2 for the table T_2 such that $\psi(\Gamma_2) = \psi_\rho(T_2)$. Since Γ_2 is a decision tree for the table T_2, for any $\bar{\delta} \in \mathrm{Row}(T_2)$ there exists a complete path $\xi(\bar{\delta})$ in tree Γ_2 such that $\bar{\delta} \in \mathrm{Row}(T_2\pi(\xi(\bar{\delta})))$, and the terminal node of the path $\xi(\bar{\delta})$ is labelled by a number $\nu_{T_2}(\bar{\delta})$. Let $\bar{\delta}_1, \bar{\delta}_2 \in \mathrm{Row}(T_2)$ and $\bar{\delta}_1 \neq \bar{\delta}_2$. Since T_2 is a diagnostic table, $\xi(\bar{\delta}_1) \neq \xi(\bar{\delta}_2)$. For every $\bar{\delta} \in \mathrm{Row}(T_2)$ the terminal node of the path $\xi(\bar{\delta})$ will be labelled by the number $\nu_{T_1}(\bar{\delta})$ instead of the number $\nu_{T_2}(\bar{\delta})$. Denote the obtained tree by Γ_1. It is clear that Γ_1 is a decision tree for the table T_1 and $\psi(\Gamma_1) = \psi(\Gamma_2)$. Therefore $\psi_\rho(T_1) \leq \psi_\rho(T_2)$. □

Proofs of Theorems 3.1–3.4

Proof (of Theorem 3.1). Let the table T contain n columns labelled by attributes $f_1, \ldots, f_n$. From Lemma 3.1 it follows that there exists a decision tree Γ for the table T such that $\psi(\Gamma) = \psi_\rho(T)$. We denote by $E(w)$ the set of numbers by which the edges issuing from certain working node w in tree Γ are labelled.

Consider the following process of transformation of the tree Γ. Each working node w in tree Γ, for which $E(w) = B$, is left untouched. Suppose $E(w) \neq B$ for some working node w in tree Γ. For every $\delta \in B \setminus E(w)$ add to the tree Γ a new terminal node w_δ and an edge d_δ issuing from the node w and entering the node w_δ. The node w_δ will be labelled by 0 and the edge d_δ by δ. All the working nodes of the tree Γ will be processed like that. As a result a new tree from Tree_ρ will be obtained. Denote this tree by G. One can show that G is a decision tree for the table T and $\psi(G) = \psi(\Gamma)$.

Let $\bar{\delta} = (\delta_1, \ldots, \delta_n)$ be an n-tuple from B^n such that $M_{\rho,\psi}(T, \bar{\delta}) = M_{\rho,\psi}(T)$. It is not difficult to prove that there exists a complete path ξ in the tree G such that $\mathrm{Alph}(\pi(\xi)) \subseteq \{(f_1, \delta_1), \ldots, (f_n, \delta_n)\}$. Taking into account that G is a decision tree for the table T and using Lemma 3.2 we obtain $T\pi(\xi) \in \mathrm{Dtab}_\rho$. Evidently, $\psi(\pi(\xi)) \leq \psi(G)$. Hence $M_{\rho,\psi}(T, \bar{\delta}) \leq \psi(G)$. Taking into account that $M_{\rho,\psi}(T) = M_{\rho,\psi}(T, \bar{\delta})$ and $\psi(G) = \psi(\Gamma) = \psi_\rho(T)$ we obtain $M_{\rho,\psi}(T) \leq \psi_\rho(T)$. □

Proof (of Theorem 3.2). From Lemma 3.1 it follows the existence of a decision tree G for the table T such that $\psi(G) = \psi_\rho(T)$. Obviously, $S(T) \leq L_t(G)$. Using Lemma 3.4 we obtain $L_t(G) \leq k^{h(G)}$. Therefore $k^{h(G)} \geq S(T)$. Bearing in mind that T is a nonempty table we obtain $S(T) > 0$ and hence $h(G) \geq \log_k S(T)$. One can easily show that $\psi(G) \geq h(G)$. Hence $\psi(G) \geq \log_k S(T)$. Taking into account that $\psi(G) = \psi_\rho(T)$ and $\psi_\rho(T) \in \mathbb{N}$ we obtain $\psi_\rho(T) \geq \lceil \log_k S(T) \rceil$. □

Proof (of Theorem 3.3). Using Lemma 3.1 we obtain that there exists a decision tree G for the table T such that $\psi(G) = \psi_\rho(T)$. By Lemma 3.3, the set $\mathrm{At}(G)$ is a test for the table T. Therefore $J(T) \leq |\mathrm{At}(G)|$. Evidently, $|\mathrm{At}(G)| \leq L_w(G)$. Therefore $J(T) \leq L_w(G)$. Using Lemma 3.4 we obtain $L_w(G) \leq (k^{h(G)} - 1)/(k-1)$. Therefore $J(T) \leq (k^{h(G)} - 1)/(k-1)$ and $k^{h(G)} \geq (k-1)J(T) + 1$. Hence $h(G) \geq \log_k((k-1)J(T)+1)$. It is easily to show that $\psi(G) \geq h(G)$. Taking into account that $\psi(G) = \psi_\rho(T)$ and $\psi_\rho(T) \in \mathbb{N}$ we obtain $\psi_\rho(T) \geq \lceil \log_k((k-1)J(T)+1) \rceil$. □

Let T be a table from Tab_ρ with $n > 1$ columns, $f \in \mathrm{At}(T)$ and $t \in B$. Denote by $T[f,t]$ the table from Tab_ρ which is obtained from the table $T(f,t)$ by removal of all columns labelled by the attribute f.

Let Γ be a finite rooted directed tree and d be an edge in the tree Γ entering the node w of the tree Γ. Let G be the subtree of the tree Γ which root is the node w. We will say that the edge d *determines the subtree G of the tree Γ*.

Proof (of Theorem 3.4). We prove by induction on n that for any nonempty table $T \in \mathrm{Tab}_\rho$ with at most n columns and for any $r \in \mathbb{Z}$ if a proof-tree for the bound $\psi_\rho(T) \geq r$ exists then the inequality $\psi_\rho(T) \geq r$ holds.

Let us show that the considered statement is true for $n = 1$. Let T contain one column, which is labelled by the attribute f, $r \in \mathbb{Z}$ and G be a proof-tree for the bound $\psi_\rho(T) \geq r$. If $r \leq 0$ then, evidently, the inequality $\psi_\rho(T) \geq r$ holds. Let $r > 0$. It is not difficult to show that the tree G consists of two nodes joined by an edge labelled by the pair of the kind (f, δ), where $\delta \in B$. Hence $T \notin \mathrm{Dtab}_\rho$

and $\psi(f) \geq r$. Using Lemma 3.1 we conclude that there exists a decision tree Γ for the table T such that $\psi(\Gamma) = \psi_\rho(T)$. Since $T \notin \mathrm{Dtab}_\rho$, at least one of the nodes of Γ is labelled by the attribute f. Hence $\psi(\Gamma) \geq \psi(f) \geq r$. Taking into account that $\psi(\Gamma) = \psi_\rho(T)$ we obtain $\psi_\rho(T) \geq r$. Thus, the considered statement is true for $n = 1$.

We assume now that the statement is true for certain $n \geq 1$. Show that it is true also for $n+1$. Let T contain $n+1$ columns, and G be a proof-tree for the bound $\psi_\rho(T) \geq r$. If $r \leq 0$ then, obviously, the inequality $\psi_\rho(T) \geq r$ holds. Let $r > 0$. Then the root of the tree G, evidently, is not a terminal node, and hence $T \notin \mathrm{Dtab}_\rho$. Using Lemma 3.5 we conclude that there exists a decision tree Γ for the table T such that $\Gamma \in \mathrm{Tree}_\rho(T)$ and $\psi(\Gamma) = \psi_\rho(T)$. Taking into account that $T \notin \mathrm{Dtab}_\rho$ and using Lemma 3.2 we conclude that the root of the tree Γ is not a terminal node. Let the root of Γ be labelled by the attribute f. Obviously, there exists an edge issuing from the root of the tree G and labelled by a pair of the kind (f, δ) where $\delta \in B$. Denote by G' the subtree of the tree G determined by this edge. One can show that G' is a proof-tree for the bound $\psi_\rho(T[f,\delta]) \geq r - \psi(f)$. Obviously, $T[f,\delta]$ is a nonempty table with at most n columns. By inductive hypothesis, $\psi_\rho(T[f,\delta]) \geq r - \psi(f)$. Obviously, $\mathrm{Row}(T(f,\delta)) \neq \emptyset$. Hence the tree Γ contains an edge issuing from the root which is labelled by the number δ. Denote by Γ' the subtree of the tree Γ determined by this edge. Taking into account that $\Gamma \in \mathrm{Tree}_\rho(T)$ it is not difficult to show that Γ' is a decision tree for the table $T[f,\delta]$. Bearing in mind that $\psi_\rho(T[f,\delta]) \geq r - \psi(f)$ we obtain $\psi(\Gamma') \geq r - \psi(f)$. Hence $\psi(\Gamma) \geq r$. From this inequality and from the choice of Γ the inequality $\psi_\rho(T) \geq r$ follows. Thus, the considered statement is proved.

We prove by induction on n that for any nonempty table $T \in \mathrm{Tab}_\rho$ with at most n columns and for any $r \in \mathbb{Z}$ if the inequality $\psi_\rho(T) \geq r$ holds then a proof-tree for the bound $\psi_\rho(T) \geq r$ exists.

Let us show that this statement is true for $n = 1$. Let T contain one column which is labelled by the attribute f. Let $r \leq 0$. Denote by G_0 the tree having only one node with no label. It is easily to see that G_0 is a proof-tree for the bound $\psi_\rho(T) \geq 0$ and hence a proof-tree for the bound $\psi_\rho(T) \geq r$. Let now $r > 0$. Then, obviously, $T \notin \mathrm{Dtab}_\rho$. Let Γ_1 be an arbitrary decision tree for the table T. Since $T \notin \mathrm{Dtab}_\rho$, the tree Γ_1 contains at least one working node. This node, evidently, is labelled by the attribute f. Therefore $\psi(\Gamma_1) \geq \psi(f)$, and hence $\psi_\rho(T) \geq \psi(f)$. It is not difficult to show that there exists a decision tree Γ_2 for the table T such that $\psi(\Gamma_2) = \psi(f)$. Hence $\psi_\rho(T) = \psi(f)$ and $\psi(f) \geq r$. Since $T \notin \mathrm{Dtab}_\rho$, there exists $\delta \in B$ such that $\mathrm{Row}(T(f,\delta)) \neq \emptyset$. Denote by G_1 the tree consisting of two non-labelled nodes and the edge joining them which is labelled by the pair (f, δ). Obviously, G_1 is a proof-tree for the bound $\psi_\rho(T) \geq \psi(f)$, and hence also a proof-tree for the bound $\psi_\rho(T) \geq r$. So the considered statement is true for $n = 1$.

Let this statement be true for certain $n \geq 1$. Let us show that it is true also for $n+1$. Let T contain $n+1$ columns, and the inequality $\psi_\rho(T) \geq r$ hold. Let $r \leq 0$. In this case the above-introduced tree G_0 is a proof-tree for the bound $\psi_\rho(T) \geq r$. Let now $r > 0$. Then, obviously, $T \notin \mathrm{Dtab}_\rho$. Let $\mathrm{At}(T) = \{f_1, \ldots, f_{n+1}\}$. It is not difficult to show that $\psi_\rho(T) \leq \psi(f_i) + \max\{\psi_\rho(T[f_i,\delta]) :$

$\delta \in B, \text{Row}(T[f_i, \delta]) \neq \emptyset\}$ for any $f_i \in \text{At}(T)$. Hence for every $f_i \in \text{At}(T)$ there exists $\delta_i \in B$ such that $\text{Row}(T[f_i, \delta_i]) \neq \emptyset$ and $\psi_\rho(T[f_i, \delta_i]) \geq \psi_\rho(T) - \psi(f_i) \geq r - \psi(f_i)$. Evidently, for any $f_i \in \text{At}(T)$ the table $T[f_i, \delta_i]$ contains at most n columns. Using the inductive hypothesis we conclude that for every $f_i \in \text{At}(T)$ there exists a proof-tree for the bound $\psi_\rho(T[f_i, \delta_i]) \geq r - \psi(f_i)$. Denote this tree by G_i. Denote by G the following labelled finite rooted directed tree: for every $f_i \in \text{At}(T)$ an edge labelled by the pair (f_i, δ_i) issues from the root of G and enters the root of G_i, and no other edges issue from the root of the tree G. Taking into account that $T \notin \text{Dtab}_\rho$ one can show that the tree G is a proof-tree for the bound $\psi_\rho(T) \geq r$. Thus, the considered statement is proved. □

3.3 Upper Bounds on Complexity of Decision Trees

Let $\rho = (F, B)$ be a signature, $|B| = k$, ψ be a weight function of the signature ρ, and T be a table from Tab_ρ. In this subsection we consider upper bounds on the value $\psi_\rho(T)$ (which is the complexity of optimal decision tree for the table T) depending on the values $N(T)$ and $M_{\rho,\psi}(T)$ defined in Sect. 3.2. The case, when the depth h is used in the capacity of decision tree complexity, is considered more explicitly.

In order to obtain the upper bounds we study the following process of decision tree construction. At each step we find a set of questions (columns, attributes) with minimal total weight satisfying the following condition: for any answers on these questions we either solve the problem (recognize the decision corresponding to the chosen row) or reduce the number of rows by half.

Bounds

Theorem 3.5. *Let ψ be a weight function of the signature ρ, and $T \in \text{Tab}_\rho$. Then*

$$\psi_\rho(T) \leq \begin{cases} 0, & \text{if } T \in \text{Dtab}_\rho \ , \\ M_{\rho,\psi}(T) \log_2 N(T), & \text{if } T \notin \text{Dtab}_\rho \ . \end{cases}$$

Corollary 3.2. *Let ψ be a weight function of the signature ρ, and T be a diagnostic table from Tab_ρ. Then*

$$\max\{M_{\rho,\psi}(T), \log_k N(T)\} \leq \psi_\rho(T) \leq \log_2 k M_{\rho,\psi}(T) \log_k N(T) \ .$$

In case the depth h is used as decision tree complexity a more precise bound is possible.

Theorem 3.6. *Let $T \in \text{Tab}_\rho$. Then*

$$h_\rho(T) \leq \begin{cases} M_{\rho,h}(T), & \text{if } M_{\rho,h}(T) \leq 1 \ , \\ 2\log_2 N(T) + M_{\rho,h}(T), & \text{if } 2 \leq M_{\rho,h}(T) \leq 3 \ , \\ M_{\rho,h}(T) \log_2 N(T) / \log_2 M_{\rho,h}(T) + M_{\rho,h}(T), & \text{if } M_{\rho,h}(T) \geq 4 \ . \end{cases}$$

Denote $C(M_{\rho,h}, N) = \{(M_{\rho,h}(T), N(T)) : T \in \text{Tab}_\rho\}$. From results of [70] the next statement follows immediately.

Theorem 3.7. *Let $\rho = (F, B)$ be a signature with infinite set F. Then*

$$C(M_{\rho,h}, N) = \{(0,0)\} \cup \{(m,n) : m, n \in \mathbb{N}, n \geq m+1\}$$

and for any pair (m,n) from $C(M_{\rho,h}, N)$ there exists a table $T(m,n)$ from Tab_ρ for which $M_{\rho,h}(T(m,n)) = m$, $N(T(m,n)) = n$ and

$$h_\rho(T(n,m)) \geq \begin{cases} m, & \text{if } 0 \leq m \leq 1 , \\ \log_2 n, & \text{if } m = 2 , \\ (m-2)\left\lfloor \frac{\log_2 n - \log_2 m}{\log_2(m-1)} \right\rfloor + m - 1, & \text{if } m \geq 3 . \end{cases}$$

It follows from Theorem 3.7 that for signature $\rho = (F, B)$ with infinite set F the bound obtained in Theorem 3.6 cannot be essentially improved.

Process $\mathcal{Y}_{\rho,\psi}$ of Decision Tree Construction. Let ψ be a weight function of the signature ρ. In this subsubsection a process $\mathcal{Y}_{\rho,\psi}$ is studied which for an arbitrary table T from Tab_ρ constructs a decision tree $\mathcal{Y}_{\rho,\psi}(T)$ for the table T. The bounds considered in Theorems 3.5 and 3.6 are the result of investigation of decision trees constructed by this process.

The set F may be uncountable and the function ψ may be non-computable. Hence generally speaking the process $\mathcal{Y}_{\rho,\psi}$ is not an algorithm but only the way of description of the tree $\mathcal{Y}_{\rho,\psi}(T)$.

Process $\mathcal{Y}_{\rho,\psi}$ includes the subprocess $\mathcal{X}_{\rho,\psi}$ which for an arbitrary nondegenerate table T from Tab_ρ constructs a decision tree $\mathcal{X}_{\rho,\psi}(T) \in \mathrm{Tree}_\rho$.

Description of the Subprocess $\mathcal{X}_{\rho,\psi}$

Let us apply the subprocess $\mathcal{X}_{\rho,\psi}$ to a nondegenerate table $T \in \mathrm{Tab}_\rho$ containing n columns which are labelled by attributes $f_1, \ldots, f_n$.

1-st Step. Let σ_i for any $i \in \{1, \ldots, n\}$ be equal to the minimal number σ from B such that

$$N(T(f_i, \sigma)) = \max\{N(T(f_i, \delta) : \delta \in B\} .$$

Denote $\bar{\sigma} = (\sigma_1, \ldots, \sigma_n)$. Choose a word $\beta \in \Omega_\rho(T)$ such that $\mathrm{Alph}(\beta) \subseteq \{(f_1, \sigma_1), \ldots, (f_n, \sigma_n)\}$, $T\beta \in \mathrm{Dtab}_\rho$ and $\psi(\beta) = M_{\rho,\psi}(T, \bar{\sigma})$. Since $T \notin \mathrm{Dtab}_\rho$, we have $\beta \neq \lambda$. Let $\beta = (f_{i(1)}, \sigma_{i(1)}) \ldots (f_{i(m)}, \sigma_{i(m)})$. Set $I_1 = \{f_{i(1)}, \ldots, f_{i(m)}\}$. Construct a tree with exactly one node labelled by the word λ. Denote the obtained tree by G_1. Proceed to the second step.

Suppose $t \geq 1$ steps have already been made and the tree G_t and the set I_t have been built.

$(t+1)$-th Step. Find the only node w in the tree G_t which is labelled by a word from $\Omega_\rho(T)$. Let w be labelled by the word α.

If $I_t = \emptyset$ then mark the node w by the number 0 instead of the word α. Denote the obtained tree by $\mathcal{X}_{\rho,\psi}(T)$. The subprocess $\mathcal{X}_{\rho,\psi}$ is completed.

Let $I_t \neq \emptyset$. Let j be the minimal number from the set $\{1, \ldots, n\}$ with the following properties: $f_j \in I_t$ and $\max\{N(T\alpha(f_j, \sigma)) : \sigma \in B \setminus \{\sigma_j\}\} \geq \max\{N(T\alpha(f_l, \sigma)) : \sigma \in B \setminus \{\sigma_l\}\}$ for any $f_l \in I_t$. Mark the node w by the attribute f_j instead of the word α . For any $\sigma \in B$ add to the tree G_t a node w_σ and an edge issuing from the node w and entering the node w_σ. Mark this edge by the number σ. If $\sigma \neq \sigma_j$ then we mark the node w_σ by the number 0. If $\sigma = \sigma_j$ then we mark the node w_σ by the word $\alpha(f_j, \sigma_j)$. Denote the obtained tree by G_{t+1}. Set $I_{t+1} = I_t \setminus \{f_j\}$. Proceed to the $(t+2)$-th step.

Example 3.9. Denote by T the table depicted in Fig. 3(a). Then the tree $\mathcal{X}_{\rho_0,h}(T)$ is depicted in Fig. 5.

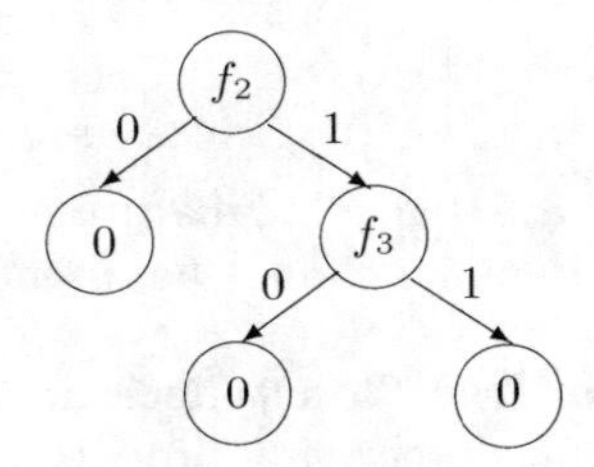

Fig. 5. Decision tree constructed by subprocess $\mathcal{X}_{\rho_0,h}$.

Description of the Process $\mathcal{Y}_{\rho,\psi}$

Let the process $\mathcal{Y}_{\rho,\psi}$ be applied to a table $T \in \text{Tab}_\rho$.

1-st Step. Let us construct a tree containing the only node w.

Let $T \in \text{Dtab}_\rho$. If T is an empty table then mark the node w by the number 0. If T is a nonempty table then the node w will be labelled by the number $\nu_T(\bar{\delta})$, where $\bar{\delta}$ is a row from $\text{Row}(T)$. Denote the obtained tree by $\mathcal{Y}_{\rho,\psi}(T)$. The process $\mathcal{Y}_{\rho,\psi}$ is completed.

Let $T \notin \text{Dtab}_\rho$. Mark the node w by the word λ and proceed to the second step.

Suppose $t \geq 1$ steps have already been made. Denote by G the tree constructed at the step t.

$(t+1)$-th Step. If no one node in the tree G is labelled by a word from $\Omega_\rho(T)$ then denote by $\mathcal{Y}_{\rho,\psi}(T)$ the tree G. The process $\mathcal{Y}_{\rho,\psi}$ is completed.

Otherwise, choose a node w in the tree G which is labelled by a word from $\Omega_\rho(T)$. Let the node w be labelled by the word α.

Let $T\alpha \in \text{Dtab}_\rho$. If $\text{Row}(T\alpha) = \emptyset$ then we mark the node w by the number 0 instead of the word α. If $\text{Row}(T\alpha) \neq \emptyset$ then instead of the word α we mark the node w by the number $\nu_T(\bar{\delta})$ where $\bar{\delta}$ is a row from $\text{Row}(T\alpha)$. Proceed to the $(t+2)$-th step.

Let $T\alpha \notin \text{Dtab}_\rho$. Construct the tree $\mathcal{X}_{\rho,\psi}(T\alpha)$ with the help of the subprocess $\mathcal{X}_{\rho,\psi}$. For any complete path ξ in the tree $\mathcal{X}_{\rho,\psi}(T)$ replace the number 0 as the label of the terminal node of this path with the word $\alpha\pi(\xi)$. Denote the obtained tree by Γ. Remove the node w from the tree G and add the tree Γ to the tree G. If there was an edge entering the node w then we join it to the root of the tree Γ. Proceed to the $(t+2)$-th step.

Example 3.10. Denote by T the table depicted in Fig. 3(a). The tree $\mathcal{Y}_{\rho_0,h}(T)$ is depicted in Fig. 3(b).

Proof of Theorem 3.5

Lemma 3.7. *Let ψ be a weight function of the signature ρ, and T be a nondegenerate table from* Tab_ρ. *Then for any complete path ξ in the tree $\mathcal{X}_{\rho,\psi}(T)$ the following statements hold:*

(a) $\psi(\pi(\xi)) \le M_{\rho,\psi}(T)$;
(b) if $T\pi(\xi)$ is a nondegenerate table then

$$N(T\pi(\xi)) \le N(T)/\max\{2, h(\pi(\xi))\} .$$

Proof. Let T contain n columns which are labelled by attributes $f_1, \ldots, f_n$. For every $i \in \{1, \ldots, n\}$ denote by σ_i the minimal number from B such that $N(T(f_i, \sigma_i)) = \max\{N(T(f_i, \sigma)) : \sigma \in B\}$. Denote $\bar{\sigma} = (\sigma_1, \ldots, \sigma_n)$. Let β be a word from $\Omega_\rho(T)$ which was chosen during the work of the subprocess $\mathcal{X}_{\rho,\psi}$ and for which $\mathrm{Alph}(\beta) \subseteq \{(f_1, \sigma_1), \ldots, (f_n, \sigma_n)\}$, $T\beta \in \mathrm{Dtab}_\rho$ and $\psi(\beta) = M_{\rho,\psi}(T, \bar{\sigma})$. Obviously, all the letters in the word β are pairwise distinct. Using this property of the word β and the description of the subprocess $\mathcal{X}_{\rho,\psi}$ we can easily conclude that there exists a complete path ξ_0 in the tree $\mathcal{X}_{\rho,\psi}(T)$ such that $\mathrm{Alph}(\pi(\xi_0)) = \mathrm{Alph}(\beta)$ and the words $\pi(\xi_0)$ and β are of the same length. Hence $T\pi(\xi_0) \in \mathrm{Dtab}_\rho$ and $\psi(\pi(\xi_0)) = \psi(\beta)$. Taking into account the choice of the word β we obtain

$$\psi(\pi(\xi_0)) = M_{\rho,\psi}(T, \bar{\sigma}) . \tag{1}$$

Let $\pi(\xi_0) = (f_{j(1)}, \sigma_{j(1)}) \ldots (f_{j(m)}, \sigma_{j(m)})$. Denote $\alpha_0 = \lambda$ and for $i = 1, \ldots, m$ denote $\alpha_i = (f_{j(1)}, \sigma_{j(1)}) \ldots (f_{j(i)}, \sigma_{j(i)})$. For $i = 1, \ldots, m$ denote by $\delta_{j(i)}$ the minimal number from the set $B \setminus \{\sigma_{j(i)}\}$ such that

$$N(T\alpha_{i-1}(f_{j(i)}, \delta_{j(i)})) = \max\{N(T\alpha_{i-1}(f_{j(i)}, \sigma)) : \sigma \in B \setminus \{\sigma_{j(i)}\}\} .$$

Let ξ be an arbitrary complete path in the tree $\mathcal{X}_{\rho,\psi}(T)$. Let $\xi = \xi_0$. Using (1) we obtain $\psi(\pi(\xi_0)) = M_{\rho,\psi}(T, \bar{\sigma}) \le M_{\rho,\psi}(T)$. Let now $\xi \ne \xi_0$. One can show that in this case there exist numbers $r \in \{1, \ldots, m\}$ and $\delta \in B$ such that $\pi(\xi) = \alpha_{r-1}(f_{j(r)}, \delta)$. Hence $\psi(\pi(\xi)) \le \psi(\pi(\xi_0))$ and $\psi(\pi(\xi)) \le M_{\rho,\psi}(T)$. Thus, the statement (a) of the lemma is proved.

Let ξ be a complete path in the tree $\mathcal{X}_{\rho,\psi}(T)$ such that $T\pi(\xi) \notin \mathrm{Dtab}_\rho$. Taking into account that $T\pi(\xi_0) \in \mathrm{Dtab}_\rho$ we obtain $\xi \ne \xi_0$. It is not difficult to show that there exist numbers $r \in \{1, \ldots, m\}$ and $\delta \in B$ such that $\delta \ne \sigma_{j(r)}$ and $\pi(\xi) = \alpha_{r-1}(f_{j(r)}, \delta)$.

We will show that $N(T\pi(\xi)) \le N(T)/2$. Obviously,

$$N(T\pi(\xi)) \le N(T(f_{j(r)}, \delta)) .$$

Taking into account the choice of the number $\sigma_{j(r)}$ we obtain

$$2N(T(f_{j(r)}, \delta)) \le N(T(f_{j(r)}, \delta)) + N(T(f_{j(r)}, \sigma_{j(r)})) .$$

Using the relation $\delta \ne \sigma_{j(r)}$ we have

$$N(T(f_{j(r)}, \delta)) + N(T(f_{j(r)}, \sigma_{j(r)})) \le N(T) .$$

Hence $N(T\pi(\xi)) \le N(T)/2$.

Obviously, $h(\pi(\xi)) = r$. Let $r \geq 2$. We will show that $N(T\pi(\xi)) \leq N(T)/r$. Bearing in mind that $\delta_{j(i+1)} \neq \sigma_{j(i+1)}$ for $i = 0, \ldots, r-2$ we obtain $N(T\alpha_{i+1}) + N(T\alpha_i(f_{j(i+1)}, \delta_{j(i+1)})) \leq N(T\alpha_i)$ for $i = 0, \ldots, r-2$. Summing these inequalities over i from 0 to $r-2$ obtain

$$N(T\alpha_{r-1}) + \sum_{i=1}^{r-2} N(T\alpha_i(f_{j(i+1)}, \delta_{j(i+1)})) \leq N(T) . \tag{2}$$

Let us show that the inequality

$$N(T\pi(\xi)) \leq N(T\alpha_i(f_{j(i+1)}, \delta_{j(i+1)})) \tag{3}$$

holds for any $i \in \{0, \ldots, r-2\}$. The inequality

$$N(T\alpha_i(f_{j(r)}, \delta)) \leq N(T\alpha_i(f_{j(i+1)}, \delta_{j(i+1)}))$$

follows from the choice of the attribute $f_{j(i+1)}$ (see the description of the subprocess $\mathcal{X}_{\rho,\psi}$) and from the definition of the number $\delta_{j(i+1)}$. The inequality $N(T\pi(\xi)) \leq N(T\alpha_i(f_{j(r)}, \delta))$ is obvious. The inequality (3) follows from the last two inequalities. The inequality $N(T\pi(\xi)) \leq N(T\alpha_{r-1})$ is obvious. From this inequality and from (2), (3) follows $rN(T\pi(\xi)) \leq N(T)$. Taking into account that $r \geq 2$ we obtain $N(T\pi(\xi)) \leq N(T)/r$. So the statement (b) of the lemma is proved. □

Using descriptions of the process $\mathcal{Y}_{\rho,\psi}$ and of the subprocess $\mathcal{X}_{\rho,\psi}$, and also Lemma 3.7 it is not difficult to prove the following statement.

Proposition 3.1. *Let ψ be a weight function of the signature ρ. Then for any table $T \in \mathrm{Tab}_\rho$ the process $\mathcal{Y}_{\rho,\psi}$ completes after realization of finite sequence of steps. The constructed tree $\mathcal{Y}_{\rho,\psi}(T)$ is a decision tree for the table T.*

Proof (of Theorem 3.5). Let $T \in \mathrm{Dtab}_\rho$. From the description of the process $\mathcal{Y}_{\rho,\psi}$ it follows that $\psi(\mathcal{Y}_{\rho,\psi}(T)) = 0$. From this equality and from Proposition 3.1 it follows that $\psi_\rho(T) \leq 0$.

Let $T \notin \mathrm{Dtab}_\rho$. Consider an arbitrary complete path ξ in the tree $\mathcal{Y}_{\rho,\psi}(T)$. From the description of the process $\mathcal{Y}_{\rho,\psi}$ and from the relation $T \notin \mathrm{Dtab}_\rho$ it follows that $\pi(\xi) = \pi(\xi_1) \ldots \pi(\xi_m)$ for certain $m \geq 1$, where ξ_1 is a complete path in the tree $\mathcal{X}_{\rho,\psi}(T)$, and if $m \geq 2$ then the path ξ_i is a complete path in the tree $\mathcal{X}_{\rho,\psi}(T\pi(\xi_1) \ldots \pi(\xi_{i-1}))$ for $i = 2, \ldots, m$.

By the assumption, $T \notin \mathrm{Dtab}_\rho$. From the description of the process $\mathcal{Y}_{\rho,\psi}$ it follows that if $m \geq 2$ then $T\pi(\xi_1) \ldots \pi(\xi_{i-1}) \notin \mathrm{Dtab}_\rho$ for $i = 2, \ldots, m$. Using the part (a) of the statement of Lemma 3.7 we obtain $\psi(\pi(\xi_1)) \leq M_{\rho,\psi}(T)$ and if $m \geq 2$ then $\psi(\pi(\xi_i)) \leq M_{\rho,\psi}(T\pi(\xi_1) \ldots \pi(\xi_{i-1}))$ for $i = 2, \ldots, m$. One can show that $M_{\rho,\psi}(T\alpha) \leq M_{\rho,\psi}(T)$ for any $\alpha \in \Omega_\rho(T)$. Therefore $\psi(\pi(\xi_i)) \leq M_{\rho,\psi}(T)$ for $i = 1, \ldots, m$. Hence

$$\psi(\pi(\xi)) = \sum_{i=1}^{m} \psi(\pi(\xi_i)) \leq mM_{\rho,\psi}(T) . \tag{4}$$

Let us show that $m \le \log_2 N(T)$. Since $T \notin \mathrm{Dtab}_\rho$, the inequality $N(T) \ge 2$ holds. Hence if $m = 1$ the considered inequality holds. Let $m \ge 2$. From the part (b) of the statement of Lemma 3.7 it follows that $N(T\pi(\xi_1)\dots\pi(\xi_{m-1})) \le N(T)/2^{m-1}$. Taking into account that $T\pi(\xi_1)\dots\pi(\xi_{m-1}) \notin \mathrm{Dtab}_\rho$ we obtain $N(T\pi(\xi_1)\dots\pi(\xi_{m-1})) \ge 2$. Hence $2^m \le N(T)$ and $m \le \log_2 N(T)$. Using (4) we have $\psi(\pi(\xi)) \le M_{\rho,\psi}(T)\log_2 N(T)$. Since ξ is an arbitrary complete path in the tree $\mathcal{Y}_{\rho,\psi}(T)$, we conclude that $\psi(\mathcal{Y}_{\rho,\psi}(T)) \le M_{\rho,\psi}(T)\log_2 N(T)$. From this inequality and from Proposition 3.1 it follows that $\psi_\rho(T) \le M_{\rho,\psi}(T)\log_2 N(T)$. □

Proof (of Corollary 3.2). The lower bounds follow from Theorem 3.1 and Corollary 3.1. The upper bound follows from Theorem 3.5 and from obvious inequality $N(T) \ge 1$. □

Proof of Theorem 3.6

Lemma 3.8. *Let $T \in \mathrm{Tab}_\rho$. Then*

a) $M_{\rho,h}(T) = 0$ if and only if $h_\rho(T) = 0$;
b) $M_{\rho,h}(T) = 1$ if and only if $h_\rho(T) = 1$;
c) if $M_{\rho,h}(T) = 1$ then there exists an attribute $f_i \in \mathrm{At}(T)$ such that $T(f_i,\delta) \in \mathrm{Dtab}_\rho$ for any $\delta \in B$.

Proof. Let $M_{\rho,h}(T) = 0$. Then, obviously, $T \in \mathrm{Dtab}_\rho$. It is easily to show that in this case $h_\rho(T) = 0$. Let $h_\rho(T) = 0$. Using Theorem 3.1 obtain $M_{\rho,h}(T) = 0$.

Let $M_{\rho,h}(T) = 1$. We will show that there exists an attribute $f_i \in \mathrm{At}(T)$ such that $T(f_i,\delta) \in \mathrm{Dtab}_\rho$ for any $\delta \in B$. Assume the contrary. Let the table T contain n columns which are labelled by attributes $f_1,\dots,f_n$, and for any $f_i \in \mathrm{At}(T)$ there exists $\delta_i \in B$ such that $T(f_i,\delta_i) \notin \mathrm{Dtab}_\rho$. Denote $\bar\delta = (\delta_1,\dots,\delta_n)$. It is easily to show that $M_{\rho,h}(T,\bar\delta) \ge 2$ which is impossible since $M_{\rho,h}(T,\bar\delta) \le M_{\rho,h}(T) = 1$. Therefore there exists $f_i \in \mathrm{At}(T)$ such that $T(f_i,\delta) \in \mathrm{Dtab}_\rho$ for any $\delta \in B$. Using this fact it is not difficult to show that there exists a decision tree Γ for the table T such that $h(\Gamma) = 1$. Hence $h_\rho(T) \le 1$. Using the equality $M_{\rho,h}(T) = 1$ and Theorem 3.1 we obtain $h_\rho(T) = 1$.

Let $h_\rho(T) = 1$. Using Theorem 3.1 obtain $M_{\rho,h}(T) \le 1$. Assume $M_{\rho,h}(T) = 0$. Then, by proved above, $h_\rho(T) = 0$ which is impossible. Hence $M_{\rho,h}(T) = 1$. □

Proof (of Theorem 3.6). For $M_{\rho,h}(T) \le 1$ the statement of the theorem follows from Lemma 3.8.

Let $M_{\rho,h}(T) \ge 2$. From this inequality it follows that $T \notin \mathrm{Dtab}_\rho$. Consider an arbitrary complete path ξ in the tree $\mathcal{Y}_{\rho,h}(T)$. From the description of the process $\mathcal{Y}_{\rho,h}$ and from the relation $T \notin \mathrm{Dtab}_\rho$ it follows that $\pi(\xi) = \pi(\xi_1)\dots\pi(\xi_m)$ for certain $m \ge 1$, where ξ_1 is a complete path in the tree $\mathcal{X}_{\rho,h}(T)$, and if $m \ge 2$ then ξ_i is a complete path in the tree $\mathcal{X}_{\rho,h}(T\pi(\xi_1)\dots\pi(\xi_{i-1}))$ for $i = 2,\dots,m$. For $i = 1,\dots,m$ denote $r_i = h(\pi(\xi_i))$. Let us estimate the value $h(\pi(\xi)) = \sum_{i=1}^m r_i$. We will show that

$$\sum_{i=1}^m r_i \le \begin{cases} 2\log_2 N(T) + M_{\rho,h}(T), & \text{if } 2 \le M_{\rho,h}(T) \le 3\ , \\ \frac{M_{\rho,h}(T)\log_2 N(T)}{\log_2 M_{\rho,h}(T)} + M_{\rho,h}(T), & \text{if } M_{\rho,h}(T) \ge 4\ . \end{cases} \qquad (5)$$

Let $m = 1$. From Lemma 3.7 it follows that $r_1 \leq M_{\rho,h}(T)$. Since $T \notin \mathrm{Dtab}_\rho$, we obtain $N(T) \geq 2$. Hence the inequality (5) holds for $m = 1$. Let $m \geq 2$. For $i = 1, \ldots, m$ denote $z_i = \max\{2, r_i\}$. By assumption, $T \notin \mathrm{Dtab}_\rho$. From the description of the process $\mathcal{Y}_{\rho,h}$ it follows that $T\pi(\xi_1)\ldots\pi(\xi_i) \notin \mathrm{Dtab}_\rho$ for $i = 1, \ldots, m-1$. Using Lemma 3.7 and the inequality $m \geq 2$ we obtain

$$N(T\pi(\xi_1)\ldots\pi(\xi_{m-1})) \leq N(T)/\prod_{i=1}^{m-1} z_i \ .$$

Taking into account that $T\pi(\xi_1)\ldots\pi(\xi_{m-1}) \notin \mathrm{Dtab}_\rho$ we obtain $N(T\pi(\xi_1)\ldots\pi(\xi_{m-1})) \geq 2$. Hence $\prod_{i=1}^{m-1} z_i \leq N(T)$. Taking the logarithm of both sides of this inequality we obtain $\sum_{i=1}^{m-1} \log_2 z_i \leq \log_2 N(T)$. From the last inequality it follows that

$$\begin{aligned} \textstyle\sum_{i=1}^{m} r_i &= r_m + \textstyle\sum_{i=1}^{m-1} (\log_2 z_i \cdot r_i / \log_2 z_i) \\ &\leq r_m + (\textstyle\sum_{i=1}^{m-1} \log_2 z_i) \max\{r_i / \log_2 z_i : i \in \{1, \ldots, m-1\}\} \\ &\leq r_m + \log_2 N(T) \max\{r_i / \log_2 z_i : i \in \{1, \ldots, m-1\}\} \ . \end{aligned} \tag{6}$$

Consider the function $q(x) = x / \log_2 \max\{2, x\}$ defined on the set of real numbers. One can show that $q(0) = 0$, $q(1) = 1$, $q(2) = 2$, $q(3) < 2$, $q(4) = 2$ and that $q(x)$ is a monotone increasing function for $x \geq 3$. Hence for any natural n the following equality holds:

$$\max\{q(i) : i \in \{0, \ldots, n\}\} = \begin{cases} 1, & \text{if } n = 1 \ , \\ 2, & \text{if } 2 \leq n \leq 3 \ , \\ n / \log_2 n, & \text{if } n \geq 4 \ . \end{cases} \tag{7}$$

Using Lemma 3.7 and the inequality $M_{\rho,h}(T\alpha) \leq M_{\rho,h}(T)$, which is true for any $\alpha \in \Omega_\rho(T)$, we obtain

$$r_i \leq M_{\rho,h}(T) \ . \tag{8}$$

From (7), (8) and from the inequality $M_{\rho,h}(T) \geq 2$ it follows that

$$\begin{aligned} \max\{r_i / \log_2 z_i : i \in \{1, \ldots, m-1\}\} &\leq \max\{q(i) : i \in \{0, \ldots, M_{\rho,h}(T)\}\} \\ &= \begin{cases} 2, & \text{if } 2 \leq M_{\rho,h}(T) \leq 3 \ , \\ M_{\rho,h}(T) / \log_2 M_{\rho,h}(T), & \text{if } M_{\rho,h}(T) \geq 4 \ . \end{cases} \end{aligned}$$

From these relations and from inequalities (6), (8) the inequality (5) follows. Taking into account that ξ is an arbitrary complete path in tree $\mathcal{Y}_{\rho,h}(T)$ and using Proposition 3.1 we conclude that the statement of the theorem holds also for $M_{\rho,h}(T) \geq 2$. □

3.4 Greedy Algorithm for Decision Tree Construction

A signature $\rho = (F, B)$ will be called *enumerated* if F is a denumerable set, elements of which are enumerated by numbers from $\mathbb{N}$, i.e. $F = \{f_i : i \in \mathbb{N}\}$. Let us fix an enumerated signature $\rho = (F, B)$, $|B| = k$.

A weight function ψ of the signature ρ will be called *computable* if there exists a general recursive function $\varphi : \mathbb{N} \to \mathbb{N} \setminus \{0\}$ such that $\psi(f_i) = \varphi(i)$ for any $i \in \mathbb{N}$.

Let ψ be a computable weight function of the signature ρ. In this subsection we consider an algorithm $\mathcal{V}_{\rho,\psi}$ which for a given table $T \in \mathrm{Tab}_\rho$ constructs a decision tree $\mathcal{V}_{\rho,\psi}(T)$ for this table. Accuracy and complexity of the algorithm $\mathcal{V}_{\rho,\psi}$ are estimated. The case, when the depth h is used in the capacity of decision tree complexity, is considered more explicitly.

The algorithm $\mathcal{V}_{\rho,\psi}$ is a greedy algorithm. As the uncertainty measure of a table we use the number of pairs of rows with different decisions. At each step we choose a question (column, attribute) which reduces the uncertainty to the greatest extent under certain constraints on the complexity (weight) of the question.

Algorithm $\mathcal{V}_{\rho,\psi}$. Define a function $R : \mathrm{Tab}_\rho \to \mathbb{N}$. For any $T \in \mathrm{Tab}_\rho$ let $R(T)$ be the number of unordered pairs of rows $\bar{\delta}_1, \bar{\delta}_2$ of the table T such that $\nu_T(\bar{\delta}_1) \neq \nu_T(\bar{\delta}_2)$.

Description of the Algorithm $\mathcal{V}_{\rho,\psi}$

Let us apply the algorithm $\mathcal{V}_{\rho,\psi}$ to a table $T \in \mathrm{Tab}_\rho$.

1-*st Step.* Construct a tree consisting of a single node w.

Let $T \in \mathrm{Dtab}_\rho$. If $\mathrm{Row}(T) = \emptyset$ then the node w will be labelled by the number 0. If $\mathrm{Row}(T) \neq \emptyset$ then the node w will be labelled by the number $\nu_T(\bar{\delta})$, where $\bar{\delta} \in \mathrm{Row}(T)$. Proceed to the second step.

Let $T \notin \mathrm{Dtab}_\rho$. Mark the node w by the word $\lambda \in \Omega_\rho(T)$ and proceed to the second step.

Suppose $t \geq 1$ steps have already been made. The tree obtained in the step t will be denoted by G.

$(t+1)$-*th Step.* If no one node of the tree G is labelled by a word from $\Omega_\rho(T)$ then we denote by $\mathcal{V}_{\rho,\psi}(T)$ the tree G. The work of the algorithm $\mathcal{V}_{\rho,\psi}$ is completed.

Otherwise we choose certain node w in the tree G which is labelled by a word from $\Omega_\rho(T)$. Let the node w be labelled by the word α.

If $T\alpha \in \mathrm{Dtab}_\rho$ then instead of the word α we mark the node w by the number $\nu_T(\bar{\delta})$, where $\bar{\delta} \in \mathrm{Row}(T\alpha)$, and proceed to the $(t+2)$-th step.

Let $T\alpha \notin \mathrm{Dtab}_\rho$. For any $f_i \in \mathrm{At}(T)$ let σ_i be the minimal number from the set B such that $R(T\alpha(f_i, \sigma_i)) = \max\{R(T\alpha(f_i, \sigma)) : \sigma \in B\}$. Set

$$I\alpha = \{f_i : f_i \in \mathrm{At}(T), R(T\alpha) > R(T\alpha(f_i, \sigma_i))\} \ .$$

For any $f_i \in I\alpha$ set $d(f_i) = \max\{\psi(f_i), R(T\alpha)/(R(T\alpha) - R(T\alpha(f_i, \sigma_i)))\}$. Let p be the minimal number from $\mathbb{N}$ for which $f_p \in I\alpha$ and $d(f_p) = \min\{d(f_i) : f_i \in I\alpha\}$. Instead of the word α we mark the node w by the attribute f_p. For every $\delta \in B$ such that $\mathrm{Row}(T\alpha(f_p, \delta)) \neq \emptyset$ add a node $w(\delta)$ to the tree G and draw an edge from the node w to the node $w(\delta)$. This edge will be labelled by the number δ, while the node $w(\delta)$ will be labelled by the word $\alpha(f_p, \delta)$. Proceed to the $(t+2)$-th step.

Note that in the description of the algorithm $\mathcal{V}_{\rho,h}$ instead of the value $d(f_i)$ we can use the value $R(T\alpha(f_i, \sigma_i))$. The output of the algorithm remains the same.

Example 3.11. Denote by T the table from Tab_{ρ_0} depicted in the Fig. 3(a). The tree $\mathcal{V}_{\rho_0,h}(T)$ is depicted in Fig. 3(b).

Consider the work of the algorithm $\mathcal{V}_{\rho,\psi}$ constructing the tree $\mathcal{V}_{\rho,\psi}(T)$. Let us show that the set $I\alpha$ (see the description of the $(t+1)$-th step of the algorithm) is not empty.

Lemma 3.9. *Let T be a nondegenerate table from Tab_ρ. Then there exists an attribute $f_i \in \mathrm{At}(T)$ such that $R(T) > \max\{R(T(f_i, \delta)) : \delta \in B\}$.*

Proof. Since $T \notin \mathrm{Dtab}_\rho$, there exist two rows $\bar{\delta}_1$ and $\bar{\delta}_2$ of the table T such that $\nu_T(\bar{\delta}_1) \neq \nu_T(\bar{\delta}_2)$. Consider an arbitrary column in which these rows are distinct. Let this column be labelled by the attribute f_i. Then, obviously, $R(T) > \max\{R(T(f_i, \delta)) : \delta \in B\}$. □

Using the description of the algorithm $\mathcal{V}_{\rho,\psi}$ and Lemma 3.9 it is not difficult to prove the following statement.

Proposition 3.2. *Let ψ be a computable weight function of the signature ρ. Then for any table $T \in \mathrm{Tab}_\rho$ the work of the algorithm $\mathcal{V}_{\rho,\psi}$ is completed in finite number of steps. The constructed tree $\mathcal{V}_{\rho,\psi}(T)$ is a decision tree for the table T.*

Bounds on Accuracy and Complexity of Algorithm $\mathcal{V}_{\rho,\psi}$. Consider upper bounds on the complexity of decision trees constructed by the algorithm $\mathcal{V}_{\rho,\psi}$.

Theorem 3.8. *Let ψ be a computable weight function of the signature ρ, and $T \in \mathrm{Tab}_\rho$. Then*

$$\psi(\mathcal{V}_{\rho,\psi}(T)) \leq \begin{cases} 0, & \text{if } T \in \mathrm{Dtab}_\rho \ , \\ M_{\rho,\psi}(T)^2 \ln R(T) + M_{\rho,\psi}(T), & \text{if } T \notin \mathrm{Dtab}_\rho \ . \end{cases}$$

Corollary 3.3. *Let ψ be a computable weight function of the signature ρ, and $T \in \mathrm{Tab}_\rho$. Then*

$$\psi(\mathcal{V}_{\rho,\psi}(T)) \leq \begin{cases} 0, & \text{if } T \in \mathrm{Dtab}_\rho \ , \\ \psi_{\rho,\psi}(T)^2 \ln R(T) + \psi_{\rho,\psi}(T), & \text{if } T \notin \mathrm{Dtab}_\rho \ . \end{cases}$$

If the depth h is taken as decision tree complexity then the following statement holds.

Theorem 3.9. *Let $T \in \mathrm{Tab}_\rho$. Then*

$$h(\mathcal{V}_{\rho,h}(T)) \leq \begin{cases} M_{\rho,h}(T), & \text{if } M_{\rho,h}(T) \leq 1 \ , \\ M_{\rho,\psi}(T)(\ln R(T) - \ln M_{\rho,\psi}(T) + 1), & \text{if } M_{\rho,h}(T) \geq 2 \ . \end{cases}$$

Corollary 3.4. *Let* $T \in \mathrm{Tab}_\rho$. *Then*

$$h(\mathcal{V}_{\rho,h}(T)) \leq \begin{cases} h_\rho(T), & \text{if } h_\rho(T) \leq 1 , \\ h_\rho(T)(\ln R(T) - \ln h_\rho(T) + 1), & \text{if } h_\rho(T) \geq 2 . \end{cases}$$

Denote

$$C(M_{\rho,h}, R) = \{(M_{\rho,h}(T), R(T)) : T \in \mathrm{Tab}_\rho\}$$

and

$$C(h_\rho, R) = \{(h_\rho(T), R(T)) : T \in \mathrm{Tab}_\rho\} .$$

The next statement follows immediately from results of [70].

Theorem 3.10. *Let* $\rho = (F, B)$ *be an enumerated signature. Then*

$$C(M_{\rho,h}, R) = C(h_\rho, R) = \{(0,0)\} \cup \{(m,r) : m, r \in \mathbb{N} \setminus \{0\}, m \leq r\}$$

and for any pair $(m,r) \in C(M_{\rho,h}, R)$ *there exists a table* $T(m,r) \in \mathrm{Tab}_\rho$ *such that* $M_{\rho,h}(T(m,r)) = h_\rho(T(m,r)) = m$, $R(T(m,r)) = r$ *and*

$$h(\mathcal{V}_{\rho,h}(T(m,r)) \geq \begin{cases} m, & \text{if } m < 2 \text{ or } r < 3m , \\ \lfloor (m-1)(\ln r - \ln 3m) \rfloor + m, & \text{if } m \geq 2 \text{ and } r \geq 3m . \end{cases}$$

From Theorem 3.10 it follows that bounds from Theorem 3.9 and Corollary 3.4 do not allow essential improvement.

From Theorem 3.10 also follows that there is no function $f : \mathbb{N} \to \mathbb{N}$ such that for any table $T \in \mathrm{Tab}_\rho$ the inequality

$$h(\mathcal{V}_{\rho,h}(T)) \leq f(h_\rho(T))$$

holds. The situation with diagnostic tables is remarkably different.

Proposition 3.3. *Let* ψ *be a computable weight function of the signature* ρ, *and* T *be a diagnostic table from* Tab_ρ. *Then*

(a) $\psi(\mathcal{V}_{\rho,\psi}(T)) \leq 2\psi_\rho(T)^3 \ln k + \psi_\rho(T)$;
(b) $h(\mathcal{V}_{\rho,h}(T)) \leq 2h_\rho(T)^2 \ln k + h_\rho(T)$.

Consider a bound on the number of steps made by algorithm $\mathcal{V}_{\rho,\psi}$ under the construction of the tree $\mathcal{V}_{\rho,\psi}(T)$. Recall that $N(T)$ denotes the number of rows in the table T.

Theorem 3.11. *Let* ψ *be a computable weight function of the signature* ρ, *and* $T \in \mathrm{Tab}_\rho$. *Then under the construction of the tree* $\mathcal{V}_{\rho,\psi}(T)$ *the algorithm* $\mathcal{V}_{\rho,\psi}$ *makes at most* $2N(T) + 2$ *steps.*

Note that the algorithm $\mathcal{V}_{\rho,\psi}$ has polynomial time complexity if there exists a polynomial algorithm which for a given $i \in \mathbb{N}$ computes the value $\psi(f_i)$. In particular, the algorithm $\mathcal{V}_{\rho,h}$ has polynomial time complexity.

Later in Sect.3.6 we will show that under some assumption on the class NP the algorithm $\mathcal{V}_{\rho,h}$ is close to best approximate polynomial algorithms for minimization of decision tree depth.

Proofs of Theorems 3.8 and 3.9. The following lemma states certain properties of the function R.

Lemma 3.10. *Let $T \in \mathrm{Tab}_\rho$. Then*

(a) for any $f_i \in \mathrm{At}(T)$, $\delta \in B$ and $\alpha \in \Omega_\rho(T)$ the inequality $R(T) - R(T(f_i,\delta)) \geq R(T\alpha) - R(T\alpha(f_i,\delta))$ holds;
(b) the equality $R(T) = 0$ holds if and only if $T \in \mathrm{Dtab}_\rho$.

Proof. Denote by D (respectively $D\alpha$) the set of unordered pairs of rows $\bar{\delta}_1, \bar{\delta}_2$ from $\mathrm{Row}(T)$ (respectively $\mathrm{Row}(T\alpha)$) such that $\nu_T(\bar{\delta}_1) \neq \nu_T(\bar{\delta}_2)$ and at least one of the rows from the pair does not belong to the set $\mathrm{Row}(T(f_i,\delta))$. One can show that $D\alpha \subseteq D$, $|D| = R(T) - R(T(f_i,\delta))$ and $|D\alpha| = R(T\alpha) - R(T\alpha(f_i,\delta))$. From these relations the statement (a) of the lemma follows.

The statement (b) of the lemma is obvious. □

The following statement characterizes properties of the tree constructed by the algorithm $\mathcal{V}_{\rho,\psi}$.

Lemma 3.11. *Let $T \in \mathrm{Tab}_\rho \setminus \mathrm{Dtab}_\rho$ and $\xi = w_1, d_1, \ldots, w_m, d_m, w_{m+1}$ be an arbitrary complete path in the tree $\mathcal{V}_{\rho,\psi}(T)$. Let for $j = 1, \ldots, m$ the node w_j be labelled by the attribute $f_{t(j)}$ and the edge d_j be labelled by the number δ_j. Let $\alpha_0 = \lambda$ and $\alpha_j = (f_{t(1)}, \delta_1) \ldots (f_{t(j)}, \delta_j)$ for $j = 1, \ldots, m$. Then for $j = 0, \ldots, m-1$ the following inequalities hold:*

$$\psi(f_{t(j+1)}) \leq M_{\rho,\psi}(T\alpha_j) \ ,$$
$$R(T\alpha_{j+1}) \leq R(T\alpha_j)(M_{\rho,\psi}(T\alpha_j) - 1)/M_{\rho,\psi}(T\alpha_j) \ .$$

Proof. Let T contain n columns which are labelled by attributes $f_{v(1)}, \ldots, f_{v(n)}$. For $j = 0, \ldots, m$ denote $M_j = M_{\rho,\psi}(T\alpha_j)$. Fix certain number $j \in \{0, \ldots, m-1\}$. For every $f_i \in \mathrm{At}(T)$ denote by σ_i the minimal number from B possessing the following property: $R(T\alpha_j(f_i,\sigma_i)) = \max\{R(T\alpha_j(f_i,\sigma)) : \sigma \in B\}$. Let β be a word from $\Omega_\rho(T)$ such that $\mathrm{Alph}(\beta) \subseteq \{(f_{v(1)}, \sigma_{v(1)}), \ldots, (f_{v(n)}, \sigma_{v(n)})\}$, $T\alpha_j\beta \in \mathrm{Dtab}_\rho$, and $\psi(\beta) = M_{\rho,\psi}(T\alpha_j, (\sigma_{v(1)}, \ldots, \sigma_{v(n)}))$. Let $\beta = (f_{l(1)}, \sigma_{l(1)}) \ldots (f_{l(r)}, \sigma_{l(r)})$. From the description of the algorithm $\mathcal{V}_{\rho,\psi}$ follows $T\alpha_j \notin \mathrm{Dtab}_\rho$. Since $T\alpha_j\beta \in \mathrm{Dtab}_\rho$, obtain

$$1 \leq r \ . \tag{9}$$

Obviously,

$$r \leq \psi(\beta) \ . \tag{10}$$

From the choice of the word β and from the definition of the value M_j follows

$$\psi(\beta) \leq M_j \ . \tag{11}$$

Since $T\alpha_j\beta \in \mathrm{Dtab}_\rho$, we have $R(T\alpha_j\beta) = 0$. Therefore

$$\begin{aligned} & R(T\alpha_j) - (R(T\alpha_j) - R(T\alpha_j(f_{l(1)}, \sigma_{l(1)}))) \\ & - (R(T\alpha_j(f_{l(1)}, \sigma_{l(1)})) - R(T\alpha_j(f_{l(1)}, \sigma_{l(1)})(f_{l(2)}, \sigma_{l(2)}))) \\ & - \ldots - (R(T\alpha_j(f_{l(1)}, \sigma_{l(1)}) \ldots (f_{l(r-1)}, \sigma_{l(r-1)})) - R(T\alpha_j\beta)) = 0 \ . \end{aligned} \tag{12}$$

Let us choose certain $q \in \{l(1), \ldots, l(r)\}$ such that

$$R(T\alpha_j(f_q, \sigma_q)) = \min\{R(T\alpha_j(f_{l(s)}, \sigma_{l(s)})) : s \in \{1, \ldots, r\}\} .$$

Using Lemma 3.10 we conclude that for $s = 2, \ldots, r$ the following inequality holds:

$$R(T\alpha_j(f_{l(1)}, \sigma_{l(1)}) \ldots (f_{l(s-1)}, \sigma_{l(s-1)})) - R(T\alpha_j(f_{l(1)}, \sigma_{l(1)}) \ldots (f_{l(s)}, \sigma_{l(s)})) \\ \leq R(T\alpha_j) - R(T\alpha_j(f_{l(s)}, \sigma_{l(s)})) .$$

The choice of q implies

$$R(T\alpha_j) - R(T\alpha_j(f_{l(s)}, \sigma_{l(s)})) \leq R(T\alpha_j) - R(T\alpha_j(f_q, \sigma_q))$$

for $s = 1, \ldots, r$. These inequalities and (12) imply

$$R(T\alpha_j) - r(R(T\alpha_j) - R(T\alpha_j(f_q, \sigma_q))) \leq 0 .$$

From this inequality and from (9) it follows that

$$R(T\alpha_j(f_q, \sigma_q)) \leq R(T\alpha_j)(r - 1)/r .$$

This last inequality together with (10) and (11) implies

$$R(T\alpha_j(f_q, \sigma_q)) \leq R(T\alpha_j)(M_j - 1)/M_j . \qquad (13)$$

Since $q \in \{l(1), \ldots, l(r)\}$, we have $\psi(f_q) \leq \psi(\beta)$. From this inequality and from (11) it follows that

$$\psi(f_q) \leq M_j . \qquad (14)$$

The description of algorithm $\mathcal{V}_{\rho,\psi}$ shows that the attribute $f_{t(j+1)}$ is defined as follows. Let

$$I\alpha_j = \{f_i : f_i \in \mathrm{At}(T), R(T\alpha_j) > R(T\alpha_j(f_i, \sigma_i))\} .$$

For every $f_i \in I\alpha_j$ denote

$$d(f_i) = \max\{\psi(f_i), R(T\alpha_j)/(R(T\alpha_j) - R(T\alpha_j(f_i, \sigma_i)))\} .$$

Let p be the minimal number from $\mathbb{N}$ such that $f_p \in I\alpha_j$ and $d(f_p) = \min\{d(f_i) : f_i \in I\alpha_j\}$. Then $f_{t(j+1)} = f_p$. Since $T\alpha_j \notin \mathrm{Dtab}_\rho$, we obtain $R(T\alpha_j) > 0$. From this inequality and from (9), (10), (11), (13) it follows that $f_q \in I\alpha_j$. One can show that for any $f_i \in I\alpha_j$ the number $d(f_i)$ is the minimum among all the numbers d such that $\psi(f_i) \leq d$ and $R(T\alpha_j(f_i, \sigma_i)) \leq R(T\alpha_j)(d - 1)/d$. From here, from (13) and from (14) it follows that $d(f_q) \leq M_j$. Therefore $d(f_{t(j+1)}) \leq M_j$. This inequality and the definition of the value $d(f_{t(j+1)})$ imply $\psi(f_{t(j+1)}) \leq M_j$ and $R(T\alpha_j(f_{t(j+1)}, \sigma_{t(j+1)})) \leq R(T\alpha_j)(M_j - 1)/M_j$. Finally, the statement of the lemma follows from these inequalities and from the choice of $\sigma_{t(j+1)}$ according to which $R(T\alpha_{j+1}) \leq R(T\alpha_j(f_{t(j+1)}, \sigma_{t(j+1)}))$. □

Proof (of Theorem 3.8). Let $T \in \mathrm{Dtab}_\rho$. It is easily to notice that in this case $\psi(\mathcal{V}_{\rho,\psi}(T)) = 0$.

Let T be a nondegenerate table containing n columns which are labelled by attributes $f_{v(1)}, \ldots, f_{v(n)}$. Consider an arbitrary complete path $\xi = w_1, d_1, \ldots, w_m, d_m, w_{m+1}$ in the tree $\mathcal{V}_{\rho,\psi}(T)$. From Proposition 3.2 it follows that the tree $\mathcal{V}_{\rho,\psi}(T)$ is a decision tree for the table T. Since $T \notin \mathrm{Dtab}_\rho$, we obtain $m \geq 1$. Let for $j = 1, \ldots, m$ the node w_j be labelled by the attribute $f_{t(j)}$ and the edge d_j be labelled by the number δ_j. Denote $\alpha_0 = \lambda$ and for $j = 1, \ldots, m$ denote $\alpha_j = (f_{t(1)}, \delta_1) \ldots (f_{t(j)}, \delta_j)$. Let us obtain the following upper bound on the value m:

$$m \leq M_{\rho,\psi}(T) \ln R(T) + 1 \ . \tag{15}$$

Since $T \notin \mathrm{Dtab}_\rho$, we obtain $R(T) \geq 1$. Obviously, $M_{\rho,\psi}(T) \geq 0$. Hence the inequality (15) holds for $m = 1$. Let $m \geq 2$. From Lemma 3.11 it follows that

$$R(T\alpha_{m-1}) \leq R(T) \prod_{j=0}^{m-2} ((M_{\rho,\psi}(T\alpha_j) - 1)/M_{\rho,\psi}(T\alpha_j)) \ . \tag{16}$$

According to the description of the algorithm $\mathcal{V}_{\rho,\psi}$ we have $T\alpha_{m-1} \notin \mathrm{Dtab}_\rho$. Therefore $R(T\alpha_{m-1}) \geq 1$. Using Lemma 3.11 we conclude that $M_{\rho,\psi}(T\alpha_j) \geq 1$ for $j = 0, \ldots, m-2$. From these inequalities and from (16) it follows that $M_{\rho,\psi}(T\alpha_j) \geq 2$ for $j = 0, \ldots, m-2$, and the next inequality holds:

$$\prod_{j=0}^{m-2} (M_{\rho,\psi}(T\alpha_j)/(M_{\rho,\psi}(T\alpha_j) - 1) \leq R(T) \ . \tag{17}$$

One can show that the inequality

$$M_{\rho,\psi}(T\alpha_j) \leq M_{\rho,\psi}(T) \tag{18}$$

holds for $j = 0, 1, \ldots, m$. From (17) and (18) it follows that

$$(m-1)\ln(1 + 1/(M_{\rho,\psi}(T) - 1)) \leq \ln R(T) \ . \tag{19}$$

The inequalities $M_{\rho,\psi}(T) \geq 2$ and $\ln(1 + 1/n) > 1/(n+1)$ (the last inequality holds for any natural n) imply

$$\ln(1 + 1/(M_{\rho,\psi}(T) - 1)) > 1/M_{\rho,\psi}(T) \ .$$

The inequality (15) follows from this inequality and from (19). From Lemma 3.11, from (15) and from (18) it follows that

$$\psi(\pi(\xi)) = \sum_{j=1}^{m} \psi(f_{t(j)}) \leq M_{\rho,\psi}(T)^2 \ln R(T) + M_{\rho,\psi}(T) \ .$$

Taking into account that ξ is an arbitrary complete path in the tree $\mathcal{V}_{\rho,\psi}(T)$ we obtain $\psi(\mathcal{V}_{\rho,\psi}(T)) \leq M_{\rho,\psi}(T)^2 \ln R(T) + M_{\rho,\psi}(T)$. □

Proof (of Corollary 3.3). If $T \in \mathrm{Dtab}_\rho$ then the considered statement follows immediately from Theorem 3.8. Let $T \notin \mathrm{Dtab}_\rho$. Using Theorem 3.1 we obtain $M_{\rho,\psi}(T) \leq \psi_\rho(T)$. From Lemma 3.10 it follows that $R(T) \geq 1$. Using these inequalities and Theorem 3.8 we conclude that considered statement holds also for the case $T \notin \mathrm{Dtab}_\rho$. □

Proof (of Theorem 3.9). Let $M_{\rho,h}(T) = 0$. Then, obviously, $T \in \mathrm{Dtab}_\rho$. Using Theorem 3.8 we obtain $h(\mathcal{V}_{\rho,h}(T)) \leq 0$.

Let $M_{\rho,h}(T) = 1$. Then, obviously, $T \notin \mathrm{Dtab}_\rho$ and $R(T) \geq 1$. Using Lemma 3.8 we conclude that there exists an attribute $f_i \in \mathrm{At}(T)$ such that $T(f_i, \delta) \in \mathrm{Dtab}_\rho$ for any $\delta \in B$. Hence $\max\{h(f_i), R(T)/(R(T) - \max\{R(T(f_i, \delta)) : \delta \in B\})\} = 1$. The obtained equality and the description of the algorithm $\mathcal{V}_{\rho,h}$ show that the root of the tree $\mathcal{V}_{\rho,h}(T)$ is labelled by certain attribute $f_j \in \mathrm{At}(T)$ such that $\max\{h(f_j), R(T)/(R(T) - \max\{R(T(f_j, \delta)) : \delta \in B\})\} = 1$. Hence $R(T(f_j, \delta)) = 0$ for any $\delta \in B$. Therefore $T(f_j, \delta) \in \mathrm{Dtab}_\rho$ for any $\delta \in B$. From here and from the description of the algorithm $\mathcal{V}_{\rho,h}$ it follows that all the nodes of tree $\mathcal{V}_{\rho,h}(T)$ except the root are terminal nodes. Therefore $h(\mathcal{V}_{\rho,h}(T)) = 1$.

Let $M_{\rho,h}(T) \geq 2$. Let $\xi = w_1, d_1, \ldots, w_m, d_m, w_{m+1}$ be a longest complete path in the tree $\mathcal{V}_{\rho,h}(T)$. Obviously, $h(\mathcal{V}_{\rho,h}(T)) = m$. Using Proposition 3.2 we conclude that $\mathcal{V}_{\rho,h}(T)$ is a decision tree for the table T. Hence $m \geq h_\rho(T)$. From this inequality, from the inequality $M_{\rho,h}(T) \geq 2$ and from Theorem 3.1 it follows that $m \geq 2$. Let for $j = 1, \ldots, m$ the node w_j be labelled by the attribute $f_{t(j)}$ and the edge d_j be labelled by the number δ_j. Denote $\alpha_0 = \lambda$ and for $j = 1, \ldots, m$ denote $\alpha_j = (f_{t(1)}, \delta_1) \ldots (f_{t(j)}, \delta_j)$. Denote $M_j = M_{\rho,h}(T\alpha_j)$ for $j = 0, \ldots, m$. Let us show that for $i = 0, \ldots, m$ the following inequality holds:

$$M_{m-i} \leq i \; . \tag{20}$$

Let Γ_i be a subtree of the tree $\mathcal{V}_{\rho,h}(T)$ defined by the edge d_{m-i}. One can show that Γ_i is a decision tree for the table $T\alpha_{m-i}$. Taking into account that ξ is a longest complete path in tree $\mathcal{V}_{\rho,h}(T)$ we obtain $h(\Gamma_i) = i$. Since Γ_i is a decision tree for the table $T\alpha_{m-i}$, the inequalities $h_\rho(T\alpha_{m-i}) \leq h(\Gamma_i) \leq i$ hold. Using Theorem 3.1 we obtain $M_{\rho,h}(T\alpha_{m-i}) \leq h_\rho(T\alpha_{m-i})$. Therefore $M_{\rho,h}(T\alpha_{m-i}) \leq i$. Thus, the inequality (20) holds. From inequality $m \geq 2$ and from Lemma 3.11 it follows that

$$R(T\alpha_{m-1}) \leq R(T) \prod_{j=0}^{m-2} ((M_j - 1)/M_j) \; . \tag{21}$$

From the description of the algorithm $\mathcal{V}_{\rho,h}$ it follows that $T\alpha_{m-1} \notin \mathrm{Dtab}_\rho$. Therefore $R(T\alpha_{m-1}) \geq 1$. The last inequality and (21) imply

$$\prod_{j=0}^{m-2} (M_j/(M_j - 1)) \leq R(T) \; . \tag{22}$$

It is not difficult to show that

$$M_j \leq M_0 \tag{23}$$

for $j = 1, \ldots, m$. From (20), (22) and (23) it follows that

$$(M_0/(M_0-1))^{m-M_0} \prod_{j=0}^{M_0-2} (M_0-j)/(M_0-j-1) \le R(T) .$$

Taking the logarithm of both sides of this inequality we obtain

$$(m - M_0)\ln(1 + 1/(M_0 - 1)) \le \ln R(T) - \ln M_0 . \tag{24}$$

Taking into account the inequality $\ln(1 + 1/n) > 1/(n + 1)$, which holds for any $n \in \mathbb{N} \setminus \{0\}$, and also $M_0 = M_{\rho,h}(T) \ge 2$ we obtain $\ln(1 + 1/(M_0 - 1)) > 1/M_0$. From this inequality and from (24) it follows that $m < M_0(\ln R(T) - \ln M_0 + 1)$. Taking into account that $m = h(\mathcal{V}_{\rho,h}(T))$ and $M_0 = M_{\rho,h}(T)$ we obtain $h(\mathcal{V}_{\rho,h}(T)) < M_{\rho,h}(T)(\ln R(T) - \ln M_{\rho,h}(T) + 1)$. □

Proof (of Corollary 3.4). Let $h_\rho(T) \le 1$. Using Theorem 3.1 obtain $M_{\rho,h}(T) \le 1$. From this inequality and from Theorem 3.9 it follows that $h(\mathcal{V}_{\rho,h}(T)) \le M_{\rho,h}(T)$. Using Theorem 3.1 we have $h(\mathcal{V}_{\rho,h}(T)) \le h_\rho(T)$.

Let $h_\rho(T) \ge 2$. Using Lemma 3.8 we obtain $M_{\rho,h}(T) \ge 2$. This inequality and Theorem 3.9 imply

$$h(\mathcal{V}_{\rho,h}(T)) \le M_{\rho,h}(T)(\ln R(T) - \ln M_{\rho,h}(T) + 1) . \tag{25}$$

Using Theorem 3.1 we have $M_{\rho,h}(T) \le h_\rho(T)$. One can show that $J(T) \le R(T)$. Using Lemma 3.1 we obtain $h_\rho(T) \le J(T)$. Hence

$$2 \le M_{\rho,h}(T) \le h_\rho(T) \le R(T) . \tag{26}$$

Let $r \in \mathbb{N}$ and $r \ge 2$. One can show that the function $x(\ln r - \ln x + 1)$ of the real variable x is a nondecreasing function on the interval $[1, r]$. Using inequalities (25) and (26) we obtain $h(\mathcal{V}_{\rho,h}(T)) \le h_\rho(T)(\ln R(T) - \ln h_\rho(T) + 1)$. □

Proof (of Proposition 3.3). Using Corollary 3.1 we obtain $\psi_\rho(T) \ge \log_k N(T)$. Obviously, $R(T) \le N(T)^2$. Therefore $\ln R(T) \le 2 \ln k \log_k N(T)$ and $\ln R(T) \le 2 \ln k \cdot \psi_\rho(T)$. Using this inequality and Corollaries 3.3 and 3.4 one can show that the inequalities (a) and (b) hold. □

Proof of Theorem 3.11. Let Γ be a finite rooted directed tree. Denote the number of nodes in the tree Γ by $L_a(\Gamma)$ and the number of terminal nodes in Γ by $L_t(\Gamma)$.

Lemma 3.12. *Let Γ be a finite rooted directed tree in which at least two edges are issuing from any nonterminal node. Then $L_a(\Gamma) \le 2L_t(\Gamma)$.*

Proof. Let us prove the statement of the lemma by induction on the value $L_a(\Gamma)$. Obviously, if $L_a(\Gamma) = 1$ then $L_a(\Gamma) \le 2L_t(\Gamma)$. One can show that there exists no tree which satisfies the conditions of the lemma and for which $L_a(\Gamma) = 2$.

Let for certain $n \ge 2$ the considered inequality hold for any tree which satisfies the conditions of the lemma and for which the number of nodes is at most n. Let a tree Γ satisfy the conditions of the lemma and $L_a(\Gamma) = n+1$. Let us show that

$L_a(\Gamma) \leq 2L_t(\Gamma)$. One can prove that there exists a nonterminal node w in the tree Γ such that each edge issuing from it enters a terminal node. Let m edges issue from w. Denote by Γ_1 the tree obtained from the tree Γ by deletion of all the edges issuing from the node w as well as all the nodes which these edges are entering. Obviously, $m \geq 2$. It is easily to see that $L_a(\Gamma_1) = L_a(\Gamma) - m$, $L_t(\Gamma) = L_t(\Gamma_1) - m + 1$ and the tree Γ_1 satisfies the condition of the lemma. By the inductive hypothesis, $L_a(\Gamma_1) \leq 2L_t(\Gamma_1)$. Therefore $L_a(\Gamma) - m \leq 2(L_t(\Gamma) - m + 1) = 2L_t(\Gamma) - 2m + 1$. Hence $L_a(\Gamma) \leq 2L_t(\Gamma) - m + 2 \leq 2L_t(\Gamma)$. □

Proof (of Theorem 3.11). Let $T \in \mathrm{Dtab}_\rho$. From the description of the algorithm $\mathcal{V}_{\rho,\psi}$ it follows that under the construction of the tree $\mathcal{V}_{\rho,\psi}(T)$ this algorithm makes exactly two steps. Therefore if $T \in \mathrm{Dtab}_\rho$ then the statement of the theorem holds.

Let $T \notin \mathrm{Dtab}_\rho$. Denote by Γ the tree $\mathcal{V}_{\rho,\psi}(T)$. Define for every node w of the tree Γ a word $\pi_\Gamma(w) \in \Omega_\rho(\Gamma)$. If w is the root of the tree Γ then $\pi_\Gamma(w) = \lambda$. Let an edge issue from a node w_1 and enter a node w_2. Let this edge be labelled by the number δ, and the node w_1 be labelled by the attribute f_i. Then $\pi_\Gamma(w_2) = \pi_\Gamma(w_1)(f_i, \delta)$.

Let w be an arbitrary nonterminal node of the tree Γ. Let the node w be labelled by the attribute f_i. From the description of the algorithm $\mathcal{V}_{\rho,\psi}$ it follows that

$$R(T\pi_\Gamma(w)) > \max\{R(T\pi_\Gamma(w)(f_i, \delta)) : \delta \in B\} .$$

Therefore the cardinality of the set $\{\delta : \delta \in B, \mathrm{Row}(T\pi_\Gamma(w)(f_i, \delta)) \neq \emptyset\}$ is at least two. Hence at least two edges are issuing from the node w. From the description of the algorithm $\mathcal{V}_{\rho,\psi}$ it follows that $\mathrm{Row}(T\pi(\xi)) \neq \emptyset$ for any complete path ξ in the tree Γ. Obviously, $\mathrm{Row}(T\pi(\xi_1)) \cap \mathrm{Row}(T\pi(\xi_2)) = \emptyset$ for any two distinct complete paths ξ_1 and ξ_2 in tree Γ. Therefore the number of complete paths in the tree Γ is at most $N(T)$. Hence $L_t(\Gamma) \leq N(T)$. Using Lemma 3.12 we obtain $L_a(\Gamma) \leq 2N(T)$. Finally, one can show that the number of steps making by the algorithm $\mathcal{V}_{\rho,\psi}$ under construction of the tree Γ is equal to $L_a(\Gamma) + 2$. □

3.5 Algorithm for Optimal Decision Tree Construction

In this subsection an algorithm is considered which for a given decision table T constructs a decision tree for the table T with minimal weighted depth (optimal decision tree). This algorithms enumerates all separable subtables of the table T and, obviously, has in general case exponential complexity depending on the number of columns in T. In Sect. 4.8 the class of all infinite information systems will be described for each of which the number of separable subtables of decision tables over the considered information system is bounded from above by a polynomial on the number of columns.

Let $\rho = (F, B)$ be an enumerated signature, $F = \{f_i : i \in \mathbb{N}\}$, $|B| = k$, and ψ be a computable weight function of the signature ρ.

Recall the notion of separable subtable given in Sect. 3.1. Let $T \in \mathrm{Tab}_\rho$. A nonempty subtable T' of the table T is called separable if there exists a word $u \in \Omega_\rho(T)$ such that $T' = Tu$. Denote by $\mathrm{Sep}(T)$ the set of all separable subtables of the table T. It is clear that $T \in \mathrm{Sep}(T)$.

Consider an algorithm $\mathcal{W}_{\rho,\psi}$ which for a given table $T \in \mathrm{Tab}_\rho$ constructs a decision tree $\mathcal{W}_{\rho,\psi}(T)$ for the table T. Let T contain n columns which, for the definiteness, are labelled by attributes $f_1, \ldots, f_n$.

Description of the Algorithm $\mathcal{W}_{\rho,\psi}$

Step 0. Construct the set $\mathrm{Sep}(T)$ and proceed to the first step.

Suppose $t \geq 0$ steps have already been made.

Step $(t+1)$. If the table T in the set $\mathrm{Sep}(T)$ is labelled by a decision tree then this decision tree is the result of the algorithm $\mathcal{W}_{\rho,\psi}$ work. Otherwise choose in the set $\mathrm{Sep}(T)$ a table D satisfying the following conditions:

a) the table D is not labelled by a decision tree;
b) either $D \in \mathrm{Dtab}_\rho$ or all separable subtables of the table D are labelled by decision trees.

If $D \in \mathrm{Dtab}_\rho$ then we mark the table D by the decision tree consisting of one node which is labelled by the number $\nu_T(\bar{\delta})$, where $\bar{\delta} \in \mathrm{Row}(D)$. Otherwise for $i \in \{1, \ldots, n\}$ denote by $E(D, i)$ the set of numbers contained in the i-th column of the table D. Denote $K(D) = \{i : i \in \{1, \ldots, n\}, |E(D, i)| \geq 2\}$. For any $i \in K(D)$ and $\delta \in E(D, i)$ denote by $\Gamma(i, \delta)$ the decision tree assigned to the table $D(f_i, \delta)$. Let $i \in K(D)$ and $E(D, i) = \{\delta_1, \ldots, \delta_r\}$. Define a decision tree Γ_i. The root of Γ_i is labelled by the attribute f_i. The root is the initial node of exactly r edges $d_1, \ldots, d_r$ which are labelled by numbers $\delta_1, \ldots, \delta_r$ respectively. The roots of the decision trees $\Gamma(i, \delta_1), \ldots, \Gamma(i, \delta_r)$ are terminal nodes of the edges $d_1, \ldots, d_r$ respectively. Mark the table D by one of the trees Γ_i, $i \in K(D)$, having minimal complexity relatively to the weight function ψ, and proceed to the $(t+2)$-th step.

It is not difficult to prove the following statement.

Theorem 3.12. *For any decision table $T \in \mathrm{Tab}_\rho$ the algorithm $\mathcal{W}_{\rho,\psi}$ constructs a decision tree $\mathcal{W}_{\rho,\psi}(T)$ for the table T such that $\psi(\mathcal{W}_{\rho,\psi}(T)) = \psi_\rho(T)$, and makes exactly $|\mathrm{Sep}(T)|+1$ steps. The time of the algorithm $\mathcal{W}_{\rho,\psi}$ work is bounded from below by $|\mathrm{Sep}(T)|$, and bounded from above by a polynomial on $|\mathrm{Sep}(T)|$, on the number of columns in the table T, and on the time for computation of weights of attributes attached to columns of the table T.*

3.6 On Complexity of Optimization Problems for Decision Trees

In this subsection two algorithmic problems connected with the computation of the minimal complexity of decision tree for a table and with construction of decision tree with minimal complexity are investigated. The solvability of these problems is shown. For the depth of decision trees the NP-hardness of problems under consideration is proved.

Moreover, in this subsection the question on accuracy of approximate polynomial algorithms for minimization of decision tree depth is discussed.

Optimization Problems. In this subsection we assume that an enumerated signature $\rho = (F, B)$ with $F = \{f_i : i \in \mathbb{N}\}$ and $|B| = k$ is fixed. Let ψ be a computable weight function of the signature ρ.

Define two algorithmic *problems of optimization of decision trees for decision tables.*

The Problem $\text{Com}(\rho, \psi)$*:* for a given table $T \in \text{Tab}_\rho$ it is required to compute the value $\psi_\rho(T)$.

The Problem $\text{Des}(\rho, \psi)$*:* for a given table $T \in \text{Tab}_\rho$ it is required to construct a decision tree Γ for the table T such that $\psi(\Gamma) = \psi_\rho(T)$.

Solvability of Problems $\text{Com}(\rho, \psi)$ and $\text{Des}(\rho, \psi)$

Proposition 3.4. *Let ψ be a computable weight function of the signature ρ. Then the problems* $\text{Com}(\rho, \psi)$ *and* $\text{Des}(\rho, \psi)$ *are solvable.*

Proof. In Sect. 3.2 a subset $\text{Tree}_\rho(T)$ of the set Tree_ρ was defined for an arbitrary table $T \in \text{Tab}_\rho$. From Lemma 3.5 it follows that this set contains at least one optimal decision tree for the table T. One can show that the set $\text{Tree}_\rho(T)$ is finite, and there exists an algorithm for enumeration of all trees from $\text{Tree}_\rho(T)$ for an arbitrary table $T \in \text{Tab}_\rho$. Using this algorithm it is not difficult to construct an algorithm which solves the problems $\text{Com}(\rho, \psi)$ and $\text{Des}(\rho, \psi)$. □

NP-Hardness of Problems $\text{Com}(\rho, h)$ and $\text{Des}(\rho, h)$. A pair $G = (V, R)$, where V is a nonempty finite set and R is a set of two-element subsets of the set V, will be called *an undirected graph without loops and multiple edges.* The elements of the set V are called *vertices*, and the elements of the set R are called *edges* of the graph G. Let $V = \{v_1, \ldots, v_n\}$. A set $W \subseteq U$ will be called *a vertex cover of the graph* G if the following conditions hold:

a) if $W = \emptyset$ then $R = \emptyset$;
b) if $R \neq \emptyset$, then for any edge $\{v_i, v_j\} \in R$ at least one of the relations $v_i \in W$ or $v_j \in W$ holds.

Denote by $\text{cv}(G)$ the minimal cardinality of vertex cover of G.

Let $r \in R$ and $r = \{v_i, v_j\}$. Denote by $\bar{\delta}(r)$ the n-tuple from $\{0, 1\}^n$ in which the i-th and j-th digits are equal to 1 while all the other digits are equal to 0.

Consider the matrix in which the set of rows coincides with the set $\{\bar{\delta}(r) : r \in R\}$ and transpose it. The so obtained matrix is called *the incidence matrix of the graph* G.

Define the table $T(G) \in \text{Tab}_\rho$ as follows. Denote by $\tilde{0}_n$ the n-tuple from $\{0, 1\}^n$ all the digits of which are equal to 0. Then the table $T(G)$ contains n columns which are labelled by attributes $f_1, \ldots, f_n$, $\text{Row}(T(G)) = \{\tilde{0}_n\} \cup \{\bar{\delta}(r) : r \in R\}$, $\nu_{T(G)}(\tilde{0}_n) = 0$ and $\nu_{T(G)}(\bar{\delta}(r)) = 1$ for any $r \in R$.

Lemma 3.13. *Let G be an undirected graph without loops and multiple edges. Then* $\text{cv}(G) = h_\rho(T(G))$.

Proof. Let G have n vertices. One can show that $\mathrm{cv}(G) = M_{\rho,h}(T(G), \tilde{0}_n)$. Therefore $M_{\rho,h}(T(G)) \geq \mathrm{cv}(G)$. Using Theorem 3.1 we obtain $h_\rho(T(G)) \geq \mathrm{cv}(G)$.

Let β be a word from $\Omega_\rho(T)$ such that $\mathrm{Alph}(\beta) \subseteq \{(f_1, 0), \ldots, (f_n, 0)\}$, $T(G)\beta \in \mathrm{Dtab}_\rho$ and $h(\beta) = M_{\rho,h}(T(G), \tilde{0}_n)$. Describe a tree Γ from Tree_ρ. Let $\beta = \lambda$. Then the tree Γ contains the only node which is labelled by the number 0. Let $\beta \neq \lambda$ and $\beta = (f_{i(1)}, 0) \ldots (f_{i(m)}, 0)$. Then Γ contains the complete path $\xi = w_1, d_1, \ldots, w_m, d_m, w_{m+1}$ in which the node w_{m+1} is labelled by the number 0 while for $j = 1, \ldots, m$ the node w_j is labelled by the attribute $f_{i(j)}$, and the edge d_j is labelled by 0. For $j = 1, \ldots, m$ an edge labelled by 1 is issuing from the node w_j. This edge enters a terminal node which is labelled by 1. The tree Γ does not contain any other nodes or edges. It is clear that Γ is a decision tree for the table $T(G)$, and $h(\Gamma) = h(\beta)$. Since $h(\beta) = \mathrm{cv}(G)$, we obtain $h_\rho(T(G)) \leq \mathrm{cv}(G)$. Hence $h_\rho(T(G)) = \mathrm{cv}(G)$. □

Proposition 3.5. *Let ρ be an enumerated signature. Then problems* $\mathrm{Com}(\rho, h)$ *and* $\mathrm{Des}(\rho, h)$ *are NP-hard.*

Proof. Define the vertex cover problem as follows: for a given undirected graph G without loops and multiple edges, represented by incidence matrix, and a number $m \in \mathbb{N}$ it is required to verify whether the inequality $\mathrm{cv}(G) \leq m$ holds. This problem is NP-complete [29].

Assume a polynomial algorithm exists for at least one of problems $\mathrm{Com}(\rho, h)$ and $\mathrm{Des}(\rho, h)$. Then using Lemma 3.13 we conclude that there exists a polynomial algorithm for the vertex cover problem. Hence the problems $\mathrm{Com}(\rho, h)$ and $\mathrm{Des}(\rho, h)$ are NP-hard. □

Note that in [29] NP-completeness is proved for certain problems slightly different from $\mathrm{Com}(\rho, h)$ and $\mathrm{Des}(\rho, h)$.

On Accuracy of Approximate Polynomial Algorithms for Problem $\mathrm{Des}(\rho, h)$. Let we have a problem which consists of the choice of a solution with minimal cost among the set of admissible solutions. Such solution is called optimal. Let r be a parameter of the problem which is a real number, and let φ be a partial function of real variable with real values. We will say that an algorithm solves the considered problem with *the multiplicative accuracy* $\varphi(r)$ if the following conditions hold:

a) if the value $\varphi(r)$ is indefinite or $\varphi(r) < 1$ then the algorithm finds an optimal solution;
b) if $\varphi(r) \geq 1$ then the cost of the obtained solution is at most the cost of optimal solution multiplied on $\varphi(r)$.

First, we consider some known results on the accuracy of solving of the set cover problem by approximate polynomial algorithms.

Let S be a set containing $N > 0$ elements, and $\mathcal{F} = \{S_1, \ldots, S_m\}$ be a family of subsets of the set S such that $S = \bigcup_{i=1}^{m} S_i$. A subfamily $\{S_{i_1}, \ldots, S_{i_t}\}$ of the family $\mathcal{F}$ will be called an *$\mathcal{F}$-cover* if $S = \bigcup_{j=1}^{t} S_{i_j}$. The problem of searching for an $\mathcal{F}$-cover with minimal cardinality is called *the set cover problem.*

In [28] it was proved that if $NP \not\subseteq DTIME(n^{O(\log_2 \log_2 n)})$ then for any ε, $0 < \varepsilon < 1$, there is no polynomial algorithm which solves the set cover problem with the multiplicative accuracy $(1-\varepsilon)\ln N$.

Similar statement holds for the problem of minimization of decision tree depth.

Proposition 3.6. *If $NP \not\subseteq DTIME(n^{O(\log_2 \log_2 n)})$ then for any ε, $0 < \varepsilon < 1$, there is no polynomial algorithm which solves the problem* $\mathrm{Des}(\rho, h)$ *with the multiplicative accuracy $(1-\varepsilon)\ln R(T)$.*

Proof. Assume the contrary: let $NP \not\subseteq DTIME(n^{O(\log_2 \log_2 n)})$, $0 < \varepsilon < 1$, and there exists a polynomial algorithm $\mathcal{A}_1$ which solves the problem $\mathrm{Des}(\rho, h)$ with the multiplicative accuracy $(1-\varepsilon)\ln R(T)$. Let us show that there exists a polynomial algorithm $\mathcal{A}_2$ which solves the set cover problem with the multiplicative accuracy $(1-\varepsilon)\ln N$. Describe the algorithm $\mathcal{A}_2$ work.

Let $S = \{a_1, \dots, a_N\}$, $\mathcal{F} = \{S_1, \dots, S_m\}$ and $S = \bigcup_{i=1}^m S_i$. If $(1-\varepsilon)\ln N < 1$ then enumerating all subfamilies of the family $\mathcal{F}$ we find an $\mathcal{F}$-cover with minimal cardinality. Let $(1-\varepsilon)\ln N \geq 1$. Then at polynomial time we can construct a decision table $T(\mathcal{F}) \in \mathrm{Tab}_\rho$ corresponding to the family $\mathcal{F}$. Denote by $\bar{\delta}_0$ the m-tuple from $\{0,1\}^m$ in which all digits are equal to 0. For $j = 1, \dots, N$ we denote by $\bar{\delta}_j$ the m-tuple $(\delta_{j1}, \dots, \delta_{jm})$ from $\{0,1\}^m$ in which $\delta_{ji} = 1$ if and only if $a_j \in S_i$, $i = 1, \dots, m$. Then the table $T(\mathcal{F})$ contains m columns which are labelled by attributes $f_1, \dots, f_m$, $\mathrm{Row}(T(\mathcal{F})) = \{\bar{\delta}_0, \bar{\delta}_1, \dots, \bar{\delta}_N\}$, $\nu_{T(\mathcal{F})}(\bar{\delta}_0) = 0$ and $\nu_{T(\mathcal{F})}(\bar{\delta}_j) = 1$ for any $j \in \{1, \dots, N\}$. One can show that $R(T(\mathcal{F})) = N$ and $h_\rho(T(\mathcal{F})) = c(\mathcal{F})$ where $c(\mathcal{F})$ is the minimal cardinality of an $\mathcal{F}$-cover.

Evidently, $(1-\varepsilon)\ln R(T(\mathcal{F})) \geq 1$. Applying the algorithm $\mathcal{A}_1$ to the table $T(\mathcal{F})$ we obtain a decision tree Γ for the table $T(\mathcal{F})$ such that

$$h(\Gamma) \leq h_\rho(T(\mathcal{F}))(1-\varepsilon)\ln R(T(\mathcal{F})) = c(\mathcal{F})(1-\varepsilon)\ln N .$$

There is a complete path in the tree Γ in which every edge is labelled by the number 0. Let $f_{i_1}, \dots, f_{i_t}$ be attributes which are labels of nodes in this path. One can show that the set $\{S_{i_1}, \dots, S_{i_t}\}$ is an $\mathcal{F}$-cover. It is clear that

$$t \leq h(\Gamma) \leq c(\mathcal{F})(1-\varepsilon)\ln N .$$

Thus, the algorithm $\mathcal{A}_2$ solves the set cover problem at polynomial time with the multiplicative accuracy $(1-\varepsilon)\ln N$ under the assumptions that $NP \not\subseteq DTIME(n^{O(\log_2 \log_2 n)})$ and $0 < \varepsilon < 1$ which, by the results from [28], is impossible. □

Algorithm $\mathcal{V}_{\rho,h}$ has polynomial time complexity. Using Corollary 3.4 we conclude that for any table $T \in \mathrm{Tab}_\rho$ with $h_\rho(T) \geq 3$ the depth of the decision tree for the table T constructed by the algorithm $\mathcal{V}_{\rho,h}$ is at most the depth of optimal decision tree for the table T multiplied on $\ln R(T)$. Therefore if $NP \not\subseteq DTIME(n^{O(\log_2 \log_2 n)})$ then the algorithm $\mathcal{V}_{\rho,h}$ is close to best (with respect to accuracy) approximate polynomial algorithms for the problem $\mathrm{Des}(\rho, h)$ solving.

It is not difficult to modify the algorithm $\mathcal{V}_{\rho,h}$ such that we obtain a polynomial algorithm $\mathcal{V}^*_{\rho,h}$ which solves the problem $\mathrm{Des}(\rho, h)$ with the multiplicative accuracy $\ln R(T)$.

For a given table $T \in \mathrm{Tab}_\rho$ the algorithm $\mathcal{V}^*_{\rho,h}$ enumerates at polynomial time all trees $\Gamma \in \mathrm{Tree}_\rho$ such that $h(\Gamma) \leq 2$, $\mathrm{At}(\Gamma) \subseteq \mathrm{At}(T)$, and terminal nodes of Γ are labelled by numbers from the set $\{\nu_T(\bar{\delta}) : \bar{\delta} \in \mathrm{Row}(T)\} \cup \{0\}$. Among these trees the algorithm $\mathcal{V}^*_{\rho,h}$ looks for decision trees for the table T. If such trees exist the algorithm $\mathcal{V}^*_{\rho,h}$ finds among them a decision tree for the table T with minimal depth. This tree is the result of the algorithm $\mathcal{V}^*_{\rho,h}$ work. If among the considered trees there are no decision trees for the table T then the algorithm $\mathcal{V}^*_{\rho,h}$ works later as the algorithm $\mathcal{V}_{\rho,h}$. In this case the result of the algorithm $\mathcal{V}^*_{\rho,h}$ work coincides with the result of the algorithm $\mathcal{V}_{\rho,h}$ work. Denote by $\mathcal{V}^*_{\rho,h}(T)$ the result of the algorithm $\mathcal{V}^*_{\rho,h}$ work.

It is clear that the algorithm $\mathcal{V}^*_{\rho,h}$ has polynomial time complexity. The following statement shows that $\mathcal{V}^*_{\rho,h}$ solves the problem $\mathrm{Des}(\rho, h)$ with the multiplicative accuracy $\ln R(T)$.

Proposition 3.7. *Let $T \in \mathrm{Tab}_\rho$. Then*

$$h(\mathcal{V}^*_{\rho,h}(T)) \leq \begin{cases} h_\rho(T), & \text{if } R(T) \leq 2 \ , \\ h_\rho(T) \ln R(T), & \text{if } R(T) \geq 3 \ . \end{cases}$$

Proof. One can show that if $h_\rho(T) \leq 2$ then $h(\mathcal{V}^*_{\rho,h}(T)) = h_\rho(T)$.

Let $R(T) \leq 2$. From Theorem 3.10 it follows that $h_\rho(T) \leq R(T)$. Therefore $h_\rho(T) \leq 2$ and $h(\mathcal{V}^*_{\rho,h}(T)) = h_\rho(T)$.

Let $R(T) \geq 3$. If $h_\rho(T) \leq 2$ then $h(\mathcal{V}^*_{\rho,h}(T)) = h_\rho(T)$, and hence $h(\mathcal{V}^*_{\rho,h}(T)) \leq h_\rho(T) \ln R(T)$. Let $h_\rho(T) \geq 3$. Then, obviously, $\mathcal{V}^*_{\rho,h}(T) = \mathcal{V}_{\rho,h}(T)$. Using Corollary 3.4 obtain $h(\mathcal{V}_{\rho,h}(T)) \leq h_\rho(T) \ln R(T)$. Hence $h(\mathcal{V}^*_{\rho,h}(T)) \leq h_\rho(T) \ln R(T)$. □

3.7 Complexity of Computation of Boolean Functions from Closed Classes

A decision table T may be interpreted as a way to define partial function $\nu_T : \mathrm{Row}(T) \to \mathbb{N}$. In this case the attribute, which is the label of i-th column, is the i-th variable of the function ν_T. Decision trees for the table T may be interpreted as algorithms for the function ν_T computation.

The unimprovable upper bounds on minimal depth of decision trees for decision tables corresponding to functions from arbitrary closed class of Boolean functions are studied in this subsection. These bounds depend on the number of variables of functions under consideration. The obtained results are of certain independent interest. Simple proofs of these results illustrates methods considered in this section. Mainly, the techniques for lower bound proving are used. The definitions, notation and results from appendix "Closed Classes of Boolean Functions" are used in this section without special notice.

Definitions and Bounds. Let $X = \{x_i : i \in \mathbb{N}\}$ be the set of variables. Denote by ρ_1 the signature $(X, \{0, 1\})$. Let C_1 be the set of all Boolean functions with variables from X containing constants 0 and 1. Let $n \geq 1$. Denote by $C_1(n)$

the set of all functions of n variables from C_1. Let $f(x_{i(1)}, \dots, x_{i(n)}) \in C_1(n)$. Associate with the function f the table $T(f)$ from Tab_{ρ_1}: the table $T(f)$ contains n columns which are labelled by variables $x_{i(1)}, \dots, x_{i(n)}$, $\mathrm{Row}(T(f)) = \{0,1\}^n$, and $\nu_{T(f)}(\bar{\delta}) = f(\bar{\delta})$ for any $\bar{\delta} \in \mathrm{Row}(T(f))$. Define the function $h : C_1 \to \mathbb{N}$. Let $f \in C_1$. If $f \equiv \mathrm{const}$ then $h(f) = 0$. If $f \not\equiv \mathrm{const}$ then $h(f) = h_{\rho_1}(T(f))$.

Example 3.12. Let $f = x_1 \vee x_2$. The table $T(f)$ is depicted in Fig. 6(a). A decision tree for the table $T(f)$ is depicted in Fig. 6(b). One can show that $h(f) = 2$.

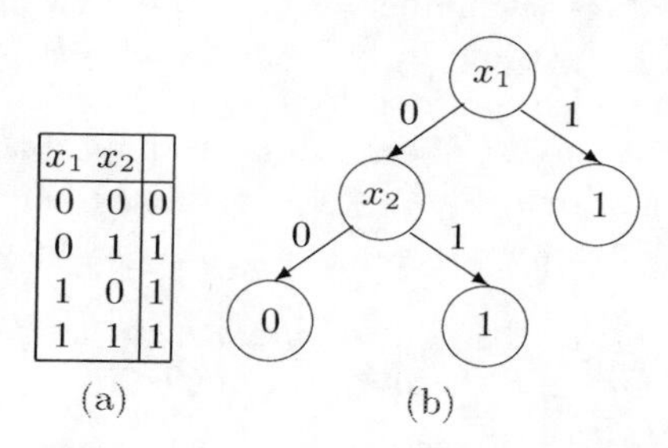

Fig. 6. Decision table $T(x_1 \vee x_2)$ and decision tree for this table.

Let V be a closed class of Boolean functions. For $n \geq 1$ denote $V(n) = V \cap C_1(n)$. Define the function $h_V : \mathbb{N} \setminus \{0\} \to \mathbb{N}$ as follows:

$$h_V(n) = \max\{h(f) : f \in V(n)\} \ .$$

The function h_V is the unimprovable upper bound on minimal depth of decision trees for tables, which correspond to functions from the class V, depending on number of variables of functions.

Theorem 3.13. *Let V be a closed class of Boolean functions, and $n \geq 1$. Then*

(a) if $V \in \{O_2, O_3, O_7\}$ then $h_V(n) = 0$;
(b) if $V \in \{O_1, O_4, O_5, O_6, O_8, O_9\}$ then $h_V(n) = 1$;
(c) if $V \in \{L_4, L_5\}$ then $h_V(n) = \begin{cases} n, & \text{if } n \text{ is odd,} \\ n-1, & \text{if } n \text{ is even;} \end{cases}$
(d) if $V \in \{D_1, D_2, D_3\}$ then $h_V(n) = \begin{cases} n, \text{ if } n \geq 3, \\ 1, \text{ if } n \leq 2; \end{cases}$
(e) if the class V coincides with neither of above-mentioned classes then $h_V(n)=n$.

Auxiliary Statements. Define the function $\mathrm{EV} : C_1 \to \mathbb{N}$. Let $f \in C_1$. If $f \equiv \mathrm{const}$ then $\mathrm{EV}(f) = 0$. If $f \not\equiv \mathrm{const}$ then $\mathrm{EV}(f)$ is the number of essential variables of the function f.

Lemma 3.14. *Let $n \geq 1$ and $f \in C_1(n)$. Then $h(f) \leq \mathrm{EV}(f) \leq n$.*

Proof. If $f \equiv \mathrm{const}$ then the statement of lemma, obviously, holds. Let $f \not\equiv \mathrm{const}$. Obviously, $\mathrm{EV}(f) \leq n$. Show that $h(f) \leq \mathrm{EV}(f)$. Let $f = f(x_{i(1)}, \dots, x_{i(n)})$. Show that for any n-tuples $\bar{\delta}, \bar{\sigma} \in \{0,1\}^n$ such that $f(\bar{\sigma}) \neq f(\bar{\delta})$ there exists a variable $x_{i(t)}$ with following properties:

a) $x_{i(t)}$ is an essential variable of the function f;
b) n-tuples $\bar{\sigma}$ and $\bar{\delta}$ differ in the t-th digit.

Let D be the set of numbers of digits in which the n-tuples $\bar{\delta}$ and $\bar{\sigma}$ differ. One can easily show that for certain $m \geq 2$ there exists a sequence $\bar{\delta}_1, \ldots, \bar{\delta}_m$ of n-tuples from $\{0,1\}^n$ in which $\bar{\delta}_1 = \bar{\delta}$, $\bar{\delta}_m = \bar{\sigma}$, for $j = 1, \ldots, m-1$ the tuples $\bar{\delta}_j$ and $\bar{\delta}_{j+1}$ differ in exactly one digit, and the number of this digit is contained in the set D. Since $f(\bar{\delta}_1) \neq f(\bar{\delta}_m)$, there exists $j \in \{1, \ldots, m-1\}$ such that $f(\bar{\delta}_j) \neq f(\bar{\delta}_{j+1})$. Let t be the number of the digit in which the tuples $\bar{\delta}_j$ and $\bar{\delta}_{j+1}$ differ. Obviously, $t \in D$ and $x_{i(t)}$ is an essential variable of the function f. Thus, the statement under consideration is proved. Using this statement it is not difficult to show that the set of essential variables of the function f is a test for the table $T(f)$. Therefore $J(T(f)) \leq \mathrm{EV}(f)$. From Lemma 3.1 it follows that $h_{\rho_1}(T(f)) \leq J(T(f))$. Therefore $h_{\rho_1}(T(f)) \leq \mathrm{EV}(f)$. Hence $h(f) \leq \mathrm{EV}(f)$. □

Corollary 3.5. *Let V be a closed class of Boolean functions and $n \geq 1$. Then the value $h_V(n)$ is definite and the inequality $h_V(n) \leq n$ holds.*

Proof. Obviously, $V(n) \neq \emptyset$. Using Lemma 3.14 obtain that the value $h_V(n)$ is definite, and the considered inequality holds. □

Let $n \geq 1$ and $\bar{\delta} \in \{0,1\}^n$. Denote by $\mathcal{O}(\bar{\delta})$ the set of all n-tuples from $\{0,1\}^n$ which differ from the n-tuple $\bar{\delta}$ in exactly one digit. Let $f \in C_1(n)$. Denote $o(f, \bar{\delta}) = \left|\{\bar{\sigma} : \bar{\sigma} \in \mathcal{O}(\bar{\delta}), f(\bar{\delta}) \neq f(\bar{\sigma})\}\right|$.

Lemma 3.15. *Let $n \geq 1$, $f \in C_1(n)$ and $\bar{\delta} \in \{0,1\}^n$. Then $h(f) \geq o(f, \bar{\delta})$.*

Proof. If $f \equiv \mathrm{const}$ then the considered inequality, obviously, holds. Let $f \not\equiv \mathrm{const}$, $f = f(x_{i(1)}, \ldots, x_{i(n)})$ and $\bar{\delta} = (\delta_1, \ldots, \delta_n)$. Show that $M_{\rho_1,h}(T(f), \bar{\delta}) \geq o(f, \bar{\delta})$. Let $M_{\rho_1,h}(T(f), \bar{\delta}) = m$. Then there exists a word $\alpha \in \Omega_{\rho_1}(T(f))$ of the length m such that $\mathrm{Alph}(\alpha) \subseteq \{(x_{i(1)}, \delta_1), \ldots, (x_{i(n)}, \delta_n)\}$ and $T(f)\alpha \in \mathrm{Dtab}_{\rho_1}$. Obviously, $\left|\mathrm{Row}(T(f)\alpha) \cap \mathcal{O}(\bar{\delta})\right| \geq n - m$. One can show that $\bar{\delta} \in \mathrm{Row}(T(f)\alpha)$. Since $T(f)\alpha \in \mathrm{Dtab}_{\rho_1}$, we obtain $n - m \leq n - o(f, \bar{\delta})$. Hence $m \geq o(f, \bar{\delta})$. Therefore $M_{\rho_1,h}(T(f), \bar{\delta}) \geq o(f, \bar{\delta})$ and $M_{\rho_1,h}(T(f)) \geq o(f, \bar{\delta})$. Using Theorem 3.1 we obtain $h_{\rho_1}(T(f)) \geq o(f, \bar{\delta})$. Hence $h(f) \geq o(f, \bar{\delta})$. □

Let $n \geq 1$ and $t \in \{0,1\}$. Denote by $\tilde{t}_n$ the n-tuple from $\{0,1\}^n$ all the digits of which are equal to t.

Define certain Boolean functions. For $n \geq 1$ denote $k_n = k_n(x_1, \ldots, x_n) = x_1 \wedge \ldots \wedge x_n$, $d_n = d_n(x_1, \ldots, x_n) = x_1 \vee \ldots \vee x_n$, $l_n = l_n(x_1, \ldots, x_n) = x_1 + \ldots + x_n \pmod 2$, $\neg l_n = \neg l_n(x_1, \ldots, x_n) = x_1 + \ldots + x_n + 1 \pmod 2$ and $\pi_n = \pi_n(x_1, \ldots, x_n) = x_1$. For $n \geq 2$ denote by $m_n = m_n(x_1, \ldots, x_n)$ the function obtained from the function l_{n-1} by insertion the unessential variable x_n. For $n \geq 3$ denote $r_n = r_n(x_1, \ldots, x_n) = (x_1 \wedge (x_2 \vee \ldots \vee x_n)) \vee (x_2 \wedge \ldots \wedge x_n)$.

Lemma 3.16. *For any $n \geq 1$ the inequalities $h(k_n) \geq n$, $h(d_n) \geq n$, $h(l_n) \geq n$, $h(\neg l_n) \geq n$ and $h(\pi_n) \geq 1$ hold. For any $n \geq 2$ the inequality $h(m_n) \geq n - 1$ holds.*

Proof. Let $n \geq 1$. Then $o(k_n, \tilde{1}_n) = n$, $o(d_n, \tilde{0}_n) = n$, $o(l_n, \tilde{0}_n) = n$, $o(\neg l_n, \tilde{0}_n) = n$, $o(\pi_n, \tilde{1}_n) = 1$. Using Lemma 3.15 we obtain $h(k_n) \geq n$, $h(d_n) \geq n$, $h(l_n) \geq n$, $h(\neg l_n) \geq n$ and $h(\pi_n) \geq 1$.

Let $n \geq 2$. Then $o(m_n, \tilde{0}_n) = n - 1$. Using Lemma 3.15 we obtain $h(m_n) \geq n - 1$. □

Lemma 3.17. *Let $n \geq 3$. Then $h(r_n) \geq n$.*

Proof. Define an $(\{x_1, \ldots, x_n\}, \rho_1)$-tree H_n. Every complete path in this tree contains exactly $n + 1$ nodes. Let $\xi = w_1, d_1, \ldots, w_n, d_n, w_{n+1}$ be an arbitrary complete path in the tree H_n, $i \in \{1, \ldots, n\}$ and let the edge d_i be labelled by the pair (x_j, σ). If $j = 1$ then $\sigma = 0$. If $j \neq 1$ and $i \neq n - 1$ then $\sigma = 1$. Let $j \neq 1$ and $i = n - 1$. If no pair $(x_1, 0)$ is present among the pairs assigned to the edges in the path from w_1 to w_{n-1} then $\sigma = 0$. Otherwise $\sigma = 1$.

We will show that the tree H_n is a proof-tree for the bound $h_{\rho_1}(T(r_n)) \geq n$. Define for an arbitrary node w of the tree H_n the word $\zeta(w) \in \Omega_{\rho_1}(T(r_n))$ according to the rule introduced in Sect. 3.2. Let w be an arbitrary terminal node in the tree H_n. One can see that $h(\zeta(w)) = n$ and $\text{Row}(T(r_n)\zeta(w)) \neq \emptyset$. Let us show that $T(r_n)\zeta(w) \notin \text{Dtab}_{\rho_1}$ for any nonterminal node w of the tree H_n. To do this it is sufficient to consider all nonterminal nodes possessing following property: each edge issuing from the node enters some terminal node of the tree H_n. Let w be one of such nodes. It is not difficult to show that there exists $i \in \{2, \ldots, n\}$ for which

$$\text{Alph}(\zeta(w)) = \{(x_1, 0)\} \cup \{(x_j, 1) : j \in \{2, \ldots, n\} \setminus \{i\}\} \tag{27}$$

or

$$\text{Alph}(\zeta(w)) = \{(x_i, 0)\} \cup \{(x_j, 1) : j \in \{2, \ldots, n\} \setminus \{i\}\} . \tag{28}$$

Denote by $\bar{\gamma}$ the n-tuple from $\{0, 1\}^n$ in which the first digit equals 0 while all other digits equal 1. Denote by $\bar{\delta}$ the n-tuple from $\{0, 1\}^n$ in which the first and the i-th digits equal 0 while all other digit equal 1. Denote by $\bar{\sigma}$ the n-tuple from $\{0, 1\}^n$ in which the i-th digit equals 0 while all other digits equal 1. Let the equality (27) hold. Then the n-tuples $\bar{\gamma}$ and $\bar{\delta}$ belong to the set $\text{Row}(T(r_n)\zeta(w))$, $\nu_{T(r_n)}(\bar{\gamma}) = 1$ and $\nu_{T(r_n)}(\bar{\delta}) = 0$. Let the equality (28) hold. Then the n-tuples $\bar{\delta}$ and $\bar{\sigma}$ belong to the set $\text{Row}(T(r_n)\zeta(w))$, $\nu_{T(r_n)}(\bar{\delta}) = 0$ and $\nu_{T(r_n)}(\bar{\sigma}) = 1$. Therefore $T(r_n)\zeta(w) \notin \text{Dtab}_{\rho_1}$. Hence the tree H_n is a proof-tree for the bound $h_{\rho_1}(T(r_n)) \geq n$. Using Theorem 3.4 we obtain $h_{\rho_1}(T(r_n)) \geq n$. Hence $h(r_n) \geq n$. □

Proof of Theorem 3.13. We obtain here certain bounds for individual closed classes.

Lemma 3.18. *Let $V \in \{O_2, O_3, O_7\}$ and $n \geq 1$. Then $h_V(n) = 0$.*

Proof. Obviously, $f \equiv \text{const}$ for any function $f \in V(n)$. Hence $h_V(n) = 0$. □

Lemma 3.19. *Let $V \in \{O_1, O_4, O_5, O_6, O_8, O_9\}$ and $n \geq 1$. Then $h_V(n) = 1$.*

Proof. One can easily see that $\pi_n \in V(n)$. Using Lemma 3.16 we obtain $h_V(n) \geq 1$. Obviously, $\mathrm{EV}(f) \leq 1$ for any function $f \in V(n)$. From Lemma 3.14 it follows that $h_V(n) \leq 1$. □

Lemma 3.20. *Let $V \in \{L_4, L_5\}$ and $n \geq 1$. Then*

$$h_V(n) = \begin{cases} n, & \text{if } n \text{ is odd} , \\ n-1, & \text{if } n \text{ is even} . \end{cases}$$

Proof. Let n be odd. Then it is easily to see that $l_n \in V(n)$. Using Lemma 3.16 and Corollary 3.5 we obtain $h_V(n) = n$.

Let n be even. One can see that $m_n \in V(n)$. Using Lemma 3.16 obtain $h_V(n) \geq n-1$. It is easily to show that $\mathrm{EV}(f) \leq n-1$ for any function f from $V(n)$. From Lemma 3.14 it follows that $h_V(n) \leq n-1$. □

Lemma 3.21. *Let $V \in \{D_1, D_2, D_3\}$ and $n \geq 1$. Then*

$$h_V(n) = \begin{cases} n, \text{ if } n \geq 3 , \\ 1, \text{ if } n \leq 2 . \end{cases}$$

Proof. Let $n \leq 2$. Obviously, $\pi_n \in V(n)$. Using Lemma 3.16 we obtain $h_V(n) \geq 1$. It is easily to show that $\mathrm{EV}(f) \leq 1$ for any function $f \in V(n)$. From Lemma 3.14 it follows that $h_V(n) \leq 1$.

Let $n \geq 3$. One can show that $r_n \in V(n)$. Using Lemma 3.17 we obtain $h(r_n) \geq n$. Hence $h_V(n) \geq n$. From Corollary 3.5 it follows that $h_V(n) = n$. □

Lemma 3.22. *Let $V \in \{S_1, L_2, L_3, P_1\}$ and $n \geq 1$. Then $h_V(n) = n$.*

Proof. It is easily to see that at least one of the functions d_n, l_n, $\neg l_n$, k_n belongs to the set $V(n)$. Using Lemma 3.16 we obtain $h_V(n) \geq n$. From Corollary 3.5 it follows that $h_V(n) = n$. □

Proof (of Theorem 3.13). Part (a) of the statement of the theorem follows from Lemma 3.18, part (b) follows from Lemma 3.19, part (c) follows from Lemma 3.20, and part (d) follows from Lemma 3.21. Let V coincide with neither of classes listed in items (a)–(d). Then it is easily to show that at least one of the following relations holds: $S_1 \subseteq V$, $L_2 \subseteq V$, $L_3 \subseteq V$, $P_1 \subseteq V$. Using Lemma 3.22 and Corollary 3.5 we obtain that $h_V(n) = n$. □

4 Local Approach to Investigation of Decision Trees

Local approach to investigation of decision trees considered in this section is based on the assumption that only attributes contained in a problem description are used by decision trees solving the problem. Bounds on complexity of decision trees are obtained and algorithms constructing decision trees are studied.

An information system $U = (A, B, F)$ consists of a set A (the universe) and a set of attributes F which are defined on A and have values from finite set B. A weight function ψ assigns a weight (a natural number) to each attribute. This weight characterizes the complexity of attribute value computation.

The notion of a problem over the information system defines as follows. We take finite number of attributes $f_1, \ldots, f_n$ from F. These attributes create a partition of the set A into classes (for each class values of the attributes are constant on elements from the class). These classes are numbered such that different classes can have the same number. The number of a class is the decision corresponding to elements of the class. For a given element a from A it is required to recognize the number of a class which contains a. Problems from various areas of applications can be represented in such form. The weight of a problem is the total weight of attributes from the problem description.

As algorithms for problem solving we consider decision trees which use attributes from the set $\{f_1, \ldots, f_n\}$. As time complexity measure we consider weighted depth of decision trees. It is clear that for each problem there exists trivial decision tree which solves this problem and which weighted depth is equal to the problem weight.

A decision table corresponds to the considered problem. The table has n columns labelling by attributes $f_1, \ldots, f_n$. Rows of this table are n-tuples of attribute $f_1, \ldots, f_n$ values corresponding to classes of partition. Each row is labelled by the number of corresponding class.

The section consists of eight subsections. First subsection contains definitions of basic notions. In the second subsection two statements are proved which allow use methods, created for decision tables, for study of problems. In the third subsection for each pair (*information system, weight function*) we investigate the behavior of local Shannon function which characterizes the growth in the worst case of minimal weighted depth of decision trees with the growth of problem weight. The criterion of solvability for problems of decision tree optimization is considered in fourth subsection. In the last four subsections the following way for construction of decision trees is considered: we construct the decision table corresponding to a problem, and then construct a decision tree for this table. In fifth subsection for an arbitrary information system we investigate the growth of the number of rows in decision tables over this system with the grows of the number of columns. In sixth subsection we study an algorithm which for a given problem constructs corresponding decision table. In sevenths and eights subsections we study approximate and exact algorithms for decision tree optimization.

4.1 Basic Notions

The notions of information system, problem, decision table corresponding to a problem, decision tree solving a problem, and weight function are considered in this subsection. We repeat here some definitions from the previous section.

Information Systems. Let A be a nonempty set, B be a finite nonempty set of integers with at least two elements, and F be a nonempty set of functions from A to B. Functions from F will be called *attributes* and the triple $U = (A, B, F)$ will be called *an information system*. In this section we will assume that different attributes from F (attributes with different names) are different as functions. If F is an infinite set then U will be called *an infinite* information system. Otherwise U will be called *a finite* information system. The signature $\rho = (F, B)$ corresponds to the information system U. Sometimes we will say that $U = (A, B, F)$ is an information system of the signature $\rho = (F, B)$.

An equation system over U is an arbitrary system of the kind

$$\{f_1(x) = \delta_1, \ldots, f_m(x) = \delta_m\}$$

where $f_1, \ldots, f_m \in F$ and $\delta_1, \ldots, \delta_m \in B$. It is possible that the considered system does not have equations. Such system will be called *empty*. The set of solutions of the empty system coincides with the set A. There is one-to-one correspondence between equation systems over U and words from the set $\Omega_\rho = \{(f, \delta) : f \in F, \delta \in B\}^*$: the word $(f_1, \delta_1) \ldots (f_m, \delta_m)$ corresponds to the considered equation system, the empty word λ corresponds to the empty equation system. For any $\alpha \in \Omega_\rho$ we denote by $\mathrm{Sol}_U(\alpha)$ the set of solutions on A of the equation system corresponding to the word α.

An information system $U = (A, B, F)$ will be called *enumerated* if $F = \{f_i : i \in \mathbb{N}\}$.

Problems and Corresponding Decision Tables. We will consider problems over the information system U. *A problem over* U is an arbitrary $(n+1)$-tuple $z = (\nu, f_1, \ldots, f_n)$ where $\nu : B^n \to \mathbb{Z}$, and $f_1, \ldots, f_n \in F$. The problem z may be interpreted as a problem of searching for the value $z(a) = \nu(f_1(a), \ldots, f_n(a))$ for an arbitrary $a \in A$. Denote $\mathrm{At}(z) = \{f_1, \ldots, f_n\}$. The tuple $(\nu, f_1, \ldots, f_n)$ is called *the description* of the problem z. The number n is called *the dimension* of the problem z and is denoted by $\dim z$. Different problems of pattern recognition, discrete optimization, fault diagnosis and computational geometry can be represented in such form. We denote by Probl_U the set of problems over U.

We denote by $T_U(z)$ the decision table of the signature $\rho = (F, B)$ satisfying the following conditions:

a) the table $T_U(z)$ contains n columns labelling by attributes $f_1, \ldots, f_n$;
b) an n-tuple $(\delta_1, \ldots, \delta_n) \in B^n$ is a row of the table $T_U(z)$ if and only if the system of equations $\{f_1(x) = \delta_1, \ldots, f_n(x) = \delta_n\}$ is compatible (has a solution) on the set A;
c) each row $(\delta_1, \ldots, \delta_n)$ of $T_U(z)$ is labelled by the decision $\nu(\delta_1, \ldots, \delta_n)$.

Denote $\mathrm{Tab}_U = \{T_U(z) : z \in \mathrm{Probl}_U\}$. Later, if an information system U is fixed, we will write sometimes $T(z)$ instead of $T_U(z)$.

Decision Trees. As algorithms for problem solving we will consider decision trees. *A decision tree over* U is a labelled finite rooted directed tree in which each terminal node is labelled by a number from $\mathbb{Z}$; each node which is not terminal (such nodes are called *working*) is labelled by an attribute from F; each edge is labelled by a number from B, and edges starting in a working node are labelled by pairwise different numbers. Denote by Tree_U the set of decision trees over U. It is clear that any decision tree over U is a decision tree of the signature ρ.

Let Γ be a decision tree over U. Denote by $\mathrm{At}(\Gamma)$ the set of attributes assigned to working nodes of Γ. *A complete path* ξ in Γ is an arbitrary path from the root to a terminal node. Let us define a word $\pi(\xi)$ from the set $\Omega_\rho(\Gamma) = \{(f, \delta) : f \in \mathrm{At}(\Gamma), \delta \in B\}^*$ associated with ξ. If there are no working nodes in ξ then $\pi(\xi) = \lambda$. Note that in this case the set $\mathrm{Sol}_U(\pi(\xi))$ coincides with the set A. Let $\xi = v_1, d_1, \ldots, v_m, d_m, v_{m+1}$ where $m > 0$, v_1 is the root, v_{m+1} is a terminal node, and v_i is the initial and v_{i+1} is the terminal node of the edge d_i for $i = 1, \ldots, m$. Let the node v_i be labelled by the attribute f_i, and the edge d_i be labelled by the number δ_i from B, $i = 1, \ldots, m$. Then $\pi(\xi) = (f_1, \delta_1) \ldots (f_m, \delta_m)$. Note that in this case the set $\mathrm{Sol}_U(\pi(\xi))$ coincides with the set of solutions on A of the equation system $\{f_1(a) = \delta_1, \ldots, f_m(a) = \delta_m\}$. Remind that $\mathrm{Path}(\Gamma)$ is the set of all complete paths in Γ.

We will say that a decision tree Γ over U *solves* a problem z over U if for each $a \in A$ there exists a complete path ξ in Γ such that $a \in \mathrm{Sol}_U(\pi(\xi))$, and the terminal node of the path ξ is labelled by the number $z(a)$.

Complexity Measures. A function $\psi : F \to \mathbb{N} \setminus \{0\}$ will be called *a weight function for* U. It is clear that any weight function for U is a weight function of the signature ρ. The value $\psi(f)$ sometimes will be called *the weight* of the attribute $f \in F$. We denote by h the weight function for which $h(f) = 1$ for any $f \in F$.

Extend a weight function ψ for U on the sets F^*, Ω_ρ, Tree_U and Probl_U.

Let $\alpha \in F^*$. If $\alpha = \lambda$ then $\psi(\alpha) = 0$. Let $\alpha \neq \lambda$ and $\alpha = f_1 \ldots f_m$. Then $\psi(\alpha) = \sum_{i=1}^m \psi(f_i)$.

Let $\beta \in \Omega_\rho$. If $\beta = \lambda$ then $\psi(\beta) = 0$. Let $\beta \neq \lambda$ and $\beta = (f_1, \delta_1) \ldots (f_m, \delta_m)$. Then $\psi(\beta) = \sum_{i=1}^m \psi(f_i)$.

For $\Gamma \in \mathrm{Tree}_U$ let $\psi(\Gamma) = \max\{\psi(\pi(\xi)) : \xi \in \mathrm{Path}(\Gamma)\}$. The value $\psi(\Gamma)$ will be called *the weighted depth* of the decision tree Γ. The value $h(\Gamma)$ will be called *the depth* of the decision tree Γ.

Let $z = (\nu, f_1, \ldots, f_n) \in P(U)$. Then $\psi(z) = \sum_{i=1}^n \psi(f_i)$. The value $\psi(z)$ will be called *the weight* of the problem z. Note that $h(z) = \dim z = n$.

Put into correspondence to the weight function ψ the function $\psi_U^l : \mathrm{Probl}_U \to \mathbb{N}$. Let $z \in \mathrm{Probl}_U$. Then $\psi_U^l(z) = \min\{\psi(\Gamma) : \Gamma \in \mathrm{Tree}_U^l(z)\}$ where $\mathrm{Tree}_U^l(z)$ is the set of all decision trees over U which solves the problem z and for which $\mathrm{At}(\Gamma) \subseteq \mathrm{At}(z)$. In other words, $\psi_U^l(z)$ is the minimal weighted depth of a decision tree over U which solves z and uses only attributes from the description of z. Such decision trees will be called *locally optimal for the problem* z.

Examples of Information Systems. We define here some information systems which will be used in examples in this and following sections.

Let m, t be natural numbers. We denote by $\mathrm{Pol}(m)$ be the set of all polynomials which have integer coefficients and depend on variables $x_1, \ldots, x_m$. We denote by $\mathrm{Pol}(m, t)$ the set of all polynomials from $\mathrm{Pol}(m)$ such that the degree of each polynomial is at most t.

We define information systems $U(\mathbb{R}, m)$, $U(\mathbb{R}, m, t)$ and $U(\mathbb{Z}, m, t)$ in the following way: $U(\mathbb{R}, m) = (\mathbb{R}^m, E, F(m))$, $U(\mathbb{R}, m, t) = (\mathbb{R}^m, E, F(m, t))$ and

$U(\mathbb{Z}, m, t) = (\mathbb{Z}^m, E, F(m, t))$ where $E = \{-1, 0, +1\}$, $F(m) = \{\text{sign}(p) : p \in \text{Pol}(m)\}$ and $F(m, t) = \{\text{sign}(p) : p \in \text{Pol}(m, t)\}$.

In the considered information systems we will not distinguish attributes which coincide as functions.

4.2 Use of Decision Tables

This subsection contains two statements which allow to use decision tables for analysis of decision trees solving problems.

Theorem 4.1. *Let $U = (A, B, F)$ be an information system, z be a problem over U, Γ be a decision tree over U, and $\text{At}(\Gamma) \subseteq \text{At}(z)$. Then the decision tree Γ solves the problem z if and only if Γ is a decision tree for the table $T_U(z)$.*

Proof. Let $z = (\nu, f_1, \ldots, f_n)$. For an arbitrary $a \in A$ denote $\bar{\delta}(a) = (f_1(a), \ldots, f_n(a))$. Denote $T = T_U(z)$. Then, obviously, $\text{Row}(T) = \{\bar{\delta}(a) : a \in A\}$ and $\nu_T(\bar{\delta}(a)) = z(a)$ for any $a \in A$. One can show that for any $a \in A$ and for any $\xi \in \text{Path}(\Gamma)$ we have $a \in \text{Sol}_U(\pi(\xi))$ if and only if $\bar{\delta}(a) \in \text{Row}(T\pi(\xi))$.

Let Γ be a decision tree for the table T, and let $a \in A$. Then $\bar{\delta}(a) \in \text{Row}(T)$ and there exists a complete path $\xi \in \text{Path}(\Gamma)$ such that $\bar{\delta}(a) \in \text{Row}(T\pi(\xi))$ and the terminal node of the path ξ is labelled by the number $\nu_T(\bar{\delta}(a))$. Obviously, $a \in \text{Sol}_U(\pi(\xi))$ and $\nu_T(\bar{\delta}(a)) = z(a)$. Therefore Γ solves the problem z.

Assume that the decision tree Γ solves the problem z. Let $\bar{\delta} \in \text{Row}(T)$. Then there exists $a \in A$ such that $\bar{\delta} = \bar{\delta}(a)$. There exists also a complete path $\xi \in \text{Path}(\Gamma)$ such that $a \in \text{Sol}_U(\pi(\xi))$, and the terminal node of ξ is labelled by the number $z(a)$. Obviously, $\bar{\delta} = \bar{\delta}(a) \in \text{Row}(T\pi(\xi))$ and $z(a) = \nu_T(\bar{\delta}(a)) = \nu_T(\bar{\delta})$. Besides, $\text{At}(\Gamma) \subseteq \text{At}(z)$. Therefore Γ is a decision tree for the table T. □

Corollary 4.1. *Let $U = (A, B, F)$ be an information system, $\rho = (F, B)$, and ψ be a weight function for U. Then $\psi_U^l(z) = \psi_\rho(T_U(z))$ for any problem z over U.*

The statement of Theorem 4.1 allows easy generalization for the case when the set $\text{At}(\Gamma)$ is not necessarily a subset of $\text{At}(z)$.

Let $\Gamma \in \text{Tree}_U$ and $z = (\nu, f_1, \ldots, f_n) \in \text{Probl}_U$. Define the problem $z \circ \Gamma \in \text{Probl}_U$. Let $m \geq n$ and let $f_1, \ldots, f_m$ be pairwise distinct attributes from F such that $\{f_1, \ldots, f_m\} = \text{At}(z) \bigcup \text{At}(\Gamma)$. Then $z \circ \Gamma = (\gamma, f_1, \ldots, f_m)$ where $\gamma : B^m \to \mathbb{Z}$ and for any m-tuple $\bar{\delta} = (\delta_1, \ldots, \delta_m) \in B^m$ the equality $\gamma(\bar{\delta}) = \nu(\delta_1, \ldots, \delta_n)$ holds.

Theorem 4.2. *Let $U = (A, B, F)$ be an information system, z be a problem over U and Γ be a decision tree over U. Then the decision tree Γ solves the problem z if and only if Γ is a decision tree for the table $T_U(z \circ \Gamma)$.*

Proof. Denote $y = z \circ \Gamma$. Obviously, $z(a) = y(a)$ for any $a \in A$. Hence the decision tree Γ solves the problem z if and only if Γ solves the problem y. From Theorem 4.1 it follows that the decision tree Γ solves the problem y if and only if Γ is a decision tree for the table $T_U(z \circ \Gamma)$. □

4.3 Local Shannon Functions

Notion of Local Shannon Function. Let $U = (A, B, F)$ be an information system and ψ be a weight function for U. Remind that for a problem $z = (\nu, f_1, \ldots, f_n)$ over U we denote by $\psi_U^l(z)$ the minimal weighted depth of a decision tree over U which solves the problem z and uses only attributes from the set $\{f_1, \ldots, f_n\}$, and we denote by $\psi(z)$ the total weight of attributes $f_1, \ldots, f_n$ from the problem z description. We will consider the relationships between the parameters $\psi_U^l(z)$ and $\psi(z)$. One can interpret the value $\psi(z)$ for the problem $z = (\nu, f_1, \ldots, f_n)$ as the weighted depth of the decision tree which solves the problem z in trivial way by computing sequentially the values of the attributes $f_1, \ldots, f_n$. So we will consider relationships between weighted depth of locally optimal and trivial decision trees. To this end we define the function $H_{U,\psi}^l : \mathbb{N} \setminus \{0\} \to \mathbb{N}$ in the following way:

$$H_{U,\psi}^l(n) = \max\{\psi_U^l(z) : z \in \mathrm{Probl}_U, \psi(z) \leq n\}$$

for any $n \in \mathbb{N} \setminus \{0\}$, where Probl_U is the set of problems over U. The value $H_{U,\psi}^l(n)$ is the unimprovable upper bound on the value $\psi_U^l(z)$ for problems $z \in \mathrm{Probl}_U$ such that $\psi(z) \leq n$. The function $H_{U,\psi}^l(n)$ will be called *the local Shannon function* for the information system U and the weight function ψ. Denote by $\mathrm{Dom}(H_{U,\psi}^l)$ the domain of the function $H_{U,\psi}^l$. It is clear that $\mathrm{Dom}(H_{U,\psi}^l) = \{n : n \in \mathbb{N}, n \geq m_0\}$ where $m_0 = \min\{\psi(f) : f \in F\}$.

Possible Types of Local Shannon Functions. We will show that for arbitrary information system U and weight function ψ for U either $H_{U,\psi}^l(n) = O(1)$, or $H_{U,\psi}^l(n) = \Theta(\log_2 n)$, or $H_{U,\psi}^l(n) = n$ for infinitely many $n \in \mathbb{N} \setminus \{0\}$.

The first type of behavior ($H_{U,\psi}^l(n) = O(1)$) realizes only for finite information systems.

The second type of behavior ($H_{U,\psi}^l(n) = \Theta(\log_2 n)$) is most interesting for us since there exist two natural numbers c_1 and c_2 such that for any problem z over U with sufficiently large value of $\psi(z)$ the inequality $\psi_U^l(z) \leq c_1 \log_2 \psi(z) + c_2$ holds.

The third type of behavior ($H_{U,\psi}^l(n) = n$ for infinitely many $n \in \mathbb{N} \setminus \{0\}$) is bad for us: for infinitely many natural n there exists a problem z over U such that $\psi_U^l(z) = \psi(z) = n$.

Thus, we must have possibility to discern the types of behavior. Now we consider the criterions of the local Shannon function $H_{U,\psi}^l$ behavior.

We will say that the information system $U = (A, B, F)$ satisfies *the condition of reduction relatively* ψ if there exists a number $m \in \mathbb{N} \setminus \{0\}$ such that for each compatible on A system of equations

$$\{f_1(x) = \delta_1, \ldots, f_r(x) = \delta_r\} ,$$

where $r \in \mathbb{N} \setminus \{0\}$, $f_1, \ldots, f_r \in F$ and $\delta_1, \ldots, \delta_r \in B$, there exists a subsystem

$$\{f_{i_1}(x) = \delta_{i_1}, \ldots, f_{i_t}(x) = \delta_{i_t}\}$$

of this system which has the same set of solutions and for which $\sum_{j=1}^t \psi(f_{i_j}) \leq m$.

In the following theorem the criterions of the local Shannon function behavior are considered.

Theorem 4.3. *Let U be an information system and ψ be a weight function for U. Then the following statements hold:*

a) if U is a finite information system then $H^l_{U,\psi}(n) = O(1)$;
b) if U is an infinite information system which satisfies the condition of reduction relatively ψ then $H^l_{U,\psi}(n) = \Theta(\log_2 n)$;
c) if U is an infinite information system which does not satisfy the condition of reduction relatively ψ then $H^l_{U,\psi}(n) = n$ for infinitely many natural n.
d) if U is an infinite information system which does not satisfy the condition of reduction relatively ψ and $\psi = h$ then $H^l_{U,\psi}(n) = n$ for any natural n.

Examples. Let m, t be natural numbers. The infinite information systems $U(\mathbb{R}, m) = (\mathbb{R}^m, E, F(m))$ and $U(\mathbb{R}, m, t) = (\mathbb{R}^m, E, F(m,t))$ were defined in Sect. 4.1. Denote $\bar{x} = (x_1, \ldots, x_m)$.

Consider an information system $U(\mathbb{R}, 1, 1)$. One can show that for any compatible system of equations over $U(\mathbb{R}, 1, 1)$ there exists an equivalent subsystem with at most two equations. Therefore $U(\mathbb{R}, 1, 1)$ satisfies the condition of reduction relatively h. Using Theorem 4.3 we obtain $H^l_{U(\mathbb{R},1,1),h}(n) = \Theta(\log_2 n)$.

Consider an information system $U(\mathbb{R}, m, t)$ such that $m \geq 2$ or $t \geq 2$. Let us show that $U(\mathbb{R}, m, t)$ does not satisfy the condition of reduction relatively h.

Let $m \geq 2$ and n be an arbitrary natural number. It is clear that there exists integers $a_1, b_1, c_1, \ldots, a_n, b_n, c_n$ such that the set of solutions on $\mathbb{R}^2$ of the inequality system

$$\{a_1 x_1 + b_1 x_2 + c_1 \leq 0, \ldots, a_n x_1 + b_n x_2 + c_1 \leq 0\}$$

is a polygon with n sides. For $i = 1, \ldots, n$ denote $f_i(\bar{x}) = \text{sign}(a_i x_1 + b_i x_2 + c_i)$. It is clear that $f_i(\bar{x}) \in F(m,t)$. One can show that on the set $\mathbb{R}^2$ the system of equations

$$\{f_1(\bar{x}) = -1, \ldots, f_n(\bar{x}) = -1\}$$

is not equivalent to any its proper subsystem. Taking into account that n is an arbitrary natural number we conclude that $U(\mathbb{R}, m, t)$ does not satisfy the condition of reduction relatively h.

Let $t \geq 2$ and n be an arbitrary natural number. Let $a_1 < b_1 < a_2 < b_2 < \ldots < a_n < b_n$ be integers. For $i = 1, \ldots, n$ denote $g_i(\bar{x}) = \text{sign}((x_1 - a_i)(x_1 - b_i))$. It is clear that $g_i(\bar{x}) \in F(m,t)$. One can show that on the set $\mathbb{R}^2$ the system of equations

$$\{g_1(\bar{x}) = +1, \ldots, g_n(\bar{x}) = +1\}$$

is not equivalent to any its proper subsystem. Taking into account that n is an arbitrary natural number we conclude that $U(\mathbb{R}, m, t)$ does not satisfy the condition of reduction relatively h.

Using Theorem 4.3 we obtain $H^l_{U(\mathbb{R},m,t),h}(n) = n$ for any natural n.

One can show that $H^l_{U(\mathbb{R},m,t),h}(n) \leq H^l_{U(\mathbb{R},m),h}(n) \leq n$ for any natural n. Therefore $H^l_{U(\mathbb{R},m),h}(n) = n$ for any natural n.

Local Shannon Functions for Finite Information Systems. Theorem 4.3 gives us some information about the behavior of local Shannon functions for infinite information systems. But for a finite information system U we have only the relation $H^l_{U,\psi}(n) = O(1)$. However finite information systems are very important for different applications.

Now we consider the behavior of the local Shannon function for an arbitrary finite information system $U = (A, B, F)$ such that $f \not\equiv$ const for any $f \in F$, and for the depth h.

A set $\{f_1, \ldots, f_n\} \subseteq F$ will be called *dependent* if $n \geq 2$ and there exist $i \in \{1, \ldots, n\}$ and $\mu : B^{n-1} \to B$ such that

$$f_i(a) = \mu(f_1(a), \ldots, f_{i-1}(a), f_{i+1}(a), \ldots, f_n(a))$$

for each $a \in A$. If the set $\{f_1, \ldots, f_n\}$ is not dependent then it will be called *independent*. We denote by $\text{in}(U)$ the maximal number of attributes in an independent subset of the set F.

A systems of equations $S = \{f_1(x) = \delta_1, \ldots, f_n(x) = \delta_n\}$ over U will be called *cancellable* if $n \geq 2$ and there exists a number $i \in \{1, \ldots, n\}$ such that the system

$$\{f_1(x) = \delta_1, \ldots, f_{i-1}(x) = \delta_{i-1}, f_{i+1}(x) = \delta_{i+1}, \ldots, f_n(x) = \delta_n\}$$

has the same set of solutions just as the system S. If the system S is not cancellable then it will be called *uncancellable*. We denote by $\text{un}(U)$ the maximal number of equations in an uncancellable compatible system over U.

One can show that

$$1 \leq \text{un}(U) \leq \text{in}(U) .$$

The values $\text{un}(U)$ and $\text{in}(U)$ will be called *the first and the second local critical points of the information system* $U = (A, B, F)$. Now we describe the behavior of the local Shannon function $H^l_{U,h}$ in terms of local critical points of U and the cardinality of the set B.

Theorem 4.4. *Let* $U = (A, B, F)$ *be a finite information system such that* $f \not\equiv$ const *for any* $f \in F$, *and* $n \in \mathbb{N} \setminus \{0\}$. *Then the following statements hold:*

a) if $n \leq \text{un}(U)$ *then* $H^l_{U,h}(n) = n$;
b) if $\text{un}(U) \leq n \leq \text{in}(U)$ *then*

$$\max\{\text{un}(U), \log_k(n+1)\} \leq H^l_{U,h}(n) \leq \min\{n, 2\text{un}(U)^2 \log_2(kn)\}$$

where $k = |B|$;
c) if $n \geq \text{in}(U)$ *then* $H^l_{U,h}(n) = H^l_{U,h}(\text{in}(U))$.

Of course, the problem of computing the values $\text{un}(U)$ and $\text{in}(U)$ for a given finite information system U is very complicated problem. But obtained results allow us to constrict essentially the class of possible types of local Shannon functions for finite information systems.

Example 4.1. Denote by A the set of all points in the plane. Consider an arbitrary straight line l, which divides the plane into positive and negative open half-planes, and the line l itself. Assign a function $f : A \to \{0, 1\}$ to the line l. The function f takes the value 1 if a point is situated on the positive half-plane, and f takes the value 0 if a point is situated on the negative half-plane or on the line l. Denote by F the set of functions which correspond to certain r mutually disjoint finite classes of parallel straight lines. Consider a finite information system $U = (A, \{0, 1\}, F)$. One can show that $\text{in}(U) = |F|$ and $\text{un}(U) \leq 2r$.

Auxiliary Statements. Let $U = (A, B, F)$ be an information system, and ψ be a weight function for U. Denote $\rho = (F, B)$. Let $|B| = k$.

Let us define the function $Q_\psi : \text{Tab}_\rho \to \mathbb{N}$. Let $T \in \text{Tab}_\rho$. If T is an empty table then $Q_\psi(T) = 0$. Let T be a nonempty table and $\bar{\delta} \in \text{Row}(T)$. Then $Q_\psi(T, \bar{\delta}) = \min\{\psi(\alpha) : \alpha \in \Omega_\rho(T), \text{Row}(T\alpha) = \{\bar{\delta}\}\}$ and $Q_\psi(T) = \max\{Q_\psi(T, \bar{\delta}) : \bar{\delta} \in \text{Row}(T)\}$. Note that $Q_\psi(T, \bar{\delta})$ is the minimal total weight of a set of columns on which the row $\bar{\delta}$ differs from all other rows. Define the function $m_\psi : \text{Tab}_\rho \to \mathbb{N}$ as follows: $m_\psi(T) = \max\{\psi(f) : f \in \text{At}(T)\}$ for any $T \in \text{Tab}_\rho$. In other words, $m_\psi(T)$ is the maximal weight of a column in the table T.

One can show that the information system U is finite if and only if there exists a natural r such that $N(T) \leq r$ for any table $T \in \text{Tab}_U$.

One can show also that U satisfies the condition of reduction relatively ψ if and only if there exists natural m such that the inequality $Q_\psi(T) \leq m$ holds for any table $T \in \text{Tab}_U$.

Later we will denote by $\dim T$ the number of columns in the table T.

Lemma 4.1. *Let $z \in \text{Probl}_U$. Then the value $\psi_U^l(z)$ is definite and the inequality $\psi_U^l(z) \leq \psi(z)$ holds.*

Proof. Let $z = (\nu, f_1, \ldots, f_n)$. Consider a decision tree Γ from Tree_U possessing following properties: from every working node of Γ exactly k edges are issuing and every complete path in the tree Γ contains exactly n working nodes. Let $\xi = v_1, d_1, \ldots, v_n, d_n, v_{n+1}$ be an arbitrary complete path in the tree Γ. Then for $j = 1, \ldots, n$ the node v_j is labelled by the attribute f_j and, if for $j = 1, \ldots, n$ the edge d_j is labelled by the number δ_j, then the node v_{n+1} is labelled by the number $\nu(\delta_1, \ldots, \delta_n)$. One can show that the decision tree Γ solves the problem z, $\text{At}(\Gamma) \subseteq \text{At}(z)$ and $\psi(\Gamma) = \psi(f_1 \ldots f_n)$. Therefore the value $\psi_U^l(z)$ is definite and the inequality $\psi_U^l(z) \leq \psi(z)$ holds. □

Define the function $\hat{Q}_{U,\psi} : \text{Tab}_U \to \mathbb{N}$. Let $T \in \text{Tab}_U$. Then

$$\hat{Q}_{U,\psi}(T) = \max\{Q_\psi(T') : T' \in \text{Tab}_U, \text{At}(T') \subseteq \text{At}(T)\} .$$

Lemma 4.2. *Let $T \in \text{Tab}_U$. Then $M_{\rho,\psi}(T) \leq 2\hat{Q}_{U,\psi}(T)$.*

Proof. Let T contain n columns labelling by attributes $f_1, \ldots, f_n$. Let $\bar{\delta} = (\delta_1, \ldots, \delta_n) \in B^n$. Let us show that $M_{\rho,\psi}(T, \bar{\delta}) \leq 2\hat{Q}_{U,\psi}(T)$.

For an arbitrary $\bar{\gamma} = (\gamma_1, \ldots, \gamma_n) \in \text{Row}(T)$ let $\beta(\bar{\gamma})$ be a word from $\Omega_\rho(T)$ such that $\text{Row}(T\beta(\bar{\gamma})) = \{\bar{\gamma}\}$ and $\psi(\beta(\bar{\gamma})) = Q_\psi(T, \bar{\gamma})$. Obviously, $\psi(f_i) \leq Q_\psi(T, \bar{\gamma}) \leq Q_\psi(T)$ for any pair $(f_i, \gamma_i) \in \text{Alph}(\beta(\bar{\gamma}))$.

Let there exist $\bar{\gamma} \in \text{Row}(T)$ such that $\text{Alph}(\beta(\bar{\gamma})) \subseteq \{(f_1, \delta_1), \ldots, (f_n, \delta_n)\}$. It is clear that $T\beta(\bar{\gamma}) \in \text{Dtab}_\rho$ and $\psi(\beta(\bar{\gamma})) \leq Q_\psi(T) \leq 2\hat{Q}_{U,\psi}(T)$. Hence $M_{\rho,\psi}(T, \bar{\delta}) \leq 2\hat{Q}_{U,\psi}(T)$.

Let the set $\text{Alph}(\beta(\bar{\gamma}))$ be not a subset of $\{(f_1, \delta_1), \ldots, (f_n, \delta_n)\}$ for any $\bar{\gamma} \in \text{Row}(T)$. Then there exists a subset $\{f_{i(1)}, \ldots, f_{i(m)}\}$ of the set $\{f_1, \ldots, f_n\}$ with following properties: $\text{Row}(T(f_{i(1)}, \delta_{i(1)}) \ldots (f_{i(m)}, \delta_{i(m)})) = \emptyset$ and $\psi(f_{i(j)}) \leq Q_\psi(T)$ for $j = 1, \ldots, m$. If $\text{Row}(T(f_{i(1)}, \delta_{i(1)})) = \emptyset$ then, evidently, $M_{\rho,\psi}(T, \bar{\delta}) \leq Q_\psi(T) \leq 2\hat{Q}_{U,\psi}(T)$. Let $\text{Row}(T(f_{i(1)}, \delta_{i(1)})) \neq \emptyset$. Then there exists a number $t \in \{1, \ldots, m-1\}$ for which $\text{Row}(T(f_{i(1)}, \delta_{i(1)}) \ldots (f_{i(t)}, \delta_{i(t)})) \neq \emptyset$ and $\text{Row}(T(f_{i(1)}, \delta_{i(1)}) \ldots (f_{i(t)}, \delta_{i(t)})(f_{i(t+1)}, \delta_{i(t+1)})) = \emptyset$. Consider the problem $z = (\nu, f_{i(1)}, \ldots, f_{i(t)})$ such that $\nu : B^t \to \{0\}$. Denote $T' = T_U(z)$ and $\bar{\delta}' = (\delta_{i(1)}, \ldots, \delta_{i(t)})$. Obviously, $T' \in \text{Tab}_U$ and $\bar{\delta}' \in \text{Row}(T')$. Let α be a word from the set $\Omega_\rho(T')$ such that $\text{Row}(T'\alpha) = \{\bar{\delta}'\}$ and $\psi(\alpha) = Q_\psi(T', \bar{\delta}')$. Bearing in mind that $\text{At}(T') \subseteq \text{At}(T)$ we obtain $\psi(\alpha) \leq Q_\psi(T') \leq \hat{Q}_{U,\psi}(T)$. One can show that $\text{Row}(T\alpha) = \text{Row}(T(f_{i(1)}, \delta_{i(1)}) \ldots (f_{i(t)}, \delta_{i(t)}))$. Denote $\kappa = \alpha(f_{i(t+1)}, \delta_{i(t+1)})$. Then $\text{Alph}(\kappa) \subseteq \{(f_1, \delta_1) \ldots (f_n, \delta_n)\}$ and $\text{Row}(T\kappa) = \emptyset$. Taking into account that $\psi(\alpha) \leq \hat{Q}_{U,\psi}(T)$ and $\psi((f_{i(t+1)}, \delta_{i(t+1)})) \leq Q_\psi(T) \leq \hat{Q}_{U,\psi}(T)$ we obtain $\psi(\kappa) \leq 2\hat{Q}_{U,\psi}(T)$. Hence $M_{\rho,\psi}(T, \bar{\delta}) \leq 2\hat{Q}_{U,\psi}(T)$. Bearing in mind that $\bar{\delta}$ is an arbitrary n-tuple from B^n obtain $M_{\rho,\psi}(T) \leq 2\hat{Q}_{U,\psi}(T)$. □

Lemma 4.3. *Let $T \in \text{Tab}_U$. Then $N(T) \leq (k \dim T)^{Q_\psi(T)}$.*

Proof. If $N(T) = 1$ then, evidently, $Q_\psi(T) = 0$ and the statement of the lemma holds. Let $N(T) > 1$. One can show that $Q_\psi(T) > 0$. Denote $m = Q_\psi(T)$. It is easily to show that for any $\bar{\delta} \in \text{Row}(T)$ there exist attributes $f_1, \ldots, f_m \in \text{At}(T)$ and numbers $\gamma_1, \ldots, \gamma_m \in B$ for which $\text{Row}(T(f_1, \gamma_1) \ldots (f_m, \gamma_m)) = \{\bar{\delta}\}$. Hence there exists a one-to-one mapping of the set $\text{Row}(T)$ onto certain set D of pairs of m-tuples of the kind $((f_1, \ldots, f_m), (\gamma_1, \ldots, \gamma_m))$ where $f_1, \ldots, f_m \in \text{At}(T)$ and $\gamma_1, \ldots, \gamma_m \in B$. Obviously, $|D| \leq (\dim T)^m k^m$. Therefore $N(T) \leq (k \dim T)^m$ where $m = Q_\psi(T)$. □

Lemma 4.4. *Let $T \in \text{Tab}_U$ and $Q_\psi(T) = n > 0$. Then the following statements hold:*

a) there exists a problem $z \in \text{Probl}_U$ such that $\psi_U^l(z) = n$ and $\psi(z) = n$;
b) if $\psi = h$ and $n > 1$ then there exists a table $T'' \in \text{Tab}_U$ such that $Q_h(T'') = n - 1$.

Proof. Let T contain r columns which are labelled by attributes $f_1, \ldots, f_r$. Let $\bar{\delta} = (\delta_1, \ldots, \delta_r) \in \text{Row}(T)$, $Q_\psi(T, \bar{\delta}) = Q_\psi(T)$ and let α be a word of minimal length from $\Omega_\rho(T)$ such that $\psi(\alpha) = Q_\psi(T, \bar{\delta})$ and $\text{Row}(T\alpha) = \{\bar{\delta}\}$. Since $\psi(\alpha) = n > 0$, we obtain $\alpha \neq \lambda$. Let $\alpha = (f_{i(1)}, \delta_{i(1)}) \ldots (f_{i(m)}, \delta_{i(m)})$. Obviously, for any $j, l \in \{1, \ldots, m\}$ if $j \neq l$ then $f_{i(j)} \neq f_{i(l)}$. Denote $\bar{\delta}' = (\delta_{i(1)}, \ldots, \delta_{i(m)})$. Let $z = (\nu, f_{i(1)}, \ldots, f_{i(m)})$ where $\nu : B^m \to \{0, 1\}$ and for any $\bar{\gamma} \in B^m$ if $\bar{\gamma} = \bar{\delta}'$ then $\nu(\bar{\gamma}) = 0$, and if $\bar{\gamma} \neq \bar{\delta}'$ then $\nu(\bar{\gamma}) = 1$. Denote $T' = T_U(z)$. Obviously, $\bar{\delta}' \in \text{Row}(T')$. One can show that $Q_\psi(T', \bar{\delta}') = n$ and $M_{\rho,\psi}(T', \bar{\delta}') = n$. Therefore $M_{\rho,\psi}(T') \geq n$. Using Theorem 3.1 we obtain $\psi_\rho(T') \geq n$. Using this inequality and Corollary 4.1 we conclude that $\psi^l_U(z) \geq n$. Obviously, $\psi(z) = n$. From Lemma 4.1 it follows that $\psi^l_U(z) = n$.

Let $\psi = h$ and $n > 1$. Then $m = n$. Denote $z' = (\nu', f_{i(1)}, \ldots, f_{i(m-1)})$, where $\nu' : B^{m-1} \to \{0\}$, $T'' = T_U(z')$ and $\bar{\delta}'' = (\delta_{i(1)}, \ldots, \delta_{i(m-1)})$. It is easily to show that $\dim T'' = m-1$ and $Q_h(T'', \bar{\delta}'') = m-1$. Therefore $Q_h(T'') = m-1 = n-1$. □

Lemma 4.5. *Let U be a finite information system. Then $H^l_{U,\psi}(n) = O(1)$.*

Proof. It is not difficult to prove that $\psi^l_U(z) \leq \sum_{f \in F} \psi(f)$ for any problem $z \in \mathrm{Probl}_U$. Therefore $H^l_{U,\psi}(n) = O(1)$. □

For $t \in \mathbb{N}$ denote $\mathrm{Spectr}_{U,\psi}(t) = \{N(T) : T \in \mathrm{Tab}_U, m_\psi(T) \leq t\}$.

Lemma 4.6. *Let U be an infinite information system satisfying the condition of reduction relatively ψ. Then there exists natural t such that the set* $\mathrm{Spectr}_{U,\psi}(t)$ *is infinite.*

Proof. Since U satisfies the condition of reduction relatively ψ, there exists natural t such that $Q_\psi(T) \leq t$ for any table $T \in \mathrm{Tab}_U$. We will show that the set $\mathrm{Spectr}_{U,\psi}(t)$ is infinite. Assume the contrary: let there exist a number $m \in \mathbb{N}$ such that for any table $T \in \mathrm{Tab}_U$ if $m_\psi(T) \leq t$ then $N(T) \leq m$. Since U is an infinite information system, there exists a table $T \in \mathrm{Tab}_U$ such that $N(T) > m$. Let $\{f_i : f_i \in \mathrm{At}(T), \psi(f_i) \leq t\} = \{f_1, \ldots, f_n\}$. Denote $z' = (\nu, f_1, \ldots, f_n)$, where $\nu : B^n \to \{0\}$, and $T' = T_U(z')$. Taking into account that $Q_\psi(T) \leq t$ one can show that $N(T') = N(T)$. Therefore $m_\psi(T') \leq t$ and $N(T') > m$. The obtained contradiction shows that the set $\mathrm{Spectr}_{U,\psi}(t)$ is infinite. □

Lemma 4.7. *Let there exists natural t such that the set* $\mathrm{Spectr}_{U,\psi}(t)$ *is infinite. Then $H^l_{U,\psi}(n) = \Omega(\log_2 n)$.*

Proof. Let $n > t$. Denote $m = \lfloor n/t \rfloor$. Since the set $\mathrm{Spectr}_{U,\psi}(t)$ is infinite, there exists a table $T \in \mathrm{Tab}_U$ such that $m_\psi(T) \leq t$ and $N(T) \geq k^m$. Let $\{f_1, \ldots, f_p\}$ be a subset of the set $\mathrm{At}(T)$ of minimal cardinality such that $|\mathrm{Row}(T(f_1, \delta_1) \ldots (f_p, \delta_p))| \leq 1$ for any $\delta_1, \ldots, \delta_p \in B$. Since $N(T) \geq k^m$, we have $p \geq m$. Denote $z = (\nu, f_1, \ldots, f_m)$ where $\nu : B^m \to \mathbb{N}$ and for any $\bar{\delta}_1, \bar{\delta}_2 \in B^m$ if $\bar{\delta}_1 \neq \bar{\delta}_2$ then $\nu(\bar{\delta}_1) \neq \nu(\bar{\delta}_2)$. Denote $T' = T(z)$. From the choice of the set $\{f_1, \ldots, f_p\}$ it follows that for any $i \in \{1, \ldots, m\}$ there exist rows $\bar{\delta}_1, \bar{\delta}_2 \in \mathrm{Row}(T')$ which differ only in i-th digit. Therefore $J(T') = m$. From Theorem 3.3 it follows that $\psi_\rho(T') \geq \log_k(m+1) \geq \log_k(n/t)$. Using Corollary 4.1 we obtain $\psi^l_U(z) = \psi_\rho(T')$. Obviously, $\psi(z) \leq mt \leq n$. Hence $H^l_{U,\psi}(n) \geq \log_k n - \log_k t$. Therefore $H^l_{U,\psi}(n) = \Omega(\log_2 n)$. □

Lemma 4.8. *Let U be an infinite information system satisfying the condition of reduction relatively ψ. Then $H^l_{U,\psi}(n) = O(\log_2 n)$.*

Proof. Since U satisfies the condition of reduction relatively ψ, there exists natural m such that $Q_\psi(T) \leq m$ for any table $T \in \mathrm{Tab}_U$. Therefore $\hat{Q}_{U,\psi}(T) \leq m$ for any table $T \in \mathrm{Tab}_U$. Let $n \in \mathrm{Dom}(H^l_{U,\psi})$ and let z be an arbitrary problem from Probl_U for which $\psi(z) \leq n$. Obviously, $n \geq 1$. Denote $T = T_U(z)$. Since $\psi(z) \leq n$, we have $\dim T \leq n$. Besides, $Q_\psi(T) \leq m$ and $\hat{Q}_{U,\psi}(T) \leq m$. Using Lemma 4.2 we obtain $M_{\rho,\psi}(T) \leq 2\hat{Q}_{U,\psi}(T) \leq 2m$. Using Lemma 4.3 obtain $N(T) \leq (k \dim T)^{Q_\psi(T)} \leq (nk)^m$. From these inequalities and from Theorem 3.5 it follows that $\psi_\rho(T) \leq 2m^2 \log_2 n + 2m^2 \log_2 k$. Using Corollary 4.1 we conclude that $\psi^l_U(z) \leq 2m^2 \log_2 n + 2m^2 \log_2 k$. Taking into account that n is an arbitrary number from the set $\text{Dom}(H^l_{U,\psi})$ and that z is an arbitrary problem from Probl_U with $\psi(z) \leq n$ we obtain $H^l_{U,\psi}(n) = O(\log_2 n)$. □

For any natural n denote

$$N_U(n) = \max\{N(T) : T \in \text{Tab}_U, \dim T \leq n\} \ .$$

Lemma 4.9. *Let U be a finite information system such that $f \not\equiv \text{const}$ for any $f \in F$, $n \in \mathbb{N} \setminus \{0\}$ and $n \leq \text{in}(U)$. Then*

$$n + 1 \leq N_U(n) \leq (kn)^{\text{un}(U)} \ .$$

Proof. One can show that $Q_h(T) \leq \text{un}(U)$ for any table $T \in \text{Tab}_U$. Using Lemma 4.3 we conclude that $N_U(n) \leq (kn)^{\text{un}(U)}$.

Let us prove that $N_U(n) \geq n + 1$. Let $\{f_1, \dots, f_{\text{in}(U)}\}$ be an independent set of attributes from F, $i \in \{1, \dots, \text{in}(U)\}$, and $z_i = (\nu, f_1, \dots, f_i)$ be the problem from Probl_U such that $\nu \equiv \{0\}$. Since $f_1 \not\equiv \text{const}$ we have $N(T(z_1)) \geq 2$. Let us prove that if $\text{in}(U) > 1$ then $N(T(z_i)) < N(T(z_{i+1}))$ for any $i \in \{1, \dots, \text{in}(U) - 1\}$. Assume the contrary: $N(T(z_i)) = N(T(z_{i+1}))$ for some $i \in \{1, \dots, \text{in}(U) - 1\}$. It is not difficult to prove that in this case there exists $\mu : B^i \to B$ such that $f_{i+1}(a) = \mu(f_1(a), \dots, f_i(a))$ for any $a \in A$, but this is impossible. Thus, $N(T(z_1)) \geq 2$ and if $\text{in}(U) > 1$ then $N(T(z_i)) < N(T(z_{i+1}))$ for any $i \in \{1, \dots, \text{in}(U) - 1\}$. Therefore $N(T(z_n)) \geq n + 1$. Since $h(z_n) = n$ we obtain $N_U(n) \geq n + 1$. □

Lemma 4.10. *Let U be a finite information system such that $f \not\equiv \text{const}$ for any $f \in F$, $n \in \mathbb{N} \setminus \{0\}$ and $n \geq \text{in}(U)$. Then*

$$H^l_{U,h}(n) = H^l_{U,h}(\text{in}(U)) \ .$$

Proof. Evidently, $H^l_{U,h}(n) \geq H^l_{U,h}(\text{in}(U))$.

Let us prove by induction on m, $m \geq \text{in}(U)$, that for any problem $z \in \text{Probl}_U$ with $h(z) = m$ there exists a problem $z' \in \text{Probl}_U$ such that $h(z') = \text{in}(U)$, $\text{At}(z') \subseteq \text{At}(z)$, and $z'(a) = z(a)$ for any $a \in A$. Evidently, the considered statement holds for $m = \text{in}(U)$. Let it hold for some m, $m \geq \text{in}(U)$. Let us show that this statement holds for $m+1$ too. Let $z = (\nu, f_1, \dots, f_{m+1}) \in \text{Probl}_U$. It is clear that the set $\{f_1, \dots, f_{m+1}\}$ is dependent set of attributes. Therefore there exist $i \in \{1, \dots, m+1\}$ and $\mu : B^m \to B$ such that

$$f_i(a) = \mu(f_1(a), \dots, f_{i-1}(a), f_{i+1}(a), \dots, f_{m+1}(a))$$

for any $a \in A$. Let $\gamma : B^m \to B$ and $\gamma(\delta_1, \ldots, \delta_m) = \nu(\delta_1, \ldots, \delta_{i-1}, \mu(\delta_1, \ldots, \delta_m), \delta_{i+1}, \ldots, \delta_m)$ for any $(\delta_1, \ldots, \delta_m) \in B^m$. Let us consider the problem $z_1 = (\gamma, f_1, \ldots, f_{i-1}, f_{i+1}, \ldots, f_{m+1})$ from Probl_U. Evidently, $h(z_1) = m$, $\mathrm{At}(z_1) \subseteq \mathrm{At}(z)$ and $z_1(a) = z(a)$ for any $a \in A$. Using the inductive hypothesis we conclude that there exists a problem $z' \in \mathrm{Probl}_U$ such that $h(z') = \mathrm{in}(U)$, $\mathrm{At}(z') \subseteq \mathrm{At}(z_1)$ and $z'(a) = z_1(a)$ for any $a \in A$. It is clear that $h(z') = \mathrm{in}(U), \mathrm{At}(z') \subseteq \mathrm{At}(z)$ and $z'(a) = z(a)$ for any $a \in A$. Thus, the considered statement holds.

Let us prove that $H^l_{U,h}(n) \leq H^l_{U,h}(\mathrm{in}(U))$. Let $z \in \mathrm{Probl}_U$ and $h(z) \leq n$. If $h(z) \leq \mathrm{in}(U)$ then $h^l_U(z) \leq H^l_{U,h}(\mathrm{in}(U))$. Let $h(z) > \mathrm{in}(U)$. Then there exists a problem $z' \in \mathrm{Probl}_U$ such that $h(z') = \mathrm{in}(U)$, $\mathrm{At}(z') \subseteq \mathrm{At}(z)$, and $z'(a) = z(a)$ for any $a \in A$. One can show that $h^l_U(z) \leq h^l_U(z')$. Evidently, $h^l_U(z') \leq H^l_{U,h}(\mathrm{in}(U))$. Therefore $h^l_U(z) \leq H^l_{U,h}(\mathrm{in}(U))$. Since z is an arbitrary problem over U such that $h(z) \leq n$ we obtain $H^l_{U,h}(n) \leq H^l_{U,h}(\mathrm{in}(U))$. Thus, $H^l_{U,h}(n) = H^l_{U,h}(\mathrm{in}(U))$. □

Proofs of Theorems 4.3 and 4.4

Proof (of Theorem 4.3). Statement a) follows from Lemma 4.5. Statement b) follows from Lemmas 4.6–4.8.

Let us prove the statement c). Since U does not satisfy the condition of reduction relatively ψ, the set $D = \{Q_\psi(T) : T \in \mathrm{Tab}_U\}$ is infinite. Let $n \in D \setminus \{0\}$ and T be a table from Tab_U for which $Q_\psi(T) = n$. Using Lemma 4.4 we conclude that there exists a problem $z \in \mathrm{Probl}_U$ such that $\psi^l_U(z) = \psi(z) = n$. Therefore $H^l_{U,\psi}(n) = n$.

Let us prove the statement d). Using the equality $\psi = h$ it is not difficult to prove that $\mathrm{Dom}(H^l_{U,\psi}) = \mathbb{N} \setminus \{0\}$. Let n be a natural number. Taking into account that U does not satisfy the condition of reduction relatively h we conclude that there exists a table $T \in \mathrm{Tab}_U$ such that $Q_h(T) \geq n$. Using Lemma 4.4 we conclude that there exists a problem $z \in \mathrm{Probl}_U$ such that $h^l_U(z) = h(z) = n$. Therefore $H^l_{U,h}(n) = n$. □

Proof (of Theorem 4.4)

a) Let $n \leq \mathrm{un}(U)$. It is clear that $H^l_{U,h}(n) \leq n$. Let us show that $H^l_{U,h}(n) \geq n$. By definition of the parameter $\mathrm{un}(U)$, there exists a compatible uncancellable system of equations over U of the kind $\{f_1(x) = \delta_1, \ldots, f_{\mathrm{un}(U)}(x) = \delta_{\mathrm{un}(U)}\}$. It is clear that the system of equations $\{f_1(x) = \delta_1, \ldots, f_n(x) = \delta_n\}$ is a compatible uncancellable system too. Let $z = (\nu, f_1, \ldots, f_n)$ where $\nu : B^n \to \{0\}$. Denote $\bar{\delta} = (\delta_1, \ldots, \delta_n)$. It is not difficult to prove that $Q_h(T(z), \bar{\delta}) = n$. Therefore $Q_h(T(z)) = n$. Using Lemma 4.4 we conclude that there exists a problem $z' \in \mathrm{Probl}_U$ such that $h(z') = n$ and $h^l_U(z') = n$. Therefore $H^l_{U,h}(n) \geq n$. Hence $H^l_{U,h}(n) = n$.
b) Let $\mathrm{un}(U) \leq n \leq \mathrm{in}(U)$. It is clear that $H^l_{U,h}(n) \leq n$. Using statement a) we obtain $H^l_{U,h}(n) \geq \mathrm{un}(U)$.
   Let us show that $H^l_{U,h}(n) \geq \log_k(n+1)$. Using Lemma 4.9 we obtain that there exists a problem $z = (\nu, f_1, \ldots, f_m)$ over U such that $m \leq n$ and

$N(T(z)) \geq n+1$. Consider the problem $z' = (\nu', f_1, \ldots, f_m)$ such that $\nu'(\bar{\delta}_1) \neq \nu'(\bar{\delta}_2)$ if $\bar{\delta}_1 \neq \bar{\delta}_2$. It is clear that $T(z')$ is a diagnostic table and $N(T(z')) \geq n+1$. Using Corollary 3.1 we obtain $h_\rho(T(z')) \geq \log_k(n+1)$. From Theorem 4.1 it follows that $h_U(z') \geq \log_k(n+1)$. Taking into account that $m \leq n$ we obtain $H^l_{U,h}(n) \geq \log_k(n+1)$.

Let us show that $H^l_{U,h}(n) \leq 2\text{un}(U)^2 \log_2(kn)$. Let $z \in \text{Probl}(U)$ and $\dim z \leq n$. Denote $T = T(z)$. From Lemma 4.9 follows that $N(T) \leq (kn)^{\text{un}(U)}$. One can show that $\hat{Q}_{U,h}(T) \leq \text{un}(U)$. Using Lemma 4.2 we obtain $M_{\rho,h}(T) \leq 2\hat{Q}_{U,h}(T)$. Therefore $M_{\rho,h}(T) \leq 2\text{un}(U)$. From the obtained inequalities and from Theorem 3.5 it follows $h_\rho(T) \leq M_{\rho,h}(T) \log_2 N(T) \leq 2\text{un}(U)^2 \log_2(kn)$. Taking into account that z is an arbitrary problem from Probl_U with $\dim z \leq n$ we obtain $H^l_{U,h}(n) \leq 2\text{un}(U)^2 \log_2(kn)$.

c) Let $n \geq \text{in}(U)$. Using Lemma 4.10 we obtain $H^l_{U,h}(n) = H^l_{U,h}(\text{in}(U))$. □

4.4 Local Optimization Problems for Decision Trees

The relationships among three algorithmic problems are considered in this section: the problem of compatibility of equation system, the problem of construction of locally optimal decision tree, and the problem of computation of locally optimal decision tree complexity.

Relationships Among Algorithmic Problems. Let $U = (A, B, F)$ be an enumerated information system, where $F = \{f_i : i \in \mathbb{N}\}$, and ψ be a computable weight function for U. Denote $\rho = (F, B)$. We formulate the following three algorithmic problems: the problem $\text{Ex}(U)$ of compatibility of equation system over U and the two problems of local optimization denoted respectively by $\text{Des}^l(U, \psi)$ and $\text{Com}^l(U, \psi)$.

The Problem $\text{Ex}(U)$ *of Compatibility of Equation System:* for a given word $\alpha \in \Omega_\rho$ it is required to determine whether the set $\text{Sol}_U(\alpha)$ is the empty set.

The Problem $\text{Des}^l(U, \psi)$*:* for a given problem $z \in \text{Probl}_U$ it is required to find a decision tree $\Gamma \in \text{Tree}_U$ which solves z and for which $\text{At}(\Gamma) \subseteq \text{At}(z)$ and $\psi(\Gamma) = \psi^l_U(z)$.

The Problem $\text{Com}^l(U, \psi)$*:* for a given problem $z \in \text{Probl}_U$ it is required to compute the value $\psi^l_U(z)$.

Theorem 4.5. *Let* $U = (A, B, F)$ *be an enumerated information system, and* ψ *be a computable weight function for* U*. Then the following statements hold:*

a) if the problem $\text{Ex}(U)$ *is solvable then problems* $\text{Com}^l(U, \psi)$ *and* $\text{Des}^l(U, \psi)$ *are also solvable;*
b) if $\text{Ex}(U)$ *is unsolvable then* $\text{Com}^l(U, \psi)$ *and* $\text{Des}^l(U, \psi)$ *are unsolvable.*

Example 4.2. Let m, t be natural numbers and $U(\mathbb{Z}, m, t), U(\mathbb{R}, m, t)$ be information systems defined in Sect. 4.1. From results of [55] it follows that there exist natural m_0 and t_0 such that for any $m \geq m_0$ and $t \geq t_0$ the problem $\text{Ex}(U(\mathbb{Z}, m, t))$ is unsolvable. From results of [207] it follows that the problem $\text{Ex}(U(\mathbb{R}, m, t))$ is solvable for any m and t.

In this subsection we consider only infinite (enumerated) information systems. It should be pointed out that for any finite information system U the problem $\mathrm{Ex}(U)$ is solvable.

Let $\rho = (F, B)$ be a signature and $\alpha \in \Omega_\rho$. The word α will be called *inconsistent* if there exist an attribute $f \in F$ and numbers $\delta, \gamma \in B$ such that $(f, \delta) \in \mathrm{Alph}(\alpha)$, $(f, \gamma) \in \mathrm{Alph}(\alpha)$ and $\delta \neq \gamma$, and *consistent* otherwise.

Proof (of Theorem 4.5). Let $U = (A, B, F)$. Denote $\rho = (F, B)$.

a) Let the problem $\mathrm{Ex}(U)$ be solvable. One can show that there exists an algorithm which for a given problem $z \in \mathrm{Probl}_U$ constructs the table $T_U(z)$. From Proposition 3.4 it follows that the problems $\mathrm{Com}(\rho, \psi)$ and $\mathrm{Des}(\rho, \psi)$ are solvable. Using Theorem 4.1 we conclude that the problems $\mathrm{Com}^l(U, \psi)$ and $\mathrm{Des}^l(U, \psi)$ are solvable too.

b) Let the problem $\mathrm{Ex}(U)$ be unsolvable. Let us show that the problem $\mathrm{Com}^l(U, \psi)$ is unsolvable. Assume the contrary. Let us show that in this case the problem $\mathrm{Ex}(U)$ is solvable. Since U is an infinite information system, there exists an attribute $f_{i(0)} \in F$ which is not constant. Let $\alpha \in \Omega_\rho$. If $\alpha = \lambda$ then, obviously, $\mathrm{Sol}_U(\alpha) \neq \emptyset$. Let now $\alpha \neq \lambda$. For inconsistent α the equality $\mathrm{Sol}_U(\alpha) = \emptyset$ is evident. Let α be a consistent word and let $\mathrm{Alph}(\alpha) = \{(f_{i(1)}, \delta_1), \ldots, (f_{i(n)}, \delta_n)\}$ where $i(1) < \ldots < i(n)$.
Let $i(0) \notin \{i(1), \ldots, i(n)\}$. For any $t \in B$ define a problem $z_t \in \mathrm{Probl}_U$ as follows: $z_t = (\nu_t, f_{i(0)}, f_{i(1)}, \ldots, f_{i(n)})$, where $\nu_t : B^{n+1} \to \{0, 1\}$ and for any $\bar{\gamma} \in B^{n+1}$ if $\bar{\gamma} = (t, \delta_1, \ldots, \delta_n)$ then $\nu_t(\bar{\gamma}) = 1$, and if $\bar{\gamma} \neq (t, \delta_1, \ldots, \delta_n)$ then $\nu_t(\bar{\gamma}) = 0$. Since the problem $\mathrm{Com}^l(U, \psi)$ is solvable, the value of $\psi_U^l(z_t)$ can be found for every $t \in B$. Taking into account that the attribute $f_{i(0)}$ is not constant one can show that $\mathrm{Sol}_U(\alpha) = \emptyset$ if and only if $\psi_U^l(z_t) = 0$ for any $t \in B$.
Let $i(0) \in \{i(1), \ldots, i(n)\}$. Define a problem $z \in \mathrm{Probl}_U$ as follows: $z = (\nu, f_{i(1)}, \ldots, f_{i(n)})$, where $\nu : B^n \to \{0, 1\}$ and for any $\bar{\gamma} \in B^n$ if $\bar{\gamma} = (\delta_1, \ldots, \delta_n)$ then $\nu(\bar{\gamma}) = 1$ and if $\bar{\gamma} \neq (\delta_1, \ldots, \delta_n)$ then $\nu(\bar{\gamma}) = 0$. From solvability of the problem $\mathrm{Com}^l(U, \psi)$ it follows the possibility of computation of the value $\psi_U^l(z)$. It is easily to show that $\mathrm{Sol}_U(\alpha) = \emptyset$ if and only if $\psi_U^l(z) = 0$. So the problem $\mathrm{Ex}(U)$ is solvable which is impossible. Therefore the problem $\mathrm{Com}^l(U, \psi)$ is unsolvable. Since ψ is a computable function, we conclude that the problem $\mathrm{Des}^l(U, \psi)$ is unsolvable too. □

4.5 Cardinality Characteristics of Test Tables

Let $\rho = (F, B)$ be a signature and $U = (A, B, F)$ be an information system.

In the first subsubsection of this subsection upper and lower bounds on the value $N(T)$ for a table $T \in \mathrm{Tab}_\rho$ are considered.

The second subsubsection is devoted to investigation of a function $N_U : \mathbb{N} \setminus \{0\} \to \mathbb{N}$ defined as follows:

$$N_U(n) = \max\{N(T) : T \in \mathrm{Tab}_U, \dim T \leq n\} ,$$

where $\dim T$ is the number of columns in the table T. This function will be used later for analysis of complexity of decision trees and also for analysis of complexity of algorithms for decision tree construction.

The third subsubsection deals with examples.

The results contained in this subsection are similar to results from [2, 182, 183]. The notions close to Vapnik-Chervonenkis dimension [7, 210] and independent dimension or I-dimension [47] are applied in the analysis of the function N_U.

Bounds on Value $N(T)$. Let T be a table from Tab_ρ with n columns labelling by attributes $f_1, \ldots, f_n$. Denote by $RO(T)$ the set of all nonempty words $f_{i(1)} \cdots f_{i(m)}$ from the set $\mathrm{At}(T)^*$ such that $i(1) < i(2) < \ldots < i(m)$. Let $\beta = f_{i(1)} \cdots f_{i(m)} \in RO(T)$. Denote $\mathrm{Row}(T, \beta) = \{(\delta_{i(1)}, \ldots, \delta_{i(m)}) : (\delta_1, \ldots, \delta_n) \in \mathrm{Row}(T)\}$. Define the value $I(T)$ as follows. If $N(T) \leq 1$ then $I(T) = 0$. Let $N(T) \geq 2$. Then $I(T)$ is the maximal number $m \in \{1, \ldots, n\}$ for which there exist a word $\beta \in RO(T)$ and sets $G_1, \ldots, G_m \subseteq B$ possessing the following properties: the length of the word β is equal to m, $|G_1| = \ldots = |G_m| = 2$ and $G_1 \times \ldots \times G_m \subseteq \mathrm{Row}(T, \beta)$.

Theorem 4.6. *Let $\rho = (F, B)$ be a signature such that F is an infinite set and $|B| = k$. Let T be a nonempty table from Tab_ρ. Then*

$$2^{I(T)} \leq N(T) \leq (k^2 \dim T)^{I(T)} .$$

Proof. Let $n, t \in \mathbb{N}$ and $n \geq 1$. Denote

$$N_\rho(n, t) = \max\{N(T) : T \in \mathrm{Tab}_\rho, \dim T \leq n, I(T) \leq t\} .$$

It is easily to notice that the value $N_\rho(n, t)$ is definite and the following inequalities hold:

$$N_\rho(n, 0) \leq 1 , \tag{29}$$

$$N_\rho(1, t) \leq k . \tag{30}$$

Let us prove that for any $n \in \mathbb{N} \setminus \{0\}$ and $t \in \mathbb{N}$ the inequality

$$N_\rho(n+1, t+1) \leq N_\rho(n, t+1) + k^2 N_\rho(n, t) \tag{31}$$

holds. It is clear that there exists a table $T \in \mathrm{Tab}_\rho$ which contains $n+1$ columns and for which $I(T) \leq t+1$ and $N(T) = N_\rho(n+1, t+1)$. Let columns of T be labelled by attributes $f_1, \ldots, f_{n+1}$. Denote $\beta = f_1 \ldots f_n$. Let $\bar{\delta} = (\delta_1, \ldots, \delta_n) \in \mathrm{Row}(T, \beta)$. Denote $\kappa(\bar{\delta}) = \{\delta : \delta \in B, (\delta_1, \ldots, \delta_n, \delta) \in \mathrm{Row}(T)\}$. Let $l, m \in B$ and $l \neq m$. Define tables $T(1)$ and $T(l, m)$ as follows: each of the tables $T(1)$ and $T(l, m)$ contains n columns labelling by attributes $f_1, \ldots, f_n$; $\mathrm{Row}(T(1)) = \{\bar{\delta} : \bar{\delta} \in \mathrm{Row}(T, \beta), |\kappa(\bar{\delta})| = 1\}$; $\mathrm{Row}(T(l, m)) = \{\bar{\delta} : \bar{\delta} \in \mathrm{Row}(T, \beta), l \in \kappa(\bar{\delta}), m \in \kappa(\bar{\delta})\}$; all rows of the considered tables are labelled by the same number 0. It is not difficult to show that

$$N(T) \leq N(T(1)) + \sum_{l,m \in B, l \neq m} N(T(l, m))$$

and $I(T(1)) \leq t+1$. Let $l, m \in B$ and $l \neq m$. Let us show that $I(T(l,m)) \leq t$. Assume the contrary. One can prove that in this case $I(T) \geq t+2$ which is impossible. Therefore $I(T(l,m)) \leq t$. Taking into account that $N(T) = N_\rho(n+1, t+1)$ we obtain $N_\rho(n+1,t+1) \leq N_\rho(n,t+1)+k^2 N_\rho(n,t)$. Hence the inequality (31) holds.

Prove now that

$$N_\rho(n,t) \leq k^{2t} n^t \qquad (32)$$

for any $n \in \mathbb{N} \setminus \{0\}$ and $t \in \mathbb{N}$. The inequality (32) will be proven by induction on $n+t$. From inequalities (29) and (30) it follows that the inequality (32) holds if $n = 1$ or $t = 0$. Therefore the inequality (32) holds if $n + t \leq 2$. Assume (32) holds if $n + t \leq r$ where $r \in \mathbb{N}$ and $r \geq 2$. Let $n' \in \mathbb{N} \setminus \{0\}$, $t' \in \mathbb{N}$ and $n' + t' = r + 1$. The inequality (32) holds if $n' = 1$ or $t' = 0$. Let $n' = n + 1$ and $t' = t+1$, where $n \in \mathbb{N} \setminus \{0\}$ and $t \in \mathbb{N}$. Using (31) and the inductive hypothesis obtain

$$N_\rho(n+1,t+1) \leq k^{2(t+1)} n^{t+1} + k^{2(t+1)} n^t \leq k^{2(t+1)} (n+1)^{t+1} .$$

Hence the inequality(32) holds.

Let T be a nonempty table from Tab_ρ. The inequality $N(T) \geq 2^{I(T)}$ is obvious. The inequality $N(T) \leq (k^2 \dim T)^{I(T)}$ follows from (32). □

Bounds on Function N_U. Let $U = (A, B, F)$ be an information system, $z = (\nu, f_1, \ldots, f_n) \in \mathrm{Probl}_U$ and $T = T_U(z)$. Let us give one more definition of the parameter $I(T)$ which is equivalent to the stated above. If $N(T) \leq 1$ then $I(T) = 0$. Let $N(T) \geq 2$. Then $I(T)$ is the maximal $m \in \{1, \ldots, n\}$ such that there exist numbers $i(1), \ldots, i(m) \in \{1, \ldots, n\}$ and sets $G_1, \ldots, G_m \subseteq B$ possessing following properties: $|G_1| = \ldots = |G_m| = 2$ and for any $\delta_1 \in G_1, \ldots, \delta_m \in G_m$ the equation system

$$\{f_{i(1)}(x) = \delta_1, \ldots, f_{i(m)}(x) = \delta_m\}$$

is compatible on the set A.

Theorem 4.7. *Let $U = (A, B, F)$ be an information system and $|B| = k$. Then the following statements hold:*

a) if there exists a constant $c \in \mathbb{N}$ such that $I(T) \leq c$ for any table $T \in \mathrm{Tab}_U$ then $N_U(n) \leq (k^2 n)^c$ for any $n \in \mathbb{N} \setminus \{0\}$;
b) if there exists no constant $c \in \mathbb{N}$ such that $I(T) \leq c$ for any table $T \in \mathrm{Tab}_U$ then $N_U(n) \geq 2^n$ for any $n \in \mathbb{N} \setminus \{0\}$.

Proof. a) Let $c \in \mathbb{N}$ and let $I(T) \leq c$ for any table $T \in \mathrm{Tab}_U$. Let $n \in \mathbb{N} \setminus \{0\}$, $T \in \mathrm{Tab}_U$ and $\dim T \leq n$. Using Theorem 4.6 we obtain $N(T) \leq (k^2 n)^c$. Hence $N_U(n) \leq (k^2 n)^c$.

b) Assume that no constant $c \in \mathbb{N}$ exists such that $I(T) \leq c$ for any table $T \in \mathrm{Tab}_U$. Let $n \in \mathbb{N} \setminus \{0\}$. Then there exists a table $T \in \mathrm{Tab}_U$ such that $I(T) \geq n$. Let $I(T) = m$. Then there exist a word $\beta = f_{i(1)} \ldots f_{i(m)} \in RO(T)$ and sets

$G_1, \ldots, G_m \subseteq B$ such that $|G_1| = \ldots = |G_m| = 2$ and $G_1 \times \ldots \times G_m \subseteq \mathrm{Row}(T, \beta)$. Define a problem $z \in \mathrm{Probl}_U$ as follows: $z = (\nu, f_{i(1)}, \ldots, f_{i(n)})$ where $\nu : B^n \to \{0\}$. Denote $T' = T_U(z)$. One can show that $\dim T' = I(T') = n$. Using Theorem 4.6 we obtain $N(T') \geq 2^n$. Therefore $N_U(n) \geq 2^n$. □

Example 4.3. Let m, t be natural numbers, and $U(\mathbb{R}, m), U(\mathbb{R}, m, t)$ be information systems defined in Sect. 4.1. We will show in Sect. 6.5 (see Note 6.2) that the function I is not bounded from above on the set of tables $\mathrm{Tab}_{U(\mathbb{R},m)}$, but it is bounded from above on the set $\mathrm{Tab}_{U(\mathbb{R},m,t)}$.

4.6 Algorithm of Decision Table Construction

In this subsection an algorithm of decision table construction is considered and their complexity characteristics are studied.

Let $U = (A, B, F)$, where $F = \{f_i : i \in \mathbb{N}\}$ and $|B| = k$, be an enumerated information system. Denote $\rho = (F, B)$. Let the problem $\mathrm{Ex}(U)$ be solvable, and $\mathcal{E}$ be an algorithm solving the problem $\mathrm{Ex}(U)$.

Description of Algorithm of Decision Table Construction. Describe an algorithm $\mathcal{IE}$ which for a given problem $z \in \mathrm{Probl}_U$ constructs the table $\mathcal{IE}(z) = T(z)$. Let $z = (\nu, f_{i(1)}, \ldots, f_{i(n)})$.

1-*st Step.* Construct the tree containing only one node. Label this node by the word λ and proceed to the second step.

Suppose t steps have already been made. Denote by D the labelled finite rooted directed tree built on the step t.

$(t+1)$-*th Step.* Let every terminal node in the tree D be labelled by n-tuple from B^n. Define the table $T \in \mathrm{Tab}_\rho$ as follows: T contains n columns labelling by attributes $f_{i(1)}, \ldots, f_{i(n)}$, the set $\mathrm{Row}(T)$ coincides with the set of n-tuples attached to terminal nodes of the tree D, and ν_T is the restriction of the mapping ν to the set $\mathrm{Row}(T)$. Set $\mathcal{IE}(z) = T$. Algorithm $\mathcal{IE}$ stops.

Let not all terminal nodes in the tree D be labelled by n-tuples from B^n. Choose a terminal node w in the tree D which is labelled by a word from Ω_ρ. Let the node w be labelled by the word α. Let the length of α be equal to n and let $\alpha = (f_{i(1)}, \delta_1) \ldots (f_{i(n)}, \delta_n)$. Instead of the word α we mark the node w by n-tuple $(\delta_1, \ldots, \delta_n)$ and proceed to the $(t+2)$-th step. Let the length r of the word α be less than n and $\alpha = (f_{i(1)}, \delta_1) \ldots (f_{i(r)}, \delta_r)$. (If $r = 0$ then $\alpha = \lambda$.) By applying k times the algorithm $\mathcal{E}$ we construct the set $B(\alpha) = \{\delta : \delta \in B, \mathrm{Sol}_U(\alpha(f_{i(r+1)}, \delta)) \neq \emptyset\}$. Erase the label α at node w. For every $\delta \in B(\alpha)$ add to the tree D a node $w(\delta)$. Draw the edge from the node w to the node $w(\delta)$ and mark the node $w(\delta)$ by the word $\alpha(f_{i(r+1)}, \delta)$. Proceed to the $(t+2)$-th step.

Complexity Parameters of Decision Table Construction Algorithm. For $z \in \mathrm{Probl}_U$ denote by $S^{\mathcal{IE}}(z)$ the number of steps made by algorithm $\mathcal{IE}$ to construct the table $\mathcal{IE}(z)$, and by $C^{\mathcal{IE}}(z)$ the number of calls of algorithm $\mathcal{E}$ by algorithm $\mathcal{IE}$ while constructing the table $\mathcal{IE}(z)$.

Consider dependencies of values $S^{\mathcal{IE}}(z)$ and $C^{\mathcal{IE}}(z)$ on the value $\dim z$ and on the value of $N(T(z))$. Note that $\dim z = \dim T(z)$.

Proposition 4.1. *Let $U = (A, B, F)$, where $F = \{f_i : i \in \mathbb{N}\}$ and $|B| = k$, be an enumerated information system, the problem* Ex(U) *be solvable, and $\mathcal{E}$ be an algorithm solving the problem* Ex(U)*. Then $\mathcal{IE}(z) = T(z)$ for any problem $z \in \mathrm{Probl}_U$, and the following inequalities hold:*

$$2 + N(T(z)) \leq S^{\mathcal{IE}}(z) \leq 2 + (\dim z + 1)N(T(z)) ,$$
$$N(T(z)) \leq C^{\mathcal{IE}}(z) \leq k(\dim z + 1)N(T(z)) .$$

Proof. Let $z \in \mathrm{Probl}_U$. Simple analysis of the algorithm $\mathcal{IE}$ shows that $\mathcal{IE}(z) = T(z)$.

Let t be the number of steps made by algorithm $\mathcal{IE}$ in process of construction of the table $\mathcal{IE}(z)$. Denote by D the tree constructed on the $(t-1)$-th step, and by $L_a(D)$ the number of nodes in the tree D. One can show that $t = L_a(D) + 2$. Obviously, the number of terminal nodes in the tree D is equal to $N(T(z))$, and every complete path in D contains exactly $\dim z + 1$ nodes. Therefore

$$N(T(z)) \leq L_a(D) \leq (\dim z + 1)N(T(z)) .$$

Hence

$$2 + N(T(z)) \leq S^{\mathcal{IE}}(z) \leq 2 + (\dim z + 1)N(T(z)) .$$

Obviously, every step of the algorithm $\mathcal{IE}$, with the exception of the first and the last, includes at most k calls of the algorithm $\mathcal{E}$. The first and the last steps of algorithm $\mathcal{IE}$ do not contain calls of algorithm $\mathcal{E}$. Hence $C^{\mathcal{IE}}(z) \leq k(\dim z + 1)N(T(z))$. Let $L_w(D)$ be the number of nonterminal nodes in D. Obviously, the number of steps containing exactly k calls of algorithm $\mathcal{E}$ by algorithm $\mathcal{IE}$ is equal to $L_w(D)$. Let $L_t(D)$ be the number of terminal nodes in the tree D. One can show that $L_t(D) \leq kL_w(D)$. Taking into account that $L_t(D) = N(T(z))$ we obtain $L_w(D) \geq N(T(z))/k$. Therefore $C^{\mathcal{IE}}(z) \geq N(T(z))$. □

Consider dependence of values $S^{\mathcal{IE}}(z)$ and $C^{\mathcal{IE}}(z)$ on $\dim z$.

Define the functions $S_U^{\mathcal{IE}} : \mathbb{N} \setminus \{0\} \to \mathbb{N}$ and $C_U^{\mathcal{IE}} : \mathbb{N} \setminus \{0\} \to \mathbb{N}$ as follows:

$$S_U^{\mathcal{IE}}(n) = \max\{S^{\mathcal{IE}}(z) : z \in \mathrm{Probl}_U, \dim z \leq n\} ,$$
$$C_U^{\mathcal{IE}}(n) = \max\{C^{\mathcal{IE}}(z) : z \in \mathrm{Probl}_U, \dim z \leq n\} .$$

From Proposition 4.1 it follows that for any $n \geq 1$ the values $S_U^{\mathcal{IE}}(n)$ and $C^{\mathcal{IE}}(z)$ are definite.

Using Theorem 4.7 and Proposition 4.1 it is not difficult to prove the following statement.

Theorem 4.8. *Let $U = (A, B, F)$, where $|B| = k$, be an enumerated information system, the problem* Ex(U) *be solvable, and $\mathcal{E}$ be an algorithm solving the problem* Ex(U)*. Then the following statements hold:*

a) *if there exists a constant* $c \in \mathbb{N}$ *such that for any table* $T \in \mathrm{Tab}_U$ *the inequality* $I(T) \le c$ *holds then* $S_U^{\mathcal{IE}}(n) \le 2 + k^{2c}n^c(n+1)$ *and* $C_U^{\mathcal{IE}}(n) \le k^{2c+1}n^c(n+1)$ *for any natural* n*;*
b) *if there does not exist a constant* $c \in \mathbb{N}$ *such that* $I(T) \le c$ *for any table* $T \in \mathrm{Tab}_U$ *then* $S_U^{\mathcal{IE}}(n) \ge 2^n$ *and* $C_U^{\mathcal{IE}}(n) \ge 2^n$ *for any natural* n*.*

Corollary 4.2. *Let* U *be an enumerated information system, the problem* $\mathrm{Ex}(U)$ *be solvable, and* $\mathcal{E}$ *be an algorithm solving the problem* $\mathrm{Ex}(U)$*. Then the following three statements are equivalent:*

a) *there exists a constant* $c \in \mathbb{N}$ *such that for any table* $T \in \mathrm{Tab}_U$ *the inequality* $I(T) \le c$ *holds;*
b) *there exists a polynomial* p_0 *such that for any problem* $z \in \mathrm{Probl}_U$ *the number of steps made by algorithm* $\mathcal{IE}$ *is bounded from above by* $p_0(\dim z)$*;*
c) *there exists a polynomial* p_1 *such that for any problem* $z \in \mathrm{Probl}_U$ *the number of calls of algorithm* $\mathcal{E}$ *by algorithm* $\mathcal{IE}$ *is bounded from above by* $p_1(\dim z)$*.*

4.7 Greedy Algorithm of Decision Tree Construction

Let $U = (A, B, F)$ be an enumerated information system, $\rho = (F, B)$, the problem $\mathrm{Ex}(U)$ be solvable, and $\mathcal{E}$ be an algorithm solving the problem $\mathrm{Ex}(U)$. Let ψ be a computable weight function for U. In this subsection we consider the algorithm $\mathcal{VE}_\psi$ which for a given problem $z \in \mathrm{Probl}_U$ constructs the decision tree $\mathcal{VE}_\psi(z)$ with following properties: the decision tree $\mathcal{VE}_\psi(z)$ solves the problem z and $\mathrm{At}(\mathcal{VE}_\psi(z)) \subseteq \mathrm{At}(z)$.

Describe the work of the algorithm $\mathcal{VE}_\psi$. Let $z \in \mathrm{Probl}_U$.

Using algorithm $\mathcal{IE}$ from Sect. 4.6 we construct the decision table $T(z)$. Next, using algorithm $\mathcal{V}_{\rho,\psi}$ from Sect. 3.4 we construct the decision tree $\mathcal{V}_{\rho,\psi}(T(z))$ for the table $T(z)$. Then $\mathcal{VE}_\psi(z) = \mathcal{V}_{\rho,\psi}(T(z))$.

Obviously, $\mathrm{At}(\mathcal{VE}_\psi(z)) \subseteq \mathrm{At}(z)$. From Theorem 4.1 it follows that the decision tree $\mathcal{VE}_\psi(z)$ solves the problem z.

Bounds on Complexity of Algorithm $\mathcal{VE}_\psi$. Complexity of algorithm $\mathcal{IE}$ which for a given problem $z \in \mathrm{Probl}_U$ constructs the decision table $T(z)$ was investigated in previous subsection. In this subsection we consider bounds on the number of steps made by algorithm $\mathcal{V}_{\rho,\psi}$ in process of the tree $\mathcal{V}_{\rho,\psi}(T(z))$ construction.

For $z \in \mathrm{Probl}_U$ we denote by $S^{\mathcal{V},\psi}(z)$ the number of steps which the algorithm $\mathcal{V}_{\rho,\psi}$ makes in the process of the tree $\mathcal{V}_{\rho,\psi}(T(z))$ construction. The function $S_U^{\mathcal{V},\psi} : \mathbb{N} \setminus \{0\} \to \mathbb{N}$ is defined as follows:

$$S_U^{\mathcal{V},\psi}(n) = \max\{S^{\mathcal{V},\psi}(z) : z \in \mathrm{Probl}_U, \dim z \le n\} \ .$$

Theorem 3.11 allows to conclude that the value $S_U^{\mathcal{V},\psi}(n)$ is definite for any $n \in \mathbb{N} \setminus \{0\}$.

Theorem 4.9. *Let* $U = (A, B, F)$ *be an enumerated information system for which* $|B| = k$*,* $\rho = (F, B)$*, the problem* $\mathrm{Ex}(U)$ *be solvable, and* ψ *be a computable weight function for* U*. Then the following statements hold:*

a) if there exists a constant $c \in \mathbb{N}$ such that $I(T) \leq c$ for any table $T \in \mathrm{Tab}_U$ then $S_U^{\mathcal{V},\psi}(n) \leq 2(k^2 n)^c + 2$ for any $n \in \mathbb{N} \setminus \{0\}$;
b) if there does not exist a constant $c \in \mathbb{N}$ such that $I(T) \leq c$ for any table $T \in \mathrm{Tab}_U$ then $S_U^{\mathcal{V},\psi}(n) \geq 2^n$ for any $n \in \mathbb{N} \setminus \{0\}$.

Proof. a) Let there exists a constant $c \in \mathbb{N}$ such that $I(T) \leq c$ for any table $T \in \mathrm{Tab}_U$. Let $n \in \mathbb{N} \setminus \{0\}, z \in \mathrm{Probl}_U$ and $\dim z \leq n$. From Theorem 3.11 it follows that $S_U^{\mathcal{V},\psi}(n) \leq 2N(T(z)) + 2$. Using Theorem 4.7 we obtain $N(T(z)) \leq (k^2 n)^c$. Therefore $S^{\mathcal{V},\psi}(z) \leq 2(k^2 n)^c + 2$. Hence $S_U^{\mathcal{V},\psi}(n) \leq 2(k^2 n)^c + 2$.
b) Let there do not exist a constant $c \in \mathbb{N}$ such that $I(T) \leq c$ for any table $T \in \mathrm{Tab}_U$. Let $n \in \mathbb{N} \setminus \{0\}$. Using Theorem 4.7 we conclude that there exists a table $T \in \mathrm{Tab}_U$ such that $\dim T \leq n$ and $N(T) \geq 2^n$. Let $z \in \mathrm{Probl}_U$, $z = (\nu, f_{i(1)}, \ldots, f_{i(m)})$ and $T = T(z)$. Obviously, $\dim z = m \leq n$. Let $\nu_1 : B^m \rightarrow \mathbb{N}$ be a mapping such that for any $\bar{\delta}_1, \bar{\delta}_2 \in B^m$ from $\bar{\delta}_1 \neq \bar{\delta}_2$ follows $\nu_1(\bar{\delta}_1) \neq \nu_1(\bar{\delta}_2)$. Denote $z_1 = (\nu_1, f_{i(1)}, \ldots, f_{i(m)})$ and $T_1 = T(z_1)$. Obviously, the tree $\mathcal{V}_{\rho,\psi}(T_1)$ contains at least 2^n terminal nodes. It was noted in the proof of Theorem 3.11 that the number of steps made by the algorithm $\mathcal{V}_{\rho,\psi}$ to construct the tree $\mathcal{V}_{\rho,\psi}(T_1)$ is equal to $L_a(\mathcal{V}_{\rho,\psi}(T_1))+2$, where $L_a(\mathcal{V}_{\rho,\psi}(T_1))$ is the number of nodes in the tree $\mathcal{V}_{\rho,\psi}(T_1)$. Therefore $S^{\mathcal{V},\psi}(z_1) \geq 2^n$. Taking into account that $\dim z_1 \leq n$ we obtain $S_U^{\mathcal{V},\psi}(n) \geq 2^n$. □

Bounds on Accuracy of Algorithm $\mathcal{VE}_\psi$. Define partial function $H_{U,\psi}^{l,\mathcal{V}} : \mathbb{N} \rightarrow \mathbb{N}$ as follows:

$$H_{U,\psi}^{l,\mathcal{V}}(n) = \max(\psi(\mathcal{VE}(z)) : z \in \mathrm{Probl}_U, \psi(z) \leq n\} \ .$$

Comparison of functions $H_{U,\psi}^{l,\mathcal{V}}$ and $H_{U,\psi}^{l}$ allows to estimate the accuracy of the algorithm $\mathcal{VE}_\psi$.

Theorem 4.10. *Let $U = (A, B, F)$ be an enumerated information system for which the problem $\mathrm{Ex}(U)$ is solvable, and ψ be a computable weight function for U. Then the following statements hold:*

a) $\mathrm{Dom}(H_{U,\psi}^{l,\mathcal{V}}) = \mathrm{Dom}(H_{U,\psi}^{l})$ and $H_{U,\psi}^{l}(n) \leq H_{U,\psi}^{l,\mathcal{V}}(n) \leq n$ for any $n \in \mathrm{Dom}(H_{U,\psi}^{l})$;
b) if U is a finite information system then $H_{U,\psi}^{l}(n) = O(1)$ and $H_{U,\psi}^{l,\mathcal{V}}(n) = O(1)$;
c) if U is an infinite information system satisfying the condition of reduction relatively to ψ then $H_{U,\psi}^{l}(n) = \Theta(\log_2 n)$ and $H_{U,\psi}^{l,\mathcal{V}}(n) = \Theta(\log_2 n)$;
d) if U is an infinite information system which does not satisfy the condition of reduction relatively to ψ then $H_{U,\psi}^{l}(n) = n$ for infinitely many natural n and $H_{U,\psi}^{l,\mathcal{V}}(n) = O(H_{U,\psi}^{l}(n)^3)$;
e) if U is an infinite information system which does not satisfy the condition of reduction relatively to ψ and $\psi = h$ then $H_{U,\psi}^{l,\mathcal{V}}(n) = H_{U,\psi}^{l}(n) = n$ for any natural n.

Proof. Let $\rho = (F, B)$ and $F = \{f_i : i \in \mathbb{N}\}$. Using Theorem 3.8 one can show that for any nonempty table T from Tab_ρ the following inequality holds:

$$\psi(\mathcal{V}_{\rho,\psi}(T)) \le 2(M_{\rho,\psi}(T))^2 \log_2 N(T) \ . \tag{33}$$

a) Let $n \notin \mathrm{Dom}(H^l_{U,\psi})$. One can show that $n \notin \mathrm{Dom}(H^{l,\mathcal{V}}_{U,\psi})$. Let $n \in \mathrm{Dom}(H^l_{U,\psi})$, $z \in \mathrm{Probl}_U$ and $\psi(z) \le n$. Denote $T = T(z)$. Using description of the algorithm $\mathcal{V}_{\rho,\psi}$ it is not difficult to show that working nodes of any complete path in the tree $\mathcal{V}_{\rho,\psi}(T)$ are labelled by pairwise distinct attributes from the set $\mathrm{At}(T)$. Therefore $\psi(\mathcal{V}_{\rho,\psi}(T)) \le n$. Hence the value $H^{l,\mathcal{V}}_{U,\psi}(n)$ is definite and $H^{l,\mathcal{V}}_{U,\psi}(n) \le n$. The inequality $H^l_{U,\psi}(n) \le H^{l,\mathcal{V}}_{U,\psi}(n)$ is obvious.

b) Let U be a finite information system, and $z \in \mathrm{Probl}_U$. Since working nodes of any complete path in the tree $\mathcal{V}_{\rho,\psi}(T(z))$ are labelled by pairwise distinct attributes, we have $\psi(\mathcal{V}_{\rho,\psi}(T(z))) \le \sum_{f_i \in F} \psi(f_i)$. Therefore $H^{l,\mathcal{V}}_{U,\psi}(n) = O(1)$. From Theorem 4.3 it follows that $H^l_{U,\psi}(n) = O(1)$.

c) Let U be an infinite information system satisfying the condition of reduction relatively to ψ. Then there exists a constant $m \in \mathbb{N}$ such that $Q_\psi(T) \le m$ for any table $T \in \mathrm{Tab}_U$. Therefore $\hat{Q}_\psi(T) \le m$ for any table $T \in \mathrm{Tab}_U$. Let $n \in \mathrm{Dom}(H^l_{U,\psi})$. Let $z \in \mathrm{Probl}_U$ and $\psi(z) \le n$. Denote $T = T(z)$. Using Lemma 4.2 obtain $M_{\rho,\psi}(T) \le 2m$. Since $\psi(z) \le n$, the inequality $\dim T \le n$ holds. Using Lemma 4.3 we obtain $N(T) \le (kn)^m$. From the inequality (33) it follows that $\psi(\mathcal{V}_{\rho,\psi}(T)) \le 8m^3 \log_2 n + 8m^3 \log_2 k$. Therefore $H^{l,\mathcal{V}}_{U,\psi}(n) = O(\log_2 n)$. From Theorem 4.3 it follows that $H^l_{U,\psi}(n) = \Theta(\log_2 n)$. Taking into account that $H^l_{U,\psi}(n) \le H^{l,\mathcal{V}}_{U,\psi}(n)$ we obtain that $H^{l,\mathcal{V}}_{U,\psi}(n) = \Theta(\log_2 n)$.

d) Let U be an infinite information system which does not satisfy the condition of reduction relatively to ψ. From Theorem 4.3 it follows that $H^l_{U,\psi}(n) = n$ for infinitely many natural n. Let $n \in \mathrm{Dom}(H^l_{U,\psi})$, $z \in \mathrm{Probl}_U$, $\psi(z) \le n$ and $z = (\nu, f_{i(1)}, \dots, f_{i(m)})$. Denote $T = T(z)$. Using Corollary 4.1 we obtain $\psi_\rho(T) = \psi^l_U(z)$. Hence $\psi_\rho(T) \le H^l_{U,\psi}(n)$. From Theorem 3.1 it follows that $M_{\rho,\psi}(T) \le \psi_\rho(T)$. Therefore

$$M_{\rho,\psi}(T) \le H^l_{U,\psi}(n) \ . \tag{34}$$

Let us show that

$$\log_2 N(T) \le (\log_2 k) H^l_{U,\psi}(n) \ . \tag{35}$$

Let $\nu_1 : B^m \to \mathbb{N}$ be a mapping such that for any $\bar{\delta}_1, \bar{\delta}_2 \in B^m$ from $\bar{\delta}_1 \ne \bar{\delta}_2$ follows $\nu_1(\bar{\delta}_1) \ne \nu_1(\bar{\delta}_2)$. Denote $z_1 = (\nu_1, f_{i(1)}, \dots, f_{i(m)})$ and $T_1 = T(z_1)$. Obviously, $\psi(z_1) \le n$. Using Corollary 4.1 obtain $\psi_\rho(T_1) \le H^l_{U,\psi}(n)$. Evidently, T_1 is a diagnostic table. Using Corollary 3.1 obtain $\log_k N(T) \le \psi_\rho(T_1)$. Evidently, $N(T) = N(T_1)$. Therefore $\log_k N(T) \le H^l_{U,\psi}(n)$. From the last inequality the inequality (35) follows. From (33)–(35) the inequality $\psi(\mathcal{V}_{\rho,\psi}(T)) \le (2 \log_2 k)(H^l_{U,\psi}(n))^3$ follows. Therefore $H^{l,\mathcal{V}}_{U,\psi}(n) = O(H^l_{U,\psi}(n)^3)$.

e) Let U be an infinite information system which does not satisfy the condition of reduction relatively to ψ and $\psi = h$. Let $n \in \mathrm{Dom}(H^l_{U,\psi})$. From the statement a) and Theorem 4.3 it follows that $H^{l,\mathcal{V}}_{U,\psi}(n) = H^l_{U,\psi}(n) = n$. □

4.8 Algorithm of Decision Tree Optimization

Let $U = (A, B, F)$ be an enumerated information system, $\rho = (F, B)$, the problem $\mathrm{Ex}(U)$ be solvable, and $\mathcal{E}$ be an algorithm solving the problem $\mathrm{Ex}(U)$. Let ψ be a computable weight function for U. In this subsection we consider the algorithm $\mathcal{WE}_\psi$ which for a given problem $z \in \mathrm{Probl}_U$ constructs the decision tree $\mathcal{WE}_\psi(z)$ with following properties: the decision tree $\mathcal{WE}_\psi(z)$ solves the problem z, $\mathrm{At}(\mathcal{WE}_\psi(z)) \subseteq \mathrm{At}(z)$ and $\psi(\mathcal{WE}_\psi(z)) = \psi^l_U(z)$.

Describe the work of the algorithm $\mathcal{WE}_\psi$. Let $z \in \mathrm{Probl}_U$.

Using algorithm $\mathcal{IE}$ from Sect. 4.6 we construct the decision table $T(z)$. Next, using algorithm $\mathcal{W}_{\rho,\psi}$ from Section 3.5 we construct the decision tree $\mathcal{W}_{\rho,\psi}(T(z))$ for the table $T(z)$. Then $\mathcal{WE}_\psi(z) = \mathcal{W}_{\rho,\psi}(T(z))$.

Obviously, $\mathrm{At}(\mathcal{WE}_\psi(z)) \subseteq \mathrm{At}(z)$. From Theorem 3.12 follows $\psi(\mathcal{WE}_\psi(z)) = \psi_\rho(T(z))$. Using Theorem 4.1 we conclude that the decision tree $\mathcal{WE}_\psi(z)$ solves the problem z and $\psi(\mathcal{WE}_\psi(z)) = \psi^l_U(z)$.

Bounds on Complexity of Algorithm $\mathcal{WE}_\psi$. Complexity of algorithm $\mathcal{IE}$ which for a given problem $z \in \mathrm{Probl}_U$ constructs the decision table $T(z)$ was investigated in Sect. 4.6. In this subsection we consider bounds on the number of steps made by algorithm $\mathcal{W}_{\rho,\psi}$ in process of the tree $\mathcal{W}_{\rho,\psi}(T(z))$ construction.

For $z \in \mathrm{Probl}_U$ we denote by $S^{\mathcal{W},\psi}(z)$ the number of steps which the algorithm $\mathcal{W}_{\rho,\psi}$ makes in process of the tree $\mathcal{W}_{\rho,\psi}(T(z))$ construction. The function $S^{\mathcal{W},\psi}_U : \mathbb{N} \setminus \{0\} \to \mathbb{N}$ is defined as follows:

$$S^{\mathcal{W},\psi}_U(n) = \max\{S^{\mathcal{W},\psi}(z) : z \in \mathrm{Probl}_U, \dim z \le n\} \ .$$

A compatible system of equations over U is called *uncancellable* if each proper subsystem of this system is not equivalent to the system.

Theorem 4.11. *Let $U = (A, B, F)$ be an enumerated information system for which $|B| = k$, $\rho = (F, B)$, the problem $\mathrm{Ex}(U)$ is solvable, and ψ be a computable weight function for U. Then the following statements hold:*

a) if the information system U satisfies the condition of reduction relatively to h, and each compatible equation system over U has an equivalent subsystem with at most r equations then $S^{\mathcal{W},\psi}_U(n) \le (nk)^r + 2$ for any natural n;
b) if the information system U does not satisfy the condition of reduction relatively to h then $S^{\mathcal{W},\psi}_U(n) \ge 2^n + 1$ for any natural n.

Proof. Let, for the definiteness, $B = \{0, \ldots, k-1\}$. Let $n \in \mathbb{N} \setminus \{0\}, m \le n$ and $z = (\nu, f_1, \ldots, f_m) \in \mathrm{Probl}_U$. From Theorem 3.12 it follows that for decision table $T(z)$ the algorithm $\mathcal{W}_{\rho,\psi}$ makes exactly $|\mathrm{Sep}(T(z))| + 1$ steps. One can show that the value $|\mathrm{Sep}(T(z))|$ coincides with the number of pairwise nonequivalent compatible subsystems of the system of equations $\{f_1(x) = 0, \ldots, f_m(x) = 0, \ldots, f_1(x) = k-1, \ldots, f_m(x) = k-1\}$ including the empty system (the set of solutions of the empty system is equal to A).

a) Let U satisfy the condition of reduction relatively to h, and each compatible equation system over U have an equivalent subsystem with at most r equations. Then $|\mathrm{Sep}(T(z))| \leq m^r k^r + 1$, and hence $S_U^{\mathcal{W},\psi}(n) \leq (nk)^r + 2$.
b) Let U do not satisfy the condition of reduction. Then there exists an uncancellable system of equations over U with at least n equations. Evidently, each subsystem of this system is uncancellable. Therefore there exists an uncancellable system over U with n equations. Denote this system by Δ. We prove that every two different subsystems Δ_1 and Δ_2 of the system Δ are nonequivalent. Assume the contrary. Then subsystems $\Delta \setminus (\Delta_1 \setminus \Delta_2)$ and $\Delta \setminus (\Delta_2 \setminus \Delta_1)$ are equivalent to Δ, and at least one of them is a proper subsystem of Δ which is impossible. Let $\Delta = \{f_1(x) = \delta_1, \ldots, f_n(x) = \delta_n\}$. Let $z = (\nu, f_1, \ldots, f_m)$ where $\nu : B^n \to \mathbb{Z}$ and $\nu(\bar{\delta}_1) \neq \nu(\bar{\delta}_2)$ if $\bar{\delta}_1 \neq \bar{\delta}_2$. Then $|\mathrm{Sep}(T(z))| \geq 2^n$. Hence $S_U^{\mathcal{W},\psi}(n) \geq 2^n + 1$. □

5 Global Approach to Investigation of Decision Trees

Global approach to investigation of decision trees considered in this section is based on the assumption that any attributes from information system can be used by decision trees solving a problem.

Remind basic notions defined in the previous section.

An information system $U = (A, B, F)$ consists of a set A (the universe) and a set of attributes F which are defined on A and have values from finite set B. In this section we will assume that different attributes from F (attributes with different names) are different as functions.

A weight function ψ assigns a weight (a natural number) to each attribute. This weight characterizes the complexity of attribute value computation.

The notion of a problem over the information system defines as follows. We take finite number of attributes $f_1, \ldots, f_n$ from F. These attributes create a partition of the set A into classes (for each class values of the attributes are constant on elements from the class). These classes are numbered such that different classes can have the same number. The number of a class is the decision corresponding to elements of the class. For a given element a from A it is required to recognize the number of a class which contains a. The weight of a problem is the total weight of attributes from the problem description.

As algorithms for problem solving we consider decision trees which use any attributes from the set F. As time complexity measure we consider weighted depth of decision trees.

The global Shannon function corresponds to each pair (*information system, weight function*) which characterizes the growth in the worst case of minimal weighted depth of decision trees with the growth of problem weight.

The section consists of five subsections. In the first subsection global Shannon functions for infinite information systems and depth are considered. In the second subsection global Shannon functions for finite two-valued information systems and depth are investigated. In the third subsection global Shannon functions for infinite information systems and arbitrary weight functions are studied. In the fourth subsection relationships among three algorithmic problems are considered: the problem of compatibility of equation system and two problems of decision tree global optimization (the problem of construction of globally optimal decision tree and the problem of computation of globally optimal decision tree complexity). In the fifth subsection all computable weight functions are described for each of which for any enumerated information system the problem of compatibility of equation system is solvable if and only if two problems of decision tree global optimization are solvable.

5.1 Global Shannon Functions for Infinite Information Systems and Depth

Let $U = (A, B, F)$ be an information system. For a problem $z = (\nu, f_1, \ldots, f_n)$ over U we denote by $h_U^g(z)$ the minimal depth of a decision tree over U which solves the problem z. We will consider the relationships between the parameters $h_U^g(z)$ and $\dim z = n$. Recall that one can interpret the value $\dim z$ for the problem z as the depth of the decision tree which solves the problem z in trivial way by computing sequentially the values of the attributes $f_1, \ldots, f_n$. We define the function $H_{U,h}^g : \mathbb{N} \setminus \{0\} \to \mathbb{N}$ in the following way:

$$H_{U,h}^g(n) = \max\{h_U^g(z) : z \in \mathrm{Probl}_U, \dim z \leq n\}$$

for any $n \in \mathbb{N} \setminus \{0\}$. The value $H_{U,h}^g(n)$ is the unimprovable upper bound on the value $h_U^g(z)$ for problems $z \in \mathrm{Probl}_U$ such that $\dim z \leq n$. The function $H_{U,h}^g$ will be called *the global Shannon function* for the information system U and the weight function h.

We will show that for an arbitrary information system U either $H_{U,h}^g(n) = O(1)$, or $H_{U,h}^g(n) = \Omega(\log_2 n)$ and $H_{U,h}^g(n) = O((\log_2 n)^{1+\varepsilon})$ for any $\varepsilon > 0$, or $H_{U,h}^g(n) = n$ for any $n \in \mathbb{N} \setminus \{0\}$.

The first type of behavior ($H_{U,h}^g(n) = O(1)$) realizes only for finite information systems. The second type of behavior ($H_{U,h}^g(n) = \Omega(\log_2 n)$ and $H_{U,h}^g(n) = O((\log_2 n)^{1+\varepsilon})$) is most interesting for us: for an arbitrary problem with large enough dimension the depth of the globally optimal decision tree is essentially less than the depth of the trivial decision tree. The third type of behavior ($H_{U,h}^g(n) = n$ for each $n \in \mathbb{N} \setminus \{0\}$) is bad for us: for problems of arbitrary dimension in the worst case the depth of the globally optimal decision tree is equal to the depth of the trivial decision tree.

Now we consider the criterions of the global Shannon function $H_{U,h}^g$ behavior. Define the notion of *I-dimension* of information system U. A finite subset $\{f_1, \ldots, f_p\}$ of the set F is called *an I-set* if there exists two-element subsets $B_1, \ldots, B_p$ of the set B such that for any $\delta_1 \in B_1, \ldots, \delta_p \in B_p$ the system of equations

$$\{f_1(x) = \delta_1, \ldots, f_p(x) = \delta_p\} \tag{36}$$

is compatible on the set A (has a solution from A). If for any natural p there exists a subset of the set F, which cardinality is equal to p and which is an

I-set, then we will say that the information system U has infinite I-dimension. Otherwise, I-dimension of U is the maximal cardinality of a subset of F, which is an I-set. Relationships between I-dimension and Vapnik-Chervonenkis dimension were discussed in [47].

Now we consider the condition of decomposition for the information system U. Let $p \in \mathbb{N}$. A nonempty subset D of the set A will be called (p, U)-set if D coincides with the set of solutions on A of a system of the kind (36) where $f_1, \ldots, f_p \in F$ and $\delta_1, \ldots, \delta_p \in B$ (we admit that among the attributes $f_1, \ldots, f_p$ there are identical ones).

We will say that the information system U satisfies *the condition of decomposition* if there exist numbers $m, t \in \mathbb{N}$ such that every $(m + 1, U)$-set is a union of t sets each of which is (m, U)-set (we admit that among the considered t sets there are identical ones). If the last statement holds we will say that the information system U satisfies *the condition of decomposition with parameters m and t.*

Theorem 5.1. *Let U be an information system. Then the following statements hold:*

a) if U is a finite information system then $H^g_{U,h}(n) = O(1)$;
b) if U is an infinite information system which has finite I-dimension and satisfies the condition of decomposition then $H^g_{U,h}(n) = \Omega(\log_2 n)$ and $H^g_{U,h}(n) = O((\log_2 n)^{1+\varepsilon})$ for any $\varepsilon > 0$;
c) if U is an infinite information system which has infinite I-dimension or does not satisfy the condition of decomposition then $H^g_{U,h}(n) = n$ for any $n \in \mathbb{N} \setminus \{0\}$.

In the following theorem bounds are considered in which instead of ε a function stands that decreases with the growth on n.

Theorem 5.2. *Let U be an infinite information system which has finite I-dimension and satisfies the condition of decomposition. Then there exists positive constant c such that $H^g_{U,h}(n) = O\left((\log_2 n)^{1+\frac{c}{\sqrt{\log_2 \log_2 n}}}\right)$.*

Now we consider an example.

Example 5.1. Let $m, t \in \mathbb{N}$. We denote by $\mathrm{Pol}(m)$ the set of all polynomials which have integer coefficients and depend on variables $x_1, \ldots, x_m$. We denote by $\mathrm{Pol}(m, t)$ the set of all polynomials from $\mathrm{Pol}(m)$ such that the degree of each polynomial is at most t. We define information systems $U(\mathbb{R}, m)$ and $U(\mathbb{R}, m, t)$ as follows: $U(\mathbb{R}, m) = (\mathbb{R}^m, E, F(m))$ and $U(\mathbb{R}, m, t) = (\mathbb{R}^m, E, F(m, t))$ where $E = \{-1, 0, +1\}$, $F(m) = \{\mathrm{sign}(p) : p \in \mathrm{Pol}(m)\}$ and $F(m, t) = \{\mathrm{sign}(p) : p \in \mathrm{Pol}(m, t)\}$. One can prove that the system $U(\mathbb{R}, m)$ has infinite I-dimension, but the system $U(\mathbb{R}, m, t)$ has finite I-dimension and satisfies the condition of decomposition.

The proofs of the theorems are divided into a sequence of lemmas. Subsubsection "Special Lemmas" consists of lemmas connected with the study of infinite information systems, which has finite I-dimension and satisfies the condition of decomposition, on the basis of notions of system saturation and covering of attributes. All the other lemmas are gathered into subsubsection "Common Lemmas".

Common Lemmas. Let $U = (A, B, F)$ be an information system, and $f_1, ..., f_n \in F$. Denote by $N_U(f_1, \ldots, f_n)$ the number of n-tuples $(\delta_1, \ldots, \delta_n) \in B^n$ for which the system of equations

$$\{f_1(x) = \delta_1, \ldots, f_n(x) = \delta_n\}$$

is compatible on A. For an arbitrary natural n let

$$N_U(n) = \max\{N_U(f_1, \ldots, f_n) : f_1, \ldots, f_n \in F\} .$$

The next statement follows immediately from Theorem 4.7.

Lemma 5.1. *Let $U = (A, B, F)$ be an information system. Then the following statements hold:*

a) if the system U has finite I-dimension, which is at most v, then $N_U(n) \leq (|B|^2 n)^v$ for any natural n;
b) if the system U has infinite I-dimension then $N_U(n) \geq 2^n$ for any natural n.

Lemma 5.2. *Let $U = (A, B, F)$ be an infinite information system with finite I-dimension. Then $N_U(n) \geq n + 1$ for any natural n.*

Proof. It is clear that $N_U(n) \geq 2$. Let us show that $N_U(n+1) \geq N_U(n) + 1$ for any natural n. It is clear that $N_U(n+1) \geq N_U(n)$ for any natural n. Assume that there exists natural n such that $N_U(n+1) = N_U(n)$. Using this equality one can prove that F is a finite set which is impossible. Therefore $N_U(n) \geq n+1$ for any natural n. □

Lemma 5.3. *Let $U = (A, B, F)$ be an information system and n be a natural number. Then there exists a problem z over U with $\dim z = n$ such that $h(\Gamma) \geq \log_{|B|} N_U(n)$ for any decision tree Γ over U which solves the problem z.*

Proof. Let $f_1, \ldots, f_n$ be such attributes from F that $N_U(f_1, \ldots, f_n) = N_U(n)$. Let $\nu : B^n \to \mathbb{N}$ and $\nu(\bar{\delta}_1) \neq \nu(\bar{\delta}_2)$ if $\bar{\delta}_1 \neq \bar{\delta}_2$. Consider the problem $z = (\nu, f_1, \ldots, f_n)$. Evidently, there exist elements $a_1, \ldots, a_{N_U(n)} \in A$ such that the values $z(a_1), \ldots, z(a_{N_U(n)})$ are mutually distinct.

Let Γ be a decision tree over U which solves the problem z. It is clear that the number of terminal nodes in Γ is at least $N_U(n)$. It is not difficult to show that the number of terminal nodes in Γ is at most $|B|^{h(\Gamma)}$. Therefore $h(\Gamma) \geq \log_{|B|} N_U(n)$. □

Lemma 5.4. *Let $U = (A, B, F)$ be an infinite information system with finite I-dimension. Then for any natural n there exists a problem z over U with $\dim z = n$ such that for each decision tree Γ over U, which solves the problem z, the inequality $h(\Gamma) \geq \log_{|B|}(n + 1)$ holds.*

Proof. The statement of the lemma follows from Lemmas 5.2 and 5.3. □

Lemma 5.5. *Let $U = (A, B, F)$ be an infinite information system for which there exists natural m such that $H^g_{U,h}(m+1) \leq m$. Then $H^g_{U,h}((m+1)^t) \leq m^t$ for any natural t.*

Proof. By induction on t we will show that $H^g_{U,h}((m+1)^t) \leq m^t$ for any $t \in \mathbb{N} \setminus \{0\}$. Evidently, this inequality holds if $t = 1$. Assume that the considered inequality holds for certain $t \geq 1$. Let us prove that $H^g_{U,h}((m+1)^{t+1}) \leq m^{t+1}$.

Using the inequality $H^g_{U,h}(m+1) \leq m$ one can show that for any problem $z \in \mathrm{Probl}_U$ with $\dim z = m+1$ there exists a decision tree Γ_z over U such that the decision tree Γ_z solves the problem z and every complete path in Γ_z has exactly m working nodes.

Let $z = (\nu, f_1, \dots, f_{(m+1)^{t+1}}) \in \mathrm{Probl}_U$. For $i = 1, \dots, (m+1)^t$ denote $z_i = (\nu_i, f_{(m+1)(i-1)+1}, \dots, f_{(m+1)(i-1)+m+1})$ where $\nu_i : B^{m+1} \to \mathbb{N}$ and $\nu_i(\bar{\delta}_1) \neq \nu_i(\bar{\delta}_2)$ for any $\bar{\delta}_1, \bar{\delta}_2 \in B^{m+1}$ such that $\bar{\delta}_1 \neq \bar{\delta}_2$. Let us describe the work of a decision tree Γ over U which simulates parallel work of the decision trees $\Gamma_{z_1}, \dots, \Gamma_{z_{(m+1)^t}}$ on arbitrary element $a \in A$ and solves the problem z. Let $j \in \{1, \dots, m\}$. Let at the j-th step decision trees $\Gamma_1, \dots, \Gamma_{(m+1)^t}$ compute values on the element a of the attributes $g_1, \dots, g_{(m+1)^t}$ respectively. According to the inductive hypothesis, $H^g_{U,h}((m+1)^t) \leq m^t$. Using this inequality we can easily show that there exists a decision tree Γ' over U which depth is at most m^t and which for a given $b \in A$ recognizes the values $g_1(b), \dots, g_{(m+1)^t}(b)$. When the decision tree Γ simulates the j-th step of work of the decision trees $\Gamma_{z_1}, \dots, \Gamma_{z_{(m+1)^t}}$ on the element a, it works like the decision tree Γ' on the element a. Evidently, the results of the work of decision trees $\Gamma_{z_1}, \dots, \Gamma_{z_{(m+1)^t}}$ on the element a are sufficient to recognize the value $z(a)$ which will be attached to the corresponding terminal node of the decision tree Γ. Thus, the decision tree Γ over U solves the problem z, and the depth of Γ is at most $m \cdot m^t = m^{t+1}$. Therefore $h^g_U(z) \leq m^{t+1}$. Since $h^g_U(z) \leq m^{t+1}$ for an arbitrary problem z from Probl_U with $\dim z = (m+1)^{t+1}$, one can show that $h^g_U(z) \leq m^{t+1}$ for any problem $z \in \mathrm{Probl}_U$ with $\dim z \leq (m+1)^{t+1}$. Therefore $H^g_{U,h}((m+1)^{t+1}) \leq m^{t+1}$. □

Lemma 5.6. *Let $U = (A, B, F)$ be an infinite information system for which there exists natural m such that $H^g_{U,h}(m+1) \leq m$. Then U has finite I-dimension and satisfies the condition of decomposition.*

Proof. Let us show that U has finite I-dimension. Assume the contrary. Using Lemmas 5.1 and 5.3 we conclude that for any natural n there exists a problem z over U with $\dim z = n$ such that $h(\Gamma) \geq \log_{|B|} 2^n = \frac{n}{\log_2 |B|}$ for any decision tree Γ over U which solves z. Therefore $H^g_{U,h}((m+1)^t) \geq \frac{(m+1)^t}{\log_2 |B|}$ for any natural t. By Lemma 5.5, $H^g_{U,h}((m+1)^t) \leq m^t$ for any natural t. It is clear that for sufficiently large t the inequality $m^t < \frac{(m+1)^t}{\log_2 |B|}$ holds, which is impossible. Therefore the information system U has finite I-dimension.

Let us show that U satisfies the condition of decomposition with parameters m and $|B|^m$. Let D be an arbitrary $(m+1, U)$-set. We show that D is a union of $|B|^m$ sets each of which is an (m, U)-set. Let D be the set of solutions on A of the system $\{f_1(x) = \delta_1, \ldots, f_{m+1}(x) = \delta_{m+1}\}$ where $f_1, \ldots, f_{m+1} \in F$ and $\delta_1, \ldots, \delta_{m+1} \in B$. Let $\nu : B^{m+1} \to \mathbb{N}$ and $\nu(\overline{\sigma}_1) \neq \nu(\overline{\sigma}_2)$ if $\overline{\sigma}_1 \neq \overline{\sigma}_2$. Consider the problem $z = (\nu, f_1, \ldots, f_{m+1})$. By assumption, there exists a decision tree Γ over U which solves this problem and for which $h(\Gamma) < m + 1$. Denote by Π the set of all complete paths of Γ in which terminal nodes are labelled by the number $\nu(\delta_1, \ldots, \delta_{m+1})$. Using definition of decision tree, which solves the problem z, one can show that D is the union of the sets of solutions on A of equation systems $\mathrm{Sol}_U(\pi(\xi))$, $\xi \in \Pi$. It is clear that each system $\mathrm{Sol}_U(\pi(\xi))$ has at most m equations. One can show that $|\Pi| \leq |B|^m$. Using these facts we conclude that D is the union of $|B|^m$ sets each of which is an (m, U)-set. Hence U satisfies the condition of decomposition with parameters m and $|B|^m$. □

Lemma 5.7. *Let* $U = (A, B, F)$ *be a finite information system. Then* $H^g_{U,h}(n) = O(1)$.

Proof. It is clear that $h^g_U(z) \leq |F|$ for any problem z over U. Hence $H^g_{U,h}(n) = O(1)$. □

Special Lemmas. Everywhere in this subsubsection we assume that some infinite information system U, which has finite I-dimension and satisfies the condition of decomposition, is fixed. Let $U = (A, B, F)$ and U satisfy the condition of decomposition with parameters m and t. It is not difficult to show that if U satisfies the condition of decomposition with parameters m and t then U satisfies the condition of decomposition with parameters $m + 1$ and t. Later we assume that $m \geq 2$. Choose a natural v such that $v \geq m$ and I-dimension of U is at most v. Set $r = \max\{v + 1, m + t + 1\}$. Denote $k = |B|$.

Notions of Saturation and Covering. Let Φ be a nonempty finite subset of the set F and $\Phi = \{f_1, \ldots, f_p\}$. Denote by $\Sigma_r(A, \Phi)$ the set of all p-tuples $(\sigma_1, \ldots, \sigma_p) \in B^p$ such that all subsystems of equation system

$$\{f_1(x) = \sigma_1, \ldots, f_p(x) = \sigma_p\}$$

that contain at most r equations are compatible on A. Define values of attributes from Φ on elements of the set $\Sigma_r(A, \Phi)$. Let $\overline{\sigma} = (\sigma_1, \ldots, \sigma_p) \in \Sigma_r(A, \Phi)$. Then $f_1(\overline{\sigma}) = \sigma_1, \ldots, f_p(\overline{\sigma}) = \sigma_p$. The information system $W = (\Sigma_r(A, \Phi), B, \Phi)$ will be called *r-saturation* of the information system $V = (A, B, \Phi)$.

Let $Y \in \{V, W\}$, z be a problem over Y, and Γ be a decision tree over Y. We will say about z and Γ that there are a problem and a decision tree *with attributes from* Φ. With a view to distinguishing of information systems V and W we will say that Γ solves z *on the set* A (for the system V) or Γ solves z *on the set* $\Sigma_r(A, \Phi)$ (for the system W).

An arbitrary equation system of the kind $\{f_{i_1}(x) = \delta_1, \ldots, f_{i_d}(x) = \delta_d\}$, where $f_{i_1}, \ldots, f_{i_d} \in \Phi$ and $\delta_1, \ldots, \delta_d \in B$, will be called *$d$-system of equations with attributes from* Φ.

Let $f_1, \ldots, f_n \in F$ and d be a natural number. Let Φ_1, Φ_2 be nonempty finite subsets of the set F, $\Phi_1 \subseteq \Phi_2$ and C be a nonempty set of d-systems of equations with attributes from Φ_1. The pair (Φ_1, C) will be called *d-covering of attributes* $f_1, \ldots, f_n$ *on the set* $\Sigma_r(A, \Phi_2)$ if the following conditions hold:

a) $f_1, \ldots, f_n \in \Phi_1$;
b) each element from $\Sigma_r(A, \Phi_2)$ is a solution of an equation system from C;
c) for each system of equations S from C there exist elements $\delta_1, \ldots, \delta_n \in B$ for which the set of solutions of S on $\Sigma_r(A, \Phi_2)$ is a subset of the set of solutions on $\Sigma_r(A, \Phi_2)$ of the equation system $\{f_1(x) = \delta_1, \ldots, f_n(x) = \delta_n\}$.

Properties of Saturations

Lemma 5.8. *Let* $\Phi = \{f_1, \ldots, f_p\}$ *be a nonempty finite subset of the set* F, d *be a natural number and* S *be a* d*-system of equations with attributes from* Φ. *Then*

a) if the system S *is compatible on* A *then* S *is compatible on* $\Sigma_r(A, \Phi)$;
b) if $d \leq r$ *then the system* S *is compatible on* A *if and only if* S *is compatible on* $\Sigma_r(A, \Phi)$.

Proof. a). Let S be compatible on A and a be an element of A which is a solution of S. It is clear that $\overline{\sigma}(a) = (f_1(a), \ldots, f_p(a))$ belongs to $\Sigma_r(A, \Phi)$ and $\overline{\sigma}(a)$ is a solution of S.

b). Let $d \leq r$, S be compatible on $\Sigma_r(A, \Phi)$, and $\overline{\sigma} = (\sigma_1, \ldots, \sigma_p)$ be an element of $\Sigma_r(A, \Phi)$ which is a solution of S. Consider the system of equations $\{f_1(x) = \sigma_1, \ldots, f_p(x) = \sigma_p\}$. By definition, each subsystem of this system with at most r equations is compatible on A. It is clear that S is a subsystem of the considered system, and contains at most r equations. Therefore S is compatible on A. □

Lemma 5.9. *Let* Φ_1 *and* Φ_2 *be nonempty finite subsets of the set* F, $\Phi_1 \subseteq \Phi_2$ *and* S *be a system of equations with attributes from* Φ_1. *If the system* S *is incompatible on the set* $\Sigma_r(A, \Phi_1)$ *then* S *is incompatible on the set* $\Sigma_r(A, \Phi_2)$.

Proof. Let, for the definiteness, $\Phi_1 = \{f_1, \ldots, f_p\}$, $\Phi_2 = \{f_1, \ldots, f_q\}$ and $S = \{f_1(x) = \delta_1, \ldots, f_n(x) = \delta_n\}$, where $n \leq p \leq q$. Let S be incompatible on $\Sigma_r(A, \Phi_1)$. Assume that S is compatible on $\Sigma_r(A, \Phi_2)$. Then there exists q-tuple $\overline{\sigma} = (\sigma_1, \ldots, \sigma_q) \in \Sigma_r(A, \Phi_2)$ such that $\sigma_1 = \delta_1, \ldots, \sigma_n = \delta_n$. By definition, all subsystems of the system of equations

$$\{f_1(x) = \sigma_1, \ldots, f_q(x) = \sigma_q\}$$

with at most r equations are compatible on A. Therefore all subsystems of the system of equations

$$\{f_1(x) = \sigma_1, \ldots, f_p(x) = \sigma_p\}$$

with at most r equations are compatible on A. Hence the p-tuple $(\sigma_1, \ldots, \sigma_p)$ belongs to the set $\Sigma_r(A, \Phi_1)$, and the system S is compatible on $\Sigma_r(A, \Phi_1)$ which is impossible. Consequently, S is incompatible on $\Sigma_r(A, \Phi_2)$. □

Lemma 5.10. *Let Φ be a nonempty finite subset of the set F. Then I-dimension of the information system $W = (\Sigma_r(A,\Phi), B, \Phi)$ is at most v.*

Proof. Assume that I-dimension of the information system W is at least $v+1$. Then there exist attributes $f_1, ..., f_{v+1} \in \Phi$ and two-element subsets $B_1, ..., B_{v+1}$ of the set B such that for any $\delta_1 \in B_1, \ldots, \delta_{v+1} \in B_{v+1}$ the system of equations

$$\{f_1(x) = \delta_1, \ldots, f_{v+1}(x) = \delta_{v+1}\}$$

is compatible on the set $\Sigma_r(A,\Phi)$. Taking into account that $v+1 \leq r$, and using Lemma 5.8 we conclude that the considered equation system is compatible on A. Consequently, I-dimension of the information system U is at least $v+1$, which is impossible. Therefore I-dimension of the information system W is at most v. □

Lemma 5.11. *Let Φ be a nonempty finite subset of the set F. Then $|\Sigma_r(A,\Phi)| \leq (k^2 |\Phi|)^v$.*

Proof. Let $\Phi = \{f_1, \ldots, f_p\}$ and $\overline{\sigma} = (\sigma_1, \ldots, \sigma_p) \in B^p$. Let us show that $\overline{\sigma} \in \Sigma_r(A,\Phi)$ if and only if the system of equations

$$S = \{f_1(x) = \sigma_1, \ldots, f_p(x) = \sigma_p\} \tag{37}$$

is compatible on $\Sigma_r(A,\Phi)$. Let $\overline{\sigma} \in \Sigma_r(A,\Phi)$. Then $f_1(\overline{\sigma}) = \sigma_1, \ldots, f_p(\overline{\sigma}) = \sigma_p$. Therefore the system S is compatible on the set $\Sigma_r(A,\Phi)$. Let now the system S be compatible on the set $\Sigma_r(A,\Phi)$ and $\overline{\delta} = (\delta_1, \ldots, \delta_p)$ be a p-tuple from $\Sigma_r(A,\Phi)$ which is a solution of S. Then $\delta_1 = \sigma_1, \ldots, \delta_p = \sigma_p$. Therefore $\overline{\sigma} \in \Sigma_r(A,\Phi)$.

Using Lemma 5.10 we conclude that I-dimension of the information system $(\Sigma_r(A,\Phi), B, \Phi)$ is at most v. From here and from Lemma 5.1 follows that the number of systems of the kind (37), which are compatible on the set $\Sigma_r(A,\Phi)$, is at most $(k^2p)^v$. Therefore $|\Sigma_r(A,\Phi)| \leq (k^2 |\Phi|)^v$. □

Lemma 5.12. *Let Φ be a nonempty finite subset of the set F, $S, S_1, \ldots, S_t$ be equation systems with attributes from Φ which are compatible on A, S contain $m+1$ equations, each of systems $S_1, \ldots, S_t$ contain m equations, and the set of solutions on A of the system S coincide with the union of sets of solutions on A of systems $S_1, \ldots, S_t$. Then the systems $S, S_1, \ldots, S_t$ are compatible on $\Sigma_r(A,\Phi)$, and the set of solutions on $\Sigma_r(A,\Phi)$ of the system S coincides with the union of sets of solutions on $\Sigma_r(A,\Phi)$ of systems $S_1, \ldots, S_t$.*

Proof. Using Lemma 5.8 and the inequality $r > m+1$ we conclude that systems of equations $S, S_1, \ldots, S_t$ are compatible on $\Sigma_r(A,\Phi)$. Let $D \in \{A, \Sigma_r(A,\Phi)\}$ and $P(D), P_1(D), \ldots, P_t(D)$ be sets of solutions on D of systems $S, S_1, \ldots, S_t$. Let, for the definiteness, $S = \{f_1(x) = \delta_1, \ldots, f_{m+1}(x) = \delta_{m+1}\}$ and for $i = 1, \ldots, t$ let $S_i = \{f_{i1}(x) = \delta_{i1}, \ldots, f_{im}(x) = \delta_{im}\}$.

It is not difficult to show that the condition $P(D) \subseteq P_1(D) \cup \ldots \cup P_t(D)$ is equivalent to the following condition: each system of equations of the kind

$$S \cup \{f_{1j_1}(x) = \sigma_1, \ldots, f_{tj_t}(x) = \sigma_t\} \ , \tag{38}$$

where $j_i \in \{1, \ldots, m\}$ and $\sigma_i \in B \setminus \{\delta_{ij_i}\}$ for $i = 1, \ldots, t$, is incompatible on D.

It is not difficult to show that the condition $P_1(D) \cup \ldots \cup P_t(D) \subseteq P(D)$ is equivalent to the following condition: each system of equations of the kind

$$S_i \cup \{f_j(x) = \sigma\} , \tag{39}$$

where $i \in \{1, \ldots, t\}$, $j \in \{1, \ldots, m+1\}$ and $\sigma \in B \setminus \{\delta_j\}$, is incompatible on D.

By assumption, $P(A) = P_1(A) \cup \ldots \cup P_t(A)$. Therefore all systems of equations of the kind (38) and (39) are incompatible on A. Each of these systems contains at most $m + t + 1$ equations. Taking into account that $m + t + 1 \leq r$ and using Lemma 5.8 we conclude that all systems of equations of the kind (38) and (39) are incompatible on $\Sigma_r(A, \Phi)$. Therefore $P(\Sigma_r(A, \Phi)) = P_1(\Sigma_r(A, \Phi)) \cup \ldots \cup P_t(\Sigma_r(A, \Phi))$. □

Lemma 5.13. *Let Φ be a nonempty finite subset of the set F, z be a problem over U with attributes from Φ, and Γ be a decision tree over U with attributes from Φ which solves z on the set $\Sigma_r(A, \Phi)$. Then Γ solves z on the set A.*

Proof. Let $\Phi = \{f_1, \ldots, f_p\}$ and $z = (\nu, f_1, \ldots, f_n)$. For an arbitrary $a \in A$ denote $\overline{\sigma}(a) = (f_1(a), \ldots, f_p(a))$. Since the system of equations $\{f_1(x) = f_1(a), \ldots, f_p(x) = f_p(a)\}$ is compatible on A, the p-tuple $\overline{\sigma}(a)$ belongs to the set $\Sigma_r(A, \Phi)$. Since $f_1, \ldots, f_n \in \Phi$, the equality $z(a) = z(\overline{\sigma}(a))$ holds. Let ξ be a complete path in the decision tree Γ. Since Γ is a decision tree with attributes from Φ, the element a is a solution of the equation system $\mathrm{Sol}_U(\pi(\xi))$ if and only if the element $\overline{\sigma}(a)$ is a solution of the system $\mathrm{Sol}_U(\pi(\xi))$. Taking into account that Γ solves z on $\Sigma_r(A, \Phi)$ we conclude that there exists a complete path ξ in Γ such that $\overline{\sigma}(a)$ is a solution of the system $\mathrm{Sol}_U(\pi(\xi))$, and the terminal node of ξ is labelled by the number $z(\overline{\sigma}(a))$. Therefore there exists a complete path ξ in Γ such that a is a solution of the system $\mathrm{Sol}_U(\pi(\xi))$, and the terminal node of ξ is labelled by the number $z(a)$. Taking into account that a is an arbitrary element from the set A we obtain Γ solves the problem z on A. □

Properties of Coverings

Lemma 5.14. *Let Φ_1 and Φ_2 be nonempty finite subsets of the set F, $\Phi_1 \subseteq \Phi_2$, d be a natural number, C be a nonempty set of d-systems of equations with attributes from Φ_1, $f_1, \ldots, f_n \in F$, and the pair (Φ_1, C) be a d-covering of attributes $f_1, \ldots, f_n$ on the set $\Sigma_r(A, \Phi_1)$. Then the pair (Φ_1, C) is a d-covering of attributes $f_1, \ldots, f_n$ on the set $\Sigma_r(A, \Phi_2)$.*

Proof. Since (Φ_1, C) is a d-covering of attributes $f_1, \ldots, f_n$ on the set $\Sigma_r(A, \Phi_1)$ and $\Phi_1 \subseteq \Phi_2$, we have $f_1, \ldots, f_n \in \Phi_2$. Let $C = \{S_1, \ldots, S_w\}$ and for $i = 1, \ldots, w$ let $S_i = \{f_{i1}(x) = \delta_{i1}, \ldots, f_{id}(x) = \delta_{id}\}$.

Let $q \in \{1, 2\}$. It is not difficult to show that any element from $\Sigma_r(A, \Phi_q)$ is a solution of some system from C if and only if each system of equations of the kind

$$\{f_{1j_1}(x) = \sigma_1, \ldots, f_{wj_w}(x) = \sigma_w\} , \tag{40}$$

where $j_i \in \{1, \ldots, d\}$ and $\sigma_i \in B \setminus \{\delta_{ij_i}\}$ for $i = 1, \ldots, w$, is incompatible on $\Sigma_r(A, \Phi_q)$. Since the pair (Φ_1, C) is a d-covering of attributes $f_1, \ldots, f_n$ on the set $\Sigma_r(A, \Phi_1)$, any element from $\Sigma_r(A, \Phi_1)$ is a solution of some system from C. Therefore all systems of the kind (40) are incompatible on $\Sigma_r(A, \Phi_1)$. Using Lemma 5.9 we conclude that all systems of the kind (40) are incompatible on $\Sigma_r(A, \Phi_2)$. Hence any element from $\Sigma_r(A, \Phi_2)$ is a solution of some system from C.

Consider an arbitrary system of equations S_i from C. Since the pair (Φ_1, C) is a d-covering of attributes $f_1, \ldots, f_n$ on the set $\Sigma_r(A, \Phi_1)$, there exist elements $\delta_1, \ldots, \delta_n \in B$ such that the set of solutions on $\Sigma_r(A, \Phi_1)$ of the system S_i is a subset of the set of solutions on $\Sigma_r(A, \Phi_1)$ of the system $S = \{f_1(x) = \delta_1, \ldots, f_n(x) = \delta_n\}$.

Let $q \in \{1, 2\}$. We denote by $P_i(q)$ and $P(q)$ the sets of solutions on $\Sigma_r(A, \Phi_q)$ of systems S_i and S respectively. It is not difficult to show that $P_i(q) \subseteq P(q)$ if and only if each system of equations of the kind

$$S_i \cup \{f_j(x) = \sigma\} \tag{41}$$

where $j \in \{1, \ldots, n\}$ and $\sigma \in B \setminus \{\delta_j\}$ is incompatible on $\Sigma_r(A, \Phi_q)$. Since $P_i(1) \subseteq P(1)$, all systems of the kind (41) are incompatible on $\Sigma_r(A, \Phi_1)$. Using Lemma 5.9 we conclude that all systems of the kind (41) are incompatible on $\Sigma_r(A, \Phi_2)$. Therefore $P_i(2) \subseteq P(2)$.

Thus, the pair (Φ_1, C) is a d-covering of attributes $f_1, \ldots, f_n$ on the set $\Sigma_r(A, \Phi_2)$. □

Lemma 5.15. *Let Φ be a nonempty finite subset of the set F, C be a nonempty set of m-systems of equations with attributes from Φ, $z = (\nu, f_1, \ldots, f_n)$ be a problem over U, and (Φ, C) be an m-covering of attributes $f_1, \ldots, f_n$ on the set $\Sigma_r(A, \Phi)$. Then there exists a decision tree Γ over U with attributes from Φ which solves the problem z on the set $\Sigma_r(A, \Phi)$ and for which $h(\Gamma) \leq rv \log_2(k^2 |\Phi|)$.*

Proof. Let, for the definiteness, $\Phi = \{f_1, ..., f_p\}$, where $p \geq n$. Let $\overline{\sigma} = (\sigma_1, ..., \sigma_p) \in \Sigma_r(A, \Phi)$. Then $z(\overline{\sigma}) = \nu(\sigma_1, \ldots, \sigma_n)$. If the function z is constant on $\Sigma_r(A, \Phi)$ then in the capacity of Γ we can take the decision tree with one node which is labelled by the value of z on p-tuples from $\Sigma_r(A, \Phi)$. It is clear that Γ solves the problem z on the set $\Sigma_r(A, \Phi)$ and $h(\Gamma) = 0$. So for the considered case the statement of the lemma is true.

Let the function z be not constant on the set $\Sigma_r(A, \Phi)$. Consider a decision tree Γ with attributes from Φ which solves z on the set $\Sigma_r(A, \Phi)$. In order to define Γ we describe its work on an arbitrary p-tuple $\overline{b}$ from $\Sigma_r(A, \Phi)$.

Define a linear ordering on the set B. For $i \in \{1, \ldots, p\}$ and $\delta \in B$ we denote by $N(i, \delta)$ the number of p-tuples from $\Sigma_r(A, \Phi)$ which in the i-th digit have the element δ. For $i = 1, \ldots, p$ we denote by δ_i the minimal element from the set B such that $N(i, \delta_i) = \max\{N(i, \delta) : \delta \in B\}$. Denote $\overline{\delta} = (\delta_1, \ldots, \delta_p)$.

First, we consider the case when $\overline{\delta} \in \Sigma_r(A, \Phi)$. Since (Φ, C) is an m-covering of attributes $f_1, \ldots, f_n$ on the set $\Sigma_r(A, \Phi)$, there exists an equation system $S \in C$ such that $\overline{\delta}$ is a solution of this system, and there exist elements $\gamma_1, \ldots, \gamma_n \in B$ for which the set of solutions on $\Sigma_r(A, \Phi)$ of the system S is a subset of the set of solutions on $\Sigma_r(A, \Phi)$ of the system $\{f_1(x) = \gamma_1, \ldots, f_n(x) = \gamma_n\}$. It is clear that the function z is constant on the set of solutions on $\Sigma_r(A,\Phi)$ of last system. Therefore the function z is constant on the set of solutions on $\Sigma_r(A,\Phi)$ of the system S.

Let $S=\{f_{i_1}(x)=\sigma_1,\ldots,f_{i_m}(x)=\sigma_m\}$. It is clear that $\sigma_1=\delta_{i_1},\ldots,\sigma_m=\delta_{i_m}$. In the considered case the work of described decision tree Γ on the p-tuple $\bar{b}$ is beginning with the computation of values of attributes $f_{i_1},\ldots,f_{i_m}$ on $\bar{b}$. As a result the p-tuple $\bar{b}$ will be localized in the set P of solutions on $\Sigma_r(A,\Phi)$ of the equation system $S'=\{f_{i_1}(x)=f_{i_1}(\bar{b}),\ldots,f_{i_m}(x)=f_{i_m}(\bar{b})\}$.

If $f_{i_1}(\bar{b})=\delta_{i_1},\ldots,f_{i_m}(\bar{b})=\delta_{i_m}$ then S' coincides with S, and the function z is constant on P. So, we know the value $z(\bar{b})$.

Let $f_{i_j}(\bar{b})\neq\delta_{i_j}$ for some $j\in\{1,\ldots,m\}$. It is clear that in this case the cardinality of the set P is at most one-half of the cardinality of the set $\Sigma_r(A,\Phi)$.

Consider now the case when $\bar{\delta}\notin\Sigma_r(A,\Phi)$. In this case the system of equations $\{f_1(x)=\delta_1,\ldots,f_p(x)=\delta_p\}$ has a subsystem with r equations $S=\{f_{i_1}(x)=\delta_{i_1},\ldots,f_{i_r}(x)=\delta_{i_r}\}$ which is incompatible on A (we admit that among the equations $f_{i_1}(x)=\delta_{i_1},\ldots,f_{i_r}(x)=\delta_{i_r}$ there are identical ones). Using Lemma 5.8 we conclude that S is incompatible on $\Sigma_r(A,\Phi)$. In the considered case the work of described decision tree Γ on the p-tuple $\bar{b}$ is beginning with the computation of values of attributes $f_{i_1},\ldots,f_{i_r}$ on $\bar{b}$. As a result the p-tuple $\bar{b}$ will be localized in the set P of solutions on $\Sigma_r(A,\Phi)$ of the equation system $S'=\{f_{i_1}(x)=f_{i_1}(\bar{b}),\ldots,f_{i_r}(x)=f_{i_r}(\bar{b})\}$. Since S is incompatible on $\Sigma_r(A,\Phi)$, there exists $j\in\{1,\ldots,m\}$ for which $f_{i_j}(\bar{b})\neq\delta_{i_j}$. Therefore the cardinality of the set P is at most one-half of the cardinality of the set $\Sigma_r(A,\Phi)$.

Later the tree Γ works similarly, but in the definition of the value $N(i,\delta)$ instead of the set $\Sigma_r(A,\Phi)$ we consider its subset P in which the p-tuple $\bar{b}$ is localized. The process described above will be called *a big step* of decision tree Γ work. During a big step we compute values of at most r attributes (note that $m<r$). As a result of a big step we either recognize the value $z(\bar{b})$ and finish the work of Γ, or reduce the cardinality of a set, in which $\bar{b}$ is localized, in at least 2 times.

Let during the work on the p-tuple $\bar{b}$ the decision tree Γ make q big steps. After the big step with the number $q-1$ we obtain a set Q in which $\bar{b}$ is localized. Since we must make additional big step in the order to recognize the value $z(\bar{b})$, $|Q|\geq 2$. It is clear that $|Q|\leq\frac{|\Sigma_r(A,\Phi)|}{2^{q-1}}$. Therefore $2^q\leq|\Sigma_r(A,\Phi)|$ and $q\leq\log_2|\Sigma_r(A,\Phi)|$. Taking into account that during each big step we compute values of at most r attributes, and $\bar{b}$ is an arbitrary p-tuple from $\Sigma_r(A,\Phi)$ we obtain $h(\Gamma)\leq r\log_2|\Sigma_r(A,\Phi)|$. From Lemma 5.11 follows that $|\Sigma_r(A,\Phi)|\leq(k^2|\Phi|)^v$. Therefore $h(\Gamma)\leq rv\log_2(k^2|\Phi|)$. □

Construction of Coverings

Lemma 5.16. *If $n\leq m$ then for any $f_1,\ldots,f_n\in F$ there exist an n-covering (Φ,C_1) of attributes $f_1,\ldots,f_n$ on $\Sigma_r(A,\Phi)$ and an m-covering (Φ,C_2) of attributes $f_1,\ldots,f_n$ on $\Sigma_r(A,\Phi)$ such that $|\Phi|=n$ and $|C_1|=|C_2|\leq(k^2n)^v$.*

Proof. Set $\Phi=\{f_1,\ldots,f_n\}$. Denote by C_1 the set of all compatible on $\Sigma_r(A,\Phi)$ systems of equations of the kind $\{f_1(x)=\delta_1,\ldots,f_n(x)=\delta_n\}$ where $\delta_1,\ldots,\delta_n\in B$.

If $n = m$ then $C_2 = C_1$. Let $n < m$. Then in each system from C_1 we repeat the last equation $m - n$ times. We denote by C_2 the obtained family of systems of equations. One can show that the pair (Φ, C_1) is an n-covering of attributes $f_1, \ldots, f_n$ on $\Sigma_r(A, \Phi)$, and the pair (Φ, C_2) is an m-covering of attributes $f_1, \ldots, f_n$ on $\Sigma_r(A, \Phi)$. It is clear that $|\Phi| = n$. From Lemma 5.10 follows that I-dimension of the information system $(\Sigma_r(A, \Phi), B, \Phi)$ is at most v. Using Lemma 5.1 we obtain $|C_1| = |C_2| \leq (k^2 n)^v$. □

Lemma 5.17. *For any $f_1, \ldots, f_{m+1} \in F$ there exists an m-covering (Φ, C) of attributes $f_1, \ldots, f_{m+1}$ on $\Sigma_r(A, \Phi)$ such that $|\Phi| \leq k^{2v}(m+1)^{v+1}t$ and $|C| \leq k^{2v}(m+1)^v t$.*

Proof. Denote by D the set of all compatible on A equation systems of the kind $\{f_1(x) = \delta_1, \ldots, f_{m+1}(x) = \delta_{m+1}\}$ where $\delta_1, \ldots, \delta_{m+1} \in B$. Let $D = \{S_1, \ldots, S_p\}$. Let $i \in \{1, \ldots, p\}$. Since the information system U satisfies the condition of decomposition with parameters m and t, there exist compatible on A m-systems of equations $S_{i1}, \ldots, S_{it}$ with attributes from F such that the set of solutions on A of the system S_i coincides with the union of sets of solutions on A of systems $S_{i1}, \ldots, S_{it}$. Denote $C = \bigcup_{i=1}^p \{S_{i1}, \ldots, S_{it}\}$. We denote by Φ the union of the set $\{f_1, \ldots, f_{m+1}\}$ and the set of all attributes which are contained in equations from systems belonging to C. Taking into account that I-dimension of the information system U is at most v and using Lemma 5.1 we obtain $p \leq k^{2v}(m+1)^v$. Therefore $|C| \leq k^{2v}(m+1)^v t$ and $|\Phi| \leq k^{2v}(m+1)^v tm + m + 1 \leq k^{2v}(m+1)^{v+1}t$.

Let us show that the pair (Φ, C) is an m-covering of attributes $f_1, \ldots, f_{m+1}$ on $\Sigma_r(A, \Phi)$. It is clear that $m+1 \leq r$. From Lemma 5.8 follows that the set of compatible on $\Sigma_r(A, \Phi)$ systems of equations of the kind $\{f_1(x) = \delta_1, \ldots, f_{m+1}(x) = \delta_{m+1}\}$, where $\delta_1, \ldots, \delta_{m+1} \in B$, coincides with D. Let $i \in \{1, \ldots, p\}$. Using Lemma 5.12 we conclude that the set of solutions on $\Sigma_r(A, \Phi)$ of equation system S_i coincides with the union of sets of solutions on $\Sigma_r(A, \Phi)$ of systems $S_{i1}, \ldots, S_{it}$. Let $\overline{\sigma} \in \Sigma_r(A, \Phi)$. Then, evidently, $\overline{\sigma}$ is a solution of a system S_{ij} for some $i \in \{1, \ldots, p\}$ and $j \in \{1, \ldots, t\}$. It is clear that the set of solutions on $\Sigma_r(A, \Phi)$ of the system S_{ij} is a subset of the set of solutions on $\Sigma_r(A, \Phi)$ of the system S_i. Therefore the pair (Φ, C) is an m-covering of attributes $f_1, \ldots, f_{m+1}$ on $\Sigma_r(A, \Phi)$. □

Lemma 5.18. *Let $n \geq m+1$, $n = (m+1)p+q$, $0 \leq q \leq m$, and $f_1, \ldots, f_n \in F$. Then there exists an $(mp+q)$-covering (Φ, C) of attributes $f_1, \ldots, f_n$ on $\Sigma_r(A, \Phi)$ such that $|\Phi| \leq ntk^{2v}(m+1)^{v+1}$ and $|C| \leq n^v t^v k^{2v(v+1)}(m+1)^{v(v+1)}$.*

Proof. Let, for the definiteness, $q > 0$. The case $q = 0$ can be considered in the same way. Divide the sequence $f_1, \ldots, f_n$ into p blocks of the length $m+1$ and one block of the length q. From Lemma 5.17 follows that for $i = 1, \ldots, p$ there exists an m-covering (Φ_i, C_i) of i-th block on $\Sigma_r(A, \Phi_i)$ such that $|\Phi_i| \leq k^{2v}(m+1)^{v+1}t$. From Lemma 5.16 follows that there exists an q-covering (Φ_{p+1}, C_{p+1}) of $(p+1)$-th block on $\Sigma_r(A, \Phi_{p+1})$ such that $|\Phi_{p+1}| \leq q$. Set $\Phi = \Phi_1 \cup \ldots \cup \Phi_p \cup \Phi_{p+1}$. Then $|\Phi| \leq pk^{2v}(m+1)^{v+1}t + q \leq nk^{2v}(m+1)^{v+1}t$. Using Lemma 5.11 we obtain

$$|\Sigma_r(A,\Phi)| \le (k^2|\Phi|)^v \le k^{2v} n^v k^{2v^2}(m+1)^{v(v+1)} t^v \\ \le n^v t^v k^{2v(v+1)}(m+1)^{v(v+1)} . \quad (42)$$

It is clear that $f_1,\dots,f_n \in \Phi$. Using Lemma 5.14 we conclude that for $i = 1,\dots,p$ the pair (Φ_i, C_i) is an m-covering of i-th block on $\Sigma_r(A,\Phi)$, and the pair (Φ_{p+1}, C_{p+1}) is an q-covering of $(p+1)$-th block on $\Sigma_r(A,\Phi)$. Let $\overline{\sigma} \in \Sigma_r(A,\Phi)$. Denote $S = \{f_1(x) = f_1(\overline{\sigma}),\dots,f_n(x) = f_n(\overline{\sigma})\}$. Divide the system of equations S into $p+1$ parts $S_1,\dots,S_{p+1}$ according to the partition of the sequence $f_1,\dots,f_n$ into blocks. Denote by P the set of solutions on $\Sigma_r(A,\Phi)$ of the system S. For $i = 1,\dots,p+1$ denote by P_i the set of solutions on $\Sigma_r(A,\Phi)$ of the system S_i. It is clear that for $i = 1,\dots,p+1$ the tuple $\overline{\sigma}$ is a solution of the system S_i. Also, it is clear that $S = S_1 \cup \ldots \cup S_{p+1}$ and $P = P_1 \cap \ldots \cap P_{p+1}$. Since for $i = 1,\dots,p+1$ the pair (Φ_i, C_i) is an m-covering or an q-covering of i-th block on $\Sigma_r(A,\Phi)$, there exists a system of equations $S'_i \in C_i$ such that $\overline{\sigma}$ is a solution of S'_i. Denote by P'_i the set of solutions on $\Sigma_r(A,\Phi)$ of the system S'_i. Using the definition of covering and the fact that $\overline{\sigma} \in P'_i$ and $\overline{\sigma} \in P_i$ it is not difficult to prove that $P'_i \subseteq P_i$. Denote $S_{\overline{\sigma}} = S'_1 \cup \ldots \cup S'_{p+1}$. Denote by P' the set of solutions on $\Sigma_r(A,\Phi)$ of the system $S_{\overline{\sigma}}$. It is clear that $P' = P'_1 \cap \ldots \cap P'_{p+1}$ and $\overline{\sigma} \in P'$. Taking into account that $P = P_1 \cap \ldots \cap P_{p+1}$ and $P'_i \subseteq P_i$ for $i = 1,\dots,p+1$ we obtain $P' \subseteq P$. Evidently, the system $S_{\overline{\sigma}}$ contains $mp+q$ equations. Denote $C = \{S_{\overline{\sigma}} : \overline{\sigma} \in \Sigma_r(A,\Phi)\}$. It is clear that the pair (Φ, C) is an $(mp+q)$-covering of attributes $f_1,\dots,f_n$ on $\Sigma_r(A,\Phi)$. Using (42) we obtain $|C| \le n^v t^v k^{2v(v+1)}(m+1)^{v(v+1)}$. □

Lemma 5.19. *Let $f_1,\dots,f_n \in F$. Then there exists an m-covering (Φ, C) of attributes $f_1,\dots,f_n$ on $\Sigma_r(A,\Phi)$ such that $|\Phi| \le n^{2(v+1)^2 \ln n}(k^2 t(m+1))^{2(v+1)^3 \ln n}$ and $|C| \le (k^2|\Phi|)^v$.*

Proof. We prove the statement of the lemma by induction on n. If $n \le m$ then by Lemma 5.16 the statement of the lemma is true. Let this statement be true for all tuples with at most $n-1$ attributes, $n-1 \ge m$. Let us show that the statement of the lemma is true for any tuple with n attributes.

Let $f_1,\dots,f_n \in F$, $n \ge m+1$ and $n = (m+1)p+q$ where $0 \le q \le m$. Using Lemma 5.18 we obtain that there exists an $(mp+q)$-covering (Φ', C') of attributes $f_1,\dots,f_n$ on $\Sigma_r(A,\Phi')$ such that $|\Phi'| \le ntk^{2v}(m+1)^{v+1}$ and $|C'| \le n^v t^v k^{2v(v+1)}(m+1)^{v(v+1)}$. Denote $d = mp+q$. Let $S \in C'$ and $S = \{f_{i_1}(x) = \delta_1,\dots,f_{i_d}(x) = \delta_d\}$. Since $n \ge m+1$, $d < n$. By inductive hypothesis, there exists an m-covering $(\Phi(S), C(S))$ of attributes $f_{i_1},\dots,f_{i_d}$ on $\Sigma_r(A,\Phi(S))$ such that $|\Phi(S)| \le d^{2(v+1)^2 \ln d}(k^2 t(m+1))^{2(v+1)^3 \ln d}$. Set $\Phi = \Phi' \cup \bigcup_{S \in C'} \Phi(S)$. Denote $c = \frac{2m}{2m+1}$. It is not difficult to show that $d \le cn$. Therefore

$$\begin{aligned} |\Phi| &\le ntk^{2v}(m+1)^{v+1} \\ &+ n^v t^v k^{2v(v+1)}(m+1)^{v(v+1)} d^{2(v+1)^2 \ln d}(k^2 t(m+1))^{2(v+1)^3 \ln d} \\ &\le n^v (k^2 t(m+1))^{(v+1)^2} d^{2(v+1)^2 \ln d}(k^2 t(m+1))^{2(v+1)^3 \ln d} \\ &\le n^{v+2(v+1)^2 \ln cn}(k^2 t(m+1))^{(v+1)^2 + 2(v+1)^3 \ln cn} . \end{aligned}$$

It is known that for any natural w the inequality $\ln(1+\frac{1}{w}) > \frac{1}{w+1}$ holds. Therefore $\ln(\frac{1}{c}) = \ln(1+\frac{1}{2m}) > \frac{1}{2m+1}$. Taking into account that $v \geq m$ we obtain $\ln(\frac{1}{c}) > \frac{1}{2m+1} > \frac{1}{2(m+1)} \geq \frac{1}{2(v+1)}$. Therefore $\ln c < -\frac{1}{2(v+1)}$. From this inequality follows that

$$v + 2(v+1)^2 \ln cn < v + 2(v+1)^2 \ln n - (v+1) < 2(v+1)^2 \ln n$$

and

$$(v+1)^2 + 2(v+1)^3 \ln cn < (v+1)^2 + 2(v+1)^3 \ln n - (v+1)^2 = 2(v+1)^3 \ln n .$$

Therefore $|\Phi| \leq n^{2(v+1)^2 \ln n}(k^2t(m+1))^{2(v+1)^3 \ln n}$.

Let $\overline{\sigma} \in \Sigma_r(A,\Phi)$. Denote $S^0 = \{f_1(x) = f_1(\overline{\sigma}), \ldots, f_n(x) = f_n(\overline{\sigma})\}$. It is clear that $\overline{\sigma}$ is a solution of the system S^0. Using Lemma 5.14 we conclude that the pair (Φ', C') is a d-covering of attributes $f_1, \ldots, f_n$ on $\Sigma_r(A,\Phi)$. Therefore there exists an equation system $S \in C'$ such that $\overline{\sigma}$ is a solution of S. Taking into account that $\overline{\sigma}$ is a solution of S^0 we conclude that the set of solutions on $\Sigma_r(A,\Phi)$ of the system S is a subset of the set of solutions on $\Sigma_r(A,\Phi)$ of the system S^0. Let $S = \{f_{i_1}(x) = \delta_1, \ldots, f_{i_d}(x) = \delta_d\}$. From Lemma 5.14 follows that the pair $(\Phi(S), C(S))$ is an m-covering of attributes $f_{i_1}, \ldots, f_{i_d}$ on $\Sigma_r(A,\Phi)$. Therefore there exists an m-system of equations $S_{\overline{\sigma}} \in C(S)$ such that $\overline{\sigma}$ is a solution of $S_{\overline{\sigma}}$. Taking into account that $\overline{\sigma}$ is a solution of S we conclude that the set of solutions on $\Sigma_r(A,\Phi)$ of the system $S_{\overline{\sigma}}$ is a subset of the set of solutions on $\Sigma_r(A,\Phi)$ of the system S. Therefore the set of solutions on $\Sigma_r(A,\Phi)$ of the system $S_{\overline{\sigma}}$ is a subset of the set of solutions on $\Sigma_r(A,\Phi)$ of the system S^0. Denote $C = \{S_{\overline{\sigma}} : \overline{\sigma} \in \Sigma_r(A,\Phi)\}$. It is clear that the pair (Φ, C) is an m-covering of attributes $f_1, \ldots, f_n$ on $\Sigma_r(A,\Phi)$. Using Lemma 5.11 we obtain $|\Sigma_r(A,\Phi)| \leq (k^2 |\Phi|)^v$. Therefore $|C| \leq (k^2 |\Phi|)^v$. □

Lemma 5.20. *Let s be a natural number. Then for any $f_1, \ldots, f_n \in F$ there exists an m-covering (Φ, C) of attributes $f_1, \ldots, f_n$ on $\Sigma_r(A,\Phi)$ such that $|\Phi| \leq 2^{c(s)(\log_2 n)^{1+\frac{1}{s}}}$ and $|C| \leq (k^2 |\Phi|)^v$ where $c(s) = 8(v+1)^{s+2}(\log_2(8ktm))^{2s-1}$.*

Proof. We prove the statement of the lemma by induction on s. We denote by $St(s)$ the statement of lemma for fixed natural number s. Let us show that $St(1)$ is true. From Lemma 5.19 follows that for any $f_1, \ldots, f_n \in F$ there exists an m-covering (Φ, C) of attributes $f_1, \ldots, f_n$ on $\Sigma_r(A,\Phi)$ such that $|\Phi| \leq n^{2(v+1)^2 \ln n}(k^2t(m+1))^{2(v+1)^3 \ln n}$ and $|C| \leq (k^2 |\Phi|)^v$. It is clear that

$$\begin{aligned} |\Phi| &\leq 2^{2(v+1)^2(\log_2 n)^2 + 2(v+1)^3 \log_2 n \log_2(k^2t(m+1))} \\ &\leq 2^{8(v+1)^3 \log_2(ktm)(\log_2 n)^2} \leq 2^{8(v+1)^3 \log_2(8ktm)(\log_2 n)^{1+\frac{1}{1}}} . \end{aligned}$$

Therefore $St(1)$ is true. Assume that $St(s)$ is true for some natural s. Let us show that $St(s+1)$ is true. We prove the statement $St(s+1)$ by induction on n.

First, we consider the case when $n \leq 2^{(\log_2 4m)^{s+1}}$. We have assumed that $St(s)$ is true. Therefore there exists an m-covering (Φ', C') of attributes $f_1, \ldots, f_n$

on $\Sigma_r(A, \Phi')$ such that $|\Phi'| \le 2^{c(s)(\log_2 n)^{1+\frac{1}{s}}}$ and $|C'| \le (k^2 |\Phi'|)^v$. Taking into account that $n \le 2^{(\log_2 4m)^{s+1}}$ we obtain

$$c(s)(\log_2 n)^{1+\frac{1}{s}} = c(s)(\log_2 n)^{\frac{1}{s(s+1)}}(\log_2 n)^{1+\frac{1}{s+1}}$$
$$\le c(s)(\log_2 4m)^{\frac{1}{s}}(\log_2 n)^{1+\frac{1}{s+1}} \le c(s+1)(\log_2 n)^{1+\frac{1}{s+1}} .$$

Hence, $|\Phi'| \le 2^{c(s+1)(\log_2 n)^{1+\frac{1}{s+1}}}$. Therefore the statement $St(s+1)$ is true if $n \le 2^{(\log_2 4m)^{s+1}}$.

Let $n > 2^{(\log_2 4m)^{s+1}}$, and the statement $St(s+1)$ be true for any sequence of attributes $f_1, \ldots, f_l \in F$ where $l \le n-1$. Let us show that the statement $St(s+1)$ is true for an arbitrary sequence of attributes $f_1, \ldots, f_n \in F$.

Set $b = 2^{\lceil (\log_2 n)^{\frac{s}{s+1}} + \log_2 m + 1 \rceil}$. It is clear that $b \ge m+1$. Let us show that $b < n$. Evidently, $\log_2 b \le (\log_2 n)^{\frac{s}{s+1}} + \log_2 4m$. Let us show that $(\log_2 n)^{\frac{s}{s+1}} + \log_2 4m < \log_2 n$. This inequality is equivalent to the inequality $\log_2 4m < \log_2 n - (\log_2 n)^{\frac{s}{s+1}}$. Since $\log_2 n - (\log_2 n)^{\frac{s}{s+1}} = (\log_2 n)^{\frac{s}{s+1}}((\log_2 n)^{\frac{1}{s+1}} - 1)$, the last inequality is equivalent to the inequality $\log_2 4m < (\log_2 n)^{\frac{s}{s+1}}((\log_2 n)^{\frac{1}{s+1}} - 1)$. Since $n > 2^{(\log_2 4m)^{s+1}}$, the inequalities

$$(\log_2 n)^{\frac{s}{s+1}}((\log_2 n)^{\frac{1}{s+1}} - 1) > (\log_2 4m)^s(\log_2 4m - 1) > \log_2 4m$$

hold. Therefore $b < n$.

Let $n = pb + q$ where $0 \le q < b$. Divide the sequence $f_1, \ldots, f_n$ into p blocks of the length b and one block of the length q. Assume, for the definiteness, that $q > 0$. The case $q = 0$ can be considered in the same way. We have assumed that the statement $St(s)$ is true. Therefore for $i = 1, \ldots, p$ there exists an m-covering (Φ_i, C_i) of i-th block on $\Sigma_r(A, \Phi_i)$ such that $|\Phi_i| \le 2^{c(s)(\log_2 b)^{1+\frac{1}{s}}}$ and $|C_i| \le (k^2 |\Phi_i|)^v$. If $q \ge m$ then there exists an m-covering (Φ_{p+1}, C_{p+1}) of $(p+1)$-th block on $\Sigma_r(A, \Phi_{p+1})$ such that $|\Phi_{p+1}| \le 2^{c(s)(\log_2 q)^{1+\frac{1}{s}}}$ and $|C_{p+1}| \le (k^2 |\Phi_{p+1}|)^v$. If $q < m$ then by Lemma 5.16 there exists an q-covering (Φ_{p+1}, C_{p+1}) of $(p+1)$-th block on $\Sigma_r(A, \Phi_{p+1})$ such that $|\Phi_{p+1}| \le q$ and $|C_{p+1}| \le (k^2 |\Phi_{p+1}|)^v$. It is clear that $|\Phi_{p+1}| \le 2^{c(s)(\log_2 b)^{1+\frac{1}{s}}}$. Set $w = m$ if $q \ge m$, and $w = q$ if $q < m$. Set $\Phi' = \Phi_1 \cup \ldots \cup \Phi_p \cup \Phi_{p+1}$. Then $|\Phi'| \le (p+1)2^{c(s)(\log_2 b)^{1+\frac{1}{s}}}$. Since $b < n$, $p + 1 = \lfloor \frac{n}{b} \rfloor + 1 \le \frac{n}{b} + 1 < \frac{2n}{b}$. Therefore

$$|\Phi'| \le \frac{2n}{b} 2^{c(s)(\log_2 b)^{1+\frac{1}{s}}} . \tag{43}$$

Using Lemma 5.11 we obtain

$$|\Sigma_r(A, \Phi')| \le (k^2 |\Phi'|)^v \le \left(k^2 \frac{2n}{b} 2^{c(s)(\log_2 b)^{1+\frac{1}{s}}}\right)^v . \tag{44}$$

From Lemma 5.14 follows that for $i = 1, \ldots, p$ the pair (Φ_i, C_i) is an m-covering of i-th block on $\Sigma_r(A, \Phi')$, and the pair (Φ_{p+1}, C_{p+1}) is an w-covering of $(p+1)$-th block on $\Sigma_r(A, \Phi')$.

Let $\overline{\sigma} \in \Sigma_r(A, \Phi')$. Denote $S' = \{f_1(x) = f_1(\overline{\sigma}), \ldots, f_n(x) = f_n(\overline{\sigma})\}$. Divide the system S' into $p+1$ parts $S'_1, \ldots, S'_{p+1}$ according to the partition of the sequence $f_1, \ldots, f_n$ into blocks. Denote by P' the set of solutions on $\Sigma_r(A, \Phi')$ of the system S'. For $i = 1, \ldots, p+1$ denote by P'_i the set of solutions on $\Sigma_r(A, \Phi')$ of the system S'_i. It is clear that for $i = 1, \ldots, p+1$ the tuple $\overline{\sigma}$ is a solution of the system S'_i. Also, it is clear that $S' = S'_1 \cup \ldots \cup S'_{p+1}$ and $P' = P'_1 \cap \ldots \cap P'_{p+1}$. Since for $i = 1, \ldots, p+1$ the pair (Φ_i, C_i) is an m-covering or an q-covering of i-th block on $\Sigma_r(A, \Phi')$, there exists a system of equations $S''_i \in C_i$ such that $\overline{\sigma}$ is a solution of S''_i. Denote by P''_i the set of solutions on $\Sigma_r(A, \Phi')$ of the system S''_i. Using the definition of covering and the fact that $\overline{\sigma} \in P''_i$ and $\overline{\sigma} \in P'_i$ it is not difficult to prove that $P''_i \subseteq P'_i$. Denote $S'_{\overline{\sigma}} = S''_1 \cup \ldots \cup S''_{p+1}$. Denote by P'' the set of solutions on $\Sigma_r(A, \Phi')$ of the system $S'_{\overline{\sigma}}$. It is clear that $P'' = P''_1 \cap \ldots \cap P''_{p+1}$. Taking into account that $P' = P'_1 \cap \ldots \cap P'_{p+1}$ and $P''_i \subseteq P'_i$ for $i = 1, \ldots, p+1$ we obtain $P'' \subseteq P'$. Denote $d = mp + w$. We see that the system $S'_{\overline{\sigma}}$ contains d equations. Denote $C' = \{S'_{\overline{\sigma}} : \overline{\sigma} \in \Sigma_r(A, \Phi')\}$. It is clear that the pair (Φ', C') is a d-covering of attributes $f_1, \ldots, f_n$ on $\Sigma_r(A, \Phi')$. Using (44) we obtain

$$|C'| \leq \left(k^2 \frac{2n}{b} 2^{c(s)(\log_2 b)^{1+\frac{1}{s}}}\right)^v . \tag{45}$$

Let $S \in C'$ and $S = \{f_{i_1}(x) = \delta_1, \ldots, f_{i_d}(x) = \delta_d\}$. Denote by $F(S)$ the sequence of attributes $f_{i_1}, \ldots, f_{i_d}$. Since $b \geq m+1$, the inequality $d \leq n-1$ holds. By inductive hypothesis, the statement $St(s+1)$ is true for d. Therefore there exists an m-covering $(\Phi(S), C(S))$ of the sequence of attributes $F(S)$ on $\Sigma_r(A, \Phi(S))$ such that $|\Phi(S)| \leq 2^{c(s+1)(\log_2 d)^{1+\frac{1}{s+1}}}$ and $|C(S)| \leq (k^2 |\Phi(S)|)^v$. It is clear that $d \leq (p+1)m$. Since $p+1 < \frac{2n}{b}$, the inequality $d \leq \frac{2nm}{b}$ holds. Therefore

$$|\Phi(S)| \leq 2^{c(s+1)\left(\log_2\left(\frac{2nm}{b}\right)\right)^{1+\frac{1}{s+1}}} . \tag{46}$$

Set $\Phi = \Phi' \cup \bigcup_{S \in C'} \Phi(S)$. From (43), (45) and (46) follows that

$$\begin{aligned}
|\Phi| &\leq \tfrac{2n}{b} 2^{c(s)(\log_2 b)^{1+\frac{1}{s}}} + \left(k^2 \tfrac{2n}{b} 2^{c(s)(\log_2 b)^{1+\frac{1}{s}}}\right)^v 2^{c(s+1)\left(\log_2\left(\frac{2nm}{b}\right)\right)^{1+\frac{1}{s+1}}} \\
&\leq \left(k^2 \tfrac{2n}{b} 2^{c(s)(\log_2 b)^{1+\frac{1}{s}}}\right)^{v+1} 2^{c(s+1)\left(\log_2\left(\frac{2nm}{b}\right)\right)^{1+\frac{1}{s+1}}} \\
&\leq 2^{(v+1)\left(2\log_2 k + 1 + \log_2 n - (\log_2 n)^{\frac{s}{s+1}} - \log_2 m - 1 + c(s)\left((\log_2 n)^{\frac{s}{s+1}} + \log_2 m + 2\right)^{\frac{s+1}{s}}\right)} \\
&\quad \times 2^{c(s+1)\left(1 + \log_2 n + \log_2 m - (\log_2 n)^{\frac{s}{s+1}} - \log_2 m - 1\right)^{1+\frac{1}{s+1}}} \\
&\leq 2^{(v+1)\left(2\log_2 k + \log_2 n + c(s)\left((\log_2 n)^{\frac{s}{s+1}} + \log_2 m + 2\right)^{\frac{s+1}{s}}\right)} \\
&\quad \times 2^{c(s+1)\left(\log_2 n - (\log_2 n)^{\frac{s}{s+1}}\right)^{1+\frac{1}{s+1}}} .
\end{aligned} \tag{47}$$

It is clear that $(\log_2 n)^{\frac{s}{s+1}} \geq 1$. Hence

$$(\log_2 n)^{\frac{s}{s+1}} + \log_2 m + 2 \leq (\log_2 n)^{\frac{s}{s+1}} \log_2 8m .$$

Also, it is clear that

$$\left(\log_2 n - (\log_2 n)^{\frac{s}{s+1}}\right)^{1+\frac{1}{s+1}} = \left(\log_2 n - (\log_2 n)^{\frac{s}{s+1}}\right)\left(\log_2 n - (\log_2 n)^{\frac{s}{s+1}}\right)^{\frac{1}{s+1}}$$
$$\leq \left(\log_2 n - (\log_2 n)^{\frac{s}{s+1}}\right)(\log_2 n)^{\frac{1}{s+1}} = (\log_2 n)^{1+\frac{1}{s+1}} - \log_2 n \ .$$

Using obtained inequalities and (47) we have

$$|\Phi| \leq 2^{(v+1)\left(2\log_2 k + \log_2 n + c(s)\log_2 n (\log_2 8m)^{\frac{s+1}{s}}\right) + c(s+1)(\log_2 n)^{1+\frac{1}{s+1}} - c(s+1)\log_2 n}$$
$$\leq 2^{(v+1)\left(2\log_2 k + 1 + c(s)(\log_2 8m)^2\right)\log_2 n + c(s+1)(\log_2 n)^{1+\frac{1}{s+1}} - c(s+1)\log_2 n} \ . \quad (48)$$

It is clear that

$$(v+1)\left(2\log_2 k + 1 + c(s)(\log_2 8m)^2\right) \leq c(s)(v+1)(\log_2 8km)^2$$
$$\leq c(s)(v+1)(\log_2 8ktm)^2 = c(s+1) \ .$$

Using (48) we obtain

$$|\Phi| \leq 2^{c(s+1)\log_2 n + c(s+1)(\log_2 n)^{1+\frac{1}{s+1}} - c(s+1)\log_2 n} = 2^{c(s+1)(\log_2 n)^{1+\frac{1}{s+1}}} \ .$$

From Lemma 5.14 follows that the pair (Φ', C') is a d-covering of attributes $f_1, \ldots, f_n$ on $\Sigma_r(A, \Phi)$, and for any $S \in C'$ the pair $(\Phi(S), C(S))$ is an m-covering of the sequence of attributes $F(S)$ on $\Sigma_r(A, \Phi)$.

Let $\overline{\sigma} \in \Sigma_r(A, \Phi)$. Then $\overline{\sigma}$ is a solution of the equation system $S^0_{\overline{\sigma}} = \{f_1(x) = f_1(\overline{\sigma}), \ldots, f_n(x) = f_n(\overline{\sigma})\}$. Denote by P^0 the set of solutions on $\Sigma_r(A, \Phi)$ of the system $S^0_{\overline{\sigma}}$. Since the pair (Φ', C') is a d-covering of attributes $f_1, \ldots, f_n$ on $\Sigma_r(A, \Phi)$, there exists a system of equations $S^1_{\overline{\sigma}} \in C'$ such that $\overline{\sigma}$ is a solution of $S^1_{\overline{\sigma}}$. Denote by P^1 the set of solutions on $\Sigma_r(A, \Phi)$ of the system $S^1_{\overline{\sigma}}$. Using the definition of d-covering it is not difficult to show that $P^1 \subseteq P^0$. Since the pair $(\Phi(S^1_{\overline{\sigma}}), C(S^1_{\overline{\sigma}}))$ is an m-covering of the sequence of attributes $F(S^1_{\overline{\sigma}})$ on $\Sigma_r(A, \Phi)$, there exists an m-system of equations $S^2_{\overline{\sigma}} \in C(S^1_{\overline{\sigma}})$ such that $\overline{\sigma}$ is a solution of $S^2_{\overline{\sigma}}$. Denote by P^2 the set of solutions on $\Sigma_r(A, \Phi)$ of the system $S^2_{\overline{\sigma}}$. Using the definition of m-covering it is not difficult to show that $P^2 \subseteq P^1$. Therefore $P^2 \subseteq P^0$. Denote $C = \{S^2_{\overline{\sigma}} : \overline{\sigma} \in \Sigma_r(A, \Phi)\}$. It is clear that the pair (Φ, C) is an m-covering of attributes $f_1, \ldots, f_n$ on $\Sigma_r(A, \Phi)$. Using Lemma 5.11 we obtain $|\Sigma_r(A, \Phi)| \leq \left(k^2 |\Phi|\right)^v$. Therefore $|C| \leq \left(k^2 |\Phi|\right)^v$. So the statement $St(s+1)$ is true for any n. Therefore the statement of the lemma is true for any s. □

Lemma 5.21. *Let* $\alpha = \sqrt{2\log_2(2(v+1)\log_2(8ktm))}$ *and* n *be natural number such that* $\log_2 \log_2 n \geq 9\alpha^2$. *Then for any* $f_1, \ldots, f_n \in F$ *there exists an* m-*covering* (Φ, C) *of attributes* $f_1, \ldots, f_n$ *on* $\Sigma_r(A, \Phi)$ *such that* $|C| \leq \left(k^2 |\Phi|\right)^v$ *and* $|\Phi| \leq 2^{(\log_2 n)^{1+\frac{2.5\alpha}{\sqrt{\log_2 \log_2 n}}}}$.

Proof. Set $s = \left\lfloor \frac{\sqrt{\log_2 \log_2 n}}{\alpha} \right\rfloor$. It is clear that s is a natural number and $s \geq 3$. From Lemma 5.20 follows that there exists an m-covering (Φ, C) of attributes

$f_1, \ldots, f_n$ on $\Sigma_r(A, \Phi)$ such that $|\Phi| \leq 2^{c(s)(\log_2 n)^{1+\frac{1}{s}}}$ and $|C| \leq (k^2 |\Phi|)^v$ where $c(s) = 8(v+1)^{s+2}(\log_2(8ktm))^{2s-1}$. Since $s \geq 3$, the inequality $c(s) \leq 2^{\alpha^2 s}$ holds. Taking into account that $\sqrt{\log_2 \log_2 n} \geq 3\alpha$ we obtain $s \geq \frac{\sqrt{\log_2 \log_2 n}}{\alpha} - 1 = \frac{\sqrt{\log_2 \log_2 n} - \alpha}{\alpha} \geq \frac{2\sqrt{\log_2 \log_2 n}}{3\alpha}$. Therefore $\frac{1}{s} \leq \frac{3\alpha}{2\sqrt{\log_2 \log_2 n}}$. Using these inequalities we obtain

$$|\Phi| \leq 2^{c(s)(\log_2 n)^{1+\frac{1}{s}}} = 2^{(\log_2 n)^{1+\frac{1}{s}+\frac{\log_2 c(s)}{\log_2 \log_2 n}}} \leq 2^{(\log_2 n)^{1+\frac{3\alpha}{2\sqrt{\log_2 \log_2 n}}+\frac{\alpha^2 s}{\log_2 \log_2 n}}}$$

$$\leq 2^{(\log_2 n)^{1+\frac{3\alpha}{2\sqrt{\log_2 \log_2 n}}+\frac{\alpha}{\sqrt{\log_2 \log_2 n}}}} = 2^{(\log_2 n)^{1+\frac{2.5\alpha}{\sqrt{\log_2 \log_2 n}}}} .$$

□

Bounds on Complexity

Lemma 5.22. *Let $s \in \mathbb{N}$, $s \geq 1$ and $d(s) = 8(v+1)^{s+3}(\log_2(8ktm))^{2s-1}$. Then for any problem $z = (\nu, f_1, \ldots, f_n)$ over U there exists a decision tree Γ over U which solves z on A and for which $h(\Gamma) \leq rd(s)(\log_2 n)^{1+\frac{1}{s}} + 1$.*

Proof. First, we consider the case when $n = 1$ and $z = (\nu, f_1)$. Let $B = \{b_1, \ldots, b_k\}$. Denote by Γ the decision tree over U which consists of the root v_0, terminal nodes $v_1, \ldots, v_k$ and edges that connect the node v_0 with the nodes $v_1, \ldots, v_k$. The node v_0 is labelled by the attribute f_1. The nodes $v_1, \ldots, v_k$ are labelled by numbers $\nu(b_1), \ldots, \nu(b_k)$ respectively. The edges connecting v_0 with $v_1, \ldots, v_k$ are labelled by numbers $b_1, \ldots, b_k$ respectively. It is clear that Γ solves the problem z on A and $h(\Gamma) = 1$. So the statement of the lemma holds if $n = 1$.

Consider now the case when $n > 1$. From Lemma 5.20 follows that there exists an m-covering (Φ, C) of attributes $f_1, \ldots, f_n$ on $\Sigma_r(A, \Phi)$ such that $|\Phi| \leq 2^{c(s)(\log_2 n)^{1+\frac{1}{s}}}$ and $|C| \leq (k^2 |\Phi|)^v$ where $c(s) = 8(v+1)^{s+2}(\log_2(8ktm))^{2s-1}$. It is clear that

$$|C| \leq 2^{vc(s)(\log_2 n)^{1+\frac{1}{s}} + \log_2 k^{2v}} . \tag{49}$$

Using Lemma 5.15 we conclude that there exists a decision tree Γ over U with attributes from Φ which solves the problem z on the set $\Sigma_r(A, \Phi)$ and for which $h(\Gamma) \leq rv \log_2(k^2 |\Phi|)$ and $L(\Gamma) \leq k^{h(\Gamma)+1}$. It is clear that $h(\Gamma) \leq rv \left(2 \log_2 k + c(s)(\log_2 n)^{1+\frac{1}{s}}\right) \leq r(v+1)c(s)(\log_2 n)^{1+\frac{1}{s}} = rd(s)(\log_2 n)^{1+\frac{1}{s}} < rd(s)(\log_2 n)^{1+\frac{1}{s}} + 1$. From Lemma 5.13 follows that the decision tree Γ solves the problem z on A. □

Lemma 5.23. *Let $\alpha = \sqrt{2 \log_2(2(v+1) \log_2(8ktm))}$ and n be natural number such that $\log_2 \log_2 n \geq 9\alpha^2$. Then for any problem $z = (\nu, f_1, \ldots, f_n)$ over U there exists a decision tree Γ over U which solves z on A and for which $h(\Gamma) \leq 3rv \log_2 k(\log_2 n)^{1+\frac{2.5\alpha}{\sqrt{\log_2 \log_2 n}}}$.*

Proof. From Lemma 5.21 follows that there exists an m-covering (Φ, C) of attributes $f_1, \ldots, f_n$ on $\Sigma_r(A, \Phi)$ such that $|\Phi| \leq 2^{(\log_2 n)^{1+\frac{2.5\alpha}{\sqrt{\log_2 \log_2 n}}}}$ and $|C| \leq \left(k^2 |\Phi|\right)^v$. It is clear that

$$|C| \le 2^{v(\log_2 n)^{1+\frac{2.5\alpha}{\sqrt{\log_2 \log_2 n}}}+\log_2 k^{2v}} . \quad (50)$$

Using Lemma 5.15 we conclude that there exists a decision tree Γ over U with attributes from Φ which solves the problem z on $\Sigma_r(A, \Phi)$ and for which $h(\Gamma) \le rv \log_2(k^2 |\Phi|)$. It is clear that $h(\Gamma) \le rv \left(2\log_2 k + (\log_2 n)^{1+\frac{2.5\alpha}{\sqrt{\log_2 \log_2 n}}} \right) \le 3rv \log_2 k (\log_2 n)^{1+\frac{2.5\alpha}{\sqrt{\log_2 \log_2 n}}}$. From Lemma 5.13 follows that the decision tree Γ solves the problem z on A. □

Proofs of Theorems

Proof (of Theorem 5.1). Statement a) of the theorem follows from Lemma 5.7. Statement b) of the theorem follows from Lemma 5.4 and Lemma 5.22 if we set $s = \lceil \frac{1}{\varepsilon} \rceil$. Statement c) of the theorem follows from Lemma 5.6. □

Proof (of Theorem 5.2). Statement of the theorem follows from Lemma 5.23. □

5.2 Global Shannon Functions for Two-Valued Finite Information Systems and Depth

An information system $U = (A, B, F)$ will be called *two-valued* if $|B| = 2$. Now we consider the behavior of the global Shannon function for an arbitrary two-valued finite information system $U = (A, B, F)$ such that $f \not\equiv \text{const}$ for any $f \in F$.

Recall that by $\text{in}(U)$ we denote the maximal number of attributes in an independent subset of the set F (see Sect. 4.3).

A problem $z \in \text{Probl}_U$ will be called *stable* if $h_U^g(z) = \dim z$. We denote by $\text{st}(U)$ the maximal dimension of a stable problem over U.

One can show that

$$1 \le \text{st}(U) \le \text{in}(U) .$$

The values $\text{st}(U)$ and $\text{in}(U)$ will be called *the first and the second global critical points of the information system* U. Now we describe the behavior of the global Shannon function $H_{U,h}^g$ in terms of global critical points of U.

Theorem 5.3. *Let $U = (A, B, F)$ be a two-valued finite information system such that $f \not\equiv \text{const}$ for any $f \in F$. Then for any natural n the following statements hold:*

a) if $n \le \text{st}(U)$ then $H_{U,h}^g(n) = n$;
b) if $\text{st}(U) < n \le \text{in}(U)$ then

$$\max\{\text{st}(U), \log_2(n+1)\} \le H_{U,h}^g(n)$$
$$\le \min\left\{n-1, 4(\text{st}(U)+1)^4(\log_2 n)^2 + 4(\text{st}(U)+1)^5 \log_2 n\right\} ;$$

c) if $n \ge \text{in}(U)$ then $H_{U,h}^g(n) = H_{U,h}^g(\text{in}(U))$.

The problem of computing the values $\mathrm{st}(U)$ and $\mathrm{in}(U)$ for given two-valued finite information system U is complicated problem. However obtained results allow us to constrict the class of possible types of global Shannon functions.

Example 5.2. Denote by A the set of all points in the plane. Consider an arbitrary straight line l, which divides the plane into positive and negative open half-planes, and the line l itself. Assign a function $f : A \to \{0, 1\}$ to the line l. The function f takes the value 1 if a point is situated on the positive half-plane, and f takes the value 0 if a point is situated on the negative half-plane or on the line l. Denote by F the set of functions which correspond to certain r mutually disjoint finite classes of parallel straight lines. Consider a finite information system $U = (A, \{0, 1\}, F)$. One can show that $\mathrm{in}(U) = |F|$ and $\mathrm{st}(U) \le 2r$.

First, we prove some auxiliary statements.

Lemma 5.24. *Let $U = (A, B, F)$ be a two-valued finite information system such that $f \not\equiv \mathrm{const}$ for any $f \in F$. Then for any natural $n \ge 2$ the following inequality holds:*

$$H^g_{U,h}(n) \le 4(\mathrm{st}(U) + 1)^4(\log_2 n)^2 + 4(\mathrm{st}(U) + 1)^5 \log_2 n \ .$$

Proof. Denote $\rho = (F, B)$ and $m = \mathrm{st}(U)$. Let, for the definiteness, $B = \{0, 1\}$. It is clear that $H^g_{U,h}(m + 1) \le m$. For an arbitrary $p \in \mathbb{N} \setminus \{0\}$ denote by ν_p the mapping of the set $\{0, 1\}^p$ into the set $\mathbb{N}$ such that $\nu_p(\bar{\delta}_1) \ne \nu_p(\bar{\delta}_2)$ for any $\bar{\delta}_1, \bar{\delta}_2 \in \{0, 1\}^p$, $\bar{\delta}_1 \ne \bar{\delta}_2$.

Let us show that $I(T) \le m$ for any table $T \in \mathrm{Tab}_U$. Assume the contrary: let there exists a table $T \in \mathrm{Tab}_U$ such that $I(T) = t > m$. Then there exist attributes $f_1, \ldots, f_t \in F$ such that the system of equations

$$\{f_1(x) = \delta_1, \ldots, f_t(x) = \delta_t\}$$

is compatible on A for any $\delta_1, \ldots, \delta_t \in \{0, 1\}$. Consider a problem $z = (\nu_t, f_1, \ldots, f_t)$. It is not difficult to prove that $h^g_U(z) \ge t$. Since $\dim z = t$, we conclude that z is a stable problem which is impossible. Therefore $I(T) \le m$ for any table $T \in \mathrm{Tab}_U$. Using Theorem 4.6 we obtain that for any problem z over U the following inequality holds:

$$N(T(z)) \le 2^{2m}(\dim z)^m \ . \tag{51}$$

Let $f_1, \ldots, f_{m+1}$ be pairwise distinct attributes from F. Denote $z(f_1, \ldots, f_{m+1}) = (\nu_{m+1}, f_1, \ldots, f_{m+1})$. From $H^g_{U,h}(m + 1) \le m$ follows the existence of a decision tree $\Gamma(f_1, \ldots, f_{m+1})$ over U such that $h(\Gamma(f_1, \ldots, f_{m+1})) \le m$ and the decision tree $\Gamma(f_1, \ldots, f_{m+1})$ solves the problem $z(f_1, \ldots, f_{m+1})$. Evidently, for any $\delta_1, \ldots, \delta_{m+1} \in \{0, 1\}$ and for any complete path ξ in the tree $\Gamma(f_1, \ldots, f_{m+1})$ either $\mathrm{Sol}_U(\pi(\xi)) \cap \mathrm{Sol}_U((f_1, \delta_1) \ldots (f_{m+1}, \delta_{m+1})) = \emptyset$ or

$$\mathrm{Sol}_U(\pi(\xi)) \subseteq \mathrm{Sol}_U((f_1, \delta_1) \ldots (f_{m+1}, \delta_{m+1})) \ .$$

Using Lemma 3.4 obtain

$$L_w(\Gamma(f_1, \ldots, f_{m+1})) \le 2^m \ . \tag{52}$$

Let $f_1, \ldots, f_q$ be pairwise distinct attributes from F and let $q \leq m$. Define a decision tree $\Gamma(f_1, \ldots, f_q)$ over U in the following way. Let every working node of the tree $\Gamma(f_1, \ldots, f_q)$ have exactly two edges issuing from it, and let every complete path in the tree $\Gamma(f_1, \ldots, f_q)$ contain exactly q working nodes. Let $\xi = v_1, d_1, \ldots, v_q, d_q, v_{q+1}$ be an arbitrary complete path in the tree $\Gamma(f_1, \ldots, f_q)$. Then for $i = 1, \ldots, q$ the node v_i is labelled by the attribute f_i, and the node v_{q+1} is labelled by the number 0.

For every problem z over U we define by induction a subset $J(z)$ of the set F. If $\dim z \leq m$ then $J(z) = \mathrm{At}(z)$. Assume that for some n, $n \geq m+1$, for any problem z' over U with $\dim z' < n$ the set $J(z')$ has already been defined. Define the set $J(z)$ for a problem $z = (\nu, f_1, \ldots, f_n)$ over U. Let $n = t(m+1) + q$, where $t \in \mathbb{N} \setminus \{0\}$ and $0 \leq q \leq m$. For $i = 1, \ldots, t$ denote $\Gamma_i = \Gamma(f_{(m+1)(i-1)+1}, \ldots, f_{(m+1)(i-1)+m+1})$. Define a decision tree Γ_{t+1} over U. If $q = 0$ then the tree Γ_{t+1} contains only the node labelled by the number 0. If $q > 0$ then $\Gamma_{t+1} = \Gamma(f_{(m+1)t+1}, \ldots, f_{(m+1)t+q})$. Define decision trees $G_1, \ldots, G_{t+1}$ from Tree_U in the following way: $G_1 = \Gamma_1$ and for $i = 1, \ldots, t$ the tree G_{i+1} is obtained from the tree G_i by replacing of every terminal node v in the tree G_i with the tree Γ_{i+1} (the edge which had entered the node v will be entered the root of the tree Γ_{i+1}). Denote by $\Gamma(z)$ the decision tree that consists of all nodes and edges of the tree G_{t+1} for each of which there exists a complete path ξ containing it and satisfying the condition $\mathrm{Sol}_U(\pi(\xi)) \neq \emptyset$. One can show that $\bigcup_{\xi \in \mathrm{Path}(\Gamma(z))} \mathrm{Sol}_U(\pi(\xi)) = A$. Denote $c = 2m/(2m+1)$. One can easily show $h(G_{t+1}) \leq mt + q \leq cn$. Therefore

$$h(\Gamma(z)) \leq cn \ . \tag{53}$$

From (52) and from the description of the tree Γ_{t+1} it follows that $|\mathrm{At}(G_{t+1})| \leq t2^m + q \leq n2^m$. Using these inequalities and the inequality (51) we conclude that the tree G_{t+1} contains at most $2^{2m}(n2^m)^m = n^m 2^{m^2+2m}$ complete paths ξ such that $\mathrm{Sol}_U(\pi(\xi)) \neq \emptyset$. Therefore

$$|\mathrm{Path}(\Gamma(z))| \leq n^m 2^{m^2+2m} \ . \tag{54}$$

We correspond to every complete path ξ in the tree $\Gamma(z)$ a problem z_ξ over U. Let $\{f_{i_1}, \ldots, f_{i_p}\}$ be the set of attributes from F attached to working nodes of the path ξ. Then $z_\xi = (\nu_p, f_{i_1}, \ldots, f_{i_p})$. From (53) it follows that for any $\xi \in \mathrm{Path}(\Gamma(z))$ the inequality

$$\dim z_\xi \leq cn \tag{55}$$

holds. Hence, by assumption, the set $J(z_\xi)$ has already been determined for any $\xi \in \mathrm{Path}(\Gamma(z))$. Set

$$J(z) = \mathrm{At}(z) \cup \left(\bigcup_{\xi \in \mathrm{Path}(\Gamma(z))} J(z_\xi) \right) \ . \tag{56}$$

For $n \in \mathbb{N} \setminus \{0\}$ denote

$$J_U(n) = \max\{|J(z)| : z \in \mathrm{Probl}_U, \dim z \leq n\} \ .$$

The inequality

$$J_U(n) \le n^{2(m+1)^2 \ln n} 2^{2(m+1)^3 \ln n} \quad (57)$$

will be proven by induction on $n \ge 1$. It is clear that if $n \le m$ then $J_U(n) \le n$. Hence for $n \le m$ the inequality (57) holds. Let for some n, $n \ge m+1$, for any n', $1 \le n' < n$, the inequality (57) hold. Let us show that it holds also for n. Let $z \in \mathrm{Probl}_U$ and $\dim z \le n$. If $\dim z < n$ then using inductive hypothesis one can show $|J(z)| \le n^{2(m+1)^2 \ln n} 2^{2(m+1)^3 \ln n}$. Let $\dim z = n$. Evidently, $1 \le \lfloor cn \rfloor < n$ and $J_U(\lfloor cn \rfloor) \ge 1$. Using (54)–(56) obtain $|J(z)| \le n + n^m 2^{m^2+2m} J_U(\lfloor cn \rfloor) \le n^m 2^{m^2+2m+1} J_U(\lfloor cn \rfloor)$. Using the inductive hypothesis obtain $|J(z)| \le n^m 2^{(m+1)^2} (\lfloor cn \rfloor)^{2(m+1)^2 \ln \lfloor cn \rfloor} 2^{2(m+1)^3 \ln \lfloor cn \rfloor} \le n^{m+2(m+1)^2 \ln cn} 2^{(m+1)^2 + 2(m+1)^3 \ln cn}$.

Using the inequality $\ln(1 + 1/r) > 1/(r+1)$ which is true for any natural r we obtain $\ln c < -1/(2m+1) < -1/2(m+1)$. Hence

$$|J(z)| \le n^{2(m+1)^2 \ln n} 2^{2(m+1)^3 \ln n} \ .$$

Taking into account that z is an arbitrary problem over U such that $\dim z \le n$ we conclude that the inequality (57) holds.

The following statement will be proven by induction in n. Let $z = (\nu, f_1, ..., f_n) \in \mathrm{Probl}_U$, $J(z) = \{f_1, \ldots, f_p\}$, $\bar{\delta} = (\delta_1, \ldots, \delta_p) \in \{0,1\}^p$, let $\alpha(z, \bar{\delta}) = (f_1, \delta_1) \ldots (f_p, \delta_p)$ and $\beta(z, \bar{\delta}) = (f_1, \delta_1) \ldots (f_n, \delta_n)$. Then there exists a word $\gamma(z, \bar{\delta})$ from the set Ω_ρ such that $\mathrm{Alph}(\gamma(z, \bar{\delta})) \subseteq \mathrm{Alph}(\alpha(z, \bar{\delta}))$, $\mathrm{Sol}_U(\gamma(z, \bar{\delta})) \subseteq \mathrm{Sol}_U(\beta(z, \bar{\delta}))$ and the length of the word $\gamma(z, \bar{\delta})$ is at most $2(m+1)$. Recall that $\Omega_\rho = \{(f, \delta) : f \in F, \delta \in \{0,1\}\}^*$ and $\mathrm{Alph}(w)$ is the set of letters from the alphabet $\{(f, \delta) : f \in F, \delta \in \{0,1\}\}$ contained in the word $w \in \Omega_\rho$.

For $n \le 2(m+1)$ this statement is true since we can take the word $\beta(z, \bar{\delta})$ in the capacity of the word $\gamma(z, \bar{\delta})$. Suppose that for certain n, $n \ge 2(m+1)+1$, the statement is true for any problem z over U with $\dim z < n$. We will show that the considered statement holds for an arbitrary problem $z = (\nu, f_1, \ldots, f_n)$ over U. Let $J(z) = \{f_1, \ldots, f_p\}$ and $\bar{\delta} = (\delta_1, \ldots, \delta_p) \in \{0,1\}^p$. One can show that $\mathrm{At}(\Gamma(z)) \subseteq J(z)$. Consider a directed path $\kappa = v_1, d_1, \ldots, v_r, d_r, v_{r+1}$ in the tree $\Gamma(z)$ starting in the root and possessing the following properties:

1) if the node v_i, $i \in \{1, \ldots, r\}$, is labelled by an attribute f_l, then the edge d_i is labelled by the number δ_l;
2) if v_{r+1} is a working node in the tree $\Gamma(z)$ which is labelled by the attribute f_l then from v_{r+1} can not issue an edge labelled by the number δ_l.

First, assume that κ is a complete path in the tree $\Gamma(z)$. Let $n = t(m+1)+q$ where $t \ge 1$ and $0 \le q \le m$. For $i = 1, \ldots, t$ denote

$$\Gamma_i = \Gamma(f_{(m+1)(i-1)+1}, \ldots, f_{(m+1)(i-1)+m+1}) \ .$$

Define a decision tree Γ_{t+1} over U. If $q = 0$ then Γ_{t+1} consists of the root labelling by 0. If $q > 0$ then $\Gamma_{t+1} = \Gamma(f_{(m+1)t+1}, \ldots, f_{(m+1)t+q})$. Define words $\beta_1, \ldots, \beta_{t+1}$. For $i = 1, \ldots, i$ let

$$\beta_i = (f_{(m+1)(i-1)+1}, \delta_{(m+1)(i-1)+1}) \ldots (f_{(m+1)(i-1)+m+1}, \delta_{(m+1)(i-1)+m+1}) \ .$$

If $q = 0$ then $\beta_{t+1} = \lambda$. If $q > 0$ then

$$\beta_{t+1} = (f_{(m+1)t+1}, \delta_{(m+1)t+1}) \ldots (f_{(m+1)t+q}, \delta_{(m+1)t+q}) \ .$$

Evidently, $\beta(z,\bar{\delta}) = \beta_1 \dots \beta_{t+1}$. One can show that the word $\pi(\kappa)$ can be represented in the form $\pi(\kappa) = \pi(\xi_1)\dots\pi(\xi_{t+1})$ where ξ_i is a complete path in the tree Γ_i, $i = 1,\dots,t+1$.

Let there exist $i \in \{1,\dots,t\}$ such that $\mathrm{Sol}_U(\beta_i) \cap \mathrm{Sol}_U(\pi(\xi_i)) = \emptyset$. Denote $\gamma = \beta_i\pi(\xi_i)$. It is clear that $\mathrm{Alph}(\gamma) \subseteq \mathrm{Alph}(\alpha(\sigma,\bar{\delta}))$ and $\mathrm{Sol}_U(\gamma) = \emptyset$. Hence $\mathrm{Sol}_U(\gamma) \subseteq \mathrm{Sol}_U(\beta(z,\bar{\delta}))$ and the length of the word γ is at most $m+1+m < 2(m+1)$. Thus, in the considered case the word γ can be taken in the capacity of the word $\gamma(z,\bar{\delta})$.

Let $\mathrm{Sol}_U(\beta_i) \cap \mathrm{Sol}_U(\pi(\xi_i)) \neq \emptyset$ for $i = 1,\dots,t$. Then, as mentioned above, we have $\mathrm{Sol}_U(\pi(\xi_i)) \subseteq \mathrm{Sol}_U(\beta_i)$ for $i = 1,\dots,t$. Evidently, $\mathrm{Sol}_U(\pi(\xi_{t+1})) = \mathrm{Sol}_U(\beta_{t+1})$ and hence

$$\mathrm{Sol}_U(\pi(\kappa)) \subseteq \mathrm{Sol}_U(\beta(z,\bar{\delta})) \ . \tag{58}$$

Consider the problem z_κ. Let $z_\kappa = (\nu_l, f_{j_1},\dots,f_{j_l})$ and $J(z_\kappa) = \{f_{j_1},\dots,f_{j_u}\}$. From (56) follows $J(z_\kappa) \subseteq J(z)$. Denote $\bar{\delta}' = (\delta_{j_1},\dots,\delta_{j_u})$. Using (55) obtain $\dim z_\kappa < n$. From this inequality and from the inductive hypothesis follows that there exists a word $\gamma(z_\kappa,\bar{\delta}') \in \Omega_\rho$ such that $\mathrm{Alph}(\gamma(z_\kappa,\bar{\delta}')) \subseteq \mathrm{Alph}(\alpha(z_\kappa,\bar{\delta}'))$, $\mathrm{Sol}_U(\gamma(z_\kappa,\bar{\delta}')) \subseteq \mathrm{Sol}_U(\beta(z_\kappa,\bar{\delta}'))$, and the length of the word $\gamma(z_\kappa,\bar{\delta}')$ is at most $2(m+1)$. It is clear that $\mathrm{Alph}(\alpha(z_\kappa,\bar{\delta}')) \subseteq \mathrm{Alph}(\alpha(z,\bar{\delta}))$ and $\mathrm{Alph}(\gamma(z_\kappa,\bar{\delta}')) \subseteq \mathrm{Alph}(\alpha(z,\bar{\delta}))$. One can easily show $\mathrm{Sol}_U(\pi(\kappa)) = \mathrm{Sol}_U(\beta(z_\kappa,\bar{\delta}'))$. Using (58) obtain $\mathrm{Sol}_U(\gamma(z_\kappa,\bar{\delta}')) \subseteq \mathrm{Sol}_U(\beta(z,\bar{\delta}))$. Hence in this case the word $\gamma(z_\kappa,\bar{\delta}')$ can be taken in the capacity of the word $\gamma(z,\bar{\delta})$.

Suppose now that the path κ is not a complete path in the tree $\Gamma(z)$. Evidently, there exists a complete path ξ in the tree $\Gamma(z)$ containing the node v_{r+1}. Consider the problem z_ξ. Let $z_\xi = (\nu_l, f_{j_1},\dots,f_{j_l})$ and $J(z_\xi) = \{f_{j_1},\dots,f_{j_u}\}$. From (56) follows $J(z_\xi) \subseteq J(z)$. Denote $\bar{\delta}' = (\delta_{j_1},\dots,\delta_{j_u})$. Recalling that the path κ is not a complete path in the tree $\Gamma(z)$ we can show $\mathrm{Sol}_U(\beta(z_\xi,\bar{\delta}')) = \emptyset$. Using (55) obtain $\dim z_\xi < n$. From this inequality and from the inductive hypothesis follows that there exists word $\gamma(z_\xi,\bar{\delta}') \in \Omega_\rho$ such that $\mathrm{Alph}(\gamma(z_\xi,\bar{\delta}')) \subseteq \mathrm{Alph}(\alpha(z_\xi,\bar{\delta}'))$, $\mathrm{Sol}_U(\gamma(z_\xi,\bar{\delta}')) \subseteq \mathrm{Sol}_U(\beta(z_\xi,\bar{\delta}'))$ and the length of the word $\gamma(z_\xi,\bar{\delta}')$ is at most $2(m+1)$. It is clear that $\mathrm{Alph}(\alpha(z_\xi,\bar{\delta}')) \subseteq \mathrm{Alph}(\alpha(z,\bar{\delta}))$. Therefore $\mathrm{Alph}(\gamma(z_\xi,\bar{\delta}')) \subseteq \mathrm{Alph}(\alpha(z,\bar{\delta}))$. From the relation $\mathrm{Sol}_U(\gamma(z_\xi,\bar{\delta}')) \subseteq \mathrm{Sol}_U(\beta(z_\xi,\bar{\delta}')) = \emptyset$ follows $\mathrm{Sol}_U(\gamma(z_\xi,\bar{\delta}')) \subseteq \mathrm{Sol}_U(\beta(z,\bar{\delta}))$. Thus, in the considered case the word $\gamma(z_\xi,\bar{\delta}')$ may be taken in the capacity of the word $\gamma(z,\bar{\delta})$.

Let $n \geq 2$. Consider an arbitrary problem z over U with $\dim \sigma \leq n$. Let $z = (\nu, f_1,\dots,f_r)$ and $J(z) = \{f_1,\dots,f_p\}$. Consider also the problem $z' = (\nu', f_1,\dots,f_p)$ where $\nu' : \{0,1\}^p \to \mathbb{Z}$ and the equality $\nu'(\bar{\delta}) = \nu(\delta_1,\dots,\delta_r)$ holds for any tuple $\bar{\delta} = (\delta_1,\dots,\delta_p) \in \{0,1\}^p$. Denote $T = T(z')$. Using (51) and (57) obtain

$$\begin{aligned} N(T) &\leq 2^{2m}(n^{2(m+1)^2 \ln n} 2^{2(m+1)^3 \ln n})^m \\ &= n^{2m(m+1)^2 \ln n} 2^{2m+2m(m+1)^3 \ln n} \\ &\leq 2^{2(m+1)^3 (\log_2 n)^2 + 2(m+1)^4 \log_2 n} \ . \end{aligned} \tag{59}$$

Let us show that

$$M_{\rho,h}(T) \leq 2(m+1) \ . \tag{60}$$

Let $\bar{\delta} = (\delta_1, \ldots, \delta_p) \in \{0,1\}^p$. Then, by proved above, there exists a word $\gamma(z, \bar{\delta})$ from the set Ω_ρ such that $\mathrm{Alph}(\gamma(z, \bar{\delta})) \subseteq \mathrm{Alph}(\alpha(z, \bar{\delta}))$, $\mathrm{Sol}_U(\gamma(z, \bar{\delta})) \subseteq \mathrm{Sol}_U(\beta(z, \bar{\delta}))$, and the length of the word $\gamma(z, \bar{\delta})$ is at most $2(m+1)$. It is clear that $\mathrm{Alph}(\gamma(z, \bar{\delta})) \subseteq \{(f_1, \delta_1), \ldots, (f_p, \delta_p)\}$. Taking into account $\mathrm{Sol}_U(\gamma(z, \bar{\delta})) \subseteq \mathrm{Sol}_U(\beta(z, \bar{\delta}))$ we can easily show $T\gamma(z, \bar{\delta}) \in \mathrm{Dtab}_\rho$. Therefore $M_{\rho,h}(T, \bar{\delta}) \leq 2(m+1)$. Recalling that $\bar{\delta}$ is an arbitrary tuple from $\{0,1\}^p$ we conclude that the inequality (60) holds. From Theorem 3.5 and from inequalities (59) and (60) it follows that $h_\rho(T) \leq M_{\rho,h}(T) \log_2 N(T) \leq 4(m+1)^4 (\log_2 n)^2 + 4(m+1)^5 \log_2 n$. Using Theorem 4.1 obtain $h^l_U(z') = h_\rho(T)$. Evidently, for any element $a \in A$ the equality $z(a) = z'(a)$ holds. Therefore $h^g_U(z) = h^g_U(z')$. It is clear that $h^g_U(z') \leq h^l_U(z')$. Thus, $h^g_U(z) \leq h_\rho(T)$. Taking into account that z is an arbitrary problem over U with $\dim z \leq n$ we obtain $H^g_{U,h}(n) \leq 4(m+1)^4 (\log_2 n)^2 + 4(m+1)^5 \log_2 n$. □

Proof (of Theorem 5.3). Let $n \geq 1$.

a). Let $n \leq \mathrm{st}(U)$. Let $z = (\nu, f_1, \ldots, f_{\mathrm{st}(U)})$ be a problem over U such that $h^g_U(z) = \mathrm{st}(U)$. Consider the problem $z' = (\nu_n, f_1, \ldots, f_n)$ over U. Assume that $h^g_U(z') < n$. One can show that in this case $h^g_U(z) \leq \mathrm{st}(U)$ which is impossible. Therefore $h^g_U(z') = n$ and $H^g_{U,h}(n) \geq n$. It is clear that $H^g_{U,h}(n) \leq n$. Thus, $H^g_{U,h}(n) = n$.

b). Let $\mathrm{st}(U) < n \leq \mathrm{in}(U)$. It is clear that $\mathrm{st}(U) \leq H^g_{U,h}(n)$. Let us show that $\log_2(n+1) \leq H^g_{U,h}(n)$.
From the inequality $n \leq \mathrm{in}(U)$ it follows that there exists an independent subset $\{f_1, \ldots, f_n\}$ of the set F. It is clear that $n \geq 2$. For $i = 1, \ldots, n$ denote $z_i = (\nu_i, f_1, \ldots, f_i)$ and $T_i = T(z_i)$. Since $f_1 \not\equiv \mathrm{const}$, $N(T_1) = 2$. Let us show that $N(T_i) < N(T_{i+1})$ for $i = 1, \ldots, n-1$. Assume the contrary: let $N(T_i) = N(T_{i+1})$ for some $i \in \{1, \ldots, n-1\}$. One can show that in this case there exists a mapping $\mu : B^i \to B$ such that $f_{i+1}(a) = \mu(f_1(a), \ldots, f_i(a))$ for any $a \in A$ which is impossible. Thus, $N(T_1) = 2$ and $N(T_i) < N(T_{i+1})$ for $i = 1, \ldots, n-1$. Therefore $N(T_n) \geq n+1$. Let Γ be an arbitrary decision tree over U which solves the problem z_n. Denote $z = z_n \circ \Gamma$ and $T = T(z_n \circ \Gamma)$. Since $N(T_n) \geq n+1$ and T_n is a diagnostic table, we have $S(T) \geq n+1$. From Theorem 3.2 it follows that $h_\rho(T) \geq \log_2(n+1)$. Using Theorem 4.2 we conclude that $h(\Gamma) \geq \log_2(n+1)$. Since Γ is an arbitrary decision tree over U which solves the problem z_n, we obtain $h^g_U(z_n) \geq \log_2(n+1)$. Using the equality $h(z_n) = n$ we conclude that $\log_2(n+1) \leq H^g_{U,h}(n)$.
It is clear that $H^g_{U,h}(n) \leq n-1$. The inequality $H^g_{U,h}(n) \leq 4(\mathrm{st}(U)+1)^4(\log_2 n)^2 + 4(\mathrm{st}(U)+1)^5 \log_2 n$ follows from Lemma 5.24.

c). Let $n \geq \mathrm{in}(U)$. Consider an arbitrary problem z over U such that $\dim z \leq n$. Let $\{f_1, \ldots, f_t\}$ be an independent subset of the set $\mathrm{At}(z)$ with maximal cardinality. It is clear that $t \leq \mathrm{in}(U)$. One can show that there exists a mapping $\nu : B^t \to \mathbb{Z}$ such that for the problem $z' = (\nu, f_1, \ldots, f_t)$ the equality $z(a) = z'(a)$ holds for any $a \in A$. It is clear that $h^g_U(z) = h^g_U(z')$. Therefore $h^g_U(z) \leq H^g_{U,h}(t) \leq H^g_{U,h}(\mathrm{in}(U))$. Since z is an arbitrary problem over U such that $\dim z \leq n$, we obtain $H^g_{U,h}(n) \leq H^g_{U,h}(\mathrm{in}(U))$. It is clear that $H^g_{U,h}(n) \geq H^g_{U,h}(\mathrm{in}(U))$. Thus, $H^g_{U,h}(n) = H^g_{U,h}(\mathrm{in}(U))$. □

5.3 Global Shannon Functions for Infinite Information Systems and Arbitrary Weight Function

Let $U = (A, B, F)$ be an information system and ψ be a weight function for U. Let $z = (\nu, f_1, \ldots, f_n)$ be a problem over U. Denote $\psi(z) = \sum_{i=1}^{n} \psi(f_i)$. We denote by $\psi_U^g(z)$ the minimal weighted depth of a decision tree over U which solves the problem z. We will consider the relationships between the parameters $\psi_U^g(z)$ and $\psi(z)$. One can interpret the value $\psi(z)$ for the problem z as the weighted depth of the decision tree which solves the problem z in trivial way by computing sequentially the values of the attributes $f_1, \ldots, f_n$. We define the function $H_{U,\psi}^g : \mathbb{N} \setminus \{0\} \to \mathbb{N}$ in the following way:

$$H_{U,\psi}^g(n) = \max\{\psi_U^g(z) : z \in \mathrm{Probl}_U, \psi(z) \leq n\}$$

for any $n \in \mathbb{N} \setminus \{0\}$. The value $H_{U,\psi}^g(n)$ is the unimprovable upper bound on the value $\psi_U^g(z)$ for problems $z \in \mathrm{Probl}_U$ such that $\psi(z) \leq n$. The function $H_{U,\psi}^g$ is called *the global Shannon function* for the information system U and the weight function ψ.

If U is a finite information system then, evidently, $\psi_U^g(z) \leq \sum_{f \in F} \psi(f)$ for any problem z over U. Therefore $H_{U,\psi}^g(n) = O(1)$ for any finite information system U.

Let U be an infinite information system. It is clear that either $H_{U,\psi}^g(n) = n$ for infinitely many natural n or $H_{U,\psi}^g(n) < n$ for sufficiently large n.

The first type of behavior ($H_{U,\psi}^g(n) = n$ for infinitely many natural n) is bad for us: for problems with arbitrarily large total weight of attributes in problem description in the worst case the weighted depth of the globally optimal decision tree is equal to the weighted depth of the trivial decision tree. The second type of behavior ($H_{U,\psi}^g(n) < n$ for sufficiently large n) is most interesting for us: for an arbitrary problem with sufficiently large total weight of attributes in the problem description the weighted depth of the globally optimal decision tree is less than the weighted depth of the trivial decision tree.

The information system U will be called *ψ-compressible* if there exists n_0 such that for any $n \geq n_0$ the inequality $H_{U,\psi}^g(n) < n$ holds.

The information systems, which are h-compressible, were investigated in Sect. 5.1: the system U is h-compressible if and only if it has finite I-dimension and satisfies the condition of decomposition

In this subsection we consider arbitrary weight functions. We describe all pairs (U, ψ) of information systems and weight functions such that the information system U is ψ-compressible. For each such pair we investigate the behavior of the global Shannon function $H_{U,\psi}^g$.

Let $f \in F$ and Γ be a decision tree over U. We will say that Γ *simulates* f if Γ solves the problem $z_f = (\nu, f)$ where $\nu(\delta) = \delta$ for any number $\delta \in B$. Let $\Phi \subseteq F$. We will say that Γ is a decision tree *with attributes from* Φ if each working node of Γ is labelled by an attribute from Φ. For $p \in \mathbb{N} \setminus \{0\}$ denote $F_1(\psi, p) = \{f : f \in F, \psi(f) \leq p\}$ and $F_2(\psi, p) = F \setminus F_1(\psi, p)$.

We will say that U is *a two-layer information system regarding to the weight function* ψ if there exists $p \in \mathbb{N} \setminus \{0\}$ such that the information system

$$(A, B, F_1(\psi, p))$$

is h-compressible, and for any attribute $f \in F_2(\psi, p)$ there exists a decision tree Γ over U with attributes from $F_1(\psi, p)$ which simulates the attribute f and for which $\psi(\Gamma) < \psi(f)$. If U is a two-layer information system regarding to ψ, and p is the minimal number, for which the considered conditions hold, then the system U will be called *a* (ψ, p)*-two-layer* information system.

Let U be a (ψ, p)-two-layer system. For any $f \in F_2(\psi, p)$ denote by $\psi_{U,p}(f)$ the minimal weighted depth of a decision tree over U with attributes from $F_1(\psi, p)$ which simulates f. Define a function $K_{U,\psi} : \mathbb{N} \setminus \{0\} \to \mathbb{N}$. If $F_2(\psi, p) = \emptyset$ then $K_{U,\psi} \equiv 0$. Let $F_2(\psi, p) \neq \emptyset$ and $q = \min\{\psi(f) : f \in F_2(\psi, p)\}$. Let $n \in \mathbb{N} \setminus \{0\}$. If $n < q$ then $K_{U,\psi}(n) = 0$. If $n \geq q$ then

$$K_{U,\psi}(n) = \max\{\psi_{U,p}(f) : f \in F_2(\psi, p), \psi(f) \leq n\} .$$

Define a function $P_{U,\psi} : \mathbb{R} \to \mathbb{R}$ as follows:

$$P_{U,\psi}(n) = \max\left\{K_{U,\psi}(n), \log_{|B|} n - \log_{|B|} p\right\} .$$

The following theorem characterizes ψ-compressible information systems.

Theorem 5.4. *Let U be an infinite information system and ψ be a weight function for U. Then the following statements hold:*

a) the system U is ψ-compressible if and only if U is a two-layer system regarding to ψ;
b) if U is ψ-compressible then for any $\varepsilon > 0$ the following equalities hold: $H^g_{U,\psi}(n) = \Omega(P_{U,\psi}(n))$ and $H^g_{U,\psi}(n) = O(P_{U,\psi}(n)^{1+\varepsilon})$.

The following theorem illustrates the variety of the function $H^g_{U,\psi}$ behavior for pairs (U, ψ) such that U is ψ-compressible information system.

Theorem 5.5. *Let $\varphi : \mathbb{N} \setminus \{0\} \to \mathbb{N} \setminus \{0\}$ be a non-decreasing function such that $\lfloor \log_2 n \rfloor + 2 \leq \varphi(n) \leq n - 3$ for any $n \geq 7$. Then there exist an infinite information system U and a weight function ψ for U such that $\varphi(n) \leq H^g_{U,\psi}(n) \leq \varphi(n) + 2$ for any $n \geq 7$.*

Proofs of Theorems 5.4 and 5.5. The proof of Theorem 5.4 is divided into a sequence of lemmas.

Lemma 5.25. *Let $U = (A, B, F)$ be an infinite information system, ψ be a weight function for U, p be a natural number, and U be a (ψ, p)-two-layer system. Then the set $F_1(\psi, p)$ is an infinite set.*

Proof. Assume the contrary: let the set $F_1(\psi, p)$ be a finite set. Since for any attribute $f \in F_2(\psi, p)$ there exists a decision tree over U with attributes from $F_1(\psi, p)$, which simulates f, and $F_1(\psi, p)$ is a finite set, the set $F_2(\psi, p)$ is a finite set. Therefore F is a finite set, which is impossible. Thus, the set $F_1(\psi, p)$ is an infinite set. □

Lemma 5.26. *Let $U = (A, B, F)$ be an infinite information system, ψ be a weight function for U, p be a natural number, and U be a (ψ, p)-two-layer system. Then for any natural n, $n \geq p$, the inequality $H^g_{U,\psi}(n) \geq \log_{|B|} n - \log_{|B|} p$ holds.*

Proof. Let $n \geq p$. Set $t = \left\lfloor \frac{n}{p} \right\rfloor$. From Lemma 5.25 follows that $F_1(\psi, p)$ is an infinite set. Using this fact it is not difficult to prove that there exist attributes $f_1, \ldots, f_t \in F_1(\psi, p)$ such that the system of equations

$$\{f_1(x) = \delta_1, \ldots, f_t(x) = \delta_t\}$$

is compatible on A for at least $t+1$ tuples $(\delta_1, \ldots, \delta_t) \in B^t$. Consider a problem $z = (\nu, f_1, \ldots, f_t)$ over U such that $\nu(\overline{\sigma}_1) \neq \nu(\overline{\sigma}_2)$ if $\overline{\sigma}_1 \neq \overline{\sigma}_2$. Let Γ be a decision tree over U which solves z. It is clear that Γ has at least $t+1$ terminal nodes. One can show that the number of terminal nodes in Γ is at most $|B|^{h(\Gamma)}$. Therefore $h(\Gamma) \geq \log_{|B|}(t+1) = \log_{|B|}(\left\lfloor \frac{n}{p} \right\rfloor + 1) \geq \log_{|B|} n - \log_{|B|} p$. It is clear that $\psi(\Gamma) \geq h(\Gamma)$. Taking into account that Γ is an arbitrary decision tree over U, which solves z, we obtain $\psi^g_U(z) \geq \log_{|B|} n - \log_{|B|} p$. It is clear that $\psi^g_U(z) \leq n$. Therefore $H^g_{U,\psi}(n) \geq \log_{|B|} n - \log_{|B|} p$. □

Let $U = (A, B, F)$ be an infinite information system, ψ be a weight function for U, and U be a ψ-compressible system. Let n_0 be the minimal natural number such that for any $n > n_0$ the inequality $H^g_{U,\psi}(n) < n$ holds. Then we will say that U is ψ-compressible *with the threshold of compressibility* n_0. For any $f \in F$ denote $z_f = (\nu, f)$ where $\nu(\delta) = \delta$ for any number $\delta \in B$.

Lemma 5.27. *Let $U = (A, B, F)$ be an infinite information system, ψ be a weight function for U, U be ψ-compressible with the threshold of compressibility n_0, and $f \in F_2(\psi, n_0)$. Then $\psi^g_U(z_f) < \psi(f)$ and there exists a decision tree Γ_f over U with attributes from $F_1(\psi, n_0)$ which solves the problem z_f (i.e. simulates f) and for which $\psi(\Gamma_f) = \psi^g_U(z_f)$.*

Proof. Taking into account that $\psi(z_f) = \psi(f) > n_0$ we obtain $\psi^g_U(z_f) < \psi(f)$.

We prove the existence of Γ_f, satisfying the statement of lemma, by induction on the value $\psi(f)$. Let $\psi(f) = n_0 + 1$ and Γ be a decision tree over U which solves z_f and for which $\psi(\Gamma) = \psi^g_U(z_f)$. By proved above, $\psi(\Gamma) \leq n_0$. Therefore Γ is a decision tree with attributes from $F_1(\psi, n_0)$, and in the capacity of Γ_f we can take the decision tree Γ.

Assume that an appropriate decision tree Γ_f exists for any attribute $f \in F$ such that $n_0 < \psi(f) \leq n$, where $n \geq n_0 + 1$. Let $f \in F$ and $\psi(f) = n + 1$. Let us prove that there exists a decision tree Γ_f satisfying the statement of lemma. Let Γ be a decision tree over U which solves z_f and for which $\psi(\Gamma) = \psi^g_U(z_f)$. It is clear that $\psi(\Gamma) \leq n$. Therefore all working nodes of Γ are labelled by attributes which weight is at most n. By inductive hypothesis, for each attribute $g \in F_2(\psi, n_0)$ used in Γ there exists a decision tree Γ_g satisfying the statement of lemma. For each attribute $g \in F_2(\psi, n_0)$ used in Γ we substitute Γ_g for g in the following way. Let v be a working node in Γ which is labelled by an attribute

$g \in F_2(\psi, n_0)$. Denote by D the set of numbers from B which are labels of edges starting in v. For each $\delta \in D$ we denote by G_δ the subtree of Γ which root is the node connected with v by the edge starting in v and labelled by δ. We remove from Γ all edged starting in v and all subtrees G_δ, $\delta \in D$, connected with v by these edges. We substitute Γ_g for the node v. For each $\delta \in D$ we substitute the subtree G_δ for each terminal node of Γ_g which is labelled by δ. Denote Γ' the decision tree obtained from Γ by change of all attributes $g \in F_2(\psi, n_0)$. It is clear that $\psi(\Gamma') \leq \psi(\Gamma)$, Γ' solves the problem z_f, and Γ' is a decision tree over U with attributes from $F_1(\psi, n_0)$. Since $\psi(\Gamma) = \psi_U^g(z_f)$, we conclude that $\psi(\Gamma') = \psi_U^g(z_f)$. Therefore we can take in the capacity of Γ_f the decision tree Γ'. □

Lemma 5.28. *Let $U = (A, B, F)$ be an infinite information system, ψ be a weight function for U, U be ψ-compressible with the threshold of compressibility n_0, and z be a problem over U. Then there exists a decision tree Γ over U with attributes from $F_1(\psi, n_0)$ which solves the problem z and for which $\psi(\Gamma) = \psi_U^g(z)$.*

Proof. Let Γ' be a decision tree over U which solves z and for which $\psi(\Gamma') = \psi_U^g(z)$. From Lemma 5.27 follows that for any attribute f from $F_2(\psi, n_0)$ there exists a decision tree Γ_f over U with attributes from $F_1(\psi, n_0)$ which solves the problem z_f (i.e. simulates f) and for which $\psi(\Gamma_f) = \psi_U^g(z_f) < \psi(f)$. For each attribute $f \in F_2(\psi, n_0)$ used in Γ' we substitute Γ_f for f in the same way as in the proof of Lemma 5.27. As a result we obtain a decision tree Γ over U with attributes from $F_1(\psi, n_0)$. One can show that Γ solves z and $\psi(\Gamma) \leq \psi(\Gamma')$. Therefore $\psi(\Gamma) = \psi_U^g(z)$. □

Lemma 5.29. *Let $U = (A, B, F)$ be an infinite information system, ψ be a weight function for U, and U be ψ-compressible with the threshold of compressibility n_0. Then $W = (A, B, F_1(\psi, n_0))$ is ψ-compressible infinite information system.*

Proof. Let $n > n_0$, $z \in \mathrm{Probl}_W$ and $\psi(z) \leq n$. From Lemma 5.28 follows that $\psi_W^g(z) = \psi_U^g(z)$. Taking into account that U is ψ-compressible with the threshold of compressibility n_0 we obtain $\psi_W^g(z) < n$. Since z is an arbitrary problem over W with $\psi(z) \leq n$, we conclude that $\psi_W^g(n) < n$. Therefore W is ψ-compressible system.

Assume that $F_1(\psi, n_0)$ is a finite set. Using Lemma 5.27 it is not difficult to prove that the set $F_2(\psi, n_0)$ is a finite set too, which is impossible since F is an infinite set. □

Lemma 5.30. *Let $U = (A, B, F)$ be an infinite information system, ψ be a weight function for U, U be ψ-compressible, and there exists natural p such that $\psi(f) \leq p$ for any $f \in F$. Then U is h-compressible information system.*

Proof. Let U be ψ-compressible with the threshold of compressibility n_0. If $p = 1$ then the statement of the lemma holds. Let $p \geq 2$. Denote $m_0 = n_0 + p$. Let us show that for any natural $n \geq 2m_0$ the following inequality holds:

$$H^g_{U,\psi}(n) \le n - \frac{n}{m_0} + 1 \ . \tag{61}$$

Since $n > p$, the value $H^g_{U,\psi}(n)$ is definite. Let z be a problem over U such that $\psi(z) \le n$. Let $\psi(z) = b$ and $z = (\nu, f_1, \ldots, f_r)$. Let us show that $\psi^g_U(z) \le n - \frac{n}{m_0} + 1$. If $b \le m_0$ then, evidently, the considered inequality holds. Let $b > m_0$. Divide the sequence $f_1, \ldots, f_r$ into blocks such that the total weight of attributes in each block (with the exception of the last block) is at least $n_0 + 1$ and at most m_0. The total weight of attributes from the last block can be less than $n_0 + 1$. Let the number of blocks for which the total weight of attributes is at least $n_0 + 1$ is equal to q. For $i = 1, \ldots, q$ we denote by y_i the total weight of attributes in i-th block. One can show that $q \ge \left\lfloor \frac{b}{m_0} \right\rfloor$. It is clear that for $i = 1, \ldots, q$ there exists a decision tree which recognizes values of attributes from i-th block and which weighted depth is at most $y_i - 1$. Using this fact it is not difficult to prove that there exists a decision tree over U which solves z and which weighted depth is at most $b - q$. Therefore $\psi^g_U(z) \le b - q \le b - \left\lfloor \frac{b}{m_0} \right\rfloor \le b - \frac{b}{m_0} + 1 \le n - \frac{n}{m_0} + 1$. Thus, $\psi^g_U(z) \le n - \frac{n}{m_0} + 1$. Since z is an arbitrary problem over U such that $\psi(z) \le n$, the inequality (61) holds.

Set $c = \frac{2m_0 - 1}{2m_0}$. One can show that if $n \ge 2m_0$ then $n - \frac{n}{m_0} + 1 \le cn$. Using the inequality (61) we conclude that if $n \ge 2m_0$ then

$$H^g_{U,\psi}(n) \le cn \ . \tag{62}$$

Denote $m = 4m_0(1 + p)$. Let us prove by induction on t that for any natural t the following inequality holds:

$$H^g_{U,\psi}(m^t) \le (m - 1)^t \ . \tag{63}$$

First, consider the case $t = 1$. Let z be a problem over U and $\psi(z) \le m$. If $\psi(z) < m$ then $\psi^g_U(z) \le m - 1$. Let $\psi(z) = m$. Since $m > n_0$, we have $\psi^g_U(z) \le m - 1$. Taking into account that z is an arbitrary problem over U with $\psi(z) \le m$ we obtain $\psi^g_U(z) \le m - 1$. Therefore the inequality (63) holds if $t = 1$.

Assume that (63) holds for some $t \ge 1$. Let us prove that (63) holds for $t + 1$. Let $z = (\nu, f_1, \ldots, f_r)$ be a problem over U and $\psi(z) \le m^{t+1}$. If $\psi(z) < m^{t+1}$ then we add attributes $f_{r+1}, \ldots, f_{r'}$ such that the total weight of attributes $f_1, \ldots, f_{r'}$ is at least $m^{t+1} - p + 1$ and at most m^{t+1}. Later on we will solve the problem of recognition of values of attributes $f_1, \ldots, f_{r'}$. If this problem will be solved, the problem z will be solved too. Divide the sequence $f_1, \ldots, f_{r'}$ into $m + 1$ blocks. Denote by y_i the total weight of attributes from i-th block. We choose the division such that $m^t - p + 1 \le y_i \le m^t$ for $i = 1, \ldots, m$. The $(m + 1)$-th block contains all the other attributes from the sequence $f_1, \ldots, f_{r'}$. Since $m > p$, we have $y_{m+1} \le (p - 1)m \le p(m - 1)$.

By inductive hypothesis, for $j = 1, \ldots, m$ there exists a decision tree Γ_j over U which recognizes values of all attributes from j-th block and for which $\psi(\Gamma_j) \le (m - 1)^t$. It is clear that $h(\Gamma_j) \le (m - 1)^t$.

Let us describe the work of a decision tree Γ over U which simulates parallel work of decision trees $\Gamma_1, \ldots, \Gamma_m$ on an element $a \in A$, and after that computes the values of attributes from $(m+1)$-th block on a. Let $i \in \{1, \dots, (m-1)^t\}$. Let at the i-th step decision trees $\Gamma_1, \dots, \Gamma_m$ compute values of attributes which total weight is equal to w_i. If $w_i < 2m_0$ then at the i-th step Γ computes the values of the considered attributes directly. If $w_i \geq 2m_0$ then we can recognize values of the considered attributes with the help of a decision tree G_i over U for which $\psi(G_i) \leq cw_i$. The existence of G_i follows from (62). In this case at the i-th step Γ works as G_i. After the simulation of $\Gamma_1, \dots, \Gamma_m$ the decision tree Γ computes the values of attributes from $(m+1)$-th block. It is clear that Γ recognizes values of attributes $f_1, \dots, f_{r'}$. Therefore $\psi_U^g(z) \leq \psi(\Gamma)$.

Since the weight of each complete path in decision trees $\Gamma_1, \dots, \Gamma_m$ is at most $(m-1)^t$, we have $\sum_{i=1}^{(m-1)^t} w_i \leq m(m-1)^t$. Denote $M = \{1, \dots, (m-1)^t\}$, $I = \{i : i \in M, w_i < 2m_0\}$ and $D = \sum_{i \in I} w_i$. Then $D < 2m_0(m-1)^t$ and

$$
\begin{aligned}
\psi(\Gamma) &\leq c\sum_{i \in M \setminus I} w_i + \sum_{i \in I} w_i + p(m-1) \\
&\leq c(m(m-1)^t - D) + D + p(m-1) \\
&= cm(m-1)^t + (1-c)D + p(m-1) \\
&\leq (1 - \tfrac{1}{2m_0})m(m-1)^t + (m-1)^t + p(m-1) \\
&= (1 - \tfrac{1}{4m_0})m(m-1)^t - \tfrac{1}{4m_0}m(m-1)^t + (m-1)^t + p(m-1) \\
&= (1 - \tfrac{1}{4m_0})m(m-1)^t - (p+1)(m-1)^t + (m-1)^t + p(m-1) \\
&\leq (1 - \tfrac{1}{4m_0})m(m-1)^t = (4m_0 - 1)(1+p)(m-1)^t \\
&= (4m_0(1+p) - p - 1)(m-1)^t < (m-1)^{t+1} .
\end{aligned}
$$

Thus, $\psi(\Gamma) < (m-1)^{t+1}$ and $\psi_U(z) < (m-1)^{t+1}$. Taking into account that z is an arbitrary problem over U such that $\psi(z) \leq m^{t+1}$ we obtain $\psi_U^g(m^{t+1}) \leq (m-1)^{t+1}$. Therefore the inequality (63) holds for any natural t.

Denote $b_t = \frac{m^t}{(m-1)^t}$. It is known that the inequality $\ln(1 + \frac{1}{s}) > \frac{1}{s+1}$ holds for any natural s. Therefore $\ln b_t = t\ln(1 + \frac{1}{m-1}) > \frac{t}{m}$. Hence the sequence $b_1, b_2, \dots$ is not bounded from above. Choose a natural t such that $\frac{m^t}{(m-1)^t} > 3p$. Denote $r = \left\lfloor \frac{m^t}{p} \right\rfloor$. Let us show that

$$
h_U(r) \leq r - 1 . \tag{64}
$$

Let z be a problem over U and $h(z) \leq r$. If $h(z) < r$ then $h_U^g(z) \leq r - 1$. Let $h(z) = r$. Then $\psi(z) \leq rp \leq m^t$. Therefore $\psi_U^g(z) \leq (m-1)^t$. It is clear that $h_U^g(z) \leq (m-1)^t$. Assume that $h_U^g(z) \geq r$. Then $(m-1)^t \geq r$ and

$$
\frac{m^t}{(m-1)^t} \leq \frac{m^t}{r} = \frac{m^t}{\left\lfloor \frac{m^t}{p} \right\rfloor} \leq \frac{m^t}{\frac{m^t}{p} - 1} \leq \frac{m^t}{\frac{m^t}{2p}} = 2p
$$

which is impossible. Therefore $h_U^g(z) \leq r - 1$. Since z is an arbitrary problem over U with $h(z) \leq r$, we conclude that the inequality (64) holds.

Assume that U is not h-compressible. In this case from Theorem 5.1 follows that $H_{U,h}^g(r) = r$ which is impossible. Thus, U is h-compressible information system. □

Lemma 5.31. *Let $U = (A, B, F)$ be an infinite information system, ψ be a weight function for U, and U be ψ-compressible. Then U is a two-layer information system regarding to the weight function ψ.*

Proof. Let U be ψ-compressible with the threshold of compressibility n_0. From Lemma 5.27 follows that for any attribute $f \in F_2(\psi, n_0)$ there exists a decision tree Γ over U with attributes from $F_1(\psi, n_0)$ which simulates f and for which $\psi(\Gamma) < \psi(f)$. From Lemma 5.29 follows that the information system $W = (A, B, F_1(\psi, n_0))$ is ψ-compressible infinite information system. Using Lemma 5.30 we conclude that W is h-compressible information system. Therefore U is a two-layer information system regarding to the weight function ψ. □

Lemma 5.32. *Let $U = (A, B, F)$ be an infinite information system, ψ be a weight function for U, and U be a two-layer information system regarding to the weight function ψ. Then U is ψ-compressible information system.*

Proof. Let U be a (ψ, p)-two-layer system. Then the information system $W = (A, B, F_1(\psi, p))$ is h-compressible information system. From Lemma 5.25 follows that $F_1(\psi, p)$ is an infinite set. Therefore W is h-compressible infinite information system. From Theorem 5.1 follows that there exist positive constants c_1 and c_2 such that $H^g_{W,h}(n) \leq c_1(\log_2 n)^2 + c_2$ for any natural n. It is clear that there exists natural m such that $p\left(c_1(\log_2 n)^2 + c_2\right) < n$ for any natural n, $n \geq m$.

Let us show that for any $n \geq m$ the following inequality holds:

$$H^g_{U,\psi}(n) < n \; . \tag{65}$$

Let $n \geq m$, $z = (\nu, f_1, \ldots, f_t) \in \text{Probl}_U$ and $\psi(z) \leq n$. Let us show that $\psi_U(z) < n$.

First, consider the case when $f_1, \ldots, f_t \in F_1(\psi, p)$. In this case $z \in \text{Probl}_W$. Evidently, $h(z) \leq n$. Therefore there exists a decision tree Γ over W which solves z and for which $h(\Gamma) \leq c_1(\log_2 n)^2 + c_2$. It is clear that Γ is a decision tree over U and $\psi(\Gamma) \leq p\left(c_1(\log_2 n)^2 + c_2\right)$. Since $n \geq m$, we have $\psi(\Gamma) < n$ and $\psi^g_U(z) < n$.

Consider now the case when there exists an attribute $f_i \in \{f_1, \ldots, f_t\}$ such that $f_i \in F_2(\psi, p)$. Then there exists a decision tree Γ_{f_i} over U with attributes from $F_1(\psi, p)$ which simulates f_i and for which $\psi(\Gamma_{f_i}) < \psi(f_i)$. Describe the work of a decision tree Γ on an element $a \in A$. At the beginning Γ works as the decision tree Γ_{f_i} and recognizes the value of $f_i(a)$. Later Γ computes values $f_1(a), \ldots, f_{i-1}(a), f_{i+1}(a), \ldots, f_t(a)$ directly and as the final result finds the value $z(a) = \nu(f_1(a), \ldots, f_t(a))$. It is clear that Γ is a decision tree over U that solves the problem z. Since $\psi(\Gamma_{f_i}) < \psi(f_i)$, we conclude that $\psi(\Gamma) < \psi(z) \leq n$. Therefore $\psi^g_U(z) < n$.

Taking into account that z is an arbitrary problem over U, for which $\psi(z) \leq n$, we conclude that the inequality (65) holds if $n \geq m$. Thus, U is ψ-compressible information system. □

Lemma 5.33. *Let $U = (A, B, F)$ be an infinite information system, ψ be a weight function for U, and U be a (ψ, p)-two-layer system. Then for any natural n, $n \geq p$, the following inequality holds:*

$$H^g_{U,\psi}(n) \geq P_{U,\psi}(n) \ .$$

Proof. From Lemma 5.25 follows that $F_1(\psi, p)$ is a nonempty set. Therefore for $n \geq p$ the value $H^g_{U,\psi}(n)$ is definite and $H^g_{U,\psi}(n) \geq 0$. From Lemma 5.26 follows that the inequality

$$H^g_{U,\psi}(n) \geq \log_{|B|} n - \log_{|B|} p \tag{66}$$

holds for any natural n, $n \geq p$. Let us show that for any $n \geq p$ the inequality

$$H^g_{U,\psi}(n) \geq K_{U,\psi}(n) \tag{67}$$

holds. If $F_2(\psi, p) = \emptyset$ then $K_{U,\psi} \equiv 0$, and (67) holds. Let $F_2(\psi, p) \neq \emptyset$ and $q = \min\{\psi(f) : f \in F_2(\psi, p)\}$. If $n < q$ then $K_{U,\psi}(n) = 0$, and (67) holds. Let $n \geq q$. Then $K_{U,\psi}(n) = \max\{\psi_{U,p}(f) : f \in F_2(\psi, p), \psi(f) \leq n\}$, where $\psi_{U,p}(f)$ is the minimal weighted depth of a decision tree over U with attributes from $F_1(\psi, p)$ which simulates f. Let $y \in F_2(\psi, p)$, $\psi(y) \leq n$ and $\psi_{U,p}(y) = K_{U,\psi}(n)$. Let $z_y = (\nu, y)$ where $\nu(\delta) = \delta$ for any $\delta \in B$. It is clear that $\psi(z_y) = \psi(y) \leq n$. From Lemma 5.27 follows that $\psi^g_U(z_y) = \psi_{U,p}(y) = K_{U,\psi}(n)$. Therefore $H^g_{U,\psi}(n) \geq K_{U,\psi}(n)$. Thus, the inequality (67) holds. By definition, $P_{U,\psi}(n) = \max\{K_{U,\psi}(n), \log_{|B|} n - \log_{|B|} p\}$. Using (66) and (67) we obtain $H^g_{U,\psi}(n) \geq P_{U,\psi}(n)$. □

Lemma 5.34. *Let $U = (A, B, F)$ be an infinite information system, ψ be a weight function for U, and U be a (ψ, p)-two-layer system. Then for any $\varepsilon > 0$ there exists positive constant c such that for any natural n, $n \geq p^2 |B|$ the following inequality holds:*

$$H^g_{U,\psi}(n) \leq c P_{U,\psi}(n)^{1+\varepsilon} \ .$$

Proof. Denote $W = (A, B, F_1(\psi, p))$. Using Lemma 5.25 we conclude that W is an h-compressible infinite information system. Let $\varepsilon > 0$. From Theorem 5.1 follows that there exist positive constant d such that for any natural n the following inequality holds:

$$H^g_{W,h}(n) \leq d \left(\log_2 n\right)^{1+\varepsilon} + 1 \ . \tag{68}$$

Let $n \in \mathbb{N} \setminus \{0\}$ and $n \geq p^2 |B|$. Let $z \in \mathrm{Probl}_U$, $\psi(z) \leq n$ and, for the definiteness, $z = (\nu, f_1, \ldots, f_m, f_{m+1}, \ldots, f_{m+t})$ where $f_1, \ldots, f_m \in F_1(\psi, p)$ and $f_{m+1}, \ldots, f_{m+t} \in F_2(\psi, p)$. It is clear that $\psi(f_{m+i}) \leq n$ for $i = 1, \ldots, t$. Therefore $\psi_{U,p}(f_{m+i}) \leq K_{U,\psi}(n)$ for $i = 1, \ldots, t$. Hence for $i = 1, \ldots, t$ there exists a decision tree Γ_i over U with attributes from $F_1(\psi, p)$ which simulates the attribute f_{m+i} and for which $h(\Gamma_i) \leq \psi(\Gamma_i) \leq K_{U,\psi}(n)$. Denote $k = |B|$. It is not difficult to prove that for $i = 1, \ldots, t$ the decision tree Γ_i has at most $k^{K_{U,\psi}(n)}$ working nodes. Therefore for $i = 1, \ldots, t$ the decision tree Γ_i uses at most $k^{K_{U,\psi}(n)}$ attributes from $F_1(\psi, p)$. Values of these attributes are sufficient for recognition of the value of the attribute f_{m+i}. It is clear that $t \leq n - m$ and $K_{U,\psi}(n) \geq 0$. Thus, for the problem z solving it is sufficient to recognize values of at most $m + (n-m)k^{K_{U,\psi}(n)} \leq nk^{K_{U,\psi}(n)}$ attributes from $F_1(\psi, p)$. Denote $u = nk^{K_{U,\psi}(n)}$. Estimate the value $\log_2 u = \log_2 n + \log_2 kK_{U,\psi}(n)$. Let us show that $\log_2 n \leq 2\log_2 k(\log_k n - \log_k p)$. This inequality is equivalent to the inequality $\log_2 n \leq 2\log_2 n - 2\log_2 p$. The last inequality is equivalent to the inequality $p^2 \leq n$ which is true. Therefore $\log_2 n \leq 2\log_2 k(\log_k n - \log_k p)$. Taking into account that $\log_k n - \log_k p \leq P_{U,\psi}(n)$ we obtain $\log_2 n \leq 2\log_2 kP_{U,\psi}(n)$. It is clear that $\log_2 kK_{U,\psi}(n) \leq \log_2 kP_{U,\psi}(n)$. Therefore $\log_2 u \leq 3\log_2 kP_{U,\psi}(n)$. Using (68) we conclude that there exists a decision tree over U with attributes from $F_1(\psi, p)$ which solves z and which depth is at most

$$d\left(3\log_2 kP_{U,\psi}(n)\right)^{1+\varepsilon} + 1 \leq \left(d\left(3\log_2 k\right)^{1+\varepsilon} + 1\right) P_{U,\psi}(n)^{1+\varepsilon} .$$

The last inequality holds since $P_{U,\psi}(n)^{1+\varepsilon} \geq 1$. Denote $c_1 = pd\left(3\log_2 k\right)^{1+\varepsilon} + p$. Then $\psi_U^g(z) \leq cP_{U,\psi}(n)^{1+\varepsilon}$. Taking into account that z is an arbitrary problem from Probl_U such that $\psi(z) \leq n$ we obtain $H_{U,\psi}^g(n) \leq cP_{U,\psi}(n)^{1+\varepsilon}$. □

Proof (of Theorem 5.4). Statement a) of the theorem follows from Lemmas 5.31 and 5.32. Statement b) of the theorem follows from Lemmas 5.33 and 5.34. □

For $i \in \mathbb{N}$ define a function $f_i : \mathbb{R} \to \{0,1\}$ in the following way:

$$f_i(a) = \begin{cases} 0, \text{ if } a < i \ , \\ 1, \text{ if } a \geq i \ , \end{cases}$$

for any $a \in \mathbb{R}$. Denote $W = (\mathbb{R}, \{0,1\}, F)$ where $F = \{f_i : i \in \mathbb{N}\}$.

For any natural n define a problem z_n over the information system W in the following way: $z_n = (\nu_n, f_1, \ldots, f_n)$ where $\nu_n : \{0,1\}^n \to \{0,1\}$. Let $\bar{\delta} = (\delta_1, \ldots, \delta_n) \in \{0,1\}^n$, $\delta_1 = \ldots = \delta_k = 1$ (if $\delta_1 = 0$ then $k = 0$) and if $k < n$ then $\delta_{k+1} = 0$. Then

$$\nu_n(\bar{\delta}) = \begin{cases} 0, \text{ if } k \text{ is even} \ , \\ 1, \text{ if } k \text{ is odd} \ . \end{cases}$$

Define a function $q : \mathbb{N} \setminus \{0\} \to \mathbb{N}$ as follows: $q(n) = \lfloor \log_2 n \rfloor + 1$ for $n \in \mathbb{N} \setminus \{0\}$.

Lemma 5.35. *For the information system $W = (\mathbb{R}, \{0,1\}, F)$ the equalities $H_{W,h}^g(n) = q(n)$ and $h_W^g(z_n) = q(n)$ hold for any natural n.*

Proof. Denote $\rho = (F, \{0,1\})$. By induction on m it is not difficult to prove that $H_{W,h}^g(2^m - 1) \leq m$ for any natural m. Hence $H_{W,h}^g(n) \leq q(n)$ for any natural n.

Let $n \geq 1$ and Γ_n be a decision tree over W such that $h(\Gamma_n) = h_W^g(z_n)$ and the decision tree Γ_n solves the problem z_n. Denote $T = T_W(z_n \circ \Gamma_n)$. It is easily to show that the set $\{f_1, \ldots, f_n\}$ is a test for the table T of minimal cardinality. Therefore $J(T) = n$. Using Theorem 3.3 we obtain $h_\rho(T) \geq \lceil \log_2(n+1) \rceil = q(n)$. Using Theorem 4.2 we conclude that the tree Γ_n is a decision tree for the table T. Therefore $h(\Gamma_n) \geq q(n)$ and $h_W^g(z_n) \geq q(n)$. Taking into account that $h(z_n) = n$ and $H_{W,h}^g(n) \leq q(n)$ obtain $H_{W,h}^g(n) = q(n)$ and $h_W^g(z_n) = q(n)$. □

In the proof of Theorem 5.5 we will use notation which were defined before Lemma 5.35.

Proof (of Theorem 5.5). It is easily to notice that for any natural n the function $q(n)$ takes the value n on exactly 2^{n-1} natural numbers $2^{n-1}, \dots, 2^n - 1$.

Denote $\mathbb{N}(7) = \{n : n \in \mathbb{N}, n \geq 7\}$. Let $n \geq 1$. Let us show that the function $\varphi(n)$ takes the value n on at most 2^{n-1} numbers from $\mathbb{N}(7)$. Let $m \in \mathbb{N}(7)$ and $\varphi(m) = n$. Then $n \geq \lfloor \log_2 m \rfloor + 2$ and $n > \log_2 m + 1$. Therefore $m < 2^{n-1}$.

Denote $D = \{\varphi(n) : n \in \mathbb{N}(7)\}$. Let $d \in D$. Let $i_1, \dots, i_m$ be all numbers from $\mathbb{N}(7)$ in ascending order on which the function φ takes the value d. Let $n(1), \dots, n(k)$ be all numbers from $\mathbb{N} \setminus \{0\}$ in ascending order on which the function q takes the value d. As we have noticed above, $m \leq k$. For any natural i let y_i be a function such that $y_i(r) = z_i(r)$ for any $r \in \mathbb{R}$. Denote $G(d) = \{y_{n(1)}, \dots, y_{n(m)}\}$. Define on the set $G(d)$ the function ψ in the following way: $\psi(y_{n(1)}) = i_1, \dots, \psi(y_{n(m)}) = i_m$. Let $j \in \{1, \dots, m\}$. Since $\varphi(i_j) = d$, $i_j \in \omega(7)$ and $\varphi(n) < n$ for any $n \in \mathbb{N}(7)$, we have $i_j > d$. Therefore $\psi(y_{n(j)}) > d$. Using Lemma 5.35 obtain

$$\psi(y_{n(j)}) > h^g_W(z_{n(j)}) \ . \tag{69}$$

Denote $G = \bigcup_{d \in D} G(d)$. The function ψ had already been defined on the set G. Extend it to the set F in the following way: $\psi(f_i) = 1$ for any $f_i \in F$. Denote $V = (\mathbb{R}, \{0,1\}, F \cup G)$.

Let Γ be a decision tree over V, z be a problem over V, and the decision tree Γ solves the problem z. Using (69) we conclude that for any attribute $y_i \in G$ there exists a decision tree Γ_i over W for which $\psi(\Gamma_i) < \psi(y_i)$ and the decision tree Γ_i solves the problem z_i. Therefore all attributes $y_i \in G$ in the tree Γ may be "replaced" by the corresponding trees Γ_i such that we obtain a decision tree Γ' over W for which $\psi(\Gamma') \leq \psi(\Gamma)$ and the decision tree Γ' solves the problem z. That implies, in particular, that $\psi^g_V(z_n) = h^g_W(z_n)$ for any $n \geq 1$.

Let $n \in \mathbb{N}(7)$. One can show that the value $H^g_{V,\psi}(n)$ is definite.

We will show that $H^g_{V,\psi}(n) \geq \varphi(n)$. Let $\varphi(n) = d$. By construction, there exists an attribute $y_i \in G$ such that $\psi(y_i) = n$ and $q(i) = d$. Consider the problem $z = (\gamma, y_i)$ where $\gamma : \{0,1\} \to \{0,1\}$ and $\gamma(\delta) = \delta$ for any $\delta \in \{0,1\}$. Evidently, $\psi(z) = n$ and $\psi^g_V(z) = \psi^g_V(z_i)$. By proved above, $\psi^g_V(z_i) = h^g_W(z_i)$. Using Lemma 5.35 obtain $h^g_W(z_i) = q(i) = d$. Hence $H^g_{V,\psi}(n) \geq \varphi(n)$.

Let us show that $H^g_{V,\psi}(n) \leq \varphi(n) + 2$. Let z be a problem over V and $\psi(z) \leq n$. For the definiteness, let $z = (\nu, f_{i(1)}, \dots, f_{i(m)}, y_{j(1)}, \dots, y_{j(s)})$. Show that $\psi^g_V(z) \leq \varphi(n) + 2$.

Let $s = 0$. Using Lemma 5.35 one can easily show that $\psi^g_V(z) \leq q(n) \leq \varphi(n) - 1$.

Let $s > 0$. Denote $k = \max\{j(1), \dots, j(s)\}$. Let

$$f_{i(1)}, \dots, f_{i(m)}, f_{i(m+1)}, \dots, f_{i(p)}$$

be pairwise different attributes from F such that

$$\{f_{i(1)}, \dots, f_{i(p)}\} = \{f_{i(1)}, \dots, f_{i(m)}\} \cup \{f_1, \dots, f_k\} \ .$$

One can show that there exists a mapping $\gamma : \{0,1\}^p \to \{0,1\}$ such that for the problem $\vartheta = (\gamma, f_{i(1)}, \ldots, f_{i(p)})$ the equality $z(r) = \vartheta(r)$ holds for any real r. Therefore $\psi_V^g(z) = \psi_V^g(\vartheta)$. Obviously, $p \leq m + k$. Using Lemma 5.35 one can show that $\psi_V^g(\vartheta) \leq q(m+k)$. Hence

$$\psi_V^g(z) \leq q(m+k) \ . \tag{70}$$

Let us evaluate the value of k. Evidently, $m + \psi(y_k) \leq n$. Let $\psi(y_k) = a$. Therefore $\varphi(a) = q(k)$. Obviously, $a \leq n - m$. Since φ is a nondecreasing function, $q(k) \leq \varphi(n-m)$. Therefore $\lfloor \log_2 k \rfloor \leq \varphi(n-m) - 1$. Hence $\log_2 k < \varphi(n-m)$ and $k < 2^{\varphi(n-m)}$. Taking into account that q is a nondecreasing function and using (70) obtain $\psi_V^g(z) \leq \lfloor \log_2(m + 2^{\varphi(n-m)}) \rfloor + 1$. Let $c = \max\{m, 2^{\varphi(n-m)}\}$. Then $\psi_V^g(z) \leq \lfloor \log_2(2c) \rfloor + 1$. Let $c = m$. Then $\psi_V^g(z) \leq \lfloor \log_2 m \rfloor + 2 \leq \lfloor \log_2 n \rfloor + 2 \leq \varphi(n)$. Let $c = 2^{\varphi(n-m)}$. Then $\psi_V^g(z) \leq \lfloor \log_2 2^{\varphi(n-m)+1} \rfloor + 1 = \varphi(n-m) + 2 \leq \varphi(n) + 2$. Thus, $H_{V,\psi}^g(n) \leq \varphi(n) + 2$. □

5.4 Global Optimization Problems for Decision Trees

Relationships among three algorithmic problems are considered in this subsection: the problem of compatibility of equation system, the problem of construction of globally optimal decision tree, and the problem of computation of globally optimal decision tree complexity.

Relationships Among Algorithmic Problems. Let $U = (A, B, F)$ be an enumerated information system where $F = \{f_i : i \in \mathbb{N}\}$, and ψ be a computable weight function for U. Denote $\rho = (F, B)$. Let us describe two algorithmic problems of global optimization.

The Problem $\mathrm{Des}^g(U, \psi)$*:* for a given problem z over U it is required to find a decision tree Γ over U which solves z and for which $\psi(\Gamma) = \psi_U^g(z)$.

The Problem $\mathrm{Com}^g(U, \psi)$*:* for a given problem z over U it is required to compute the value $\psi_U^g(z)$.

Theorem 5.6. *Let $U = (A, B, F)$ be an enumerated information system, and ψ be a computable weight function for U. Then the following statements hold:*

a) if the problem $\mathrm{Ex}(U)$ *is unsolvable then the problems* $\mathrm{Com}^g(U, \psi)$ *and* $\mathrm{Des}^g(U, \psi)$ *are unsolvable;*
c) if the problem $\mathrm{Ex}(U)$ *is solvable then either both problems* $\mathrm{Com}^g(U, \psi)$ *and* $\mathrm{Des}^g(U, \psi)$ *are solvable or both of them are unsolvable.*

In the next subsection we will consider computable weight functions ψ for each of which there is no enumerated information system V such that the problem $\mathrm{Ex}(V)$ is solvable but the problems $\mathrm{Com}^g(V, \psi)$ and $\mathrm{Des}^g(V, \psi)$ are unsolvable.

Proof of Theorem 5.6. Let $U = (A, B, F)$ be an enumerated information system, $|B| = k$, $F = \{f_i : i \in \mathbb{N}\}$, and ψ be a computable weight function for U. Denote $\rho = (F, B)$.

Lemma 5.36. *Let Γ be a decision tree over U. Then the following statements hold:*

a) $\psi(\Gamma) \geq h(\Gamma)$;
b) $\psi(\Gamma) \geq \max\{\psi(f_i) : f_i \in \mathrm{At}(\Gamma)\}$;
c) $\psi(\Gamma) \geq 0$;
d) $\psi(\Gamma) = 0$ *if and only if the tree Γ consists of the only node.*

Proof. Let $\alpha \in F^*$. One can show that $\psi(\alpha) \geq h(\alpha)$, $\psi(\alpha) \geq \psi(f_i)$ for any letter f_i of the word α, $\psi(\alpha) \geq 0$, and $\psi(\alpha) = 0$ if and only if $\alpha = \lambda$. Using these relations it is not difficult to prove the statements of the lemma. □

Corollary 5.1. *Let z be a problem over U. Then $\psi_U^g(z) = 0$ if and only if* $z \equiv \mathrm{const}$.

Define an algorithmic problem $R(U)$.

Problem $R(U)$: for a given problem z over U and a decision tree Γ over U it is required to recognize whether the decision tree Γ solves the problem z.

Lemma 5.37. *The problem $R(U)$ is solvable if and only if the problem* $\mathrm{Ex}(U)$ *is solvable.*

Proof. Let the problem $\mathrm{Ex}(U)$ be solvable. Using Theorem 4.2 one can show that the problem $R(U)$ is solvable.

Let the problem $R(U)$ be solvable. Let us show that the problem $\mathrm{Ex}(U)$ is solvable. Let $\alpha \in \Omega_\rho$. If $\alpha = \lambda$ then, evidently, $\mathrm{Sol}_U(\alpha) \neq \emptyset$. Let $\alpha \neq \lambda$ and $\alpha = (f_{i(1)}, \delta_1) \dots (f_{i(n)}, \delta_n)$. Denote $z = (\nu, f_{i(1)}, \dots, f_{i(n)})$ where $\nu : B^n \to \{0, 1\}$ and for any tuple $\bar{\gamma} \in B^n$ the equality $\nu(\bar{\gamma}) = 1$ holds if and only if $\bar{\gamma} = (\delta_1, \dots, \delta_n)$. Denote by Γ the decision tree containing the only node. Let this node be labelled by the number 0. It is easily to notice that the decision tree Γ solves the problem z if and only if $\mathrm{Sol}_U(\alpha) = \emptyset$. Taking into account that the problem $R(U)$ is solvable we conclude that the problem $\mathrm{Ex}(U)$ is solvable. □

Lemma 5.38. *If problem* $\mathrm{Des}^g(U, \psi)$ *is solvable then problem* $\mathrm{Com}^g(U, \psi)$ *is solvable too.*

Proof. Let the problem $\mathrm{Des}^g(U, \psi)$ be solvable. Obviously, the function ψ is computable on the set of decision trees over U. Hence the problem $\mathrm{Com}^g(U, \psi)$ is solvable. □

Lemma 5.39. *If the problems* $\mathrm{Ex}(U)$ *and* $\mathrm{Com}^g(U, \psi)$ *are solvable then the problem* $\mathrm{Des}^g(U, \psi)$ *is also solvable.*

Proof. One can show that there exists an algorithm which consequently enumerates all decision trees over U.

Let the problems $\text{Ex}(U)$ and $\text{Com}^g(U, \psi)$ be solvable. Using Lemma 5.37 we conclude that the problem $R(U)$ is solvable. Evidently, the function ψ is computable on the set of decision trees over U.

Describe an algorithm which solves the problem $\text{Des}^g(U, \psi)$. Let z be a problem over U. First, we compute the value $\psi_U^g(z)$ with the help of algorithm solving the problem $\text{Com}^g(U, \psi)$. Then we will consequently scan the decision trees over U, using the algorithm for enumeration of the decision trees over U, the algorithm for computation of the function ψ and the algorithm for solution of the problem $R(U)$, until find a decision tree Γ such that $\psi(\Gamma) = \psi_U^g(z)$, and the decision tree Γ solves the problem z. This tree is a solution of the problem $\text{Des}^g(U, \psi)$ for the problem z. □

Proof (of Theorem 5.6). Let $\rho = (F, B)$ and $F = \{f_i : i \in \mathbb{N}\}$.

a) Let the problem $\text{Ex}(U)$ be unsolvable. Let us show that the problem $\text{Com}^g(U, \psi)$ is unsolvable. Assume the contrary. Let us prove that in this case the problem $\text{Ex}(U)$ is solvable. It is clear that there exists an attribute $f_{i(0)} \in F$ such that $f_{i(0)} \not\equiv \text{const}$. Let $\alpha \in \Omega_\rho$. If $\alpha = \lambda$ then, obviously, $\text{Sol}_U(\alpha) \neq \emptyset$. Let $\alpha \neq \lambda$ and $\alpha = (f_{i(1)}, \delta_1) \dots (f_{i(n)}, \delta_n)$. For any $r \in B$ consider a problem $z_r = (\nu_r, f_{i(0)}, f_{i(1)}, \dots, f_{i(n)})$ over U where $\nu_r : B^{n+1} \to \{0, 1\}$ and for any $\bar{\gamma} \in B^{n+1}$ if $\bar{\gamma} = (r, \delta_1, \dots, \delta_n)$ then $\nu_r(\bar{\gamma}) = 1$ and if $\bar{\gamma} \neq (r, \delta_1, \dots, \delta_n)$ then $\nu_r(\bar{\gamma}) = 0$. Find for any $r \in B$ the value $\psi_U^g(z_r)$ using the solvability of the problem $\text{Com}^g(U, \psi)$. Taking into account that $f_{i(0)} \not\equiv \text{const}$ and using Corollary 5.1 one can easily show that $\text{Sol}_U(\alpha) = \emptyset$ if and only if $\psi_U^g(z_r) = 0$ for any $r \in B$. So the problem $\text{Ex}(U)$ is solvable which contradicts the assumption. Therefore the problem $\text{Com}^g(U, \psi)$ is unsolvable. Using Lemma 5.38 we conclude that the problem $\text{Des}^g(U, \psi)$ is unsolvable.

b) Let the problem $\text{Ex}(U)$ be solvable. Using Lemmas 5.38 and 5.39 we conclude that the problems $\text{Des}^g(U, \psi)$ and $\text{Com}^g(U, \psi)$ either both are solvable or both are unsolvable. □

5.5 Proper Weight Functions

Definitions and Main Result. Let $\rho = (F, B)$ be an enumerated signature, $|B| = k$ and $F = \{f_i : i \in \mathbb{N}\}$. Let, for the definiteness, $B = \{0, \dots, k-1\}$.

A computable weight function ψ of the signature ρ will be called *proper* if for any information system $U = (A, B, F)$, such that the problem $\text{Ex}(U)$ is solvable, the problems $\text{Com}^g(U, \psi)$ and $\text{Des}^g(U, \psi)$ are also solvable.

In this subsection we consider a criterion for a computable weight function to be proper.

Let ψ be a weight function of the signature ρ. For $i \in \mathbb{N}$ denote $\mathbb{N}_\psi(i) = \{j : j \in \mathbb{N}, \psi(f_j) = i\}$. Define a partial function $K_\psi : \mathbb{N} \to \mathbb{N}$ as follows. Let $i \in \mathbb{N}$. If $\mathbb{N}_\psi(i)$ is a finite set then $K_\psi(i) = |\mathbb{N}_\psi(i)|$. If $\mathbb{N}_\psi(i)$ is an infinite set then the value of $K_\psi(i)$ is indefinite. Denote by $\text{Dom}(K_\psi)$ the domain of K_ψ.

Theorem 5.7. *Let ψ be a computable weight function of the signature ρ. Then ψ is a proper weight function if and only if K_ψ is a general recursive function.*

Consider examples of proper weight functions of the signature ρ. Let $\varphi : \mathbb{N} \to \mathbb{N} \setminus \{0\}$ be a general recursive nondecreasing function which is unbounded from above. Then the weight function ψ^φ such that $\psi^\varphi(f_i) = \varphi(i)$ for any $i \in \mathbb{N}$ is a proper weight function.

Proof of Theorem 5.7. The proof of Theorem 5.7 is divided into a sequence of lemmas.

Let ψ be a computable weight function of the signature ρ. Obviously, the weight function ψ satisfies exactly one of the three following conditions:

(a) K_ψ is a general recursive function;
(b) $\mathrm{Dom}(K_\psi) = \mathbb{N}$ and the function K_ψ is not a general recursive function;
(c) $\mathrm{Dom}(K_\psi) \neq \mathbb{N}$.

Lemma 5.40. *Let ψ satisfy the condition (a). Then the weight function ψ is proper.*

Proof. Let $U = (A, B, F)$ be an information system for which the problem $\mathrm{Ex}(U)$ is solvable. Using Lemma 5.37 we conclude that the problem $R(U)$ is solvable.

Taking into account that the function ψ satisfies the condition (a) one can show that there exists an algorithm which for a given number $r \in \mathbb{N}$ constructs the set $\{f_i : f_i \in F, \psi(f_i) \leq r\}$. Using this fact it is not difficult to prove that there exists an algorithm which for a given number $r \in \mathbb{N}$ and a finite nonempty subset M of the set $\mathbb{Z}$ constructs the set $\mathrm{Tree}(r, M)$ of all decision trees Γ over U satisfying the following conditions: $h(\Gamma) \leq r$, $\psi(f_i) \leq r$ for any attribute $f_i \in \mathrm{At}(\Gamma)$, and each terminal node of the tree Γ is labelled by a number from the set M.

Let z be a problem over U and $z = (\nu, f_{i(1)}, \ldots, f_{i(n)})$. Denote by $M(z)$ the range of values of the mapping ν. Let us show that the set $\mathrm{Tree}(\psi(z), M(z))$ contains a tree which is a solution of the problem $\mathrm{Des}^g(U, \psi)$ for the problem z. It is easily to notice that there exists a decision tree Γ over U which is a solution of the problem $\mathrm{Des}^g(U, \psi)$ for the problem z and in which all terminal nodes are labelled by numbers from the set $M(z)$. It is clear that $\psi(\Gamma) \leq \psi(z)$. Using this inequality obtain $h(\Gamma) \leq \psi(z)$ and $\psi(f_i) \leq \psi(z)$ for any attribute $f_i \in \mathrm{At}(\Gamma)$. Therefore $\Gamma \in \mathrm{Tree}(\psi(z), M(z))$.

Describe an algorithm which solves the problem $\mathrm{Des}^g(U, \psi)$. Let z be a problem over U. Compute the value $\psi(z)$ and construct the set $M(z)$. Construct the set $\mathrm{Tree}(\psi(z), M(z))$. With the help of algorithm which solves the problem $R(U)$ we find a decision tree $\Gamma \in \mathrm{Tree}(\psi(z), M(z))$ such that the tree Γ solves the problem z and $\psi(\Gamma) = \min\{\psi(G) : G \in \mathrm{Tree}(\psi(z), M(z)), G \in \mathrm{Tree}_U^g(z)\}$ where $\mathrm{Tree}_U^g(z)$ is the set of all decision trees over U solving z. The tree Γ is a solution of the problem $\mathrm{Des}^g(U, \psi)$ for the problem z. So the problem $\mathrm{Des}^g(U, \psi)$ is solvable. Using Lemma 5.38 we conclude that the problem $\mathrm{Com}^g(U, \psi)$ is also solvable. Taking into account that U is an arbitrary information system of the signature ρ such that the problem $\mathrm{Ex}(U)$ is solvable obtain ψ is proper weight function. □

Lemma 5.41. *Let ψ satisfy the condition (b). Then the weight function ψ is not proper.*

Proof. Define a function $\gamma : \mathbb{N} \to \mathbb{N}$ possessing the following property: for any $i \in \mathbb{N}$ there exists $t \in \mathbb{N}$ such that

$$\{\gamma(0), \ldots, \gamma(i)\} = \{0, \ldots, t\} . \tag{71}$$

Let $\gamma(0) = 0$. Assume that the values $\gamma(0), \ldots, \gamma(i)$ have been already defined and the equality (71) holds. Define the value $\gamma(i+1)$. For $j = 0, \ldots, t$ compute the values $m(j) = \min\{\psi(f_k) : p \in \mathbb{N}, p \leq i, \gamma(p) = j\}$. If there exists a number $l \in \{0, \ldots, t\}$ such that $\psi(f_{i+1}) < m(l)$, and the inequality $\psi(f_{i+1}) \geq m(j)$ holds for any $j \in \{0, \ldots, t\}$ such that $j < l$ then $\gamma(i+1) = l$. If $\psi(f_{i+1}) \geq m(j)$ for $j = 0, \ldots, t$ then $\gamma(i+1) = t + 1$. It is not difficult to show that γ is a general recursive function, and $\{\gamma(i) : i \in \mathbb{N}\} = \mathbb{N}$.

Define an information system $U = (\mathbb{N}, B, F)$ of the signature ρ as follows: let for any $i, j \in \mathbb{N}$

$$f_i(j) = \begin{cases} 1, \text{ if } j = \gamma(i) \ , \\ 0, \text{ if } j \neq \gamma(i) \ . \end{cases}$$

Let us show that the problem $\mathrm{Ex}(U)$ is solvable. Let $\alpha \in \Omega_\rho$. If $\alpha = \lambda$ then, evidently, $\mathrm{Sol}_U(\alpha) \neq \emptyset$. Let $\alpha \neq \lambda$ and $\alpha = (f_{i(1)}, \delta_1) \ldots (f_{i(n)}, \delta_n)$. One can show that $\mathrm{Sol}_U(\alpha) = \emptyset$ if and only if there exists numbers $s, p \in \{1, \ldots, n\}$ such that one of the following two conditions holds:

a) $\gamma(i(s)) = \gamma(i(p))$ and $\delta_s \neq \delta_p$;
b) $\gamma(i(s)) \neq \gamma(i(p))$ and $\delta_s = \delta_p = 1$.

Taking into account that the function γ is computable we conclude that the problem $\mathrm{Ex}(U)$ is solvable.

Define a function $r : \mathbb{N} \to \mathbb{N}$ in the following way: for any $i \in \mathbb{N}$ let $r(i) = \min\{j : j \in \mathbb{N}, \gamma(j) = i\}$. One can show that r is a general recursive function. For $i \in \mathbb{N}$ define a problem z_i over U in the following way: $z_i = (\nu_i, f_{r(i)})$ where $\nu_i : B \to B$ and $\nu_i(\delta) = \delta$ for any $\delta \in B$.

Let $j \in \mathbb{N}$. Define a decision tree Γ_j over U as follows. The root of the tree Γ_j is labelled by the attribute f_j. For any $\delta \in \{0, 1\}$ there exists an edge d_δ issuing from the root of Γ_j and entering the node v_δ. Both the edge d_δ and the node v_δ are labelled by the number δ. There are no other edges or nodes in the tree Γ_j. A decision tree Γ over U will be called *a tree of the kind* Γ_j if $\Gamma = \Gamma_j$ or the tree Γ_j can be obtained from the tree Γ by removal certain terminal nodes and the edges which enter them. Let Γ be a tree of the kind Γ_j. One can easily notice that the decision tree Γ solves the problem z_i if and only if $\gamma(j) = i$.

Let $i \in \mathbb{N}$ and let Γ be a decision tree over U which is a solution of the problem $\mathrm{Des}^g(U, \psi)$ for the problem z_i. Let us show that there exists $j \in \mathbb{N}$ such that the tree Γ is a tree of the kind Γ_j. Obviously, there exists a complete path ξ in the tree Γ such that $i \in \mathrm{Sol}_U(\pi(\xi))$. The terminal node of this path is labelled by the number 1. Therefore $\mathrm{Sol}_U(\pi(\xi)) = \{i\}$. Based on this equality one can easily show that the word $\pi(\xi)$ contains the letter $(f_j, 1)$ where $\gamma(j) = i$. Assume that $\pi(\xi) \neq (f_j, 1)$. Then $\psi(\Gamma) > \psi(\Gamma_j)$ which is impossible since the decision tree Γ_j solves the problem z_i. Therefore $\pi(\xi) = (f_j, 1)$. Hence the root of Γ is labelled by the attribute f_j. Based on the obvious inequality $\psi(\Gamma) \leq \psi(\Gamma_j)$, one can easily show that the tree Γ is a tree of the kind Γ_j. Thus, for any $i \in \mathbb{N}$ the following equality holds:

$$\psi_U^g(z_i) = \min\{\psi(f_j) : j \in \mathbb{N}, \gamma(j) = i\} . \tag{72}$$

Let us show that the problem $\mathrm{Des}^g(U, \psi)$ is unsolvable. Assume the contrary. Let us show that in this case the function K_ψ is computable.

Define a function $\varphi : \mathbb{N} \to \mathbb{N}$ in the following way: $\varphi(i) = \psi_U^g(z_i)$ for any $i \in \mathbb{N}$. Using Lemma 5.38 we conclude that the problem $\mathrm{Com}^g(U, \psi)$ is solvable. Taking into account that r is a general recursive function we conclude that φ is also a general recursive function. From the equality (72) it follows that $\varphi(i) = \min\{\psi(f_j) : j \in \mathbb{N}, \gamma(j) = i\}$ for any $i \in \mathbb{N}$. Using the definition of the function γ obtain φ is a nondecreasing function and $\varphi(0) = \min\{\psi(f_j) : j \in \mathbb{N}\}$. Taking into account that $\mathrm{Dom}(K_\psi) = \mathbb{N}$ one can easily show that φ is an unbounded above function. From these properties of the function φ follows that there exists an algorithm which for a given $t \in \mathbb{N}$, $t \geq \varphi(0)$, finds the maximal number $i \in \mathbb{N}$ such that $\varphi(i) \leq t$.

Describe an algorithm which computes the function K_ψ. Let $t \in \mathbb{N}$. Compute the value $K_\psi(t)$. If $t < \varphi(0)$ then $K_\psi(t) = 0$. Let $t \geq \varphi(0)$. Find the maximal $i \in \mathbb{N}$ such that $\varphi(i) \leq t$. Find a tree Γ which is a solution of the problem $\mathrm{Des}^g(U, \psi)$ for the problem z_i. Let Γ be a tree of the kind Γ_j. One can show that if $p > j$ then $\psi(f_p) \neq t$ for any $p \in \mathbb{N}$. Therefore $K_\psi(t) = |\{p : p \in \mathbb{N}, p \leq j, \psi(f_p) = t\}|$. Thus, the function K_ψ is computable which is impossible. Therefore the problem $\mathrm{Des}^g(U, \psi)$ is unsolvable, and the function ψ is not proper. □

Lemma 5.42. *Let ψ satisfy the condition (c). Then the weight function ψ is not proper.*

Proof. Let m be a minimal number from set $\mathbb{N} \setminus \mathrm{Dom}(K_\psi)$. One can show that there exists a general recursive function $r : \mathbb{N} \to \mathbb{N}$ such that $\{r(i) : i \in \mathbb{N}\} = \{j : j \in \mathbb{N}, \psi(f_j) = m\}$ and $r(i) < r(i+1)$ for any $i \in \mathbb{N}$. Evidently, there exists an algorithm which for a given $j \in \mathbb{N}$, such that $\psi(f_j) = m$, finds $i \in \mathbb{N}$ such that $j = r(i)$.

For any $i \in \mathbb{N}$ define the functions α, β_i, γ_i and ε from $\mathbb{N}$ to $\{0, 1\}$. Let $j \in \mathbb{N}$. Then

$$\alpha(j) = \begin{cases} 1, \text{ if } j \text{ is even} , \\ 0, \text{ if } j \text{ is odd} , \end{cases}$$

$$\beta_i(j) = \begin{cases} 0, \text{ if } j \neq 2i \text{ and } j \neq 2i+1 , \\ 1, \text{ if } j = 2i \text{ or } j = 2i+1 , \end{cases}$$

$$\gamma_i(j) = \begin{cases} 0, \text{ if } j \neq 2i+1 , \\ 1, \text{ if } j = 2i+1 , \end{cases}$$

$$\varepsilon(j) = 0 .$$

Let $g : \mathbb{N} \to \mathbb{N}$ be a general recursive function with nonrecursive range.

Define an information system $U = (\mathbb{N}, B, F)$ of the signature ρ: for any $j \in \mathbb{N}$

a) if $\psi(f_j) \neq m$ then $f_j = \varepsilon$;
b) if $j = r(0)$ then $f_j = \alpha$;
c) if $j = r(2i+1)$ for some $i \in \mathbb{N}$ then $f_j = \beta_i$;
d) if $j = r(2i+2)$ for some $i \in \mathbb{N}$ then $f_j = \gamma_{g(i)}$.

Let us show that the problem $\mathrm{Ex}(U)$ is solvable. Let $\mu \in \Omega_\rho$. If $\mu = \lambda$ then, obviously, $\mathrm{Sol}_U(\mu) \neq \emptyset$. Let $\mu \neq \lambda$ and $\mu = (f_{i(1)}, \delta_1) \dots (f_{i(n)}, \delta_n)$. Denote $S = \{f_{i(1)}(x) = \delta_1, \dots, f_{i(n)}(x) = \delta_n\}$. One can show that the equation system S is incompatible on $\mathbb{N}$ if and only if it contains a subsystem of at least one of the following kinds:

- $\{f_i(x) = \delta\}$ where $i \in \mathbb{N}$ and $\delta \in B \setminus \{0, 1\}$;
- $\{f_i(x) = 0, f_i(x) = 1\}$ where $i \in \mathbb{N}$;
- $\{\varepsilon(x) = 1\}$;
- $\{\alpha(x) = 0, \gamma_i(x) = 1\}$ where $i \in \mathbb{N}$;
- $\{\gamma_i(x) = 1, \gamma_j(x) = 1\}$ where $i, j \in \mathbb{N}$ and $i \neq j$;
- $\{\beta_i(x) = 1, \beta_j(x) = 1\}$ where $i, j \in \mathbb{N}$ and $i \neq j$;
- $\{\gamma_i(x) = 1, \beta_j(x) = 1\}$ where $i, j \in \mathbb{N}$ and $i \neq j$;
- $\{\gamma_i(x) = 1, \beta_i(x) = 0\}$ where $i \in \mathbb{N}$;
- $\{\alpha(x) = 1, \beta_i(x) = 1, \gamma_i(x) = 0\}$ where $i \in \mathbb{N}$.

Thus, the problem $\mathrm{Ex}(U)$ is solvable.

Let $i \in \mathbb{N}$. Denote $z_i = (\nu_i, f_{r(0)}, f_{r(2i+1)})$ where $\nu_i : B^2 \to \{0, 1\}$ and $\nu_i(\delta_1, \delta_2) = 1$ if and only if $\delta_1 = \delta_2 = 1$. One can show that

$$z_i(j) = \gamma_i(j) \tag{73}$$

for any $j \in \mathbb{N}$. Let Γ be a decision tree over U which is a solution of the problem $\mathrm{Des}^g(U, \psi)$ for the problem z_i. From the equality (73) follows that $z_i \not\equiv \mathrm{const}$. Hence there exists an attribute $f_j \in \mathrm{At}(\Gamma)$ such that $\psi(f_j) = m$. Using Lemma 5.36 obtain $\psi(\Gamma) \geq m$.

Assume $\psi(\Gamma) = m$. Bearing in mind the relation $f_j \in \mathrm{At}(\Gamma)$ obtain $h(\Gamma) = 1$ and see that the root of Γ is labelled by the attribute f_j. Using the equality (73) one can show that $f_j = \gamma_i$.

Let for certain $j \in \mathbb{N}$ the equality $f_j = \gamma_i$ hold. Then one can easily see that $\psi_U^g(z_i) = m$.

Thus, $\psi_U^g(z_i) = m$ if and only if the number i belongs to the set of values of the function g.

Assume that the problem $\mathrm{Com}^g(U, \psi)$ is solvable. Then the range of the function g is recursive which is impossible. Thus, the problem $\mathrm{Com}^g(U, \psi)$ is unsolvable. Therefore the function ψ is not proper. □

Proof (of Theorem 5.7). The statement of the theorem follows from Lemmas 5.40–5.42. □

6 Decision Trees over Quasilinear Information Systems

The notion of linear information system was introduced in Sect. 2. Each problem over linear information system can be represented in the following form. We take finite number of hyperplanes in the space $\mathbb{R}^n$. These hyperplanes divide the space into domains. These domains are numbered such that different domains can have the same number. For a given point of the space it is required to recognize the number of a domain which contains the point. Decision trees over the considered information system used attributes of the kind $\text{sign}\,(\sum_{i=1}^{n} a_i x_i + a_{n+1})$. This attribute allows to recognize the position of a point relatively the hyperplane defined by the equality $\sum_{i=1}^{n} a_i x_i + a_{n+1} = 0$.

Quasilinear information systems is simple and useful generalization of linear information systems: instead of attributes of the kind $\text{sign}\,(\sum_{i=1}^{n} a_i x_i + a_{n+1})$ we consider attributes of the kind $\text{sign}\,(\sum_{i=1}^{n} a_i \varphi_i(x) + a_{n+1})$ where $\varphi_1, \ldots, \varphi_n$ are functions from a set A to $\mathbb{R}$. Upper bounds on complexity of decision trees and algorithms for construction of decision trees over quasilinear information systems are considered in the section.

This section consists of five subsections. The first subsection contains main definitions and results. In the second subsection preliminary lemmas and in the third subsection principal lemmas are proved. The forth subsection contains proofs of main theorems of the section. In the fifth subsection for some pairs (*information system, weight function*) the behavior of global Shannon functions is studied.

6.1 Bounds on Complexity and Algorithms for Construction of Decision Trees over Quasilinear Information Systems

In this subsection we define the notion of quasilinear information system and formulate main results of the section.

Quasilinear Information Systems. We will call a set K *a numerical ring with unity* if $K \subseteq \mathbb{R}$, $1 \in K$ and for any $a, b \in K$ the relations $a + b \in K$, $a \cdot b \in K$ and $-a \in K$ hold. For instance, $\mathbb{R}$, $\mathbb{Q}$, $\mathbb{Z}$ and $\{a + b\sqrt{2} : a, b \in \mathbb{Z}\}$ are numerical rings with unity. Let K be a numerical ring with unity, A be a nonempty set, and $\varphi_1, \ldots, \varphi_n$ be functions from A to $\mathbb{R}$. Denote

$$F(A, K, \varphi_1, \ldots, \varphi_n) = \left\{ \text{sign} \left(\sum_{i=1}^{n} a_i \varphi_i(x) + a_{n+1} \right) : a_1, \ldots, a_{n+1} \in K \right\} .$$

The information system $(A, \{-1, 0, +1\}, F(A, K, \varphi_1, \ldots, \varphi_n))$ will be denoted by $U(A, K, \varphi_1, \ldots, \varphi_n)$ and will be called *a quasilinear information system*.

Let $f \in F(A, K, \varphi_1, \ldots, \varphi_n)$ and $f = \text{sign}\,(\sum_{i=1}^{n} a_i \varphi_i(x) + a_{n+1})$. We define the parameter $r(f)$ of the attribute f as follows. If $(a_1, \ldots, a_{n+1}) = (0, \ldots, 0)$ then $r(f) = 0$. Otherwise

$$r(f) = \max\{0, \max\{\log_2 |a_i| : i \in \{1, \ldots, n+1\}, a_i \neq 0\}\} .$$

For a problem $z = (\nu, f_1, \ldots, f_k)$ over $U(A, K, \varphi_1, \ldots, \varphi_n)$ denote $r(z) = \max\{r(f_i) : i = 1, \ldots, k\}$. For a decision tree Γ over $U(A, K, \varphi_1, \ldots, \varphi_n)$ denote $r(\Gamma) = \max\{r(f) : f \in \text{At}(\Gamma)\}$ (if $\text{At}(\Gamma) = \emptyset$ then $r(\Gamma) = 0$).

Note 6.1. In contrast to Sects. 4 and 5 we assume in this section that different attributes (attributes with different names) can coincide as functions. For example, different attributes $\text{sign}(x + 1)$ and $\text{sign}(2x + 2)$ coincide as functions.

Main Results

Theorem 6.1. *Let $U = U(A, K, \varphi_1, \ldots, \varphi_n)$ be a quasilinear information system. Then for any problem z over U there exists a decision tree Γ over U which solves z and for which*

$$h(\Gamma) \leq (2(n+2)^3 \log_2(\dim z + 2n + 2))/(\log_2(n+2)) \tag{74}$$

and

$$r(\Gamma) \leq 2(n+1)^2(r(z) + 1 + \log_2(n+1)) \ . \tag{75}$$

Theorem 6.2. *Let $U = U(A, \mathbb{Z}, \varphi_1, \ldots, \varphi_n)$ be a quasilinear information system. Then there exists an algorithm which for a given problem z over U constructs a decision tree Γ over U such that Γ solves the problem z and the inequalities (74) and (75) hold.*

The proofs of Theorems 6.1 and 6.2 are contained in Sect. 6.4. The preliminary lemmas are proven in Sect. 6.2 and the principal ones on which the proofs of the theorems are based can be found in Sect. 6.3. In Sect. 6.5 some corollaries of Theorem 6.1 are considered.

6.2 Preliminary Lemmas

In this subsection certain statements are considered relating mainly to the theory of linear inequalities.

Introduce some notation which will be used in the present section. Let K be a numerical ring with unity. For $n \geq 1$ denote $L_n(K) = \{\sum_{i=1}^n a_i x_i + a_{n+1} : a_i \in K, 1 \leq i \leq n+1\}$ and denote $S_n(K) = \{\text{sign}(g) : g \in L_n(K)\}$. Functions from the sets $L_n(K)$ and $S_n(K)$ are defined on $\mathbb{R}^n$. Extend the mapping r on the set $L_n(K)$ in the following way: $r(\sum_{i=1}^n a_i x_i + a_{n+1}) = r(\text{sign}(\sum_{i=1}^n a_i x_i + a_{n+1}))$. The symbol $'$ will be used for notation of the operation of matrix transposition.

Let $f_1, \ldots, f_k \in L_n(\mathbb{R})$ and $f_j(\bar{x}) = \sum_{i=1}^n a_{ji} x_i + a_{jn+1} = q_j(\bar{x}) + a_{jn+1}$, $1 \leq j \leq k$. The maximal number of linearly independent (over $\mathbb{R}$) functions in the set $\{q_1, \ldots, q_k\}$ will be called *the rank* of each of the following systems:

$$\{f_1(\bar{x}) = 0, \ldots, f_k(\bar{x}) = 0\} \ , \tag{76}$$

$$\{f_1(\bar{x}) \geq 0, \ldots, f_k(\bar{x}) \geq 0\} \ , \tag{77}$$

$$\{f_1(\bar{x}) > 0, \ldots, f_k(\bar{x}) > 0\} \ . \tag{78}$$

Lemma 6.1. ([46], pp. 83, 85, 88). *Let V_1 be the set of solutions of the compatible system (76) of the rank t, V_2 be the set of solutions of the system*

$$\{q_1(\bar{x}) = 0, \ldots, q_k(\bar{x}) = 0\} \tag{79}$$

corresponding to the system (76), and $\bar{y}_0 \in V_1$. Then $V_1 = \{\bar{y}_0 + \bar{y} : \bar{y} \in V_2\}$ and the maximal number of linearly independent elements in the set V_2 is equal to $n - t$.

Lemma 6.2. ([46], pp. 57, 75, 83, 84). *Let $a_{ij} \in \mathbb{R}$, $1 \leq i \leq n$, $1 \leq j \leq n+1$, and let the elements $(a_{11}, \ldots, a_{1n}), \ldots, (a_{n1}, \ldots, a_{nn})$ be linearly independent. Then*

a) the system

$$\begin{cases} a_{11}x_1 + \ldots + a_{1n}x_n = a_{1n+1} \\ \ldots\ldots\ldots\ldots\ldots\ldots\ldots\ldots \\ a_{n1}x_1 + \ldots + a_{nn}x_n = a_{nn+1} \end{cases}$$

has the unique solution $(d_1/d_0, \ldots, d_n/d_0)$ *where* $d_0 = \begin{vmatrix} a_{11} \ldots a_{1n} \\ \ldots\ldots\ldots \\ a_{n1} \ldots a_{nn} \end{vmatrix}$, $d_0 \neq 0$,

and d_j, $1 \leq j \leq n$, *is an* n*-th order determinant obtained from the determinant* d_0 *by substitution the column* $(a_{1n+1}, \ldots, a_{nn+1})'$ *for the* j*-th column of* d_0*;*

b) if M_j, $1 \leq j \leq n$, *is an* $(n-1)$*-th order determinant obtained from the determinant* d_0 *by deletion of the last row and the* j*-th column then the element* $(M_1, -M_2, M_3, -M_4, \ldots, (-1)^{n-1}M_n)$ *is a solution of the system*

$$\begin{cases} a_{11}x_1 + \ldots + a_{1n}x_n = 0 \\ \ldots\ldots\ldots\ldots\ldots\ldots\ldots\ldots \\ a_{n-11}x_1 + \ldots + a_{n-1n}x_n = 0 \end{cases},$$

any other solution of this system is proportional to it, and $\sum_{i=1}^{n} |M_i| > 0$.

Any two equation systems or inequality systems will be called *equivalent* if their sets of solutions coincide.

Lemma 6.3. *Any compatible system (76) of the rank* t *contains a subsystem of the rank* t *which is equivalent to (76) and consists of* t *equations. Any incompatible system (76) of the rank* t *contains an incompatible subsystem with* $t+1$ *equations.*

Proof. Consider the system (76). Denote

$$A = \begin{vmatrix} a_{11} \ldots a_{1n} \\ \ldots\ldots\ldots \\ a_{k1} \ldots a_{kn} \end{vmatrix},$$

$$\bar{A} = \begin{vmatrix} a_{11} \ldots a_{1n} a_{1n+1} \\ \ldots\ldots\ldots\ldots \\ a_{k1} \ldots a_{kn} a_{kn+1} \end{vmatrix}.$$

We denote by $\mathrm{rank}(A)$ (respectively by $\mathrm{rank}(\bar{A})$) the maximal number of linearly independent rows of the matrix A (respectively $\bar{A}$). It is known ([46], pp. 74, 78) that the system (76) is compatible if and only if $\mathrm{rank}(\bar{A}) = \mathrm{rank}(A)$, and is incompatible if and only if $\mathrm{rank}(\bar{A}) = \mathrm{rank}(A) + 1$ (the Kronecker-Capelli theorem).

Let the system (76) of the rank t be compatible. Then it follows from the Kronecker-Capelli theorem that the maximal number of linearly independent functions in the set $\{f_1, \ldots, f_k\}$ is equal to t. Let for $i_1, \ldots, i_t \in \{1, \ldots, k\}$ the functions $f_{i_1}, \ldots, f_{i_t}$ be linearly independent. One can show that in this case the equation system $\{f_{i_1}(\bar{x}) = 0, \ldots, f_{i_t}(\bar{x}) = 0\}$ is equivalent to the equation system (76).

Let the equation system (76) of the rank t be incompatible. Then it follows from the Kronecker-Capelli theorem that there exist numbers $i_1, \ldots, i_{t+1} \in \{1, \ldots, k\}$ such that the functions $f_{i_1}, \ldots, f_{i_{t+1}}$ are linearly independent. Using the Kronecker-Capelli theorem one can show that the equation system $\{f_{i_1}(\bar{x}) = 0, \ldots, f_{i_{t+1}}(\bar{x}) = 0\}$ is incompatible. □

A plane of the space $\mathbb{R}^n$ is the set of solutions of a compatible on $\mathbb{R}^n$ equation system of the kind (76) where $f_1, \ldots, f_k \in L_n(\mathbb{R})$. If t is the rank of this equation system then the number $n-t$ is called *the dimension* of the plane. A set $V \subseteq \mathbb{R}^n$ will be called *t-dimensional set* if there exists a t-dimensional plane of the space $\mathbb{R}^n$ containing V, and if $t \geq 1$ then there does not exist an $(t-1)$-dimensional plane containing V.

Lemma 6.4. *A finite set $\{\bar{b}_1, \ldots, \bar{b}_m\}$, $m \geq 2$, of elements of the space $\mathbb{R}^n$ is of the dimension t if and only if the maximal number of linearly independent elements in the set $\{\bar{b}_1 - \bar{b}_2, \ldots, \bar{b}_1 - \bar{b}_m\}$ is equal to t.*

Proof. Let the maximal number of linearly independent elements in the set $\{\bar{b}_1 - \bar{b}_2, \ldots, \bar{b}_1 - \bar{b}_m\}$ be equal to t.

Assume that $t \neq 0$. Consider the equation systems

$$\left\{ \begin{array}{l} b_{11}x_1 + \ldots + b_{1n}x_n + x_{n+1} = 0 \\ \ldots\ldots\ldots\ldots\ldots\ldots\ldots\ldots \\ b_{m1}x_1 + \ldots + b_{mn}x_n + x_{n+1} = 0 \end{array} \right. , \tag{80}$$

$$\left\{ \begin{array}{l} (b_{11} - b_{21})x_1 + \ldots + (b_{1n} - b_{2n})x_n = 0 \\ \ldots\ldots\ldots\ldots\ldots\ldots\ldots\ldots \\ (b_{11} - b_{m1})x_1 + \ldots + (b_{1n} - b_{mn})x_n = 0 \end{array} \right. \tag{81}$$

where $(b_{j1}, \ldots, b_{jn}) = \bar{b}_j$, $1 \leq j \leq m$. Let the set $\{\bar{b}_1, \ldots, \bar{b}_m\}$ be contained in a plane of the space $\mathbb{R}^n$ and let this plane be the set of solutions of the system (76). Then the element $(a_{j1}, \ldots, a_{jn}, a_{jn+1})$, $1 \leq j \leq k$, is a solution of the equation system (80), while the element $(a_{j1}, \ldots, a_{jn})$ is a solution of the system (81). By assumption, the rank of the system (81) is equal to t. From this and from Lemma 6.1 follows that the maximal number of linearly independent solutions of the system (81) is equal to $n - t$. Hence the rank of the system (76) is at most $n - t$. Therefore there is no an $(t-1)$-dimensional plane containing the set $\{\bar{b}_1, \ldots, \bar{b}_m\}$.

Let us show that the set $\{\bar{b}_1, \ldots, \bar{b}_m\}$ is contained in an t-dimensional plane. It is clear that this statement holds if $t = n$. Assume that $t \neq n$. Then by Lemma 6.1 there exist $n - t$ linearly independent solutions

$$(c_{11}, \ldots, c_{1n}), \ldots, (c_{n-t1}, \ldots, c_{n-tn})$$

of the system (81). Consider the equation system

$$\left\{ \begin{array}{l} c_{11}x_1 + \ldots + c_{1n}x_n = \sum_{i=1}^{n} b_{1i}c_{1i} \\ \ldots\ldots\ldots\ldots\ldots\ldots\ldots\ldots \\ c_{n-t1}x_1 + \ldots + c_{n-tn}x_n = \sum_{i=1}^{n} b_{1i}c_{n-ti} \end{array} \right. .$$

It is easily to notice that the rank of this system is equal to $n-t$, and the element $\bar{b}_j$, $1 \leq j \leq m$, is a solution of it. Hence the set $\{\bar{b}_1, \ldots, \bar{b}_m\}$ is contained in a t-dimensional plane. Therefore the dimension of the set $\{\bar{b}_1, \ldots, \bar{b}_m\}$ is equal to t.

Let the dimension of the set $\{\bar{b}_1, \ldots, \bar{b}_m\}$ be equal to t. Assume that the maximal number of linearly independent elements in the set $\{\bar{b}_1 - \bar{b}_2, \ldots, \bar{b}_1 - \bar{b}_m\}$ is equal to t_1 and $t_1 \neq t$. Then, by proved above, the dimension of the set $\{\bar{b}_1, \ldots, \bar{b}_m\}$ is equal to t_1 which is impossible. Therefore the maximal number of linearly independent elements in the set $\{\bar{b}_1 - \bar{b}_2, \ldots, \bar{b}_1 - \bar{b}_m\}$ is equal to t. □

Lemma 6.5. *Let $\{\bar{b}_1, \ldots, \bar{b}_n\}$, $n \geq 2$, be an $(n-1)$-dimensional subset of $\mathbb{R}^n$ and let each of two $(n-1)$-dimensional planes in the space $\mathbb{R}^n$, defined by equations $a_1x_1 + \ldots + a_nx_n + a_{n+1} = 0$ and $c_1x_1 + \ldots + c_nx_n + c_{n+1} = 0$, contain the elements $\bar{b}_1, \ldots, \bar{b}_n$. Then there exists a number $p \in \mathbb{R}$ such that $(a_1, \ldots, a_{n+1}) = p(c_1, \ldots, c_{n+1})$.*

Proof. Let $\bar{b}_j = (b_{j1}, \ldots, b_{jn})$, $1 \leq j \leq n$. The equation systems obtained from (80) and (81) by the substitution n for m will be denoted by (80a) and (81a) respectively. It follows from Lemma 6.4 that the rank of the system (81a) is equal to $n-1$. Since the elements $(a_1, \ldots, a_n)$ and $(c_1, \ldots, c_n)$ are solutions of the system (81a), from Lemma 6.2 follows that there exists a number $p \in \mathbb{R}$ such that $(a_1, \ldots . a_n) = p(c_1, \ldots, c_n)$. From the fact that $(a_1, \ldots, a_{n+1})$ is a solution of the system (80a) it follows $a_{n+1} = -\sum_{i=1}^{n} a_i b_{1i}$. In the same way we obtain $c_{n+1} = -\sum_{i=1}^{n} c_i b_{1i}$. Therefore $a_{n+1} = pc_{n+1}$, and $(a_1, \ldots, a_{n+1}) = p(c_1, \ldots, c_{n+1})$. □

Lemma 6.6. ([15], p. 92) *If the set of solutions of the compatible system (77) with a nonzero rank is bounded then the rank of the system (77) is equal to n.*

A set of the kind $\{\sum_{i=1}^{k} p_i \bar{a}_i : p_i \in \mathbb{R}, p_i \geq 0, 1 \leq i \leq k, \sum_{i=1}^{k} p_i = 1\}$, where $\bar{a}_1, \ldots, \bar{a}_k \in \mathbb{R}^n$ and $k < \infty$, is called *a finitely generated centroid in the space* $\mathbb{R}^n$. We will say that the centroid is *generated* by the set of elements $\{\bar{a}_1, \ldots, \bar{a}_k\}$ which are *generating elements* of the centroid. An element $\bar{a}_j$, $j \in \{1, \ldots, k\}$, will be called *a vertex* of the centroid if the centroid generated by the set $\{\bar{a}_1, \ldots, \bar{a}_k\} \setminus \{\bar{a}_j\}$ does not contain the element $\bar{a}_j$. A finite set $V \subset \mathbb{R}^n$ all elements of which are vertices of the centroid generated by it will be called *centroidally independent.*

Lemma 6.7. ([15], pp. 197) *A finitely generated centroid V of the space $\mathbb{R}^n$ has unique centroidally independent set of generating elements.*

A solution of the system (77) of the rank $t > 0$ will be called *nodal solution* if the inequalities of the system (77), which are equalities on this solution, form a system of the rank t.

Lemma 6.8. ([15], pp. 194, 196) *If the set V of solutions of the compatible system (77) is bounded then the set of nodal solutions of it is finite and centroidally independent. The centroid generated by this set coincides with V.*

The set of solutions of an inequality of the kind $f(\bar{x}) \geq 0$, where $f(\bar{x})$ is a function from $L_n(\mathbb{R})$ which is not constant, will be called *a half-space of the space* $\mathbb{R}^n$. The plane defined by the equations $f(\bar{x}) = 0$ will be called *the boundary* plane of this half-space.

Let V be a finitely generated n-dimensional centroid in $\mathbb{R}^n$. *An extreme support* of the centroid U is any half-space of the space $\mathbb{R}^n$ containing the set V which boundary plane includes at least n vertices of the centroid U.

Lemma 6.9. ([15], pp. 197) *Any finitely generated n-dimensional centroid U of the space $\mathbb{R}^n$ coincides with the intersection of its extreme supports.*

Lemma 6.10. ([15], pp. 124) *Let $f \in L_n(\mathbb{R})$ and let the inequality $f(\bar{x}) \geq 0$ be a consequence of the compatible system (77). Then there exist nonnegative numbers $p_1, \ldots, p_{k+1} \in \mathbb{R}$ such that $f(\bar{a}) = \sum_{i=1}^{k} p_i f_i(\bar{a}) + p_{k+1}$ for any $\bar{a} \in \mathbb{R}^n$.*

The inequality $f_j(\bar{x}) \geq 0$, $j \in \{1, \ldots, k\}$, from the compatible system (77) will be called *unstable* if the inequality $f_j(\bar{x}) \leq 0$ is a consequence of the system (77), and *stable* otherwise.

Lemma 6.11. *If in the system (77) the function f_j is not constant for any $j \in \{1, \ldots, k\}$ then the set of solutions of this system is n-dimensional if and only if the system (78), corresponding to the system (77), is compatible.*

Proof. Let the set V of solutions of the system (77) be n-dimensional. We will prove that for any $j \in \{1, \ldots, k\}$ there exists an element $\bar{a}_j \in V$ such that $f_j(\bar{a}_j) > 0$. Assume that a number $j \in \{1, \ldots, k\}$ exists such that the equality $f_j(\bar{b}) = 0$ holds for any $\bar{b} \in V$. Then, since the function f_j is not constant on $\mathbb{R}^n$, the dimension of the set V is at most $n - 1$ which is impossible. Set $\bar{a} = (1/k) \sum_{i=1}^{k} \bar{a}_i$. Then $f_j(\bar{a}) = (1/k) \sum_{i=1}^{k} f_j(\bar{a}_i) \geq f_j(\bar{a}_j)/k > 0$ for any $j \in \{1, \ldots, k\}$. Hence the system (78) corresponding to (77) is compatible.

It is known ([15], pp. 306, 311) that if the system (77) is compatible, and for any $j \in \{1, \ldots, k\}$ the function f_j is not constant then the set of solutions of this system is n-dimensional if and only if the system (77) does not contain unstable inequalities.

Let the system (78) corresponding to (77) be compatible. Then the system (77) is compatible and does not contain unstable inequalities. Besides, by the assumption the function f_j, $1 \leq j \leq k$, is not constant. Therefore the set of solutions of the system (77) is n-dimensional. □

The system (77) will be called *stably compatible* if the system (78) corresponding to it is compatible.

A compatible inequality system will be called *uncancellable* if the deletion of any inequality from it implies a change in the set of solutions of the system.

Lemma 6.12. ([15], p. 288) *The stably compatible system (77) is uncancellable if and only if the system (78) corresponding to it is uncancellable.*

Lemma 6.13. ([15], p. 288) *If the system (78) is compatible and uncancellable then for any $j \in \{1, \ldots, k\}$ the system obtained from (78) by substitution of the equation $f_j(\bar{x}) = 0$ for the inequality $f_j(\bar{x}) > 0$ is compatible.*

Lemma 6.14. ([15], p. 111) *The incompatible system (78) of the rank t contains an incompatible subsystem of the rank t consisting of $t+1$ inequalities.*

Lemma 6.15. *Let P be the set of solutions of the compatible system (76) of rank t. Then there exists a one-to-one mapping ρ from P onto $\mathbb{R}^{n-t}$ with following properties:*

1°) *for any function $f \in L_n(\mathbb{R})$ there exists a function $f^* \in L_{n-t}(\mathbb{R})$ such that $f(\bar{a}) = f^*(\rho(\bar{a}))$ for any $\bar{a} \in P$;*
2°) *if $\bar{a}_0, \bar{a}_1, \ldots, \bar{a}_k \in P$ and $\lambda_1, \ldots, \lambda_k$ are nonnegative numbers from $\mathbb{R}$ such that $\sum_{i=1}^k \lambda_i = 1$ then $\bar{a}_0 = \sum_{i=1}^k \lambda_i \bar{a}_i$ if and only if $\rho(\bar{a}_0) = \sum_{i=1}^k \lambda_i \rho(\bar{a}_i)$;*
3°) *a finite nonempty set $\{\bar{a}_1, \ldots, \bar{a}_k\} \subset P$ is m-dimensional in $\mathbb{R}^n$ if and only if the set $\{\rho(\bar{a}_1), \ldots, \rho(\bar{a}_k)\}$ is m-dimensional in $\mathbb{R}^{n-t}$.*

Proof. Let $t = n$. Then from Lemma 6.1 follows that the system (76) has the unique solution. One can show that in this case the statement of the lemma is true.

Let $t < n$. In this case from Lemma 6.1 follows that there exist linearly independent solutions $\bar{y}_1, \ldots, \bar{y}_{n-t}$ of the system (79), corresponding to the system (76), and a solution $\bar{y}_0$ of the system (76) such that

$$P = \{\sum_{i=1}^{n-t} \alpha_i \bar{y}_i + \bar{y}_0 : \alpha_1, \ldots, \alpha_{n-t} \in \mathbb{R}\} . \tag{82}$$

Define a mapping $\rho : P \to \mathbb{R}^{n-t}$. From (82) follows that if $\bar{a} \in P$ then there exists a tuple $(\alpha_1, \ldots, \alpha_{n-t}) \in \mathbb{R}^{n-t}$ such that $\bar{a} = \sum_{i=1}^{n-t} \alpha_i \bar{y}_i + \bar{y}_0$. In this case set $\rho(\bar{a}) = (\alpha_1, \ldots, \alpha_{n-t})$. From the linear independence of the elements $\bar{y}_1, \ldots, \bar{y}_{n-t}$ and from (82) it follows that the mapping ρ is a one-to-one mapping from P onto $\mathbb{R}^{n-t}$. Let $f \in L_n(\mathbb{R})$ and $f(\bar{x}) = \sum_{i=1}^n b_i x_i + b_{n+1} = q(\bar{x}) + b_{n+1}$. Consider the function $f^* \in L_{n-t}(\mathbb{R})$ where $f^*(\bar{x}) = \sum_{i=1}^{n-t} q(\bar{y}_i)x_i + q(\bar{y}_0) + b_{n+1}$. Let $\bar{a} \in P$ and $\bar{a} = \sum_{i=1}^{n-t} \alpha_i \bar{y}_i + \bar{y}_0$. Then $f(\bar{a}) = q(\sum_{i=1}^{n-t} \alpha_i \bar{y}_i + \bar{y}_0) + b_{n+1} = \sum_{i=1}^{n-t} \alpha_i q(\bar{y}_i) + q(\bar{y}_0) + b_{n+1} = f^*(\rho(\bar{a}))$. Hence the mapping ρ has the property 1°.

Let $\bar{a}_0, \bar{a}_1, \ldots, \bar{a}_k \in P$, $\bar{a}_j = \sum_{i=1}^{n-t} \alpha_{ji} \bar{y}_i + \bar{y}_0$, $0 \le j \le k$, and $\lambda_1, \ldots, \lambda_k$ be nonnegative numbers from $\mathbb{R}$ such that $\sum_{i=1}^k \lambda_i = 1$. Let $\rho(\bar{a}_0) = \sum_{i=1}^k \lambda_i \rho(\bar{a}_i)$. Then $\bar{a}_0 = \sum_{i=1}^k \lambda_i \bar{a}_i$. Let now $\bar{a}_0 = \sum_{i=1}^k \lambda_i \bar{a}_i$. From linear independence of the elements $\bar{y}_1, \ldots, \bar{y}_{n-t}$ follows $\alpha_{0i} = \sum_{j=1}^k \lambda_j \alpha_{ji}$, $1 \le i \le n-t$. Therefore $\rho(\bar{a}_0) = \sum_{i=1}^k \lambda_i \rho(\bar{a}_i)$. Hence the mapping ρ has the property 2°.

Let $\bar{a}_1, \ldots, \bar{a}_m$ be arbitrary elements from P and $m \ge 2$. From linear independence of the elements $\bar{y}_1, \ldots, \bar{y}_{n-t}$ it follows that the elements $\bar{a}_1 - \bar{a}_2, \ldots, \bar{a}_1 - \bar{a}_m$ are linearly independent if and only if the elements $\rho(\bar{a}_1) - \rho(\bar{a}_2), \ldots, \rho(\bar{a}_1) - \rho(\bar{a}_m)$ are linearly independent. Using this fact and Lemma 6.4 we conclude that the mapping ρ has the property 3°. □

Let W be a finitely generated centroid and let V be the set of its vertices. A set $D \subseteq V^{t+1}$ will be called *a vertex $(t+1)$-covering of the set W* if D possesses the following properties:

1) if $(\bar{v}_1, \ldots, \bar{v}_{t+1}) \in D$ then the set $\{\bar{v}_1, \ldots, \bar{v}_{t+1}\}$ is t-dimensional;
2) for any element $\bar{a} \in W$ there exist a tuple $(\bar{v}_1, \ldots, \bar{v}_{t+1}) \in D$ such that $\bar{a}$ is contained in the centroid generated by the set $\{\bar{v}_1, \ldots, \bar{v}_{t+1}\}$.

Lemma 6.16. *If the set of solutions of the system (77) is an n-dimensional, $n \geq 1$, finitely generated centroid in the space $\mathbb{R}^n$ then it has a vertex $(n+1)$-covering which cardinality is at most k^{n-1}.*

Proof. We prove the statement of the lemma by induction on n. One can show that any one-dimensional finitely generated centroid in $\mathbb{R}^1$ has exactly two vertices. Hence the statement of the lemma is true for $n = 1$.

Let $n \geq 2$ and let the statement of the lemma hold for the spaces of the dimension $1, \ldots, n-1$. Let us show that it also hods for the space of dimension n. Let W be the set of solutions of the system (77), and W be an n-dimensional finitely generated centroid in $\mathbb{R}^n$. Denote by V the set of the vertices of the centroid W. Choose an uncancellable subsystem of the system (77) such that the set of the solutions of it coincides with W. Without loss of generality we can assume that the system (77) is uncancellable. From the fact that the system (77) is compatible and uncancellable it follows for any $j \in \{1, \ldots, k\}$ that the function f_j is not constant on $\mathbb{R}^n$. Let $i \in \{1, \ldots, k\}$. Denote by P_i the set of solutions on $\mathbb{R}^n$ of the system

$$\{f_i(\bar{x}) = 0\} \ . \tag{83}$$

Since the function f_i is not constant on $\mathbb{R}^n$, the rank of the system (83) is equal to 1. Consider a one-to-one mapping ζ from P_i onto $\mathbb{R}^{n-1}$ having the properties $1° - 3°$ from Lemma 6.15. From the property $1°$ it follows that for any $j \in \{1, \ldots, k\}$ there exists a function $f_j^* \in L_{n-1}(\mathbb{R})$ such that $f_j(\bar{a}) = f_j^*(\zeta(\bar{a}))$ for any $\bar{a} \in P_i$. Denote by W_i the set of solutions on $\mathbb{R}^n$ of the system

$$\begin{aligned} &\{f_1(\bar{x}) \geq 0, \ldots, f_{i-1}(\bar{x}) \geq 0, f_i(\bar{x}) \geq 0, \\ &- f_i(\bar{x}) \geq 0, f_{i+1}(\bar{x}) \geq 0, \ldots, f_k(\bar{x}) \geq 0\} \ , \end{aligned} \tag{84}$$

and let W_i^* be the set of solutions on $\mathbb{R}^{n-1}$ of the system

$$\{f_1^*(\bar{x}) \geq 0, \ldots, f_{i-1}^*(\bar{x}) \geq 0, f_{i+1}^*(\bar{x}) \geq 0, \ldots, f_k^*(\bar{x}) \geq 0\} \ . \tag{85}$$

Let us show that the set W_i^* is an $(n-1)$-dimensional finitely generated centroid in the space $\mathbb{R}^{n-1}$.

First, verify that the dimension of the set W_i^* is equal to $n-1$. From the fact that the function f_j is not constant on $\mathbb{R}^n$ for any $j \in \{1, \ldots, k\}$, from the fact that the set W is n-dimensional and from Lemma 6.11 follows that the system (78) corresponding to (77) is compatible on $\mathbb{R}^n$. From here, from the fact that the system (77) is uncancellable and from Lemma 6.12 follows that the system (78) is uncancellable. Using the fact that the system (78) is compatible and uncancellable, and using Lemma 6.13 we conclude that the system

$$\{f_1(\bar{x}) > 0, \ldots, f_{i-1}(\bar{x}) > 0, f_i(\bar{x}) = 0, f_{i+1}(\bar{x}) > 0, \ldots, f_k(\bar{x}) > 0\} \tag{86}$$

is compatible on $\mathbb{R}^n$. From the compatibility of this system and from the choice of the functions $f_j^*, 1 \leq j \leq k$, follows that the system

$$\{f_1^*(\bar{x}) > 0, \ldots, f_{i-1}^*(\bar{x}) > 0, f_{i+1}^*(\bar{x}) > 0, \ldots, f_k^*(\bar{x}) > 0\} \quad (87)$$

is compatible on $\mathbb{R}^{n-1}$. From here follows that the system (85) is compatible. Denote by (85a) and (87a) the systems obtained from the systems (85) and (87) by deletion of the inequalities in which functions f_j^* are constant on $\mathbb{R}^{n-1}$. From the compatibility of systems (85) and (87) follows that the set of solutions of (85a) coincides with the set of solutions of the system (85), and the set of solutions of (87a) coincides with the set of solutions of (87). Since the system (87a) is compatible and the functions in the inequalities of this system are not constant on $\mathbb{R}^{n-1}$, it follows from Lemma 6.11 that the set of solutions of (85a) on $\mathbb{R}^{n-1}$ is $(n-1)$-dimensional. Hence the set W_i^* is $(n-1)$-dimensional.

Verify now that the set W_i^* is a finitely generated centroid. First, let us show that the set W_i is a finitely generated centroid. From the compatibility of the system (86) on $\mathbb{R}^n$ it follows that the set W_i is a nonempty set. From the fact that the set W is a finitely generated centroid follows that the set W_i is bounded. Since the system (84) is compatible and the set of its solutions is bounded, from Lemma 6.8 follows that the set W_i is a finitely generated centroid. Denote by V_i the set of vertices of the centroid W_i. Denote $V_i^* = \{\zeta(\bar{v}) : \bar{v} \in V_i\}$.

Let us show that the centroid generated by the set V_i^* coincides with the set W_i^*. From the choice of the functions $f_j^*, 1 \leq j \leq k$, follows that $V_i^* \subseteq W_i^*$. Using this fact one can show that the centroid generated by the set V_i^* is a subset of the set W_i^*. We will demonstrate that any element $\bar{a}$ from W_i^* is contained in this centroid. Denote by ζ^{-1} the inverse mapping for the mapping ζ. From the choice of the functions f_j^*, $1 \leq j \leq k$, it follows that the element $\zeta^{-1}(\bar{a})$ is contained in the centroid W_i. From this and from the property 2° of the mapping ζ follows that the element $\bar{a}$ is contained in the centroid generated by the set V_i^*. Hence W_i^* is an $(n-1)$-dimensional finitely generated centroid in the space $\mathbb{R}^{n-1}$, and V_i^* is the set of its generating elements.

By the inductive hypothesis, for the centroid W_i^* there exists an vertex n-covering the cardinality of which is at most $(k-1)^{n-2}$. Denote this covering by D_i^*. Denote $D_i = \{(\zeta^{-1}(\bar{v}_1), \ldots, \zeta^{-1}(\bar{v}_n)) : (\bar{v}_1, \ldots, \bar{v}_n) \in D_i^*\}$. Let us show that the set D_i is a vertex n-covering of the centroid W_i. By definition, the set V_i is centroidally independent. From here and from the property 2° of the mapping ζ it follows that the set V_i^* is centroidally independent. Therefore the set V_i^* is a centroidally independent set of generating elements of the centroid W_i^*. Using Lemma 6.7 we conclude that the set V_i^* is the set of all vertices of the centroid W_i^*. From here, from the choice of the set D_i^*, from the equality $V_i = \{\zeta^{-1}(\bar{v}) : \bar{v} \in V_i^*\}$ and from properties 2° and 3° of the mapping ζ follows that the set D_i is a vertex n-covering of the centroid W_i. Note that

$$|D_i| \leq (k-1)^{n-2} \ . \quad (88)$$

Choose a vertex $\bar{v}^0$ of the centroid W. Let, for the definiteness, the vertex $\bar{v}^0$ do not belong to the planes $P_1, \ldots, P_m$, and belong to the planes $P_{m+1}, \ldots, P_k$.

For any $i \in \{m+1, \ldots, k\}$ choose a vertex $\bar{v}^i$ of the centroid W which does not belong to the plane P_i. Such a vertex exists since the set W is n-dimensional. Denote

$$D = \left(\bigcup_{i=1}^{m} \{(\bar{v}^0, \bar{v}_1, \ldots, \bar{v}_n) : (\bar{v}_1, \ldots, \bar{v}_n) \in D_i\}\right) \cup \left(\bigcup_{i=m+1}^{k} \{(\bar{v}^i, \bar{v}_1, \ldots, \bar{v}_n) : (\bar{v}_1, \ldots, \bar{v}_n) \in D_i\}\right) .$$

Let us show that the set D is a vertex $(n+1)$-covering of the centroid W. Let $(\bar{v}_1, \ldots, \bar{v}_{n+1})$ be an arbitrary element of the set D. We will demonstrate that the set $\{\bar{v}_1, \ldots, \bar{v}_{n+1}\}$ is n-dimensional. Let, for the definiteness, $\bar{v}_1 = \bar{v}^0$, $(\bar{v}_2, \ldots, \bar{v}_{n+1}) \in D_1$ and $f_1(\bar{x}) = \sum_{i=1}^{n} a_{1i}x_i + a_{1n+1} = q_1(\bar{x}) + a_{1n+1}$. Then, by definition of the set D_1, the elements $\bar{v}_2, \ldots, \bar{v}_{n+1}$ are solutions of the equation $f_1(\bar{x}) = 0$. Hence the elements $\bar{v}_2 - \bar{v}_3, \ldots, \bar{v}_2 - \bar{v}_{n+1}$ are solutions of the equation $q_1(\bar{x}) = 0$.

The set $\{\bar{v}_2, \ldots, \bar{v}_{n+1}\}$, by definition of D_1, is $(n-1)$-dimensional. From here and from Lemma 6.4 follows that the elements $\bar{v}_2 - \bar{v}_3, \ldots, \bar{v}_2 - \bar{v}_{n+1}$ are linearly independent. Assume that the elements $\bar{v}_2 - \bar{v}^0, \bar{v}_2 - \bar{v}_3, \ldots, \bar{v}_2 - \bar{v}_{n+1}$ are linearly dependent. Then the element $\bar{v}_2 - \bar{v}^0$ and hence the element $\bar{v}^0 - \bar{v}_2$ are solutions of the equation $q_1(\bar{x}) = 0$. Therefore the element $\bar{v}^0 - \bar{v}_2 + \bar{v}_2 = \bar{v}^0$ is a solution of the equation $f_1(\bar{x}) = 0$ which is impossible since the element $\bar{v}^0$ does not belong to the plane P_1. Therefore the elements $\bar{v}_2 - \bar{v}^0, \bar{v}_2 - \bar{v}_3, \ldots, \bar{v}_2 - \bar{v}_{n+1}$ are linearly independent. From here and from Lemma 6.4 follows that the set $\{\bar{v}^0, \bar{v}_2, \ldots, \bar{v}_{n+1}\}$ is n-dimensional. Let us show that for any element $\bar{a} \in W$ there exists a tuple $(\bar{v}_1, \ldots, \bar{v}_{n+1}) \in D$ such that $\bar{a}$ belongs to the centroid generated by the set $\{\bar{v}_1, \ldots, \bar{v}_{n+1}\}$. Let $\bar{a} \in W$ and $\bar{a} \notin P_{m+1} \cup \ldots \cup P_k$. Then $\bar{a} \neq \bar{v}^0$. From here and from Lemma 6.4 follows that the set $\{\bar{a}, \bar{v}^0\}$ is one-dimensional. Therefore functions $g_1, \ldots, g_{n-1} \in L_n(\mathbb{R})$ exist such that P is the set of solutions on $\mathbb{R}^n$ of the system $\{g_1(\bar{x}) = 0, \ldots, g_{n-1}(\bar{x}) = 0\}$, and P is an one-dimensional plane containing the elements $\bar{a}$ and $\bar{v}^0$. It is easily to show that the set of solutions on $\mathbb{R}^n$ of the system

$$\begin{gathered}\{g_1(\bar{x}) \geq 0, -g_1(\bar{x}) \geq 0, \ldots, g_{n-1}(\bar{x}) \geq 0, -g_{n-1}(\bar{x}) \geq 0, \\ f_1(\bar{x}) \geq 0, \ldots, f_k(\bar{x}) \geq 0\}\end{gathered} \tag{89}$$

is a nonempty bounded set. From here and from Lemma 6.8 follows that the set W, which is the set of solutions of the system (89) on $\mathbb{R}^n$, is a finitely generated centroid. Since W is contained in the one-dimensional plane P and $\bar{a}, \bar{v}^0 \in W$, we conclude that W is one-dimensional set. Using Lemma 6.4 one can show that any one-dimensional finitely generated centroid has exactly two vertices. From the compatibility of (77), from the fact that the set W is bounded and from Lemmas 6.7 and 6.8 follows that $\bar{v}^0$ is a nodal solution of the system (77). From the fact that W is a bounded set and from Lemma 6.6 follows that the rank of the system (77) is equal to n. Hence the rank of (89) is also equal to n. Therefore the element $\bar{v}^0$ is a nodal solution of the system (89). Using Lemmas 6.7 and 6.8 we conclude that $\bar{v}^0$ is a vertex of the centroid W. Let $\bar{v}$

be a vertex of the centroid W which is different from $\bar{v}^0$. From Lemmas 6.7 and 6.8 and from the fact that W is bounded follows that $\bar{v}$ is a nodal solution of the system (89). Taking into account that the rank of the system (89) is equal to n we conclude that there exists $i_0 \in \{1, \ldots, k\}$ for which $f_{i_0}(\bar{v}) = 0$. We see that $i_0 \in \{1, \ldots, m\}$ since otherwise $\bar{a} \in P_{m+1} \cup \ldots \cup P_k$ which contradicts the assumption. Therefore there exists $i_0 \in \{1, \ldots, m\}$ such that $\bar{v} \in W_{i_0}$. From the properties of the set D_{i_0} it follows that there exists a tuple $(\bar{v}_1, \ldots, \bar{v}_n) \in D_{i_0}$ such that the element $\bar{v}$ belongs to the centroid generated by the set $\{\bar{v}_1, \ldots, \bar{v}_n\}$. Since the element $\bar{a}$ is contained in the centroid generated by the set $\{\bar{v}^0, \bar{v}\}$, we obtain that $\bar{a}$ is contained in the centroid generated by the set $\{\bar{v}^0, \bar{v}_1, \ldots, \bar{v}_n\}$. Finally, $(\bar{v}^0, \bar{v}_1, \ldots, \bar{v}_n) \in D$ by the definition of D.

The case $\bar{a} \in P_i$, $i \in \{m+1, \ldots, k\}$, can be considered in the same way, but instead of $\bar{v}^0$ we must take $\bar{v}^i$. Hence the set D is a vertex $(n+1)$-covering of the centroid W. From the definition of the set D and from (88) follows that the cardinality of the set D is at most k^{n-1}. □

6.3 Principal Lemmas

In this subsection four lemmas will be proved which are used immediately in the proof of Theorem 6.1.

Lemma 6.17. *Any incompatible equation system of the kind*

$$\{f_1(\bar{x}) = \delta_1, \ldots, f_k(\bar{x}) = \delta_k\} \tag{90}$$

where $f_1, \ldots, f_k \in S_n(\mathbb{R})$ *and* $\delta_1, \ldots, \delta_k \in \{-1, 0, +1\}$ *contains an incompatible subsystem with at most* $n+1$ *equations.*

Proof. Let, for definiteness, $\delta_1 = \delta_2 = \ldots = \delta_m = 0$, $\delta_{m+1} = \ldots = \delta_{m+p} = -1$ and $\delta_{m+p+1} = \ldots = \delta_k = +1$. Let the equality $f_j(\bar{x}) = \text{sign}(g_j(\bar{x}))$, where $g_j \in L_n(\mathbb{R})$, hold for any $j \in \{1, \ldots, k\}$. Let the rank of the system

$$\{g_1(\bar{x}) = 0, \ldots, g_m(\bar{x}) = 0\} \tag{91}$$

be equal to t. Assume that the system (91) is incompatible. Then from Lemma 6.3 follows that the system (91) contains an incompatible subsystem with at most $t+1$ equations. Hence the system (90) also contains an incompatible subsystem with at most $t+1$ equations. Since $t \leq n$, the statement of the lemma holds in the considered case. Assume that the system (91) is compatible. From Lemma 6.3 follows that in this case (91) contains a subsystem of the rank t

$$\{g_{j_1}(\bar{x}) = 0, \ldots, g_{j_t}(\bar{x}) = 0\} \tag{92}$$

which contains t equations and is equivalent to the system (91). Denote by P the set of solutions of the system (92). From Lemma 6.15 follows that there exists one-to-one mapping ζ from P onto $\mathbb{R}^{n-t}$ and the functions $g_1^*, \ldots, g_k^* \in L_{n-t}(\mathbb{R})$ such that $g_j(\bar{a}) = g_j^*(\zeta(\bar{a}))$ for any $\bar{a} \in P$ and $j \in \{1, \ldots, k\}$. From the incompatibility of the system (90) and from the choice of the functions g_j^*, $1 \leq j \leq$

k, follows that the system $\{-g^*_{m+1}(\bar{x}) > 0, \ldots, -g^*_{m+p}(\bar{x}) > 0, g^*_{m+p+1}(\bar{x}) > 0, \ldots, g^*_k(\bar{x}) > 0\}$ is incompatible on $\mathbb{R}^{n-t}$. From Lemma 6.14 follows that this system contains an incompatible subsystem with at most $n - t + 1$ inequalities. Let this system be of the kind $\{-g^*_{i_1}(\bar{x}) > 0, \ldots, -g^*_{i_s}(\bar{x}) > 0, g^*_{i_{s+1}}(\bar{x}) > 0, \ldots, g^*_{i_r}(\bar{x}) > 0\}$ where $r \leq n - t + 1$. We can then easily see that the subsystem $\{f_{j_1}(\bar{x}) = 0, \ldots, f_{j_t}(\bar{x}) = 0, f_{i_1}(\bar{x}) = -1, \ldots, f_{i_s}(\bar{x}) = -1, f_{i_{s+1}}(\bar{x}) = +1, \ldots, f_{i_r}(\bar{x}) = +1\}$ of the system (90) containing at most $n + 1$ equations is incompatible on $\mathbb{R}^n$.

□

Lemma 6.18. *Let $f_1, \ldots, f_k \in S_n(\mathbb{R})$. Then there exist at most $2k^n + 1$ pairwise different tuples $(\delta_1, \ldots, \delta_k) \in \{-1, 0, +1\}^k$ for each of which the equation system $\{f_1(\bar{x}) = \delta_1, \ldots, f_k(\bar{x}) = \delta_k\}$ is compatible on $\mathbb{R}^n$.*

Proof. For arbitrary $f_1, \ldots, f_k \in S_n(\mathbb{R})$ denote by $N(f_1, \ldots, f_k)$ the number of different tuples $(\delta_1, \ldots, \delta_k) \in \{-1, 0, +1\}^k$ such that the equation system $\{f_1(\bar{x}) = \delta_1, \ldots, f_k(\bar{x}) = \delta_k\}$ is compatible on $\mathbb{R}^n$. Set

$$N(n, k) = \max\{N(f_1, \ldots, f_k) : f_1, \ldots, f_k \in S_n(\mathbb{R})\} \ .$$

Let us show that for any natural n and k the inequality

$$N(n, k) \leq 2k^n + 1 \tag{93}$$

holds. First, verify that for any natural n and k the inequality

$$N(n + 1, k + 1) \leq N(n + 1, k) + 2N(n, k) \tag{94}$$

holds. From the definition of the value $N(n + 1, k + 1)$ follows that functions $f_1, \ldots, f_{k+1} \in S_{n+1}(\mathbb{R})$ exist such that $N(f_1, \ldots, f_{k+1}) = N(n + 1, k + 1)$. Obviously, among the functions $f_1, \ldots, f_{k+1}$ there exists a function which is not constant on $\mathbb{R}^{n+1}$. Without loss of generality, we can assume that the function f_{k+1} is not constant.

For any $\bar{\delta} = (\delta_1, \ldots, \delta_k) \in \{-1, 0, +1\}^k$ and $\sigma \in \{-1, 0, +1\}$ denote by $B(\bar{\delta})$ the system of equations $\{f_1(\bar{x}) = \delta_1, \ldots, f_k(\bar{x}) = \delta_k\}$, and by $B_\sigma(\bar{\delta})$ denote the system of equations $\{f_1(\bar{x}) = \delta_1, \ldots, f_k(\bar{x}) = \delta_k, f_{k+1}(\bar{x}) = \sigma\}$. Let $B \in \{B(\bar{\delta}) : \bar{\delta} \in \{-1, 0, +1\}^k\} \cup \{B_\sigma(\bar{\delta}) : \bar{\delta} \in \{-1, 0, +1\}^k, \sigma \in \{-1, 0, +1\}\}$. Set

$$C(B) = \begin{cases} 0, & \text{if the system } B \text{ is incompatible on } \mathbb{R}^{n+1} \ , \\ 1, & \text{otherwise} \ . \end{cases}$$

Let $\bar{\delta} \in \{-1, 0, +1\}^k$. Let us show that if $C(B_0(\bar{\delta})) + C(B_2(\bar{\delta})) \geq 1$ then $C(B(\bar{\delta})) = 1$, and if $C(B_0(\bar{\delta})) + C(B_2(\bar{\delta})) = 2$ then $C(B_1(\bar{\delta})) = 1$. Let, for instance, $C(B_0(\bar{\delta})) = 1$, and $\bar{a}$ be a solution of the system $B_0(\bar{\delta})$. Then the element $\bar{a}$ is a solution of the system $B(\bar{\delta})$, and hence $C(B(\bar{\delta})) = 1$. Let $C(B_0(\bar{\delta})) + C(B_2(\bar{\delta})) = 2$, $\bar{a}_0$ be a solution of the system $B_0(\bar{\delta})$ and $\bar{a}_2$ be a solution of the system $B_2(\bar{\delta})$. Let $f_{k+1}(\bar{x}) = \text{sign}(g(\bar{x}))$ where $g \in L_{n+1}(\mathbb{R})$. Then the element $-\bar{a}_0(g(\bar{a}_2)/(g(\bar{a}_0) - g(\bar{a}_2))) + \bar{a}_2(1 + g(\bar{a}_2)/(g(\bar{a}_0) - g(\bar{a}_2)))$ is a solution of the system $B_1(\bar{\delta})$. Therefore $C(B_1(\bar{\delta})) = 1$.

From the obtained relations it follows that for any $\bar{\delta} \in \{-1, 0, +1\}^k$ the inequality

$$\sum_{\sigma \in \{-1,0,+1\}} C(B_\sigma(\bar{\delta})) \leq C(B(\bar{\delta})) + 2C(B_1(\bar{\delta}))$$

holds. Therefore

$$\begin{aligned} &\textstyle\sum_{\bar{\delta} \in \{-1,0,+1\}^k} \sum_{\sigma \in \{-1,0,+1\}} C(B_\sigma(\bar{\delta})) \\ &\leq \textstyle\sum_{\bar{\delta} \in \{-1,0,+1\}^k} C(B(\bar{\delta})) + 2 \sum_{\bar{\delta} \in \{-1,0,+1\}^k} C(B_1(\bar{\delta})) \ . \end{aligned} \tag{95}$$

From the choice of functions $f_1, \ldots, f_{k+1}$ follows that

$$\sum_{\bar{\delta} \in \{-1,0,+1\}^k} \sum_{\sigma \in \{-1,0,+1\}} C(B_\sigma(\bar{\delta})) = N(n+1, k+1) \ . \tag{96}$$

One can easily see that $\sum_{\bar{\delta} \in \{-1,0,+1\}^k} C(B(\bar{\delta})) = N(f_1, \ldots, f_k)$. Hence

$$\sum_{\bar{\delta} \in \{-1,0,+1\}^k} C(B(\bar{\delta})) \leq N(n+1, k) \ . \tag{97}$$

Let us show that

$$\sum_{\bar{\delta} \in \{-1,0,+1\}^k} C(B_1(\bar{\delta})) \leq N(n, k) \ . \tag{98}$$

Let for any $j \in \{1, \ldots, k+1\}$ the equality $f_j(\bar{x}) = \text{sign}(g_j(\bar{x}))$ holds where $g_j \in L_{n+1}(\mathbb{R})$. Denote by P the set of solutions of the system $\{g_{k+1}(\bar{x}) = 0\}$. Since the function f_{k+1} is not constant on $\mathbb{R}^{n+1}$, the rank of this system is equal to 1. Consider a mapping $\zeta : P \to \mathbb{R}^n$ satisfying the conditions of Lemma 6.15. From the property 1° of the mapping ζ follows that for any $j \in \{1, \ldots, k\}$ there exists a function $g_j^* \in L_n(\mathbb{R})$ such that $g_j(\bar{a}) = g_j^*(\zeta(\bar{a}))$ for any $\bar{a} \in P$. Let $f_j^* = \text{sign}(g_j^*)$,$1 \leq j \leq k$. Then $\sum_{\bar{\delta} \in \{-1,0,+1\}^k} C(B_1(\bar{\delta})) = N(f_1^*, \ldots, f_k^*)$. Since $N(f_1^*, \ldots, f_k^*) \leq N(n, k)$, (98) holds. From (95)–(98) follows (94). Let us prove the inequality (93) by induction on the value $n + k$. One can show that for any natural n and k the following equalities hold:

$$N(1, k) = 2k + 1 \ , \tag{99}$$

$$N(n, 1) = 3 \ . \tag{100}$$

Hence (93) is true for $n + k \leq 3$. Let $t \geq 4$. Assume that (93) holds for any natural n and k such that $n + k < t$. Let n^* and k^* be arbitrary natural numbers such that $n^* + k^* = t$. Let us show that (93) also holds for n^* and k^*. From (99) and (100) follows that if $n^* = 1$ or $k^* = 1$ then (93) holds. Therefore we can assume that $n^* = n + 1$ and $k^* = k + 1$ for some natural n and k. From (94) and from the inductive hypothesis follows that $N(n+1, k+1) \leq 2k^{n+1} + 1 + 4k^n + 2 \leq 2(k^{n+1} + (n+1)k^n + 1) + 1 \leq 2(k+1)^{n+1} + 1$. Thus, the inequality (93) holds. The statement of the lemma follows from (93). □

Let $U = (A, B, F)$ be an information system, Γ be a decision tree over U and z be a problem over U. Let us define a partial function $\Gamma : A \to \mathbb{Z}$ (later we will say that the decision tree Γ *realizes* this function). Let $a \in A$. If there is no a complete path ξ in Γ such that $a \in \mathrm{Sol}_U(\pi(\xi))$ then the value $\Gamma(a)$ is indefinite. Let ξ be a complete path in Γ such that $a \in \mathrm{Sol}_U(\pi(\xi))$, and let the terminal node of ξ be labelled by the number m. Then $\Gamma(a) = m$. It is clear that Γ solves z if and only if $\Gamma(a) = z(a)$ for any $a \in A$.

Denote by C_n the set of solutions on $\mathbb{R}^n$ of the inequality system

$$\{x_1 + 2 > 0, 2 - x_1 > 0, \ldots, x_n + 2 > 0, 2 - x_n > 0\} .$$

Let $U = U(\mathbb{R}^n, K, x_1, \ldots, x_n)$ be a quasilinear information system. Denote $U^\kappa = U(C_{n+1}, K, x_1, \ldots, x_{n+1})$. Put into correspondence to any attribute $f \in F(\mathbb{R}^n, K, x_1, \ldots, x_n)$ an attribute $f^\kappa \in F(C_{n+1}, K, x_1, \ldots, x_{n+1})$ in the following way: if $f = \mathrm{sign}(a_1x_1 + \ldots + a_nx_n + a_{n+1})$ then $f^\kappa = \mathrm{sign}(a_1x_1 + \ldots + a_nx_n + a_{n+1}x_{n+1})$. Put into correspondence to any problem z over U a problem z^κ over U^κ in the following way: if $z = (\nu, f_1, \ldots, f_k)$ then $z^\kappa = (\nu, f_1^\kappa, \ldots, f_k^\kappa)$.

Lemma 6.19. *Let $U = U(\mathbb{R}^n, K, x_1, \ldots, x_n)$ be a quasilinear information system, $U^\kappa = U(C_{n+1}, K, x_1, \ldots, x_{n+1})$, z be a problem over U, and Γ_1 be a decision tree over U^κ which solves the problem z^κ. Then there exists a decision tree Γ over U which solves the problem z and for which $h(\Gamma) \le h(\Gamma_1) + 2n$ and $r(\Gamma) \le r(\Gamma_1) + 1$.*

Proof. Denote by $\Gamma_{m,\delta}$, $1 \le m \le n+1$, $\delta \in \{-1, 1\}$, a decision tree over U which is obtained from the tree Γ_1 in the following way: if a working node w of the tree Γ_1 is labelled by an attribute $\mathrm{sign}(a_1x_1 + \ldots a_{n+1}x_{n+1} + a_{n+2})$ then the node w in the tree $\Gamma_{m,\delta}$ is labelled by the attribute $\mathrm{sign}(a_1x_1 + \ldots a_nx_n + a_{n+1} + a_{n+2})$ if $m = n+1$, or by the attribute $\mathrm{sign}(a_1x_1 + \ldots + a_{m-1}x_{m-1} + (a_m + \delta a_{n+2})x_m + a_{m+1}x_{m+1} + \ldots + a_nx_n + a_{n+1})$ if $m \ne n+1$.

Define a function $q : \mathbb{R}^n \to \mathbb{N}$. Let $(a_1, \ldots, a_n) \in \mathbb{R}^n$, $a_{n+1} = 1$ and let m be the minimal number from $\{1, \ldots, n+1\}$ such that $|a_m| = \max\{|a_i| : 1 \le i \le n+1\}$. Then $q(a_1, \ldots, a_n) = 2^m 3^{\mathrm{sign}(a_m)+1}$. We can easily see that there exists a decision tree Γ_0 over U which realizes the function q and for which $h(\Gamma_0) = 2n$ and $r(\Gamma_0) = 0$.

A decision tree Γ over U will be obtained from the tree Γ_0 and the trees $\Gamma_{m,\delta}$ in the following way. The root of Γ is the root of the tree Γ_0. Each terminal node of the tree Γ_0 is replaced by the root of a decision tree. Let w be a terminal node of the tree Γ_0 and w be labelled by the number $2^m 3^\sigma$. Then the node w is replaced by the root of the tree $\Gamma_{m,\sigma-1}$. One can show that

$$h(\Gamma) \le h(\Gamma_1) + 2n , \tag{101}$$

$$r(\Gamma) \le r(\Gamma_1) + 1 . \tag{102}$$

Let us show that the decision tree Γ solves the problem z. Define a mapping $\kappa : \mathbb{R}^n \to C_{n+1}$ in the following way: if $\bar{a} = (a_1, \ldots, a_n) \in \mathbb{R}^n$, $a_{n+1} = 1$ and $v = \max\{|a_i| : 1 \le i \le n+1\}$ then $\kappa(\bar{a}) = (a_1/v, \ldots, a_n/v, 1/v)$. It is not difficult to see that

$$z(\bar{a}) = z^{\kappa}(\kappa(\bar{a})) \tag{103}$$

for any $\bar{a} \in \mathbb{R}^n$. One can show that for any $\bar{a} \in \mathbb{R}^n$ if the value $\Gamma_1(\kappa(\bar{a}))$ is definite then the value $\Gamma(\bar{a})$ is also definite and the equality

$$\Gamma(\bar{a}) = \Gamma_1(\kappa(\bar{a})) \tag{104}$$

holds. Since the decision tree Γ_1 solves the problem z^{κ}, for any $\bar{b} \in C_{n+1}$ the value $\Gamma_1(\bar{b})$ is definite and the equality $\Gamma_1(\bar{b}) = z^{\kappa}(\bar{b})$ holds. Using this fact and the fact that $\kappa(\bar{a}) \in C_{n+1}$ for any $\bar{a} \in \mathbb{R}^n$ we conclude that $\Gamma_1(\kappa(\bar{a})) = z^{\kappa}(\kappa(\bar{a}))$ for any $\bar{a} \in \mathbb{R}^n$. From here and from (103) and (104) follows that $\Gamma(\bar{a}) = z(\bar{a})$ for any $\bar{a} \in \mathbb{R}^n$. Using this fact and (101), (102) obtain the statement of the lemma. □

Let K be a numerical ring with unity and let $W \subseteq \mathbb{R}^n$. A finite set of functions $\mathcal{F} \subset S_n(K)$ will be called *a functional (m, K)-covering of the set W* if for any $\bar{a} \in W$ there exist functions $f_1, \dots, f_m \in \mathcal{F}$ and numbers $\sigma_1, \dots, \sigma_m \in \{-1, 0, +1\}$ such that $\bar{a}$ is a solution of the equation system

$$\{f_1(\bar{x}) = \sigma_1, \dots, f_m(\bar{x}) = \sigma_m\}$$

and the set of solutions of this system on $\mathbb{R}^n$ is a subset of W.

Lemma 6.20. *Let K be a numerical ring with unity, $f_1, \dots, f_k \in S_n(K)$, $\delta_1, \dots, \delta_k \in \{-1, 0, +1\}$ and W be the nonempty set of solutions on C_n of the equation system $\{f_1(\bar{x}) = \delta_1, \dots, f_k(\bar{x}) = \delta_k\}$. Then there exists a functional $(n+1, K)$-covering $\mathcal{F}$ of the set W such that $|\mathcal{F}| \leq (n+1)(k+2n)^{n-1}$ and $\max\{r(f) : f \in \mathcal{F}\} \leq 2n^2(\log_2 n + 1 + \max\{r(f_j) : j = 1, \dots, k\}) - 1$.*

Proof. Let $f_j = \mathrm{sign}(g_j^0)$, where $1 \leq j \leq k$ and $g_j^0 \in L_n(K)$. Let, for the definiteness, $\delta_1 = \dots = \delta_m = 0, \delta_{m+1} = \dots = \delta_p = +1$ and $\delta_{p+1} = \dots = \delta_k = -1$. Let $g_j = g_j^0$ for $j = 1, \dots, p$, $g_j = -g_j^0$ for $j = p+1, \dots, k$, $g_j = x_{j-k} + 2$ for $j = k+1, \dots, k+n$ and $g_j = 2 - x_{j-k-n}$ for $j = k+n+1, \dots, k+2n$. Then the set of solutions on $\mathbb{R}^n$ of the system

$$S_0 = \{g_1(\bar{x}) = 0, \dots, g_m(\bar{x}) = 0, g_{m+1}(\bar{x}) > 0, \dots, g_{k+2n}(\bar{x}) > 0\}$$

coincides with the set W. Let U be the set of solutions on $\mathbb{R}^n$ of the system

$$\begin{gathered}\{g_1(\bar{x}) \geq 0, -g_1(\bar{x}) \geq 0, \dots, g_m(\bar{x}) \geq 0, -g_m(\bar{x}) \geq 0, \\ g_{m+1}(\bar{x}) \geq 0, g_{m+2}(\bar{x}) \geq 0, \dots, g_{k+2n}(\bar{x}) \geq 0\} \ .\end{gathered} \tag{105}$$

Denote by P the set of solutions on $\mathbb{R}^n$ of the system

$$\{g_1(\bar{x}) = 0, \dots, g_m(\bar{x}) = 0\} \ . \tag{106}$$

Since the set W is nonempty, P is also nonempty. Let the rank of the system (106) be equal to t.

If $t = n$ then from the compatibility of the system (106) and from Lemma 6.1 follows that the set P is one-element. Since the set W is nonempty and $W \subseteq P$, we have $W = P$. From the compatibility of the system (106) and from Lemma 6.3 follows that there exists of a subsystem of (106) of the rank n which is equivalent to (106) and contains exactly n equations. Let, for the definiteness, it be of the kind $\{g_1(\bar{x}) = 0, \ldots, g_n(\bar{x}) = 0\}$. Then the set of solutions on $\mathbb{R}^n$ of the system $\{f_1(\bar{x}) = 0, \ldots, f_n(\bar{x}) = 0\}$ is equal to W. Using this fact one can show that the set $\mathcal{F} = \{f_1, \ldots, f_n\}$ is a functional $(n+1, K)$-covering of the set W, and this covering satisfies the conditions of the lemma. Therefore the statement of the lemma is true if $t = n$.

Let $t < n$. Consider a mapping $\zeta : P \to \mathbb{R}^{n-t}$ possessing the properties 1°–3° from Lemma 6.15. From the property 1° of the mapping ζ follows that for any $j \in \{m+1, \ldots, k+2n\}$ there exists a function $g_j^* \in L_{n-t}(\mathbb{R})$ such that $g_j(\bar{a}) = g_j^*(\zeta(\bar{a}))$ for any element $\bar{a} \in P$. Denote by U^* the set of solutions on $\mathbb{R}^{n-t}$ of the system

$$\{g_{m+1}^*(\bar{x}) \geq 0, \ldots, g_{k+2n}^*(\bar{x}) \geq 0\} \ . \tag{107}$$

One can show that U is a bounded nonempty set. From here and from Lemma 6.8 it follows that U is a finitely generated centroid. Denote by V the set of vertices of the centroid U. Denote $V^* = \{\zeta(\bar{v}) : \bar{v} \in V\}$. In the same way as in the proof of Lemma 6.16 we can show that U^* is a finitely generated centroid and V^* is the set of vertices of the centroid U^*.

From the compatibility on $\mathbb{R}^n$ of the system S_0 and from the choice of the functions g_j^*, $m+1 \leq j \leq k+2n$, it follows that the system

$$\{g_{m+1}^*(\bar{x}) > 0, \ldots, g_{k+2n}^*(\bar{x}) > 0\} \tag{108}$$

is compatible on $\mathbb{R}^{n-t}$. The systems obtained from (107) and (108) by the deletion of the inequalities in which functions g_j^* are constant on $\mathbb{R}^{n-t}$ will be denoted by (107a) and (108a). From the compatibility of the systems (107) and (108) it follows that the set of solutions of the system (107a) coincides with the set of solutions of the system (107), and the set of solutions of the system (108a) coincides with the set of solutions of the system (108). Functions from inequalities of the system (108a) are not constant on $\mathbb{R}^{n-t}$. Therefore from compatibility of (108a) and from Lemma 6.11 follows that the set of solutions on $\mathbb{R}^{n-t}$ of the system (107a) is $(n-t)$-dimensional. Hence U^* is an $(n-t)$-dimensional finitely generated centroid in the space $\mathbb{R}^{n-t}$. From here and from Lemma 6.16 follows that there exists a vertex $(n-t+1)$-covering D^* of the set U^* which cardinality is at most $(k+2n-m)^{n-t-1}$. Since the rank of the system (106) is equal to t, we have $m \geq t$. Therefore the cardinality of the set D^* is at most $(k+2n-t)^{n-t-1}$.

Let ζ^{-1} be the mapping which is inverse to the mapping ζ and $D = \{(\zeta^{-1}(\bar{v}_1), \ldots, \zeta^{-1}(\bar{v}_{n-t+1})) : (\bar{v}_1, \ldots, \bar{v}_{n-t+1}) \in D^*\}$. In the same way as in the proof of Lemma 6.16 we can show that the set D is a vertex $(n-t+1)$-covering of the centroid U. Notice that

$$|D| \leq (k+2n-t)^{n-t-1} \ . \tag{109}$$

Let $a = 2^{\max\{1,\max\{r(f_j):j=1,\ldots,k\}\}}$ and let $\bar{v}$ be an arbitrary vertex of the centroid U. Let us show that $\bar{v} = (d_1/d_0, \ldots, d_n/d_0)$ where $d_j \in K$, $|d_j| \le a^n n!$ for any $j \in \{0, \ldots, n\}$ and $d_0 \ne 0$. From the boundedness of the set U and from Lemmas 6.7 and 6.8 follows that $\bar{v}$ is a nodal solution of the system (105). Since the set U is bounded, from Lemma 6.6 follows that the rank of the system (105) is equal to n. Therefore there exist numbers $i_1, \ldots, i_n \in \{1, \ldots, k+2n\}$ such that the element $\bar{v}$ is a solution of the system $\{g_{i_1}(\bar{x}) = 0, \ldots, g_{i_n}(\bar{x}) = 0\}$, and the rank of this system is equal to n. From Lemma 6.2 follows that $\bar{v} = (d_1/d_0, \ldots, d_n/d_0)$ where d_j is an n-th order determinant constructed from coefficients of the functions $g_{i_1}, \ldots, g_{i_n}$, $0 \le j \le n$, and $d_0 \ne 0$. Coefficients of the functions $g_{i_1}, \ldots, g_{i_n}$ are numbers from the ring K, and the absolute value of each coefficient is at most a. Therefore $d_j \in K$ and $|d_j| \le a^n n!$ for any $j \in \{0, \ldots, n\}$. Consider an arbitrary tuple $(\bar{v}_1, \ldots, \bar{v}_{n-t+1}) \in D$. Since the set D is a vertex $(n-t+1)$-covering of the centroid U, the dimension of the set $\{\bar{v}_1, \ldots, \bar{v}_{n-t+1}\}$ is equal to $n-t$. Let $\Delta \subseteq \mathbb{R}^n$ and $\Delta = \{(0, \ldots, 0), (1, 0, \ldots, 0), \ldots, (0, \ldots, 0, 1)\}$. From Lemma 6.4 follows that the dimension of the set Δ is equal to n. If $t > 0$ we will add to the set $\{\bar{v}_1, \ldots, \bar{v}_{n-t+1}\}$ elements $\bar{u}_1, \ldots, \bar{u}_t \in \Delta$ and obtain the set $\{\bar{v}_1, \ldots, \bar{v}_{n-t+1}, \bar{u}_1, \ldots, \bar{u}_t\}$ of the dimension n. Let us show that such elements $\bar{u}_1, \ldots, \bar{u}_t \in \Delta$ exist. First, we show that there exists an element $\bar{u}_1 \in \Delta$ such that the dimension of the set $\{\bar{v}_1, \ldots, \bar{v}_{n-t+1}, \bar{u}_1\}$ is equal to $n-t+1$. From Lemma 6.4 follows that for any $\bar{u} \in \Delta$ the dimension of the set $\{\bar{v}_1, \ldots, \bar{v}_{n-t+1}, \bar{u}\}$ is either $n-t$ or $n-t+1$. Assume that for any $\bar{u} \in \Delta$ the dimension of the set $\{\bar{v}_1, \ldots, \bar{v}_{n-t+1}, \bar{u}\}$ is equal to $n-t$. Using Lemma 6.4 we conclude that in this case Δ is contained in the same $(n-t)$-dimensional plane as the set $\{\bar{v}_1, \ldots, \bar{v}_{n-t+1}\}$. Then the dimension of the set Δ is less than n which is impossible. Hence there exists an element $\bar{u}_1 \in \Delta$ such that the dimension of the set $\{\bar{v}_1, \ldots, \bar{v}_{n-t+1}, \bar{u}_1\}$ is equal to $n-t+1$. In the similar way we can prove that there exist elements $\bar{u}_2, \ldots, \bar{u}_t \in \Delta$ such that the dimension of the set $\{\bar{v}_1, \ldots, \bar{v}_{n-t+1}, \bar{u}_1, \ldots, \bar{u}_t\}$ is equal to n.

Denote

$$D(\bar{v}_1, \ldots, \bar{v}_{n-t+1}) = \{(\bar{v}_2, \bar{v}_3 \ldots, \bar{v}_{n-t+1}, \bar{u}_1, \ldots, \bar{u}_t),$$
$$(\bar{v}_1, \bar{v}_3, \ldots, \bar{v}_{n-t+1}, \bar{u}_1, \ldots, \bar{u}_t), \ldots, (\bar{v}_1, \bar{v}_2, \ldots, \bar{v}_{n-t}, \bar{u}_1, \ldots, \bar{u}_t)\} .$$

Correspond to each tuple from the set $D(\bar{v}_1, \ldots, \bar{v}_{n-t+1})$ a function from $S_n(K)$ which is not constant on P and which is equal to 1 on all elements from the tuple. Consider, for instance, the tuple $(\bar{v}_1, \bar{v}_2, \ldots, \bar{v}_{n-t}, \bar{u}_1, \ldots, \bar{u}_t)$. By proved above, for $j = 1, \ldots, n-t$ the equality $\bar{v}_j = (d_{j1}/d_{j0}, \ldots, d_{jn}/d_{j0})$ holds where $d_{ji} \in K$,

$$|d_{ji}| \le a^n n! \ , \tag{110}$$

$0 \le i \le n$, and $d_{j0} \ne 0$. One can show that the element $\bar{u}_j$, $1 \le j \le t$, can be represented in the form

$$\bar{u}_j = (d_{n-t+j1}/d_{n-t+j0}, \ldots, d_{n-t+jn}/d_{n-t+j0})$$

where $d_{n-t+ji} \in K$,

$$|d_{n-t+ji}| \le a^n n! \tag{111}$$

for any $i \in \{0, \ldots, n\}$ and $d_{n-t+j0} \ne 0$. Consider the equation systems

$$\left\{\begin{array}{l}(d_{11}/d_{10})x_1 + \ldots + (d_{1n}/d_{10})x_n + x_{n+1} = 0 \\ \ldots\ldots\ldots\ldots\ldots\ldots\ldots\ldots\ldots\ldots \\ (d_{n1}/d_{n0})x_1 + \ldots + (d_{nn}/d_{n0})x_n + x_{n+1} = 0 \ ,\end{array}\right. \quad (112)$$

$$\left\{\begin{array}{l}(d_{11}/d_{10} - d_{21}/d_{20})x_1 + \ldots + (d_{1n}/d_{10} - d_{2n}/d_{20})x_n = 0 \\ \ldots\ldots\ldots\ldots\ldots\ldots\ldots\ldots\ldots\ldots \\ (d_{11}/d_{10} - d_{n1}/d_{n0})x_1 + \ldots + (d_{1n}/d_{10} - d_{nn}/d_{n0})x_n = 0 \ ,\end{array}\right. \quad (113)$$

$$\left\{\begin{array}{l}(d_{11}d_{20} - d_{21}d_{10})x_1 + \ldots + (d_{1n}d_{20} - d_{2n}d_{10})x_n = 0 \\ \ldots\ldots\ldots\ldots\ldots\ldots\ldots\ldots\ldots\ldots \\ (d_{11}d_{n0} - d_{n1}d_{10})x_1 + \ldots + (d_{1n}d_{n0} - d_{nn}d_{10})x_n = 0 \ .\end{array}\right. \quad (114)$$

Since the set $\{\bar{v}_1, \ldots, \bar{v}_{n-t+1}, \bar{u}_1, \ldots, \bar{u}_t\}$ is n-dimensional, from Lemma 6.4 follows that the set $\{\bar{v}_1, \ldots, \bar{v}_{n-t}, \bar{u}_1, \ldots, \bar{u}_t\}$ is $(n-1)$-dimensional. From here and from Lemma 6.4 it follows that the rank of the system (113) is equal to $n-1$. It is clear that the system (114) is equivalent to the system (113), and the rank of (114) is also $n-1$. From here and from Lemma 6.2 it follows that there exist determinants $M_1, \ldots, M_n$ of the order $n-1$ which are constructed from coefficients of equations of the system (114) and which satisfy the following conditions: $\sum_{i=1}^{n} |M_i| > 0$ and the element $(M_1, -M_2, M_3, -M_4, \ldots, (-1)^{n-1}M_n)$ is a solution of the system (114). From (110) and (111) follows that for any $j \in \{1, \ldots, n\}$ the following inequality holds:

$$|M_j| \le 2^{n-1}a^{2n(n-1)}(n!)^{2(n-1)}(n-1)! \ . \quad (115)$$

One can show that the element

$$(M_1d_{10}, -M_2d_{10}, \ldots, (-1)^{n-1}M_nd_{10}, -\sum_{i=1}^{n}(-1)^{i-1}M_id_{1i}) \quad (116)$$

is a solution of the system (112). Denote by g a function from $L_n(K)$ such that

$$g(\bar{x}) = \sum_{i=1}^{n}(-1)^{i-1}M_id_{10}x_i - \sum_{i=1}^{n}(-1)^{i-1}M_id_{1i} \ .$$

Since the element (116) is a solution of the system (112), we have $g(\bar{v}_1) = \ldots = g(\bar{v}_{n-t}) = g(\bar{u}_1) = \ldots = g(\bar{u}_t) = 0$. Since $\sum_{i=1}^{n} |M_i| > 0$ and $d_{10} \neq 0$, the function g is not constant on $\mathbb{R}^n$. Let us show that $g(\bar{v}_{n-t+1}) \neq 0$. Assume that $g(\bar{v}_{n-t+1}) = 0$. Then an n-dimensional set $\{\bar{v}_1, \ldots, \bar{v}_{n-t+1}, \bar{u}_1, \ldots, \bar{u}_t\}$ is contained in the set of solutions of the equation $g(\bar{x}) = 0$, which is an $(n-1)$-dimensional plane, but it is impossible. Correspond to the tuple $(\bar{v}_1, \ldots, \bar{v}_{n-t}, \bar{u}_1, \ldots, \bar{u}_t) \in D(\bar{v}_1, \ldots, \bar{v}_{n-t+1})$ the function $q = \text{sign}(g)$. This function is not constant on P since $q(\bar{v}_1) = \ldots = q(\bar{v}_{n-t}) = 0$ and $q(\bar{v}_{n-t+1}) \neq 0$. From (110), (111), (115) and from the definition of the value a follows that

$$r(q) \le 2n^2(\log_2 n + 1 + \max\{r(f_j) : j = 1, \ldots, k\}) - 1 \ . \quad (117)$$

To each of other tuples from the set $D(\bar{v}_1, \ldots, \bar{v}_{n-t+1})$ we correspond in the same way a function from $S_n(K)$ which is not constant on P and equals to 0 on all elements from the tuple. Denote by $\mathcal{F}(\bar{v}_1, \ldots, \bar{v}_{n-t+1})$ the set of functions from $S_n(K)$ corresponded to tuples of $D(\bar{v}_1, \ldots, \bar{v}_{n-t+1})$. From Lemma 6.3 follows that there exists a subsystem of the system (106) which is equivalent to the system (106) and contains t equations. Let this subsystem be of the kind $\{g_1(\bar{x}) = 0, \ldots, g_t(\bar{x}) = 0\}$. Denote

$$\mathcal{F} = \{f_1, \ldots, f_t\} \cup \left(\bigcup_{(\bar{v}_1, \ldots, \bar{v}_{n-t+1}) \in D} \mathcal{F}(\bar{v}_1, \ldots, \bar{v}_{n-t+1}) \right) .$$

From (109) follows that

$$|\mathcal{F}| \leq (n+1)(k+2n)^{n-1} . \tag{118}$$

From (117) follows that

$$\max\{r(q) : q \in \mathcal{F}\} \leq 2n^2(\log_2 n + 1 + \max\{r(f_j) : j = 1, \ldots, k\}) - 1 . \tag{119}$$

Let us show that the set $\mathcal{F}$ is a functional $(n+1, K)$-covering of the set W. Let $\bar{a}_0 \in W$. Since the set D is a vertex $(n-t+1)$-covering of the set U, there exists a tuple $(\bar{v}_1, \ldots, \bar{v}_{n-t+1}) \in D$ such that the element $\bar{a}_0$ belongs to the centroid generated by the set $\{\bar{v}_1, \ldots, \bar{v}_{n-t+1}\}$. Let $\mathcal{F}(\bar{v}_1, \ldots, \bar{v}_{n-t+1}) = \{q_1, \ldots, q_{n-t+1}\}$ and $q_j = \text{sign}(h_j)$, $1 \leq j \leq n-t+1$, where $h_j \in L_n(K)$. Let, for the definiteness, the inequality $h_j(\bar{v}_i) \geq 0$ hold for any $j \in \{1, \ldots, n-t+1\}$ and $i \in \{1, \ldots, n-t+1\}$. From the property 1° of the mapping ζ follows that there exists a function $h_j^* \in L_{n-t}(\mathbb{R})$, $1 \leq j \leq n-t+1$, such that $h_j(\bar{a}) = h_j^*(\zeta(\bar{a}))$ for any $\bar{a} \in P$.

Consider the centroid U_0^* generated by the set $\{\zeta(\bar{v}_1), \ldots, \zeta(\bar{v}_{n-t+1})\}$. From the property 2° of the mapping ζ and from the definition of the set D follows that the elements $\zeta(\bar{v}_1), \ldots, \zeta(\bar{v}_{n-t+1})$ are centroidally independent. From here and from Lemma 6.7 follows that the set $\{\zeta(\bar{v}_1), \ldots, \zeta(\bar{v}_{n-t+1})\}$ coincides with the set of vertices of the centroid U_0^*. Since the set $\{\bar{v}_1, \ldots, \bar{v}_{n-t+1}\}$ is $(n-t)$-dimensional, from the property 3° of the mapping ζ follows that the dimension of the set $\{\zeta(\bar{v}_1), \ldots, \zeta(\bar{v}_{n-t+1})\}$ is equal to $n-t$. Therefore the dimension of the centroid U_0^* is also equals to $n-t$. Since the functions $q_1, \ldots, q_{n-t+1}$ are not constant on P, the functions $h_1^*, \ldots, h_{n-t+1}^*$ are not constant on $\mathbb{R}^{n-t}$. From the properties of the functions q_j, $1 \leq j \leq n-t+1$, and from the $(n-t)$-dimensionality of U_0^* we conclude that the half-space defined by the inequality $h_j^*(\bar{x}) \geq 0$, $1 \leq j \leq n-t+1$, is an extreme support of the centroid U_0^*. Let us show that the centroid U_0^* has no other extreme supports. Consider an arbitrary extreme support of the centroid U_0^*. Let it be given by the inequality $h(\bar{x}) \geq 0$ where h is a function from $L_{n-t}(\mathbb{R})$ which is not constant on $\mathbb{R}^{n-t}$. Then there exist $n-t$ vertices of the centroid U_0^* contained in the $(n-t-1)$-dimensional plane defined by the equation $h(\bar{x}) = 0$. Let, for the definiteness, this plane contains the vertices $\zeta(\bar{v}_1), \ldots, \zeta(\bar{v}_{n-t})$. Besides, from the choice of the functions q_j, $1 \leq j \leq n-t+1$, follows that for some $j \in \{1, \ldots, n-t+1\}$ the vertices $\zeta(\bar{v}_1), \ldots, \zeta(\bar{v}_{n-t})$ are contained in the plane defined by the equation $h_j^*(\bar{x}) = 0$.

Since the set $\{\zeta(\bar{v}_1), \ldots, \zeta(\bar{v}_{n-t+1})\}$ is $(n-t)$-dimensional, from Lemma 6.4 follows that the set $\{\zeta(\bar{v}_1), \ldots, \zeta(\bar{v}_{n-t})\}$ is $(n-t-1)$-dimensional. Using Lemma 6.5 we conclude that there exists $c \in \mathbb{R}$ such that $h_j^*(\bar{a}) = c \cdot h(\bar{a})$ for any $\bar{a} \in \mathbb{R}^{n-t}$. Hence the extreme support, defined by the inequality $h(\bar{x}) \geq 0$, coincides with the extreme support defined by the inequality $h_j^*(\bar{x}) \geq 0$. Therefore the set of half-spaces defined by the inequalities $h_1^*(\bar{x}) \geq 0, \ldots, h_{n-t+1}^*(\bar{x}) \geq 0$ coincides with the set of the extreme supports of the centroid U_0^*. From here, from Lemma 6.9 and from the $(n-t)$-dimensionality of the centroid U_0^* follows that the set of solutions of the system $\{h_1^*(\bar{x}) \geq 0, \ldots, h_{n-t+1}^*(\bar{x}) \geq 0\}$ on $\mathbb{R}^{n-t}$ coincides with U_0^*. From here, from the choice of the functions $h_j^*, 1 \leq j \leq n-t+1$, and from the properties of the mapping ζ follows that the set of solutions on P of the system $\{h_1(\bar{x}) \geq 0, \ldots, h_{n-t+1}(\bar{x}) \geq 0\}$ coincides with the centroid U_0 generated by the set $\{\bar{v}_1, \ldots, \bar{v}_{n-t+1}\}$. Therefore the set of solutions on $\mathbb{R}^n$ of the system

$$\{g_1(\bar{x}) = 0, \ldots, g_t(\bar{x}) = 0, h_1(\bar{x}) \geq 0, \ldots, h_{n-t+1}(\bar{x}) \geq 0\} \tag{120}$$

coincides with U_0 and is a subset of the set U.

Let $\eta_j = g_j$ for $j = 1, \ldots, t$ and $\eta_{t+j} = h_j$ for $j = 1, \ldots, n-t+1$. Let, for the definiteness, $\eta_{t+1}(\bar{a}_0) = \ldots = \eta_{t+p}(\bar{a}_0) = 0$ and $\eta_{t+p+1}(\bar{a}_0) > 0, \ldots, \eta_{n+1}(\bar{a}_0) > 0$. Denote by W_0 the set of solutions on $\mathbb{R}^n$ of the system $\{\eta_1(\bar{x}) = 0, \ldots, \eta_{t+p}(\bar{x}) = 0, \eta_{t+p+1}(\bar{x}) > 0, \ldots, \eta_{n+1}(\bar{x}) > 0\}$. Evidently, $\bar{a}_0 \in W_0$. Let us show that $W_0 \subseteq W$. It is sufficient to show that for any $\bar{a}_1 \in W_0$ and for any $j \in \{m+1, \ldots, k+2n\}$ the inequality $g_j(\bar{a}_1) > 0$ holds. Evidently, the set of solutions of the system

$$\{\eta_1(\bar{x}) = 0, \ldots, \eta_{t+p}(\bar{x}) = 0, \eta_{t+p+1}(\bar{x}) \geq 0, \ldots, \eta_{n+1}(\bar{x}) \geq 0\} \tag{121}$$

is a subset of the set of solutions of the system (120). Hence the set of solutions of the system (121) is a subset of the set U. Therefore for any $j \in \{m+1, \ldots, k+2n\}$ the inequality $g_j(\bar{x}) \geq 0$ is a consequence of the system (121). Let $j \in \{m+1, \ldots, k+2n\}$. From Lemma 6.10 follows that there exist numbers $\lambda_1, \ldots, \lambda_{t+p} \in \mathbb{R}$ and the nonnegative numbers $\lambda_{t+p+1}, \ldots, \lambda_{n+1}, \lambda_0 \in \mathbb{R}$ such that $g_j(\bar{x}) = \sum_{i=1}^{n+1} \lambda_i \eta_i(\bar{x}) + \lambda_0$. Let $\bar{a}_1 \in W_0$. Then $\sum_{i=1}^{t+p} \lambda_i \eta_i(\bar{a}_1) = 0$, and if there exists at least one $i \in \{0, t+p+1, t+p+2, \ldots, n+1\}$ such that $\lambda_i > 0$ then $g_j(\bar{a}_1) > 0$. Assume that $\lambda_i = 0$ for any $i \in \{0, t+p+1, t+p+2, \ldots, n+1\}$. Then $g_j(\bar{a}_1) = 0$ for any $\bar{a}_1 \in W_0$. This contradicts to the fact that $g_j(\bar{a}_0) > 0$ and $\bar{a}_0 \in W_0$. Therefore $W_0 \subseteq W$. Hence the element $\bar{a}_0$ belongs to the set of solutions on $\mathbb{R}^n$ of the system $\{f_1(\bar{x}) = 0, \ldots, f_t(\bar{x}) = 0, q_1(\bar{x}) = 0, \ldots, q_p(\bar{x}) = 0, q_{p+1}(\bar{x}) = +1, \ldots, q_{n-t+1}(\bar{x}) = +1\}$, and the set of solutions of this system is a subset of the set W. Thus, $\mathcal{F}$ is a functional $(n+1, K)$-covering of the set W. From here, from (118) and from (119) follows the statement of lemma. □

6.4 Proofs of Theorems

This subsection contains proofs of Theorems 6.1 and 6.2.

Proof (of Theorem 6.1). Let $U = U(A, K, \varphi_1, \ldots, \varphi_n)$ be a quasilinear information system, z be a problem over U, $\dim z = k$, $z = (\nu, f_1, \ldots, f_k)$ and $r(z) = t$.

Denote $V = U(\mathbb{R}^n, K, x_1, \dots, x_n)$. Correspond to arbitrary attribute $f \in F(A, K, \varphi_1, \dots, \varphi_n)$ attribute $f^\tau \in F(\mathbb{R}^n, K, x_1, \dots, x_n)$. Let $f = \text{sign}(a_1\varphi_1 + \dots + a_n\varphi_n + a_{n+1})$. Then $f^\tau = \text{sign}(a_1x_1 + \dots + a_nx_n + a_{n+1})$. Denote $z^\tau = (\nu, f_1^\tau, \dots, f_k^\tau)$.

Denote $Y = U(C_{n+1}, K, x_1, \dots, x_{n+1})$ and $z^{\tau\kappa} = (\nu, f_1^{\tau\kappa}, \dots, f_k^{\tau\kappa})$. Corresponding definitions are presented before Lemma 6.19. Note that

$$f^{\tau\kappa} = \text{sign}(a_1x_1 + \dots + a_nx_n + a_{n+1}x_{n+1})$$

for the considered attribute $f = \text{sign}(a_1\varphi_1 + \dots + a_n\varphi_n + a_{n+1})$.

First, we consider decision trees for the problem $z^{\tau\kappa}$ over Y. Obtained results will then be applied consequently to the problem z^τ over V and to the problem z over U.

For $j = 1, \dots, k$ set $g_j = f_j^{\tau\kappa}$. Denote by $H(g_1, \dots, g_k)$ the set of all tuples $(\delta_1, \dots, \delta_k) \in \{-1, 0, +1\}^k$ for each of which the equation system $\{g_1(\bar{x}) = \delta_1, \dots, g_k(\bar{x}) = \delta_k\}$ is compatible on C_{n+1}. Let

$$H(g_1, \dots, g_k) = \{(\delta_{11}, \dots, \delta_{1k}), \dots, (\delta_{m1}, \dots, \delta_{mk})\} \ .$$

Let us denote by W_j, $1 \le j \le m$, the set of solutions on C_{n+1} of the equation system $\{g_1(\bar{x}) = \delta_{j1}, \dots, g_k(\bar{x}) = \delta_{jk}\}$. It is easily to see that $r(g_j) = r(f_j)$ for any $j \in \{1, \dots, k\}$. Therefore $\max\{r(g_j) : j = 1, \dots, k\} = t$. From here and from Lemma 6.20 follows that there exists a functional $(n+2, K)$-covering $\mathcal{F}_j$ of the set W_j, $1 \le j \le m$, such that

$$|\mathcal{F}_j| \le (n+2)(k+2n+2)^n \ , \tag{122}$$
$$\max\{r(f) : f \in \mathcal{F}_j\} \le 2(n+1)^2(t+1+\log_2(n+1)) - 1 \ . \tag{123}$$

Let $j \in \{1, \dots, m\}$ and $\mathcal{F}_j = \{q_1, \dots, q_l\}$ where $q_i \in S_{n+1}(K)$, $1 \le i \le l$. Let us denote $\mathcal{F}_j^0 = \{q_1, \dots, q_l\}$. Set $\mathcal{F} = \{g_1, \dots, g_k\} \cup \left(\bigcup_{j=1}^m \mathcal{F}_j^0\right)$. Denote $p = |\mathcal{F}|$. Since $m \le 2k^{n+1} + 1$ according to Lemma 6.18, using (122) we obtain

$$\begin{aligned} p &\le k + (2k^{n+1} + 1)(n+2)(k+2n+2)^n \\ &\le (k+2n+2)^{2(n+1)} - 1 \ . \end{aligned} \tag{124}$$

Let $\mathcal{F} = \{g_1, \dots, g_p\}$ and let $\nu_1 : \{-1, 0, +1\}^p \to \mathbb{Z}$ be a mapping such that $\nu_1(\delta_1, \dots, \delta_p) = \nu(\delta_1, \dots, \delta_k)$ for any $(\delta_1, \dots, \delta_p) \in \{-1, 0, +1\}^p$. Denote $z_1 = (\nu_1, g_1, \dots, g_p)$. It is clear that for any $\bar{a} \in C_{n+1}$ the equality

$$z^{\tau\kappa}(\bar{a}) = z_1(\bar{a}) \tag{125}$$

holds. Denote $T = T_Y(z_1)$ and $\rho = (F(C_{n+1}, K, x_1, \dots, x_{n+1}), \{-1, 0, +1\})$. (The definition of the decision table $T_Y(z_1)$ can be found in Sect. 4.2. The definitions of the parameters $N(T)$ and $M_{\rho,h}(T)$ can be found in Sect. 3.2.) From (124) and from Lemma 6.18 follows that

$$N(T) \le 2(k+2n+2)^{2(n+1)^2} \ . \tag{126}$$

To verify the inequality

$$M_{\rho,h}(T) \leq n + 2 \tag{127}$$

it is sufficiently to show that for any tuple $(\delta_1, \ldots, \delta_p) \in \{-1, 0, +1\}^p$ there exist numbers $i_1, \ldots, i_q \in \{1, \ldots, p\}$ such that $q \leq n + 2$ and $T(g_{i_1}, \delta_{i_1}) \ldots (g_{i_q}, \delta_{i_q}) \in \mathrm{Dtab}_\rho$. Verify this statement. Let $(\delta_1, \ldots, \delta_p) \in \{-1, 0, +1\}^p$ and the equation system

$$\{g_1(\bar{x}) = \delta_1, \ldots, g_p(\bar{x}) = \delta_p\} \tag{128}$$

be incompatible on C_{n+1}. Then the equation system

$$\begin{aligned} &\{g_1(\bar{x}) = \delta_1, \ldots, g_p(\bar{x}) = \delta_p, \mathrm{sign}(x_1 + 2) = 1, \\ &\mathrm{sign}(2 - x_1) = 1, \ldots, \mathrm{sign}(x_{n+1} + 2) = 1, \mathrm{sign}(2 - x_{n+1}) = 1\} \end{aligned} \tag{129}$$

is incompatible on $\mathbb{R}^{n+1}$. From Lemma 6.17 follows that the system (129) contains a subsystem which consists of at most $n+2$ equations and is incompatible on $\mathbb{R}^{n+1}$. By deletion from this subsystem the equations from the last $2n + 2$ equations of the system (129) we obtain an incompatible on C_{n+1} subsystem of the system (128) containing at most $n+2$ equations. Let this subsystem be of the kind $\{g_{i_1}(\bar{x}) = \delta_{i_1}, .., g_{i_q}(\bar{x}) = \delta_{i_q}\}$ where $q \leq n + 2$ and $i_1, \ldots, i_q \in \{1, \ldots, p\}$. Then $\mathrm{Row}(T(g_{i_1}, \delta_{i_1}) \ldots (g_{i_q}, \delta_{i_q})) = \emptyset$ and hence $T(g_{i_1}, \delta_{i_1}) \ldots (g_{i_q}, \delta_{i_q}) \in \mathrm{Dtab}_\rho$. Let the system (128) be compatible on C_{n+1}, and let $\bar{a}_0$ be a solutions of this system from C_{n+1}. Then for some $j \in \{1, \ldots, m\}$ we have $\bar{a}_0 \in W_j$. Let, for the definiteness, $\bar{a}_0 \in W_1$. Then, by definition of the set $\mathcal{F}_1$, there exist functions $g_{i_1}, \ldots, g_{i_{n+2}} \in \mathcal{F}_1$ and numbers $\sigma_1, \ldots, \sigma_{n+2} \in \{-1, 0, +1\}$ such that the element $\bar{a}_0$ is a solution of the equation system

$$\{g_{i_1}(\bar{x}) = \sigma_1, \ldots, g_{i_{n+2}}(\bar{x}) = \sigma_{n+2}\} \ ,$$

and the set W of solutions of this system on $\mathbb{R}^{n+1}$ is a subset of the set W_1. Since $g_{i_1}, \ldots, g_{i_{n+2}} \in \mathcal{F}$, we have $\sigma_1 = \delta_{i_1}, \ldots, \sigma_{n+2} = \delta_{i_{n+2}}$. It is clear that the function $z^{T\kappa}(\bar{x})$ is constant on the set W_1. From here and from (125) follows that the function $z_1(\bar{x})$ is constant on the set W_1. Hence the function $z_1(\bar{x})$ is constant on the set W. Using this fact it is not difficult to show that $T(g_{i_1}, \delta_{i_1}) \ldots (g_{i_{n+2}}, \delta_{i_{n+2}}) \in \mathrm{Dtab}_\rho$. Thus, the inequality (127) holds.

Let $a, b \in \mathbb{N}$, $a \geq 4$, $b \geq 1$, $T^* \in \mathrm{Tab}_\rho$, $M_{\rho,h}(T^*) \leq a$ and $N_\rho(T^*) \leq b$. Then using Theorem 3.6 one can show that

$$h_\rho(T^*) \leq a + (a \log_2 b)/\log_2 a \ .$$

From here and from (126) and (127) follows

$$\begin{aligned} h_\rho(T) &\leq (n + 3) + (n + 3)(1 + 2(n + 1)^2 \log_2(k + 2n + 2))/\log_2(n + 3) \\ &\leq 2(n + 2)^3 \log_2(k + 2n + 2)/\log_2(n + 2) - 2n \ . \end{aligned} \tag{130}$$

Let Γ_1 be a decision tree for the table T such that

$$h(\Gamma_1) = h_\rho(T) \ . \tag{131}$$

Using the inequality (123) we obtain

$$r(\Gamma_1) \leq 2(n+1)^2(t+1+\log_2(n+1)) - 1 \ . \tag{132}$$

From Theorem 4.1 and from (125) follows that the decision tree Γ_1 over Y solves the problems z_1 and $z^{\tau\kappa}$. From here and from Lemma 6.19 follows that there exists a decision tree Γ_2 over V which solves the problem z^τ over V and for which $h(\Gamma_2) \leq h(\Gamma_1) + 2n$ and $r(\Gamma_2) \leq r(\Gamma_1) + 1$. From the two last inequalities, from (130)–(132) and from the equalities $k = \dim z$ and $t = r(z)$ follows

$$h(\Gamma_2) \leq 2(n+2)^3 \log_2(\dim z + 2n + 2)/\log_2(n+2)$$

and

$$r(\Gamma_2) \leq 2(n+1)^2(r(z) + 1 + \log_2(n+1)) \ .$$

For each working node w of the tree Γ_2 instead of an attribute $\text{sign}(a_1x_1 + \ldots + a_nx_n + a_{n+1})$ attached to it we mark w by the attribute $\text{sign}(a_1\varphi_1 + \ldots + a_n\varphi_n + a_{n+1})$. Denote the obtained decision tree by Γ. It is clear that Γ is a decision tree over U, $h(\Gamma) = h(\Gamma_2)$ and $r(\Gamma) = r(\Gamma_2)$. One can show that Γ solves the problem z. □

Proof (of Theorem 6.2). Describe an algorithm which for a given problem z over U constructs a decision tree Γ over U such that the decision tree Γ solves the problem z, and the inequalities (74) and (75) hold.

Denote $V = U(\mathbb{R}^n, \mathbb{Z}, x_1, \ldots, x_n)$ and $\rho = (F(\mathbb{R}^n, \mathbb{Z}, x_1, \ldots, x_n), \{-1, 0, +1\})$. We correspond to an arbitrary attribute $f \in F(A, \mathbb{Z}, \varphi_1, \ldots, \varphi_n)$ an attribute $f^\tau \in F(\mathbb{R}^n, \mathbb{Z}, x_1, \ldots, x_n)$. Let $f = \text{sign}(a_1\varphi_1 + \ldots + a_n\varphi_n + a_{n+1})$. Then $f^\tau = \text{sign}(a_1x_1 + \ldots + a_nx_n + a_{n+1})$.

Let $z = (\nu, g_1, \ldots, g_k)$. For $i = 1, \ldots, k$ denote $f_i = g_i^\tau$. Find a number $q \in \mathbb{N}$ such that $r(z) = \log_2 q$. Construct the set $B = \{\text{sign}(a_1x_1 + \ldots + a_nx_n + a_{n+1}) \in S_n(\mathbb{Z}) : |a_i| \leq (2q(n+1))^{2(n+1)^2}, 1 \leq i \leq n+1\}$. Denote $t = 2(n+1)^2(r(z) + 1 + \log_2(n+1))$. Evidently, $B = \{f : f \in S_n(\mathbb{Z}), r(f) \leq t\}$. Let $B = \{f_1, \ldots, f_m\}$. Denote $z_1 = (\nu, f_1, \ldots, f_k)$ and $z^* = (\nu^*, f_1, \ldots, f_m)$ where $\nu^* : \{-1, 0, +1\}^m \to \mathbb{Z}$ and $\nu^*(\delta_1, \ldots, \delta_m) = \nu(\delta_1, \ldots, \delta_k)$ for any tuple $(\delta_1, \ldots, \delta_m) \in \{-1, 0, +1\}^m$.

Theory of the algebra $(\mathbb{R}; x + y, x \cdot y, 0, 1)$ is solvable [207]. Using this fact one can show that there exists an algorithm which for a given problem y over V constructs the decision table $T_V(y)$. Using this algorithm construct the decision table $T^* = T_V(z^*)$. As in the proof of Proposition 3.4 we can show that there exists an algorithm which for a given decision table $T \in \text{Tab}_\rho$ constructs a decision tree Γ_0 for the table T such that $h(\Gamma_0) = h_\rho(T)$. Using this algorithm construct a decision tree Γ_1 for the table T^* such that $h(\Gamma_1) = h_\rho(T^*)$. From Theorem 6.1 follows that there exists a decision tree G over V which solves the problem z_1 over V and for which the inequalities

$$h(G) \leq (2(n+2)^3 \log_2(\dim z_1 + 2n + 2))/(\log_2(n+2))$$

and

$$r(G) \leq 2(n+1)^2(r(z_1) + 1 + \log_2(n+1))$$

hold. Evidently, the decision tree G solves the problem z^*, and $\mathrm{At}(G) \subseteq B = \mathrm{At}(z^*)$. From Theorem 4.1 follows that G is a decision tree for the table T^*. Hence $h(\Gamma_1) \leq h(G)$. Therefore the inequality (74) holds if in the capacity of Γ we take Γ_1. Since $\mathrm{At}(\Gamma_1) \subseteq B$, the inequality (75) holds if in the capacity of Γ we take Γ_1. From Theorem 4.1 follows that the decision tree Γ_1 solves the problem z^*. Hence Γ_1 solves also the problem z_1.

For each working node w of the tree Γ_1 instead of an attribute $\mathrm{sign}(a_1x_1 + \ldots + a_nx_n + a_{n+1})$ attached to it we mark w by the attribute $\mathrm{sign}(a_1\varphi_1 + \ldots + a_n\varphi_n + a_{n+1})$. Denote the obtained decision tree by Γ. It is clear that Γ is a decision tree over U, $h(\Gamma) = h(\Gamma_1)$ and $r(\Gamma) = r(\Gamma_1)$. One can show that Γ solves the problem z. It is clear that (74) and (75) hold for Γ. □

6.5 Global Shannon Functions for Some Information Systems and Weight Functions

This subsection is devoted to the study of global Shannon functions for pairs (*information system, weight function*) belonging to the three families. The first of them consists of pairs of the kind (U, h), where U is a quasilinear information system. The second one contains pairs (U, ψ), where U is a quasilinear information system of the kind $U(\mathbb{R}, K, x_1, \ldots, x_m)$ and ψ is a weight function for U. The third family consists of pairs (U, h) in which U is either a system with attributes of the kind $\mathrm{sign}(p)$ where p is a polynomial of variables $x_1, \ldots, x_m$ with integer coefficients, or U is a system with attributes of the kind $\mathrm{sign}(q)$ where q is a polynomial of variables $x_1, \ldots, x_m$ with integer coefficients which degree is at most t.

Study of Pairs from First Family

Theorem 6.3. *Let $U = U(A, K, \varphi_1, \ldots, \varphi_m)$ be a quasilinear information system. Then the following statements hold:*

(a) if the set of functions $F(A, K, \varphi_1, \ldots, \varphi_m)$ is finite (i.e. contains only finite number of pairwise different functions) then $H^g_{U,h}(n) = O(1)$;
(b) if the set of functions $U(A, K, \varphi_1, \ldots, \varphi_m)$ is infinite then $H^g_{U,h}(n) = \Theta(\log_2 n)$.

Proof. a) Let the set of functions $F(A, K, \varphi_1, \ldots, \varphi_m)$ be finite. Then from Theorem 5.1 follows that $H^g_{U,h}(n) = O(1)$.
b) Let the set of the functions $F(A, K, \varphi_1, \ldots, \varphi_m)$ be infinite. From Theorem 6.1 follows that $H^g_{U,h}(n) = O(\log_2 n)$. Using Theorem 5.1 we conclude that $H^g_{U,h}(n) = \Omega(\log_2 n)$. Therefore $H^g_{U,h}(n) = \Theta(\log_2 n)$. □

Study of Pairs from Second Family. Let $U = U(\mathbb{R}, K, x_1, \ldots, x_m)$ be a quasilinear information system and ψ be a weight function for U. Two attributes $f, g \in F = F(\mathbb{R}, K, x_1, \ldots, x_m)$ will be called $\mathbb{R}$-*equivalent* if there exist a number $c \in \mathbb{R} \setminus \{0\}$ and numbers $a_1, \ldots, a_{m+1} \in K$ such that $f = \mathrm{sign}(a_1x_1 + \ldots + a_mx_m + a_{m+1})$ and $g = \mathrm{sign}(ca_1x_1 + \ldots + ca_mx_m + ca_{m+1})$. If $c > 0$ then we will say that f and g are *positively* $\mathbb{R}$-equivalent. If $c < 0$ then we will say that f and g are *negatively* $\mathbb{R}$-equivalent. The weight function ψ will be called $\mathbb{R}$-*bounded* if there exists a number $t \in \mathbb{N}$ with the following property: for any attribute $f \in F$ there exists an attribute $g \in F$ such that $\psi(g) \le t$ and the attributes f and g are $\mathbb{R}$-equivalent.

Theorem 6.4. *Let $U = U(\mathbb{R}, K, x_1, \ldots, x_m)$ be a quasilinear information system and ψ be a weight function for U. Then the following statements hold:*

(a) if the weight function ψ is $\mathbb{R}$-bounded then $H^g_{U,\psi}(n) = \Theta(\log_2 n)$;
(b) if the weight function ψ is not $\mathbb{R}$-bounded then $H^g_{U,\psi}(n) = n$ for infinitely many natural n.

Proof. It is clear that $\mathrm{Dom}(H^g_{U,\psi}) = \{n : n \in \mathbb{N}, n \ge \min\{\psi(f) : f \in F\}\}$.

a) Let the weight function ψ be $\mathbb{R}$-bounded, $t \in \mathbb{N}$ and for any attribute $f \in F$ there exists an attribute $g \in F$ such that $\psi(g) \le t$ and the attributes f and g are $\mathbb{R}$-equivalent.
Let $n \in \mathrm{Dom}(H^g_{U,\psi})$, z be a problem over U and $\psi(z) \le n$. Evidently, $\dim \sigma \le n$. From Theorem 6.1 follows that there exists a decision tree Γ over U which solves the problem z and for which $h(\Gamma) \le 2(m+2)^3 \log_2(n + 2m + 2)$. Let us denote by Γ^* the decision tree obtained from the tree Γ by replacement of attributes attached to working nodes and numbers attached to edges of Γ according to the following conditions. If some working node v of the tree Γ is labelled by an attribute f then we replace it by an attribute $g \in F$ such that $\psi(g) \le t$ and the attributes f and g are $\mathbb{R}$-equivalent. If f and g are negatively $\mathbb{R}$-equivalent then for every edge issuing from the node v substitute the number $-\delta$ for the number δ attached to it. If f and g are positively $\mathbb{R}$-equivalent then the numbers attached to the edges issuing from v will be left untouched. Evidently, the decision tree Γ^* solves the problem z, $h(\Gamma^*) = h(\Gamma)$ and $\psi(\Gamma^*) \le t \cdot h(\Gamma^*)$. Set $c_1 = t2(m+2)^3$ and $c_2 = t2(m+2)^3 \log_2(2m+3)$. Then $\psi(\Gamma^*) \le c_1 \log_2 n + c_2$ and hence $\psi^g_U(z) \le c_1 \log_2 n + c_2$. Since n is an arbitrary number from the set $\mathrm{Dom}(H^g_{U,\psi})$ and z is an arbitrary problem over U with $\psi(z) \le n$, we conclude that for any $n \in \mathrm{Dom}(H^g_{U,\psi})$ the inequality

$$H^g_{U,\psi}(n) \le c_1 \log_2 n + c_2$$

holds. From this inequality follows that U is ψ-compressible information system. It is clear that F contains an infinite set of pairwise distinct functions. Using these facts and Theorem 5.4 we obtain $H^g_{U,\psi}(n) = \Omega(\log_2 n)$. Thus, $H^g_{U,\psi}(n) = \Theta(\log_2 n)$.

b) Let the weight function ψ be not $\mathbb{R}$-bounded. To show that $H^g_{U,\psi}(n) = n$ for infinitely many natural n it is sufficiently to show that for any $p \in \mathbb{N}$ there exists $n \in \mathbb{N}$ such that $n \ge p$, $n \in \mathrm{Dom}(H^g_{U,\psi})$ and $H^g_{U,\psi}(n) \ge n$. Let $p \in \mathbb{N}$. Since the weight function ψ is not $\mathbb{R}$-bounded, there exists an attribute $f \in F$ possessing the following properties: $f \not\equiv \mathrm{const}$, $\psi(f) \ge p$ and for any attribute $g \in F$ which is $\mathbb{R}$-equivalent to the attribute f the inequality $\psi(g) \ge \psi(f)$ holds. Set $n = \psi(f)$. Evidently, $n \in \mathrm{Dom}(H^g_{U,\psi})$.

Define a function $\varphi : \mathbb{R}^m \to \{\{-1\}, \{+1\}, \{-1, +1\}\}$ in the following way. Let $\bar{a} \in \mathbb{R}^m$. Then

$$\varphi(\bar{a}) = \begin{cases} \{-1\}, & \text{if } f(\bar{a}) = -1 \ , \\ \{-1, +1\}, & \text{if } f(\bar{a}) = 0 \ , \\ \{+1\}, & \text{if } f(\bar{a}) = +1 \ . \end{cases}$$

Let $z = (\nu, f)$, where $\nu : \{-1, 0, +1\} \to \mathbb{Z}$, $\nu(-1) = -1$, $\nu(0) = -1$ and $\nu(1) = +1$, Γ be a decision tree over U and the decision tree Γ solve the problem z. Let $\mathrm{At}(\Gamma) = \{f_1, \ldots, f_k\}$. Obviously, there exists a mapping $\gamma : \{-1, 0, +1\}^k \to \mathbb{Z}$ such that the equality

$$\gamma(f_1(\bar{a}), \ldots, f_k(\bar{a})) = z(\bar{a})$$

holds for any $\bar{a} \in \mathbb{R}^m$. Evidently, $z(\bar{a}) \in \varphi(\bar{a})$ for any $\bar{a} \in \mathbb{R}^m$ and hence $\gamma(f_1(\bar{a}), \ldots, f_k(\bar{a})) \in \varphi(\bar{a})$ for any $\bar{a} \in \mathbb{R}^m$. From here and from the main result (Theorem 1) of [67] follows that for some $i \in \{1, \ldots, k\}$ the attributes f and f_i are $\mathbb{R}$-equivalent. Therefore $\psi(f_i) \geq n$. Since $f_i \in \mathrm{At}(\Gamma)$, we have $\psi(\Gamma) \geq n$. Taking into account that Γ is an arbitrary decision tree over U which solves the problem z we obtain $\psi_U^g(z) \geq n$. From this inequality and from the obvious equality $\psi(z) = n$ follows that $H_{U,\psi}^g(n) \geq n$. Therefore $H_{U,\psi}^g(n) = n$ for infinitely many natural n. □

Study of Pairs from Third Family. Remind some definitions from Sect. 4.1. Let m, t be natural numbers. We denote by $\mathrm{Pol}(m)$ the set of all polynomials which have integer coefficients and depend on variables $x_1, \ldots, x_m$. We denote by $\mathrm{Pol}(m, t)$ the set of all polynomials from $\mathrm{Pol}(m)$ such that the degree of each polynomial is at most t.

Define information systems $U(\mathbb{R}, m)$ and $U(\mathbb{R}, m, t)$ as follows: $U(\mathbb{R}, m) = (\mathbb{R}^m, E, F(m))$ and $U(\mathbb{R}, m, t) = (\mathbb{R}^m, E, F(m, t))$ where $E = \{-1, 0, +1\}$, $F(m) = \{\mathrm{sign}(p) : p \in \mathrm{Pol}(m)\}$ and $F(m, t) = \{\mathrm{sign}(p) : p \in \mathrm{Pol}(m, t)\}$.

Theorem 6.5. *Let m, t be natural numbers. Then the following statements hold:*

(a) $H^g_{U(\mathbb{R},m),h}(n) = n$ for any natural n;
(b) $H^g_{U(\mathbb{R},m,t),h}(n) = \Theta(\log_2 n)$.

Proof. One can show that for any natural n there exist polynomials $\pi_1, \ldots, \pi_n$ which have integer coefficients and depend on variable x_1, and for which for any $\delta_1, \ldots, \delta_n \in \{-1, +1\}$ the equation system

$$\{\mathrm{sign}(\pi_1(x_1)) = \delta_1, \ldots, \mathrm{sign}(\pi_n(x_1)) = \delta_n\}$$

is compatible on $\mathbb{R}$. From this fact follows that the information system $U(\mathbb{R}, m)$ has infinite I-dimension. It is clear that $U(\mathbb{R}, m)$ is an infinite information system. Using Theorem 5.1 obtain $H^g_{U(\mathbb{R},m),h}(n) = n$ for any natural n.

b) Consider the quasilinear information system $U = U(\mathbb{R}^m, \mathbb{Z}, \varphi_1, ..., \varphi_k)$ where $\{\varphi_1, \ldots, \varphi_k\} = \{x_1^{l_1} \cdot \ldots \cdot x_m^{l_m} : l_i \in \mathbb{N}, 1 \leq i \leq m, 1 \leq \sum_{i=1}^m l_i \leq t\}$. Denote $F = U(\mathbb{R}^m, \mathbb{Z}, \varphi_1, \ldots, \varphi_k)$. Obviously, $F = F(m,t) = \{\text{sign}(p) : p \in \text{Pol}(m,t)\}$. Using this equality one can show that $H^g_{U(\mathbb{R},m,t),h}(n) = H^g_{U,h}(n)$ for any natural n. Evidently, the set F contains infinite set of pairwise different attributes. From Theorem 6.3 follows that $H^g_{U,h}(n) = \Theta(\log_2 n)$. Therefore $H^g_{U(\mathbb{R},m,t),h}(n) = \Theta(\log_2 n)$. □

Note 6.2. Let m, t be natural numbers. In the proof of Theorem 6.5 it was shown that the information system $U(\mathbb{R}, m)$ has infinite I-dimension. Therefore the function I is not bounded from above on the set $\text{Tab}_{U(\mathbb{R},m)}$. From Theorems 5.1 and 6.5 follows that the information system $U(\mathbb{R}, m, t)$ has finite I-dimension. Therefore the function I is bounded from above on the set $\text{Tab}_{U(\mathbb{R},m,t)}$.

7 Classes of Problems over Quasilinear Information Systems

In this section six classes of problems over quasilinear information systems (three classes of discrete optimization problems and three classes of recognition and sorting problems) are considered. For each class examples and corollaries of Theorem 6.1 are given. Some definitions and notation from Sect. 6 are used without special stipulations.

7.1 Definitions and Auxiliary Statement

Let $U = U(A, K, \varphi_1, \ldots, \varphi_m)$ be a quasilinear information system. A pair (A, ϕ) where ϕ is a function from A to a finite subset of the set $\mathbb{Z}$ will be called *a problem over the set* A. The problem (A, ϕ) may be interpreted as a problem of searching for the value $\phi(a)$ for arbitrary $a \in A$. Let $k \in \mathbb{N}$, $k \geq 1$, and $t \in \mathbb{R}$, $t \geq 0$. The problem (A, ϕ) will be called (m, k, t)*-problem over* U if there exists a problem z over U such that $\phi(a) = z(a)$ for each $a \in A$, $\dim z \leq k$ and $r(z) \leq t$. Let $\phi(a) = z(a)$ for each $a \in A$ and $z = (\nu, f_1, \ldots, f_p)$. Then the set $\{f_1, \ldots, f_p\}$ will be called *a separating set* for the problem (A, ϕ). We will say that a decision tree Γ over U *solves* the problem (A, ϕ) if Γ solves the problem z.

Denote

$$L(A, K, \varphi_1, \ldots, \varphi_m) = \{\textstyle\sum_{i=1}^m d_i\varphi_i(x) + d_{m+1} : d_1, \ldots, d_{m+1} \in K\} \ ,$$
$$F(A, K, \varphi_1, \ldots, \varphi_m) = \{\text{sign}(g) : g \in L(A, K, \varphi_1, \ldots, \varphi_m)\} \ .$$

Let $g \in L(A, K, \varphi_1, \ldots, \varphi_m)$ and $g = \sum_{i=1}^m d_i\varphi_i(x) + d_{m+1}$. We define parameters $r(g)$ and $r(\text{sign}(g))$ of the functions g and $\text{sign}(g)$ as follows. If $(d_1, \ldots, d_{m+1}) = (0, \ldots, 0)$ then $r(g) = r(\text{sign}(g)) = 0$. Otherwise

$$r(g) = r(\text{sign}(g)) = \max\{0, \max\{\log_2 |d_i| : i \in \{1, \ldots, m+1\}, d_i \neq 0\}\} \ .$$

In what follows we will assume that elements of the set $\{-1, 1\}^n$, of the set $\{-1, 0, +1\}^n$, of the set Π_n of all n-degree permutations, and of the set $\{0, 1\}^n$ are enumerated by numbers from 1 to 2^n, by numbers from 1 to 3^n, by numbers from 1 to $n!$, and by numbers from 1 to 2^n respectively.

The following statement allows to proof for a problem over the set A that it is a problem over the information system U.

Proposition 7.1. *Let $U = U(A, K, \varphi_1, \ldots, \varphi_n)$ be a quasilinear information system, (A, ϕ) be a problem over A, D be a finite subset of the set $\mathbb{Z}$ such that $\{\phi(a) : a \in A\} \subseteq D$, $\{f_1, \ldots, f_k\}$ be a nonempty finite subset of the set $F = F(A, K, \varphi_1, \ldots, \varphi_n)$ such that for any $i \in D$ the set $W(i) = \{a : a \in A, \phi(a) = i\}$ is the union of the sets of solutions on A of some equation systems of the kind $\{f_{i_1}(x) = \sigma_1, \ldots, f_{i_m}(x) = \sigma_m\}$ where $i_1, \ldots, i_m \in \{1, \ldots, k\}$ and $\sigma_1, \ldots, \sigma_m \in \{-1, 0, +1\}$. Then the set $\{f_1, \ldots, f_k\}$ is a separating set of the problem (A, ϕ) and the problem (A, ϕ) is a problem over U.*

Proof. For any $\bar{\delta} = (\delta_1, \ldots, \delta_k) \in \{-1, 0, +1\}^k$ denote by $A(\bar{\delta})$ the set of solutions on A of the equation system $\{f_1(x) = \delta_1, \ldots, f_k(x) = \delta_k\}$. Let C be the equation system $\{f_{i_1}(x) = \sigma_1, \ldots, f_{i_m}(x) = \sigma_m\}$, where $i_1, \ldots, i_m \in \{1, \ldots, k\}$ and $\sigma_1, \ldots, \sigma_m \in \{-1, 0, +1\}$. Denote $H(C) = \{(\delta_1, \ldots, \delta_k) : (\delta_1, \ldots, \delta_k) \in \{-1, 0, +1\}^k, \delta_{i_1} = \sigma_1, \ldots, \delta_{i_m} = \sigma_m\}$. One can show that the set of solutions on A of the system C coincides with the set $\bigcup_{\bar{\delta} \in H(C)} A(\bar{\delta})$.

Let $i \in D$ and the set $W(i)$ be the union of sets of solutions on A of t equation systems of the kind C. Denote these systems by $C_1, \ldots, C_t$, and denote $H(i) = \bigcup_{j=1}^{t} H(C_j)$. It is clear that $W(i) = \bigcup_{\bar{\delta} \in H(i)} A(\bar{\delta})$.

Define a mapping $\nu : \{-1, 0, +1\}^k \to \mathbb{Z}$ in the following way. Let $\bar{\delta} \in \{-1, 0, +1\}^k$. If $\bar{\delta} \notin \bigcup_{i \in D} H(i)$ then $\nu(\bar{\delta}) = 1$, but if $\bar{\delta} \in \bigcup_{i \in D} H(i)$ then $\nu(\bar{\delta}) = \min\{j : j \in D, \bar{\delta} \in H(j)\}$.

Let us show that $\phi(a) = \nu(f_1(a), \ldots, f_k(a))$ for any $a \in A$. Let $\bar{\delta} = (f_1(a), \ldots, f_k(a))$ and $\phi(a) = i$. Then $a \in W(i)$. Therefore there exists a tuple $\bar{\sigma} \in H(i)$ such that $a \in A(\bar{\sigma})$. Obviously, $\bar{\sigma} = \bar{\delta}$. Therefore $\bar{\delta} \in H(i)$. Assume that there exists a number $j \in D$ such that $j \neq i$ and $\bar{\delta} \in H(j)$. Then $\phi(a) = j$ which is impossible. Hence $\{j : j \in D, \bar{\delta} \in H(j)\} = \{i\}$, $\nu(\bar{\delta}) = i$ and $\nu(f_1(a), \ldots, f_k(a)) = i$. Thus, the set $\{f_1, \ldots, f_k\}$ is a separating set for the problem (A, ϕ) and the problem (A, ϕ) is a problem over U.

□

7.2 Problems of Discrete Optimization

In this subsection three classes of discrete optimization problems are considered.

Problems of Unconditional Optimization. Let $k \in \mathbb{N} \setminus \{0\}$, $t \in \mathbb{R}$, $t \geq 0$, and $g_1, \ldots, g_k$ be functions from $L(A, K, \varphi_1, \ldots, \varphi_m)$ such that $r(g_j) \leq t$ for $j = 1, \ldots, k$.

Problem 7.1. (Unconditional optimization of values of functions $g_1, \ldots, g_k$ on an element of the set A.) For a given $a \in A$ it is required to find the minimal number $i \in \{1, \ldots, k\}$ such that $g_i(a) = \min\{g_j(a) : 1 \leq j \leq k\}$.

For this problem set $D = \{1, \ldots, k\}$. Let $i \in D$. One can show that for this problem the set $W(i)$ is the union of sets of solutions on A of all equation systems of the kind $\{\text{sign}(g_i(x) - g_j(x)) = \delta_j : 1 \leq j \leq k, j \neq i\}$ where $\delta_j = -1$

for $j \in \{1, \ldots, i-1\}$ and $\delta_j \in \{-1, 0\}$ for $j \in \{i+1, \ldots, k\}$. Using Proposition 7.1 we conclude that the set $\{\text{sign}(g_i(x) - g_j(x)) : i, j \in \{1, \ldots, k\}, i \neq j\}$ is a separating set for this problem, and the considered problem is $(m, k^2, t+1)$-problem over the information system $U(A, K, \varphi_1, \ldots, \varphi_m)$.

Example 7.1. (n-City traveling salesman problem.) Let $n \in \mathbb{N}$, $n \geq 4$, and let G_n be the complete undirected graph with n nodes. Assume that edges in G_n are enumerated by numbers from 1 to $n(n-1)/2$, and Hamiltonian circuits in G_n are enumerated by numbers from 1 to $(n-1)!/2$. Let a number $a_i \in \mathbb{R}$ be attached to the i-th edge, $i = 1, \ldots, n(n-1)/2$. We will interpret the number a_i as the length of the i-th edge. It is required to find the minimal number of a Hamiltonian circuit in G_n which has the minimal length. For each $j \in \{1, \ldots, (n-1)!/2\}$ we will associate with the j-th Hamiltonian circuit the function $g_j(\bar{x}) = \sum_{i=1}^{n(n-1)/2} \delta_{ji} x_i$ where $\delta_{ji} = 1$ if the i-th edge is contained in the j-th Hamiltonian circuit, and $\delta_{ji} = 0$ otherwise. Obviously, the considered problem is the problem of unconditional optimization of values of functions $g_1, \ldots, g_{(n-1)!/2}$ on an element of the set $\mathbb{R}^{n(n-1)/2}$. Therefore the set $\{\text{sign}(g_i(\bar{x}) - g_j(\bar{x})) : i, j \in \{1, \ldots, (n-1)!/2\}, i \neq j\}$ is a separating set for the n-city traveling salesman problem, and this problem is $(n(n-1)/2, ((n-1)!/2)^2, 0)$-problem over the information system $U = U(\mathbb{R}^{n(n-1)/2}, \mathbb{Z}, x_1, \ldots, x_{n(n-1)/2})$. From Theorem 6.1 follows that there exists a decision tree Γ over U which solves the n-city traveling salesman problem, $n \geq 4$, and for which $h(\Gamma) \leq n^7/2$ and $r(\Gamma) \leq n^4 \log_2 n$.

Example 7.2. (n-Dimensional quadratic assignment problem.) Let $n \in \mathbb{N}$ and $n \geq 2$. For given $a_{ij}, b_{ij} \in \mathbb{R}$, $1 \leq i, j \leq n$, it is required to find the minimal number of n-degree permutation π which minimizes the value $\sum_{i=1}^{n} \sum_{j=1}^{n} a_{ij} b_{\pi(i)\pi(j)}$. Obviously, this problem is the problem of unconditional optimization of values of functions from the set $\{\sum_{i=1}^{n} \sum_{j=1}^{n} x_{ij} y_{\pi(i)\pi(j)} : \pi \in \Pi_n\}$ on an element of the set $\mathbb{R}^{2n^2}$. Hence the set $\{\text{sign}(\sum_{i=1}^{n} \sum_{j=1}^{n} x_{ij} y_{\pi(i)\pi(j)} - \sum_{i=1}^{n} \sum_{j=1}^{n} x_{ij} y_{\tau(i)\tau(j)}) : \pi, \tau \in \Pi_n, \pi \neq \tau\}$ is a separating set for this problem, and the considered problem is $(n^4, (n!)^2, 0)$-problem over the information system $U = U(\mathbb{R}^{2n^2}, \mathbb{Z}, x_{11}y_{11}, \ldots, x_{nn}y_{nn})$. From Theorem 6.1 follows that there exists a decision tree Γ over U which solves the n-dimensional quadratic assignment problem and for which $h(\Gamma) \leq 3n(n^4+2)^3$ and $r(\Gamma) \leq 2(n^4+1)^2 \log_2(2n^4+2)$.

Problems of Unconditional Optimization of Absolute Values. Let $k \in \mathbb{N} \setminus \{0\}$, $t \in \mathbb{R}$, $t \geq 0$, and $g_1, \ldots, g_k$ be functions from $L(A, K, \varphi_1, \ldots, \varphi_m)$ such that $r(g_j) \leq t$ for $j = 1, \ldots, k$.

Problem 7.2. (Unconditional optimization of absolute values of functions $g_1, \ldots, g_k$ on an element of the set A.) For a given $a \in A$ it is required to find the minimal number $i \in \{1, \ldots, k\}$ such that $|g_i(a)| = \min\{|g_j(a)| : 1 \leq j \leq k\}$.

For this problem set $D = \{1, \ldots, k\}$. Let $i \in D$. One can show that $|g_i(a)| < |g_j(a)|$ if and only if $(g_i(a) + g_j(a))(g_i(a) - g_j(a)) < 0$, and $|g_i(a)| = |g_j(a)|$ if and only if $(g_i(a) + g_j(a))(g_i(a) - g_j(a)) = 0$. From here follows that the set $W(i)$ coincides with the union of the sets of solutions on A of all equation systems of the kind $\{\text{sign}(g_i(x) + g_j(x)) = \delta_{j1}, \text{sign}(g_i(x) - g_j(x)) = \delta_{j2} : 1 \leq j \leq k, i \neq j\}$, where $(\delta_{j1}, \delta_{j2}) \in \{(-1, +1), (+1, -1)\}$ for $j \in \{1, \ldots, i-1\}$ and $(\delta_{j1}, \delta_{j2}) \in \{-1, 0, +1\}^2 \setminus \{(-1, -1), (+1, +1)\}$ for $j \in \{i+1, \ldots, k\}$. Using Proposition 7.1 we conclude that the set $\{\text{sign}(g_i(x) + g_j(x)), \text{sign}(g_i(x) - g_j(x)) : i, j \in \{1, \ldots, k\}, i \neq j\}$ is a separating set for the considered problem, and this problem is $(m, 2k^2, t+1)$-problem over the information system $U(A, K, \varphi_1, \ldots, \varphi_m)$.

Example 7.3. (n-Stone problem.) Let $n \geq 1$. For a given tuple $(a_1, \ldots, a_n) \in \mathbb{R}^n$ it is required to find the minimal number of a tuple $(\delta_1, \ldots, \delta_n) \in \{-1, 1\}^n$ which minimizes the value of $|\sum_{i=1}^n \delta_i a_i|$. Obviously, this problem is the problem of unconditional optimization of absolute values of functions from the set $\{\sum_{i=1}^n \delta_i x_i : (\delta_1, \ldots, \delta_n) \in \{-1, 1\}^n\}$ on an element of the set $\mathbb{R}^n$. Therefore the set $\{\text{sign}(\sum_{i=1}^n \delta_i x_i) : (\delta_1, \ldots, \delta_n) \in \{-2, 0, 2\}^n\}$ and hence the set $\{\text{sign}(\sum_{i=1}^n \delta_i x_i) : (\delta_1, \ldots, \delta_n) \in \{-1, 0, 1\}^n\}$ are separating sets for the considered problem, and this problem is $(n, 3^n, 0)$-problem over the information system $U = U(\mathbb{R}^n, \mathbb{Z}, x_1, \ldots, x_n)$. From Theorem 6.1 follows that there exists a decision tree Γ over U which solves the n-stone problem and for which $h(\Gamma) \leq 4(n+2)^4/\log_2(n+2)$ and $r(\Gamma) \leq 2(n+1)^2 \log_2(2n+2)$.

Problems of Conditional Optimization. Let $k, p \in \mathbb{N} \setminus \{0\}$, $t \in \mathbb{R}, t \geq 0$, $D \subseteq \mathbb{R}$, $D \neq \emptyset$ and $g_1, \ldots, g_k$ be functions from $L(A, K, \varphi_1, \ldots, \varphi_m)$ such that $r(g_j) \leq t$ for $j = 1, \ldots, k$.

Problem 7.3. (Conditional optimization of values of functions $g_1, \ldots, g_k$ on an element of the set A with p restrictions from $A \times D$.) For a given tuple $(a_0, a_1, ..., a_p, d_1, \ldots, d_p) \in A^{p+1} \times D^p$ it is required to find the minimal number $i \in \{1, \ldots, k\}$ such that $g_i(a_1) \leq d_1, \ldots, g_i(a_p) \leq d_p$ and $g_i(a_0) = \max\{g_j(a_0) : g_j(a_1) \leq d_1, \ldots, g_j(a_p) \leq d_p, j \in \{1, \ldots, k\}\}$ or to show that such i does not exist. (In the last case let $k+1$ be the solution of the problem.)

For this problem set $D = \{1, \ldots, k+1\}$. The variables with values from A will be denoted by $x_0, x_1, \ldots, x_p$ and the variables with values from D will be denoted by $y_1, \ldots, y_p$. One can show that the set $W(k+1)$ coincides with the union of the sets of solutions on $A^{p+1} \times D^p$ of all equation systems of the kind $\{\text{sign}(g_1(x_{i_1}) - y_{i_1}) = +1, \ldots, \text{sign}(g_k(x_{i_k}) - y_{i_k}) = +1\}$ where $1 \leq i_j \leq p$ for any j, $1 \leq j \leq k$. Let $i \in \{1, \ldots, k\}$. It is not difficult to see that the set $W(i)$ coincides with the union of the sets of solutions on $A^{p+1} \times B^p$ of all equation systems of the kind $\{\text{sign}(g_i(x_0) - g_j(x_0)) = \delta_j : j \in C \setminus \{i\}\} \cup \left(\bigcup_{j \in C} \{\text{sign}(g_j(x_l) - y_l) = \sigma_{jl} : 1 \leq l \leq p\}\right) \cup \{\text{sign}(g_j(x_{i_j}) - y_{i_j}) = +1 : j \in \{1, \ldots, k\} \setminus C\}$, where $C \subseteq \{1, \ldots, k\}, i \in C$; for $j \in C \setminus \{i\}$, if $j < i$, then $\delta_j = +1$, but if $j > i$, then $\delta_j \in \{0, +1\}$; $\sigma_{jl} \in \{-1, 0\}$ for $j \in C$ and $l \in \{1, \ldots, p\}$; $1 \leq i_j \leq p$ for $j \in \{1, \ldots, k\} \setminus C$. Using Proposition 7.1 we conclude that the set $\{\text{sign}(g_i(x_0) - g_j(x_0)) : 1 \leq i, j \leq k\} \cup (\bigcup_{j=1}^p \{\text{sign}(g_i(x_j) - y_j) : 1 \leq i \leq k\})$ is a separating set for the considered problem, and this problem is $(p + m(p+1), pk + k^2, t+1)$-problem over the information system $U(A^{p+1} \times D^p, K, \varphi_1(x_0), \ldots, \varphi_m(x_0), \ldots, \varphi_1(x_p), \ldots, \varphi_m(x_p), y_1, \ldots, y_p)$.

Example 7.4. (Problem on 0-1-knapsack with n objects.) Let $n \in \mathbb{N} \setminus \{0\}$. For a given tuple $(a_1, \ldots, a_{2n+1}) \in \mathbb{Z}^{2n+1}$ it is required to find the minimal number of a tuple $(\delta_1, \ldots, \delta_n) \in \{0,1\}^n$ which maximizes the value $\sum_{i=1}^{n} \delta_i a_i$ under the condition $\sum_{i=1}^{n} \delta_i a_{n+i} \leq a_{2n+1}$. This is the problem of conditional optimization of values of functions from the set $\{\sum_{i=1}^{n} \delta_i x_i : (\delta_1, \ldots, \delta_n) \in \{0,1\}^n\}$ on an element of the set $\mathbb{Z}^n$ with one restriction from $\mathbb{Z}^n \times \mathbb{Z}$. The set $\{\text{sign}(\sum_{i=1}^{n} \delta_i x_i) : (\delta_1, \ldots, \delta_n) \in \{-1,0,1\}^n\} \cup \{\text{sign}(\sum_{i=1}^{n} \delta_i x_{n+i} - x_{2n+1}) : (\delta_1, \ldots, \delta_n) \in \{0,1\}^n\}$ is a separating set for the considered problem, and this problem is $(2n+1, 3^n + 2^n, 0)$-problem over the information system $U = U(\mathbb{Z}^{2n+1}, \mathbb{Z}, x_1, \ldots, x_{2n+1})$. From Theorem 6.1 follows that there exists a decision tree Γ over U which solves the problem on 0-1-knapsack with n objects and for which $h(\Gamma) \leq 2(2n+3)^4 / \log_2(2n+3)$ and $r(\Gamma) \leq 2(2n+2)^2 \log_2(4n+4)$.

7.3 Problems of Recognition and Sorting

In this subsection three classes of problems of recognition and sorting are considered.

Problems of Recognition of Values of Functions. Let $k \in \mathbb{N} \setminus \{0\}$, $f_1, \ldots, f_k \in F(A, K, \varphi_1, \ldots, \varphi_m)$ and $r(f_j) \leq t$ for any j, $1 \leq j \leq k$.

Problem 7.4. (Recognition of values of functions $f_1, \ldots, f_k$ on an element of the set A.) For a given $a \in A$ it is required to find the number of the tuple $(f_1(a), \ldots, f_k(a))$.

For this problem set $D = \{1, \ldots, 3^k\}$. Let $i \in D$ and let $(\delta_1, \ldots, \delta_k)$ be the tuple with the number i from $\{-1, 0, +1\}^k$. One can show that for the considered problem the set $W(i)$ coincides with the set of solutions on A of the equation system $\{f_1(x) = \delta_1, \ldots, f_k(x) = \delta_k\}$. Using Proposition 7.1 we conclude that the set $\{f_1, \ldots, f_k\}$ is a separating set for this problem, and the considered problem is (n, k, t)-problem over the information system $U(A, K, \varphi_1, \ldots, \varphi_m)$.

Example 7.5. (Recognition of a threshold Boolean function depending on n variables.) Let $n \in \mathbb{N} \setminus \{0\}$. For a given tuple $(a_1, \ldots, a_{n+1}) \in \mathbb{R}^{n+1}$ it is required to find the value $\text{sign}(\sum_{i=1}^{n} \delta_i a_i - a_{n+1})$ for each tuple $(\delta_1, \ldots, \delta_n) \in \{0,1\}^n$. Obviously, this problem is the problem of recognition of values of functions $\text{sign}(\sum_{i=1}^{n} \delta_i x_i - x_{n+1})$, $(\delta_1, \ldots, \delta_n) \in \{0,1\}^n$, on an element of the set $\mathbb{R}^{n+1}$. Hence the set $\{\text{sign}(\sum_{i=1}^{n} \delta_i x_i - x_{n+1}) : (\delta_1, \ldots, \delta_n) \in \{0,1\}^n\}$ is a separating set for this problem, and the considered problem is $(n+1, 2^n, 0)$-problem over the information system $U = U(\mathbb{R}^{n+1}, \mathbb{Z}, x_1, \ldots, x_{n+1})$. From Theorem 6.1 follows that there exists a decision tree Γ over U which solves the problem of recognition of a threshold Boolean function depending on n variables and for which $h(\Gamma) \leq 2(n+3)^4 / \log_2(n+3)$ and $r(\Gamma) \leq 2(n+2)^2 \log_2(2n+4)$.

Problems of Recognition of Belonging. Let we have $m \geq 1$ equation systems $C_1, \ldots, C_m$ such that $C_j = \{f_{j1}(x) = \delta_{j1}, \ldots, f_{jp_j}(x) = \delta_{jp_j}\}$ for any j, $1 \leq j \leq m$, and $f_{ji} \in F(A, K, \varphi_1, \ldots, \varphi_m)$, $\delta_{ji} \in \{-1, 0, +1\}$ for any i, $1 \leq i \leq p_j$.

Problem 7.5. (Recognition of belonging of an element from A to the union of sets of solutions on A of equation systems $C_1, \dots, C_m$.) For a given element $a \in A$ it is required to recognize whether a is a solution of at least one of the systems $C_1, \dots, C_m$. (If yes then the response is equal to 1, otherwise the response is equal to 2.)

For this problem set $D = \{1, 2\}$. One can show that for this problem the set $W(1)$ coincides with the union of the sets of solutions on A of the systems $C_1, \dots, C_m$, and the set $W(2)$ coincides with the union of the sets of solutions on A of all equation systems of the kind $\{f_{1i_1}(x) = \sigma_1, \dots, f_{mi_m}(x) = \sigma_m\}$ where $1 \leq i_j \leq p_j$, $\sigma_j \in \{-1, 0, +1\}$, and $\sigma_j \neq \delta_{ji_j}$ for any j, $1 \leq j \leq m$. From Proposition 7.1 follows that the set $F = \{f_{11}, \dots, f_{1p_1}, \dots, f_{m1}, \dots, f_{mp_m}\}$ is a separating set for this problem, and the considered problem is $(n, |F|, l)$-problem over the information system $U(A, K, \varphi_1, \dots, \varphi_m)$ where $l = \max\{r(f) : f \in F\}$.

Example 7.6. ($((n+1)\times m)$-Dimensional problem of "0-1-integer programming".) Let $n, m \in \mathbb{N} \setminus \{0\}$. For given $a_{ij} \in \mathbb{Z}$, $1 \leq i \leq m$, $1 \leq j \leq n+1$, it is required to recognize whether exists a tuple $(\delta_1, \dots, \delta_n) \in \{0,1\}^n$ such that $\sum_{j=1}^{n} a_{ij}\delta_j = a_{in+1}$ for any i, $1 \leq i \leq m$. Obviously, this problem is the problem of recognition of belonging of an element from $\mathbb{Z}^{(n+1)m}$ to the union of sets of solutions on $\mathbb{Z}^{(n+1)m}$ of all equation systems of the kind $\{\text{sign}(\sum_{j=1}^{n} x_{ij}\delta_j - x_{in+1}) = 0 : 1 \leq i \leq m\}$, where $(\delta_1, \dots, \delta_n) \in \{0,1\}^n$. Hence the set $\{\text{sign}(\sum_{j=1}^{n} x_{ij}\delta_j - x_{in+1}) : 1 \leq i \leq m, (\delta_1, \dots, \delta_n) \in \{0,1\}^n\}$ is a separating set for this problem, and the considered problem is $((n+1)m, m2^n, 0)$-problem over the information system $U = U(\mathbb{Z}^{(n+1)m}, \mathbb{Z}, x_{11}, \dots, x_{mn+1})$. From Theorem 6.1 follows that there exists a decision tree Γ over U which solves the $((n+1) \times m)$-dimensional problem of "0-1-integer programming" and for which $h(\Gamma) \leq 2(nm+m+2)^4/\log_2(nm+m+2)$ and $r(\Gamma) \leq 2(nm+m+1)^2\log_2(2nm+2m+2)$.

Example 7.7. (Problem on (n,d,k,t)-system of polynomial inequalities.) Let $n, d, k \in \mathbb{N} \setminus \{0\}$, $t \in \mathbb{R}$, $t \geq 0$ and let $\text{Pol}(n,d)$ be the set of all polynomials with integer coefficients which depend on variables $x_1, \dots, x_n$ and which degree is at most d. Let $g_1, \dots, g_k \in \text{Pol}(n,d)$ and $r(g_j) \leq t$ for any j, $1 \leq j \leq k$. For a given $\bar{a} \in \mathbb{R}^n$ it is required to recognize whether $\bar{a}$ is a solution of the inequality system $\{g_1(\bar{x}) \geq 0, \dots, g_k(\bar{x}) \geq 0\}$. Obviously, this problem is the problem of recognition of belonging of an element from $\mathbb{R}^n$ to the union of sets of solutions on $\mathbb{R}^n$ of all equation systems of the kind $\{\text{sign}(g_1(\bar{x})) = \delta_1, \dots, \text{sign}(g_k(\bar{x})) = \delta_k\}$, where $(\delta_1, \dots, \delta_k) \in \{0, +1\}^k$. Hence the set $\{\text{sign}(g_1(\bar{x})), \dots, \text{sign}(g_k(\bar{x}))\}$ is a separating set for this problem and the considered problem is $\left(\sum_{i=1}^{d}\binom{n+i-1}{i}, k, t\right)$-problem over the information system $U = U(\mathbb{R}^n, \mathbb{Z}, x_1, \dots, x_n^d)$. From Theorem 6.1 follows that there exists a decision tree Γ over U which solves the problem on (n,d,k,t)-system of polynomial inequalities and for which $h(\Gamma) \leq 2(d(n+d)^d+2)^3\log_2(k+2d(n+d)^d+2)$ and $r(\Gamma) \leq 2(d(n+d)^d+1)^2(t+1+\log_2(d(n+d)^d+1))$.

Problems of Sorting. Let $k \in \mathbb{N} \setminus \{0\}$, $t \in \mathbb{R}$, $t > 0$, and $g_1, \dots, g_k$ be functions from $L(A, K, \varphi_1, \dots, \varphi_m)$ such that $r(g_j) \leq t$ for $j = 1, \dots, k$.

Problem 7.6. (Sorting of values of functions $g_1, \ldots, g_k$ on an element from A.) For a given $a \in A$ it is required to find the minimal number of an k-degree permutation π such that $g_{\pi(1)}(a) \leq g_{\pi(2)}(a) \leq \ldots \leq g_{\pi(k)}(a)$.

For this problem set $D = \{1, \ldots, k!\}$. Denote by π_j the k-degree permutation with the number j where $1 \leq j \leq k!$. Let $i \in D$. One can show that for the considered problem the set $W(i)$ coincides with the union of the sets of solutions on A of all equation systems of the kind $\{\text{sign}(g_{\pi_i(n)}(x) - g_{\pi_i(n+1)}(x)) = \delta_n : 1 \leq n \leq k-1\} \cup \{\text{sign}(g_{\pi_j(n_j)}(x) - g_{\pi_j(n_j+1)}(x)) = +1 : 1 \leq j \leq i-1\}$, where $(\delta_1, \ldots, \delta_{k-1}) \in \{-1, 0\}^{k-1}$ and $1 \leq n_j \leq k-1$ for any j, $1 \leq j \leq i-1$. From Proposition 7.1 follows that the set $\{\text{sign}(g_i(x) - g_j(x)) : i, j \in \{1, \ldots, k\}\}$ is a separating set of this problem, and the considered problem is $(n, k^2, t+1)$-problem over the information system $U(A, K, \varphi_1, \ldots, \varphi_m)$.

Example 7.8. (Problem of sorting of values of functions from set $\{\sum_{j=1}^{n} \delta_j \sin jx: (\delta_1, \ldots, \delta_n) \in \{0, 1, \ldots, q\}^n\}$ on an element from the set $\mathbb{R}$). Obviously, the set $\{\text{sign}(\sum_{j=1}^{n} \delta_j \sin jx) : (\delta_1, \ldots, \delta_n) \in \{-q, \ldots, -1, 0, 1, \ldots, q\}^n\}$ is a separating set for this problem and the considered problem is $(n, (2q+1)^n, \log_2 q)$-problem over the information system $U = U(\mathbb{R}, \mathbb{Z}, \sin x, \sin 2x, \ldots, \sin nx)$. From Theorem 6.1 follows that there exists a decision tree Γ over U which solves the problem of sorting of values of the functions from the set $\{\sum_{j=1}^{n} \delta_j \sin jx : (\delta_1, \ldots, \delta_n) \in \{0, 1, \ldots, q\}^n\}$ on an element from the set $\mathbb{R}$ and for which $h(\Gamma) \leq 2(n+2)^4 \log_2(2q+1)/\log_2(n+2)$ and $r(\Gamma) \leq 2(n+1)^2 \log_2(2q(n+1))$.

8 On Depth of Acyclic Programs in Basis $\{x + y, x - y, 1; \text{sign}(x)\}$

In this section relationships between depth of deterministic and nondeterministic acyclic programs in the basis $B_0 = \{x+y, x-y, 1; \text{sign}(x)\}$ are considered. Proof of the main result of this section is based on Theorem 6.1 and is an example of the application of methods of decision tree theory to analysis of algorithms which are not decision trees.

8.1 Main Definitions and Result

Letters from the alphabet $X = \{x_i : i \in \mathbb{N}\}$ will be called *input variables*, while letters from the alphabet $Y = \{y_i : i \in \mathbb{N}\}$ will be called *working variables*.

A program in the basis B_0 is a labelled finite directed graph which has nodes of the following four kinds:

a) the only node without entering edges called the node "input";
b) the only node without issuing edges called the node "output";
c) functional nodes of the kinds $y_j := 1, y_j := z_l + z_k$ and $y_j := z_l - z_k$ where $z_l, z_k \in X \cup Y$;
d) predicate nodes of the kind $\text{sign}(y_j)$.

Each edge issuing from a predicate node is labelled by a number from the set $\{-1, 0, +1\}$. The other edges are not labelled.

Further we assume that in expressions assigned to nodes of a program at least one input variable and hence at least one working variable are present.

A program in the basis B_0 will be called *acyclic* if it contains no directed cycles. A program will be called *deterministic* if it satisfies the following conditions: the node "input" and each functional node have exactly one issuing edge, and edges issuing from a predicate node are labelled by pairwise different numbers. If a program is not deterministic we will call it *nondeterministic*.

Let P be an acyclic program in the basis B_0 with the input variables $x_1, \ldots, x_n$ and the working variables $y_1, \ldots, y_t$.

A complete path in P is an arbitrary directed path from the node "input" to the node "output" in the program P. Let $\xi = v_1, d_1, \ldots, v_m, d_m, v_{m+1}$ be a complete path in the program P. Define the set of elements from $\mathbb{Q}^n$ accepted by the complete path ξ. For $i = 1, \ldots, m$ we will attach to the node v_i of the path ξ a tuple $\bar{\beta}_i = (\beta_{i1}, \ldots, \beta_{it})$ composed of functions from the set $L(\mathbb{Q}^n, \mathbb{Z}, x_1, \ldots, x_n) = \{\sum_{i=1}^{n} b_i x_i + b_{n+1} : b_1, \ldots, b_{n+1} \in \mathbb{Z}\}$. Let $\bar{\beta}_1 = (0, \ldots, 0)$. Let the tuples $\bar{\beta}_1, \ldots, \bar{\beta}_{i-1}$, where $2 \leq i \leq m$, be already defined. If v_i is a predicate node then $\bar{\beta}_i = \bar{\beta}_{i-1}$. Let v_i be a functional node and let, for the definiteness, the node v_i be of the kind $y_j := x_l + y_p$. Then $\bar{\beta}_i = (\beta_{i-11}, \ldots, \beta_{i-1j-1}, x_l + \beta_{i-1p}, \beta_{i-1j+1}, \ldots, \beta_{i-1t})$. For another kinds of functional nodes the tuple $\bar{\beta}_i$ is defined in the same way.

Let the nodes $v_{i_1}, \ldots, v_{i_k}$ be all predicate nodes in the complete path ξ. Let $k > 0$, let the nodes $v_{i_1}, \ldots, v_{i_k}$ be of the kind $\mathrm{sign}(y_{j_1}), \ldots, \mathrm{sign}(y_{j_k})$, and let the edges $d_{i_1}, \ldots, d_{i_k}$ be labelled by the numbers $\delta_1, \ldots, \delta_k$. Denote $F(\xi) = \{\beta_{i_1 j_1}, \ldots, \beta_{i_k j_k}\}$ and $\mathrm{Sol}(\xi)$ the set of solutions on $\mathbb{Q}^n$ of the equation system

$$S_\xi = \{\mathrm{sign}(\beta_{i_1 j_1}(\bar{x})) = \delta_1, \ldots, \mathrm{sign}(\beta_{i_k j_k}(\bar{x})) = \delta_k\} .$$

If $k = 0$ then $F(\xi) = \emptyset$ and $\mathrm{Sol}(\xi) = \mathbb{Q}^n$. The set $\mathrm{Sol}(\xi)$ will be called *the set of elements from* $\mathbb{Q}^n$ *accepted by the complete path* ξ. The set of all complete paths in the program P will be denoted by $\mathrm{Path}(P)$. Evidently, $\mathrm{Path}(P) \neq \emptyset$. Denote $\mathrm{Rec}(P) = \bigcup_{\xi \in \mathrm{Path}(P)} \mathrm{Sol}(\xi)$. We will say that *the program* P *recognizes the set* $\mathrm{Rec}(P)$. Denote $F(P) = \bigcup_{\xi \in \mathrm{Path}(P)} F(\xi)$.

Denote by $h(\xi)$ the number of functional and predicate nodes in a complete path ξ. The value $h(P) = \max\{h(\xi) : \xi \in \mathrm{Path}(P)\}$ will be called *the depth* of the program P.

Acyclic programs P_1 and P_2 in the basis B_0 will be called *equivalent* if the sets of input variables of P_1 and P_2 coincide, and the equality $\mathrm{Rec}(P_1) = \mathrm{Rec}(P_2)$ holds.

Theorem 8.1. *For each nondeterministic acyclic program P_1 in the basis B_0 with n input variables there exists a deterministic acyclic program P_2 in the basis B_0 which is equivalent to P_1 and for which the following inequality holds:*

$$h(P_2) \leq 8(n+2)^7 (h(P_1) + 2)^2 .$$

Example 8.1. (Problem of partition of n numbers.) The set of tuples $(q_1, \ldots, q_n)$ from $\mathbb{Q}^n$ for each of which there exists a tuple $(\sigma_1, \ldots, \sigma_n) \in \{-1, 1\}^n$ such that $\sum_{i=1}^{n} \sigma_i q_i = 0$ will be denoted by W_n. The problem of recognition of belonging of a tuple from $\mathbb{Q}^n$ to the set W_n is known as the problem of partition of n numbers. Fig. 7 represents a nondeterministic acyclic program P_n in the basis B_0 with input variables $x_1, \ldots, x_n$ and working variable y_1 for which $\mathrm{Rec}(P_n) = W_n$ and $h(P_n) = n + 1$. Using Theorem 8.1 we conclude that there exists a deterministic acyclic program in the basis B_0 which recognizes the set W_n and for which the depth is at most $8(n + 3)^9$.

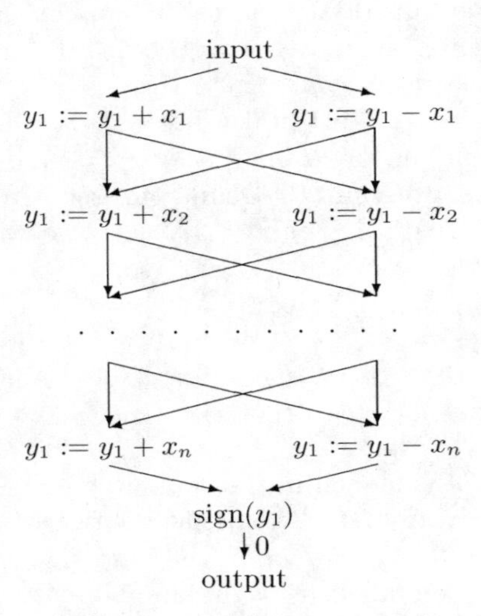

Fig. 7. Nondeterministic program P_n in basis B_0 which solves problem of partition of n numbers.

8.2 Proof of Theorem 8.1

A functional program in the basis B_0 is a deterministic acyclic program in the basis B_0 without predicate nodes. Let P be a functional program in the basis B_0 with the input variables $x_1, \ldots, x_n$ and the working variables $y_1, \ldots, y_t$. Evidently, P contains the only complete path ξ. Let $\xi = v_1, d_1, \ldots, v_m, d_m, v_{m+1}$ and let $(\beta_{m1}, \ldots, \beta_{mt})$ be the tuple of functions from the set $L(\mathbb{Q}^n, \mathbb{Z}, x_1, \ldots, x_n)$ attached to the node v_m of this path. We will say that the program P *realizes functions* $\beta_{m1}, \ldots, \beta_{mt}$ *in the working variables* $y_1, \ldots, y_t$ *respectively.*

Lemma 8.1. *Let* $n, t \in \mathbb{N}$, $n \geq 1$, $f \in L(\mathbb{Q}^n, \mathbb{Z}, x_1, \ldots, x_n)$ *and* $r(f) \leq t$. *Then there exists a functional program* P *in the basis* B_0 *with the input variables* $x_1, \ldots, x_n$ *which realizes the function* f *in a working variable and for which the following inequality holds:*

$$h(P) \leq 2(n + 1)(t + 1) \ .$$

Proof. Let $f(\bar{x}) = \sum_{i=1}^{n} a_i x_i + a_{n+1}$. It is not difficult to construct a functional program P_1 in the basis B_0 which realizes the functions $0, 2^0, 2^1, ..., 2^t, 2^0 x_1, 2^1 x_1, \ldots, 2^t x_1, \ldots, 2^0 x_n, 2^1 x_n, \ldots, 2^t x_n$ in certain working variables and for which $h(P_1) \leq (n + 1)(t + 1)$.

Later, using the program P_1 it is not difficult to construct a functional program P_2 in the basis B_0 with the input variables $x_1, \ldots, x_n$ which realizes the functions $|a_1|x_1, \ldots, |a_n|x_n, |a_{n+1}|$ in certain working variables and for which $h(P_2) \leq h(P_1) + (n+1)t$.

At last, using the program P_1 one can construct a functional program P in the basis B_0 with the input variables $x_1, \ldots, x_n$ which realizes the function f in some working variable and for which $h(P) \leq h(P_2)+(n+1) \leq 2(n+1)(t+1)$. □

Proof (of Theorem 8.1). Let P_1 be a nondeterministic acyclic program in the basis B_0 with the input variables $x_1, \ldots, x_n$. One can show that if $\mathrm{Rec}(P_1) = \mathbb{Q}^n$ then the statement of the theorem holds. Let $\mathrm{Rec}(P_1) \neq \mathbb{Q}^n$. Consider the problem of recognition of belonging of an element $\bar{q} \in \mathbb{Q}^n$ to the set $\mathrm{Rec}(P_1)$ (if $\bar{q} \in \mathrm{Rec}(P_1)$ then the response is equal to 1, otherwise the response is equal to 2). Evidently, this problem is the problem of recognition of belonging of an element from $\mathbb{Q}^n$ to the union of the sets of solutions of the equation systems S_ξ, $\xi \in \mathrm{Path}(P_1)$ (see Problem 7.5). By proved in Sect. 7, the considered problem is $(n, |F(P_1)|, l)$-problem over the information system $U = U(\mathbb{Q}^n, \mathbb{Z}, x_1, \ldots, x_n)$ where $l = \max\{r(f) : f \in F(P_1)\}$.

Estimate values l and $|F(P_1)|$. Let $f_1, f_2 \in L(\mathbb{Q}^n, \mathbb{Z}, x_1, \ldots, x_n)$. Then, obviously,

$$\max\{r(f_1 - f_2), r(f_1 + f_2)\} \leq 1 + \max\{r(f_1), r(f_2)\} \ .$$

Using this inequality one can show that

$$l \leq h(P_1) \ . \tag{133}$$

From (133) follows

$$|F(P_1)| \leq (2^{h(P_1)+1} + 1)^{n+1} \ . \tag{134}$$

Using (133) and (134) we conclude that the problem of recognition of belonging of an element $\bar{q} \in \mathbb{Q}^n$ to the set $\mathrm{Rec}(P_1)$ is $(n, (2^{h(P_1)+1}+1)^{n+1}, h(P_1))$-problem over the information system U. From here and from Theorem 6.1 follows that there exists a decision tree Γ over U which solves the considered problem and for which

$$h(\Gamma) \leq 2(n+2)^4(h(P_1)+2)/\log_2(n+2) \ , \tag{135}$$

$$r(\Gamma) \leq 2(n+1)^2(1 + h(P_1) + \log_2(n+1)) \ . \tag{136}$$

Transform the decision tree Γ into a program P_2 in the basis B_0. We add to the tree Γ the node "input" and the edge issuing from the node "input" and entering the root of Γ. Then we identify all terminal nodes of the tree Γ which are labelled by the number 1 and replace this node by the node "output". After that we delete all nodes and edges such that there are no directed paths from the node "input" to the node "output" passing through them. Denote by G the obtained graph.

Every node v of the graph G, which is not neither "input" nor "output", will be replaced by a labelled graph G_v which is constructed in the following way. Let the attribute $\mathrm{sign}(\sum_{i=1}^n a_i x_i + a_{n+1})$ be assigned to the node v. From

(136) and from Lemma 8.1 follows that there exists a functional program P_v in the basis B_0 with the input variables $x_1, \ldots, x_n$ which realizes the function $\sum_{i=1}^{n} a_i x_i + a_{n+1}$ in certain working variable $y_{j(v)}$ and for which

$$h(P_v) \leq 2(n+1)(\lceil 2(n+1)^2(1 + h(P_1) + \log_2(n+1)) \rceil + 1) . \tag{137}$$

Denote by u_1 the node of the program P_v to which the edge enters, issuing from the node "input" of P_v. Denote by G_v the graph obtained from P_v by deletion of the node "input" and the edge, issuing from it, and by substitution of the node $\text{sign}(y_{j(v)})$ for the node "output". Denote this node $\text{sign}(y_{j(v)})$ by u_2.

The replacement of the node v by the graph G_v is made in the following way: we delete the node v and connect the edge, entering v, to the node u_1 of the graph G_v; all edges, issued from the node v, issue now from the node u_2 of the graph G_v.

We assume that the sets of working variables of the programs P_{v_1} and P_{v_2} for distinct nodes v_1 and v_2 are disjoint. Denote by P_2 the graph obtained from G by replacement of each node v, which is not neither "input" nor "output", by the graphs G_v.

One can show that P_2 is a deterministic acyclic program in the basis B_0 such that $\text{Rec}(P_2) = \text{Rec}(P_1)$. Taking into account the sets of input variables of the programs P_2 and P_1 coincide we conclude that the programs P_2 and P_1 are equivalent. From (135) and (137) follows that

$$\begin{aligned} h(P_2) &\leq (2(n+2)^4(h(P_1)+2)/\log_2(n+2)) \\ &\times (2(n+1)(\lceil 2(n+1)^2(1 + h(P_1) + \log_2(n+1)) \rceil + 1) + 1) \\ &\leq 8(n+2)^7(h(P_1)+2)^2 . \end{aligned}$$

□

9 Regular Language Word Recognition

In this section we consider the problem of recognition of words of fixed length in a regular language. The word under consideration can be interpreted as a description of certain screen image in the following way: the i-th letter of the word encodes the color of the i-th screen cell. In this case a decision tree which recognizes some words may be interpreted as an algorithm for the recognition of images which are defined by considered words. We obtain a classification of all regular languages depending on the growth of minimal depth of decision trees for language word recognition with the growth of the word length.

Let L be a regular language, n be a natural number and $L(n)$ be the set of all words from L whose length is equal to n. Let $L(n) \neq \emptyset$. A two-person game can be associated with the set $L(n)$. The first player choose a word from $L(n)$. The second player must recognize this word. For this purpose he can ask questions to the first player: he can choose a number i from the set $\{1, \ldots, n\}$ and ask what is the i-th letter of the word. Strategies of the second player are represented in the form of decision trees. We denote by $h_L(n)$ the minimal depth of a decision tree which recognizes words from $L(n)$ and uses only attributes each of which recognizes an i-th letter of a word, $i \in \{1,\dots,n\}$. If $L(n)=\emptyset$ then $h_L(n)=0$. The graph of the function $h_L(n)$ may have saw-tooth form. Therefore we study a smoothed analog of the function $h_L(n)$: we consider the function

$$H_L(n)=\max\{h_L(m): m\le n\} .$$

In this section we show that either $H_L(n)=O(1)$, or $H_L(n)=\Theta(\log_2 n)$, or $H_L(n)=\Theta(n)$.

9.1 Main Definitions and Results

Problem of Recognition of Words. Let $k\in \mathbb{N}$, $k\ge 2$ and E_k=$\{0,1,\dots,k-1\}$. By $(E_k)^*$ we denote the set of all finite words over the alphabet E_k, including the empty word λ. Let L be a regular language over the alphabet E_k. For any natural n we denote by $L(n)$ the set of all words from L for which the length is equal to n. Let us assume that $L(n)\ne\emptyset$. For $i\in\{1,\dots,n\}$ we define a function $l_i: L(n)\to E_k$ as follows: $l_i(\delta_1\dots\delta_n)=\delta_i$ for any $\delta_1\dots\delta_n\in L(n)$. Let us consider an information system $U(L,n)=(L(n),E_k,\{l_1,\dots,l_n\})$ and a problem $z_{L,n}=(\nu,l_1,\dots,l_n)$ over $U(L,n)$ such that $\nu(\bar\delta_1)\ne\nu(\bar\delta_2)$ for any $\bar\delta_1,\bar\delta_2\in E_k^n$, $\bar\delta_1\ne\bar\delta_2$. The problem $z_{L,n}$ will be called *the problem of recognition of words from* $L(n)$. We denote by $h_L(n)$ the minimal depth of a decision tree over $U(L,n)$ which solves the problem of recognition of words from $L(n)$. Denote by $T(z_{L,n})$ the decision table $T_{U(L,n)}(z_{L,n})$ corresponding to the problem $z_{L,n}$. Denote $\rho(L,n)=(\{l_1,\dots,l_n\},E_k)$. From Theorem 4.1 follows that

$$h_L(n)=h_{\rho(L,n)}(T(z_{L,n})) . \tag{138}$$

If $L(n)=\emptyset$ then $h_L(n)=0$.

In this section we will consider the behavior of the function $H_L:\mathbb{N}\setminus\{0\}\to\mathbb{N}$ which is defined as follows. Let $n\in\mathbb{N}\setminus\{0\}$. Then

$$H_L(n)=\max\{h_L(m): m\in\mathbb{N}\setminus\{0\}, m\le n\} .$$

Example 9.1. Let L be the regular language which is generated by the source represented in Fig. 9(b). Let us consider the problem $z_{L,4}=(\nu,l_1,l_2,l_3,l_4)$ of recognition of words from $L(4)=\{0001,0011,0111,1111\}$. Let $\nu(0,0,0,1)=1$, $\nu(0,0,1,1)=2$, $\nu(0,1,1,1)=3$ and $\nu(1,1,1,1)=4$. The decision table $T(z_{L,4})$ is represented in Fig. 8(a). The decision tree in Fig. 8(b) solves the problem of recognition of words from $L(4)$. Note that instead of numbers of words the terminal nodes in this tree are labelled by words. The depth of the considered decision tree is equal to 2. Using Theorem 3.2 we obtain $h_L(4)=2$.

A-Sources. *An A-source over the alphabet* E_k is a triple $I=(G,q_0,Q)$ where G is a directed graph, possibly with multiple edges and loops, in which each edge is labelled by a number from E_k and any two different edges starting in a node are labelled by pairwise different numbers; q_0 is a node of G and Q is some nonempty set of the graph G nodes.

A path of the source I is an arbitrary sequence $\xi=v_1,d_1,\dots,v_m,d_m,v_{m+1}$ of nodes and edges of G such that v_i is the initial and v_{i+1} is the terminal node

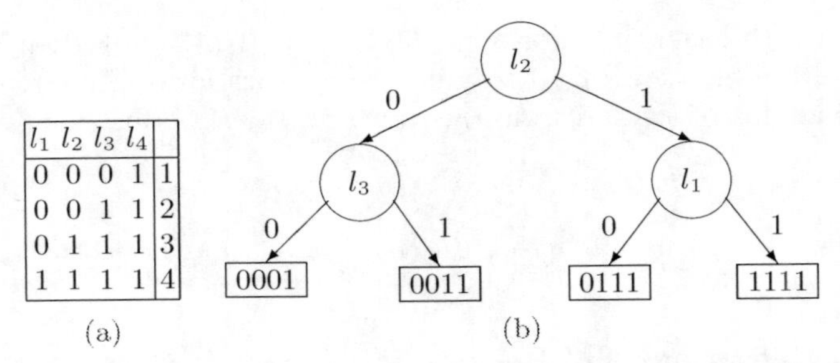

Fig. 8. Decision table and decision tree for problem of recognition of words from set $\{0001, 0011, 0111, 1111\}$.

of the edge d_i for $i = 1, \ldots, m$. Now we define a word gen(ξ) from E_k^* in the following way: if $m = 0$ then $\text{gen}(\xi) = \lambda$. Let $m > 0$, and let δ_j be the number assigned to the edge d_j, $j = 1, \ldots, m$. Then $\text{gen}(\xi) = \delta_1 \ldots \delta_m$. We will say that the path ξ *generates* the word gen(ξ). Note that different paths which start in the same node generate different words.

We denote by Path(I) the set of all paths of the source I each of which starts in the node q_0 and finishes in a node from Q. Let

$$\text{Gen}(I) = \{\text{gen}(\xi) : \xi \in \text{Path}(I)\} \ .$$

We will say that the source I *generates* the language Gen(I). It is well known that Gen(I) is a regular language.

The A-source I will be called *everywhere defined over the alphabet* E_k if each node of G is the initial node of exactly k edges which are labelled by pairwise different numbers from E_k. The A-source I will be called *reduced* if for each node of G there exists a path from Path(I) which contains this node. It is known [53] that for each regular language over the alphabet E_k there exists an everywhere defined over the alphabet E_k A-source which generates this language. Therefore for each nonempty regular language there exists a reduced A-source which generates this language. Further we will assume that a considered regular language is nonempty and it is given by reduced A-source which generates this language.

Types of Reduced A-Sources. Let $I = (G, q_0, Q)$ be a reduced A-source over the alphabet E_k. A path of the source I will be called *elementary* if nodes of this path are pairwise different. A path of the source I will be called *a cycle of the source* I if there is at least one edge in this path, and the first node of this path is equal to the last node of this path. A cycle of the source I will be called *elementary* if nodes of this cycle, with the exception of the last node, are pairwise different. As usual, paths and cycles of the source I will be considered sometimes as subgraphs of the graph G. We will say that these subgraphs *are generated* by corresponding paths and cycles.

The source I will be called *simple* if each two different (as subgraphs) elementary cycles of the source I do not have common nodes. Let I be a simple source and ξ be a path of the source I. The number of different (as subgraphs) elementary cycles of the source I, which have common nodes with ξ, will be denoted by $\mathrm{cl}(\xi)$ and will be called *the cyclic length of the path* ξ. The value

$$\mathrm{cl}(I) = \max\{\mathrm{cl}(\xi) : \xi \in \mathrm{Path}(I)\}$$

will be called *the cyclic length of the source* I.

Let I be a simple source, C be an elementary cycle of the source I, and v be a node of the cycle C. Beginning with the node v the cycle C generates an infinite periodic word over the alphabet E_k. This word will be denoted by $\mathrm{W}(I, C, v)$. We denote by $r(I, C, v)$ the minimal period of the word $\mathrm{W}(I, C, v)$. The source I will be called *dependent* if there exist two different (as subgraphs) elementary cycles C_1 and C_2 of the source I, nodes v_1 and v_2 of the cycles C_1 and C_2 respectively, and the path π of the source I from v_1 to v_2 which satisfy the following conditions: $\mathrm{W}(I, C_1, v_1) = \mathrm{W}(I, C_2, v_2)$ and the length of the path π is a number divisible by $r(I, C_1, v_1)$. If the source I is not dependent then it will be called *independent*.

Main Result and Examples

Theorem 9.1. *Let L be a nonempty regular language and I be a reduced A-source which generates the language L. Then the following statements hold:*

a) if I is an independent simple source and $\mathrm{cl}(I) \leq 1$ then $H_L(n) = O(1)$;
b) if I is an independent simple source and $\mathrm{cl}(I) \geq 2$ then $H_L(n) = \Theta(\log_2 n)$;
c) if I is not independent simple source then $H_L(n) = \Theta(n)$.

In the following figures containing a source $I = (G, q_0, Q)$ the node q_0 will be labelled by $+$ and each node from Q will be labelled by $*$.

Example 9.2. Let I_1 be the source in Fig. 9(a) and L_1 be the regular language which is generated by I_1. The source I_1 is an independent simple A-source with $\mathrm{cl}(I_1) = 1$. One can show that $H_{L_1}(n) = 0$ for each $n \in \mathbb{N} \setminus \{0\}$.

Example 9.3. Let I_2 be the source in Fig. 9(b) and L_2 be the regular language which is generated by I_2. The source I_2 is an independent simple A-source with $\mathrm{cl}(I_2) = 2$. One can show that $H_{L_2}(n) = \lceil \log_2 n \rceil$ for each $n \in \mathbb{N} \setminus \{0\}$.

Example 9.4. Let I_3 be the source in Fig. 10(a) and L_3 be the regular language which is generated by I_3. The source I_3 is a dependent simple A-source with $\mathrm{cl}(I_3) = 2$. One can show that $H_{L_3}(n) = n - 1$ for each $n \in \mathbb{N} \setminus \{0\}$.

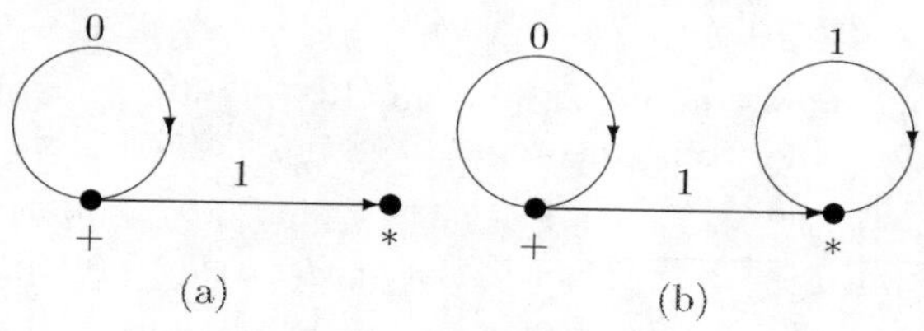

Fig. 9. Independent simple A-source I_1 with $cl(I_1) = 1$ and independent simple A-source I_2 with $cl(I_2) = 2$.

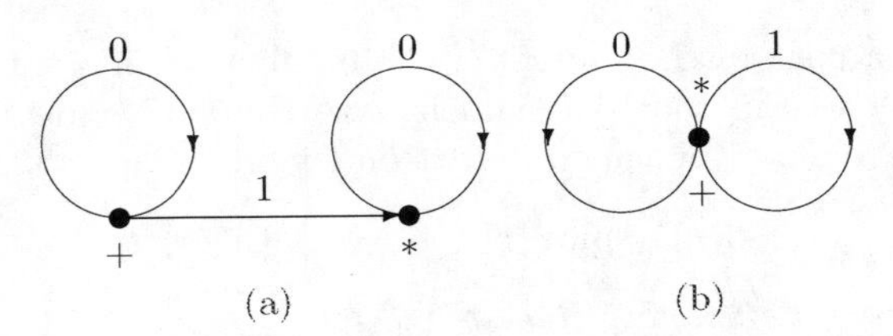

Fig. 10. Dependent simple A-source I_3 and A-source I_4 which is not simple.

Example 9.5. Let I_4 be the source in Fig. 10(b) and L_4 be the regular language which is generated by I_4. The source I_4 is not a simple A-source. One can show that $H_{L_4}(n) = n$ for each $n \in \mathbb{N} \setminus \{0\}$.

9.2 Some Bounds on $h_L(n)$

In this subsection we consider some simple corollaries of statements from Sect. 3 which are adapted for study of the problem of word recognition.

Let L be a regular language over the alphabet E_k. Now for any $n \geq 1$ we define a parameter $M_L(n)$ of the language L. Denote by $E_k^*(n)$ the set of all words of the length n over the alphabet E_k. Let $\alpha = \alpha_1 \ldots \alpha_n$ be a word from $E_k^*(n)$ and $J \subseteq \{1, \ldots, n\}$. Denote $L(\alpha, J) = \{\beta_1 \ldots \beta_n \in L(n) : \beta_j = \alpha_j, j \in J\}$ (if $J = \emptyset$ then $L(\alpha, J) = L(n)$) and $M_L(n, \alpha) = \min\{|J| : J \subseteq \{1, \ldots, n\}, |L(\alpha, J)| \leq 1\}$. Then $M_L(n) = \max\{M_L(n, \alpha) : \alpha \in E_k^*(n)\}$. It is clear that $M_L(n) = 0$ if $L(n) = \emptyset$, and

$$M_L(n) = M_{\rho(L,n),h}(T(z_{L,n})) \tag{139}$$

if $L(n) \neq \emptyset$.

Next statement follows immediately from (138), (139) and from Theorem 3.1.

Proposition 9.1. *Let L be a regular language and n be a natural number. Then $h_L(n) \geq M_L(n)$.*

Next statement follows immediately from Theorem 3.2, from obvious equality $|L(n)| = S(T(z_{L,n}))$ and from (138).

Proposition 9.2. *Let L be a regular language over the alphabet E_k, n be a natural number, and $L(n) \neq \emptyset$. Then $h_L(n) \geq \lceil \log_k |L(n)| \rceil$.*

The following statement can be easily proved by induction on $|L(n)|$.

Proposition 9.3. *Let L be a regular language over the alphabet E_k, n be a natural number, and $L(n) \neq \emptyset$. Then $h_L(n) \leq |L(n)| - 1$.*

Next statement follows immediately from (138), (139), from obvious equality $|L(n)| = N(T(z_{L,n}))$ and from Theorem 3.5.

Proposition 9.4. *Let L be a regular language over the alphabet E_k, n be a natural number, and $L(n) \neq \emptyset$. Then $h_L(n) \leq M_L(n) \log_2 |L(n)|$.*

Note that Propositions 9.1, 9.2, 9.3 and 9.4 are formulated for regular languages but they are true for arbitrary languages over finite alphabets.

9.3 Proof of Theorem 9.1

Auxiliary statements and proof of Theorem 9.1 are presented in this subsection.

Lemma 9.1. *Let L be a nonempty regular language, and I be a reduced A-source over the alphabet E_k which generates the language L and which is not simple. Then $H_L(n) = \Theta(n)$.*

Proof. Let $I = (G, q_0, Q)$ and let C_1, C_2 be different (as subgraphs) elementary cycles of the source I which have a common node v. Since I is a reduced source, it contains a path ξ_1 from the node q_0 to the node v and a path ξ_2 from the node v to a node $q_1 \in Q$. Let the length of the path ξ_1 be equal to a, the length of the path ξ_2 be equal to b, the length of the cycle C_1 be equal to c and the length of the cycle C_2 be equal to d. Consider the sequence of numbers $n_i = a+b+i \cdot c \cdot d$, $i = 0, 1, \ldots$. Let $i \in \mathbb{N}$. We can obtain a path of the length $c \cdot d$ from the node v to the node v by passage d times along the cycle C_1 or by passage c times along the cycle C_2. Using this fact one can show that $|L(n_i)| \geq 2^i$. From Proposition 9.2 follows that $h_L(n_i) \geq i/\log_2 k = (n_i - a - b)/(c \cdot d \cdot \log_2 k)$. Therefore $h_L(n_i) \geq n_i/c_1 - n_1$, where $c_1 = n_1 \cdot \log_2 k$. Let $n \geq n_1$ and let i be the maximal number from $\mathbb{N}$ such that $n \geq n_i$. Evidently, $n - n_i \leq n_1$. Hence $H_L(n) \geq h_L(n_i) \geq (n - n_1)/c_1 - n_1 \geq n/c_1 - c_2$, where $c_2 = n_1 + 1$. Therefore $H_L(n) \geq n/2c_1$ for large enough n. The inequality $H_L(n) \leq n$ is obvious. Thus, $H_L(n) = \Theta(n)$. □

Let $I = (G, q_0, Q)$ be a simple A-source and let $\xi \in \text{Path}(I)$. Denote by $G(\xi)$ the subgraph of the graph G generated by the path ξ. Define an equivalence relation on the set Path(I). We will say that two paths ξ_1 and ξ_2 from Path(I) are *equivalent* if $G(\xi_1) = G(\xi_2)$. Evidently, for this equivalence relation there are only finite number of equivalence classes. These classes will be called *I-classes.*

Note one important property of paths in simple source. Let $\xi \in \text{Path}(I)$, $\xi = v_1, d_1, \ldots, v_m, d_m, v_{m+1}$, $1 \leq i < j \leq m+1$ and the nodes v_i, v_j belong to an elementary cycle C of the source I. Since I is a simple source it is easily to show that the nodes $v_{i+1}, \ldots, v_{j-1}$ and the edges $d_i, \ldots, d_{j-1}$ belong to the cycle C.

Lemma 9.2. *Let L be a nonempty regular language and I be a simple reduced A-source over the alphabet E_k which generates the language L. Then the following statements hold:*

a) if $\text{cl}(I) \leq 1$ then there exists a constant $c_1 \geq 1$ such that for any $n \geq 1$ the inequality $|L(n)| \leq c_1$ holds;
b) if $\text{cl}(I) \geq 2$ then there exists a constant $c_2 \geq 1$ such that for any $n \geq 1$ the inequality $|L(n)| \leq c_2 \cdot n^{\text{cl}(I)}$ holds.

Proof. Let $I = (G, q_0, Q)$.

a). Let $\text{cl}(I) \leq 1$, $n \geq 1$, B be some I-class and $\xi \in B$. It is clear that the subgraph $G(\xi)$ contains at most one elementary cycle. Using this fact and taking into account that ξ is a path in a simple source one can show that in the class B at most one path of the length n is contained. Denote by c_1 the number of I-classes. Then $|L(n)| \leq c_1$.

b) Let $\mathrm{cl}(I) \geq 2$, $n \geq 1$, B be some I-class and $\xi \in B$. Let the subgraph $G(\xi)$ contain exactly m different (as subgraphs) elementary cycles $C_1, \ldots, C_m$, which are enumerated here in the same order as they are met in the path ξ. It is clear that $m \leq \mathrm{cl}(I)$. Taking into account that ξ is a path in simple source one can show that the set of words generated by paths from B is of the kind

$$\{\alpha_1 \beta_1^{i_1} \ldots \alpha_m \beta_m^{i_m} \alpha_{m+1} : i_1, \ldots, i_m \in \mathbb{N} \setminus \{0\}\} ,$$

where $\alpha_1, \ldots, \alpha_{m+1}$ are some words over the alphabet E_k and β_j, $1 \leq j \leq m$, is the word generated by single passage along the cycle C_j beginning with a certain node. The notation α^i designates the word $\alpha \ldots \alpha$ where the word α repeats i times. It is clear that each word of the length n generated by a path from B is of the kind $\alpha_1 \beta_1^{i_1} \ldots \alpha_m \beta_m^{i_m} \alpha_{m+1}$ where $i_1, \ldots, i_m \in \{1, \ldots, n\}$. It is not difficult to see that the number of such words is bounded from above by the value $n^m \leq n^{\mathrm{cl}(I)}$. Thus, $|L(n)| \leq c_2 \cdot n^{\mathrm{cl}(I)}$ where c_2 is the number of I-classes. □

Note that proofs of Lemmas 9.1 and 9.2 are close to proofs of some statements from [208].

Lemma 9.3. *Let L be a nonempty regular language and I be a simple reduced A-source over the alphabet E_k which generates the language L. Then the following statements hold:*

a) if $\mathrm{cl}(I) \leq 1$ then $H_L(n) = O(1)$;
b) if $\mathrm{cl}(I) \geq 2$ then $H_L(n) = \Omega(\log_2 n)$.

Proof. Let $I = (G, q_0, Q)$.

a). Let $\mathrm{cl}(I) \leq 1$. From Lemma 9.2 follows that there exists a constant $c_1 \geq 1$ such that for any $n \geq 1$ the inequality $|L(n)| \leq c_1$ holds. Using Proposition 9.3 we obtain that for any $n \geq 1$ the inequality $h_L(n) \leq c_1 - 1$ holds. Therefore $H_L(n) = O(1)$.

b). Let $\mathrm{cl}(I) \geq 2$. One can show that there exists a path $\xi \in \mathrm{Path}(I)$ such that the subgraph $G(\xi)$ contains $m \geq 2$ different (as subgraphs) elementary cycles. As it has been already mentioned in the proof of Lemma 9.2, the set of words generated by paths from $\mathrm{Path}(I)$, which are equivalent to the path ξ, is of the kind $\{\alpha_1 \beta_1^{i_1} \ldots \alpha_m \beta_m^{i_m} \alpha_{m+1} : i_1, \ldots, i_m \in \mathbb{N} \setminus \{0\}\}$, where $\alpha_1, \beta_1, \ldots \alpha_m, \beta_m, \alpha_{m+1}$ are certain words over the alphabet E_k. Let a be the length of the word β_1, b be the length of the word β_2 and c be the length of the word $\alpha_1 \alpha_2 \ldots \alpha_{m+1} \beta_3 \ldots \beta_m$. It is clear that $a > 0$ and $b > 0$. Consider the sequence of numbers $n_j = a \cdot b \cdot j + c$, $j = 2, 3, \ldots$. Let $j \in \mathbb{N}$, $j \geq 2$ and $t \in \{1, \ldots, j-1\}$. Then the word $\alpha_1 \beta_1^{i_1} \ldots \alpha_m \beta_m^{i_m} \alpha_{m+1}$ where $i_1 = bt$, $i_2 = a(j-t)$ and $i_3 = \ldots = i_m = 1$ belongs to $L(n_j)$. Hence $|L(n_j)| \geq j - 1$. From Proposition 9.2 follows that $h_L(n_j) \geq \log_2(j - 1)/\log_2 k = \log_2((n_j - c - a \cdot b)/a \cdot b)/\log_2 k$. Let $n_j \geq 2 \cdot (c + a \cdot b)$. Then $h_L(n_j) \geq \log_2(n_j/2 \cdot a \cdot b)/\log_2 k = (\log_2 n_j)/c_1 - c_2'$ where $c_1 = \log_2 k$ and $c_2' = (\log_2 2 \cdot a \cdot b)/\log_2 k$. Let j_0 be the minimal number from $\mathbb{N}$ such that $n_{j_0} \geq 2 \cdot (c + a \cdot b)$. Let $n \in \mathbb{N}, n \geq n_{j_0}$ and let j be the maximal number from $\mathbb{N}$ such that $n \geq n_j$. It is clear that $n - n_j \leq a \cdot b$. Hence $H_L(n) \geq h_L(n_j) \geq \log_2(n - a \cdot b)/c_1 - c_2' \geq \log_2(n/2)/c_1 - c_2' = \log_2 n/c_1 - c_2$, where $c_2 = 1/c_1 + c_2'$. Therefore $H_L(n) \geq \log_2 n/2c_1$ for large enough n. Thus, $H_L(n) = \Omega(\log_2 n)$. □

Lemma 9.4. *Let L be a nonempty regular language and I be a dependent simple reduced A-source over the alphabet E_k which generates the language L. Then $H_L(n) = \Theta(n)$.*

Proof. Let $I = (G, q_0, Q)$. Since I is a dependent source then in the source I there exist different (as subgraphs) elementary cycles C_1 and C_2, nodes v_1 and v_2 belonging to cycles C_1 and C_2 respectively, and a path ξ of the source I from the node v_1 to the node v_2 such that $\mathrm{W}(I, C_1, v_1) = \mathrm{W}(I, C_2, v_2)$ and the length of the path ξ is a number divisible by $r(I, C_1, v_1)$. Let the length of the cycle C_1 be equal to a, the length of the cycle C_2 be equal to b, and the length of the path ξ be equal to c. Let $t = a \cdot b \cdot c$. Denote by β the word generated by passage the path of the length t along the cycle C_1 beginning with the node v_1. Since I is a reduced source then there exist a path ξ_1 from the node q_0 to the node v_1 and a path ξ_2 from the node v_2 to a node $q_1 \in Q$. Let $i, j \in \mathbb{N}$. Consider a path from the set $\mathrm{Path}(I)$. This path will be described as follows. First, we pass the path ξ_1 from the node q_0 to the node v_1 (denote by α_1 the word generated by this path). Then we pass along the cycle C_1 the path of the length $i \cdot t$ beginning with the node v_1 (this path generates the word β^i and finishes in the node v_1). After that we pass the path ξ and then the path of the length $t - c$ along the cycle C_2 beginning with the node v_2 (denote by γ the generated word, and denote by v_3 the terminal node of this path which belongs to C_2). Beginning with the node v_3 we pass along the cycle C_2 the path of the length $j \cdot t$ (evidently, the word β^j is generated and the path finishes in the node v_3). Finally, we pass along the cycle C_2 the path from the node v_3 to the node v_2 and the path ξ_2 (denote by α_2 the generated word). Thus, we obtain that for any $i, j \in \mathbb{N}$ the word $\alpha_1 \beta^i \gamma \beta^j \alpha_2$ belongs to the language L.

Denote by p_i the length of the word α_i, $i \in \{1, 2\}$. Let $m \geq 1$ and $i \in \{0, \ldots, m\}$. Denote $\delta_i = \alpha_1 \beta^i \gamma \beta^{m-i} \alpha_2$ and $n_m = p_1 + p_2 + t \cdot (m + 1)$. It is clear that $\delta_i \in L(n_m), 0 \leq i \leq m$. Denote $\sigma = \alpha_1 \beta^{m+1} \alpha_2$. Obtain a lower bound on the value $M_L(n_m, \sigma) = \min\{|J| : J \subseteq \{1, \ldots, n_m\}, |L(\sigma, J)| \leq 1\}$. Let $J \subseteq \{1, \ldots, n_m\}$ and $|L(\sigma, J)| \leq 1$. For $i \in \{0, \ldots, m\}$ denote $B_i = \{p_1 + i \cdot t + 1, \ldots, p_1 + i \cdot t + t\}$. It is clear that for any $i \in \{0, \ldots, m\}$ if $J \cap B_i = \emptyset$ then $\delta_i \in L(\sigma, J)$. Assume that $|J| \leq m - 1$. Then for two distinct $i_1, i_2 \in \{0, \ldots, m\}$ the words δ_{i_1} and δ_{i_2} belong to the set $L(\sigma, J)$ which is impossible. Therefore $|J| \geq m$. From this inequality follows that $M_L(n_m, \sigma) \geq m$ and hence $M_L(n_m) \geq m$. Using Proposition 9.1 we obtain $h_L(n_m) \geq M_L(n_m) \geq m$. Therefore $h_L(n_m) \geq n_m/c_1 - c_2'$ where $c_1 = t$ and $c_2' = (p_1 + p_2 + t)/t$.

Let $n \geq n_1$ and let m be the maximal number from $\mathbb{N}$ such that $n \geq n_m$. It is clear that $n - n_m \leq t$. Hence $H_L(n) \geq h_L(n_m) \geq (n - t)/c_1 - c_2' \geq n/c_1 - c_2$ where $c_2 = 1 + c_2'$. Therefore $H_L(n) \geq n/2c_1$ for large enough n. The inequality $H_L(n) \leq n$ is obvious. Thus, $H_L(n) = \Theta(n)$. □

Lemma 9.5. *Let L be a nonempty regular language, I be an independent simple reduced A-source over the alphabet E_k which generates the language L, and let* $\mathrm{cl}(I) \geq 2$. *Then* $H_L(n) = O(\log_2 n)$.

Proof. Let $I = (G, q_0, Q)$, d be the number of nodes in the graph G and $t = 4 \cdot d$.

Let C_1 be an elementary cycle of the source I, a be the number of nodes in C_1, v_1 be a node of the cycle C_1, β be the word of the length a generated by passage along the cycle C_1 beginning with the node v_1, and let $r = r(I, C_1, v_1)$. We will show that for any $j \in \mathbb{N}$ there is no path ξ of the length $(t + j) \cdot a$ from the node v_1 to a node not belonging to the cycle C_1 which generates a word of the kind $\sigma = \gamma\beta^t$. Assume the contrary. Let for some $j \in \mathbb{N}$ there exists a path ξ of the length $(t + j) \cdot a$ from the node v_1 to a node not belonging to the cycle C_1 which generates a word of the kind $\sigma = \gamma\beta^t$.

The terminal part of the path ξ which generates the suffix β^t of the word σ will be denoted by π. We will show that no node from the path π can belong to the cycle C_1. Assume the contrary. Let some node v of the path π belong to the cycle C_1. Since I is a simple source all the nodes and edges preceding v in the path ξ also belong to the cycle C_1. Hence $\gamma = \beta^j$. If an A-source contains two paths which start in the same node and generate the same word then these paths are the same. Using this fact one can show that all nodes of the path π following the node v also belong to the cycle C_1 which is impossible. Therefore the path π contains no nodes from the cycle C_1.

It is clear that the length of the path π is equal to $t \cdot a = 4 \cdot d \cdot a$. If we assume that in the process of the path π passage each elementary cycle of the source I will be passed less than $2 \cdot a$ times then the length of the path π is at most $3 \cdot d \cdot a$ which is impossible. Let C_2 be an elementary cycle of the source I which is passed at least $2 \cdot a$ times in the process of the path π passage. It is clear that the elementary cycles C_1 and C_2 are different (as subgraphs). Let b be the number of nodes in the cycle C_2. Then we obtain that the word β^t contains a certain segment φ of the length $2 \cdot a \cdot b$ generated both by passage $2 \cdot b$ times along the cycle C_1 and by passage $2 \cdot a$ times along the cycle C_2. Hence the cycles C_1 and C_2 beginning with some nodes generate the same infinite periodic word over the alphabet E_k with the minimal period r. Divide the word σ into segments of the length a. Since the length of the word φ is equal to $2 \cdot a \cdot b$ then φ contains at least one such segment. Let v_2 be the node of the cycle C_2 such that beginning with this node the first segment generates which is wholly contained in φ. Evidently, the length of the path ξ part from v_1 to v_2 is a number divisible by a and hence it is a number divisible by r. Besides, β is a prefix of the infinite word generated by the cycle C_2 beginning with the node v_2. Since a is a number divisible by r, a is a period of this infinite word. Therefore the infinite word generated by the cycle C_2 beginning with the node v_2 is of the kind $\beta\beta\ldots$. Hence $\mathrm{W}(I, C_1, v_1) = \mathrm{W}(I, C_2, v_2)$. Therefore the considered source is a dependent one which is impossible. Thus, for any $j \in \mathbb{N}$ there is no path ξ of the length $(t + j) \cdot a$ from the node v_1 to a node not belonging to the cycle C_1 which generates a word of the kind $\sigma = \gamma\beta^t$. From here follows that if a path κ, which starts in the node v_1, generates the word of the kind $\sigma = \gamma\beta^t$ of the length $i \cdot a$,

$i \geq t$, then this word is β^i. Really, the initial and the terminal nodes of the path κ belong to the cycle C_1. Taking into account that I is a simple source we obtain that all the nodes and edges of the path κ belong to the cycle C_1. Therefore the word generated by the path κ is β^i and the path κ is uniquely determined by the initial node and the suffix β^t of the word generated by it.

Consider a path ξ of the source I. Let the subgraph $G(\xi)$ contain exactly m different elementary cycles $C_1, \ldots, C_m$, and the passage of the path ξ consists of the sequential passage of the paths $\xi_1, \kappa_1, \xi_2, \ldots, \xi_m, \kappa_m, \xi_{m+1}$. The path ξ_j is an elementary path from v_j to v_{j+1}, $1 \leq j \leq m+1$. The path κ_j consists of the passage i_j times along the elementary cycle C_j beginning with the node v_{j+1} belonging to the cycle C_j, $1 \leq j \leq m$. Here the paths ξ_j and κ_j have the only common node v_{j+1}, $1 \leq j \leq m$. Let α_j be the word generated by the elementary path ξ_j, $1 \leq j \leq m+1$, and let β_j be the word generated by single passage of the cycle C_j beginning with the node v_{j+1}, $1 \leq j \leq m$. In this case the path ξ generates the word

$$\gamma = \alpha_1 \beta_1^{i_1} \alpha_2 \ldots \alpha_m \beta_m^{i_m} \alpha_{m+1} .$$

Let us mark some letters in the word γ as follows. All the letters from the subwords $\alpha_1, \alpha_2, \ldots, \alpha_{m+1}$ will be marked. For $j = 1, \ldots, m$, if $i_j < t$ then all the letters from the subword $\beta_j^{i_j}$ will be marked, and if $i_j \geq t$, then all the letters from the suffix β_j^t of the subword $\beta_j^{i_j}$ will be marked. It is clear that at most $d(t+1)$ letters will be marked. One can show that the marked letters and the initial node v_1 of the path ξ determine uniquely the path ξ and the word γ generated by it. Really, the initial node of the path ξ_1 and the word α_1 determine uniquely the path ξ_1 and the initial node of the paths κ_1 and ξ_2. The initial node of the path κ_1 and the word $\beta_j^{i_j}$ in the case $i_j < t$ or the word β_j^t in the case $i_j \geq t$ determine uniquely the path κ_1 etc.

Let $n \geq 1$ and let $L(n) \neq \emptyset$. Find an upper bound on the value $M_L(n)$. Let σ be a word of the length n in the alphabet E_k. Let $\sigma \in L(n)$ and let the word σ be generated by the path $\xi \in \text{Path}(I)$. It is known that the initial node of this path is q_0. Therefore this path and the word generated by it are uniquely determined by at most $d(t+1)$ letters of the word σ. Using this fact it is not difficult to show that $M_L(n, \sigma) \leq d(t+1)$. Let now $\sigma \notin L(n)$. Let there exist a path ξ of the source I which starts in the node q_0 and generates the word σ (it is clear that the terminal node of the path ξ does not belong to the set Q). Then the path ξ and the word σ generated by it are determined uniquely by at most $d(t+1)$ letters of the word σ. Using this fact it is not difficult to show that $M_L(n, \sigma) \leq d(t+1)$. Let the source I contain no path which starts in the node q_0 and generates the word σ. Let ξ be the path of the source I which starts in the node q_0 and generates the prefix γ of the word σ with maximal length. Then there exist at most $d(t+1)$ letters of the word γ which determine uniquely both the path ξ and the word γ. Add to these letters the letter from the word σ, following the prefix γ. Evidently, the set $L(n)$ contains no words in which marked in the word σ letters could be found in the same places. Using this fact one can show that $M_L(n, \sigma) \leq d(t+1)+1$. Hence $M_L(n) \leq d(t+1)+1$. Denote $s = d(t+1)+1$. Using Lemma 9.2 we obtain that there exists a constant $m \geq 1$, depending only on the source I and such that $|L(n)| \leq m \cdot n^{\mathrm{cl}(I)}$. From Proposition 9.4 follows that $h_L(n) \leq M_L(n) \cdot \log_2 |L(n)| \leq s \cdot (\mathrm{cl}(I) \cdot \log_2 n + \log_2 m) = c_1 \cdot \log_2 n + c_2$, where $c_1 = s \cdot \mathrm{cl}(I)$ and $c_2 = s \cdot \log_2 m$. Evidently, in the case $L(n) = \emptyset$ the inequality $h_L(n) \leq c_1 \cdot \log_2 n + c_2$ also holds. Since the considered inequality holds for any $n \geq 1$, we conclude that the inequality $H_L(n) \leq c_1 \cdot \log_2 n + c_2$ holds for any $n \geq 1$. Therefore $H_L(n) = 2c_1 \log_2 n$ for large enough n. Thus, $H_L(n) = O(\log_2 n)$. □

Proof (of Theorem 9.1). The statement of theorem follows immediately from Lemmas 9.1, 9.3, 9.4 and 9.5. □

10 Diagnosis of Constant Faults in Circuits

Different lines of investigation of applications of decision trees to the constant fault diagnosis in combinatorial circuits are studied in this section. Faults under consideration are represented in the form of Boolean constants on some inputs of the circuit gates. The diagnosis problem consists in the recognition of the function realized by the circuit with a fixed tuple of constant faults from given set of tuples. For this problem solving we use decision trees. Each attribute of a decision tree consists in observation of output of the circuit at the inputs of which a binary tuple is given.

This section contains five subsections. In the first subsection basic notions are defined. In the second subsection the complexity of decision trees for diagnosis of arbitrary and specially constructed circuits is considered. In the third subsection the complexity of algorithms for construction of decision trees for diagnosis of faults is studied. In the fourth subsection so-called iteration-free circuits are investigated. In the fifth subsection an approach to circuit construction and diagnosis is considered.

Definitions, notation and results from appendix "Closed Classes of Boolean Functions" are used in this section without special notice.

10.1 Basic Notions

The notions of combinatorial circuit, set of tuples of constant faults and diagnosis problem are defined in this subsection.

Combinatorial Circuits. *A basis* is an arbitrary nonempty finite set of Boolean functions. Let B be a basis.

A combinatorial circuit in the basis B (a circuit in the basis B) is a labelled finite directed graph without directed cycles and, possibly, with multiple edges which has nodes of the three types: inputs, gates and outputs.

Nodes of the *input* type have no entering edges, each input is labelled by a variable, and distinct inputs are labelled by distinct variables. Every circuit has at least one input.

Each node of the *gate* type is labelled by a function from the set B. Let v be a gate and let a function g depending on t variables be attached to it. If $t = 0$

(this is the case when g is one of the constants 0 or 1) then the node v has no entering edges. If $t > 0$ then the node v has exactly t entering edges which are labelled by numbers $1, \ldots, t$ respectively. Every circuit has at least one gate.

Each node of the *output* type has exactly one entering edge which issues from a gate. Let v be an output. Nothing is attached to v, and v has no issuing edges. We will consider only circuits which have exactly one output.

Let S be a circuit in basis B which has n inputs labelled by variables $x_1, \ldots, x_n$. Let us correspond to each node v in the circuit S a Boolean function f_v depending on variables $x_1, \ldots, x_n$. If v is an input of S labelled by the variable x_i then $f_v = x_i$. If v is a gate labelled by a constant $c \in \{0, 1\}$ then $f_v = c$. Let v be a gate labelled by a function g depending on $t > 0$ variables. For $i = 1, \ldots, t$ let the edge d_i, labelled by the number i, issue from a node v_i and enter the node v. Then $f_v = g(f_{v_1}, \ldots, f_{v_t})$. If v is an output of the circuit S and an edge, issuing from a node u enters the node v, then $f_v = f_u$. The function corresponding to the output of the circuit S will be denoted by f_S. We will say that the circuit S *realizes* the function f_S.

Denote by $L(S)$ the number of gates in the circuit S. The value $L(S)$ characterizes the complexity of the circuit S.

Denote by $\mathrm{Circ}(B)$ the set of circuits in the basis B. Denote $\mathcal{F}(B) = \{f_S : S \in \mathrm{Circ}(B)\}$. One can show that $\mathcal{F}(B) = [B] \setminus \{0, 1\}$ where $[B]$ is the closure of the set B relatively to operation of substitution and operations of insertion and deletion of unessential variable.

Set of Tuples of Constant Faults on Inputs of Gates. Let S be a circuit in basis B. Edges entering gates of the circuit S will be called *inputs of gates.* Let the circuit S have m gate inputs. The circuit S will be called *degenerate* if $m = 0$ and *nondegenerate* if $m > 0$. Let S be a nondegenerate circuit. Later we will assume that the gate inputs in the circuit S are enumerated by numbers from 1 to m. Thus, each edge entering a gate has a sequential number in the circuit besides the number attached to it and corresponding to the gate.

We will consider the faults in the circuit S which consist in appearance of Boolean constants on gate inputs. Each fault of such kind is defined by *a tuple of constant faults on inputs of gates of the circuit* S which is an arbitrary m-tuple of the kind $\bar{w} = (w_1, \ldots, w_m) \in \{0, 1, 2\}^m$. If $w_i = 2$ then the i-th gate input in the circuit S operates properly. If $w_i \neq 2$ then the i-th gate input in the circuit S is faulty and realizes the constant w_i.

Define a circuit $S(\bar{w})$ in the basis $B \cup \{0, 1\}$ which will be interpreted as the result of action of the tuple of faults $\bar{w}$ on the circuit S. Let us overlook all gate inputs in the circuit S. Let $i \in \{1, \ldots, m\}$. If $w_i = 2$ then the i-th gate input will be left without changes. Let $w_i \neq 2$ and the i-th gate input is the edge d issuing from the node v_1 and entering the node v_2. Add to the circuit S new gate $v(w_i)$ which is labelled by the constant w_i. Instead of the node v_1 connect the edge d to the node $v(w_i)$.

A set of tuples of constant faults on inputs of gates of the circuit S is a subset W of the set $\{0, 1, 2\}^m$ containing the tuple $(2, \ldots, 2)$. Denote $\mathrm{Circ}(S, W) = \{S(\bar{w}) : \bar{w} \in W\}$. Note that $S((2, \ldots, 2)) = S$.

Problem of Diagnosis. Let S be a nondegenerate circuit in the basis B with n inputs and m gate inputs, and let W be a set of tuples of constant faults on gate inputs of the circuit S. *The diagnosis problem for the circuit S relative to the faults from W*: for a given circuit $S' \in \text{Circ}(S, W)$ it is required to recognize the function realized by the circuit S'. To solve this problem we will use decision trees in which the computation of the value of each attribute consists in observation of output of the circuit S' at the inputs of which a tuple from the set $\{0,1\}^n$ is given.

Define the diagnosis problem for the circuit S relative to the faults from the set W as a problem over corresponding information system. With each $\bar{\delta} \in \{0,1\}^n$ we associate the function $\bar{\delta} : \text{Circ}(S, W) \to \{0,1\}$ such that $\bar{\delta}(S') = f_{S'}(\bar{\delta})$ for any $S' \in \text{Circ}(S, W)$. Let us consider an information system $U(S, W) = (\text{Circ}(S, W), \{0,1\}, \{0,1\}^n)$ and a problem $z_{S,W} = (\nu, \bar{\delta}_1, \ldots, \bar{\delta}_{2^n})$ over $U(S, W)$ where $\{\bar{\delta}_1, \ldots, \bar{\delta}_{2^n}\} = \{0,1\}^n$ and $\nu(\bar{\sigma}_1) \neq \nu(\bar{\sigma}_2)$ for any $\bar{\sigma}_1, \bar{\sigma}_2 \in \{0,1\}^{2^n}$ such that $\bar{\sigma}_1 \neq \bar{\sigma}_2$. The problem $z_{S,W}$ is a formalization of the notion of the diagnosis problem for the circuit S relative to the faults from the set W. Denote $\rho(S, W) = (\{0,1\}^n, \{0,1\})$. Note that $U(S, W)$ is an information system of the signature $\rho(S, W)$.

The mapping ν from $z_{S,W}$ numbers all Boolean functions of n variables. The solution of the problem $z_{S,W}$ for a circuit $S' \in \text{Circ}(S, W)$ is the number of the function $f_{S'}$ realizing by the circuit S'. In some cases it will be convenient for us instead of the number of the function $f_{S'}$ use a formula which realizes a function equal to $f_{S'}$. As in appendix, two Boolean functions is called *equal* if one of them can be obtained from the other by operations of insertion and deletion of unessential variables.

Later, we will often consider the set $\{0,1,2\}^m$ of all possible tuples of constant faults on inputs of gates of the circuit S. Denote $U(S) = U(S, \{0,1,2\}^m)$, $\rho(S) = \rho(S, \{0,1,2\}^m)$, $z_S = z_{S,\{0,1,2\}^m}$ and $h(S) = h^g_{U(S)}(z_S)$. Evidently, $h^g_{U(S)}(z_S) = h^l_{U(S)}(z_S)$. It is clear that $h(S)$ is the minimal depth of a decision tree over $U(S)$ solving the diagnostic problem for the circuit S relative to the faults from the set $\{0,1,2\}^m$. If S is a degenerate circuit then $h(S) = 0$.

Example 10.1. Let S be a circuit with one gate represented in Fig. 11(a). For this circuit $L(S) = 1$. We will admit all possible tuples of constant faults on gate inputs. The circuit S (possible, with faults) realizes a function from the set $\{x \wedge y, x, y, 0, 1\}$. In Fig. 11(b) one can see corresponding decision table $T_{U(S)}(z_S)$ and in Fig. 11(c) one can see a decision tree which solves the diagnosis problem for S. Note that in the table and in the tree we use functions as labels instead of its numbers. The depth of the considered decision tree is equal to 3. Using Theorem 3.2 we obtain $h(S) = 3$.

10.2 Complexity of Algorithms for Diagnosis

In this subsection the complexity of decision trees for diagnosis of arbitrary and specially constructed circuits is considered.

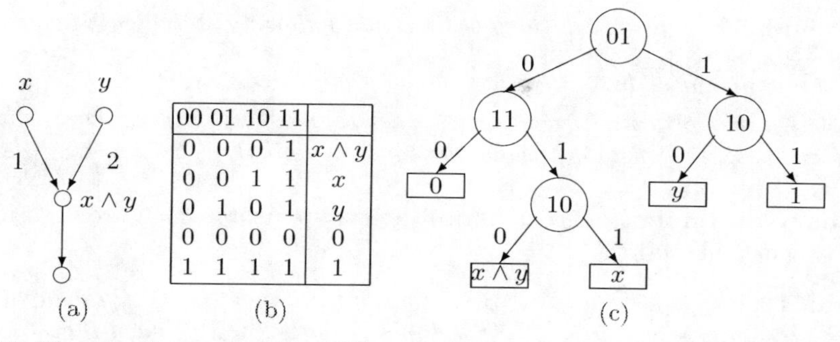

00	01	10	11	
0	0	0	1	$x \wedge y$
0	0	1	1	x
0	1	0	1	y
0	0	0	0	0
1	1	1	1	1

Fig. 11. The circuit, the decision table and the decision tree for this circuit diagnosis problem.

Arbitrary Circuits. The first line of investigation comprises the study of the complexity of fault diagnosis algorithms for arbitrary circuits in the basis B. Let us consider for this purpose the function $h_B^{(1)}$ which characterizes the worst-case dependency of $h(S)$ on $L(S)$ on the set $\mathrm{Circ}(B)$ of circuits. The function $h_B^{(1)}$ will be defined in the following way:

$$h_B^{(1)}(n) = \max\{h(S) : S \in \mathrm{Circ}(B), L(S) \leq n\} \ .$$

The basis B will be called *primitive* if at least one of the following conditions holds:

a) every function from B is either a disjunction $x_1 \vee \ldots \vee x_n$ or a constant;
b) every function from B is either a conjunction $x_1 \wedge \ldots \wedge x_n$ or a constant;
c) every function from B is either a linear function $x_1 + \ldots + x_n + c \pmod 2$, $c \in \{0, 1\}$, or a constant.

Theorem 10.1. *For any basis B the following statements hold:*

a) if B is a primitive basis then $h_B^{(1)}(n) = O(n)$;
b) b) if B is a non-primitive basis then $\log_2 h_B^{(1)}(n) = \Omega(n^{1/2})$.

Specially Constructed Circuits. As opposed to the first one, the second line of research explores complexity of diagnostic algorithms for circuits which are not arbitrary but chosen as the best from the point of view of solution of the diagnosis problem for the circuits, realizing the Boolean functions given as formulas over B. Let $\Phi(B)$ be the set of formulas over the basis B. For a formula $\varphi \in \Phi(B)$ we will denote by $L(\varphi)$ the number of functional symbols in φ. Let φ realize a function which does not belong to the set $\{0, 1\}$. Set $h(\varphi) = \min h(S)$, where the minimum is taken over all possible combinatorial circuits S (not necessarily in the basis B) which realize the same function as the formula φ. If φ realizes a function from the set $\{0, 1\}$ then $h(\varphi) = 0$. We will study the behavior of a function $h_B^{(2)}$ which characterizes the worst-case dependency of $h(\varphi)$ on $L(\varphi)$ on the set of formulas over B and is defined as follows:

$$h_B^{(2)}(n) = \max\{h(\varphi) : \varphi \in \Phi(B), L(\varphi) \leq n\} \ .$$

Theorem 10.2. *For an arbitrary basis B the following statements hold:*

a) if B is a primitive basis then $h_B^{(2)}(n) = O(n)$;
b) if B is a non-primitive basis then the equality $\log_2 h_B^{(2)}(n) = \Omega(n^c)$ holds for certain positive constant c which depends only on B.

Auxiliary Statements. Let us prove three statements which are used in this and following subsections.

Lemma 10.1. *Let B be a basis, S be a circuit in the basis B with n inputs and $m > 0$ gate inputs, $\bar{w}_0, \bar{w}_1, \ldots, \bar{w}_r$ be tuples from $\{0,1,2\}^m$ such that $f_{S(\bar{w}_0)} \neq f_{S(\bar{w}_i)}$ for $i = 1, \ldots, r$, and p be the minimal cardinality of a set of tuples from $\{0,1\}^n$ on which the function $f_{S(\bar{w}_0)}$ is distinct from all functions $f_{S(\bar{w}_1)}, \ldots, f_{S(\bar{w}_r)}$. Then $h(S) \geq p$.*

Proof. Let $T = T(z_S)$ be the decision table corresponding to the problem z_S and let $\bar{\gamma}$ be the tuple of values of the function $f_{S(\bar{w}_0)}$ on elements of the set $\{0,1\}^n$. One can show that $M_{\rho(S),h}(T, \bar{\gamma}) \geq p$ and hence $M_{\rho(S),h}(T) \geq p$. Using Theorem 3.1 obtain $h_{\rho(S)}(T) \geq p$. From Theorem 4.1 follows $h^l_{U(S)}(z_S) = h_{\rho(S)}(T)$. Since $h(S) = h^l_{U(S)}(z_S)$, we conclude $h(S) \geq p$. □

Next statement is a simple corollary of a result from [209].

Lemma 10.2. *Let B_1 and B_2 be bases such that $[B_1] \subseteq [B_2]$. Then there exist constants $c_1, c_2 \geq 1$ with the following property: for any formula φ_1 over B_1 there exists a formula φ_2 over B_2 which realizes the function equal to the function realized by the formula φ_1 and for which $L(\varphi_2) \leq c_1(L(\varphi_1))^{c_2}$.*

Proof. Let φ be a formula over some basis. Denote by $\Lambda(\varphi)$ the number of symbols of variables and constants in the formula φ. It has been proven in [209] that there exist constants $d_1, d_2 \geq 1$ such that for any formula φ_1 over B_1 there exists a formula φ_2 over B_2 which realizes a function equal to the function realized by the formula φ_1 and for which $\Lambda(\varphi_2) \leq d_1(\Lambda(\varphi_1))^{d_2}$.

Let φ_1 be a formula over B_1. Denote by φ_2 a formula over B_2, satisfying the following conditions:

(a) φ_2 realizes a function which is equal to the function realized by the formula φ_1;
(b) $\Lambda(\varphi_2) \leq d_1(\Lambda(\varphi_1))^{d_2}$;
(c) φ_2 has minimal number of functional symbols among all formulas satisfying the conditions (a) and (b).

The formula φ_2 can be represented in a natural way as a finite rooted directed tree D nodes of which are labelled by symbols of variables or functions. Denote by Λ_0 the number of nodes in D which are labelled by symbols of variables or constants. Let Λ_1 be the number of nodes in D which are labelled by symbols of functions of one variable. Denote by Λ_2 the number of nodes in D which are labelled by symbols of functions of two and more variables. It is clear that the number of terminal nodes in the tree D is equal to Λ_0. Using Lemma 3.12 one can show that $\Lambda_0 + \Lambda_2 \le 2\Lambda_0$ and hence $\Lambda_2 \le \Lambda_0$. Since the formula φ_2 satisfies the condition (c), the tree D does not contain three nodes consecutively linked by edges such that they are labelled by symbols of monadic functions. Using this fact one can show $\Lambda_1 \le 2(\Lambda_0 + \Lambda_2) \le 4\Lambda_0$. It is clear that $\Lambda(\varphi_2) = \Lambda_0$ and $L(\varphi_2) \le \Lambda_0 + \Lambda_1 + \Lambda_2$. Therefore $L(\varphi_2) \le 6\Lambda_0 = 6\Lambda(\varphi_2)$. Denote by p the maximal number of variables in functions from B_1. Evidently, $\Lambda(\varphi_1) \le (p+1)L(\varphi_1)$. Taking into account that φ_2 satisfies the condition (b) obtain $L(\varphi_2) \le 6\Lambda(\varphi_2) \le 6d_1(\Lambda(\varphi_1))^{d_2} \le 6d_1((p+1)L(\varphi_1))^{d_2} = 6d_1(p+1)^{d_2}(L(\varphi_1))^{d_2}$. Set $c_1 = 6d_1(p+1)^{d_2}$ and $c_2 = d_2$. Then $L(\varphi_2) \le c_1(L(\varphi_1))^{c_2}$. Since φ_1 is an arbitrary formula over B_1, the statement of the lemma holds. □

Lemma 10.3. *Let B be a non-primitive basis. Then there exist the functions $\psi_1(x,y,z)$ and $\psi_2(x,y,z)$ in $[B]$ such that $\psi_1(x,y,1) = x \vee y$ and $\psi_2(x,y,0) = x \wedge y$.*

Proof. From the results contained in appendix follows that at least one of the relations $F_2^\infty \subseteq [B]$, $D_2 \subseteq [B]$ and $F_6^\infty \subseteq [B]$ holds where $F_2^\infty = [\{x \vee (y \wedge z)\}]$, $D_2 = [\{(x \wedge y) \vee (x \wedge z) \vee (y \wedge z)\}]$ and $F_6^\infty = [\{x \wedge (y \vee z)\}]$. If $F_2^\infty \subseteq [B]$ then we can take functions $x \vee (y \wedge z)$ and $z \vee (x \wedge y)$ in the capacity of the functions ψ_1 and ψ_2. If $D_2 \subseteq [B]$ then in the capacity of the functions ψ_1 and ψ_2 we can take one and the same function $(x \wedge y) \vee (x \wedge z) \vee (y \wedge z)$. If $F_6^\infty \subseteq [B]$ then we can take the functions $z \wedge (x \vee y)$ and $x \wedge (y \vee z)$ in the capacity of the functions ψ_1 and ψ_2. □

Proofs of Theorems 10.1 and 10.2. The following statement characterizes relationships between the functions $h_B^{(1)}$ and $h_B^{(2)}$.

Lemma 10.4. *Let B be a basis. Then for any $n \in \mathbb{N} \setminus \{0\}$ the values of $h_B^{(1)}(n)$ and $h_B^{(2)}(n)$ are definite and the inequality $h_B^{(2)}(n) \le h_B^{(1)}(n)$ holds.*

Proof. Let $n \in \mathbb{N} \setminus \{0\}$. It is clear that $\{S : S \in \mathrm{Circ}(B), L(S) \le n\}$ is a nonempty set. Denote by p the maximal number of variables in functions from B. Let S be a circuit in the basis B such that $L(S) \le n$. Assume that S is a nondegenerate circuit. It is clear that the number of inputs of the circuit S which are connected by edges with gates is at most pn. Using this fact one can show that $h(S) \le 2^{pn}$. If S is a degenerate circuit then $h(S) = 0 < 2^{pn}$. Taking into account that S is an arbitrary circuit in the basis B with $L(S) \le n$ we obtain that the value $h_B^{(1)}(n)$ is definite.

It is clear that $\{\varphi : \varphi \in \Phi(B), L(\varphi) \le n\}$ is a nonempty set. Let $\varphi \in \Phi(B)$ and $L(\varphi) \le n$. If the formula φ realizes a function belonging to the set $\{0,1\}$ then $h(\varphi) = 0 \le h_B^{(1)}(n)$. Let φ realize a function which does not belong to $\{0,1\}$. It is not difficult to construct a circuit S in the basis B which realizes the same function as the formula φ and for which $L(S) = L(\varphi) \le n$. It is clear that $h(\varphi) \le h(S) \le h_B^{(1)}(n)$. Since φ is an arbitrary formula over B for which $L(\varphi) \le n$, we conclude that the value $h_B^{(2)}(n)$ is definite and that the inequality $h_B^{(2)}(n) \le h_B^{(1)}(n)$ holds. □

Let us study the behavior of the function $h_B^{(1)}$ for an arbitrary primitive basis B.

Lemma 10.5. *Let B be a primitive basis. Then $h_B^{(1)}(n) = O(n)$.*

Proof. Since B is a primitive basis, at least one of the relations $[B] \subseteq S_6$, $[B] \subseteq P_6$, $[B] \subseteq L_1$ holds where $S_6 = [\{x \vee y, 0, 1\}]$, $P_6 = [\{x \wedge y, 0, 1\}]$ and $L_1 = [\{x + y (\mathrm{mod}\ 2), 1\}]$.

Consider the case $[B] \subseteq L_1$. Let $n \in \mathbb{N} \setminus \{0\}$ and let S be a circuit from $\mathrm{Circ}(B)$ with $L(S) \leq n$. Assume that S is a nondegenerate circuit. Let the circuit S have exactly r inputs labelled by variables $x_1, \ldots, x_r$ respectively. Denote by m the number of gate inputs in the circuit S. Let exactly t inputs of the circuit S be linked by edges to gates, and let these inputs be labelled by variables $x_{i_1}, \ldots, x_{i_t}$ (possibly, $t = 0$). One can show that any circuit S' from $\mathrm{Circ}(S, \{0, 1, 2\}^m)$ realizes a function of the kind $(d_1 \wedge x_1) + \ldots + (d_r \wedge x_r) + d_0 (\mathrm{mod}\ 2)$ where $d_j \in \{0, 1\}$, $0 \leq j \leq r$. It is clear that $d_j = 0$ for any $j \in \{1, \ldots, r\} \setminus \{i_1, \ldots, i_t\}$.

Let us describe the work of a decision tree solving the problem z_S which is the diagnosis problem for the circuit S relative to the faults from the set $\{0, 1, 2\}^m$. Let $S' \in \mathrm{Circ}(S, \{0, 1, 2\}^m)$. Give on the inputs of the circuit S' the tuple consisting of zeros. We obtain on the output of the circuit S' the value d_0. For each $j \in \{1, \ldots, t\}$ give some tuple on inputs of the circuit S'. Let $j \in \{1, \ldots, t\}$. Give the unity on the input of the circuit S' labelled by the variable x_{i_j}, and give zeros on the other inputs of the circuit. We obtain value $d_{i_j} + d_0 (\mathrm{mod}\ 2)$ on the output of the circuit. Thus, after the giving on the inputs of the circuit S' of the considered $t + 1$ tuples the coefficients $d_1, \ldots, d_r, d_0$ of the formula $(d_1 \wedge x_1) + \ldots + (d_r \wedge x_r) + d_0 (\mathrm{mod}\ 2)$ will be recognized. Hence the considered decision tree solves the problem z_S, and the depth of this decision tree is at most $t + 1$. Therefore $h(S) \leq t + 1$. Denote by p the maximal number of variables in functions from B. It is clear that $t \leq pn$. Set $c_1 = p + 1$. Then $h(S) \leq c_1 n$. If S is a degenerate circuit then $h(S) = 0 < c_1 n$. Taking into account that S is an arbitrary circuit in the basis B with $L(S) \leq n$ we obtain $h_B^{(1)}(n) \leq c_1 n$. Therefore $h_B^{(1)}(n) = O(n)$.

The cases $[B] \subseteq S_6$ and $[B] \subseteq P_6$ can be considered in the same way. □

Let us study the behavior of the functions $h_B^{(1)}$ and $h_B^{(2)}$ for a non-primitive basis B.

Lemma 10.6. *Let B be a non-primitive basis. Then following statements hold:*

a) $\log_2 h_B^{(1)} = \Omega(n^{1/2})$;
b) there exists a constant $c \geq 0$ such that $\log_2 h_B^{(2)} = \Omega(n^c)$.

Proof. Using Lemma 10.3 obtain that there exist the functions $\psi_1(x, y, z)$ and $\psi_2(x, y, z) \in [B]$ such that $\psi_1(x, y, 1) = x \vee y$ and $\psi_2(x, y, 0) = x \wedge y$. Denote $B_1 = \{\psi_1, \psi_2\}$. Let $r \in \mathbb{N} \setminus \{0\}$. It is easily to show that there exists a formula φ_r over B_1 with the following properties: φ_r realizes the function

$$g_r = g_r(x_1, \ldots, x_r, y_1, \ldots, y_r, z_1, \ldots, z_r, t_1, \ldots, t_r, u_0, u_1)$$

such that $g_r(x_1,\ldots,x_r,y_1,\ldots,y_r,z_1,\ldots,z_r,t_1,\ldots,t_r,0,1) = \bigvee_{j=1}^{r}((x_1 \vee y_1) \wedge \ldots \wedge (x_{j-1} \vee y_{j-1}) \wedge (x_j \wedge y_j) \wedge (x_{j+1} \vee y_{j+1}) \wedge \ldots \wedge (x_r \vee y_r)) \vee \bigwedge_{i=1}^{r}((x_i \wedge z_i) \vee (y_i \wedge t_i))$ and the inequality $L(\varphi_r) \leq 6r^2$ holds.

Let S be a circuit in an arbitrary basis realizing a function equal to the function g_r. Let us show that $h(S) \geq 2^r$. Let the circuit S have exactly n inputs. One can show that the variables $x_1,\ldots,x_r,y_1,\ldots,y_r,z_1,\ldots,z_r,t_1,\ldots,t_r$ are essential variables of the function g_r. Therefore the circuit S has inputs labelled by these variables. If the variable u_i, $i \in \{0,1\}$, is an essential variable of the function g_r then S has an input labelled by this variable. Thus, the circuit S realizes a function $f_S(x_1,\ldots,x_r,y_1,\ldots,y_r,z_1,\ldots,z_r,t_1,\ldots,t_r,\ldots)$ depending on $m \geq 4r$ variables and equal to the function g_r. Let $\bar{\sigma} = (\sigma_1,\ldots,\sigma_{2r}) \in \{0,1\}^{2r}$. Define a tuple $w(\bar{\sigma})$ of constant faults on inputs of the circuit S. For $i = 1,\ldots,r$ the constant σ_i is realized on all gate inputs connected to the input z_i of the circuit S. For $i = 1,\ldots,r$ the constant σ_{r+i} is realized on all gate inputs connected to the input t_i of the circuit S. For $i = 0,1$ the constant i is realized on all gate inputs connected to the input u_i of the circuit S (if such input exists). The other gate inputs in the circuit S operate correctly. Denote by $f_{\bar{\sigma}}$ the function realized by the circuit $S(w(\bar{\sigma}))$. It is clear that the function $f_{\bar{\sigma}}$ is obtained from the function f_S by substitution of the constants $\sigma_1,\ldots,\sigma_{2r}$ for the variables $z_1,\ldots,z_r,t_1,\ldots,t_r$ and of the constant i for the variable u_i (if S contains the input u_i), $i = 0,1$. Denote by Σ the set of all possible tuples $\bar{\sigma} = (\sigma_1,\ldots,\sigma_{2r}) \in \{0,1\}^{2r}$ such that $\sigma_i + \sigma_{r+i} = 1$ for $i = 1,\ldots,r$. Denote by $\bar{0}$ the tuple $(0,\ldots,0)$ from $\{0,1\}^{2r}$. Let $\bar{\sigma} = (\sigma_1,\ldots,\sigma_{2r}) \in \Sigma$. One can show that the functions $f_{\bar{0}}$ and $f_{\bar{\sigma}}$ differ on a tuple $(\delta_1,\ldots,\delta_m) \in \{0,1\}^m$ if and only if the equalities $\delta_1 = \sigma_1,\ldots,\delta_{2r} = \sigma_{2r}$ hold. It is clear that $|\Sigma| = 2^r$. Hence the minimal cardinality of the set of tuples from $\{0,1\}^m$ on which the function $f_{\bar{0}}$ differs from all functions $f_{\bar{\sigma}}$, $\bar{\sigma} \in \Sigma$, is equal to 2^r. Using Lemma 10.1 obtain $h(S) \geq 2^r$.

a) Using the relation $B_1 \subset [B]$ one can show that there exists a constant $d_1 \geq 1$ with the following property: for any $r \in \mathbb{N} \setminus \{0\}$ there exists a circuit S_r in the basis B which realizes the function g_r and for which $L(S_r) \leq d_1 r^2$. Set $c_1 = 4d_1$. Let $n \in \mathbb{N}$ and $n \geq c_1$. Denote $r = \lfloor (n/d_1)^{1/2} \rfloor$. One can show that $L(S_r) \leq d_1 r^2 \leq n$ and $h(S_r) \geq 2^r = 2^{\lfloor (n/d_1)^{1/2} \rfloor} \geq 2^{n^{1/2}/c_1}$. Hence $h_B^{(1)}(n) \geq 2^{n^{1/2}/c_1}$ and $\log_2 h_B^{(1)} = \Omega(n^{1/2})$.

b) b) It is clear that $[B_1] \subseteq [B]$. Using Lemma 10.2 we conclude that there exist constants $d_2, d_3 \geq 1$ with the following property: for any $r \in \mathbb{N} \setminus \{0\}$ there exists a formula π_r over B which realizes a function equal to the function g_r and for which $L(\pi_r) \leq d_2 r^{d_3}$. Set $c_2 = 2^{d_3} d_2$ and $c = 1/d_3$. Let $n \in \mathbb{N}$ and $n \geq c_2$. Denote $r = \lfloor (n/d_2)^{1/d_3} \rfloor$. One can show that $L(\pi_r) \leq d_2 r^{d_3} \leq n$ and $h(\pi_r) \geq 2^r = 2^{\lfloor (n/d_2)^{1/d_3} \rfloor} \geq 2^{n^c/c_2}$. Hence $h_B^{(2)}(n) \geq 2^{n^c/c_2}$ and $\log_2 h_B^{(1)} = \Omega(n^c)$. □

Proof (of Theorem 10.1). The statement of the theorem follows from Lemmas 10.5 and 10.6. □

Proof (of Theorem 10.2). The statement of the theorem follows from Lemmas 10.4, 10.5 and 10.6. □

10.3 Complexity of Construction of Algorithms for Diagnosis

The third line of research is to study the complexity of algorithms for construction of decision trees for diagnosis problem.

A basis B will be called *degenerate* if $B \subseteq \{0,1\}$, and *nondegenerate* otherwise. Let B be a nondegenerate basis. Define an algorithmic problem $\mathrm{Con}(B)$.

The problem $\mathrm{Con}(B)$: for a given circuit S from $\mathrm{Circ}(B)$ and a given set W of tuples of constant faults on inputs of gates of the circuit S it is required to construct a decision tree which solves the diagnosis problem for the circuit S relative to the faults from W.

Note that there exists a decision tree which solves the diagnosis problem for the circuit S relative to the faults from W and the number of nodes in which is at most $2|W| - 1$.

Theorem 10.3. *Let B be a nondegenerate basis. Then the following statements hold:*

a) if B is a primitive basis then there exists an algorithm which solves the problem $\mathrm{Con}(B)$ *with polynomial time complexity;*
b) if B is a non-primitive basis then the problem $\mathrm{Con}(B)$ *is NP-hard.*

Proof of Theorem 10.3. Study the complexity of the problem $\mathrm{Con}(B)$ solving for a nondegenerate primitive basis B.

Lemma 10.7. *Let B be a nondegenerate primitive basis. Then there exists an algorithm with polynomial time complexity which solves the problem* $\mathrm{Con}(B)$.

Proof. Since the basis B is primitive, at least one of the relations $[B] \subseteq S_6$, $[B] \subseteq P_6$, $[B] \subseteq L_1$ holds.

Consider the case $[B] \subseteq S_6$. Let S be a circuit in the basis B with n inputs labelled by variables $x_1, \ldots, x_n$, and let W be a set of tuples of constant faults on gate inputs of the circuit S. One can show that any circuit $S' \in \mathrm{Circ}(S, W)$ realizes a function of the kind

$$(c_1 \wedge x_1) \vee \ldots \vee (c_n \wedge x_n) \vee c_0 \qquad (140)$$

where $c_0, c_1, \ldots, c_n \in \{0,1\}$. Denote by $\bar{\delta}_0$ the n-tuple $(0, \ldots, 0)$. For $i = 1, \ldots, n$ denote by $\bar{\delta}_i$ the n-tuple $(0, \ldots, 0, 1, 0, \ldots, 0)$ in which 1 is in the i-th digit. One can show that for the recognition of the coefficients $c_0, c_1, \ldots, c_n$ in the formula (140) which realizes the function $f_{S'}$ it is sufficient to know the values of the function $f_{S'}$ on n-tuples $\bar{\delta}_0, \bar{\delta}_1, \ldots, \bar{\delta}_n$. In fact, $f_{S'}(\bar{\delta}_0) = c_0$. If $c_0 = 1$ then $f_{S'} \equiv 1$ and we may set $c_1 = \ldots = c_n = 1$. If $c_0 = 0$ then $f_{S'}(\bar{\delta}_i) = c_i$ for $i = 1, \ldots, n$.

Let us show that there exists an algorithm which solves the problem $\mathrm{Con}(B)$ with polynomial time complexity. Let us describe the work of this algorithm for a circuit S and a set W of tuples of faults. Define a problem z over the information system $U = U(S, W)$ in the following way: $z = (\nu, \bar{\delta}_0, \bar{\delta}_1, \ldots, \bar{\delta}_n)$ where $\nu : \{0,1\}^{n+1} \to \mathbb{N}$ and for any $\bar{\alpha} \in \{0,1\}^{n+1}$ the value of $\nu(\bar{\alpha})$ is a number from $\mathbb{N}$, the binary representation of which is the n-tuple $\bar{\alpha}$. Denote

$T = T_U(z)$. By simulation of work of the circuit $S(\bar{w})$ on the n-tuple $\bar{\delta}_i$ for any $\bar{w} \in W$ and $i \in \{0, 1, \ldots, n\}$ we can construct the table T in polynomial time. Using the algorithm $\mathcal{V}_{\rho(S),h}$, which has polynomial time complexity, we can construct the decision tree $\mathcal{V}_{\rho(S),h}(T)$ for the table T. (The description of the algorithm $\mathcal{V}_{\rho(S),h}$ can be found in Sect. 3.4.) Each terminal node of the tree $\mathcal{V}_{\rho(S),h}(T)$ is labelled by the number $\nu(\bar{\alpha})$ of an n-tuple $\bar{\alpha} \in \{0,1\}^{n+1}$. Using this number we can construct in polynomial time a formula of the kind (140) which realizes a function f such that $(f(\bar{\delta}_0), f(\bar{\delta}_1), \ldots, f(\bar{\delta}_n)) = \bar{\alpha}$. For each terminal node of the tree $\mathcal{V}_{\rho(S),h}(T)$ replace the number attached to it by the corresponding formula. Denote by Γ the obtained decision tree. Using Theorem 4.1 one can show that the decision tree Γ solves the problem $z_{S,W}$.

The cases $[B] \subseteq P_6$ and $[B] \subseteq L_1$ can be considered in the same way. □

Study the complexity of the problem Con(B) solving for a nondegenerate basis B which is not primitive.

Lemma 10.8. *Let B be a nondegenerate basis which is not primitive. Then the problem* Con(B) *is NP-hard.*

Proof. Using Lemma 10.3 one can show that there exist circuits S_1 and S_2 in the basis B which realizes functions $\psi_1(x, y, z)$ and $\psi_2(x, y, z)$ such that $\psi_1(x, y, 1) = x \vee y$ and $\psi_2(x, y, 0) = x \wedge y$.

Assume that there exists an algorithm which solves the problem Con(B) and has polynomial time complexity. Let us show that in this case there exists an algorithm which has polynomial time complexity and solves the vertex cover problem which is NP-complete (see Sect. 3.6). In this problem for a given undirected graph G without loops and multiple edges and a number $m \in \mathbb{N}$ it is required to establish whether the inequality cv$(G) \leq m$ holds where cv(G) is the minimal cardinality of vertex cover of the graph G. Let $G = (V, R)$ where $V = \{v_1, \ldots, v_n\}$ is the set of vertices of the graph G and $R = \{\{v_{i_1}, v_{j_1}\}, \ldots, \{v_{i_t}, v_{j_t}\}\}$ is the set of edges of the graph G. If $m \geq n$ then the inequality cv$(G) \leq m$ holds. If $m = 0$ then the inequality cv$(G) \leq m$ holds if and only if $R = \emptyset$. Assume $0 < m < n$. Consider the Boolean function $\psi_G(x_1, \ldots, x_n) = (x_{i_1} \vee x_{j_1}) \wedge \ldots \wedge (x_{i_t} \vee x_{j_t})$. It is clear that inequality cv$(G) \leq m$ holds if and only if the function ψ_G has the value 1 on an n-tuple from $\{0,1\}^n$ containing at most m numbers 1. Define a function $\psi_n^{m+1} : \{0,1\}^n \to \{0,1\}$ in the following way: $\psi_n^{m+1}(x_1, \ldots, x_n) = 1$ if and only if $\sum_{i=1}^{n} x_i \geq m + 1$.

Using the circuits S_1 and S_2 it is easily to construct in polynomial time a circuit S in the basis B with n inputs, and two tuples $\bar{w}_1$, $\bar{w}_2$ of constant faults on gate inputs of the circuit S such that the circuit $S(\bar{w}_1)$ realizes the function $\psi_n^{m+1}(x_1, \ldots, x_n)$ and the circuit $S(\bar{w}_2)$ realizes the function $\psi_G(x_1, \ldots, x_n) \vee \psi_n^{m+1}(x_1, \ldots, x_n)$. It is clear that $f_{S(\bar{w}_1)} \neq f_{S(\bar{w}_2)}$ if and only if the inequality cv$(G) \leq m$ holds. Denote $W = \{\bar{w}_1, \bar{w}_2, (2, \ldots, 2)\}$. Apply the algorithm, which solves the problem Con(B) and has polynomial time complexity, to the circuit S and to the set W. As a result we obtain a decision tree Γ over $U(S, W))$ which solves the problem $z_{S,W}$. Construct the circuits $S(\bar{w}_1)$ and $S(\bar{w}_2)$ and apply the decision tree Γ to them. The results of the decision tree Γ work on circuits $S(\bar{w}_1)$

and $S(\bar{w}_2)$ will be different if and only if the inequality $\mathrm{cv}(G) \leq m$ holds. Note that the processes of construction of circuits $S(\bar{w}_1)$ and $S(\bar{w}_2)$ and of applying to them the decision tree Γ can be carried out in polynomial time.

Thus, if an algorithm exists which solves the problem Con(B) with polynomial time complexity then an algorithm exists which solves the vertex cover problem in polynomial time. Hence the problem Con(B) is NP-hard. □

Proof (of Theorem 10.3). The statement of the theorem follows from Lemmas 10.7 and 10.8. □

10.4 Diagnosis of Iteration-Free Circuits

From the point of view of the solution of the diagnosis problem for arbitrary tuples of constant faults on inputs of gates of arbitrary circuits only primitive bases seem to be admissible. The extension of the set of such bases is possible by the substantial restriction on the class of the circuits under consideration. The fourth line of research is the study of the complexity of fault diagnosis algorithms for iteration-free circuits.

Bounds on Complexity of Algorithms for Diagnosis. Let B be a basis. A circuit in the basis B is called *iteration-free* if each node (input or gate) of it has at most one issuing edge. Let us denote by $\mathrm{Circ}^1(B)$ the set of iteration-free circuits in the basis B with only one output. Let us consider the function $h_B^{(3)}$ which characterizes the worst-case dependency of $h(S)$ on $L(S)$ for circuits from $\mathrm{Circ}^1(B)$ and is defined in the following way:

$$h_B^{(3)}(n) = \max\{h(S) : S \in \mathrm{Circ}^1(B), L(S) \leq n\} .$$

Let us call a Boolean function $f(x_1, \ldots, x_n)$ *quasimonotone* if there exist numbers $\sigma_1, \ldots, \sigma_n \in \{0, 1\}$ and a monotone Boolean function $g(x_1, \ldots, x_n)$ such that $f(x_1, \ldots, x_n) = g(x_1^{\sigma_1}, \ldots, x_n^{\sigma_n})$ where $x^\sigma = x$ if $\sigma = 1$, and $x^\sigma = \neg x$ if $\sigma = 0$.

The basis B will be called *quasiprimitive* if at least one of the following conditions is true:

a) all functions from B are linear functions or constants;
b) all functions from B are quasimonotone functions.

The class of the quasiprimitive bases is rather large: we will show (see Proposition 10.1) that for any basis B_1 there exists a quasiprimitive basis B_2 such that $[B_1] = [B_2]$.

Theorem 10.4. *Let B be a basis. Then the following statements holds:*

a) if B is a quasiprimitive basis then $h_B^{(3)}(n) = O(n)$;
b) if B is not a quasiprimitive basis then $\log_2 h_B^{(3)}(n) = \Omega(n)$.

Quasiprimitive Bases. The following statement characterizes the class of quasiprimitive bases.

Proposition 10.1. *For any basis B_1 there exists a quasiprimitive basis B_2 such that $[B_1] = [B_2]$.*

Proof. In appendix for each closed class U of Boolean functions a basis B_U is considered such that $U = [B_U]$. One can show that for all classes U, with the exception of the classes C_2, C_3 and C_4, the basis B_U is quasiprimitive. Denote $B'_{C_2} = \{x \wedge y, x \vee \neg y\}$, $B'_{C_3} = \{x \vee y, x \wedge \neg y\}$ and $B'_{C_4} = \{x \vee y, x \wedge y, x \vee (y \wedge \neg z)\}$. From Lemma 18 of [217] follows $C_2 = [B'_{C_2}]$ and $C_3 = [B'_{C_3}]$. From Lemma 17 of [217] follows that the equality $C_4 = [\{x \vee y, x \wedge y, g\}]$ holds for any nonmonotone α-function g. One can show that the function $x \vee (y \wedge \neg z)$ is a nonmonotone α-function. Hence $C_4 = [B'_{C_4}]$. Evidently, bases B'_{C_2}, B'_{C_3} and B'_{C_4} are quasiprimitive. Thus, for any closed class U of Boolean functions there exists a quasiprimitive basis B such that $U = [B]$.

Let B_1 be a basis. Denote $U = [B_1]$. By proved above, for the closed class U there exists a quasiprimitive basis B_2 such that $U = [B_2]$. □

The following statement characterizes iteration-free circuits in a basis which consists of quasimonotone functions.

Lemma 10.9. *Let B be a nondegenerate basis consisting of quasimonotone functions, and let S be an iteration-free circuit in the basis B with n inputs and $m > 0$ gate inputs. Then there exist tuples $\bar{\sigma}_0$ and $\bar{\sigma}_1$ from $\{0,1\}^n$ such that for any circuit $S' \in \mathrm{Circ}(S, \{0,1,2\}^m)$ if $f_{S'} \not\equiv \mathrm{const}$ then $f_{S'}(\bar{\sigma}_0) = 0$ and $f_{S'}(\bar{\sigma}_1) = 1$.*

Proof. We prove the statement of the lemma by induction on the parameter $t = L(S)$. Consider an arbitrary nondegenerate circuit $S \in \mathrm{Circ}^1(B)$ for which $L(S) = 1$. Let the circuit S have n inputs labelled by variables $x_1, \ldots, x_n$. Let the only gate v in the circuit S be labelled by the function g of m variables. Assume, for the definiteness, that the inputs of the gate v are connected to the inputs of the circuit S labelled by the variables $x_1, \ldots, x_m$. Since g is a quasimonotone function, there exists a monotone function q of m variables and the numbers $\delta_1, \ldots, \delta_m \in \{0,1\}$ such that $g(x_1, \ldots, x_m) = q(x_1^{\delta_1}, \ldots, x_m^{\delta_m})$. Define two tuples $\bar{\sigma}_0$ and $\bar{\sigma}_1$ from $\{0,1\}^n$ in the following way: $\bar{\sigma}_0 = (\neg\delta_1, \ldots, \neg\delta_m, 0, \ldots, 0)$ and $\bar{\sigma}_1 = (\delta_1, \ldots, \delta_m, 0, \ldots, 0)$. Let $S' \in \mathrm{Circ}(S, \{0,1,2\}^m)$. It is clear that a function equal to the function $f_{S'}$ can be obtained from the function $g(x_1, \ldots, x_m)$ by substitution of constants from $\{0,1\}$ for some variables. Taking into account that q is a monotone function one can show that either $f_{S'} \equiv \mathrm{const}$ or $f_{S'}(\bar{\sigma}_0) = 0$ and $f_{S'}(\bar{\sigma}_1) = 1$. Therefore the statement of the lemma holds if $t = 1$.

Assume that the statement of lemma is true for some $t \geq 1$. Show that this statement also holds for $t + 1$. Let $S \in \mathrm{Circ}^1(B)$, $L(S) = t + 1$, the circuit S have $m > 0$ gate inputs, and the output of the circuit S be linked by an edge to a gate v labelled by a function g. If $g \in \{0,1\}$ then $f_{S'} \equiv \mathrm{const}$ for any $S' \in \mathrm{Circ}(S, \{0,1,2\}^m)$ and the statement of the lemma holds for the circuit S. Let $g \notin \{0,1\}$ and edges $d_1, \ldots, d$, which issue from nodes $v_1, \ldots, v_r$, enter the node v. For $i = 1, \ldots, r$ let S_i be a subcircuit of S consisting of nodes and edges such that each of them is contained in a directed path terminating in v_i. Let

$i \in \{1,\ldots,r\}$. If the subcircuit S_i does not contain inputs of S then a constant is realized on the i-th input of the gate v for any circuit $S' \in \mathrm{Circ}(S, \{0,1,2\}^m)$. If the subcircuit S_i does not contain gates then it consists of one input of the circuit S. Let the subcircuit S_i contain at least one input of the circuit S and at least one gate. Add to the subcircuit S_i a node u and an edge d issuing from the node v_i and entering the node u. Denote the obtained graph by G_i. It is clear that G_i is a nondegenerate circuit from $\mathrm{Circ}^1(B)$, and $L(G_i) \leq t$. By the inductive hypothesis, for the circuit G_i there exist tuples $\bar{\sigma}_0^i$ and $\bar{\sigma}_1^i$, satisfying the statement of the lemma. Since g is a quasimonotone function, there exist a monotone function $q(x_1,\ldots,x_r)$ and numbers $\delta_1,\ldots,\delta_r \in \{0,1\}$ such that $g(x_1,\ldots,x_r) = q(x_1^{\delta_1},\ldots,x_r^{\delta_r})$. Define a tuple $\bar{\sigma}_0$ of values of inputs of S. Let $i \in \{1,\ldots,r\}$. If the subcircuit S_i consists of the only input of the circuit S then we give on this input the number $\neg\delta_i$. Let the subcircuit S_i contain at least one input of the circuit S and at least one gate. Then we give on the inputs of the circuit S, included into the subcircuit S_i, the tuple $\bar{\sigma}_0^i$ if $\delta_i = 1$, and the tuple $\bar{\sigma}_1^i$ if $\delta_i = 0$. We give zeros on the inputs of the circuit S not belonging to any subcircuit S_i, $i \in \{1,\ldots,r\}$. Define a tuple $\bar{\sigma}_1$ of values of inputs of the circuit S. Let $i \in \{1,\ldots,r\}$. If the subcircuit S_i consists of the only input of the circuit S then we give on this input the number δ_i. Let the subcircuit S_i contain at least one input of the circuit S and at least one gate. Then we give on the inputs of the circuit S, included into the subcircuit S_i, the tuple $\bar{\sigma}_0^i$ if $\delta_i = 0$, and the tuple $\bar{\sigma}_1^i$ if $\delta_i = 1$. We give zeros on the inputs of the circuit S not belonging to any subcircuit S_i, $i \in \{1,\ldots,r\}$. Let $S' \in \mathrm{Circ}(S, \{0,1,2\}^m)$ and let, for the definiteness, for any $i \in \{1,\ldots,p\}$ on the i-th input of the gate v the function equal to the constant γ_i is realized, while for any $i \in \{p+1,\ldots,r\}$ on the i-th input of the gate v a function is realized which is not constant. Then $f_{S'}(\bar{\sigma}_0) = q(\gamma_1^{\delta_1},\ldots,\gamma_p^{\delta_p},0,\ldots,0)$ and $f_{S'}(\bar{\sigma}_1) = q(\gamma_1^{\delta_1},\ldots,\gamma_p^{\delta_p},1,\ldots,1)$. Since q is a monotone function, either $f_{S'} \equiv \mathrm{const}$, or $f_{S'}(\bar{\sigma}_0) = 0$ and $f_{S'}(\bar{\sigma}_1) = 1$. Hence the statement of the lemma also holds for $t+1$. □

Study the behavior of the function $h_B^{(3)}$ for a quasiprimitive basis B.

Lemma 10.10. *Let B be a quasiprimitive basis. Then $h_B^{(3)}(n) = O(n)$.*

Proof. Let all functions from B be linear. Then the basis B is primitive. Using Theorem 10.1 we conclude that $h_B^{(1)}(n) = O(n)$. It is clear that $h_B^{(3)}(n) \leq h_B^{(1)}(n)$ for any natural n. Hence $h_B^{(3)}(n) = O(n)$.

Let all functions from B be quasimonotone. If B is a degenerate basis then all functions from B are linear, and for the basis B the statement of the lemma is true. Let us assume now that the basis B is nondegenerate. Let g be a function from B depending on $r > 0$ variables, and let S_g be an iteration-free circuit in the basis B with r inputs labelled by variables $x_1,\ldots,x_r$ and with one gate labelled by the function g. For $i = 1,\ldots,r$ let the i-th gate input be linked to the input of the circuit which is labelled by the variable x_i. One can show that there exists a decision tree Γ_g which solves the problem z_{S_g} and for which the result of the work of Γ for any circuit $S' \in \mathrm{Circ}(S_g, \{0,1,2\}^r)$ consists of

a) a formula $\varphi = g(\alpha_1, \ldots, \alpha_r)$ realizing a function equal to $f_{S'}$, where $\alpha_i = x_i$ if x_i is an essential variable of the function $f_{S'}$, and α_i is a constant from $\{0, 1\}$ if x_i is not an essential variable of the function $f_{S'}$;
b) for any essential variable x_i of the function $f_{S'}$ a tuple $(\beta_1, \ldots, \beta_{i-1}, \beta_{i+1}, \ldots, \beta_r) \in \{0,1\}^{r-1}$ such that $f_{S'}(\beta_1, \ldots, \beta_{i-1}, 0, \beta_{i+1}, \ldots, \beta_r) \neq f_{S'}(\beta_1, \ldots, \beta_{i-1}, 1, \beta_{i+1}, \ldots, \beta_r)$.

Let $c = \max\{h(\Gamma_g) : g \in B \setminus \{0, 1\}\}$. Set $c_1 = \max\{1, c\}$. By induction on the parameter $t = L(S)$ we will prove that for any nondegenerate iteration-free circuit S in B there exists a decision tree Γ which solves the problem z_S and for which $h(\Gamma) \leq c_1 L(S)$. If the circuit S has m gate inputs then for any circuit $S' \in \mathrm{Circ}(S, \{0, 1, 2\}^m)$ the result of Γ work is a formula realizing a function equal to the function $f_{S'}$. Let $S \in \mathrm{Circ}^1(B)$, $L(S) = 1$, the only gate of the circuit S be labelled by a function g of r variables, and the i-th gate input be linked to the input of the circuit S which is labelled by a variable x_{j_i}, $i = 1, \ldots, r$. For the problem z_S solving we will use the decision tree Γ_g modified in the following way: we give tuples, generated by the decision tree Γ_g, on the inputs of the circuit S linked to the gate while on the other inputs of the circuit S we give zeros. In each formula attached to a terminal node of the decision tree Γ_g the variables $x_1, \ldots, x_r$ are replaced by the variables $x_{j_1}, \ldots, x_{j_r}$ respectively. One can show that the modified in such a way decision tree Γ_g solves the problem z_S. The depth of this decision tree is at most c_1. Hence if $t = 1$ then the considered statement holds.

Assume that this statement holds for some $t \geq 1$. Let us show that it holds for $t + 1$. Let $S \in \mathrm{Circ}^1(B)$, $L(S) = t + 1$, the circuit S have $m > 0$ gate inputs and the output of the circuit S be linked by an edge to a gate v which is labelled by a function g. Let $g \in \{0, 1\}$. Then $f_{S'} \equiv g$ for any circuit $S' \in \mathrm{Circ}(S, \{0, 1, 2\}^m)$. Denote by Γ the decision tree consisting of only one node labelled by the formula g. It is clear that Γ solves the problem z_S and $h(\Gamma) = 0 < c_1 L(S)$. Therefore the considered statement holds for the circuit S. Assume now that $g \notin \{0, 1\}$. Let the edges $d_1, \ldots, d_r$, issuing from nodes $v_1, \ldots, v_r$ respectively, enter the gate v. For $i = 1, \ldots, r$ let S_i be a subcircuit of S consisting of those nodes and edges of the circuit S everyone of which is included into a directed path terminating in v_i. Let $i \in \{1, \ldots, r\}$. If the subcircuit S_i does not contain the inputs of the circuit S then a constant is realized on the i-th input of the gate v for any circuit $S' \in \mathrm{Circ}(S, \{0, 1, 2\}^m)$. If the subcircuit S_i does not contain gates, it consists of an only input of the circuit S. Let the subcircuit S_i contain at least one input of the circuit S and at least one gate. Add to the subcircuit S_i a node u and an edge d issuing from the node v_i and entering the node u. Denote by G_i the obtained graph. Evidently, G_i is a nondegenerate circuit from $\mathrm{Circ}^1(B)$ for which $L(G_i) \leq t$. Let G_i have m_i gate inputs. By inductive hypothesis, there exists a decision tree Γ_i which solves the problem z_{G_i} and for which $h(\Gamma_i) \leq c_1 L(G_i)$. Using Lemma 10.9 obtain that there exist input tuples $\bar{\sigma}_0^i$ and $\bar{\sigma}_1^i$ of the circuit G_i such that for any circuit $G' \in \mathrm{Circ}(G_i, \{0, 1, 2\}^{m_i})$ if $f_{G'} \not\equiv \mathrm{const}$ then $f_{G'}(\bar{\sigma}_0^i) = 0$ and $f_{G'}(\bar{\sigma}_1^i) = 1$.

Let us describe the work of a decision tree Γ which solves the problem z_S. Let $S' \in \mathrm{Circ}(S, \{0, 1, 2\}^m)$. First, the decision tree Γ simulates in some way the work of the decision tree Γ_g. The gate v is analyzed at that. Let the decision tree Γ_g generates the tuple $(\delta_1, \ldots, \delta_r)$. Let $i \in \{1, \ldots, r\}$. If the subcircuit S_i consists of an only input of the circuit S then we give the number δ_i on this input. Let the subcircuit S_i contain at least one input of the circuit S and at least one gate. Then we give on the inputs of the circuit S', belonging to the subcircuit S_i, the tuple $\bar{\sigma}_0^i$ if $\delta_i = 0$, and the tuple $\bar{\sigma}_1^i$ if $\delta_i = 1$. We give zeros on the inputs of the circuit S' not belonging to any subcircuit S_i, $i \in \{1, \ldots, r\}$. Let the result of the work of the decision tree Γ_g be a formula $g(\alpha_1, \ldots, \alpha_r)$. Let, for the definiteness, $\alpha_i = x_i$ for $i = 1, \ldots, q$ and α_i be a constant from $\{0, 1\}$ for $i = q + 1, \ldots, r$. Assume that for $i = 1, \ldots, p$ the subcircuit S_i contains at least one input of the circuit S and at least one gate, and for $i = p + 1, \ldots, q$ the subcircuit S_i consists of one input of the circuit S which is labelled by the variable x_{l_i}.

Later, the decision tree Γ consequently simulates the work of the decision trees $\Gamma_1, \ldots, \Gamma_p$. The subcircuits $S_1, \ldots, S_p$ are analyzed at that. Let $i \in \{1, \ldots, p\}$. Describe the work of the decision tree Γ when it simulates the decision tree Γ_i. Let $(\beta_1, \ldots, \beta_{i-1}, \beta_{i+1}, \ldots, \beta_r)$ be a tuple constructed by the decision tree Γ_g for an essential variable x_i of the function $g'(x_1, \ldots, x_r) = g(\alpha_1, \ldots, \alpha_r)$. For this tuple $g'(\beta_1, \ldots, \beta_{i-1}, 0, \beta_{i+1}, \ldots, \beta_r) \neq g'(\beta_1, \ldots, \beta_{i-1}, 1, \beta_{i+1}, \ldots, \beta_r)$. Let $j \in \{1, \ldots, r\} \setminus \{i\}$. If the subcircuit S_j consists of the only input of the circuit S then we give the number β_j on this input. Let the subcircuit S_j contain at least one input of the circuit S and at least one gate. On the inputs of the circuit S', belonging to the subcircuit S_j, we give the tuple $\bar{\sigma}_0^j$ if $\beta_j = 0$, and the tuple $\bar{\sigma}_1^j$ if $\beta_j = 1$. On the inputs of S', belonging to the subcircuit S_i, we give the tuples generated by the decision tree Γ_i. We give zeros on the inputs of the circuit S' not belonging to any subcircuit S_j, $j \in \{1, \ldots, r\}$. If $g'(\beta_1, \ldots, \beta_{i-1}, 0, \beta_{i+1}, \ldots, \beta_r) = 1$ then the value of the output of the circuit S' is inverted before it is used by the decision tree Γ_i. Let the result of the work of the decision tree Γ_i be a formula φ_i.

After the construction of formulas $\varphi_1, \ldots, \varphi_p$ in the process of the simulation by the decision tree Γ of the work of decision trees $\Gamma_1, \ldots, \Gamma_p$, the work of the decision tree Γ is over. The formula $g(\varphi_1, \ldots, \varphi_p, x_{l_{p+1}}, \ldots, x_{l_q}, \alpha_{q+1}, \ldots, \alpha_r)$ will be obtained as the result. One can show that this formula realizes a function equal to the function $f_{S'}$. Denote by I the set of all $i \in \{1, \ldots, r\}$ such that the subcircuit S_i contains at least one input of the circuit S and at least one gate. It is clear that $h(\Gamma) \leq h(\Gamma_g) + \sum_{i \in I} h(\Gamma_i) \leq c_1 + c_1 \sum_{i \in I} L(G_i) \leq c_1 L(S)$. Hence the considered statement holds also for $t + 1$.

Let $n \in \mathbb{N} \setminus \{0\}$, $S \in \mathrm{Circ}^1(B)$ and $L(S) \leq n$. If S is a degenerate circuit then $h(S) = 0 < c_1 n$. If S is a nondegenerate circuit then $h(S) \leq c_1 L(S) \leq c_1 n$. Hence $h_B^{(3)}(n) \leq c_1 n$ and $h_B^{(3)}(n) = O(n)$. □

Non-quasiprimitive Bases. In this subsubsection the behavior of the function $h_B^{(3)}$ for a non-quasiprimitive basis B is studied. The subsubsection contains also the proof of Theorem 10.4.

Let us prove some auxiliary statements.

Lemma 10.11. *Let B be a non-quasiprimitive basis. Then there exists iteration-free circuit $S_\neg$ in the basis B and tuples $\bar{w}_1$ and $\bar{w}_2$ of constant faults on gate inputs of the circuit $S_\neg$ such that circuit $S_\neg(\bar{w}_1)$ realizes a function equal to x, and the circuit $S_\neg(\bar{w}_2)$ realizes a function equal to $\neg x$.*

Proof. Since the basis B is non-quasiprimitive then there exists a function $g(x_1, \ldots, x_n) \in B$ which is not quasimonotone. Let us show that there exist a number $i \in \{1, \ldots, n\}$ and two tuples $(\alpha_1, \ldots, \alpha_{i-1}, \alpha_{i+1}, \ldots, \alpha_n)$ and $(\beta_1, \ldots, \beta_{i-1}, \beta_{i+1}, \ldots, \beta_n)$ from $\{0,1\}^{n-1}$ such that

$$\begin{aligned} g(\alpha_1, \ldots, \alpha_{i-1}, 0, \alpha_{i+1}, \ldots, \alpha_n) &= 0 \ , \\ g(\alpha_1, \ldots, \alpha_{i-1}, 1, \alpha_{i+1}, \ldots, \alpha_n) &= 1 \ , \\ g(\beta_1, \ldots, \beta_{i-1}, 0, \beta_{i+1}, \ldots, \beta_n) &= 1 \ , \\ g(\beta_1, \ldots, \beta_{i-1}, 1, \beta_{i+1}, \ldots, \beta_n) &= 0 \ . \end{aligned} \tag{141}$$

Assume the contrary. In this case for any $i \in \{1, \ldots, n\}$ there exists a number $\sigma_i \in \{0,1\}$ such that for any tuple $(\gamma_1, \ldots, \gamma_{i-1}, \gamma_{i+1}, \ldots, \gamma_n) \in \{0,1\}^{n-1}$ the inequality $g(\gamma_1, \ldots, \gamma_{i-1}, \sigma_i, \gamma_{i+1}, \ldots, \gamma_n) \geq g(\gamma_1, \ldots, \gamma_{i-1}, \neg\sigma_i, \gamma_{i+1}, \ldots, \gamma_n)$ holds. One can show that the function $g(x_1^{\sigma_1}, \ldots, x_n^{\sigma_n})$ is monotone. Therefore $g(x_1, \ldots, x_n)$ is a quasimonotone function which is impossible. Thus, for an $i \in \{1, \ldots, n\}$ and some tuples

$$(\alpha_1, \ldots, \alpha_{i-1}, \alpha_{i+1}, \ldots, \alpha_n), (\beta_1, \ldots, \beta_{i-1}, \beta_{i+1}, \ldots, \beta_n) \in \{0,1\}^{n-1}$$

the equalities (141) hold. Using these equalities one can construct an iteration-free circuit $S_\neg$ in the basis B, containing one gate labelled by the function g, and the tuples $\bar{w}_1$ and $\bar{w}_2$ of constant faults on inputs of this gate such that the circuit $S_\neg(\bar{w}_1)$ realizes a function equal to x, while the circuit $S_\neg(\bar{w}_2)$ realizes a function equal to $\neg x$. □

The proof of the following lemma is similar to the proof of Lemma 11 from [217].

Lemma 10.12. *Let B be a non-quasiprimitive basis. Then there exist an iteration-free circuit $S_\wedge$ in the basis B and a tuple $\bar{w}$ of constant faults on gate inputs of the circuit $S_\wedge$ such that the circuit $S_\wedge(\bar{w})$ realizes a function equal to the function $x \wedge y$.*

Proof. Since B is non-quasiprimitive, it contains a nonlinear function $q(x_1,\ldots,x_n)$. It is well known (see Theorem 6 with the consequent note in [217]) that the function q has the unique representation in the form of a Zhegalkin polynomial:

$$q(x_1, \ldots, x_n) = \sum_{\{i_1,\ldots,i_s\} \subseteq \{1,\ldots,n\}} a_{i_1\ldots i_s} \cdot x_{i_1} \cdot \ldots \cdot x_{i_s} (\mathrm{mod}\ 2) \ ,$$

where $\cdot$ is the usual multiplication and $a_{i_1\ldots i_s} \in \{0,1\}$, $\{i_1, \ldots, i_s\} \subseteq \{1, \ldots, n\}$. Since q is nonlinear, the Zhegalkin polynomial for q contains a term with at least two variables. Without loss of generality, we can assume that x_1 and x_2 are among these variables. Then we can transform the considered polynomial as follows:

$$q(x_1,\ldots,x_n) = x_1 \cdot x_2 \cdot q_1(x_3,\ldots,x_n) + x_1 \cdot q_2(x_3,\ldots,x_n) + x_2 \cdot q_3(x_3,\ldots,x_n) + q_4(x_3,\ldots,x_n) (\mathrm{mod}\ 2) .$$

From uniqueness of the Zhegalkin polynomial follows that $q_1(x_3,\ldots,x_n) \not\equiv 0$. Let $\alpha_3,\ldots,\alpha_n$ be numbers from $\{0,1\}$ such that $q_1(\alpha_3,\ldots,\alpha_n) = 1$. Then $\chi(x_1,x_2) = q(x_1,x_2,\alpha_3,\ldots,\alpha_n) = x_1 \cdot x_2 + \alpha \cdot x_1 + \beta \cdot x_2 + \gamma (\mathrm{mod}\ 2)$ where $\alpha,\beta,\gamma \in \{0,1\}$. Consider the function $\psi(x_1,x_2) = \chi(x_1 + \beta(\mathrm{mod}\ 2), x_2 + \alpha(\mathrm{mod}\ 2)) + \gamma + \alpha \cdot \beta(\mathrm{mod}\ 2)$. One can show that $\psi(x_1,x_2) = x_1 \cdot x_2 = x_1 \wedge x_2$. Denote $\sigma_1 = \neg\beta, \sigma_2 = \neg\alpha$ and $\sigma_3 = \neg(\gamma + \alpha \cdot \beta(\mathrm{mod}\ 2))$. Then

$$(q(x_1^{\sigma_1}, x_2^{\sigma_2}, \alpha_3,\ldots,\alpha_n))^{\sigma_3} = x_1 \wedge x_2 . \tag{142}$$

From Lemma 10.11 follows that there exist a circuit $S_\neg \in \mathrm{Circ}^1(B)$ and a tuple $\bar{u}$ of constant faults on gate inputs of the circuit $S_\neg$ such that the circuit $S_\neg(\bar{u})$ realizes a function equal to the function $\neg x$. From the equality (142) follows that based on the function q and on the circuit $S_\neg$ we can construct an iteration-free circuit $S_\wedge$ in the basis B and a tuple $\bar{w}$ of constant faults on gate inputs of the circuit $S_\wedge$ such that the circuit $S_\wedge(\bar{w})$ realizes a function equal to the function $x \wedge y$. □

Lemma 10.13. *Let B be a non-quasiprimitive basis. Then* $\log_2 h_B^{(3)}(n) = \Omega(n)$.

Proof. Using Lemma 10.11 we conclude that there exist a circuit $S_\neg \in \mathrm{Circ}^1(B)$ and tuples $\bar{w}_1$ and $\bar{w}_2$ of constant faults on gate inputs of the circuit $S_\neg$ such that the circuit $S_\neg(\bar{w}_1)$ realizes a function equal to the function x, while the circuit $S_\neg(\bar{w}_2)$ realizes a function equal to the function $\neg x$. Using Lemma 10.12 we obtain that there exist a circuit $S_\wedge \in \mathrm{Circ}^1(B)$ and a tuple $\bar{w}$ of constant faults on gate inputs of the circuit $S_\wedge$ such that the circuit $S_\wedge(\bar{w})$ realizes a function equal to the function $x \wedge y$.

Using the circuits $S_\neg$ and $S_\wedge$ one can construct for any $r \in \mathbb{N} \setminus \{0\}$ an iteration-free circuit S_r in the basis B with the following properties:

a) for any $\bar{\sigma} = (\sigma_1,\ldots,\sigma_r) \in \{0,1\}^r$ there exists a tuple $u(\bar{\sigma})$ of constant faults on gate inputs of the circuit S_r such that the circuit $S_r(u(\bar{\sigma}))$ realizes a function equal to the function $x_1^{\sigma_1} \wedge \ldots \wedge x_r^{\sigma_r}$;
b) there exists a tuple $\bar{v}$ of constant faults on gate inputs of the circuit S_r such that the circuit $S_r(\bar{v})$ realizes a function equal to the function 0;
c) $L(S_r) \le dr$, where $d = L(S_\neg) + L(S_\wedge)$.

Let us show that $h(S_r) \ge 2^r$. Let the circuit S_r have exactly m inputs labelled by variables $x_1,\ldots,x_m$. The function realized by the circuit $S_r(\bar{v})$ will be denoted by f, and for any tuple $\bar{\sigma} \in \{0,1\}^r$ the function realized by the circuit $S_r(u(\bar{\sigma}))$ will be denoted by $f_{\bar{\sigma}}$. Let $\bar{\sigma} = (\sigma_1,\ldots,\sigma_r) \in \{0,1\}^r$. One can show that the functions f and $f_{\bar{\sigma}}$ have different values on a tuple $(\delta_1,\ldots,\delta_m) \in \{0,1\}^m$ if and only if $\delta_1 = \sigma_1, \ldots, \delta_r = \sigma_r$. Hence the minimal cardinality of a set of tuples from $\{0,1\}^m$, on which the function f is different from any function $f_{\bar{\sigma}}$, $\bar{\sigma} \in \{0,1\}^r$, is equal to 2^r. Using Lemma 10.1 we obtain $h(S_r) \geq 2^r$.

Set $c_1 = 2d$. Let $n \geq c_1$. Denote $r = \lfloor n/d \rfloor$. One can show that $L(S_r) \leq dr \leq n$ and $h(S_r) \geq 2^r = 2^{\lfloor n/d \rfloor} \geq 2^{n/c_1}$. Hence $h_B^{(3)}(n) \geq 2^{n/c_1}$ and $\log_2 h_B^{(3)}(n) = \Omega(n)$.

□

Proof (of Theorem 10.4). Statement of the theorem follows from Lemmas 10.10 and 10.13.

□

10.5 Approach to Circuit Construction and Diagnosis

The fifth line of research deals with the approach to the circuit construction and to the effective diagnosis of faults based on the results obtained for the iteration-free circuits. From Proposition 10.1 and Lemma 10.2 follows that for each basis B_1 there exists a quasiprimitive basis B_2 with the following properties:

a) $[B_1] = [B_2]$, i.e. the set of functions realized by circuits in the basis B_2 coincides with the set of functions realized by circuits in the basis B_1;
b) there exists a polynomial p such that for any formula φ_1 over B_1 there exists a formula φ_2 over B_2 which realizes the function equal to that realized by φ_1, and such that $L(\varphi_2) \leq p(L(\varphi_1))$.

The considered approach to the circuit construction and fault diagnosis consists in the following. Let φ_1 be a formula over B_1 realizing certain function $f, f \notin \{0,1\}$, and let us construct the formula φ_2 over B_2 realizing the function equal to f and satisfying the inequality $L(\varphi_2) \leq p(L(\varphi_1))$. Next a circuit S in the basis B_2 is constructed (according to the formula φ_2) realizing the function f, satisfying the equality $L(S) = L(\varphi_2)$ and the condition that from each gate of the circuit S at most one edge issues. In addition to the usual work mode of the circuit S there exists the diagnostic mode in which the inputs of the circuit S are "split" so that it becomes the iteration-free circuit $\tilde{S}$. From Theorem 10.4 follows that the inequalities $h(\tilde{S}) \leq cL(S) \leq cp(L(\varphi_1))$, where c is a constant depending only on the basis B_2, hold for the circuit $\tilde{S}$.

11 Decision Trees for (1, 2)-Bayesian Networks

Bayesian Networks (BN for short) are convenient tool for representation of joint probability distribution of variables [39, 163]. Some of these variables are hidden (unobservable). Using values of open (observable) variables and information about probability distribution from BN we can draw some conclusions about values of hidden variables.

In this section we study time complexity of decision trees which compute values of all open variables from BN. We consider (1, 2)-BN in which each node has at most 1 entering edge, and each variable has at most 2 values. For an arbitrary (1, 2)-BN we obtain lower and upper bounds on minimal depth of decision tree that differ not more than by a factor of 4, and can be computed by an algorithm which has polynomial time complexity. The number of nodes in considered decision trees can grow as exponential on number of open variables in BN. We develop an polynomial algorithm for simulation of the work of decision trees which depth lies between the obtained bounds.

The investigation of decision trees for computation of all open variable values may be helpful for BN use. Let the process of computation of open variable values be rather expensive (it may be connected with use of remote sensors, carrying out of experiments, etc.), there exist a decision tree whose depth is essentially less than the number of open variables, and there exist an efficient algorithm for simulation of the decision tree work. Then it is appropriate to use this decision tree instead of the sequential computation of all open variable values.

This section consists of six subsections. The first subsection contains definitions of main notions. The second one contains preliminary bounds on complexity of decision trees for (1, 2)-BN. In the third subsection we consider a decomposition of an arbitrary (1, 2)-BN into a forest of so-called monotone increasing BN and a forest of so-called monotone decreasing BN. The fourth subsection is devoted to the study of monotone increasing and monotone decreasing BN. In the fifth subsection bounds on minimal depth of decision trees for (1, 2)-BN are obtained, and in the sixth subsection an algorithm for simulation of the work of decision tree which depth is close to minimal is considered.

11.1 Main Notions

In this subsection we consider notions of (1, 2)-BN, decision tree for BN and abridged description of BN.

(1,2)-BN. (1, 2)-*BN* is a directed acyclic graph in which each node has at most one entering edge. Nodes of the graph are labelled by pairwise different variables which have values from the set $\{0,1\}$. Sometimes we will not distinguish nodes and variables. A probability distribution is attached to each node (variable) y. This is probability distribution $\Pr(y = a)$, $a \in \{0,1\}$, if y has no entering edges, or probability distribution $\Pr(y = a \mid x = b)$, $a, b \in \{0,1\}$, if there is an edge from x to y. Some variables are *hidden* or *unobservable*. The other variables are *open* or *observable*. Further we will consider only (1, 2)-BN.

Let S be BN with n variables $v_1, \ldots, v_n$. With each n-tuple $\bar{a} = (a_1, \ldots, a_n) \in \{0,1\}^n$ of variable $v_1, \ldots, v_n$ values we will associate the probability of its appearance. This probability is equal to $p(\bar{a}) = p_1(\bar{a}) \cdot \ldots \cdot p_n(\bar{a})$ where $p_i(\bar{a}) = \Pr(v_i = a_i)$ if v_i has no entering edges, and $p_i(\bar{a}) = \Pr(v_i = a_i \mid v_j = a_j)$ if there is an edge from v_j to v_i. We will assume that the n-tuple $\bar{a}$ of variable values is *realizable* if $p(\bar{a}) \neq 0$.

Let $v_1, \ldots, v_m$ be open variables and $v_{m+1}, \ldots, v_n$ be hidden variables. The m-tuple $(b_1, \ldots, b_m)$ of open variable values will be called *realizable* if there exists a realizable n-tuple $(a_1, \ldots, a_n)$ of variable $v_1, \ldots, v_n$ values such that $b_1 = a_1$, ..., $b_m = a_m$. Denote by $\mathcal{R}(S)$ the set of all realizable m-tuples of open variable values.

We will associate with BN S the following *problem* $\mathcal{P}(S)$: for a given m-tuple from $\mathcal{R}(S)$ we must recognize it. To this end we can ask questions about values of open variables.

Of course, we can formulate the problem $\mathcal{P}(S)$ as a problem over an appropriate information system. However, in this case we must code m-tuples from $\mathcal{R}(S)$ (solutions of the problem $\mathcal{P}(S)$) by integers which is not convenient for the proofs. So, we will consider original formulation of the problem $\mathcal{P}(S)$. For this problem solving we will use decision trees in which each terminal node is labelled by an m-tuple from $\mathcal{R}(S)$.

Decision Trees for BN. *A decision tree over* BN S is a finite rooted directed tree in which each terminal node is labelled by an m-tuple from $\mathcal{R}(S)$; each node which is not terminal (such nodes are called *working*) is labelled by an open variable from S; there are exactly two edges starting in a working node, and these edges are labelled by numbers 0 and 1 respectively.

Let Γ be a decision tree over S, $\bar{b} = (b_1, \ldots, b_m) \in \mathcal{R}(S)$ and ξ be a directed path from the root to a terminal node of Γ. We will say that the path ξ *accepts* the m-tuple $\bar{b}$ if the following conditions hold. If there are no working nodes in ξ then ξ accepts $\bar{b}$. Otherwise ξ accepts $\bar{b}$ if and only if for an arbitrary working node of ξ if this node is labelled by v_i then the edge of ξ starting in the node is labelled by b_i. It is clear that Γ has exactly one directed path from the root to a terminal node which accepts $\bar{b}$.

We will say that the decision tree Γ *solves* the problem $\mathcal{P}(S)$ if for any $\bar{b} \in \mathcal{R}(S)$ the terminal node of the path which accepts $\bar{b}$ is labelled by $\bar{b}$.

As time complexity measure for decision trees we will consider the depth of decision tree. Denote by $h(S)$ the minimal depth of a decision tree over S which solves the problem $\mathcal{P}(S)$.

As space complexity measure we will consider the number of nodes in decision tree. Denote by $L(\Gamma)$ the number of nodes in decision tree Γ. Denote by $L(S)$ the minimal number of nodes in a decision tree over S which solves the problem $\mathcal{P}(S)$.

Abridged Description of BN. Let S be BN. The set of decision trees over S which solve the problem $\mathcal{P}(S)$ is defined completely by the set $\mathcal{R}(S)$ and by the list of open variables which are given in the right order. The belonging to the set $\mathcal{R}(S)$ is defined completely by values of variables for which the probabilities (from probability distributions attached to nodes) are positive.

For an arbitrary node y of S we change the distribution attached to y in the following way. If y has no entering edges then, instead of probability distribution $\Pr(y = a)$, $a \in \{0, 1\}$, we attach to y the set $V = \{a : a \in \{0, 1\}, \Pr(y = a) \neq 0\}$. If there is an edge from x to y then instead of probability distribution $\Pr(y = a \mid x = b)$, $a, b \in \{0, 1\}$, we attach to y a pair of sets V_0, V_1 where $V_0 = \{a : a \in \{0, 1\}, \Pr(y = a \mid x = 0) \neq 0\}$ and $V_1 = \{a : a \in \{0, 1\}, \Pr(y = a \mid x = 1) \neq 0\}$. The obtained abridged description of BN will be called BN too. Further we will consider only such BN.

11.2 Preliminary Bounds on Complexity of Decision Trees for (1, 2)-BN

First, we consider bounds on minimal depth and minimal number of nodes in decision trees depending on the number of open variables in (1, 2)-BN.

Lemma 11.1. *For any* $(1,2)$-*BN* S *the following equality holds:*

$$L(S) = 2|\mathcal{R}(S)| - 1 \ .$$

Proof. Consider an arbitrary decision tree over S. One can show that the number of working nodes in this tree is equal to the number of terminal nodes minus 1. It is clear that each decision tree which solves the problem $\mathcal{P}(S)$ must have at least $|\mathcal{R}(S)|$ terminal nodes. Therefore $L(S) \geq 2|\mathcal{R}(S)| - 1$. By induction on $|\mathcal{R}(S)|$, it is not difficult to prove that $L(S) \leq 2|\mathcal{R}(S)| - 1$. □

Lemma 11.2. *For any* $(1,2)$-*BN* S *the following inequality holds:*

$$h(S) \geq \lceil \log_2 |\mathcal{R}(S)| \rceil \ .$$

Proof. Let Γ be a decision tree which solves the problem $\mathcal{P}(S)$, and for which $h(\Gamma) = h(S)$. It is clear that Γ must have at least $|\mathcal{R}(S)|$ terminal nodes. One can show that Γ has at most $2^{h(\Gamma)}$ terminal nodes. Therefore $h(\Gamma) \geq \lceil \log_2 |\mathcal{R}(S)| \rceil$ and $h(S) \geq \lceil \log_2 |\mathcal{R}(S)| \rceil$. □

Proposition 11.1. *Let* S *be* $(1,2)$-*BN with* m *open variables. Then*

$$0 \leq h(S) \leq m \ , \qquad 1 \leq L(S) \leq 2^{m+1} - 1 \ .$$

Both lower and upper bounds on $h(S)$ *and* $L(S)$ *are unimprovable in general case.*

Proof. It is clear that $h(S) \geq 0$ and $L(S) \geq 1$. Consider a decision tree which computes sequentially values of all open variables of S. It is clear that the depth of this tree is equal to m. Therefore $h(S) \leq m$. Evidently, $|\mathcal{R}(S)| \leq 2^m$. Using Lemma 11.1 we conclude that $L(S) \leq 2^{m+1} - 1$.

Now we prove that the considered bounds are unimprovable.

Let S_1 be BN such that S_1 is a directed path with m nodes, all nodes are open, the set $\{0\}$ is attached to the root, and the pair $\{0\}, \{0,1\}$ is attached to an arbitrary node which is not root. It is clear that $\mathcal{R}(S_1) = \{(0,0,\ldots,0)\}$. Therefore there exists a decision tree with exactly one node which solves the problem $\mathcal{P}(S_1)$. Using this fact we obtain that $h(S_1) = 0$ and $L(S_1) = 1$. Thus, lower bounds are unimprovable.

Let S_2 be BN such that S_2 is a directed path with m nodes, all nodes are open, the set $\{0,1\}$ is attached to the root, and the pair $\{0,1\}, \{0,1\}$ is attached to an arbitrary node which is not root. It is clear that $\mathcal{R}(S_2) = \{0,1\}^m$. From Lemmas 11.1 and 11.2 follows that $h(S_2) \geq m$ and $L(S_2) \geq 2^{m+1} - 1$. Thus, upper bounds are unimprovable too. □

More precise bounds can be found if instead of the set of open variables we will consider such its subsets that the values of all open variables can be found if we know values of variables from considered subset. Let S be BN with m open variables $v_1, \ldots, v_m$. A set $B \subseteq \{v_1, \ldots, v_m\}$ will be called *a basis* of S if each two different m-tuples from $\mathcal{R}(S)$ are differed in a digit corresponding to a variable from B. Denote by $b(S)$ the minimal cardinality of a basis of B. Now we consider bounds on minimal depth and minimal number of nodes in decision trees depending on the parameter $b(S)$.

Proposition 11.2. *Let S be $(1,2)$-BN. Then*

$$\lceil \log_2(b(S)+1) \rceil \leq h(S) \leq b(S) \ , \qquad 2b(S)+1 \leq L(S) \leq 2^{b(S)+1} - 1 \ .$$

Both lower and upper bounds on $h(S)$ and $L(S)$ are unimprovable in general case.

Proof. By induction on $|\mathcal{R}(S)|$ it is not difficult to prove that $b(S) \leq |\mathcal{R}(S)| - 1$ and therefore $|\mathcal{R}(S)| \geq b(S) + 1$. Using Lemmas 11.1 and 11.2 we obtain that $h(S) \geq \lceil \log_2(b(S)+1) \rceil$ and $L(S) \geq 2b(S) + 1$.

Let B be a basis of S such that $|B| = b(S)$. It is not difficult to construct a decision tree over S which solves the problem $\mathcal{P}(S)$ by sequential computation of values of all variables from B. It is clear that the depth of this tree is equal to $b(S)$. Therefore $h(S) \leq b(S)$. One can show that $|\mathcal{R}(S)| \leq 2^{b(S)}$. Using Lemma 11.1 we obtain that $L(S) \leq 2^{b(S)+1} - 1$.

Now we show that the considered bounds are unimprovable.

Let S_3 be a BN such that S_3 is a directed path with m nodes, all nodes are open, the set $\{0, 1\}$ is attached to the root, and the pair $\{0\}, \{0, 1\}$ is attached to an arbitrary node which is not root. One can show that $\mathcal{R}(S_3) = \{(0, \ldots, 0), (1, 0, \ldots, 0), \ldots, (1, \ldots, 1, 0), (1, \ldots, 1)\}$, $|\mathcal{R}(S_3)| = m+1$ and $b(S) = m$. From Lemma 11.1 follows that $L(S_3) = 2b(S_3) + 1$. It is not difficult to prove (see Sect. 11.4) that $h(S_3) \leq \lceil \log_2(b(S_3) + 1) \rceil$. Thus, lower bounds are unimprovable.

Consider BN S_2 which was described in the proof of Proposition 11.1. It is clear that $b(S_2) = m$ and $\mathcal{R}(S_2) = \{0, 1\}^m$. Using Lemmas 11.1 and 11.2 we obtain that $h(S_2) \geq b(S_2)$ and $L(S_2) \geq 2^{b(S_2)+1} - 1$. Thus, upper bounds are unimprovable too.

□

From the considered propositions follows that lower and upper unimprovable bounds on minimal depth of decision trees depending on number of open variables are essentially distinct (as constant and linear function). The same situation is with bounds depending on the parameter $b(S)$. These bounds behave as logarithm and linear function. So, if we want to have more precise bounds, we must consider bounds which depend essentially on the structure of BN.

From Proposition 11.1 follows also that the minimal number of nodes in decision trees can grow as exponential on the number of open variables. So, instead of construction of decision tree we must try to simulate its work by algorithms that have polynomial time complexity.

11.3 Decomposition of (1,2)-BN

In this subsection we consider a process of decomposition of an arbitrary $(1,2)$-BN into a forest of so-called monotone increasing BN and a forest of so-called monotone decreasing BN.

Structure of (1,2)-BN. Let S be an $(1,2)$-BN. Graph S consists of connected components. Denote the set of this components by $F_1(S)$.

Lemma 11.3. *Let S be an $(1,2)$-BN and $P \in F_1(S)$. Then P is a rooted directed tree.*

Proof. Since P is an acyclic graph it has a directed path of maximal length. It is clear that the initial node of this path has no entering edges. Therefore P has a node without entering edges. Show that such node is unique. Suppose that x and y are different nodes of P each of which has no entering edges. Since P is a connected graph, there is an undirected path in P from x to y. It is clear that this path has a node with two entering edges which is impossible. Therefore P has exactly one node v without entering edges. Each other node has exactly one entering edge. So, the number of edges in P is equal to the number of nodes minus one. Taking into account that P is a connected graph we obtain that P is a tree.

Let x be an arbitrary node of P which is not equal to v. Consider an arbitrary path in P from v to x with pairwise different edges. Assume that this path is not directed. Then the path has a node with two entering edges which is impossible. Hence P is a directed tree with the root v. □

An $(1,2)$-BN which is a rooted directed tree will be called *tree-like* BN. Thus, an arbitrary $(1,2)$-BN S is a forest of tree-like BN. One can prove the following statement.

Lemma 11.4. *Let S be an $(1,2)$-BN. Then*

$$h(S) = \sum_{P \in F_1(S)} h(P) .$$

It is clear that the solution of the problem $\mathcal{P}(S)$ is the union of solutions of problems $\mathcal{P}(P)$, $P \in F_1(S)$.

From Tree-Like BN to Forest of Reduced BN. Consider an arbitrary tree-like BN $P \in F_1(S)$. Describe a process of transformation of P. As a result we obtain a forest of so-called reduced BN. *A reduced* BN is a tree-like BN in which the set $\{0,1\}$ is attached to the root, and the pair $\{0,1\},\{1\}$ or the pair $\{0\},\{0,1\}$ is attached to an arbitrary node which is not root.

First, we consider some informal reasons. Let P be not reduced BN. If the set $\{0\}$ or the set $\{1\}$ is attached to the root v of BN P then v is a constant. Variables, which are constants, will be removed from P. Let x, y be nodes of P, there be an edge from x to y, and the pair of sets V_0, V_1 be attached to y.

Let x be a constant a. If $|V_a| = 1$ then y is a constant. If $|V_a| = 2$ then y is an "independent" variable (it can be equal to 0 or to 1 independently of values of other variables). "Independent" variables will become roots of reduced BN. Let now x be not constant. We consider all possible 7 pairs V_0, V_1 which are different from $\{0,1\},\{1\}$ and $\{0\},\{0,1\}$. If V_0, V_1 is $\{0\},\{0\}$ or $\{1\},\{1\}$ then y is a constant. If V_0, V_1 is $\{0,1\},\{0,1\}$ then y is an "independent" variable. If V_0, V_1 is $\{0\},\{1\}$ or $\{1\},\{0\}$ then $y = x$ or $y = \neg x$ ($\neg x$ is the negation of x). Nodes (variables) which are equal to the same variable or to its negation will be identified. If V_0, V_1 is $\{0,1\},\{0\}$ or $\{1\},\{0,1\}$ then we replace y by $\neg y$ and pass to the pair $\{0,1\},\{1\}$ or to the pair $\{0\},\{0,1\}$. So we see a way to transform P into a forest of reduced BN.

Now we describe the process of transformation in detail. In the beginning of the process we create a list C_P of all variables from P. We will use this list for recording of current information on variables. The process consists of two phases. First phase of transformation begins from the root of P and finishes in terminal nodes. During each step we treat a node, predecessor of which is already treated (if there is an edge from x to y then x is called *predecessor* of y, and y is called *successor* of x).

If the set $\{0,1\}$ is attached to the root v of BN P then the root does not change. If the set $\{0\}$ or the set $\{1\}$ is attached to the root then we mark the root as 0- or 1-*constant* node respectively. We equate the variable v in the list C_P to 0 or to 1 respectively.

Let a node y be not treated, a node x be treated and there be an edge from x to y. Let the pair of sets V_0, V_1 be attached to y.

Let x be a-constant node, $a \in \{0,1\}$. If $|V_a| = 2$ then we mark the node y as *"independent"* node, and instead of the pair of sets V_0, V_1 we attach to y the set $\{0,1\}$. If $|V_a| = \{b\}$, then we mark the node y as b-constant node, and equate the variable y in the list C_P to constant b.

Let x be not constant node.

If $V_0 = V_1 = \{0\}$ or $V_0 = V_1 = \{1\}$ then we mark y as 0- or 1-constant node respectively. We equate the variable y in the list C_P to 0 or to 1 respectively.

If $V_0 = V_1 = \{0,1\}$ then we mark y as "independent" node, and instead of the pair of sets V_0, V_1 we attach to y the set $\{0,1\}$.

If $V_0 = \{0\}$ and $V_1 = \{1\}$ then instead of the variable y we label the considered node by the variable x. We equate the variable y in the list C_P to the variable x.

If $V_0 = \{1\}$ and $V_1 = \{0\}$ then instead of the variable y we label the considered node by the variable x. We equate the variable y in the list C_P to $\neg x$. Instead of the pair of sets V_0, V_1 we attach to the considered node the pair $\{0\},\{1\}$. For each successor of the node we instead of a pair W_0, W_1, attached to the successor, attach to it the pair W_1, W_0.

If $V_0 = \{0,1\}$ and $V_1 = \{0\}$ then instead of the variable y we label the considered node by the variable $\neg y$. We equate the variable y in the list C_P to $\neg\neg y$ (here $\neg y$ is new variable, and y is equal to the negation of this variable). Instead of the pair of sets V_0, V_1 we attach to the considered node the pair $\{0,1\},\{1\}$. For each successor of the node we instead of a pair W_0, W_1, attached to the successor, attach to it the pair W_1, W_0.

If $V_0 = \{1\}$ and $V_1 = \{0,1\}$ then instead of the variable y we label the considered node by the variable $\neg y$. We equate the variable y in the list C_P to $\neg\neg y$. Instead of the pair of sets V_0, V_1 we attach to the considered node the pair $\{0\}, \{0,1\}$. For each successor of the node we instead of a pair W_0, W_1, attached to the successor, attach to it the pair W_1, W_0.

If $V_0 = \{0,1\}, V_1 = \{1\}$ or $V_0 = \{0\}$, $V_1 = \{0,1\}$ then the considered node does not change.

We assume that for any variable x from P variables x and $\neg x$ have the same state: if x is open then $\neg x$ is open, if x is hidden then $\neg x$ is hidden.

After all nodes have been treated, the second phase of transformation is beginning.

We remove each constant node with each edge which is incident to it.

We remove each edge which enters to an "independent" node.

After the first phase of transformation different nodes may be labelled by the same variable. It is clear that the subgraph generated by these nodes is a directed tree with root.

For each variable z which is met more than one time we carry out the following operations:

1. We identify all nodes which are labelled by z (in other words we collapse all nodes marked with z to a single node).
2. We remove all edges each of which starts in the obtained node and enters to it.
3. We attach to the obtained node the set or the pair of sets which was attached to the root of identified tree.
4. If z is an open variable or all identified nodes were labelled in P by hidden variables, then we label the obtained node by z.
5. Otherwise we choose an open variable x which was label in P of an identified node. If the equality $x = z$ belongs to C_P then we label the obtained node by x, instead of z put x in right parts of equalities from C_P, remove the equality $x = x$ from C_P, add variable x to C_P, and equate the variable z in the list C_P to x. If the equality $x = \neg z$ belongs to C_P then we label the obtained node by $\neg x$, instead of z put $\neg x$ in right parts of equalities from C_P, and equate the variable z in the list C_P to $\neg x$.

As a result we obtain a forest which consists of reduced BN. Denote the set of BN from this forest by $F_2(P)$. Denote $F_2(S) = \bigcup_{P \in F_1(S)} F_2(P)$.

It is not difficult to prove the following statement.

Lemma 11.5. *Let S be an $(1,2)$-BN and $P \in F_1(S)$. Then*

$$h(P) = \sum_{Q \in F_2(P)} h(Q) .$$

If we know solutions of problems $\mathcal{P}(Q)$, $Q \in F_2(P)$, we can easily restore the solution of the problem $\mathcal{P}(P)$:

1. If in the list C_P an open variable is not equated to constant or other variable, or negation of other variable then the value of this variable was found during the process of problem $\mathcal{P}(Q)$, $Q \in F_2(P)$, solving.

2. If in the list C_P an open variable is equated to a constant then it is equal to this constant.
3. If in the list C_P an open variable is equated to other variable or to negation of other variable then the value of the other variable was found during the process of problem $\mathcal{P}(Q)$, $Q \in F_2(P)$, solving, and the value of the considered variable is equal to the value of the other variable or to negation of it respectively.

From Reduced BN to Forest of Monotone Increasing BN and Forest of Monotone Decreasing BN. Let Q be one of a reduced BN from $F_2(S)$. A node of Q will be called 0-*node* if it is the root, or the pair $\{0\}, \{0,1\}$ is attached to this node. A node of Q will be called 1-*node* if the pair $\{0,1\}, \{1\}$ is attached to this node.

Remove from BN Q all 1-nodes. As a result we obtain a forest which consists of some directed trees with root. In each such tree instead of a set or a pair of sets we attach to the root the set $\{0,1\}$. As a result we obtain the forest of tree-like BN. Each of these BN satisfies the following conditions: the set $\{0,1\}$ is attached to the root; the pair $\{0\}, \{0,1\}$ is attached to an arbitrary node which is not root. Such BN will be called *monotone increasing* BN. Denote the set of monotone increasing BN from the considered forest by $F_3(Q)$. Denote $F_3(S) = \bigcup_{Q \in F_2(S)} F_3(Q)$.

One can consider a tree-like BN as *a poset* (partially ordered set) in which the root is the maximal element. Therefore we can say about nodes which are greater or less than given node.

If the value of some node (variable) v of monotone increasing BN is equal to 0 then the value of each node which is less than v is equal to 0. If the value of v is equal to 1 then the value of each node which is greater than v is equal to 1.

Remove from BN Q all 0-nodes. As a result we obtain a forest which consists of some directed trees with root. In each such tree instead of a set or a pair of sets we attach to the root the set $\{0,1\}$. As a result we obtain the forest of tree-like BN. Each of these BN satisfies the following conditions: the set $\{0,1\}$ is attached to the root; the pair $\{0,1\}, \{1\}$ is attached to an arbitrary node which is not root. Such BN will be called *monotone decreasing* BN. Denote the set of monotone decreasing BN from the considered forest by $F_4(Q)$. Denote $F_4(S) = \bigcup_{Q \in F_2(S)} F_4(Q)$.

If the value of some node (variable) v of monotone decreasing BN is equal to 1 then the value of each node which is less than v is equal to 1. If the value of v is equal to 0 then the value of each node which is greater than v is equal to 0.

Lemma 11.6. *Let S be an $(1,2)$-BN and $Q \in F_2(S)$. Then*

$$\max\{\sum_{U \in F_3(Q)} h(U), \sum_{W \in F_4(Q)} h(W)\} \le h(Q) \le \sum_{U \in F_3(Q)} h(U) + \sum_{W \in F_4(Q)} h(W) .$$

Proof. If we know solutions of problems $\mathcal{P}(U)$, $U \in F_3(Q)$, and $\mathcal{P}(W)$, $W \in F_4(Q)$, we can easily restore the solution of the problem $\mathcal{P}(Q)$: it is equal to the union of solutions of considered problems. Therefore $h(Q) \leq \sum_{U \in F_3(Q)} h(U) + \sum_{W \in F_4(Q)} h(W)$.

Let us consider the following values of variables from Q: each 1-node (variable) is equal to 1; for any $U \in F_3(Q)$ as tuple of values of variables from U we take an arbitrary realizable for U tuple of values. One can show that the considered tuple of values of variables from Q is realizable for Q. Using this fact one can show that $h(Q) \geq \sum_{U \in F_3(Q)} h(U)$. The inequality $h(Q) \geq \sum_{W \in F_4(Q)} h(W)$ can be proved similarly. □

11.4 Analysis of Monotone Increasing and Monotone Decreasing BN

In this subsection we consider bounds on minimal depth of decision trees for monotone increasing and monotone decreasing BN, algorithms for computation of these bounds and algorithms for simulation of the work of decision trees.

Open Monotone Increasing BN. Let U be a monotone increasing BN from $F_3(S)$. Remove from this BN all hidden nodes with incident edges. If a pair of open nodes x, y was connected in U by a directed path from x to y containing hidden nodes only then we add an edge from x to y. As a result we obtain a directed tree with root. If a pair of sets is attached to the root of the tree then instead of this pair we attach the set $\{0, 1\}$ to the root. We denote the obtained BN by $U^{(1)}$. It is clear that $U^{(1)}$ is a monotone increasing BN in which all nodes are open. Such BN will be called *open monotone increasing* BN. One can prove the following statement.

Lemma 11.7. *Let S be an* $(1, 2)$*-BN and* $U \in F_3(S)$. *Then*

$$h(U) = h(U^{(1)}) \ .$$

Note that the solution of the problem $\mathcal{P}(U)$ coincides with the solution of the problem $\mathcal{P}(U^{(1)})$. Note also that the problem $\mathcal{P}(U^{(1)})$ is equivalent to the problem of deciphering of monotone 0-1 function on $U^{(1)}$. Such function f is defined on nodes of $U^{(1)}$, has values from $\{0, 1\}$, and if there is an edge from x to y then $f(x) \geq f(y)$.

Open Monotone Increasing Chains. Let Y be an open monotone increasing BN such that Y is a directed path with t nodes. Such BN will be called *open monotone increasing chain.*

Now we describe a dichotomous algorithm $\mathcal{A}_1$ which for a given open monotone increasing chain Y simulates the work of a decision tree which solves the problem $\mathcal{P}(Y)$. Let Y has t nodes labelled by variables $v_1, \ldots, v_t$, where v_1 is the root of Y, and v_{i+1} is a successor of v_i, $i = 1, \ldots, t-1$.

Description of the Algorithm $\mathcal{A}_1$

Compute the value of v_p where $p = \lceil t/2 \rceil$.

If $v_p = 1$ then $v_1 = \ldots = v_{p-1} = 1$. Compute the value of $v_{p+\lceil (t-p)/2 \rceil}$, and so on.

If $v_p = 0$ then $v_{p+1} = \ldots = v_t = 0$. Compute the value of $v_{\lceil (p-1)/2 \rceil}$, and so on.

It is clear that the algorithm $\mathcal{A}_1$ has polynomial time complexity.

Lemma 11.8. *Let Y be an open monotone increasing chain with t nodes. Then the algorithm $\mathcal{A}_1$ computes values of at most $\lceil \log_2(t+1) \rceil$ variables from Y.*

Proof. By induction on m it is not difficult to prove that if $t \le 2^m - 1$ then $\mathcal{A}_1$ computes values of at most m variables from Y. From this fact the statement of the lemma follows. □

Lemma 11.9. *Let Y be an open monotone increasing chain with t nodes. Then*

$$h(Y) = \lceil \log_2(t+1) \rceil \ .$$

Proof. One can show that $|\mathcal{R}(Y)| = t+1$. From Lemma 11.2 follows that $h(Y) \ge \lceil \log_2(t+1) \rceil$. Using Lemma 11.8 we conclude that $h(Y) = \lceil \log_2(t+1) \rceil$. □

From Open Monotone Increasing BN to Compressed BN. Consider the open monotone increasing BN $U^{(1)}$. The number of edges starting in a node will be called *the degree* of the node. Remove from $U^{(1)}$ all nodes (with the exception of the root) which degree is equal to 1. If a pair of remaining nodes x, y was connected in $U^{(1)}$ by a directed path from x to y containing nodes of degree 1 only then we add an edge from x to y. Denote the obtained open monotone increasing BN by $U^{(2)}$. In this BN each node which is not root or terminal has at least two edges starting in it. Such BN will be called *compressed.*

We attach to each edge of $U^{(2)}$ a number. Let d be an edge from x to y in $U^{(2)}$. If there is edge from x to y in $U^{(1)}$ then we attach 0 to d. Let there be no edge from x to y in $U^{(1)}$, and there be a directed path from x to y containing nodes of degree 1 only. Let this path have exactly t nodes of degree 1. Then we attach the number $\lceil \log_2(t+1) \rceil$ to the edge d.

A set of edges from $U^{(2)}$ will be called *independent* if any two different edges from the set do not belong to the same directed path in $U^{(2)}$. *The weight* of a set of edges is the sum of numbers attached to edges from the set. Denote by $M(U^{(2)})$ the maximal weight of an independent set of edges from $U^{(2)}$.

Later we will consider a polynomial algorithm $\mathcal{A}_4$ that for a given compressed BN $U^{(2)}$ simulates the work of a decision tree which solves the problem $\mathcal{P}(U^{(2)})$ and which depth is equal to $h(U^{(2)})$.

Now we describe an algorithm $\mathcal{A}_2$ that for a given open monotone increasing BN $U^{(1)}$ simulates the work of a decision tree which solves the problem $\mathcal{P}(U^{(1)})$.

Description of the Algorithm $\mathcal{A}_2$

Using algorithm $\mathcal{A}_4$ we solve the problem $\mathcal{P}(U^{(2)})$. Further we construct the set J of edges from $U^{(2)}$ each of which satisfies the following conditions:

1. The edge starts in a node (variable) which value is equal to 1.
2. The edge enters to a node which value is equal to 0.
3. Positive number is attached to the edge.

Let $r \in J$ and r start in x and enter to y. Then there is a directed path in $U^{(1)}$ from x to y containing nodes of degree 1 only (with the exception of x and y). We remove nodes x and y from this path, and denote the obtained path by X_r. Let this path contain t nodes of degree 1. Then we transform it in a natural way into the open monotone increasing chain Y_r with t nodes. Using the algorithm $\mathcal{A}_1$ we solve the problem $\mathcal{P}(Y_r)$ for each $r \in J$.

Since algorithms $\mathcal{A}_1$ and $\mathcal{A}_4$ have polynomial time complexity, algorithm $\mathcal{A}_2$ has polynomial time complexity too.

Lemma 11.10. *Let S be an $(1,2)$-BN and $U \in F_3(S)$. Then for open monotone increasing BN $U^{(1)}$ the algorithm $\mathcal{A}_2$ simulates the work of a decision tree which solves the problem $\mathcal{P}(U^{(1)})$ and which depth is at most $h(U^{(2)}) + M(U^{(2)})$.*

Proof. It is clear that after the problem $\mathcal{P}(U^{(2)})$ solving we do not know in BN $U^{(1)}$ only values of nodes (variables) of degree 1 which belong to paths X_r, $r \in J$. Therefore when the problems $\mathcal{P}(Y_r)$, $r \in J$, will be solved we will know values of all open variables from $U^{(1)}$. Thus, the algorithm $\mathcal{A}_2$ simulates the work of a decision tree which solves the problem $\mathcal{P}(U^{(1)})$.

It is clear that J is an independent set of edges. Let w be the weight of J. Using Lemma 11.8 we conclude that during the solving of problems $\mathcal{P}(Y_r)$, $r \in J$, the algorithm $\mathcal{A}_2$ computes values of at most w variables from $U^{(1)}$. Evidently, $w \leq M(U^{(2)})$. Algorithm $\mathcal{A}_4$ simulates the work of a decision tree which depth is equal to $h(U^{(2)})$. Therefore the algorithm $\mathcal{A}_2$ simulates the work of a decision tree the depth of which is at most $h(U^{(2)}) + M(U^{(2)})$. □

Lemma 11.11. *Let S be an $(1,2)$-BN and $U \in F_3(S)$. Then*

$$\max\{h(U^{(2)}), M(U^{(2)})\} \leq h(U^{(1)}) \leq h(U^{(2)}) + M(U^{(2)}) .$$

Proof. At first we show that

$$h(U^{(1)}) \geq h(U^{(2)}) . \tag{143}$$

Consider a decision tree Γ which solves the problem $\mathcal{P}(U^{(1)})$ and which depth is equal to $h(U^{(1)})$. Transform this tree as follows:

1. From each tuple, which is label of a terminal node, we remove values of all variables which do not belong to $U^{(2)}$.
2. If a working node is labelled by a variable y which does not belong to $U^{(2)}$, then instead of y we label the considered node by minimal (relatively the poset $U^{(1)}$) variable from $U^{(2)}$ which is greater than y.

One can show that the obtained decision tree over $U^{(2)}$ solves the problem $\mathcal{P}(U^{(2)})$, and has the same depth as Γ. Therefore the equality (143) holds.

Now we show that

$$h(U^{(1)}) \geq M(U^{(2)}) . \tag{144}$$

Let $J = \{d_1, \ldots, d_s\}$ be an independent set of edges from $U^{(2)}$ such that the weight of J is equal to $M(U^{(2)})$, and a positive number is attached to each edge from J. For $i = 1, \ldots, s$ we denote by X_i the directed path in $U^{(1)}$ which corresponds to the edge d_i and consists of nodes of degree 1 only. For $i = 1, \ldots, s$ we denote by Y_i the open monotone increasing chain which corresponds to the path X_i in a natural way. Let D be the set of variables from $U^{(1)}$ which do not belong to paths $X_1, \ldots, X_s$.

One can show that there exists such realizable tuple of values of variables from D that for $i = 1, \ldots, s$ the value of node from which the edge d_i starts is equal to 1, and the value of node to which d_i enters is equal to 0. For $i = 1, \ldots, s$ we give to variables from X_i values from an arbitrary tuple belonging to $\mathcal{R}(Y_i)$. As a result we obtain a tuple of values of all variables from $U^{(1)}$. One can show that this tuple belongs to $\mathcal{R}(U^{(1)})$. Using this fact it is not difficult to prove that

$$h(U^{(1)}) \geq h(Y_1) + \ldots + h(Y_s) \ .$$

From Lemma 11.9 and from the choose of the set J follows that $h(Y_1) + \ldots + h(Y_s) = M(U^{(2)})$. Therefore the inequality (144) holds.

From (143), (144) and from Lemma 11.10 follows the statement of the lemma.

□

Analysis of Compressed BN. Let K be an open monotone increasing BN. A node of K will be called *preterminal* if it is not terminal, and each successor of this node is a terminal node. Denote by $L_t(K)$ the number of terminal nodes in K, and by $L_p(K)$ we denote the number of preterminal nodes in K.

Open monotone increasing BN will be called *proper* BN if it has at least one nonterminal node, and the degree of each nonterminal node is at least 2.

Now we describe an algorithm $\mathcal{A}_3$ which for a given proper open monotone increasing BN K simulates the work of a decision tree which solves the problem $\mathcal{P}(K)$. During the work of this algorithm some nodes which values are already known will be marked as treated. Before the beginning of the algorithm work there are no marked nodes.

Description of the Algorithm $\mathcal{A}_3$

(*) We choose an untreated node v such that each successor of this node is a treated or a terminal node (using description of the algorithm one can show that such node exists).

If there is a treated successor of v which value is equal to 1, then $v = 1$. We compute values of all successors of v which are terminal nodes.

Let there be no a treated successor of v which value is equal to 1. Then we compute the value of v. If $v = 1$ then we compute values of all successors of v which are terminal nodes. If $v = 0$ then values of all successors of v which are terminal nodes are equal to 0.

We mark v and all successors of v which are terminal nodes as treated nodes.

If all nodes in K are marked as treated we finish the work of algorithm. If there is node in K which is not marked as treated we return to (*).

It is clear that the algorithm $\mathcal{A}_3$ simulates the work of a decision tree which solves the problem $\mathcal{P}(K)$. One can show that this algorithm has polynomial time complexity.

Each node v of BN K is the root of a subtree of the tree K. We denote this subtree by K_v. The depth of K_v (the maximal length of a directed path from the root to a terminal node of K_v) will be called *the depth* of the considered node v.

Lemma 11.12. *Let K be an arbitrary proper open monotone increasing BN. Then for each nonterminal node v of K the algorithm $\mathcal{A}_3$ computes values of at most $L_t(K_v) + L_p(K_v)$ variables from K_v if $v = 1$, and values of at most $L_t(K_v) + L_p(K_v) - 1$ variables from K_v if $v = 0$.*

Proof. We prove the statement of lemma by induction on the depth of node. Note that the depth of a nonterminal node is at least 1.

Let v be a node which depth is equal to 1. It is clear that v is a preterminal node. Let v have r successors. Then $L_t(K_v) + L_p(K_v) = r + 1$. Evidently, the considered algorithm computes values of at most $r+1$ variables from K_v if $v = 1$, and value of one variable from K_v if $v = 0$.

Hence for each node from K which depth is equal to 1 the statement of lemma is true.

Assume that the statement is true for any node from K which depth is at most $q, q \geq 1$. Let v be an arbitrary node from K which depth is equal to $q+1$. We show that the considered statement is true for v too. Let v have exactly r successors $v_1, \ldots, v_r$ which are roots of subtrees $K_{v_1}, \ldots, K_{v_r}$. It is clear that $r \geq 2$. For the definiteness, let for $i = 1, \ldots, s$ graph K_{v_i} have more than 1 node, and for $i = s+1, \ldots, r$ graph K_{v_i} have exactly 1 node (i.e. $v_{s+1}, \ldots, v_r$ are terminal nodes). It is clear that $s \geq 1$, and it is possible that $s = r$.

It is clear that the depth of the node v_i, $i = 1, \ldots, s$, is at most q. Using inductive hypothesis we conclude that for $i = 1, \ldots, s$ the considered algorithm computes values of at most $L_t(K_{v_i}) + L_p(K_{v_i})$ variables from K_{v_i} if $v_i = 1$, and values of at most $L_t(K_{v_i}) + L_p(K_{v_i}) - 1$ variables from K_{v_i} if $v_i = 0$.

Denote $z = \sum_{i=1}^{s}(L_t(K_i) + L_p(K_i))$. One can show that $L_t(K_v) + L_p(K_v) = z + r - s$.

Consider the following phase of the algorithm $\mathcal{A}_3$ work: nodes $v_1, \ldots, v_s$ are treated, nodes $v, v_{s+1}, \ldots, v_r$ are not treated, and we choose the node v.

If there is $i \in \{1, \ldots, s\}$ such that $v_i = 1$, then $v = 1$. We compute values of variables $v_{s+1}, \ldots, v_r$ (in this case algorithm computes values of at most $z+r-s$ variables from K_v).

Let $v_1 = \ldots = v_s = 0$. Then we compute the value of v. If $v = 1$ then we compute values of variables $v_{s+1}, \ldots, v_r$ (in this case algorithm computes values of at most $z - s + r - s + 1 \leq z + r - s$ variables from K_v). If $v = 0$ then $v_{s+1} = \ldots = v_r = 0$ (in this case algorithm computes values of at most $z - s + 1 \leq z + r - s - 1$ variables from K_v).

We see that the statement of lemma is true for v too. □

Describe an algorithm $\mathcal{A}_4$ that for a given compressed BN $U^{(2)}$ simulates the work of a decision tree which solves the problem $\mathcal{P}(U^{(2)})$.

Description of the Algorithm $\mathcal{A}_4$

If $U^{(2)}$ has at most 2 nodes then we compute values of all variables from $U^{(2)}$.

Let $U^{(2)}$ have at least 3 nodes. If $U^{(2)}$ is proper then we apply the algorithm $\mathcal{A}_3$ to BN $U^{(2)}$.

Let $U^{(2)}$ be not proper. Denote by u the root of $U^{(2)}$. Then there is exactly one edge which starts in u. This edge enters to the root of a subtree K. Denote by v the root of K. Instead of the pair $\{0\}, \{0,1\}$ we attach to v the set $\{0,1\}$. Denote by K^* the obtained BN. It is clear that K^* is proper BN. We apply the algorithm $\mathcal{A}_3$ to BN K^*. If $v = 0$ then we compute the value of u. If $v = 1$ then $u = 1$.

Since the algorithm $\mathcal{A}_3$ has polynomial time complexity, the algorithm $\mathcal{A}_4$ has polynomial time complexity too.

Lemma 11.13. *Let S be an $(1,2)$-BN and $U \in F_3(S)$. Then for compressed BN $U^{(2)}$ the algorithm $\mathcal{A}_4$ simulates the work of a decision tree which solves the problem $\mathcal{P}(U^{(2)})$ and which depth is at most $L_t(U^{(2)}) + L_p(U^{(2)})$.*

Proof. It is clear that $\mathcal{A}_4$ simulates the work of a decision tree which solves the problem $\mathcal{P}(U^{(2)})$. Denote this decision tree by Γ, and denote $z = L_t(U^{(2)}) + L_p(U^{(2)})$.

If $U^{(2)}$ has at most 2 nodes then, evidently, $h(\Gamma) \leq z$.

Let $U^{(2)}$ have at least three node. Let $U^{(2)}$ be proper BN. Using Lemma 11.12 we conclude that $h(\Gamma) \leq z$. Let $U^{(2)}$ be not proper. It is clear that $z = L_t(K^*) + L_p(K^*)$ (we use here notation from algorithm $\mathcal{A}_4$ description). Using Lemma 11.12 we conclude that the algorithm $\mathcal{A}_4$ computes values of at most $z - 1$ variables from K^* and value of u if $v = 0$. If $v = 1$ then $\mathcal{A}_4$ computes only values of at most z variables from K^*. Therefore $h(\Gamma) \leq z$. □

Lemma 11.14. *Let S be an $(1,2)$-BN and $U \in F_3(S)$. Then*

$$h(U^{(2)}) = L_t(U^{(2)}) + L_p(U^{(2)}) \ .$$

Proof. From Lemma 11.13 follows that $h(U^{(2)}) \leq L_t(U^{(2)}) + L_p(U^{(2)})$.

Now we show that $h(U^{(2)}) \geq L_t(U^{(2)}) + L_p(U^{(2)})$. Let Γ be a decision tree which solves the problem $\mathcal{P}(U^{(2)})$ and for which $h(\Gamma) = h(U^{(2)})$. Consider the following realizable tuple $\bar{b}$ of values of variables from $U^{(2)}$: all terminal nodes (variables) are equal to 0, and all nonterminal nodes are equal to 1. Consider a path ξ from the root to a terminal node of Γ which accepts $\bar{b}$. We show that all terminal and preterminal nodes (variables) of $U^{(2)}$ must be computed on this path.

Assume that some terminal variable is not computed on the path. Instead of 0 we write 1 to the digit corresponding to this variable in the tuple $\bar{b}$. Denote the obtained tuple by $\bar{b}'$. It is clear that $\bar{b}'$ is realizable, and Γ can not distinguish tuples $\bar{b}$ and $\bar{b}'$ which is impossible.

Assume that some preterminal variable is not computed on the path. Instead of 1 we write 0 to the digit corresponding to this variable in the tuple $\bar{b}$. Denote the obtained tuple by $\bar{b}''$. It is clear that $\bar{b}''$ is realizable, and Γ can not distinguish tuples $\bar{b}$ and $\bar{b}''$ which is impossible.

Therefore ξ has at least $L_t(U^{(2)}) + L_p(U^{(2)})$ working nodes. Taking into account the choose of Γ, we obtain $h(U^{(2)}) \geq L_t(U^{(2)}) + L_p(U^{(2)})$. □

From Lemmas 11.13 and 11.14 follows

Corollary 11.1. *Let S be an $(1,2)$-BN and $U \in F_3(S)$. Then for compressed BN $U^{(2)}$ the algorithm $\mathcal{A}_4$ simulates the work of a decision tree which solves the problem $\mathcal{P}(U^{(2)})$ and which depth is equal to $h(U^{(2)})$.*

From Lemmas 11.7, 11.11 and 11.14 follows

Corollary 11.2. *Let S be an $(1,2)$-BN and $U \in F_3(S)$. Then*

$$\max\{L_t(U^{(2)}) + L_p(U^{(2)}), M(U^{(2)})\} \leq h(U) \leq L_t(U^{(2)}) + L_p(U^{(2)}) + M(U^{(2)}) .$$

It is clear that parameters $L_t(U^{(2)})$ and $L_p(U^{(2)})$ can be computed by an algorithm which has polynomial time complexity. In the next subsubsection a polynomial algorithm for parameter $M(U^{(2)})$ computation is considered.

Algorithm for Computation of $M(U^{(2)})$. Each node of BN $U^{(2)}$ is the root of a subtree. The depth of this subtree will be called *the depth* of the considered node.

Beginning with nodes which depth is equal to 0 we will attach to each node a number that is equal to maximal weight of independent set of edges in subtree corresponding to the node. We attach the number 0 to each node which depth is equal to 0. Let for $r, r \geq 0$, each node which depth is equal to r have an attached number. Let v be a node which depth is equal to $r+1$. Let edges $d_1, \ldots, d_t$ start in v and enter nodes $v_1, \ldots, v_t$. Let numbers $w_1, \ldots, w_t$ be attached to edges $d_1, \ldots, d_t$, and numbers $u_1, \ldots, u_t$ be already attached to nodes $v_1, \ldots, v_t$. Then we attach to the node v the number

$$\max\{w_1, u_1\} + \ldots + \max\{w_t, u_t\} .$$

After the finishing of the algorithm work a number will be attached to the root of $U^{(2)}$. One can show that this number is equal to $M(U^{(2)})$.

It is clear that the considered algorithm has polynomial time complexity.

From Monotone Decreasing to Monotone Increasing BN. Let W be a monotone decreasing BN from $F_4(S)$. If instead of variables we label nodes of W by negations of variables we obtain a monotone increasing BN. This allows us to use results received earlier. To this end we transform each node of W as follows:

1. If the node is labelled by a variable x then instead of x we label this node by variable $\neg x$.
2. If the node is not root then instead of the pair of sets $\{0,1\}$, $\{1\}$ we attach to this node the pair of sets $\{0\}$, $\{0,1\}$.
3. The state of the node (it is hidden or open) does not change.

We denote the obtained BN by $\overline{W}$. It is not difficult to prove the following statement.

Lemma 11.15. *Let S be an $(1,2)$-BN and $W \in F_4(S)$. Then $\overline{W}$ is a monotone increasing BN such that*

$$h(\overline{W}) = h(W) \ .$$

From this lemma and from Corollary 11.2 follows

Corollary 11.3. *Let S be an $(1,2)$-BN and $W \in F_4(S)$. Then*

$$\max\{L_t(\overline{W}^{(2)}) + L_p(\overline{W}^{(2)}), M(\overline{W}^{(2)})\} \leq h(W)$$
$$\leq L_t(\overline{W}^{(2)}) + L_p(\overline{W}^{(2)}) + M(\overline{W}^{(2)}) \ .$$

If we know the solution of the problem $\mathcal{P}(\overline{W})$ we can easily restore the solution of the problem $\mathcal{P}(W)$: values of open variables from W are equal to negation of values of corresponding open variables from $\overline{W}$.

11.5 Bounds on Minimal Depth of Decision Trees for (1,2)-BN

Let S be $(1,2)$-BN. Then S is a forest which consists of tree-like BN (see Sect. 11.3). We have denoted the set of tree-like BN from this forest by $F_1(S)$.

As it was described in Sect. 11.3, an arbitrary tree-like BN $P \in F_1(S)$ may be transformed into a forest of reduced BN. We have denoted the set of reduced BN from this forest by $F_2(P)$, and denoted $F_2(S) = \bigcup_{P \in F_1(S)} F_2(P)$.

As it was described in Sect. 11.3, an arbitrary reduced BN $Q \in F_2(S)$ may be transformed into a forest of monotone increasing BN and a forest of monotone decreasing BN. We have denoted the set of monotone increasing BN from the first forest by $F_3(Q)$, and the set of monotone decreasing BN from the second forest by $F_4(Q)$. We have denoted $F_3(S) = \bigcup_{Q \in F_2(S)} F_3(Q)$ and $F_4(S) = \bigcup_{Q \in F_2(S)} F_4(Q)$.

As it was described in Sect. 11.4, an arbitrary monotone increasing BN $U \in F_3(S)$ may be transformed into an open monotone increasing BN $U^{(1)}$, and BN $U^{(1)}$ may be transformed into a compressed BN $U^{(2)}$ in which numbers are attached to edges. Denote

$$A(S) = \sum_{U \in F_3(S)} \left(L_t(U^{(2)}) + L_p(U^{(2)})\right) \ , \qquad B(S) = \sum_{U \in F_3(S)} M(U^{(2)}) \ .$$

The parameters $M(U^{(2)})$, $L_t(U^{(2)})$ and $L_p(U^{(2)})$ were defined in Sect. 11.4.

As it was described in Sect. 11.4, an arbitrary monotone decreasing BN $W \in F_4(S)$ may be transformed into a monotone increasing BN $\overline{W}$. Denote

$$C(S) = \sum_{W \in F_4(S)} \left(L_t(\overline{W}^{(2)}) + L_p(\overline{W}^{(2)})\right) \ , \qquad D(S) = \sum_{W \in F_4(S)} M(\overline{W}^{(2)}) \ .$$

Theorem 11.1. *For an arbitrary $(1,2)$-BN S the following inequalities hold:*

$$\max\{A(S), B(S), C(S), D(S)\} \leq h(S) \leq A(S) + B(S) + C(S) + D(S) \ .$$

Proof. One can show that $|F_2(S)| = \sum_{P \in F_1(S)} |F_2(P)|$. From this fact and from Lemmas 11.4 and 11.5 follows that

$$h(S) = \sum_{Q \in F_2(S)} h(Q) . \tag{145}$$

From Lemma 11.6 follows that for an arbitrary BN $Q \in F_2(S)$

$$\begin{aligned} \max\left\{\sum_{U \in F_3(Q)} h(U), \sum_{W \in F_4(Q)} h(W)\right\} &\leq h(Q) \\ \leq \sum_{U \in F_3(Q)} h(U) + \sum_{W \in F_4(Q)} h(W) . \end{aligned} \tag{146}$$

One can show $|F_3(S)| = \sum_{Q \in F_2(S)} |F_3(Q)|$ and $|F_4(S)| = \sum_{Q \in F_2(S)} |F_4(Q)|$. From this fact and from (145) and (146) follows that

$$\begin{aligned} \max\left\{\sum_{U \in F_3(S)} h(U), \sum_{W \in F_4(S)} h(W)\right\} &\leq h(S) \\ \leq \sum_{U \in F_3(S)} h(U) + \sum_{W \in F_4(S)} h(W) . \end{aligned} \tag{147}$$

Using Corollary 11.2 we obtain

$$\max\{A(S), B(S)\} \leq \sum_{U \in F_3(S)} h(U) \leq A(S) + B(S) . \tag{148}$$

Using Corollary 11.3 we obtain

$$\max\{C(S), D(S)\} \leq \sum_{W \in F_4(S)} h(W) \leq C(S) + D(S) . \tag{149}$$

The statement of the theorem follows from (147), (148) and (149). □

It is clear that upper bound from Theorem 11.1 is at most four lower bounds from Theorem 11.1. Therefore lower and upper bounds from Theorem 11.1 differ not more than by a factor of 4.

One can show that there is a polynomial algorithm which for a given (1, 2)-BN S constructs sets $F_3(S)$ and $F_4(S)$. It is clear that there is a polynomial algorithm which for given $U \in F_3(S)$ and $W \in F_4(S)$ constructs compressed BN $U^{(2)}$ and $\overline{W}^{(2)}$. The parameters $L_t(U^{(2)})$, $L_p(U^{(2)})$, $L_t(\overline{W}^{(2)})$ and $L_p(\overline{W}^{(2)})$, evidently, can be computed by an algorithm which has polynomial complexity. From results of Sect. 11.4 follows that there is an algorithm which has polynomial complexity, and computes parameters $M(U^{(2)})$ and $M(\overline{W}^{(2)})$. Hence bounds from Theorem 11.1 can be computed by an algorithm with polynomial time complexity.

11.6 Algorithm for Simulation of Decision Tree Work

Now we describe an algorithm $\mathcal{A}$ which for a given (1, 2)-BN S simulates the work of a decision tree which solves the problem $\mathcal{P}(S)$ and which depth lies between bounds from Theorem 11.1. The work of the algorithm $\mathcal{A}$ consists of three phases.

During the first phase we decompose the BN S. During the second phase we simulate the work of decision trees for obtained open monotone increasing BN. During the third phase we restore the solution of the problem $\mathcal{P}(S)$. Note that we compute values of variables from S only during the second phase.

Description of the Algorithm $\mathcal{A}$

1. *Decomposition of* $(1,2)$*-BN* S. At first, as it was described in Sect. 11.3, we construct the set $F_1(S)$ of tree-like BN. Then for each $P \in F_1(S)$ we construct (see details in Sect. 11.3) the list C_P of all variables from P and the set $F_2(P)$ of reduced BN. Also we construct the set $F_2(S) = \bigcup_{P \in F_1(S)} F_2(P)$.
   Further, as it was described in Sect. 11.3, we construct for each BN $Q \in F_2(S)$ the set $F_3(Q)$ of monotone increasing BN and the set $F_4(Q)$ of monotone decreasing BN. Next we construct the sets $F_3(S) = \bigcup_{Q \in F_2(S)} F_3(Q)$ and $F_4(S) = \bigcup_{Q \in F_2(S)} F_4(Q)$. For each monotone increasing BN $U \in F_3(S)$ we construct open monotone increasing BN $U^{(1)}$ (see Sect. 11.4) and compressed BN $U^{(2)}$ (see Sect. 11.4). For each monotone decreasing BN $W \in F_4(S)$ we construct monotone increasing BN $\overline{W}$ (see Sect. 11.4), open monotone increasing BN $\overline{W}^{(1)}$ and compressed BN $\overline{W}^{(2)}$.
2. *Simulation of Decision Tree Work for Open Monotone Increasing BN.* Using algorithm $\mathcal{A}_2$ (see Sect. 11.4) we simulate for each $U \in F_3(S)$ the work of a decision tree which solves the problem $\mathcal{P}(U^{(1)})$. Using algorithm $\mathcal{A}_2$ we simulate for each $W \in F_4(S)$ the work of a decision tree which solves the problem $\mathcal{P}(\overline{W}^{(1)})$.
   Some variables of considered BN may be negations of variables from S. In this case we compute values of corresponding variables from S and use negations of these values.
3. *Restoration of Problem* $\mathcal{P}(S)$ *solution.* It is clear that the solution of $\mathcal{P}(U)$ coincides with the solution of $\mathcal{P}(U^{(1)})$ for any $U \in F_3(S)$. The solution of $\mathcal{P}(\overline{W})$ coincides with the solution of $\mathcal{P}(\overline{W}^{(1)})$ for any $W \in F_4(S)$. For each $W \in F_4(S)$ the solution of $\mathcal{P}(W)$ is digit by digit negation of the solution of $\mathcal{P}(\overline{W})$.
   For each $Q \in F_2(S)$ the solution of $\mathcal{P}(Q)$ is the union of solutions of $\mathcal{P}(U)$, $U \in F_3(Q)$, and $\mathcal{P}(W)$, $W \in F_4(Q)$.
   For each $P \in F_1(S)$ we can obtain the solution of $\mathcal{P}(P)$ using solutions of $\mathcal{P}(Q)$, $Q \in F_2(P)$, and information from the list C_P as it was described in the end of Sect. 11.3.
   The solution of $\mathcal{P}(S)$ is the union of solutions of $\mathcal{P}(P)$, $P \in F_1(S)$.

Theorem 11.2. *The algorithm $\mathcal{A}$ has polynomial time complexity. For any* $(1,2)$*-BN S the algorithm $\mathcal{A}$ simulates the work of a decision tree which solves the problem $\mathcal{P}(S)$ and which depth is at most* $A(B) + B(S) + C(S) + D(S)$.

Proof. One can show that the first and the third phases of the algorithm $\mathcal{A}$ work may be realized during polynomial time. Since the algorithm $\mathcal{A}_2$ has polynomial time complexity, the second phase may be realized during polynomial time too.

It is not difficult to see that $\mathcal{A}$ simulates the work of a decision tree which solves the problem $\mathcal{P}(S)$. Using Lemma 11.10 we obtain that the depth of this decision tree is bounded from above by

$$\sum_{U \in F_3(S)} (h(U^{(2)}) + M(U^{(2)})) + \sum_{W \in F_4(S)} (h(\overline{W}^{(2)}) + M(\overline{W}^{(2)})) .$$

From here and from Lemma 11.14 follows that the depth of the considered decision tree is at most $A(B) + B(S) + C(S) + D(S)$. □

Conclusion

We have considered bounds on weighted depth of decision trees over finite and infinite information systems, algorithms for decision tree construction, and some examples of mathematical applications of decision tree theory and rough set theory. We did not say anything about such an important subject as relationships between time complexity of deterministic and nondeterministic decision trees. However, it was possible to see that the considered fragments of the decision tree theory have interesting mathematical problems and, partially, can be useful in practice.

In Sect. 6 it was shown that for some NP-complete problems of fixed dimension there exist linear decision trees with small depth which use attributes with small coefficients. Unfortunately, such trees may have large number of nodes. This is the general problem for decision tree theory: let us have a decision tree with small depth, then we must be able to describe effectively the work of this decision tree.

We did not consider this complicated problem especially. Note, however, that this general problem was solved for some discrete optimization problems (but not for NP-complete problems) in [54, 62–65, 69, 71] and also for some problems of pattern recognition (recognition of words from regular languages generated by independent simple A-sources, Sect. 9), fault diagnosis (diagnosis of constant faults in iteration-free circuits in quasiprimitive bases, Sect. 10), and probabilistic reasoning (recognition of values of all open variables in $(1, 2)$-Bayesian networks, Sect. 11).

Acknowledgments

The author wishes to express his deep gratitude to the late Al.A. Markov and S.V. Yablonskii, and also to thank O.B. Lupanov, Z. Pawlak, A. Skowron and Ju.I. Zhuravlev who influenced greatly the author's views on the subject of the present investigation.

The author is greatly indebted to J.F. Peters, S. Ramanna, S.P. Salmin and A. Skowron for the various help during the preparation of the text of this monograph.

Different stages of this investigation were partially supported by Russian Foundation for Basic Research, Program "Universities of Russia", Program "Fundamental Investigations in Mathematics" of Ministry of Higher Education of Russian Federation, Russian Federal Program "Integration of Science and Higher Education", Polish State Committee for Scientific Research, and Intel, Inc.

References

1. Ahlswede, R., Wegener, I.: Suchprobleme. B.G. Teubner, Stuttgart, 1979
2. Alexeyev, V.E.: On entropy of two-dimensional fragmentary closed languages. Combinatorial-Algebraic Methods and its Application. Edited by Al.A. Markov. Gorky University Publishers, Gorky (1987) 5–13 (in Russian)
3. Angluin, D.: Queries and concept learning. Machine Learning **2**(4) (1988) 319–342
4. Armstrong, D.B.: On finding of nearly minimal set of fault detection tests for combinatorial logic nets. IEEE Trans. on Elec. Comp. EC-**15**(1) (1966) 66–73
5. Bazan, J., Nguyen, H. Son, Nguyen, S. Hoa, Synak, P., Wróblewski, J.: Rough set algorithms in classification problems. Rough Set Methods and Applications: New Developments in Knowledge Discovery in Information Systems (Studies in Fuzziness and Soft Computing **56**). Edited by L. Polkowski, T.Y. Lin and S. Tsumoto. Phisica-Verlag. A Springer-Verlag Company (2000) 48–88
6. Ben-Or, M.: Lower bounds for algebraic computation trees. Proceedings of 15th ACM Annual Symp. on Theory of Comput. (1983) 80–86
7. Blumer, A., Ehrenfeucht, A., Haussler, D., Warmuth, M.: Learnability and the Vapnik-Chervonenkis dimension. J. ACM **36**(4) (1989) 929–965
8. Bondarenko, V.A.: Non-polynomial lower bound on complexity of traveling salesman problem in one class of algorithms. Automation and Telemechanics 9 (1983) 45–50 (in Russian)
9. Bondarenko, V.A.: Complexity bounds for combinatorial optimization problems in one class of algorithms. Russian Academy of Sciences Doklady **328**(1) (1993) 22–24 (in Russian)
10. Bondarenko, V.A., Yurov, S.V.: About a polyhedron of cubic graphs. Fundamenta Informaticae **25** (1996) 35–38
11. Breiman, L., Friedman, J.H., Olshen, R.A., Stone, C.J.: Classification and Regression Trees. Chapman and Hall, New York, 1984
12. Brodley, C.E., Utgoff, P.E.: Multivariate decision trees. Machine Learning **19** (1995) 45–77
13. Buntine, W.: Learning classification trees. Statistics and Computing **2** (1992) 63–73
14. Chegis, I.A., Yablonskii, S.V.: Logical methods of electric circuit control. Trudy MIAN SSSR **51** (1958) 270–360 (in Russian)
15. Chernikov, S.N.: Linear Inequalities. Nauka Publishers, Moscow, 1968 (in Russian)
16. Chikalov, I.V.: On decision trees with minimal average depth. Proceedings of the First International Conference on Rough Sets and Current Trends in Computing. Warsaw, Poland. Lecture Notes in Artificial Intelligence **1424**, Springer-Verlag (1998) 506–512
17. Chikalov, I.V.: Bounds on average weighted depth of decision trees depending only on entropy. Proceedings of the Seventh International Conference on Information Processing and Management of Uncertainty in Knowledge-based Systems, Vol. 2. Paris, France (1998) 1190–1194

18. Chikalov, I.V.: On average time complexity of decision trees and branching programs. Fundamenta Informaticae **39** (1999) 337–357
19. Chikalov, I.V.: Algorithm for constructing of decision trees with minimal average depth. Proceedings of the Eighth International Conference on Information Processing and Management of Uncertainty in Knowledge-based Systems, Vol. 1. Madrid, Spain (2000) 376–379
20. Chikalov, I.V.: Algorithm for constructing of decision trees with minimal number of nodes. Proceedings of the Second International Conference on Rough Sets and Current Trends in Computing. Banff, Canada (2000) 107–111
21. Dietterich, T.G., Shavlik, J.W. (Eds.): Readings in Machine Learning. Morgan Kaufmann, 1990
22. Dobkin, D., Lipton, R.J.: Multidimensional searching problems. SIAM J. Comput. **5**(2) (1976) 181–186
23. Dobkin, D., Lipton, R.J.: A lower bound of $(1/2)n^2$ on linear search programs for the knapsack problem. J. Comput. Syst. Sci. **16** (1978) 413–417
24. Dobkin, D., Lipton, R.J.: On the complexity of computations under varying sets of primitives. J. Comput. Syst. Sci. **18** (1979) 86–91
25. Duda, R.O., Hart, P.E., Stork, D.G.: Pattern Recognition. Wiley, New York, 2000
26. Dudina, Ju.V., Knyazev, A.N.: On complexity of recognition of words from languages generated by context-free grammars with one nonterminal symbol. Bulletin of Nizhny Novgorod State University. Mathematical Simulation and Optimal Control 2 (1998) 214–223 (in Russian)
27. Eldred, B.D.: Test routines based on symbolic logic statements. J. ACM **6**(1) (1959) 33–36
28. Feige, U.: A threshold of $\ln n$ for approximating set cover (Preliminary version). Proceedings of 28th Annual ACM Symposium on the Theory of Computing (1996) 314–318
29. Garey, M.R., Jonson, D.S.: Computers and Intractability. A Guide to the Theory of NP-Completeness. W.N. Freeman and Company, San Francisco, 1979
30. Goldman, R.S., Chipulis, V.P.: Diagnosis of iteration-free combinatorial circuits. Discrete Analysis **14**. Edited by Ju.I. Zhuravlev. Nauka Publishers, Novosibirsk (1969) 3–15 (in Russian)
31. Grigoriev, D., Karpinski, M., Vorobjov, N.: Improved lower bound on testing membership to a polyhedron by algebraic decision trees. Proceedings IEEE FOCS (1995) 258–265
32. Hastie, T., Tibshirani, R., Friedman, J.: The Elements of Statistical Learning: Data Mining, Inference, and Prediction. Springer-Verlag, Berlin, 2001
33. Hegedüs, T: Generalized teaching dimensions and the query complexity of learning. Proceedings of the 8th Annual ACM Conference on Computational Learning Theory. Santa Cruz, USA. ACM, New York (1995) 108–117
34. Hellerstein, L., Pillaipakkamnatt, K., Raghavan, V.V., Wilkins, D.: How many queries are needed to learn? J. ACM **43**(5) (1996) 840–862
35. Humby, E.: Programs from Decision Tables. Macdonald, London and American Elsevier, New York, 1973
36. Imam, I.F., Michalski, R.S.: Learning decision trees from decision rules: a method and initial results from a comparative study. Journal of Intelligent Information Systems **2** (1993) 279–304
37. Inuiguchi, M., Tsumoto, S., Hirano S. (Eds.): Rough Set Theory and Granular Computing (Studies in Fuzziness and Soft Computing **125**). Phisica-Verlag. A Springer-Verlag Company, 2003

38. Johnson, D.S.: Approximation algorithms for combinatorial problems. J. Comput. System Sci. **9** (1974) 256–278
39. Jordan, M. I. (Ed.): Learning in Graphical Models. MIT Press, 1999
40. Karavai, M.F.: Diagnosis of tree-like circuits in arbitrary basis. Automation and Telemechanics 1 (1973) 173–181 (in Russian)
41. Knyazev, A.N.: On recognition of words from languages generated by linear grammars with one nonterminal symbol. Proceedings of the First International Conference on Rough Sets and Current Trends in Computing. Warsaw, Poland. Lecture Notes in Artificial Intelligence **1424**, Springer-Verlag (1998) 111–114
42. Knyazev, A.N.: On recognition of words from languages generated by 1-context-free grammars. Proceedings of the Twelfth International Conference Problems of Theoretical Cybernetics, Part 1. Nizhny Novgorod, Russia (1999) 96 (in Russian)
43. Knyazev, A.N.: On recognition of words from languages generated by context-free grammars with one nonterminal symbol. Proceedings of the Eighth International Conference on Information Processing and Management of Uncertainty in Knowledge-based Systems, Vol. 1. Madrid, Spain (2000) 1945–1948
44. Komorowski, J., Pawlak, Z., Polkowski, L., Skowron, A.: Rough sets. A tutorial. Rough-Fuzzy Hybridization: A New Trend in Decision-Making. Edited by S.K. Pal and A. Skowron, Springer-Verlag, Singapore (1999) 3–98
45. Kospanov, E.Sh.: On algorithm for construction of simple enough tests. Discrete Analysis **8**. Nauka Publishers, Novosibirsk (1966) 43–47 (in Russian)
46. Kurosh, A.G.: Higher Algebra, 11-th ed. Nauka Publishers, Moscow, 1975 (in Russian)
47. Laskowski, M.C.: Vapnik-Chervonenkis classes of definable sets. J. London Math. Society **45** (1992) 377–384
48. Liu, H., Motoda, H.: Feature Selection for Knowledge Discovery and Data Mining. Kluwer Academic Publishers, Boston, 1998
49. Liu, H., Motoda, H. (Eds.): Feature Extraction, Construction and Selection: A Data Mining Approach. Kluwer Academic Publishers, Boston, 1998
50. Loh, W.-Y., Shih, Y.-S.: Split selection methods for classification trees. Statistica Sinica **7** (1997) 815–840
51. Lund, C., Yannakakis, M.: On the hardness of approximating minimization problems. J. ACM **41**(5) (1994) 960–981
52. Madatyan, Ch.A.: Complete test for iteration-free contact circuits. Problems of Cybernetics **23**. Edited by S.V. Yablonskii. Nauka Publishers, Moscow (1970) 103–118 (in Russian)
53. Markov Al.A.: Introduction into Coding Theory. Nauka Publishers, Moscow, 1982
54. Markov, Al.A.: Circuit complexity of discrete optimization. Discrete Mathematics **4**(3) (1992) 29–46 (in Russian)
55. Matiyasevich, Ju.V.: Diophantinity of enumerable sets. Academy of Sciences Doklady **191**(2) (1970) 279–382 (in Russian)
56. Meyer auf der Heide, F.: A polynomial linear search algorithm for the n-dimensional knapsack problem. J. ACM **31**(3) (1984) 668–676
57. Meyer auf der Heide, F.: Fast algorithms for n-dimensional restrictions of hard problems. J. ACM **35**(3) (1988) 740–747
58. Michalski, R.S.: Discovering classification rules using variable-valued logic system VL1. Proceedings of the Third International Joint Conference on Artificial Intelligence. Stanford, USA (1973) 162–172
59. Moore, E.F.: Gedanken-experiments on sequential machines. Automata Studies. Edited by C. Shannon and J. McCarty. Princeton University Press (1956) 129–153

60. Moravek, J.: On the complexity of discrete programming problems. Appl. Mat. **14**(6) (1969) 442–474
61. Moravek, J.: A localization problems in geometry and complexity of discrete programming. Kybernetika **8**(6) (1972) 498–516
62. Morzhakov, N.M.: On relationship between complexity of a set description and complexity of problem of linear form minimization on this set. Combinatorial-Algebraic Methods in Applied Mathematics. Edited by Al.A. Markov. Gorky University Publishers, Gorky (1985) 83–98 (in Russian)
63. Morzhakov, N.M.: Bounds on complexity of construction of finite subsets of the set $\mathbb{R}^n$. Combinatorial-Algebraic Methods in Applied Mathematics. Edited by Al.A. Markov. Gorky University Publishers, Gorky (1986) 84–106 (in Russian)
64. Morzhakov, N.M.: On possibilities of compression of finite subsets of the set $\mathbb{R}^n$. Combinatorial-Algebraic and Probabilistic Methods in Applied Mathematics. Edited by Al.A. Markov. Gorky University Publishers, Gorky (1988) 22–33 (in Russian)
65. Morzhakov, N.M.: On complexity of discrete extremal problem solving in the class of circuit algorithms. Mathematical Problems of Cybernetics **6**. Edited by S.V. Yablonskii. Nauka Publishers, Moscow (1996) 215–238 (in Russian)
66. Moshkov, M.Ju.: Problems of consequence in some subalgebras of real function algebras. Combinatorial-Algebraic Methods in Applied Mathematics. Edited by Al.A. Markov. Gorky University Publishers, Gorky (1979) 70–81 (in Russian)
67. Moshkov, M.Ju.: About uniqueness of uncancellable tests for recognition problems with linear decision rules. Combinatorial-Algebraic Methods in Applied Mathematics. Edited by Al.A. Markov. Gorky University Publishers, Gorky (1981) 97–109 (in Russian)
68. Moshkov, M.Ju.: On conditional tests. Academy of Sciences Doklady **265**(3) (1982) 550–552 (in Russian); English translation: Sov. Phys. Dokl. **27** (1982) 528–530
69. Moshkov, M.Ju.: Test approach to extremal combinatorial problems. Ph.D. thesis. Gorky University (1982) (in Russian)
70. Moshkov, M.Ju.: Conditional tests. Problems of Cybernetics **40**. Edited by S.V. Yablonskii. Nauka Publishers, Moscow (1983) 131–170 (in Russian)
71. Moshkov, M.Ju.: On problem of linear form minimization on finite set. Combinatorial-Algebraic Methods in Applied Mathematics. Edited by Al.A. Markov. Gorky University Publishers, Gorky (1985) 98–119 (in Russian)
72. Moshkov, M.Ju.: Conditional tests for diagnosis of constant faults in combinatorial circuits. Proceedings of Eights All-Union Conference Problems of Theoretical Cybernetics, Part 2. Gorky, USSR (1988) 50 (in Russian)
73. Moshkov, M.Ju.: On relationship of depth of deterministic and nondeterministic acyclic programs in the basis $\{x + y, x - y, 1; \text{sign } x\}$. Mathematical Problems in Computation Theory, Banach Center Publications **21**. PWN, Polish Scientific Publishers, Warsaw (1988) 523–529 (in Russian)
74. Moshkov, M.Ju.: On depth of conditional tests for tables from closed classes. Combinatorial-Algebraic and Probabilistic Methods of Discrete Analysis. Edited by Al.A. Markov. Gorky University Publishers, Gorky (1989) 78–86 (in Russian)
75. Moshkov, M.Ju.: On minimization of object complexity. Proceedings of Workshop on Discrete Mathematics and its Applications. Moscow State Universiry Publishers, Moscow (1989) 156–161 (in Russian)
76. Moshkov, M.Ju.: On complexity of algorithms for construction of tests for diagnosis constant faults on inputs of combinatorial circuits. Proceedings of the Ninth All-Union Conference Problems of Theoretical Cybernetics, Part I(1). Volgograd, Russia (1990) 81 (in Russian)

77. Moshkov, M.Ju.: Decision trees with quasilinear checks. Trudy IM SO RAN **27** (1994) 108–141 (in Russian)
78. Moshkov, M.Ju.: Optimization problems for decision trees. Fundamenta Informaticae **21** (1994) 391–401
79. Moshkov, M.Ju.: Decision Trees. Theory and Applications. Nizhny Novgorod University Publishers, Nizhny Novgorod, 1994 (in Russian)
80. Moshkov, M.Ju.: About the depth of decision trees computing Boolean functions. Fundamenta Informaticae **22** (1995) 203–215
81. Moshkov, M.Ju.: Two approaches to investigation of deterministic and nondeterministic decision tree complexity. Proceedings of the World Conference on the Fundamentals of AI. Paris, France (1995) 275–280
82. Moshkov, M.Ju.: Complexity of decision trees for regular language word recognition. Preproceedings of the Second International Conference Developments in Language Theory. Magdeburg, Germany (1995)
83. Moshkov, M.Ju.: Comparative analysis of complexity of deterministic and nondeterministic decision trees. Local Approach. Actual Problems of Modern Mathematics **1**. NII MIOO NGU Publishers, Novosibirsk (1995) 109–113 (in Russian)
84. Moshkov, M.Ju.: Comparative analysis of deterministic and nondeterministic decision tree complexity. Global approach. Fundamenta Informaticae **25** (1996) 201–214
85. Moshkov, M.Ju.: Lower bounds on time complexity of deterministic conditional tests. Discrete Mathematics **8**(3) (1996) 98–110 (in Russian)
86. Moshkov, M.Ju.: On the depth of decision trees over arbitrary check system. Proceedings of the Eleventh International Conference Problems of Theoretical Cybernetics. Uljanovsk, Russia (1996) 146–147 (in Russian)
87. Moshkov, M.Ju.: On the depth of decision trees over infinite information systems. Proceedings of the Congress Information Processing and Management of Uncertainty in Knowledge-based Systems. Granada, Spain (1996) 885–886
88. Moshkov, M.Ju.: On global Shannon functions of two-valued information systems. Proceedings of the Fourth International Workshop on Rough Sets, Fuzzy Sets and Machine Discovery. Tokyo, Japan (1996) 142–143
89. Moshkov, M.Ju.: Bounds on the depth of decision trees that compute Boolean functions, Russian Academy of Sciences Doklady **350**(1) (1996) 22–24 (in Russian); English translation: Dokl. Math. **54**(2) (1996) 662-664.
90. Moshkov, M.Ju.: Local and global approaches to comparative analysis of complexity of deterministic and nondeterministic decision trees. Actual Problems of Modern Mathematics **2**. NII MIOO NGU Publishers, Novosibirsk (1996) 110–118 (in Russian)
91. Moshkov, M.Ju.: Diagnosis of constant faults of circuits. Proceedings of the Fourth International Workshop on Rough Sets, Fuzzy Sets and Machine Discovery. Tokyo, Japan (1996) 325–327
92. Moshkov, M.Ju.: Some bounds on minimal decision tree depth. Fundamenta Informaticae **27** (1996) 197–203
93. Moshkov, M.Ju.: Unimprovable upper bounds on complexity of decision trees over information systems. Foundations of Computing and Decision Sciences **21**(4) (1996) 219–231
94. Moshkov, M.Ju.: Optimization of decision trees. Intellectual Systems **1**(1-4) (1996) 199–204 (in Russian)
95. Moshkov, M.Ju.: On complexity of decision trees over infinite information systems. Proceedings of the Third Joint Conference on Information Systems. USA, Duke University (1997) 353–354

96. Moshkov, M.Ju.: Comparative analysis of time complexity of deterministic and nondeterministic tree-programs. Actual Problems of Modern Mathematics **3**. NII MIOO NGU Publishers, Novosibirsk (1997) 117–124 (in Russian)
97. Moshkov, M.Ju.: Algorithms for constructing of decision trees. Proceedings of the First European Symposium Principles of Data Mining and Knowledge Discovery. Trondheim, Norway. Lecture Notes in Artificial Intelligence **1263**, Springer-Verlag (1997) 335–342
98. Moshkov, M.Ju.: Unimprovable upper bounds on time complexity of decision trees. Fundamenta Informaticae **31**(2) (1997) 157–184
99. Moshkov, M.Ju.: Rough analysis of tree-programs. Proceedings of the Fifth European Congress on Intelligent Techniques and Soft Computing. Aachen, Germany (1997) 231–235
100. Moshkov, M.Ju.: Complexity of deterministic and nondeterministic decision trees for regular language word recognition. Proceedings of the Third International Conference Developments in Language Theory. Thessaloniki, Greece (1997) 343–349
101. Moshkov, M.Ju.: Some relationships between decision trees and decision rule systems. Proceedings of the First International Conference on Rough Sets and Current Trends in Computing. Warsaw, Poland. Lecture Notes in Artificial Intelligence **1424**, Springer-Verlag (1998) 499–505
102. Moshkov, M.Ju.: On time complexity of decision trees. Rough Sets in Knowledge Discovery **1**. Methodology and Applications (Studies in Fuzziness and Soft Computing **18**). Edited by L. Polkowski and A. Skowron. Phisica-Verlag. A Springer-Verlag Company (1998) 160–191
103. Moshkov, M.Ju.: On the depth of decision trees. Russian Academy of Sciences Doklady **358**(1) (1998) 26 (in Russian)
104. Moshkov, M.Ju.: On time complexity of decision trees. Proceedings of International Siberian Conference on Operations Research. Novosibirsk, Russia (1998) 28–31 (in Russian)
105. Moshkov, M.Ju.: Bounds on depth of decision trees over finite two-valued check systems. Mathematical Problems of Cybernetics **7**. Edited by S.V. Yablonskii. Nauka Publishers, Moscow (1998) 162–168 (in Russian)
106. Moshkov, M.Ju.: Local approach to construction of decision trees. Rough-Fuzzy Hybridization: A New Trend in Decision-Making. Edited by S.K. Pal and A. Skowron, Springer-Verlag, Singapore (1999) 163–176
107. Moshkov, M.Ju.: On complexity of deterministic and nondeterministic decision trees. Proceedings of the Twelfth International Conference Problems of Theoretical Cybernetics, Part 2. Nizhny Novgorod, Russia (1999) 164 (in Russian)
108. Moshkov, M.Ju.: Time complexity of decision trees. Proceedings of the Ninth Interstates Workshop Design and Complexity of Control Systems. Nizhny Novgorod, Russia (1999) 52–62 (in Russian)
109. Moshkov, M.Ju.: Deterministic and nondeterministic decision trees for rough computing. Fundamenta Informaticae **41**(3) (2000) 301–311
110. Moshkov, M.Ju.: Decision trees for regular language word recognition. Fundamenta Informaticae **41**(4) (2000) 449–461
111. Moshkov, M.Ju.: About papers of R.G. Nigmatullin on approximate algorithms for solving of discrete extremal problems. Discrete Analysis and Operations Research **7**(1) (2000) 6–17 (in Russian)
112. Moshkov, M.Ju.: Classification of infinite information systems. Proceedings of the Second International Conference on Rough Sets and Current Trends in Computing. Banff, Canada (2000) 167–171

113. Moshkov, M.Ju.: On time and space complexity of deterministic and nondeterministic decision trees. Proceedings of the Eighth International Conference Information Processing and Management of Uncertainty in Knowledge-based Systems, Vol. 3. Madrid, Spain (2000) 1932–1936
114. Moshkov, M.Ju.: On complexity of decision trees over infinite check systems. Proceedings of the Fourth International Conference on Discrete Models in Control System Theory. Krasnovidovo, Russia (2000) 83–86 (in Russian)
115. Moshkov, M.Ju.: Diagnosis of constant faults in circuits. Mathematical Problems of Cybernetics **9**. Edited by O.B. Lupanov. Nauka Publishers, Moscow (2000) 79–100 (in Russian)
116. Moshkov, M.Ju.: Elements of Mathematical Theory of Tests with Applications to Problems of Discrete Optimization. Nizhny Novgorod University Publishers, Nizhny Novgorod, 2001 (in Russian)
117. Moshkov, M.Ju.: On space and time complexity of decision trees. Discrete Mathematics and its Applications. Collection of Lectures for Youth Scientific Schools on Discrete Mathematics and its Applications, Vol. 2. Center for Applied Investigations of Faculty of Mathematics and Mechanics, Moscow State University, Moscow (2001) 35–40 (in Russian)
118. Moshkov, M.Ju.: Classification of infinite check systems depending on complexity of decision trees and decision rule systems. Proceedings of the Eleventh Interstates Workshop Design and Complexity of Control Systems, Part 1. Nizhny Novgorod, Russia (2001) 109–116 (in Russian)
119. Moshkov, M.Ju.: Test theory and problems of machine learning. Proceedings of the International School-Seminar on Discrete Mathematics and Mathematical Cybernetics. Ratmino, Russia (2001) 6–10
120. Moshkov, M.Ju.: On transformation of decision rule systems into decision trees. Proceedings of the Seventh International Workshop Discrete Mathematics and its Applications, Part 1. Moscow, Russia (2001) 21–26 (in Russian)
121. Moshkov, M.Ju.: On deciphering of monotone 0-1 function defined on tree with root. Proceedings of the Twelfth International Workshop Design and Complexity of Control Systems, Part 2. Penza, Russia (2001) 157–160 (in Russian)
122. Moshkov, M.Ju.: On decision trees for (1,2)-Bayesian networks. Fundamenta Informaticae **50**(1) (2002) 57–76
123. Moshkov, M.Ju.: On compressible information systems. Proceedings of the Third International Conference on Rough Sets and Current Trends in Computing. Penn State Great Valley, USA. Lecture Notes in Artificial Intelligence **2475**, Springer-Verlag (2002) 156–160
124. Moshkov, M.Ju.: On closed classes of machine learning problems. Proceedings of the Thirteenth International Conference Problems of Theoretical Cybernetics, Part 2. Kazan, Russia (2002) 128 (in Russian)
125. Moshkov, M.Ju.: Greedy algorithm for set cover in context of knowledge discovery problems. Proceedings of the International Workshop on Rough Sets in Knowledge Discovery and Soft Computing (ETAPS 2003 Satellite Event). Warsaw, Poland. Electronic Notes in Theoretical Computer Science **82**(4) (2003) http://www.elsevier.nl/locate/entcs/volume82.html
126. Moshkov, M.Ju.: Approximate algorithm for minimization of decision tree depth. Proceedings of the Ninth International Conference Rough Sets, Fuzzy Sets, Data Mining, and Granular Computing. Chongqing, China. Lecture Notes in Computer Science **2639**, Springer-Verlag (2003) 611–614

127. Moshkov, M.Ju.: Classification of infinite information systems depending on complexity of decision trees and decision rule systems. Fundamenta Informaticae **54**(4) (2003) 345–368
128. Moshkov, M.Ju.: Compressible infinite information systems. Fundamenta Informaticae **55**(1) (2003) 51–61
129. Moshkov, M.Ju., Chikalov, I.V.: On the average depth of decision trees over information systems. Proceedings of the Fourth European Congress on Intelligent Techniques and Soft Computing, Vol. 1. Aachen, Germany (1996) 220–222
130. Moshkov, M.Ju., Chikalov, I.V.: Upper bound on average depth of decision trees over information systems. Proceedings of the Fourth International Workshop on Rough Sets, Fuzzy Sets and Machine Discovery. Tokyo, Japan (1996) 139–141
131. Moshkov, M.Ju., Chikalov, I.V.: Bounds on average weighted depth of decision trees. Fundamenta Informaticae **31**(2) (1997) 145–156
132. Moshkov, M.Ju., Chikalov, I.V.: Bounds on average depth of decision trees. Proceedings of the Fifth European Congress on Intelligent Techniques and Soft Computing. Aachen, Germany (1997) 226–230
133. Moshkov, M.Ju., Chikalov, I.V.: On effective algorithms for construction of decision trees. Proceedings of the Twelfth International Conference Problems of Theoretical Cybernetics, Part 2. Nizhny Novgorod, Russia (1999) 165 (in Russian)
134. Moshkov, M.Ju., Chikalov, I.V.: On algorithm for constructing of decision trees with minimal depth. Fundamenta Informaticae **41**(3) (2000) 295–299
135. Moshkov, M.Ju., Chikalov, I.V.: On complexity of construction of minimal tests and minimal conditional tests for some class of problems. Proceedings of the Thirteenth International Workshop Design and Complexity of Control Systems, Part 2. Penza, Russia (2002) 165–168 (in Russian)
136. Moshkov, M.Ju., Chikalov, I.V.: Sequential optimization of decision trees relatively different complexity measures. Proceedings of the Sixth International Conference Soft Computing and Distributed Processing. Rzeszow, Poland (2002) 53–56
137. Moshkov, M.Ju., Moshkova, A.M.: Optimal bases for some closed classes of Boolean functions. Proceedings of the Fifth European Congress on Intelligent Techniques and Soft Computing. Aachen, Germany (1997) 1643–1647
138. Moshkova, A.M.: Diagnosis of retaining faults of combinatorial circuits, Bulletin of Nizhny Novgorod State University. Mathematical Simulation and Optimal Control 2 (1998) 204–233 (in Russian)
139. Moshkova, A.M.: On diagnosis of retaining faults in circuits. Proceedings of the First International Conference on Rough Sets and Current Trends in Computing. Warsaw, Poland. Lecture Notes in Artificial Intelligence **1424**, Springer-Verlag (1998) 513–516
140. Moshkova, A.M.: On time complexity of "retaining" fault diagnosis in circuits. Proceedings of the Eighth International Conference on Information Processing and Management of Uncertainty in Knowledge-based Systems, Vol. 1. Madrid, Spain (2000) 372–375
141. Müller, W., Wysotzki, F.: Automatic construction of decision trees for classification. Annals of Operations Research **52** (1994) 231–247
142. Murthy, S.K., Kasif, S., Salzberg, S.: A system for induction of oblique decision trees. Journal of Artificial Intelligence Research **2** (1994) 1–33
143. Nguyen, H. Son: From optimal hyperplanes to optimal decision trees. Fundamenta Informaticae **34**(1-2) (1998) 145–174

144. Nguyen, H. Son: On efficient handling of continuous attributes in large data bases. Fundamenta Informaticae **48**(1) (2001) 61–81
145. Nguyen, H. Son, Nguyen, H. Hoa: Discretization methods in data mining. Rough Sets in Knowledge Discovery **1**. Methodology and Applications (Studies in Fuzziness and Soft Computing **18**). Edited by L. Polkowski and A. Skowron. Phisica-Verlag. A Springer-Verlag Company (1998) 451–482
146. Nguyen, S. Hoa, Nguyen, H. Son: Pattern extraction from data. Fundamenta Informaticae **34**(1-2) (1998) 129–144
147. Nguyen, H. Son, Slezak, D.: Approximate reducts and association rules – correspondence and complexity results. Proceedings of the Seventh International Workshop on Rough Sets, Fuzzy Sets, Data Mining, and Granular-Soft Computing. Yamaguchi, Japan. Lecture Notes in Artificial Intelligence **1711**, Springer-Verlag (1999) 137–145
148. Nigmatullin, R.G.: Method of steepest descent in problems on cover. Memoirs of Symposium Problems of Precision and Efficiency of Computing Algorithms **5**. Kiev, USSR (1969) 116–126 (in Russian)
149. Okolnishnikova, E.A.: Lower bounds on complexity of realization of characteristic functions of binary codes by branching programs. Methods of Discrete Analysis **51**. Edited by A.D. Korshunov. IM SO AN USSR Publishers, Novosibirsk (1991) 61–83 (in Russian)
150. Pal, S.K., Polkowski, L., Skowron, A. (Eds.): Rough-Neural Computing. Techniques for Computing with Words. Springer Verlag series in Cognitive Technologies, Berlin, 2003
151. Parchomenko, P.P. : Theory of questionnaires, Automation and Telemechanics 4 (1970) 140–159 (in Russian)
152. Parchomenko, P.P., Sogomonyan, E.S.: Fundamentals of Technical Diagnosis. Energoizdat Publishers, Moscow, 1981 (in Russian)
153. Pawlak, Z.: Information Systems – Theoretical Foundations. PWN, Warsaw, 1981 (in Polish)
154. Pawlak, Z.: Rough sets. International J. Comp. Inform. Science **11** (1982) 341–356
155. Pawlak, Z.: Rough classification. Report of the Computing Center of the Polish Academy of Sciences **506** (1983)
156. Pawlak, Z.: Rough sets and fuzzy sets. Fuzzy Sets and Systems **17** (1985) 99–102
157. Pawlak, Z.: Rough sets and decision tables. Lecture Notes in Computer Science **208**, Springer-Verlag (1985) 186–196
158. Pawlak, Z.: On rough dependency of attributes in information systems. Bull. Polish Acad. Sci. Tech. **33** (1985) 551–599
159. Pawlak, Z.: On decision tables. Bull. Polish Acad. Sci. Tech. **34** (1986) 553–572
160. Pawlak, Z.: Rough logic. Bull. Polish Acad. Sci. Tech. **35** (1987) 253–258
161. Pawlak, Z.: Decision tables – a rough set approach. Bull. of EATCS **33** (1987) 85–96
162. Pawlak, Z.: Rough Sets – Theoretical Aspects of Reasoning about Data. Kluwer Academic Publishers, Dordrecht, Boston, London, 1991
163. Pearl, J.: Probabilistic Inference in Intelligent Systems, Morgan Kaufman, 1988
164. Peters, J.F., Skowron, A., Stepaniuk, J., Ramanna, S.: Towards an ontology of approximate reason. Fundamenta Informaticae **51**(1-2) (2002) 157–173
165. Peters, J.F., Skowron, A., Synak, P., Ramanna, S.: Rough sets and information granulation. Proceedings of the Tenth International Fuzzy Systems Association World Congress. Istanbul, Turkey. Lecture Notes in Computer Science **2715**, Springer-Verlag (2003) 370–377

166. Picard, C.F.: Theorie des Questionnaires. Gauthier-Villars, Paris, 1965
167. Picard, C.F.: Graphes et Questionnaires, Vol. 1, Vol. 2. Gauthier-Villars, Paris, 1972
168. Polkowski, L.: Rough Sets. Mathematical Foundations (Advances in Soft Computing). Physica-Verlag, Heidelberg, 2002
169. Polkowski, L., Lin, T.Y., Tsumoto, S. (Eds.): Rough Set Methods and Applications: New Developments in Knowledge Discovery in Information Systems (Studies in Fuzziness and Soft Computing **56**). Phisica-Verlag. A Springer-Verlag Company, 2000
170. Pollack, S.L.: Decision Tables: Theory and Practice. J. Wiley & Sons Inc., 1971
171. Post, E.: Introduction to a general theory of elementary propositions. Amer. J. Math. **43** (1921) 163-185
172. Post, E. : Two-valued iterative systems of mathematical logic. Annals of Math. Studies **5**, Princeton Univ. Press, Princeton-London, 1941
173. Preparata, F.P., Shamos, M.I.: Computational Geometry: An Introduction. Springer-Verlag, 1985
174. Quinlan, J.R.: Discovering rules by induction from large collections of examples. Experts Systems in the Microelectronic Age. Edited by D. Michie. Edinburg University Press (1979)
175. Quinlan, J.R.: Induction of decision trees. Machine Learning **1**(1) (1986) 81–106
176. Quinlan, J.R.: Generating production rules from decision trees. Proc. of the Tenth Int. Joint Conf. on AI (1987) 304–307
177. Quinlan, J.R.: C4.5: Programs for Machine Learning. Morgan Kaufmann, San Mateo, 1993
178. Redkin, N.P.: Reliability and Diagnosis of Circuits, Moscow University Publishers, Moscow, 1992 (in Russian)
179. Rissanen, J.: Modeling by shortest data description. Automatica **14** (1978) 465–471
180. Roth, J.P. : Diagnosis of automata failures: a calculus and method. Journal Research and Development (1966) 278–291
181. Sapozhenko, A.A.: On a proof of upper bound on complexity of minimal disjunctive normal form for almost all functions. Proceedings of the First All-Union Conference Problems of Theoretical Cybernetics. Novosibirsk, USSR (1969) 103
182. Sauer, N.: On the density of families of sets. J. of Combinatorial Theory (A) **13** (1972) 145–147
183. Shelah, S.: A combinatorial problem; stability and order for models and theories in infinitary languages. Pacific J. of Mathematics **41** (1972) 241–261
184. Shevtchenko, V.I.: On complexity of diagnosis of one type of faults in combinatorial circuits by conditional tests. Combinatorial-Algebraic and Probabilistic Methods in Applied Mathematics. Edited by Al.A. Markov. Gorky University Publishers, Gorky (1988) 86–97 (in Russian)
185. Shevtchenko, V.I.: On complexity of diagnosis of faults of the type "$\oplus$" in combinatorial circuits. Combinatorial-Algebraic and Probabilistic Methods of Discrete Analysis. Edited by Al.A. Markov, Gorky University Publishers, Gorky, (1989) 129–140 (in Russian)
186. Shevtchenko, V.I.: On complexity of diagnosis of faults of types "0", "1", "&" and "$\vee$" in combinatorial circuits. Combinatorial-Algebraic and Probabilistic Methods and its Application. Edited by Al.A. Markov. Gorky University Publishers, Gorky (1990) 125–150 (in Russian)

187. Shevtchenko, V.I.: On depth of conditional tests for diagnosis of "negation" type faults in circuits. Siberian Journal on Operations Research **1** (1994) 63–74 (in Russian)
188. Shevtchenko, V.I.: On the depth of decision trees for diagnosing faults in circuits. Soft Computing (Third International Workshop on Rough Sets and Soft Computing). The Society for Computer Simulation, San Diego (1995) 200–203
189. Shevtchenko, V.I.: On the depth of decision trees for control faults in circuits. Proceedings of the Fourth International Workshop on Rough Sets, Fuzzy Sets and Machine Discovery. Tokyo, Japan (1996) 328–330
190. Shevtchenko, V.I.: On complexity of conditional tests for diagnosis of circuits. Intellectual Systems **1**(1–4) (1996) 247–251 (in Russian)
191. Shevtchenko, V.I.: On the depth of decision trees for diagnosing of nonelementary faults in circuits. Proceedings of the First International Conference on Rough Sets and Current Trends in Computing. Warsaw, Poland. Lecture Notes in Artificial Intelligence **1424**, Springer-Verlag (1998) 517–520
192. Shevtchenko, V.I., Moshkov, M.Ju., Moshkova, A.M.: Effective methods for diagnosis of faults in circuits. Proceedings of the Eleventh Interstates Workshop Design and Complexity of Control Systems, Part 2. Nizhny Novgorod, Russia (2001) 228-238 (in Russian)
193. Skowron, A.: Rough sets in KDD. Proceedings of the 16-th World Computer Congress (IFIP'2000). Beijing, China (2000) 1–14
194. Skowron, A., Pal, S.K. (Eds.): Special issue "Rough sets, pattern recognition and data mining". Pattern Recognition Letters **24**(6) (2003) 829–933
195. Skowron, A., Pawlak, Z., Komorowski, J., Polkowski, L.: A rough set perspective on data and knowledge. Handbook of KDD. Edited by W. Kloesgen and J. Żytkow. Oxford University Press (2002) 134–149
196. Skowron, A., Polkowski, L.: Synthesis of decision systems from data tables. Rough Sets and Data Mining: Analysis for Imprecise Data. Edited by T.Y. Lin and N. Cercone. Kluwer Academic Publishers, Boston (1997) 259–300
197. Skowron, A., Rauszer, C.: The discernibility matrices and functions in information systems. Intelligent Decision Support. Handbook of Applications and Advances of the Rough Set Theory. Edited by R. Slowinski. Kluwer Academic Publishers, Dordrecht, Boston, London (1992) 331–362
198. Skowron, A., Swiniarski, R.: Information granulation and pattern recognition. In Rough-Neural Computing. Techniques for Computing with Words. Edited by S.K. Pal, L. Polkowski and A. Skowron. Springer Verlag series in Cognitive Technologies, Berlin (2003) 599–636
199. Slezak, D.: Approximate decision reducts. Ph.D. thesis. Warsaw University (2002) (in Polish)
200. Slezak, D.: Approximate Markov boundaries and bayesian networks: Rough set approach. Rough Set Theory and Granular Computing (Studies in Fuzziness and Soft Computing **125**). Edited by M. Inuiguchi, S. Tsumoto and S. Hirano. Phisica-Verlag. A Springer-Verlag Company (2003) 109–121
201. Slowinski, R. (Ed.): Intelligent Decision Support. Handbook of Applications and Advances of the Rough Set Theory. Kluwer Academic Publishers, Dordrecht, Boston, London, 1992
202. Soloviev, N.A.: On certain property of tables with uncancellable tests of equal length. Discrete Analysis **12**. Edited by Ju.I. Zhuravlev. Nauka Publishers, Novosibirsk (1968) 91–95 (in Russian)
203. Soloviev, N.A.: Tests (Theory, Construction, Applications). Nauka Publishers, Novosibirsk, 1978 (in Russian)

204. Steele, J.M., Yao, A.C.: Lower bounds for algebraic decision trees. J. of Algorithms **3** (1982) 1–8
205. Swiniarski, R., Skowron, A.: Rough set methods in feature selection and recognition. Pattern Recognition Letters **24** (2003) 833–849
206. Tarasova, V.P.: Opponent Strategy Method in Optimal Search Problems. Moscow University Publishers, Moscow, 1988 (in Russian)
207. Tarski, A.: Arithmetical classes and types of mathematical systems, Mathematical aspects of arithmetical classes and types, Arithmetical classes and types of Boolean algebras, Arithmetical classes and types of algebraically closed and real closed fields. Bull. Amer. Math. Soc. **55** (1949) 63–64
208. Ufnarovskii, V. A.: Criterion of growth of graphs and algebras defined by words. Mathematical Notes **31**(3) (1982) 465–472 (in Russian)
209. Ugolnikov, A.B.: On depth and polynomial equivalence of formulas for closed classes of binary logic. Mathematical Notes **42**(4) (1987) 603–612 (in Russian)
210. Vapnik, V.N., Chervonenkis, A.Ya.: On the uniform convergence of relative frequencies of events to their probabilities. Theory of Probability and its Applications **16**(2) (1971) 264–280
211. Vasilevsky, M.P.: On recognition of faults of automata. Cybernetics 4 (1973) 98–108 (in Russian)
212. Vasilevsky, M.P.: On deciphering of automata, Cybernetics 2, (1974) 19–23 (in Russian)
213. Wegener, I.: The Complexity of Boolean Functions. John Wiley and Sons, and B.G. Teubner, Stuttgart, 1987
214. Yablonskii, S.V.: Tests. Encyklopaedia Kybernetiki. Edited by V.M. Glushkov. Main Editorial Staff of Ukrainian Soviet Encyklopaedia, Kiev (1975) 431–432 (in Russian)
215. Yablonskii, S.V.: Some problems of reliability and diagnosis in control systems. Mathematical Problems of Cybernetics **1**. Edited by S.V. Yablonskii. Nauka Publishers, Moscow (1988) 5–25 (in Russian)
216. Yablonskii, S.V., Chegis, I.A.: On tests for electric circuits. UMN **10**(4) (1955) 182–184 (in Russian)
217. Yablonskii, S.V., Gavrilov, G.P., Kudriavtzev, V.B.: Functions of Algebra of Logic and Classes of Post. Nauka Publishers, Moscow, 1966 (in Russian)
218. Yao, A.: Algebraic decision trees and Euler characteristics. Proceedings IEEE FOCS (1992) 268–277
219. Yao, A.: Decision tree complexity and Betti numbers. Proceedings ACM STOC (1994) 615–624
220. Zhuravlev, Ju.I.: On a class of partial Boolean functions. Discrete Analysis **2**. Edited by Ju.I. Zhuravlev. IM SO AN USSR Publishers, Novosibirsk (1964) 23–27 (in Russian)

Appendix. Closed Classes of Boolean Functions

The structure of all classes of Boolean functions closed relatively the operation of substitution has been described by E. Post in [171, 172]. In [217] S.V. Yablonskii, G.P. Gavrilov and V.B. Kudriavtzev considered the structure of all classes of Boolean functions closed relatively the operation of substitution and the operations of insertion and deletion of unessential variable. Appendix contains the description of this structure which is slightly different from the Post structure.

Some Definitions and Notation

A function $f(x_1,\ldots,x_n)$ with variables defined on the set $E_2 = \{0,1\}$ and with values from E_2 will be called *a Boolean function*. The constants 0 and 1 also are Boolean functions.

Let U be a set of Boolean functions, $f(x_1,\ldots,x_n)$ be a function from U, and g_i be either function from U or variable, $i = 1,\ldots,n$. We will say that the function $f(g_1,\ldots,g_n)$ is obtained from functions from U *by operation of substitution*.

Let $f(x_1,\ldots,x_n)$ be a Boolean function. A variable x_i of the function f will be called *essential* if there exist two n-tuples $\bar{\delta}$ and $\bar{\sigma}$ from E_2^n which differ only in the i-th digit and for which $f(\bar{\delta}) \neq f(\bar{\sigma})$. Variables of the function f which are not essential will be called *unessential*. Let x_j be an unessential variable of the function f and $g(x_1,\ldots,x_{j-1},x_{j+1},\ldots,x_n) = f(x_1,\ldots,x_{j-1},0,x_{j+1},\ldots,x_n)$. We will say that the function g is obtained from f *by operation of deletion of unessential variable*. We will say that the function f is obtained from g *by operation of insertion of unessential variable*.

Let U be a nonempty set of Boolean functions. We denote by $[U]$ the closure of the set U relatively the operation of substitution and the operations of insertion and deletion of unessential variable. The set U will be called *a closed class* if $U = [U]$.

The notion of *a formula over* U will be defined inductively in the following way:

a) The expression $f(x_1,\ldots,x_n)$, where $f(x_1,\ldots,x_n)$ is a function from U, is a formula over U.
b) Let $f(x_1,\ldots,x_n)$ be a function from U and $\varphi_1,\ldots,\varphi_n$ be expressions which are either formulas over U or variables. Then the expression $f(\varphi_1,\ldots,\varphi_n)$ is a formula over U.

A Boolean function corresponds in natural way to any formula over U. We will say that the formula *realizes* this Boolean function.

Denote by $[U]_1$ the closure of the set U relatively the operation of substitution. One can show that $[U]_1$ coincides with the set of functions realized by formulas over U. Denote by $[U]_2$ the closure of the set $[U]_1$ relatively operations of insertion and deletion of unessential variable. One can show that $[U] = [U]_2$.

The operation of negation will be denoted by the symbol $\neg$. Let $n \in \mathbb{N} \setminus \{0\}$ and $t \in E_2$. Denote by $\tilde{t}_n$ the n-tuple from E_2^n all the digits of which are equal to t. Let f be a Boolean function depending on n variables. The function f will be called *α-function* if $f(\tilde{t}_n) = t$ for any $t \in E_2$, *β-function* if $f(\tilde{t}_n) = 1$ for any $t \in E_2$, and *γ-function* if $f(\tilde{t}_n) = 0$ for any $t \in E_2$.

A function f will be called *a linear* function if $f = c_0 + c_1x_1 + \ldots + c_nx_n \pmod 2$, where $c_i \in E_2$, $0 \leq i \leq n$. A function f will be called *a self-dual* function if $f(x_1,\ldots,x_n) = \neg f(\neg x_1,\ldots,\neg x_n)$. A function f will be called *a monotone* function if for any n-tuples $\bar{\delta} = (\delta_1,\ldots,\delta_n)$ and $\bar{\sigma} = (\sigma_1,\ldots,\sigma_n)$ from E_2^n such that $\delta_i \leq \sigma_i$, $1 \leq i \leq n$, the inequality $f(\bar{\delta}) \leq f(\bar{\sigma})$ holds.

Let $\mu \in \mathbb{N}$ and $\mu \geq 2$. We will say that a function $f(x_1,\ldots,x_n)$ satisfies *the condition* $\langle a^\mu \rangle$ if for any μ tuples from E_2^n on which f has the value 0 there exists a number $j \in \{1, \ldots, n\}$ such that in each of the considered tuples the j-th digit is equal to 0. We will say that the function f satisfies *the condition* $\langle a^\infty \rangle$ if there exists a number $j \in \{1, \ldots, n\}$ such that in any n-tuple from E_2^n on which f has the value 0 the j-th digit is equal to 0. We will say that the function f satisfies *the condition* $\langle A^\mu \rangle$ if for any μ tuples from E_2^n on which f has the value 1 there exists a number $j \in \{1, \ldots, n\}$ such that in each of the considered tuples the j-th digit is equal to 1. We will say that a function f satisfies *the condition* $\langle A^\infty \rangle$ if there exists a number $j \in \{1, \ldots, n\}$ such that in any n-tuple from E_2^n on which f has the value 1 the j-th digit is equal to 1. The constant 1, by definition, satisfies the condition $\langle a^\infty \rangle$ and does not satisfy the condition $\langle A^2 \rangle$. The constant 0, by definition, satisfies the condition $\langle A^\infty \rangle$ and does not satisfy the condition $\langle a^2 \rangle$.

Let $\mu \in \mathbb{N}$ and $\mu \geq 2$. Denote

$$h_\mu = \bigvee_{i=1}^{\mu+1} (x_1 \wedge x_2 \wedge \ldots \wedge x_{i-1} \wedge x_{i+1} \wedge \ldots \wedge x_{\mu+1})$$

and

$$h_\mu^* = \bigwedge_{i=1}^{\mu+1} (x_1 \vee x_2 \vee \ldots \vee x_{i-1} \vee x_{i+1} \vee \ldots \vee x_{\mu+1}) \ .$$

Description of All Closed Classes of Boolean Functions

In this subsection all closed classes of Boolean functions are listed. For each class the Post notation is given, the description of functions contained in the considered class is presented, and a finite set of Boolean functions is given such that its closure relatively the operation of substitution and the operations of insertion and deletion of unessential variable is equal to this class.

As in [217] two Boolean functions is called *equal* if one of them can be obtained from the other by operations of insertion and deletion of unessential variable.

The inclusion diagram for closed classes of Boolean functions [217] is depicted in Fig. 12. Two points in this diagram corresponding to certain classes U and V are connected by an edge if the class V is immediately included into the class U (there are no intermediate classes between U and V). The point corresponding to the class U is placed at that above the point corresponding to the class V.

1. The class $O_1 = [\{x\}]$. This class consists of all functions equal to the function x, and all functions obtained from them by renaming of variables without identification.
2. The class $O_2 = [\{1\}]$. This class consists of all functions equal to the function 1.
3. The class $O_3 = [\{0\}]$. This class consists of all functions equal to the function 0.
4. The class $O_4 = [\{\neg x\}]$. This class consists of all functions equal to functions x or $\neg x$, and all functions obtained from them by renaming of variables without identification.

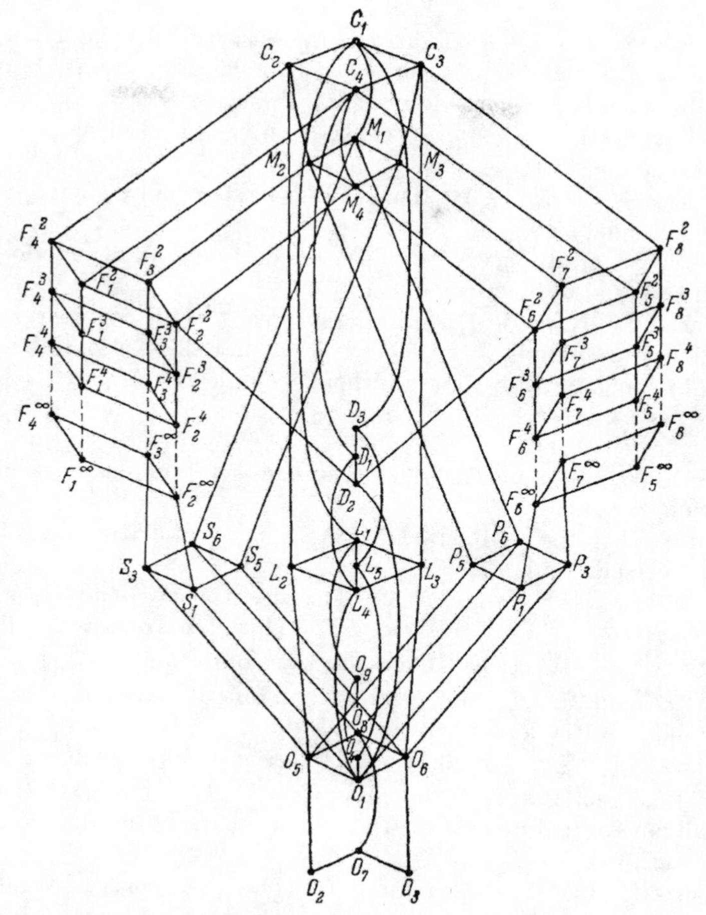

Fig. 12. Inclusion diagram for closed classes of Boolean functions.

5. The class $O_5 = [\{x, 1\}]$. This class consists of all functions equal to functions 1 or x, and all functions obtained from them by renaming of variables without identification.
6. The class $O_6 = [\{x, 0\}]$. This class consists of all functions equal to functions 0 or x, and all functions obtained from them by renaming of variables without identification.
7. The class $O_7 = [\{0, 1\}]$. This class consists of all functions equal to functions 0 or 1.
8. The class $O_8 = [\{x, 0, 1\}]$. This class consists of all functions equal to functions 0, 1 or x, and all functions obtained from them by renaming of variables without identification.
9. The class $O_9 = [\{\neg x, 0\}]$. This class consists of all functions equal to functions 0, 1, $\neg x$, or x, and all functions obtained from them by renaming of variables without identification.

10. The class $S_1 = [\{x \vee y\}]$. This class consists of all disjunctions (that is functions of the kind $\bigvee_{i=1}^{n} x_i$, $n = 1, 2, \ldots$, and all functions obtained from them by renaming of variables without identification).
11. The class $S_3 = [\{x \vee y, 1\}]$. This class consists of all disjunctions, and all functions equal to 1.
12. The class $S_5 = [\{x \vee y, 0\}]$. This class consists of all disjunctions, and all functions equal to 0.
13. The class $S_6 = [\{x \vee y, 0, 1\}]$. This class consists of all disjunctions, and all functions equal to functions 0 or 1.
14. The class $P_1 = [\{x \wedge y\}]$. This class consists of all conjunctions (that is functions of the kind $\bigwedge_{i=1}^{n} x_i$, $n = 1, 2, \ldots$, and all functions obtained from them by renaming of variables without identification).
15. The class $P_3 = [\{x \wedge y, 0\}]$. This class consists of all conjunctions, and all functions equal to 0.
16. The class $P_5 = [\{x \wedge y, 1\}]$. This class consists of all conjunctions, and all functions equal to 1.
17. The class $P_6 = [\{x \wedge y, 0, 1\}]$. This class consists of all conjunctions, and all functions equal to 0 or 1.
18. The class $L_1 = [\{x+y(\text{mod } 2), 1\}]$. This class consists of all linear functions.
19. The class $L_2 = [\{x + y + 1(\text{mod } 2)\}]$. This class consists of all linear α-functions and β-functions (that is functions of the kind $\sum_{i=1}^{2k} x_i + 1(\text{mod } 2)$, $\sum_{i=1}^{2l+1} x_i(\text{mod } 2)$, $k, l = 0, 1, 2, \ldots$, and all functions obtained from them by renaming of variables without identification).
20. The class $L_3 = [\{x + y(\text{mod } 2)\}]$. This class consists of all linear α-functions and γ-functions (that is the functions of the kind $\sum_{i=1}^{l} x_i(\text{mod } 2)$, l=0, 1, 2, ..., and all functions obtained from them by renaming of variables without identification).
21. The class $L_4 = [\{x + y + z(\text{mod } 2)\}]$. This class consists of all linear α-functions (that is functions of the kind $\sum_{i=1}^{2l+1} x_i(\text{mod } 2)$, $l = 0, 1, 2, \ldots$, and all functions obtained from them by renaming of variables without identification).
22. The class $L_5 = [\{x+y+z+1(\text{mod } 2)\}]$. This class consists of all linear self-dual functions (functions of the kind $\sum_{i=1}^{2l+1} x_i + 1(\text{mod } 2), \sum_{i=1}^{2l+1} x_i(\text{mod } 2)$, $l = 0, 1, 2, \ldots$, and all functions obtained from them by renaming of variables without identification).
23. The class $D_2 = [\{(x \wedge y) \vee (x \wedge z) \vee (y \wedge z)\}]$. This class consists of all self-dual monotone functions.
24. The class $D_1 = [\{(x \wedge y) \vee (x \wedge \neg z) \vee (y \wedge \neg z)\}]$. This class consists of all self-dual α-functions.
25. The class $D_3 = [\{(x \wedge \neg y) \vee (x \wedge \neg z) \vee (\neg y \wedge \neg z)\}]$. This class consists of all self-dual functions.
26. The class $A_1 = M_1 = [\{x \wedge y, x \vee y, 0, 1\}]$. This class consists of all monotone functions.
27. The class $A_2 = M_2 = [\{x \wedge y, x \vee y, 1\}]$. This class consists of all monotone α-functions and β-functions.

28. The class $A_3 = M_3 = [\{x \wedge y, x \vee y, 0\}]$. This class consists of all monotone α-functions and γ-functions.
29. The class $A_4 = M_4 = [\{x \wedge y, x \vee y\}]$. This class consists of all monotone α-functions.
30. The class $C_1 = [\{\neg(x \wedge y)\}]$. This class consists of all Boolean functions.
31. The class $C_2 = [\{x \vee y, x + y + 1 (\bmod 2)\}]$. This class consists of all α-functions and β-functions.
32. The class $C_3 = [\{x \wedge y, x + y (\bmod 2)\}]$. This class consists of all α-functions and γ-functions.
33. The class $C_4 = [\{x \vee y, x \wedge (y + z + 1 (\bmod 2))\}]$. This class consists of all α-functions.
34. The class $F_1^\mu = [\{x \vee (y \wedge \neg z), h_\mu^*\}]$, $\mu = 2, 3, \ldots$. This class consists of all α-functions satisfying the condition $\langle a^\mu \rangle$.
35. The class F_2^μ, $\mu = 2, 3, \ldots$, where $F_2^\mu = [\{x \vee (y \wedge z), h_2^*\}]$ if $\mu = 2$, and $F_2^\mu = [\{h_\mu^*\}]$ if $\mu \geq 3$. This class consists of all monotone α-functions satisfying the condition $\langle a^\mu \rangle$.
36. The class $F_3^\mu = [\{1, h_\mu^*\}]$, $\mu = 2, 3, \ldots$. This class consists of all monotone functions satisfying the condition $\langle a^\mu \rangle$.
37. The class $F_4^\mu = [\{x \vee \neg y, h_\mu^*\}]$, $\mu = 2, 3, \ldots$. This class consists of all functions satisfying the condition $\langle a^\mu \rangle$.
38. The class $F_5^\mu = [\{x \wedge (y \vee \neg z), h_\mu\}]$, $\mu = 2, 3, \ldots$. This class consists of all α-functions satisfying the condition $\langle A^\mu \rangle$.
39. The class F_6^μ, $\mu = 2, 3, \ldots$, where $F_6^\mu = [\{x \wedge (y \vee z), h_2\}]$ if $\mu = 2$, and $F_6^\mu = [\{h_\mu\}]$ if $\mu \geq 3$. This class consists of all monotone α-functions satisfying the condition $\langle A^\mu \rangle$.
40. The class $F_7^\mu = [\{0, h_\mu\}]$, $\mu = 2, 3, \ldots$. This class consists of all monotone functions satisfying the condition $\langle A^\mu \rangle$.
41. The class $F_8^\mu = [\{x \wedge \neg y, h_\mu\}]$, $\mu = 2, 3, \ldots$. This class consists of all functions satisfying the condition $\langle A^\mu \rangle$.
42. The class $F_1^\infty = [\{x \vee (y \wedge \neg z)\}]$. This class consists of all α-functions satisfying the condition $\langle a^\infty \rangle$.
43. The class $F_2^\infty = [\{x \vee (y \wedge z)\}]$. This class consists of all monotone α-functions satisfying the condition $\langle a^\infty \rangle$.
44. The class $F_3^\infty = [\{1, x \vee (y \wedge z)\}]$. This class consists of all monotone functions satisfying the condition $\langle a^\infty \rangle$.
45. The class $F_4^\infty = [\{x \vee \neg y\}]$. This class consists of all functions satisfying the condition $\langle a^\infty \rangle$.
46. The class $F_5^\infty = [\{x \wedge (y \vee \neg z)\}]$. This class consists of all α-functions satisfying the condition $\langle A^\infty \rangle$.
47. The class $F_6^\infty = [\{x \wedge (y \vee z)\}]$. This class consists of all monotone α-functions satisfying the condition $\langle A^\infty \rangle$.
48. The class $F_7^\infty = [\{0, x \wedge (y \vee z)\}]$. This class consists of all monotone functions satisfying the condition $\langle A^\infty \rangle$.
49. The class $F_8^\infty = [\{x \wedge \neg y\}]$. This class consists of all functions satisfying the condition $\langle A^\infty \rangle$.

Author Index

Lecture Notes in Computer Science

For information about Vols. 1–3382

please contact your bookseller or Springer

Vol. 3525: A.E. Abdallah, C.B. Jones, J.W. Sanders (Eds.), Communicating Sequential Processes. XIV, 321 pages. 2005.

Vol. 3517: H.S. Baird, D.P. Lopresti (Eds.), Human Interactive Proofs. IX, 143 pages. 2005.

Vol. 3510: T. Braun, G. Carle, Y. Koucheryavy, V. Tsaoussidis (Eds.), Wired/Wireless Internet Communications. XIV, 366 pages. 2005.

Vol. 3508: P. Bresciani, P. Giorgini, B. Henderson-Sellers, G. Low, M. Winikoff (Eds.), Agent-Oriented Information Systems II. X, 227 pages. 2005. (Subseries LNAI).

Vol. 3503: S.E. Nikoletseas (Ed.), Experimental and Efficient Algorithms. XV, 624 pages. 2005.

Vol. 3501: B. Kégl, G. Lapalme (Eds.), Advances in Artificial Intelligence. XV, 458 pages. 2005. (Subseries LNAI).

Vol. 3500: S. Miyano, J. Mesirov, S. Kasif, S. Istrail, P. Pevzner, M. Waterman (Eds.), Research in Computational Molecular Biology. XVII, 632 pages. 2005. (Subseries LNBI).

Vol. 3498: J. Wang, X. Liao, Z. Yi (Eds.), Advances in Neural Networks – ISNN 2005, Part III. L, 1077 pages. 2005.

Vol. 3497: J. Wang, X. Liao, Z. Yi (Eds.), Advances in Neural Networks – ISNN 2005, Part II. L, 947 pages. 2005.

Vol. 3496: J. Wang, X. Liao, Z. Yi (Eds.), Advances in Neural Networks – ISNN 2005, Part II. L, 1055 pages. 2005.

Vol. 3495: P. Kantor, G. Muresan, F. Roberts, D.D. Zeng, F.-Y. Wang, H. Chen, R.C. Merkle (Eds.), Intelligence and Security Informatics. XVIII, 674 pages. 2005.

Vol. 3494: R. Cramer (Ed.), Advances in Cryptology – EUROCRYPT 2005. XIV, 576 pages. 2005.

Vol. 3492: P. Blache, E. Stabler, J. Busquets, R. Moot (Eds.), Logical Aspects of Computational Linguistics. X, 363 pages. 2005. (Subseries LNAI).

Vol. 3489: G.T. Heineman, J.A. Stafford, H.W. Schmidt, K. Wallnau, C. Szyperski, I. Crnkovic (Eds.), Component-Based Software Engineering. XI, 358 pages. 2005.

Vol. 3488: M.-S. Hacid, N.V. Murray, Z.W. Raś, S. Tsumoto (Eds.), Foundations of Intelligent Systems. XIII, 700 pages. 2005. (Subseries LNAI).

Vol. 3475: N. Guelfi (Ed.), Rapid Integration of Software Engineering Techniques. X, 145 pages. 2005.

Vol. 3467: J. Giesl (Ed.), Term Rewriting and Applications. XIII, 517 pages. 2005.

Vol. 3465: M. Bernardo, A. Bogliolo (Eds.), Formal Methods for Mobile Computing. VII, 271 pages. 2005.

Vol. 3463: M. Dal Cin, M. Kaâniche, A. Pataricza (Eds.), Dependable Computing - EDCC 2005. XVI, 472 pages. 2005.

Vol. 3462: R. Boutaba, K. Almeroth, R. Puigjaner, S. Shen, J.P. Black (Eds.), NETWORKING 2005. XXX, 1483 pages. 2005.

Vol. 3461: P. Urzyczyn (Ed.), Typed Lambda Calculi and Applications. XI, 433 pages. 2005.

Vol. 3459: R. Kimmel, N.A. Sochen, J. Weickert (Eds.), Scale Space and PDE Methods in Computer Vision. XI, 634 pages. 2005.

Vol. 3456: H. Rust, Operational Semantics for Timed Systems. XII, 223 pages. 2005.

Vol. 3455: H. Treharne, S. King, M. Henson, S. Schneider (Eds.), ZB 2005: Formal Specification and Development in Z and B. XV, 493 pages. 2005.

Vol. 3454: J.-M. Jacquet, G.P. Picco (Eds.), Coordination Models and Languages. X, 299 pages. 2005.

Vol. 3453: L. Zhou, B.C. Ooi, X. Meng (Eds.), Database Systems for Advanced Applications. XXVII, 929 pages. 2005.

Vol. 3452: F. Baader, A. Voronkov (Eds.), Logic for Programming, Artificial Intelligence, and Reasoning. XI, 562 pages. 2005. (Subseries LNAI).

Vol. 3450: D. Hutter, M. Ullmann (Eds.), Security in Pervasive Computing. XI, 239 pages. 2005.

Vol. 3449: F. Rothlauf, J. Branke, S. Cagnoni, D.W. Corne, R. Drechsler, Y. Jin, P. Machado, E. Marchiori, J. Romero, G.D. Smith, G. Squillero (Eds.), Applications of Evolutionary Computing. XX, 631 pages. 2005.

Vol. 3448: G.R. Raidl, J. Gottlieb (Eds.), Evolutionary Computation in Combinatorial Optimization. XI, 271 pages. 2005.

Vol. 3447: M. Keijzer, A. Tettamanzi, P. Collet, J.v. Hemert, M. Tomassini (Eds.), Genetic Programming. XIII, 382 pages. 2005.

Vol. 3444: M. Sagiv (Ed.), Programming Languages and Systems. XIII, 439 pages. 2005.

Vol. 3443: R. Bodik (Ed.), Compiler Construction. XI, 305 pages. 2005.

Vol. 3442: M. Cerioli (Ed.), Fundamental Approaches to Software Engineering. XIII, 373 pages. 2005.

Vol. 3441: V. Sassone (Ed.), Foundations of Software Science and Computational Structures. XVIII, 521 pages. 2005.

Vol. 3440: N. Halbwachs, L.D. Zuck (Eds.), Tools and Algorithms for the Construction and Analysis of Systems. XVII, 588 pages. 2005.

Vol. 3439: R.H. Deng, F. Bao, H. Pang, J. Zhou (Eds.), Information Security Practice and Experience. XII, 424 pages. 2005.

Vol. 3437: T. Gschwind, C. Mascolo (Eds.), Software Engineering and Middleware. X, 245 pages. 2005.

Vol. 3436: B. Bouyssounouse, J. Sifakis (Eds.), Embedded Systems Design. XV, 492 pages. 2005.

Vol. 3434: L. Brun, M. Vento (Eds.), Graph-Based Representations in Pattern Recognition. XII, 384 pages. 2005.

Vol. 3433: S. Bhalla (Ed.), Databases in Networked Information Systems. VII, 319 pages. 2005.

Vol. 3432: M. Beigl, P. Lukowicz (Eds.), Systems Aspects in Organic and Pervasive Computing - ARCS 2005. X, 265 pages. 2005.

Vol. 3431: C. Dovrolis (Ed.), Passive and Active Network Measurement. XII, 374 pages. 2005.

Vol. 3429: E. Andres, G. Damiand, P. Lienhardt (Eds.), Discrete Geometry for Computer Imagery. X, 428 pages. 2005.

Vol. 3427: G. Kotsis, O. Spaniol (Eds.), Wireless Systems and Mobility in Next Generation Internet. VIII, 249 pages. 2005.

Vol. 3423: J.L. Fiadeiro, P.D. Mosses, F. Orejas (Eds.), Recent Trends in Algebraic Development Techniques. VIII, 271 pages. 2005.

Vol. 3422: R.T. Mittermeir (Ed.), From Computer Literacy to Informatics Fundamentals. X, 203 pages. 2005.

Vol. 3421: P. Lorenz, P. Dini (Eds.), Networking - ICN 2005, Part II. XXXV, 1153 pages. 2005.

Vol. 3420: P. Lorenz, P. Dini (Eds.), Networking - ICN 2005, Part I. XXXV, 933 pages. 2005.

Vol. 3419: B. Faltings, A. Petcu, F. Fages, F. Rossi (Eds.), Constraint Satisfaction and Constraint Logic Programming. X, 217 pages. 2005. (Subseries LNAI).

Vol. 3418: U. Brandes, T. Erlebach (Eds.), Network Analysis. XII, 471 pages. 2005.

Vol. 3416: M. Böhlen, J. Gamper, W. Polasek, M.A. Wimmer (Eds.), E-Government: Towards Electronic Democracy. XIII, 311 pages. 2005. (Subseries LNAI).

Vol. 3415: P. Davidsson, B. Logan, K. Takadama (Eds.), Multi-Agent and Multi-Agent-Based Simulation. X, 265 pages. 2005. (Subseries LNAI).

Vol. 3414: M. Morari, L. Thiele (Eds.), Hybrid Systems: Computation and Control. XII, 684 pages. 2005.

Vol. 3412: X. Franch, D. Port (Eds.), COTS-Based Software Systems. XVI, 312 pages. 2005.

Vol. 3411: S.H. Myaeng, M. Zhou, K.-F. Wong, H.-J. Zhang (Eds.), Information Retrieval Technology. XIII, 337 pages. 2005.

Vol. 3410: C.A. Coello Coello, A. Hernández Aguirre, E. Zitzler (Eds.), Evolutionary Multi-Criterion Optimization. XVI, 912 pages. 2005.

Vol. 3409: N. Guelfi, G. Reggio, A. Romanovsky (Eds.), Scientific Engineering of Distributed Java Applications. X, 127 pages. 2005.

Vol. 3408: D.E. Losada, J.M. Fernández-Luna (Eds.), Advances in Information Retrieval. XVII, 572 pages. 2005.

Vol. 3407: Z. Liu, K. Araki (Eds.), Theoretical Aspects of Computing - ICTAC 2004. XIV, 562 pages. 2005.

Vol. 3406: A. Gelbukh (Ed.), Computational Linguistics and Intelligent Text Processing. XVII, 829 pages. 2005.

Vol. 3404: V. Diekert, B. Durand (Eds.), STACS 2005. XVI, 706 pages. 2005.

Vol. 3403: B. Ganter, R. Godin (Eds.), Formal Concept Analysis. XI, 419 pages. 2005. (Subseries LNAI).

Vol. 3402: M. Daydé, J.J. Dongarra, V. Hernández, J.M.L.M. Palma (Eds.), High Performance Computing for Computational Science - VECPAR 2004. XI, 732 pages. 2005.

Vol. 3401: Z. Li, L.G. Vulkov, J. Waśniewski (Eds.), Numerical Analysis and Its Applications. XIII, 630 pages. 2005.

Vol. 3400: J.F. Peters, A. Skowron (Eds.), Transactions on Rough Sets III. IX, 461 pages. 2005.

Vol. 3399: Y. Zhang, K. Tanaka, J.X. Yu, S. Wang, M. Li (Eds.), Web Technologies Research and Development - APWeb 2005. XXII, 1082 pages. 2005.

Vol. 3398: D.-K. Baik (Ed.), Systems Modeling and Simulation: Theory and Applications. XIV, 733 pages. 2005. (Subseries LNAI).

Vol. 3397: T.G. Kim (Ed.), Artificial Intelligence and Simulation. XV, 711 pages. 2005. (Subseries LNAI).

Vol. 3396: R.M. van Eijk, M.-P. Huget, F. Dignum (Eds.), Agent Communication. X, 261 pages. 2005. (Subseries LNAI).

Vol. 3395: J. Grabowski, B. Nielsen (Eds.), Formal Approaches to Software Testing. X, 225 pages. 2005.

Vol. 3394: D. Kudenko, D. Kazakov, E. Alonso (Eds.), Adaptive Agents and Multi-Agent Systems II. VIII, 313 pages. 2005. (Subseries LNAI).

Vol. 3393: H.-J. Kreowski, U. Montanari, F. Orejas, G. Rozenberg, G. Taentzer (Eds.), Formal Methods in Software and Systems Modeling. XXVII, 413 pages. 2005.

Vol. 3392: D. Seipel, M. Hanus, U. Geske, O. Bartenstein (Eds.), Applications of Declarative Programming and Knowledge Management. X, 309 pages. 2005. (Subseries LNAI).

Vol. 3391: C. Kim (Ed.), Information Networking. XVII, 936 pages. 2005.

Vol. 3390: R. Choren, A. Garcia, C. Lucena, A. Romanovsky (Eds.), Software Engineering for Multi-Agent Systems III. XII, 291 pages. 2005.

Vol. 3389: P. Van Roy (Ed.), Multiparadigm Programming in Mozart/Oz. XV, 329 pages. 2005.

Vol. 3388: J. Lagergren (Ed.), Comparative Genomics. VII, 133 pages. 2005. (Subseries LNBI).

Vol. 3387: J. Cardoso, A. Sheth (Eds.), Semantic Web Services and Web Process Composition. VIII, 147 pages. 2005.

Vol. 3386: S. Vaudenay (Ed.), Public Key Cryptography - PKC 2005. IX, 436 pages. 2005.

Vol. 3385: R. Cousot (Ed.), Verification, Model Checking, and Abstract Interpretation. XII, 483 pages. 2005.

Vol. 3383: J. Pach (Ed.), Graph Drawing. XII, 536 pages. 2005.